2000.

Biopsy Pathology and Cytology of the Cervix

Biopsy Pathology Series

Titles in the series

1. Biopsy Pathology of the Small Intestine
   F.D. Lee and P.G. Toner
3. Brain Biopsy
   J.H. Adams, D.I. Graham and D. Doyle
4. Biopsy Pathology of the Lymphoreticular System
   D.H. Wright and P.G. Isaacson
6. Biopsy Pathology of Bone and Bone Marrow
   B. Frisch, S.M. Lewis, R. Burkhardt and R. Bartl
7. Biopsy Pathology of the Breast
   J. Sloane
9. Biopsy Pathology of the Bronchi
   E.M. McDowell and T.F. Beals
10. Biopsy Pathology in Colorectal Disease
    I.C. Talbot and A.B. Price
12. Biopsy Pathology of the Liver (2nd edn)
    R.S. Patrick and J.O'D. McGee
13. Biopsy Pathology of the Pulmonary Vasculature
    C.A. Wagenvoort and W.J. Mooi
14. Biopsy Pathology of the Endometrium
    C.H. Buckley and H. Fox
15. Biopsy Pathology of Muscle (2nd edn)
    M. Swash and M.S. Schwartz
16. Biopsy Pathology of the Skin
    N. Kirkham
17. Biopsy Pathology of Melanocytic Disorders
    W.J. Mooi and T. Krausz
18. Biopsy Pathology of the Thyroid and Parathyroid
    O. Ljungberg
19. Biopsy Pathology of the Oesophagus, Stomach and Duodenum (2nd edn)
    D.W. Day and M.F. Dixon
20. Biopsy Pathology of the Prostate
    D.G. Bostwick and P.A. Dundore
21. Biopsy Pathology of the Eye and Ocular Adnexa
    N.A. Rao
22. Biopsy Pathology of the Oral Tissues
    E.W. Odell and P.R. Morgan
23. Biopsy Pathology and Cytology of the Cervix (2nd edn)
    D.V. Coleman and D.M.D. Evans

BIOPSY PATHOLOGY AND CYTOLOGY OF THE CERVIX

Second Edition

Dulcie V. Coleman MD FRCPath FIAC
Professor and Head of Department of Cytopathology and Cytogenetics,
St Mary's Hospital, London, UK

and

David M. D. Evans MD FRCP FRCPath
formerly Consultant Histopathologist, Llandough Hospital, South Glamorgan;
Honorary Clinical Teacher, College of Medicine, University of Wales, Cardiff,
UK

with the assistance of

Robert Baker BA FI BMS CT(IAC)
Department of Cytopathology and Cytogenetics,
St Mary's Hospital, London, UK

A member of the Hodder Headline Group
LONDON · SYDNEY · AUCKLAND
Co-published in the USA by Oxford University Press Inc., New York

First published in Great Britain in 1988
Second edition published in 1999 by
Arnold, a member of the Hodder Headline Group,
338 Euston Road, London NW1 3BH

http://www.arnoldpublishers.com

Co-published in the United States of America by
Oxford University Press Inc.,
198 Madison Avenue, New York, NY10016
Oxford is a registered trademark of Oxford University Press

British Library Cataloguing in Publication Data
A catalogue record for this book is available from the British Library

Library of Congress Cataloging-in-Publication Data
A catalog record for this book is available from the Library of Congress

ISBN 0 340 74087 6

1 2 3 4 5 6 7 8 9 10

Typeset by Type Study, Scarborough, North Yorkshire
Printed and bound in Italy

Contents

PREFACE TO THE SECOND EDITION

The first edition of *Biopsy Pathology and Cytology of the Cervix* was published ten years ago and has been reprinted several times to meet continuing demand for the book .

The time has now come to bring the book up to date and a second edition has now been prepared with this in mind. The authors have made extensive revision to the text so that it more accurately reflects the current concepts of cervical disease. They have also added new material to encompass the most recent developments in the field of gynaecological histopathology and cytology.

Many of the key changes to be found in the second edition have been made in response to requests from readers. A new chapter has been added to address management issues relevant to the provision of a cervical cytology service. It reflects the need for quality standards in cytology and describes the procedures that need to be incorporated into laboratory practice to ensure screening is effective.

Another new element of the book is the introduction of colour photographs of the cytological findings. Colour illustrations have been introduced selectively wherever there is a need to demonstrate subtle morphological details of the cells such as irregular dispersion of nuclear chromatin or delicate cytoplasmic vacuolation; these features are less well displayed in monochrome photographs than in colour. Colour is also used wherever there are overlapping clusters of cells to assist with the interpretation of three-dimensional structures. However, monochrome photographs have been retained wherever patterns of cells or tissue architecture is of prime importance as these are best illustrated in black and white photos.

Readers who are familiar with the first edition will note that that there are several sections in the book where the text has been expanded. Particular attention has been focused on laser cone biopsy or large loop excision biopsy (LLETZ) which were only just introduced as the first edition went to press. Other sections which have been expanded or revised include the review of terminology for reporting cervical neoplasia embodying more recently recognized entities, current concepts of glandular neoplasia and the cytodiagnosis of endocervical brush specimens. In addition, the section relating to the role of human papillomaviruses in cervical neoplasia has been updated in the light of current knowledge of the carcinogenic process.

Biopsy Pathology and Cytology of the Cervix is a book with a unique practical value for the pathologist. It is a handy reference manual which can be used at the microscope for the interpretation of histological and cytological material. It is also a book with an educational function as it describes the pathological basis for disease and the collection and preparation of specimens. This new edition ensures that the book is modern, comprehensive and up to date. The authors hope that both experienced pathologists and pathologists in training will continue to find it useful in clinical practice and that the book helps them to enhance their knowledge and understanding of pathological lesions of the cervix.

D.V. Coleman
D.M.D. Evans

1
INTRODUCTION

1.1 HISTORICAL NOTE

The techniques of biopsy and cytology were first described in the early part of the 19th century (Bennett, 1845; Marmy, 1846; Sedillot, 1846; Vogel, 1847). Microscopic analysis of exfoliated cells from sputum and urine was quickly acknowledged to be a safe, accurate and convenient method of establishing the presence of malignancy and cytological specimens were soon recognized as a useful source of material for diagnosis of malignant and other diseases. In contrast, biopsy techniques were slow to be accepted into clinical practice, as tissue processing techniques were not well developed and many pathologists found the specimens difficult to interpret (Ewing, 1915). As tissue processing improved and pathologists gained in experience in the interpretation of histological sections, confidence in the technique grew and biopsy soon replaced cytology as the method of choice for the investigation of a patient with a clinically suspicious lesion. By the turn of the century, most pathologists had abandoned cytology and some even denied it a place among their diagnostic methods (Bamforth and Osborn, 1958).

1.2 DEVELOPMENT OF CERVICAL BIOPSY AND THE CONCEPT OF CARCINOMA *IN SITU*

Biopsy of the cervix was used toward the end of the 19th century for the preoperative diagnosis of cervical cancer (Ruge and Veit, 1878; Friedlander, 1885; Cullen, 1900). At first, wedge biopsies were taken with scalpel and forceps to establish the benign or malignant nature of clinically obvious lesions of the cervix. It soon became apparent that only advanced cancers which were already infiltrating the surrounding tissue could be detected by this approach and methods of detecting cancerous lesions at their earliest stages were sought.

Interest in this approach was stimulated by the observations of several Austrian pathologists. Schauenstein (1908), Rubin (1910) and Schottlander and Kermauner (1912) described structural changes in the epithelium at the periphery of invasive cancer, which they considered to be a manifestation of early cervical cancer. The normal epithelium at this site was replaced throughout its thickness by abnormal cells, a pattern which was subsequently designated carcinoma *in situ*.

Several gynaecologists including Hinselmann (1925) and Schiller (1928) were also convinced of the malignant potential of carcinoma *in situ* and developed techniques for the diagnosis of these lesions. Concern about the need to identify and treat early cancer led Hinselmann to develop the colposcope with which he could inspect the cervix at a magnification of 20 times. In his search for early carcinomas, he supplemented colposcopic inspection with colposcopically directed cervical biopsy and by careful correlation was able to identify a profusion of changes in the cervical epithelium (Hinselmann, 1925). Unfortunately, Hinselmann's achievements were not widely known and, for many years, colposcopy was practised in only a few centres outside Germany.

With the introduction of cervical cytology in the 1950s, new biopsy techniques were developed to identify microscopic foci of carcinoma *in situ* in the cervical epithelium of women with abnormal smears. Multiple blind punch biopsies and quadrant biopsies were taken but these proved unreliable methods of localizing the lesions and were soon abandoned in favour of cold knife conization of the cervix. This operation, which was first advocated by Thornton *et al.* (1954) and Scott and Reagan (1956), combined accurate diagnosis and effective treatment in one operation and still has a firm place in the management of the women with an abnormal smear today.

The risk of complication following cone biopsy stimulated many gynaecologists to seek safer methods of diagnosing and treating cervical lesions. Reappraisal of the work of Hinselmann showed that colposcopically directed biopsy was a useful method of detecting precancerous changes and this technique therefore assumed an important role in the diagnosis of cervical neoplasia and rapidly gained acceptance in modern clinical practice.

More recently, the technique of large loop excision of the transformation zone (LLETZ) was shown to have distinct advantages over colposcopic biopsy in that the risk of complications was small and more material was available for histological examination. Consequently, in the United Kingdom, LLETZ is currently the preferred method of sampling the cervices of women with an abnormal cervical smear.

The extensive use of cervical biopsy in the latter half of this century has revealed a spectrum of changes in the cervical epithelium (dysplasia and carcinoma *in situ*), some of which are now considered to be precursors of invasive squamous cancer. They are currently collectively described as cervical intraepithelial neoplasia (CIN) and have been extensively studied by cell pathologists and cell biologists alike in an attempt to understand the mechanism of cervical carcinogenesis. The histological and cytological appearance and clinical significance of these lesions are discussed in detail in Chapter 10.

1.3 DEVELOPMENT OF CERVICAL CYTOLOGY

As described in the preceding section, cytological techniques were quickly accepted into clinical practice in the 19th century and were widely used for over 50 years before they were superseded by biopsy as a routine diagnostic tool for the diagnosis of malignancy. Lionel Beale, Professor of Medicine at King's College, London, was at the forefront of the development of cytology in the United Kingdom, and Donaldson (1853) promoted cytological techniques in the United States. Beale (1854) was the first to illustrate the cells of uterine carcinoma in the urine of a woman with advanced cancer.

At the turn of the century, interest in cytological diagnosis declined as biopsy techniques became more acceptable. It was against this background of declining interest that George Papanicolaou presented his paper at the 3rd Race Betterment Conference, Battlecreek, Michigan, in which he reported that tumour cells could be found in vaginal smears from women with cervical cancer (Papanicolaou, 1928). He tentatively suggested that cytological examination of vaginal secretions could provide a new opportunity for diagnosing cancer, but his paper was received with little enthusiasm. The prevailing opinion at the time, as expressed by Dr James Ewing, one of the outstanding American pathologists of the day, was that, since the uterine cervix was accessible to diagnostic exploration by biopsy, which is a relatively simple procedure, 'the use of cytologic examination appeared to be superfluous' (Carmichael, 1973).

Fifteen years later, Papanicolaou published his monograph *Diagnosis of uterine cancer by the vaginal smear* (Papanicolaou and Traut, 1943). By then the climate of opinion had changed. The concept of carcinoma *in situ* of the cervix had been developed and was widely accepted by gynaecologists and pathologists. Papanicolaou's new technique appeared to provide a unique opportunity for detecting cervical cancer at its earliest preinvasive stage.

Figure 1.1 Dr George Papanicolaou (1883–1962).

Confirmation of Papanicolaou's finding by other workers (Ayre, 1944; Jones *et al.*, 1945; Meigs *et al.*, 1945) reinforced this view, and the first screening clinic was opened in 1945 in Massachusetts (McSweeney and McKay, 1948).

The driving force behind the expansion of medical and technical expertise in cytodiagnosis was the prospect of eliminating cervical cancer by the detection and removal of precancerous lesions. The hurried introduction of population screening programmes, both in the United States and Europe meant that no initial prospective studies to establish the validity of this hypothesis were carried out. In consequence, the usefulness of the diagnostic test has long been in doubt. Only recently, after many years of uncertainty, is evidence emerging that population screening is a positive force in the control of cervical cancer. Significant reduction in mortality and morbidity has been reported in countries with comprehensive and systematic screening programmes such as British Columbia, Iceland, Denmark, Sweden, Finland and some areas of Scotland (Miller, 1981; Boyes *et al.*, 1982; Johannesson *et al.*, 1982; Patterson *et al.*, 1985; Hakama, 1978, 1982; Duguid *et al.*, 1985; Macgregor, 1971; Macgregor and Teper, 1978). In view of this, many countries which hitherto only offered cervical screening on an ad hoc basis, are now considering the introduction of organized screening programmes in order to ensure that all women at risk benefit from the protection afforded by a cervical smear test.

1.4 APPLICATION OF CERVICAL CYTOLOGY

Cervical cytology is a simple, safe, non-invasive method of detecting precancerous changes in the cervix, which is acceptable to most women. It is therefore a particularly appropriate tool for the screening of apparently healthy women for evidence of CIN and this is its main use today. The majority of smears submitted to the pathology laboratory for diagnosis are taken as a preventive measure from asymptomatic women participating in a national screening programme. Unfortunately the test is not uniformly acceptable to all social groups and women most at risk frequently fail to avail themselves of the screening service. Nevertheless, in those communities where intensive effort has been made to ensure that women at greatest risk are screened, a remarkable fall in the incidence of invasive cervical cancer has been observed and a significant reduction in mortality from this disease has been recorded (Draper and Cook, 1983; Worth, 1984; Day, 1984).

Cervical cytology has another important application in clinical practice. It is a valuable diagnostic tool for the investigation of women with clinical symptoms and signs of cervical cancer and for the follow-up of women who have been treated for preinvasive or invasive cancer. The high degree of reliability of a positive smear report has convinced gynaecologists and pathologists alike of the importance of further investigation of all women with an abnormal smear to determine the significance of the cytological findings.

1.5 INDICATION FOR BIOPSY AND THE PLACE OF COLPOSCOPY

Before the introduction of cytology, any abnormality of the cervical mucosa, however small, was a target for biopsy in an endeavour to detect early cancerous change. As a result, many benign lesions were biopsied unnecessarily and some *in situ* and occult invasive cancers were missed. The introduction of cervical screening has modified the indications for biopsy.

At the present time, the most frequent indication for biopsy is an abnormal smear report. Some of the abnormal smears will have been taken from apparently healthy women participating in the cervical cancer screening programme; others will be taken from women referred for gynaecological investigation with symptoms and signs suggestive of cervical cancer.

It is important to remember that biopsy is still indicated for women with a negative smear but with symptoms and signs suggestive of cervical cancer. A cervical smear from an advanced, fungating or ulcerating cancer may contain only blood, inflammatory cells and necrotic debris and is therefore not always a reliable guide to the presence of malignancy. Macgregor (1971) has shown that 10% of cervical smears taken from patients with invasive cancer are negative so that whenever an invasive cancer is suspected, the decision to biopsy the cervix must be made on clinical grounds alone.

The types of biopsy taken from women who have an abnormal smear has varied greatly over the past 20 years. Before the introduction of colposcopy, wedge biopsy, cold knife conization of the cervix and even amputation of the cervix was the rule. The introduction of colposcopy into routine gynaecological practice has led to a more conservative approach. This technique which was undervalued as a clinical tool for many years permits the gynaecologist to view a magnified image of the cervix so that preinvasive and early invasive cancers invisible to the naked eye can be identified and the site and extent determined with a reasonable degree of accuracy. Thus, biopsy can be tailored according to the lesion and the large wedge and cone biopsies which were common 20 years ago have been largely replaced by punch biopsy and the much smaller cones obtained by laser excision or large loop diathermy excision of the transformation zone (LLETZ).

REFERENCES

Ayre, J. E. (1944) A simple office test for uterine cancer diagnosis. *Can. Med. Assoc. J.*, **51**, 17.

Bamforth, J. and Osborn, G. R. (1958) Diagnosis from cells. *J. Clin. Pathol.*, **11**, 473–482.

Beale, L. S. (1854) Morbid growths, in *The Microscope and its Application to Clinical Medicine*, Samuel Highley, London, Chapter 11, pp. 160–168.

Bennett, J. H. (1845) Introductory address to a course of lectures on histology and the use of the microscope. *Lancet*, **i**, 517–522.

Boyes, D. A., Morrison, R., Knox, E. G., Draper, G. J. and Miller, A. B. (1982) A cohort study of cervical cancer screening in British Columbia. *Clin. Invest. Med.*, **5**, 1–29.

Carmichael, D. E. (1973) Diagnosis by the vaginal smear, in *The Papanicolaou Smear; Life of George N. Papanicolaou*, Charles C. Thomas Publishers, Springfield, Illinois, pp. 5–61.

Cullen, T. S. (1900) The removal and examination of uterine tissue for diagnostic purposes, in *Cancer of the Uterus*, Henry Kimpton, London, Chapter 3, pp. 27–39.

Day, N. E. (1984) Effect of cervical cancer screening in Scandinavia. *Obstet. Gynecol.*, **63**, 714–718.

Donaldson, F. (1853) The practical application of the microscope to the diagnosis of cancer. *Am. J. Med. Sci.*, **25**, 43–70.

Draper, G. J. and Cook, G. A. (1983) Changing patterns of cervical cancer rates. *Br. Med. J.*, **287**, 510–512.

Duguid, H. L. D., Duncan, L. D. and Currie, J. (1985) Screening for cervical intraepithelial neoplasia in Dundee and Angus 1962–81 and its relation with invasive cancer, *Lancet*, **2**, 1053–1056.

Ewing, J. (1915) The incision of tumours for diagnosis. *N. York Med. J.*, **102**, 10–14.

Friedlander, C. (1885) *The Use of the Microscope in Clinical and Pathological Examinations* (2nd edn. translated by H. C. Coe), D. Appleton and Co., New York, 99–100.

Hakama, M. (1978) Mass screening for cervical cancer in Finland, in *Report of a UICC International Workshop Toronto* (ed. A. B. Miller) pp. 93–107. UICC Tech. Rep. Sv40, 9080, 682.

Hakama, M. (1982) Trends in incidence of cervical cancer in the Nordic countries, in *Trends in Cancer Incidence* (ed. K. Magnus), Hemisphere Publishing Corporation, New York, pp. 279–291.

Hinselmann, H. (1925) Verbesserung der Inspection öglichkeiten von Vulva, Vagina und Portio. Munch. *Med. Wochenschr.*, **72**, 17–33.

Johannesson, G., Geirsson, G., Day, N. and Tulinus, H. (1982) Screening for cancer of the uterine cervix in Iceland 1965–1978 *Acta Obstet. Gynecol. Scand.*, **61**, 199–203.

Jones, C. A., Neustaedter, T. and Mackenzie, L. (1945) The value of vaginal smears in the diagnosis of early malignancy. *Am. J. Obstet. Gynecol.*, **49**, 159–168.

Macgregor, J. E. (1971) The consequences of comprehensive screening for cervical cancer 1959–1969, in *The Early Diagnosis of Cancer of the Cervix* (ed. J. M. Roggott), University of Hull, pp. 82–90.

Macgregor, J. E. and Teper, S. (1978) Mortality from carcinoma of cervix in Britain. *Lancet*, **2**, 774–776.

Marmy, J. (1846) The utility of microscope observations in the diagnosis of cancer. *Revue Medicochirurgicale de Paris*, **1**, 215.

McSweeney, D. J. and McKay, D. (1948) Uterine cancer: its detection by simple screening methods. *N. Engl. J. Med.*, **238**, 867–870.

Meigs, J. V., Graham, R. M., Fremont-Smith, M., Kapnick,I. and Rawson, R. (1945) The value of the vaginal smear in the diagnosis of uterine cancer. *Surg. Gynecol. Obstet.*, **81**, 337–345.

Miller, A. B. (1981) An evaluation of population screening for cervical cancer, in *Advances In Clinical Cytology* (eds L. G. Koss and D. V. Coleman), Butterworths, London, Chapter 3, pp. 64–89.

Papanicolaou, G. N. (1928) New cancer diagnosis, in *Proceedings of Third Race Betterment Conference*. Battle Creek, Race Betterment Foundation, p. 528.

Papanicolaou, G. N. and Traut, H. F. (1943) *Diagnosis of Uterine Cancer by the Vaginal Smear*, The Commonwealth Fund, New York, May 1943.

Patterson, F., Bjorkholm, E. and Naslund, I. (1985) Evaluation of screening for cervical cancer in Sweden. Trends in incidence and mortality 1985–1980. *Int. J. Epidemiol.*, **14**, 521–527.

Rubin, I. C. (1910) The pathological diagnosis of incipient carcinoma of the uterus. *Am. J. Obstet.*, **62**, 668–676.

Ruge, C. and Veit, J. (1878) Zur Pathologie der Vaginalportion. *Zeitschr. fur Geburtsch. u Gynak.*, **2**, 415.

Schauenstein, W. (1908) Histologische Untersuchungen über atypisches Plattenepithel an der Portio und an der Innenfläche der *Cervix uteri. Arch. Gynaek*, **85**, 576–616.

Schiller, W. (1928) Über Frühstadien des Portiocarcinoms und ihre Diagnose. *Arch. Gynaek.*, **133**, 211–283.

Schottlander, J. and Kermauner, F. (1912) *Zur Kenntis des Uteruskarzinoms; monographische Studie über Morphologie, Entwichlung, Waschstum nebst Beitragezl zue Klinik der Erkrankung*, S. Karger, Berlin.

Scott, R. B. and Reagan, J. W. (1956) Diagnostic cervical biopsy techniques for study of early cancer; value of cold knife conisation procedure. *J. Am. Med. Assoc.*, **160**, 343–347.

Sedillot, C. E. (1846) *Recherches sur cancer*, Strasbourg.

Thornton, W. N., Waters, L. N., Pearce, L. S., Wilson,

L. A. and Nodes, T. M. (1954) Carcinoma *in situ*; value of cold knife cone biopsy. *Obstet. Gynecol.*, **3**, 587–594.

Vogel, J. (1847) Malignant heterologous tumours, in *Pathological Anatomy of the Human Body*. (Translated from the German with additions by G. E. Day), H. Ballière, London, Chapter 5, pp. 261–269.

Worth, A. J. (1984) The Walton report and its subsequent impact on cervical cancer screening programmes in Canada. *Obstet. Gynecol.*, **63**, 135–139.

2

Cytological methods

2.1 SPECIMEN COLLECTION

Several methods have been described for collecting material from the cervix for cytological study, each of which has its advantages and limitations. The technique which is most widely used is the cervical scrape taken with the Ayre spatula or a modification of it. Other methods include vaginal aspiration and endocervical brushings.

2.1.1 Cervical scrape

Numerous studies have shown that this is the most efficient single method of detecting premalignant and clinically unsuspected invasive cancers of the cervix. An excellent illustrated account of the technique has been prepared by Macgregor (1981) updated by Wolfendale (1989) and published by the British Society for Clinical Cytology. Copies of this video and booklet should be available in every pathology laboratory providing a cytology service, for distribution to clinicians or nurses carrying out the procedure.

Briefly, the cervical scrape is obtained under direct vision with the vaginal speculum in position. The Ayre spatula is inserted into the os and the spatula gently but firmly rotated through 360° (Figure 2.1(a, b)). The cervical mucus and cellular material on the spatula is spread evenly across a glass slide, the frosted end of which has been previously labelled in pencil with the patient's name. The slide must be fixed immediately by immersing it in a solution of 95% ethanol. It must be left in the fixative for 30 minutes, after which it can be safely stored dry at room temperature. Fixing the smear with spray fixative or a carbowax spirit-based fixative also produces satisfactory results. *On no account should the smear be allowed to dry before fixation as this impairs the staining properties of the cells.*

The cervical smear, when correctly taken, will provide a sample of the surface epithelial layers from the whole of the transformation zone. It is most important that this area be sampled as it is at this site that intraepithelial neoplasia and invasive cancer is most likely to develop. One of the limitations of cervical sampling is that the smear provides little guidance to the location or extent of a focus of neoplastic change in the cervix. As the changes of intraepithelial and early invasive cancers of the cervix are rarely obvious to the naked eye, methods for defining the lesions for biopsy and treatment have been devised. These are described in Chapter 3.

The efficiency of sampling depends largely

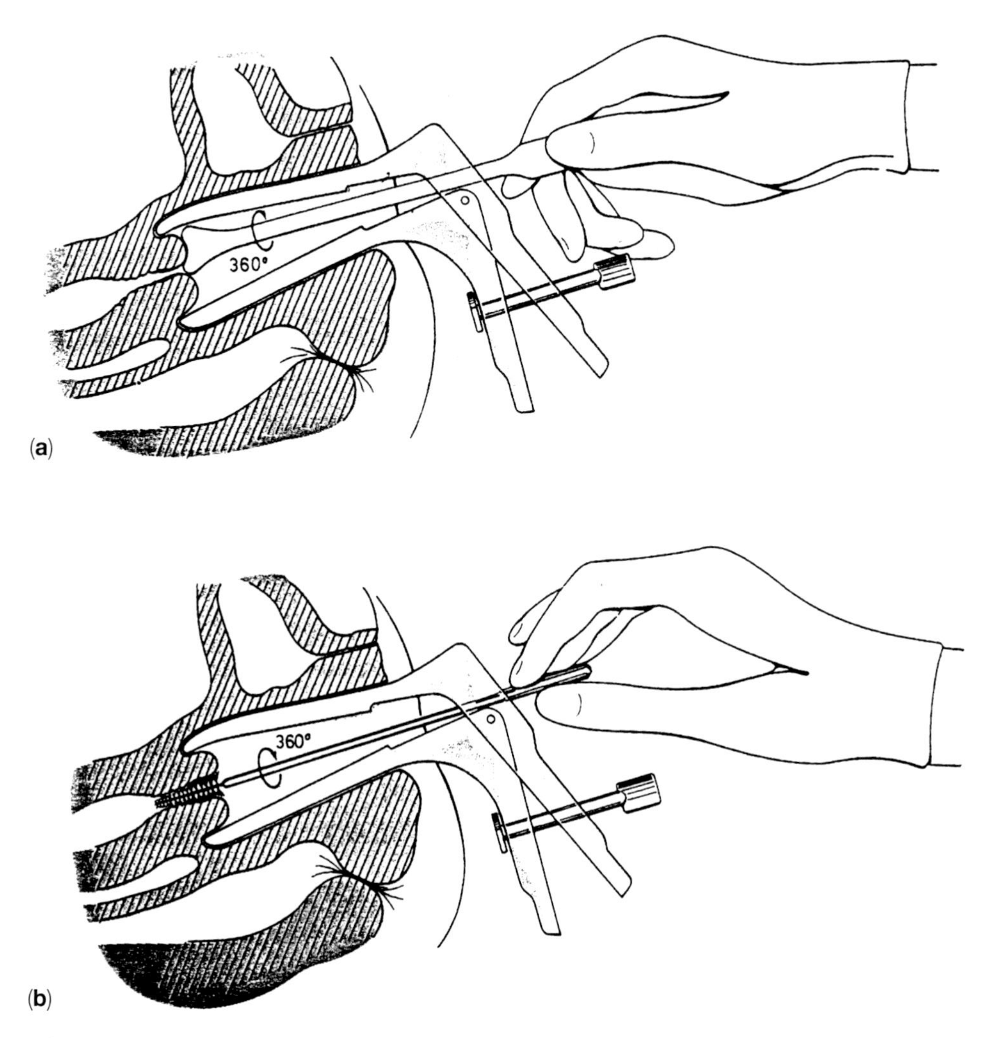

Figure 2.1 (**a**) Illustration showing the correct position of the cervical spatula in the external os. The spatula is rotated through 360° while in close contact with the cervical epithelium at the squamocolumnar junction. (**b**) Illustration showing the correct position of the endocervical brush in the cervical canal. The brush is rotated through 360° while in close contact with the cervical epithelium at the squamocolumnar junction.

on the skill with which the specimens are taken. Ten per cent of smears submitted for routine cytological examination are considered to be unsatisfactory, due to inadequate sampling technique. Other common faults are air drying or improper fixation, both of which will impair the quality of staining. Scanty smears or heavily blood-stained specimens are also unsuitable for a reliable assessment. Great care must be exercised to ensure that the smear is actually taken from the squamocolumnar junction, otherwise the sample will not be representative of the changes that are present in the transformation zone.

The quality of the cervical scrape is also dependent on the instrument used to take the sample. The spatula selected should fit the shape of the cervix (Figure 2.2). Clearly no single design will be appropriate for all cervices and smear takers should ensure that they have several different sampling devices to hand, in addition to the Ayre spatula, when arranging a screening clinic. The Aylesbury spatula which was devised and tested by Wolfendale *et al.* (1987) is a particularly useful model. The single elongated prong extends high into the endocervical canal while the broad part of the blade allows a full sweep of the endocervix. The rounded end of the Aylesbury or Ayre spatula can be used to obtain samples from women with large open patulous cervices.

2.1.2 Vaginal aspirate

This method of obtaining samples of cervical cells is rarely used in clinical practice. Aspiration from the vaginal pool by means of a pipette was the method of smear taking developed by Papanicolaou and Traut (1943). This approach was originally considered to be an effective method of detecting neoplasms arising in the body of the uterus as well as in the cervix. However, careful evaluation of the technique has shown that it is inefficient for this purpose (Macgregor *et al.*, 1966) and it has now very largely fallen into disuse. The main value of vaginal aspiration is to provide a specimen in the rare cases where the cervix cannot be visualized for anatomical reasons.

2.1.3 Sampling the endocervical canal

In older women, sampling the transformation zone may be difficult as the squamocolumnar junction tends to recede into the endocervical canal with age. Similar difficulties in sampling the squamocolumnar junction may be encountered if the os is stenosed after surgery or is congenitally narrowed. Special spatulae with a long point can be used on these occasions (Pistofidis *et al.*, 1984; Wolfendale *et al.*, 1987) to ensure that samples are obtained from the appropriate site. Alternatively, an endocervical brush or endocervical swab can be used for this purpose (Figure 2.1(b)). Trimbos and Arentz

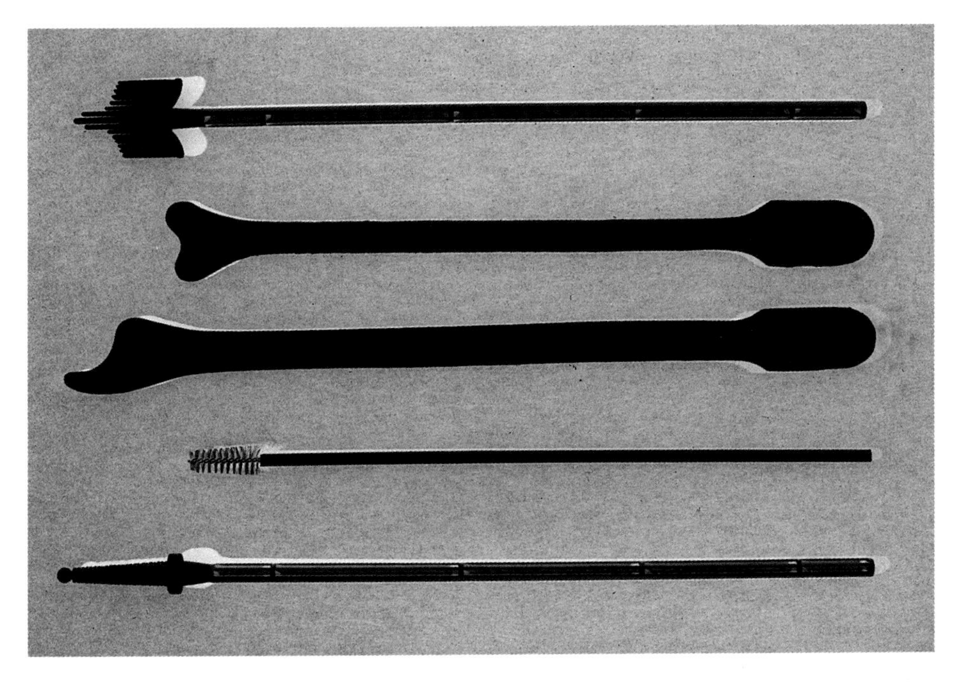

Figure 2.2 A range of different cervical spatulas and brushes that have been developed for taking cervical smears.

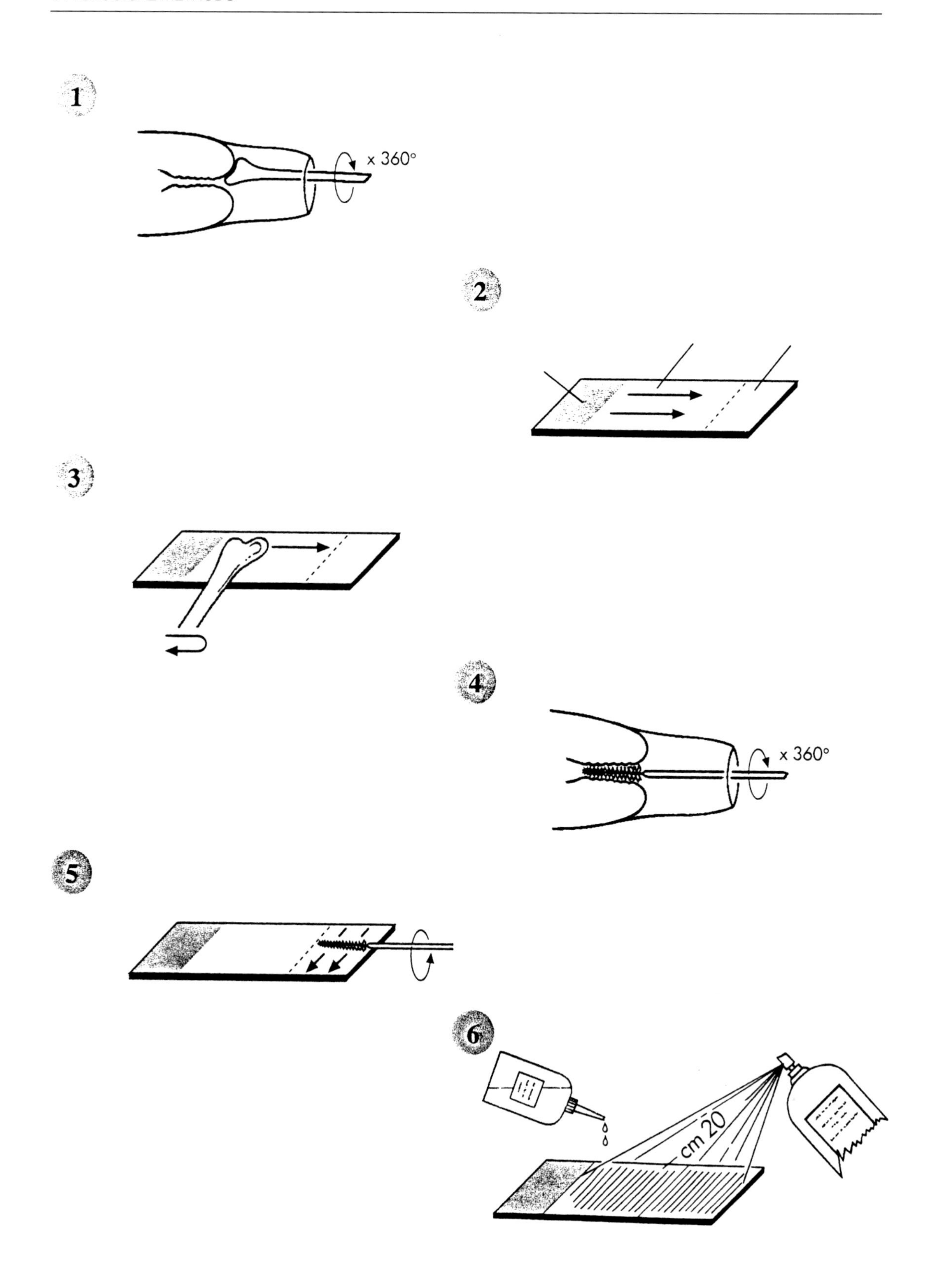
1
x 360°
2
3
4
x 360°
5
6
cm 20

(1986) and Selvaggi (1989) have shown that the cytobrush is superior to the cotton swab as a method of collecting endocervical cells although this is not a universal finding (Schettino *et al.*, 1993). A delicate nylon brush developed by Professor Nils Stormby in the 1970s or a modification of it is widely used in clinical practice. Wietzman *et al.* (1988), reporting on 128 women who had previously had colposcopically directed endocervical curettage found that the endocervical brush was significantly more sensitive than curettage in detecting endocervical neoplasia, as well as being less costly and less painful. Van Erp *et al.* (1988), investigating 130 high-risk patients found that the cytobrush reduced the number of inadequate and false-negative smears.

Studies have shown that neoplastic lesions of the cervix may be missed if the endocervical brush is used on its own (Buntinx *et al.*, 1991) and the endocervical brush should always be used in combination with an Ayre or Aylesbury spatula when specimens are taken for screening purposes. The ectocervix should be sampled first followed by endocervical sampling. Both samples may be smeared on the same slide providing the smear is not allowed to dry before fixation (Figure 2.3).

2.1.4 Vaginal vault smear

One of the key uses of cytological techniques is to monitor the progress of patients who have been treated for cervical neoplasia by hysterectomy with amputation of the cervix. A smear from the vaginal vault may detect residual cancer or recurrence even though a lesion is not visible to the naked eye. The rounded end of the Ayre spatula can be used for this purpose and the material obtained is spread and fixed exactly as for a cervical smear.

2.1.5 Upper-third vaginal wall scrape

This technique is a useful method of evaluating the hormonal response of the vaginal epithelium. Scrapings are taken with the rounded end of an Ayre spatula and a smear prepared in the usual way. A quick staining method has been described for the study of these smears (Shorr, 1941).

2.1.6 Cytological surveillance following laser ablation

Hughes *et al.* (1992) studied 856 women with CIN 3 who had been treated by CO_2 laser or cold coagulation between 9 and 30 months earlier. They carried out a randomized study of the ability of a range of cervical cytological sampling devices to reveal neoplasia in women found on punch biopsy at their follow-up visit to have a focus of recurrent or residual CIN. Dyskariotic cells were seen in 10% of smears taken with the Ayre spatula, 8.3% of smears taken with the Rocket sampler, 4% of smears taken with the Aylesbury spatula, and none of the smears taken with the Multispatula. The Cytobrush did not substantially improve the detection rate of dyskariosis in the absence of cervical stenosis or endocervical neoplasia. They concluded that surveillance should be by cytology using the Ayre spatula and by colposcopy with directed punch biopsy.

2.1.7 Assessment of diagnostic yield of different sampling devices used in cervical screening

Buntinx and Brouwers (1996) compared the yield of histological and cytological

Figure 2.3 Preparation of a combined cervical spatula and cervical brush smear. It is important to smear both sides of the spatula on the slide, and to roll the endocervical brush through 360° to ensure that all the material on the spatula and brush is transferred to the slide.

abnormalities for a range of different cervical sampling devices. Data from 29 papers were included in this study. Meta-analysis of randomized and quasi-randomized studies revealed no substantial differences in the yield of mild dysplasia or worse between the Ayre spatula, the Cytobrush or the cotton swab when used alone. There were also no substantial differences in the yield of cytological abnormalities between the extended tip spatula, the Ayre spatula combined with the cotton swab, or the Cervix brush. However, the Ayre spatula, Cytobrush or cotton swab when used alone performed significantly worse than the combination of any spatula plus the Cytobrush or cotton swab. This last combination is recommended for cervical screening.

2.2 STAINING PROCEDURES

Cervical smears are received in the laboratory ready-fixed and can be stored indefinitely at room temperature. By convention, a modified Papanicolaou stain is used to stain the slide. After staining, the epithelial cells should exhibit a clearly defined nuclear chromatin pattern, differential cytoplasmic counterstaining and cytoplasmic transparency. The stain is designed to distinguish between the keratinized and non-keratinized epithelial cells in the smear. Squames which contain keratin or keratin precursors assume a delicate pink colour, whereas the cytoplasm of the non-keratinized cells is blue. As the pattern of staining depends on the pH of the smear and on excellent fixation, it is unsafe to place too much emphasis on the cytoplasmic staining pattern of the cells for the purposes of diagnosis. Moreover, hormonal evaluation should never be attempted on a cervical smear as the cervical epithelium is much less responsive to oestrogen than the upper third of the vagina. The formula of the Papanicolaou stain, its characteristics and instructions for its preparation have been described in detail by Bales and Durfee (1979). Staining protocols have been documented by Proctor (1989) and Boon and Suurmeijer (1991).

2.3 ASSESSMENT OF THE CERVICAL SMEAR

Two broad categories of cervical smear have been defined:

1. Those which are deemed adequate for reliable assessment and reporting.
2. Those which are considered to be inadequate for reporting (Evans *et al.*, 1986).

There are few scientific studies of smear adequacy and valid measures of adequacy are hard to find. Guidelines have been prepared but these must be interpreted with care and due reference to the particular clinical circumstances of each case.

Four elements which affect adequacy have been defined: (i) accurate and complete patient identification data; (ii) pertinent clinical information; (iii) technical interpretability; and (iv) cellular composition and evidence of sampling of the transformation zone. These elements are discussed in Section 6.6.

2.4 SMEAR REPORTING, TERMINOLOGY AND CLINICAL MANAGEMENT

It is not always appreciated that the preparation of a cytology report requires just as much care as the preparation of a surgical pathology report. It must be accurate and concise and contain information which is of clinical relevance. It is generally agreed that communication between pathologists and clinicians is best if the report is written in a narrative form, as shorthand reports are open to misinterpretation. For this reason, numerical classification systems, including that of Papanicolaou (1954), are strongly discouraged. The report should contain a brief description of the findings in precise cytological terms which are widely understood. This may be followed where appropriate by prediction of the probable histological diagnosis suggested by the smear pattern and a recommendation for management. The recommended terminology for reporting is given in a document prepared by the British Society for Clinical Cytology (Evans *et al.*, 1986).

In Table 2.1, we have set out some of the terms used to describe normal and abnormal epithelial cells in the smear, together with the corresponding histological changes suggested by the cytological findings. The term 'dyskaryosis' (Gr. *dys* = abnormal, *karyon* = nucleus or nut) is used to describe the nuclear changes in epithelial cells suggestive of intraepithelial neoplasia of the cervix. When dyskaryotic cells are seen in the smear, an effort should be made to deduce the grade of CIN in the cervix from the severity of the dyskaryosis (Sections 10.3–10.5). 'Malignant cells' or 'evidence of malignancy' should be used only when a confident diagnosis of invasive cancer can be made. The report should indicate the type of lesion suspected as this is helpful to the clinician for determining the management of the patient.

Table 2.1 Interpretation of smear results (based on terminology recommended by British Society for Clinical Cytology)

Cytology report	*Explanation*	*Action*
Inadequate	Insufficient cellular material Inadequate fixation Smear consisting mainly of blood or inflammatory cell exudate Little or no material to suggest that the transformation zone has been sampled	Repeat smear
Negative	Normal. Includes simple inflammatory changes, including a mild polymorph exudate	Routine recall
Borderline changes with or without HPV change	Cellular appearances that cannot be described as normal. Smears in which there is doubt as to whether the nuclear changes are inflammatory or dyskaryosis	Repeat smear at 6 months. Consider for colposcopy if changes persist
Mild dyskaryosis with or without HPV change	Cellular appearances consistent with origin from CIN (mild dysplasia)	Repeat smear at 6 months. Consider for colposcopy if changes persist
Moderate dyskaryosis with or without HPV change	Cellular appearances consistent with origin from CIN 2 (moderate dysplasia)	Refer for colposcopy
Severe dyskaryosis with or without HPV change	Cellular appearances consistent with origin from CIN 3 (severe dysplasia/carcinoma *in situ*)	Refer for colposcopy
Severe dyskaryosis/ ? invasive carcinoma	Cellular appearances consistent with origin from CIN 3, but with additional features which suggest the possibility of invasive cancer	Refer for colposcopy
Glandular neoplasia or suspicion of glandular neoplasia	Cellular appearances suggesting pre-cancer or cancer in the cervical canal or the endometrium	Refer for colposcopy

Source: *Practical Guides for General Practice (14): Cervical Screening* (1992) Austoker, J. and McPherson, A. Cancer Research Campaign, Oxford Medical Publications.

The term 'borderline' is recommended to describe the changes in epithelial cells which differ from the norm but which cannot confidently be reported as dyskaryotic or malignant (Evans *et al.*, 1986). In the past such equivocal cells have tended to be described as 'atypical' or 'suspicious' or 'abnormal' or 'positive' or 'Papanicolaou Class III' which are terms commonly applied to tumour cells. In an effort to achieve consistency of reporting 'borderline' changes, the British Society for Clinical Cytology have prepared guidelines for their recognition in cervical smears (National Coordinating Network Working Party Report, 1994). The guidelines contain an interesting set of histograms which illustrate the variation in frequency of reporting this category of cells between laboratories. The proportion of smears showing borderline changes varies inversely with the proportion showing mild dyskariosis. In order to ensure that the proportion of smears assigned to the 'borderline' category is kept to a minimum, the guidelines contain examples of the morphological changes which fall into this category.

'Borderline' cells may show nuclear enlargement and irregularity, abnormal chromatin structure, hyperchromasia – all of which are features of dyskariosis but to a lesser degree. Cells showing borderline changes are most frequently found in the presence of human papilloma virus infection and chronic cervicitis and are discussed further in Sections 7.2.2, 7.13.2 and 10.7.2. Women whose smears contain cells showing borderline changes should be managed conservatively in the first instance, with repeat smears at 6-monthly intervals until the smear pattern returns to normal or there is clear evidence of dyskariosis. Colposcopy is indicated only if borderline changes persist for a year.

The need for standardization of cervical smears reports has been recognized world-wide. It has provided the stimulus for a group of American cytologists to devise a system of reporting which it was hoped would have world-wide appeal. The Bethesda system (so called because the meeting to discuss the preparation of a unified system of reporting was sponsored by the National Institutes of Health, Bethesda, Maryland) is shown in abbreviated form in Table 2.2. The full report of the Bethesda system can be found in *Acta Cytologica*, 1993, **37**, 115–124.

The Bethesda system appears to be more complicated than the British System with which it can be equated. The most striking difference is the method of classifying squamous intraepithelial neoplasia. In the British system, three grades of cervical intraepithelial neoplasia are recognized (CIN 1, 2 and 3). In the Bethesda system two grades are recognized: low-grade squamous intraepithelial lesions (LSIL) which incorporate both human papillomavirus changes and CIN 1; and high-grade squamous intraepithelial lesions (HSIL) encompassing CIN 2 and CIN 3. The decision by the Bethesda group to adopt a two-grade system is based on the requirement for treatment. In general, conservative managment by cytological surveillance is appropriate for LSIL in the first instance whereas HSIL requires colposcopy and biopsy. The Bethesda system has been criticized on the grounds that a two-grade system is insufficiently flexible and may lead to overtreatment of lesions which in time may have regressed.

Another difference between the UK terminology and the Bethesda system relates to the classification of epithelial cells with undetermined nuclear changes. In the UK such cells are classified as 'borderline'. In the Bethesda system the same cells are designated as 'atypical squamous (or glandular) cells of uncertain significance' (ASCUS or AGCUS), qualified if possible as to whether a reactive or neoplastic process is favoured. The Bethesda system also eliminates the distinction between HPV changes and CIN 1. Sherman *et al.* (1992) observe that, owing to the wide spectrum of changes included under ASCUS, an explanatory note and/or recommendation is desirable for the management of the patient with this diagnosis.

In order to facilitate comparability of cervical screening internationally a system of

Table 2.2 Bethesda system

Adequacy of the specimen (see pp. 87 and 88)
  Satisfactory for evaluation
  Satisfactory for evaluation but limited by . . . (specify reason)
  Unsatisfactory for evaluation . . . (specify reason)

General categorization (optional)
  Within normal limits
  Benign cellular changes: see descriptive diagnosis
  Epithelial cell abnormality: see descriptive diagnosis

Descriptive diagnoses
  BENIGN CELLULAR CHANGES
    Infection
      Trichomonas vaginalis
      Fungal organisms morphologically consistent with *Candida* sp.
      Predominance of coccobacilli consistent with shift in vaginal flora
      Bacteria morphologically consistent with *Actinomyces* spp.
      Cellular changes associated with herpes simplex virus
      Other*
    Reactive changes
      Reactive cellular changes associated with:
      Inflammation (includes typical repair)
      Atrophy with inflammation ('atrophic vaginitis')
      Radiation
      Intrauterine contraceptive device (IUD)
      Other

Epithelial cell abnormalities
  SQUAMOUS CELL
    Atypical squamous cells of undetermined significance: qualify
    Low-grade squamous intraepithelial lesion (LSIL) encompassing: HPV** mild dysplasia/CIN 1
    High-grade squamous intraepithelial lesion (HSIL) encompassing: moderate and severe dysplasia, CIS/CIN 2 and CIN 3
  GLANDULAR CELL
    Endometrial cells, cytologically benign in a post-menopausal woman
    Atypical glandular cells of undetermined significance: qualify
    Endocervical adenocarcinoma
    Endometrial adenocarcinoma
    Extrauterine adenocarcinoma
    Adenocarcinoma, NOS

Other malignant neoplasms: Specify

Hormonal evaluation (applies to vaginal smears only)
  Hormonal pattern compatible with age and history
  Hormonal pattern incompatible with age and history: specify
  Hormonal evaluation not possible due to: specify

* Atypical squamous or glandular cells of undetermined significance should be further qualified, if possible, as to whether a premalignant/malignant process is favoured.
** Cellular changes of human papillomavirus (HPV) (previously termed *koilocytosis, koilocytotic atypia* and *condylomatous atypia*) are included in the category of LSIL.

Table 2.3 Protocol for the management of smears

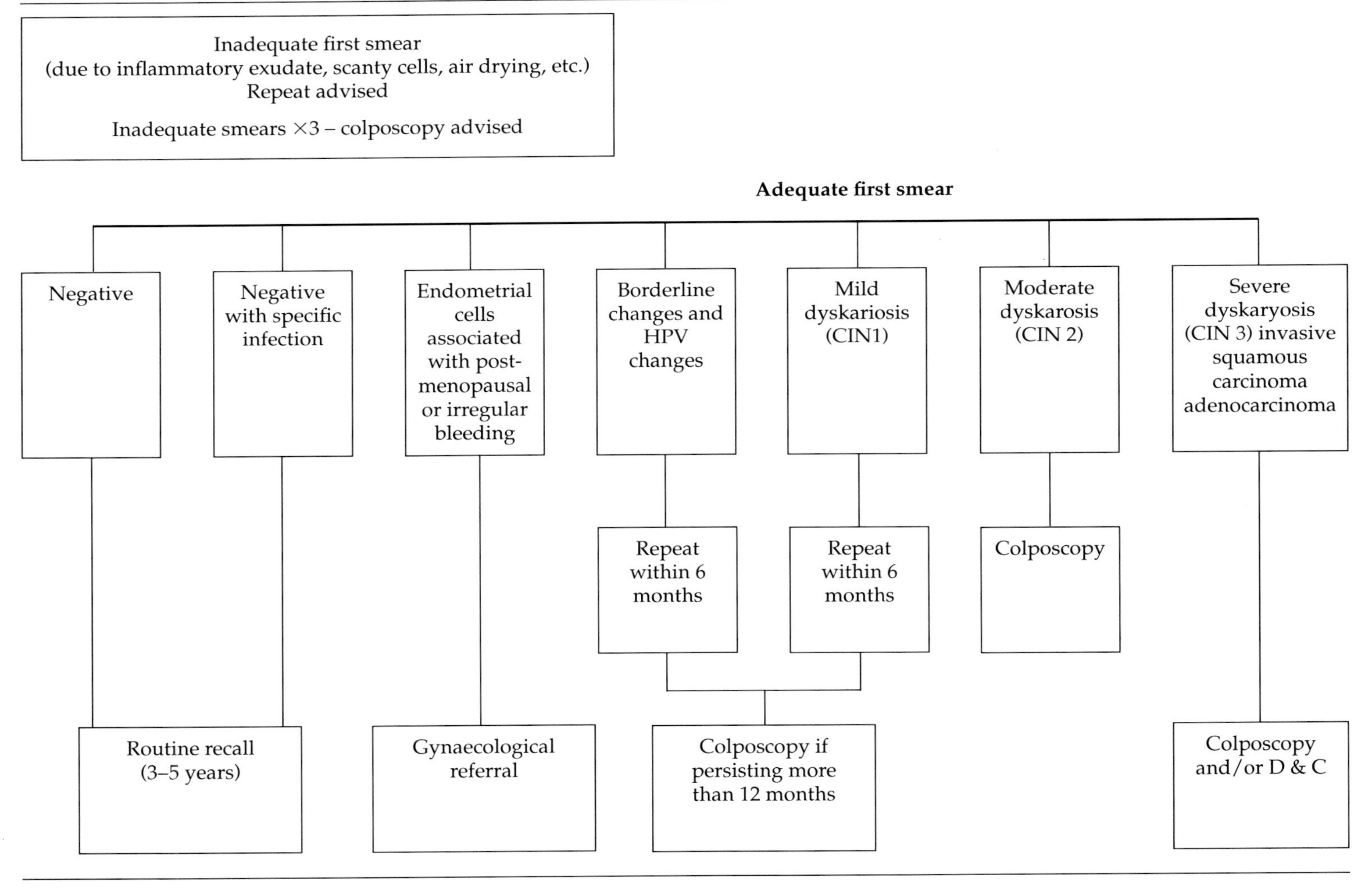

The guidelines apply to women under the age of 35. Older women with abnormal smear merit closer surveillance or earlier referral for coloscopy due to their increased risk of cervical cancer. Severe dyskaryosis at any stage merits immediate referral to coposcopy, as does any clinical indication of carcinoma.

equivalent terminology has been devised whereby the terminology used in each European country is equated with that of Bethesda and the UK system. Equivalent Terminology tables have been included in the European Guidelines for Quality Assurance in Cervical Cancer Screening (Coleman *et al.*, 1993).

There are at present no international guidelines on the management of women with abnormal smears. Management is based on the current concept of the way in which cervical cancer develops. Cervical cancer is a multistage disease comprising a preinvasive stage (CIN), microinvasive stage and frank invasive stage. Progression from CIN to invasive cancer may take many years and the risk of progression varies with the grade of CIN. Management protocols are based on the concept that the risk of progression is least in women with CIN I and greatest in women with CIN 3. However, the risk for any individual woman is not known. Each case should be managed individually on the basis of age, clinical and cytological history and degree of dyskariosis. A protocol which is widely accepted in the UK is shown in Table 2.3. Women with a mildly dyskariotic smear are advised to have a repeat smear in 6 months and those with moderate dyskariosis or severe dyskariosis suggestive of CIN 2 or CIN 3 or invasive cancer should be referred for colposcopy without delay. Colposcopy is recommended for women with borderline changes or mild dyskariosis whose lesions persist for 12 months or longer. Immediate colposcopy is advised for older women with an abnormal smear of any grade. It is also essential if there is a clinical suspicion of cervical cancer even if the cervical smear is normal.

Although the aim of the cervical screening programme is to detect cervical neoplasia, other diagnoses can be made from the smear. These include a small number of specific infections such as *Trichomonas vaginalis*, monilia, herpes simplex virus, hormonal imbalance and malignancy elsewhere in the genital tract. Evidence of these conditions should be included in the report.

No matter how skilful the cytopathologist may be, there will always be a small number of cases where a definite diagnosis cannot be made on the basis of the cytological findings. This may be due to inadequate sampling, or the presence of an unusual tumour. In such cases, a description of the smear is all that can be given and a repeat sample can be requested if it is thought that this will further the diagnosis. If there is any doubt about the quality of the smear, it must be reported as unsatisfactory and a repeat smear requested. It is useful for the clinician if the cytologist indicates in the report precisely why a repeat smear is needed, e.g. because of air drying, heavy blood-staining, etc., so that steps can be taken to avoid these problems when the second smear is taken (Figures 2.4 (a–c) and 2.5).

2.4.1 Validity of smear report

The degree of confidence that can be placed in a smear report has been a topic of concern for many years. The interpretation of cervical smears is a subjective exercise and diversity of opinion on the significance of minor variations of colour, size and shape of the cells in the smears is inevitable. The fact that there may be observer variation emphasizes the need for standards against which cytologists can measure their performance. Several professional bodies have addressed this need and methodology for measuring laboratory performance has been defined in National Publications (see references). Factors influencing laboratory performance and laboratory standards are discussed elsewhere (Chapter 17).

Because of the subjective nature of cytology it is usual to assess the validity of a smear report by comparing it with a 'gold standard' which in most cases is taken to be the histology report. One measure of validity that has been assessed in this way is the specificity of cervical screening. The specificity of cytology was demonstrated by Reagan and Hamonic (1958) who correlated the cytological and histological findings in 127 cases who were reported on

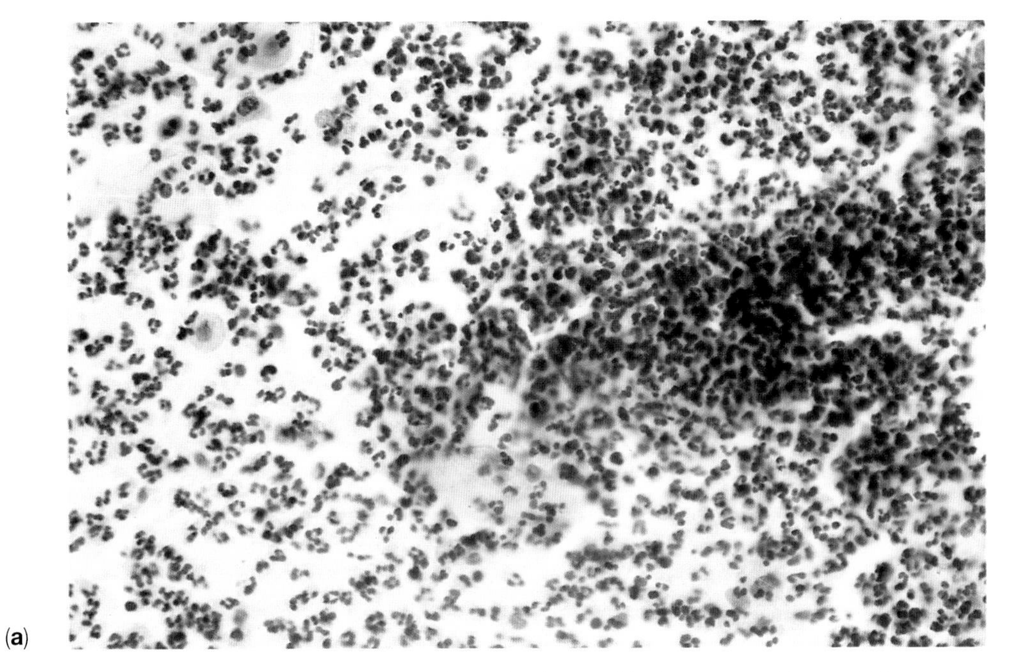

(a)

Figure 2.4 (**a–c**) Inadequate smears. (**a**) Epithelial cells obscured by leucocytes (Pap, ×400).

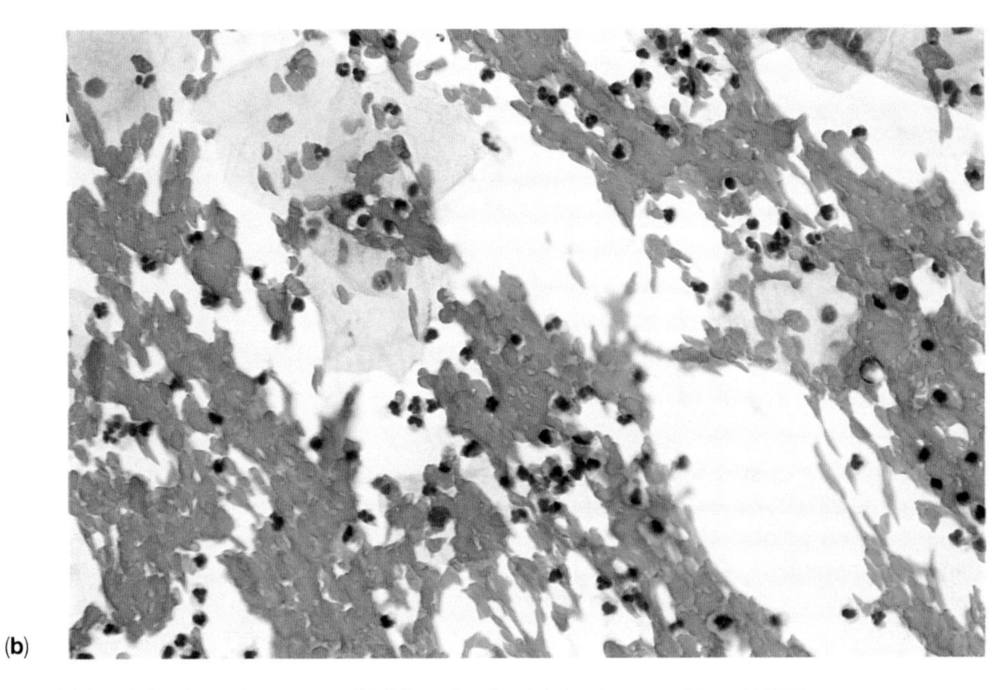

(b)

Figure 2.4 (**a–c**) Inadequate smears. (**b**) A heavily bloodstained smear (Pap, ×400).

(**c**) 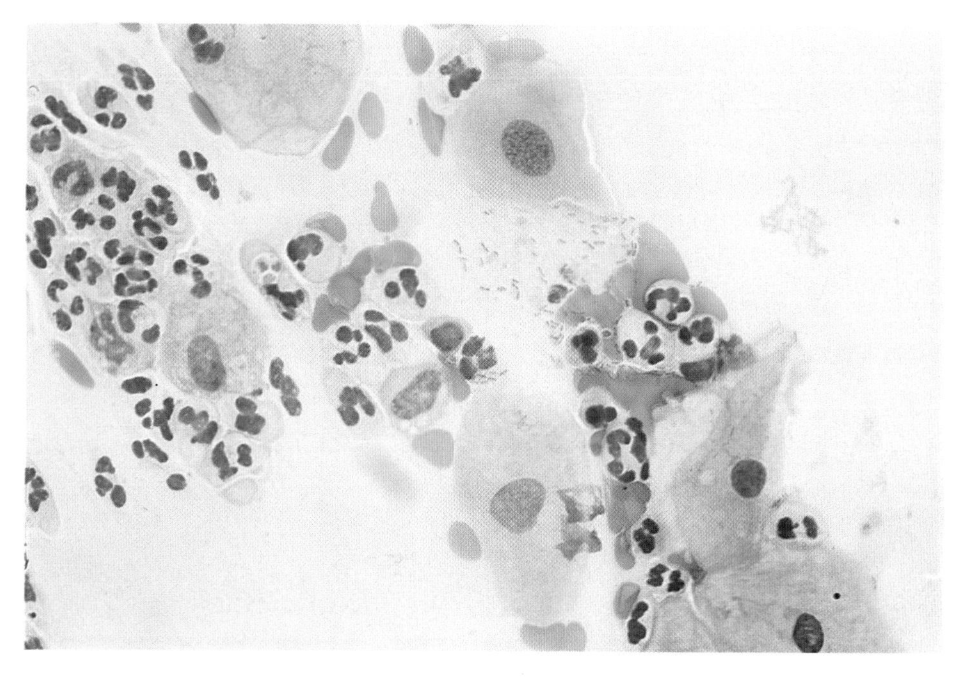

Figure 2.4 (**a–c**) Inadequate smears. (**c**) An inadequately fixed smear, with pale swollen nuclei, lacking nuclear detail (Pap, ×630).

cytological grounds to have carcinoma *in situ*. Correlation was observed in 120 cases (94.5%). In six cases the histology revealed invasive carcinoma, and dysplasia was reported in one case. In a number of cases in this study, correlation was achieved only when additional sections were cut from the block. Although the specificity of cytology for the diagnosis of high-grade cervical lesions (CIN 3) is impressive this is not the case for low-grade lesions. Recent reports show that cervical biopsy of women with a cytodiagnosis of borderline dyskariosis or mild dyskariosis is not infrequently negative, raising a number of issues with regard to correct management of women with a low-grade cervical smear (Jones *et al.*, 1996).

Another measurement of the validity of a smear report is sensitivity. Sensitivity measures the ability of a test to detect correctly diseased individuals in the population under investigation. The sensitivity of cervical screening is difficult to determine, as false-negative smears are not immediately apparent and have a variety of causes. False-negative reports may be due to sampling failure or to screening error. A small proportion of errors may be due to failure of the tumour to exfoliate. This was clearly demonstrated by Yule (1973) who recalled 14 437 women taking part in a population screening programme for a second smear within 3 months of taking the first. Twenty-nine additional positive cases were detected on the second examination giving a false-negative rate of 16.9%. In 16 of the 29 cases, review of the initial smear revealed dyskariotic cells that had been missed at the first screening. The remaining 13 cases were probably due to sampling error or failure of the tumour to exfoliate.

A potentially important cause of erroneous reporting has been described by Anderson and Hartley (1979). These authors pointed out that follow-up cytology and colposcopy of patients

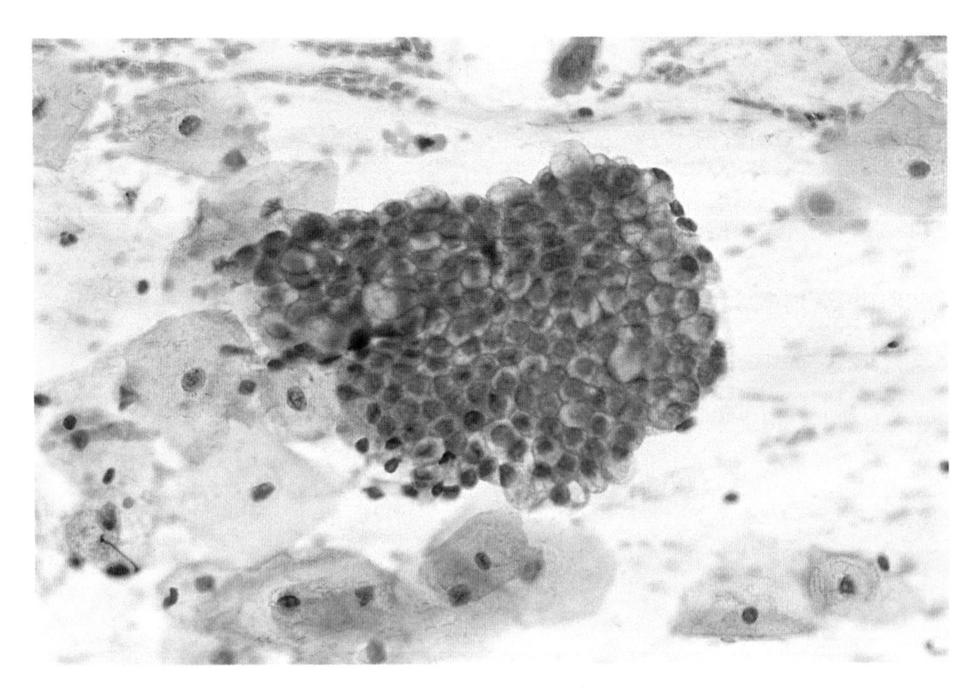

Figure 2.5 Well-displayed superficial squames and endocervical cells in an adequate cervical smear (Pap, ×200).

whose CIN has been treated by electrocautery or laser therapy may be misleading if the tissue has not been destroyed to a sufficient depth (estimated at 4 mm) to eliminate all the tumour. Residual CIN in the depth of the crypts may be viable, giving rise to an invasive cancer. This would not be apparent from the smear until the tumour breaks through the surface.

Errors of cytodiagnosis may also occur after cone biopsy due to cervical stenosis. In such cases, the smear report may not reflect residual or recurrent neoplasia in the cervix.

Attention has recently been drawn to the biopsy findings in women who have been reported to have 'inflammatory smears'. These are smears which contain numerous polymorphs and show cellular atypia reflecting degenerative or regenerative changes consistent with acute or chronic cervicitis. Unfortunately many of the cellular features which characterize inflammatory atypia are similar to those found in CIN and the cytological distinction between the two conditions can be quite difficult. Moreover, colposcopic biopsy has shown that a small proportion (less than 10%) of women with inflammatory atypia have a focus of CIN (Ostergard and Condos, 1971; Slater and Duke, 1985; Walker *et al.*, 1986; Kaminski *et al.*, 1989). This has led to a call for colposcopy of all women with an 'inflammatory smear'. A more conservative management policy is recommended for this group of women in the UK in an effort to avoid inducing alarm and anxiety in these women whose risk of cancer is extremely low.

We have also noted that smears taken from an area of leukoplakia may not always reflect the presence of concomitant CIN. The smears contain numerous highly keratinized anucleate cells and dyskeratocytes; dyskaryotic cells from the deeper layers of the epithelium are not always seen (Figure 9.21(a,b)). Colposcopy is advised in these cases.

There have been many attempts to measure

the false-negative rate for cervical cytology and the rates vary according to the method used. It has been reported to be as low as 7% and as high as 30%. Further discussion of the factors which affect the sensitivity and specificity of cervical screening is contained in Chapter 17 together with information about the steps that need to be taken to ensure that the screening programme is as efficient and effective as possible.

2.5 SPECIAL TECHNIQUES

The indication for special tests on cervical smears are few indeed. However, when they are needed they represent a special problem for the pathologist. Unless the investigation is anticipated, the test has to be performed directly on the original Papanicolaou-stained slide. Even if special staining is anticipated and multiple smears are taken from the cervix to provide additional material, it is unlikely that the smears will be identical. An alternative approach is to prepare a suspension of cells from the cervical scrape and make multiple cytocentrifuge preparations for immunocytochemical staining.

Within these limitations, transmission electron microscopy and immunocytochemical techniques have been used successfully to resolve specific diagnostic problems of cervical cytology. Scanning electron microscopy, fluorescence microscopy and cytometry (including DNA ploidy studies) have also been used but have a limited role at present in diagnostic cytology or cervical screening (Murphy *et al.*, 1975; Ito and Kudo, 1982). DNA hybridization of cervical scrapes has assumed an important place as a research tool for the detection of human papillomaviruses in the cervix. It has been suggested that it might replace the Papanicolaou smear as a screening test for cervical cancer (Walboomers *et al.*, 1994; Cuzick *et al.*, 1995). There are also those who advocate its use for the triage of women with low-grade cervical lesions (borderline dyskariosis or mild dyskariosis) to identify those at greatest risk of cervical cancer (see also p. 250).

2.5.1 Transmission electron microscopy (TEM)

Transmission electron microscopy is occasionally indicated to verify the presence of virus infection of the cervix. Cells infected with herpes simplex virus or human papillomaviruses may resemble tumour cells and the differential diagnosis can be quite difficult. A technique has been developed for retrospective studies of mounted cytological material, whereby fixed and stained cells identified in the Papanicolaou smear by light microscopy can be re-examined in the electron microscope for the presence of virus particles. A full description of this technique has been published by Smith and Coleman (1984). Briefly, a photograph is taken of the abnormal cells in the Papanicolaou smear. The cell is marked with a ring, drawn with a diamond pencil on the under surface of the slide. The coverslip is removed and the smear post-fixed in osmium tetroxide and restained with uranyl acetate.

After dehydration, the cell is embedded in epoxy resin by placing an open-ended polyethylene embedding capsule in position over the cell (Figure 2.6), and filling it with resin which is allowed to polymerize. The slide and the capsule are then immersed in liquid nitrogen. The difference in the coefficient of expansion of glass and epoxy resin is such that the cells or tissue cleave off the glass leaving the cell embedded in the surface of the resin block. The block is trimmed and ultrathin sections of the cell cut for examination in the electron microscope.

The major use for this technique has been to establish the diagnosis of virus infection retrospectively, when fresh specimens and conventional viral studies are not available. Virus particles are clearly visible after reprocessing of the alcohol-fixed Papanicolaou-stained smear by this method (see Figures 7.31 and 7.43). The different virus types can be

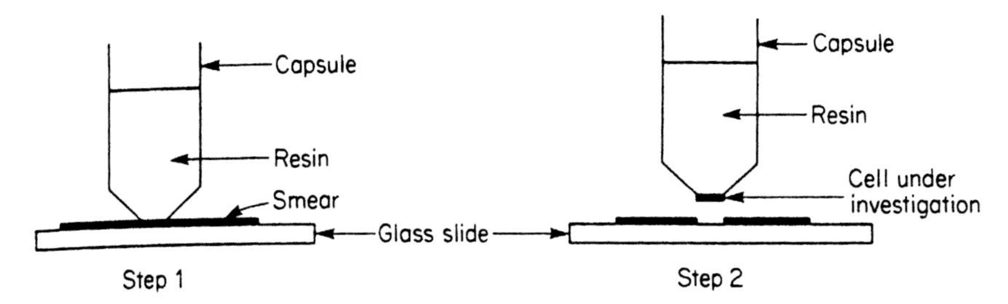

Figure 2.6 Preparation of cytological or histological specimen for electron microscopy. An open-ended embedding capsule is fixed over the cell under investigation (step 1). The slide and embedding capsule are plunged into liquid nitrogen. The cells cleave off the glass and are now embedded on the surface of the resin block (step 2).

recognized by their morphology and particle size (Smith and Coleman, 1984). Unfortunately, the fine structure of the cytoplasm and nucleus of the cell is poorly preserved by alcohol fixation. If ultrastructural studies of cell organelles are required, the freshly taken smear must be fixed in phosphate-buffered glutaraldehyde with a pH of 7.4 for 15 minutes before routine staining by the Papanicolaou method and processing for electron microscopy.

2.5.2 Immunocytochemical staining

Immunocytochemical techniques provide the pathologist with the opportunity to study simultaneously the morphology of a cell and

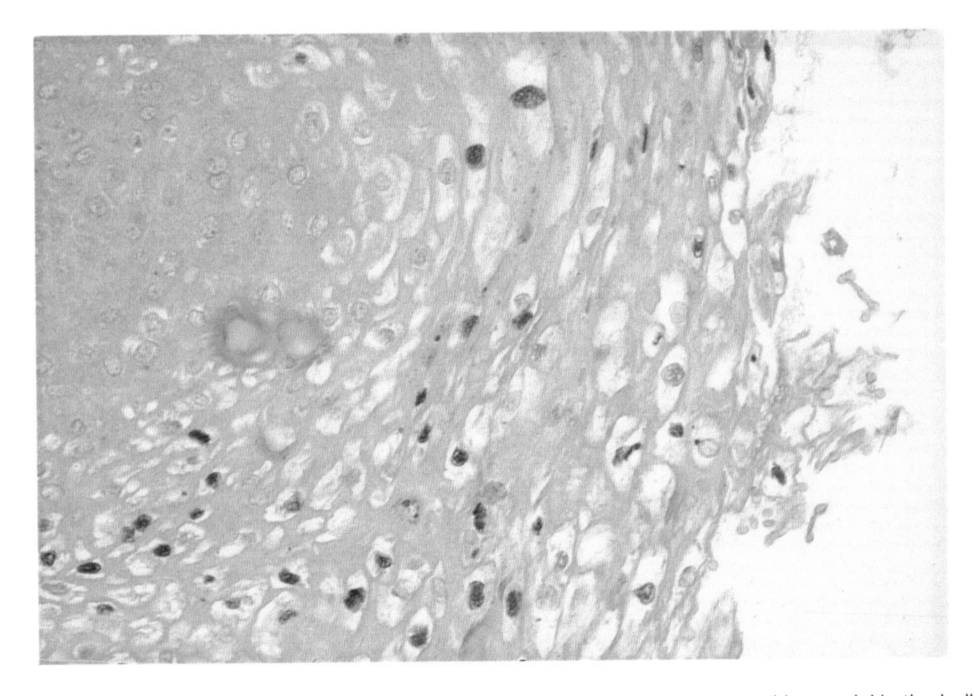

Figure 2.7 Cervical biopsy stained for papillomavirus antigen. Note the antigen-positive nuclei in the koilocytes in the surface layers of the epithelium (alkaline phosphatase, ×630).

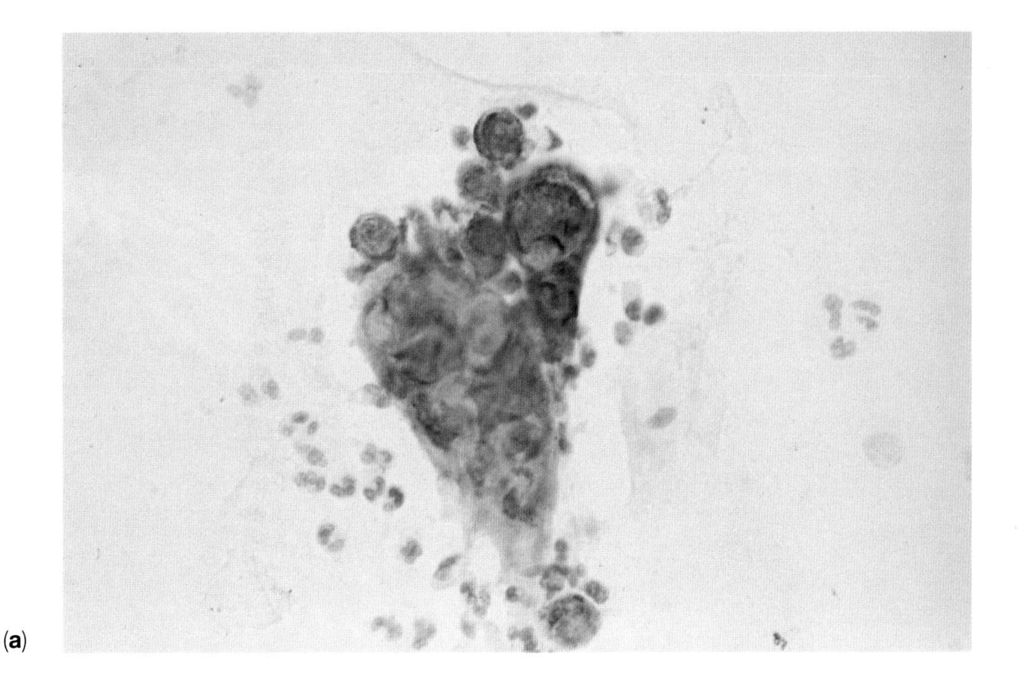

(**a**)

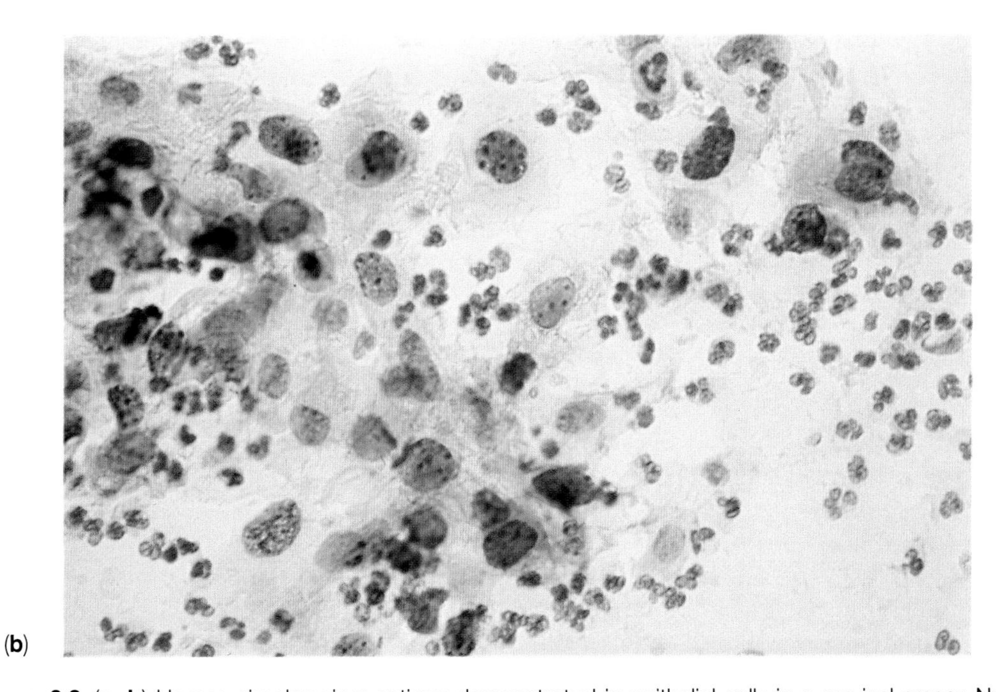

(**b**)

Figure 2.8 (**a, b**) Herpes simplex virus antigen demonstrated in epithelial cells in a cervical smear. Note the infected cells in (**b**) do not show the characteristic multi-nucleation which make diagnosis possible from a Papanicolaou-stained smear (Courtesy of Dr Ghiradini, Ospedale Mirandola, Italy) (both alkaline phosphatase, ×630).

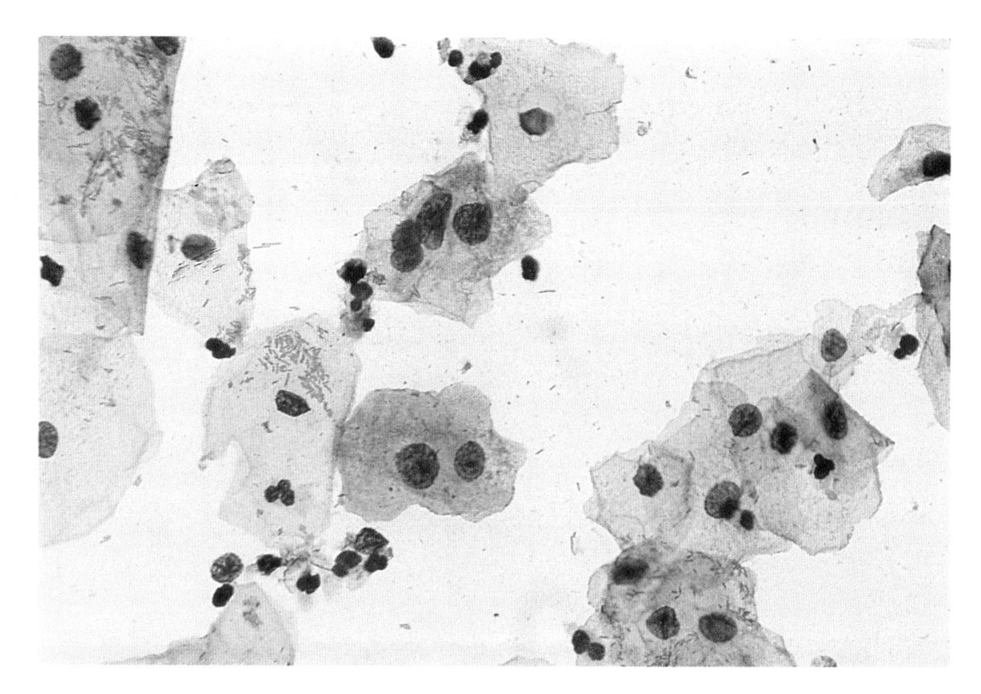

Figure 2.9 Cervical smear stained for epithelial membrane antigen. Note positive staining of cytoplasm of dyskaryotic cell (alkaline phosphatase, ×400).

aspects of its function and antigenic expression. Immunofluorescence and immunoperoxidase techniques have been used to detect the presence of viral antigen in epithelial cells of cervical smears. Human papillomavirus (Figure 2.7) and herpes simplex virus (Figure 2.8 (a,b)) have both been demonstrated in cervical smears and biopsies by immunocytochemical methods (Gupta *et al.*, 1983; Walker *et al.*, 1983).

Immunoperoxidase techniques have also been applied to smears to study the distribution of tumour marker substances (Figure 2.9) to determine whether staining for tumour antigens would be of value in discriminating between benign and malignant cells. Were a marker specific for malignancy to be found, we would have a powerful new tool which could provide the basis for automated analysis of smears (Fray *et al.*, 1984; Moncrieff *et al.*, 1984; Valkova *et al.*, 1984). Early studies of the distribution of epithelial markers for human milk fat globulin, cytokeratin and carcinoembryonic antigen have been unrewarding as far as grading cervical neoplasia is concerned as the antigens are expressed by neoplastic as well as non-neoplastic cells (Fray *et al.*, 1984; Moncreiff *et al.*, 1984). More recent studies of the distribution of markers of cell proliferation and cell adhesion molecules have been more rewarding. Vessey *et al.* (1995) showed that loss of E cadherin expression may represent one of the abnormalities underlying loss of cell polarity and differentiation which characterize CIN and invasive cervical cancer. Integrin expression which is concerned with epithelial cell/basement membrane interaction has been shown to alter in cervical neoplasia (Hughes *et al.*, 1994).

Many of the problems associated with the immunoperoxidase staining of cervical smears are similar to those encountered when staining histological sections. As with histology, the technique demands careful selection of controls, absolute specificity of antisera and complete blocking of endogenous enzyme activity (Heyderman, 1979). One problem

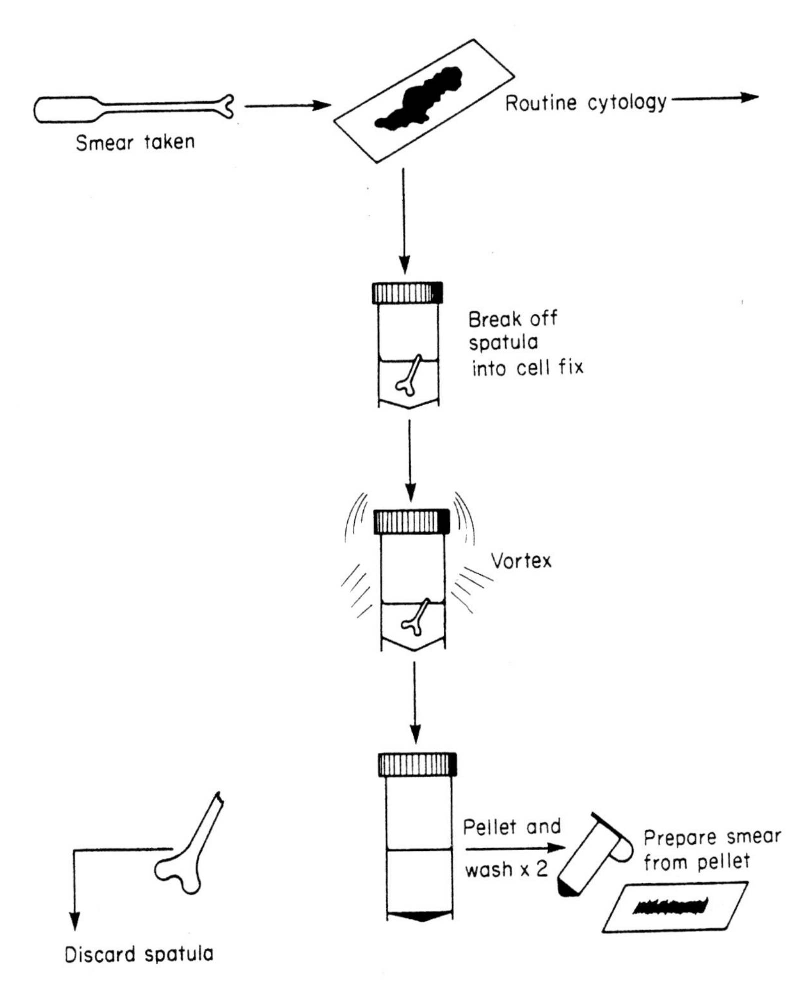

Figure 2.10 Diagram to show stages in the preparation of a cervical scrape for immunocytochemical staining.

however, is peculiar to cervical smears. In these specimens, the presence of mucus, pus and cell clusters results in non-specific staining. The problem can be resolved by preparing a single cell suspension of the material (Figure 2.10). After the scrape is taken, the tip of the Ayre spatula is broken off and placed in a cell fix solution. The clumps of cells in the solution are broken up by repeated aspiration through a fine needle and syringe. The suspension is centrifuged and monolayers prepared from the deposits which can be stained by standard immunocytochemical techniques (Moncrieff *et al.*, 1984; Valkova *et al.*, 1984).

2.5.3 DNA hybridization techniques

The interest in viral oncogenic agents particularly the human papillomavirus has provided a stimulus for the development of DNA hybridization techniques which can be used on cervical smears. Several hybridization techniques have been applied to cervical scrape material including Southern blotting, dot blotting (filter hybridization) and *in situ* hybridization. The principles of all three approaches are fundamentally the same in that they exploit the ability of complementary DNA sequences in single strands of DNA or RNA to pair with

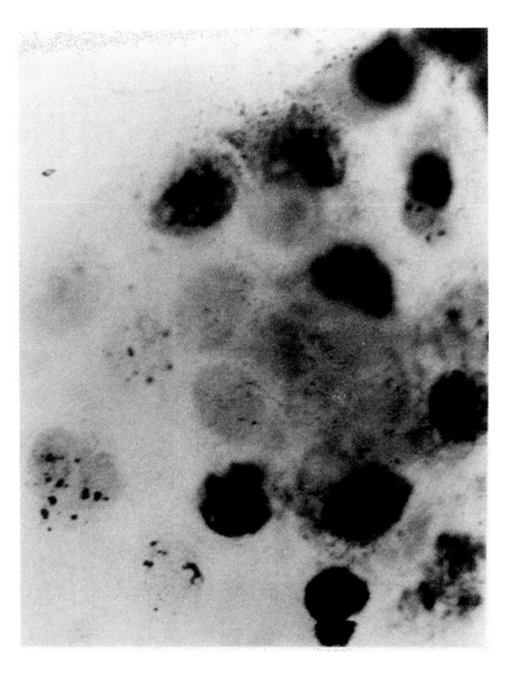
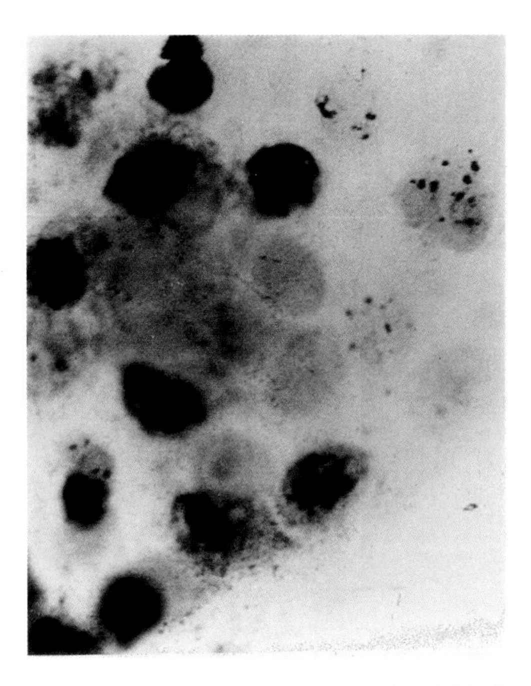

Figure 2.11 *In situ* hybridization of a cervical smear using probes for HPV DNA. Diffuse staining of nuclei indicates episomal virus, punctate signal indicates integration of virus. (Photographs courtesy of Dr C.S. Herrington, University of Liverpool).

each other to form a double helix. However, the techniques differ in that *in situ* hybridization preserves the morphology of the cells so that cellular localization of the signal is possible, whereas with Southern blotting or dot blotting it is not. Although the sensitivity of *in situ* hybridization is less than that of the dot blot technique, sensitivity can be enhanced by prior amplification of the DNA sequences under investigation in the smear using the polymerase chain reaction (PCR).

In practice, single strands of DNA or RNA are labelled with a radioactive, enzymatic or fluorescent label and used to probe the DNA or RNA of the test sample which has also been rendered single stranded by a denaturation process. The presence of complementary sequences in the test sample can be recognized by the demonstration of a labelled product (or hybrid) in this test sample.

DNA hybridization techniques have had a limited application in cytology. They have been used mainly for the detection of human papillomavirus DNA. *In situ* hybridization techniques have been applied to cervical smears by Herrington to demonstrate integrated and episomal HPVDNA in cervical epithelial cells (Cooper *et al.*, 1991) (Figure 2.11). Dot blotting has been used to determine virus prevalence in the cervix using DNA extracted from cells which have been harvested from the cervical scrape (Wickenden *et al.* 1987) (Figure 2.12). In view of the known association of HPV with cervical carcinoma, Cuzick *et al.* (1995) have suggested that HPV testing be used routinely to complement cervical cytology as a method of identifying women at risk of cervical cancer. A critical evaluation of this approach is needed (see also Section 10.6.4).

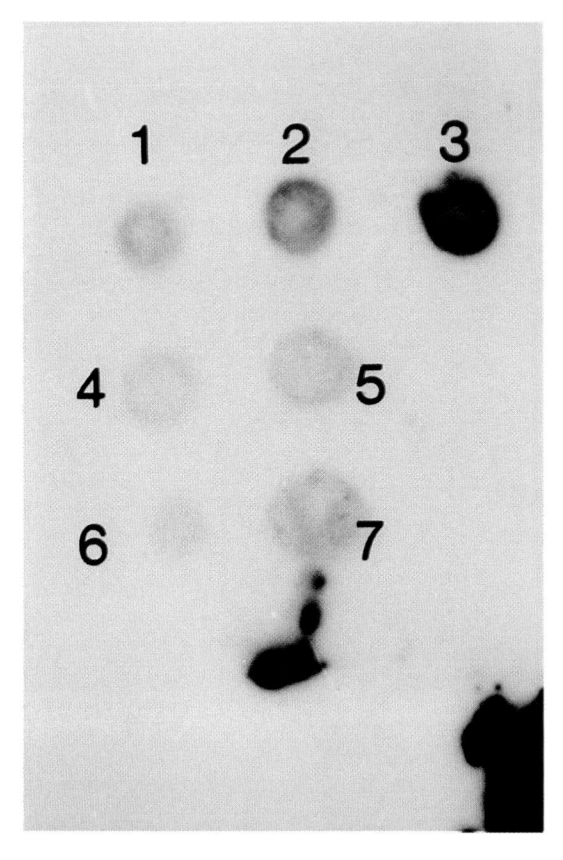

Figure 2.12 Dot blot hybridization. DNA has been extracted from cells in a cervical scrape and blotted onto a filter. Hybridization with probe for HPV DNA reveals positive signals in dot blots numbers 2 and 3.

REFERENCES

Anderson, M. C. and Hartley, R. B. (1979) Cervical crypt involvement by intraepithelial neoplasia. *Obstet. Gynaecol. Surv.*, **34**, 852–853.

Bales, C. E. and Durfee, G. R. (1979) Cytological techniques, in *Diagnostic Cytology and its Histopathologic Bases*, 3rd edn (ed. L. Koss), J.B. Lippincott Company, Philadelphia, Chapter 33, pp. 1211–1220.

Beckman, A. M., Myerson, D., Daling, J., Kiviat, N. B., Fenoglio, C. M. and McDougall, J. K. (1985) Detection and localization of human papillomavirus DNA in human genital condylomas by *in situ* hybridisation with biotinylated probes. *J. Med. Virol.*, **16**, 265–273.

Boon, M. E. and Suurmeijer, A. J. H. (1991) Papanicolaou staining method. In *The Pap Smear.* Coulomb Press, Leyden, Chapter 16.

Buntinx, F., Boon, M. E., Beck, S., Knotterus, J. A. and Essed, G. G. M. (1991) Comparison of cytobrush sampling, spatula sampling and cytobrush–spatula sampling of the uterine cervix. *Acta Cytologica*, **35**, 64–68.

Buntinx, F. and Brouwers, M. (1996) Relation between sampling device and detection of abnormality in cervical smears: a meta-analysis of randomized and quasi randomized studies. *Br. Med. J.*, **313**, 1285–1290.

Coleman, D., Day, N., Douglas, G., Farmery, E., Lynge, E., Phillip, J. and Segnan, N. (1993) European Guidelines for quality assurance in cervical cancer screening. *Eur. J. Cancer*, **29A**(suppl. 4), S1–S38.

Cooper, K., Herrington, C. S., Stickland, J. F., Evans, M. F. and Gee, J. O'D. (1991) Episomal and integrated HPV in cervical neoplasia demonstrated by non isotopic *in situ* hybridisation. *J. Clin. Pathol.*, **44**, 990–996.

Covell, J. L. and Frierson, H. F. (1992) Intraepithelial neoplasia mimicking microinvasive squamous cell carcinoma in endocervical brushings. *Diagnostic Cytopathol.*, **8**, 18–22.

Cuzick, J., Szarewski, A., Terry, G. *et al.* (1995) Human papilloma virus testing in primary screening, *Lancet*, **345**, 1533–1536.

Evans, D. M. D., Hudson, E. A., Brown, C. L., Boddington, M. M., Hughes, H. C., Mackenzie, E. F. D. and Marshall, T. (1986) Terminology in gynaecological cytopathology: report of the working party of the British Society for Clinical Cytology. *J. Clin. Pathol.*, **39**, 933–944.

Evans, D. M. D., Hudson, E. A., Brown, C. L., Boddington, M. M., Hughes, H. C. and Mackenzie, E. (1987) Management of women with abnormal cervical smears: Supplement to Terminology in Gynaecological Cytopathology. *J. Clin. Pathol.*, **40**, 530–531.

Fray, R. E., Husain, O. A. N, To, A. C. W, Watts, K. C. *et al.* (1984) The value of immunohistochemical markers in the diagnosis of cervical neoplasia. *Br. J. Obstet. Gynaecol.*, **91**, 1037–1041.

Gupta, J. W., Gupta, P. K., Shah, K. V. and Kelly, D. P. (1983) Distribution of human papillomavirus antigen in cervicovaginal smears and cervical tissues. *Int. J. Gynecol. Pathol.*, **2**, 160–170.

Heyderman, E. A. (1979) Immunoperoxidase techniques in histopathology: applications, methods and controls. *J. Clin. Pathol.*, **32**, 971–978.

Hughes, D. E., Rebello, G. and Al Nafussi, A. (1994) Integrin expression in squamous neoplasia of the cervix. *J. Pathol.*, **173**, 97–104.

Hughes, R. G., Haddad, N. G., Smart, G. E., Colquhoun, M., McGoogan, E., McIntyre, C. O., and Prescott, R. J. (1992) The cytological detection of persistent cervical intraepithelial neoplasia after local ablative treatment: a comparison of sampling devices. *Br. J. Obstet. Gynaecol.*, **99**, 498–502.

Ito, E. and Kudo, R. (1982) Scanning electron microscopy of normal cells, dyskaryotic cells and malignant cells exfoliated from the uterine cervix. *Acta Cytol.*, **26**, 457–465.

Jones, M. H., Jenkins, D. and Singer, A. (1996) Regular audit of colposcopic biopsies in women with a mildly dyskaryotic or borderline cervical smear results in fewer cases of CIN III. *Cytopathology*, **7**(1), 17–24.

Kaminski, P. F., Stevens, C. W. and Wheelock, J. B. (1989) Squamous atypia on cytology. The influence of age. *J. Reprod. Med.*, **34**, 617–620.

MacGregor, J. E., Fraser, M. E. and Mann, E. M. F. (1966) The cytopipette in the diagnosis of early cervical carcinoma. *Lancet*, **i**, 252–256.

Mitchell, H. and Medley, G. (1993) Cellular differences between true negative and false negative Papanicolaou smears. *Cytopathology*, **4**, 285–290.

Moncrieff, D., Ormerod, M. G. and Coleman, D. V. (1984) Tumour marker studies of cervical smears: potential for automation. *Acta Cytol.*, **28**, 407–410.

Murphy, J. F., Allen, J. M., Jordan, J. A. and William, A. E. (1975) Scanning electron microscopy of normal and abnormal exfoliated cervical squamous cells. *Br. J. Obstet. Gynaecol.*, **82**, 44–51.

National Coordinating Network (National cervical cancer screening programme) British Society for Clinical Cytology and Royal College of Pathologists working party (1994) Borderline nuclear changes in cervical smears: guidelines on their recognition and managment. *J. Clin. Pathol.*, **47**, 481–492.

National Health Service Cervical Screening Programme, Publ. No. 3. (1996) Quality Assurance Guidelines for the Cervical Programme: Report of a working Party convened by the NHS Cervical Screening Programme.

Ostergard, D. R. and Gondos, B. (1971) The incidence of false negative cervical cytology as determined by colposcopically directed biopsies. *Acta Cytol.*, **15**, 292–293.

Papanicolaou, G. N. and Traut, H. F. (1943) *Diagnosis of Uterine Cancer by the Vaginal Smear.* The Commonwealth Fund, New York.

Papanicolaou, G. N. (1954) *Atlas of Exfoliative Cytology.* Harvard University Press, Cambridge, Mass.

Pistofidis, G. A., House, F. R., Moir Shepherd, J. and Vale, J. C. (1984) The multispatula: a new dimension in sampling the cervix. *Lancet*, **i**, 1214–1215.

Proctor, D. T. (1989) Staining techniques, in *Clinical Cytotechnology* (eds D. V. Coleman and P. A. Chapman), Butterworths, London, Chapter 6.

Reagan, J. W. and Hamonic, M. J. (1958) The cellular pathology of carcinoma *in situ*. *Cancer*, **9**, 385–402.

Scherman, M. E., Schiffman, M. H., Erozan, Y. S., Wacholder, S. and Kurman, R. K. (1992) The Bethesda System. A proposal for reporting abnormal cervical smears based on the reproducibility of cytopathologic diagnoses. *Arch. Pathol. Lab. Med.*, **110**, 1155–1158.

Schettino, F., Sideri, M., Cangini, L., Candiani, M., Zannoni, E., Maggi, R. and Ferrari, A. (1993) Endocervical detection of CIN. Cytobrush vs cotton. *Eur. J. Gynaecol. Oncol.*, **14**, 234–236.

Selvaggi, S. M. (1989) Spatula/cytobrush vs spatula/cotton swab detection of cervical condylomatous lesions. *J. Reprod. Med.*, **34**, 629–633.

Shorr, E. (1941) New techniques for staining vaginal smears: III, a simple differential stain. *Science*, **94**, 545–546.

Slater, D. and Duke, E. (1985) Cervical smear policy. *Lancet*, **ii**, 1305.

Smith, J. and Coleman, D. V. (1984) Electron microscopy of cells showing viral cytopathic effects in Papanicolaou smears. *Acta Cytol.*, **27**, 605–613.

The Bethesda System for reporting cervical/vaginal cytologic diagnoses (1993) *Acta Cytologica*, **37**, 115-124.

Trimbos, J. B. and Arentz, N. P. W. (1986) The efficiency of the cytobrush versus the cotton swab for the collection of endocervical cells in cervical smears. *Acta Cytol.*, **30**, 261–263.

Valkova, B., Ormerod, M. G., Moncrieff, D. and Coleman, D. V. (1984) Epithelial membrane antigen in cells from the uterine cervix. *J. Clin. Pathol.*, **37**, 984–989.

Van Erp, E. J., Blaschek-Lut, C. H., Arentz, N. P. and Trimbos, J. B. (1988) Performance of the cytobrush in patients at risk of cervical pathology: does it add anything to the wooden spatula? *Eur. J. Gynaecol. Oncol.*, **9**, 456–460.

Vessey, C. J., Wilding, J., Folarin, N. *et al.* (1995) Altered expression and function of E cadherin in cervical intraepithelial neoplasia and invasive squamous cell carcinoma. *J. Pathol.*, **176**, 151–159.

Walboomers, J. M. M., Husman. A., Van den Brule, A. J. C., Snijders, P. J. F. and Meijner, C. J. L. M. (1994) Detection of genital human papillomavirus infections: critical review of methods and prevalence studies in relation to cervical cancer. In *Human Papillomavirus Biology and Immunology* (eds P. L. Slen and M. Stanley), Oxford University Press, Oxford, pp. 41–71.

Walker, E., Dodgson, J. and Duncan, I. (1986) Does mild atypia in a cervical smear warrant further investigation? *Lancet*, **ii**, 672–673.

Walker, P. G., Singer, A., Dyson, J. L., Shah, D. V., To, A. A. and Coleman, D. V. (1983) The prevalence of human papillomavirus antigen in patients with cervical intraepithelial neoplasia. *Br. J. Cancer*, **48**, 99–101.

Weitzman, G. A., Koritonen, M. O., Reeves K. O., Irwin, J. F., Carlter, T. S. and Kaufman, R. H. (1988) Endocervical brush cytology. An alternative to endocervical curettage. *J. Reprod. Med.*, **33**, 677–683.

Wickenden, C., Malcolm, A. D. B., Byrne, M., Smith, C., Anderson, M. C. and Coleman, D. V. (1987) Prevalence of HPVDNA and viral copy number in cervical scrapes from women with normal and abnormal cervices. *J. Pathol.*, **153**, 127–135.

Wilbanks, G. D., Ikomi, E., Prado, R. B. and Richart, R. M. (1968) An evaluation of a one-slide cervical cytology method for the detection of cervical intraepithelial neoplasia. *Acta Cytol.*, **12**, 157–158.

Wolfendale, M. R., Howe-Guest, R., Usherwood, M. McD. and Draper, G. J. (1987) Controlled trial of a new cervical spatula. *Br. Med. J.*, **294**, 33–35.

Yule, R. (1973) The prevention of cancer of the cervix by cytological screening of the population, in *Cancer of the Uterine Cervix* (ed. E. C. Easson), W. B. Saunders, London, pp. 11–25.

3

DEFINING THE LESION FOR BIOPSY

Intraepithelial neoplasia rarely produces changes in the cervix which are visible to the naked eye. Thus, speculum examination of a woman with an abnormal smear may not reveal the source of the abnormal cells and the gynaecologist will have difficulty in deciding where to target the biopsy. For this reason, several techniques have been developed which can be used to identify the site for biopsy. These include Schiller's iodine test, colposcopy and membrane cytology and cervicography.

3.1 SCHILLER'S IODINE TEST

This test is defined by the reaction obtained when an iodine solution is painted on to the portio vaginalis of the cervix and the vaginal mucosa. Normal squamous epithelium, which is glycogen-rich, stains brown, whereas epithelium that contains little or no glycogen does not take up the stain. Cancerous areas are usually devoid of glycogen and do not express the stain so that the surgeon can direct the biopsy appropriately to the iodine-negative areas.

Schiller (1933) recommended this test as a useful screen for cancer in a busy out-patient setting. Unfortunately, experience has shown that the staining patterns are not a totally reliable guide, as immature metaplastic epithelium atrophic and glandular epithelia do not contain glycogen and do not stain with iodine whereas a few cancers which are rich in glycogen stain a deep brown. Richart (1964) has shown that 72% of carcinomas *in situ* and 57% of dysplasias do not take up the stain when iodine is applied to the cervix.

The solutions traditionally available for the test are: (1) Schiller's solution (1 g of pure iodine, 2 g of potassium chloride and 300 g of water), and (2) Lugol's solution (5% iodine and 10% potassium chloride in water). Alcoholic solution of iodine must not be used as this spoils the tissue for histological study. Other stains have been applied to the cervix in an attempt to develop a more accurate test (toluidine blue, acridine orange fluorescence and tetracycline fluorescence) but none has proved to be of substantial value in clinical practice.

American gynaecologists frequently refer to the epithelium they have biopsied as Schiller-positive, indicating that the tissue does not react with iodine. The terminology used to describe the test differs in Europe and the United States and is described in Table 3.1.

Table 3.1 Description of results of Schiller test

Country	*Iodine staining of glycogen-positive cells*	*No glycogen present No staining seen*
America	Schiller negative	Schiller positive
Europe	Iodine positive	Iodine negative
Recommended terminology	Schiller dark	Schiller light

3.2 COLPOSCOPY

The colposcope consists of a stereoscopic binocular microscope, equipped with a central illuminating device of high intensity which is used to inspect the cervix, vagina and vulva. When the instrument is positioned, 6–7 cm from the introitus, the portio vaginalis of the cervix and the outer third of the endocervical canal can be seen and examined at a magnification of 4 to 40 times (Figure 3.1).

Colposcopic examination of a woman with an abnormal smear is directed primarily at the transformation zone (Section 5.2), as it is within this region that the neoplastic change occurs. A particular advantage of colposcopic investigation is that biopsies can be taken from abnormal epithelium under direct vision without anaesthesia so that the histological diagnosis can be made and the site and extent of the lesion defined before treatment.

It has been observed that abnormalities of the transformation zone can be identified more easily in the colposcope if the cervix is first gently moistened with a saline solution to render it translucent. This procedure reveals subtle changes in contour, colour and opacity of the surface and in the subepithelial vasculature, which may escape the naked eye. Correlative studies have shown that the visual images observed with the colposcope are determined by the structure of the epithelium and the subepithelial vascular network. Colposcopic diagnosis is based on careful evaluation of these patterns and an experienced colposcopist can make a distinction between normal and abnormal epithelium with a high degree of accuracy.

An alternative method of revealing subtle changes in the topography and angioarchitecture of the cervical epithelium is to apply a dilute solution (5%) of acetic acid to the cervix. This technique was first described by Hinselmann (1925) and is widely used in the United Kingdom (Figure 3.2 (a,b)). The pattern of reflected light from areas of cervix affected by CIN changes in the presence of acetic acid and the abnormal epithelium appears opaque or 'aceto-white'. Biopsy studies have shown that aceto-white epithelium is found in areas where the cells are densely packed and the nuclei are crowded and it is therefore a useful guide to the presence of CIN. It is not, however, specific for neoplasia as aceto-white epithelium may

Table 3.2 Abnormal colposcopic findings

1. Atypical transformation zone
   (a) Whiteness of epithelium after application of acetic acid
   (b) Changes in angioarchitecture (e.g. mosaic pattern, punctuation, irregular vessels)
   (c) Keratosis (syn: leucoplakia)
   (d) Slight surface irregularity
2. Changes suggestive of invasive cancer
3. Other colposcopic findings, e.g. condyloma, erosio vera, trichomoniasis, herpes genitalis

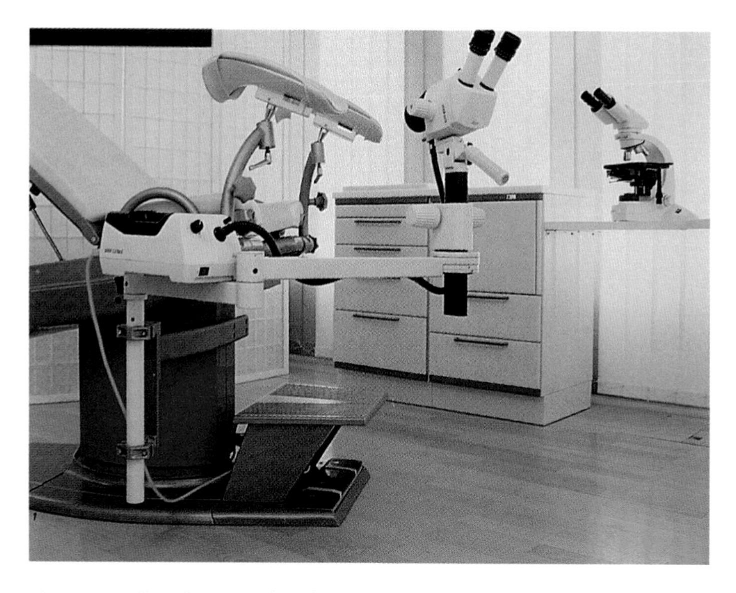

Figure 3.1 Leica colposcope fitted to a patient's chair in the colposcopy clinic.

also be seen in an area of immature metaplasia or in association with regenerating epithelium, human papillomavirus infection, a congenital transformation zone or cervicitis (Anderson *et al.*, 1992a).

Some of the colposcopic changes which indicate an underlying epithelial atypia in the region of the transformation zone are listed in Table 3.2. To the trained colposcopist certain characteristics of aceto-white epithelium indicate a significant lesion. The intensity of the whiteness and its duration, as well as the sharpness of the borders of the lesion, are all taken into account by the colposcopist when assessing a lesion. Moreover, the severity of the epithelial abnormality can be assessed from the regularity of the vascular pattern, the calibre of the vessels and the distance between them. The latter increases from about 200 μm in CIN 1 to 450–500 μm in CIN 3 (Anderson *et al.*, 1992b). The vascular proliferation associated with CIN may produce a punctate pattern if the new vessels are viewed 'end on' by the colposcopist. If they branch at the surface they will form a mosaic pattern (Figure 3.3 (a,b)).

Colposcopists have developed a system of grading the changes in the atypical transformation zone to assist with their interpretation and prediction of the histological findings. Generally speaking, the higher the grade the more severe the epithelial abnormality. The colposcopic appearances can also be recorded by colour photographs of the magnified cervix. Colpophotography provides an accurate record of the changes in the transformation zone after the application of acetic acid and is of great value for teaching purposes, for correlating with membrane cytology (see below) and for surveillance of patients over the course of several visits. Colpophotography is not generally considered necessary for routine clinical records (Anderson *et al.*, 1992c). However, it must be emphasized that the predicted colposcopic diagnosis and the histological diagnosis do not always correspond and biopsy is essential for a definitive diagnosis.

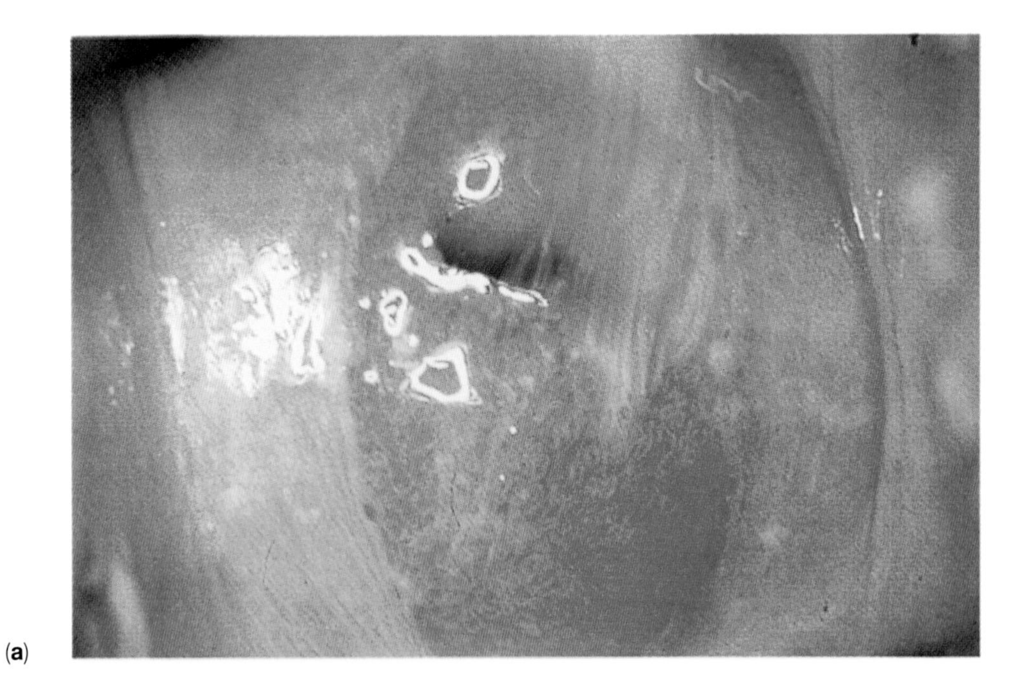

(**a**)

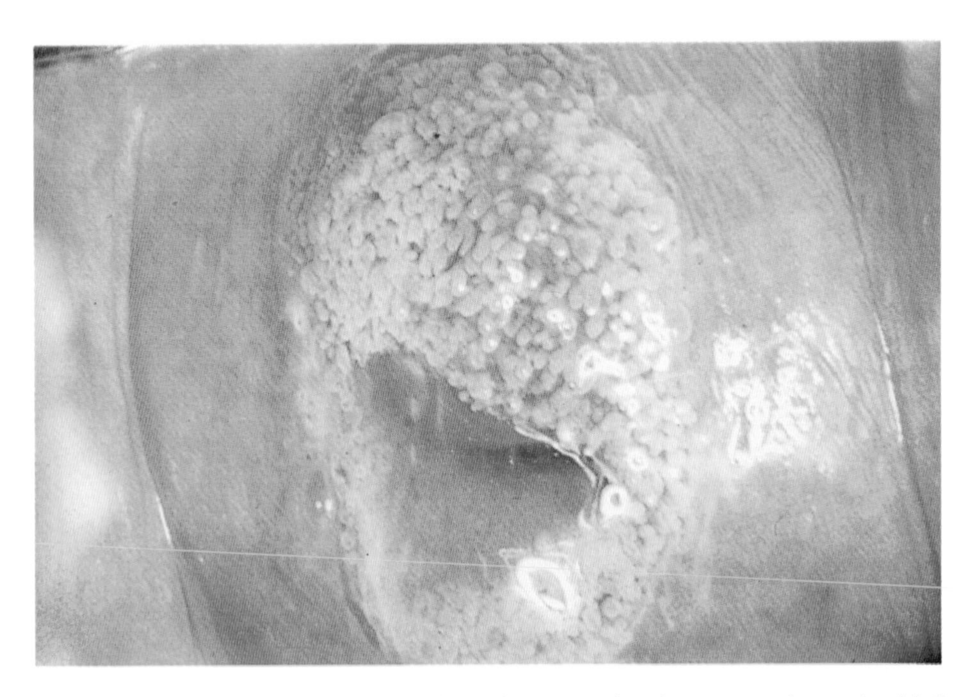

(**b**)

Figure 3.2 (**a**, **b**) The normal cervix, as seen through a colposcope, showing an area of ectropion (**a**), the villous structure and extent of which are demonstrated more clearly after application of 3% acetic acid (**b**).

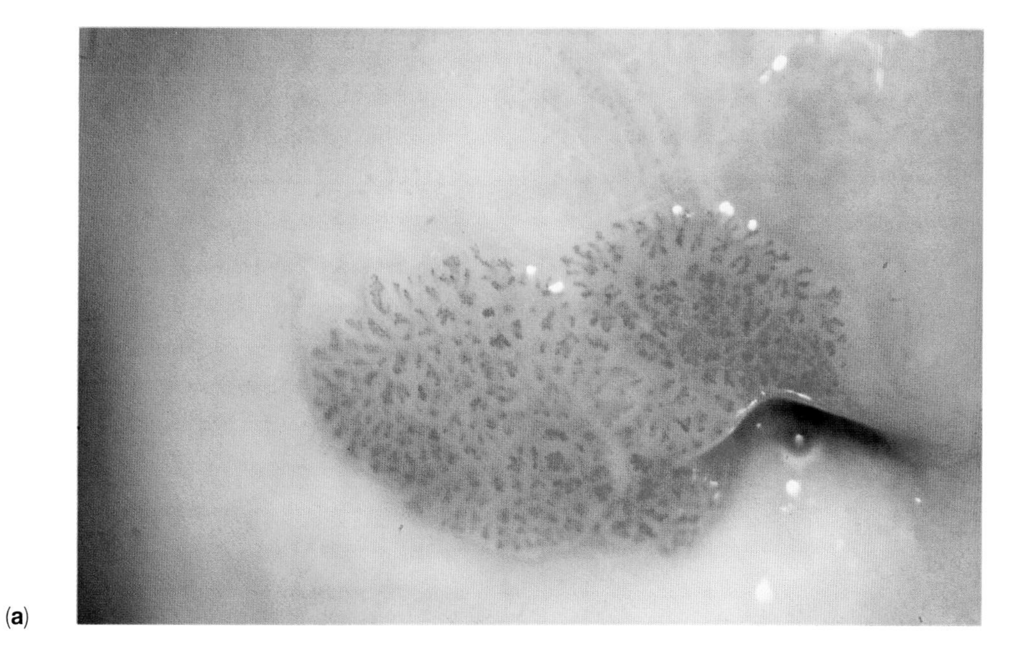

(**a**)

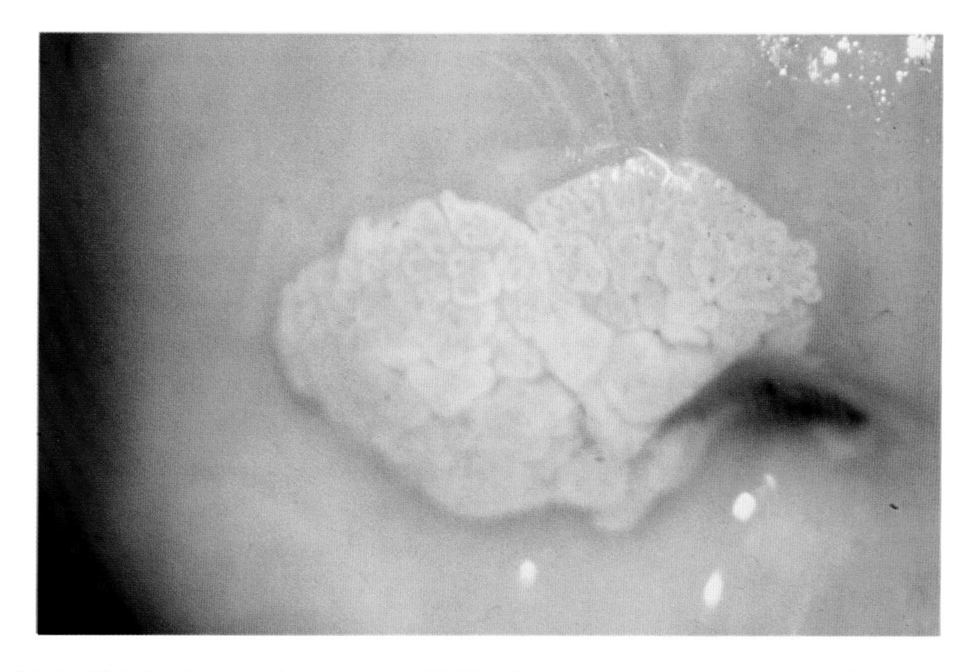

(**b**)

Figure 3.3 (**a, b**) Colpophotographs of an area of CIN before and after the application of 3% acetic acid. Note the presence of punctation and mosaicism of the vessels consistent with neoplastic change of the epithelium.

Although the main use of colposcopy is to identify the source of the abnormal cells in women presenting with an abnormal smear, it has other roles to play in the management of these patients. Colposcopic assessment of the extent of the CIN lesion ensures that the most appropriate treatment, e.g. cone biopsy or laser ablation, is adopted for each patient. Moreover, colposcopy is of value in distinguishing between CIN and benign conditions such as infections of the cervix which can, on occasion, give rise to a false-positive cervical report. It is also useful for the follow-up of women who have received treatment for CIN or invasive cancer and for monitoring women who have been exposed to diethyl stilboestrol *in utero*. Thus, colposcopy permits the treatment of each patient according to her individual needs and, for this reason, it has an important role to play in gynaecological practice. Its main limitation is its inability to bring lesions situated high in the endocervical canal into view. When such lesions are suspected, an endocervical brush or swab and/or curettage (Section 2.1.3) or cone biopsy are required to provide diagnostic cells or tissue. An endocervical membrane cytology applicator has also been devised (Evans, 1969).

3.3 MEMBRANE CYTOLOGY

This technique is used to map the cervix in order to locate a focus of CIN in the woman with an abnormal smear (Evans, 1967; Evans *et al.*, 1969). Although it has been largely superseded by colposcopy it still has a role to play when colposcopy is not available. It is also a method of obtaining sheets of cells on a membrane, e.g. for scanning electron microscopy (Murphy *et al.*, 1975), and DNA *in situ* hybridization (Wickenden *et al.*, 1987).

The basic principles of the technique are evident from Figure 3.4. A suitable membrane e.g. Macrofol (Bayer), Nucleopore (GEC) or a Millipore filter is applied briefly but firmly to the ectocervix using a simple applicator. The membrane is suitably marked to ensure correct orientation once the specimen is taken. The superficial cells of the cervical mucosa adhere to the membrane, thus providing a mirror image of their distribution on the cervix. By marking the centre of the anterior lip of the cervix with gentian violet immediately before collection a mark is produced on the membrane which provides a guide to the position of the external os. Aerosol spray fixative is applied to the membrane while it is still moist. The approximate site of the external os is marked by puncturing the posterocentral point of the gentian violet mark with a needle, as the mark itself disappears during processing. The membrane is stained by the Papanicolaou schedule while enclosed in a large tissue-processing capsule. It is then mounted like a tissue section on a large glass slide. A two-dimensional map can be prepared as a guide to the site and extent of the CIN. The information assists the surgeon to determine the extent of the cone biopsy required to

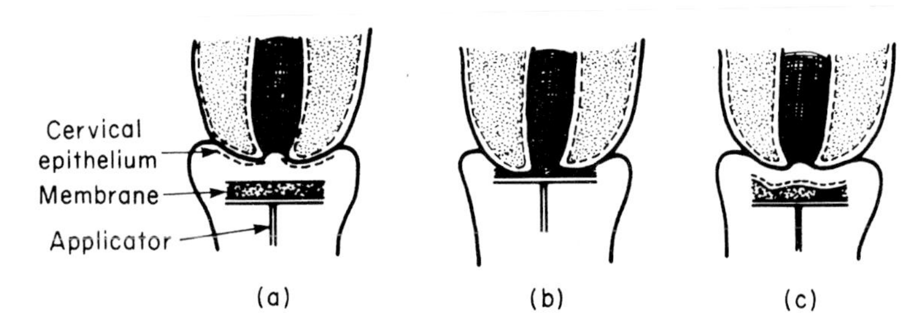

Figure 3.4 Membrane cytology. The dotted line in (**a**) represents surface epithelial cells. As the membrane is pressed against the cervix (as in (**b**)) many of the surface cells adhere to it, as represented in (**c**).

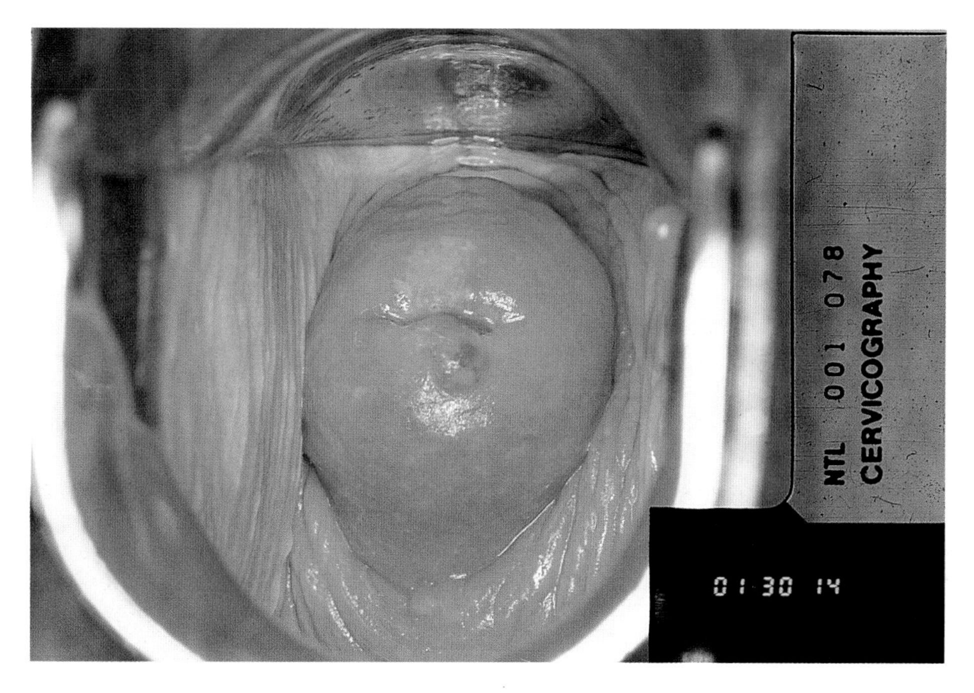

Figure 3.5 Photographic image of the cervix (cervicogram) showing wart virus changes (courtesy of Prof. A. Singer).

achieve complete removal of the lesion. The technique can be modified to map the endocervix (Evans, 1969) and the squamocolumnar junction.

3.4 CERVICOGRAPHY

This technique was developed by Stafl in 1981. It involves the use of a camera (a cerviscope) and the preparation of a permanent record of the appearances of the ectocervix in the form of a photographic image or cervicogram (Figure 3.5). The cervicograms are projected onto a screen and viewed at a greatly enlarged magnification by a trained medical practitioner or nurse. Since the taking of the cervicogram and its interpretation is much less demanding in terms of technical skill and specialist time than conventional colposcopy it has been suggested that cervicography should be used as a screening tool for CIN.

Several studies have been carried out to determine the place of cervicography in clinical practice (Szarewski *et al.*, 1991; Ceccini *et al.*, 1993; Coibion *et al.*, 1994). Most studies show that although cervicography is a sensitive method of detecting cervical neoplasia its specificity is low and overdiagnosis of CIN is common.

REFERENCES

Anderson, M. C., Jordan, J. A., Morse, A. R. and Sharp, F. (1992a) *A Text and Atlas of Integrated Colposcopy*. Chapman & Hall, London, p. 82.

Anderson, M. C., Jordan, J. A, Morse, A. R. and Sharp, F. (1992b) *A Text and Atlas of Integrated Colposcopy*. Chapman & Hall, London, p. 84.

Anderson, M. C., Jordan, J. A., Morse, A. R. and Sharp, F. (1992c) *A Text and Atlas of Integrated Colposcopy*. Chapman & Hall, London, p. 67.

Cecchini, S., Iossa, A., Bonardi, R. *et al.* (1992) Evaluation of the sensitivity of cervicography on a consecutive colposcopic series. *Tumori*, **78**, 211–213.

Cecchini, S., Bonardi, R., Mazotta, A. *et al.* (1993) Testing cervicography and cerviscopy as screening tests for cervical cancer. *Tumori*, **79**, 22–25.

Coibion, M., Autier, P., Vandam, P. *et al.* (1994) Is there a role for cervicography in the detection of premalignant lesions of the cervix? *Br. J. Cancer*, **70**, 1–4.

Evans, D. M. D. (1967) Cytological method for assessing the topography of neoplastic cells on the ectocervix. *Lancet*, **ii**, 972.

Evans, D. M. D. (1969) Cytological method for assessing the topography of neoplastic change in the endocervical canal. *Lancet*, **ii**, 574–575.

Evans, D. M. D., McCormack, J., Sanerkin, N. G., Ponsford, P. and Jones, J. (1969) A membrane cytologic technique for assessing the extent of ectocervical carcinoma using the Tenovus applicator. *Acta Cytol.*, **13**, 119–121.

Hinselmann, H. (1925) Verbesserung der Inspectionsmoglichkeiten von Vulva, Vagina und Portio. *Munch. Med. Wochenschr*, **72**, 1733.

Murphy, J. F., Allen, J. M., Jordan, J. A. and Williams, A. E. (1975) Scanning electron microscopy of normal and abnormal exfoliated cervical squamous cells. *Br. J. Obstet. Gynaecol.*, **82**, 44–51.

Richart, R. M. (1964) The correlation of Schiller positive areas on the exposed portion of the cervix with intraepithelial neoplasia. *Am. J. Obstet. Gynecol.*, **90**, 697–701.

Schiller, W. (1933) Early diagnosis of carcinoma of the cervix. *Surg. Gynecol. Obstet.*, **56**, 210–222.

Stafl, A. (1981) Cervicography: a new method for cervical cancer detection. *Am. J. Obstet. Gynecol.*, **139**, 815–825.

Szarewski, A., Cuzick, J., Edwards, R., Butler, B. and Singer, A. (1991) The use of cervicography in a primary screening service. *Br. J. Obstet. Gynaecol.*, **98**, 313–317.

Wickenden, C., Malcolm, A. D. B. and Coleman, D. V. (1987) DNA hybridisation of cervical tissue, in *CRC Critical Reviews in Clinical Laboratory Science, vol. 25*, CRC Press Inc, USA, pp. 1–18.

4

HISTOLOGICAL METHODS AND TYPES OF BIOPSY

4.1 TYPES OF SPECIMEN AND FIXATION

Punch biopsy, wedge (segmental) biopsy, cone biopsy, endocervical curettage, laser biopsy and large loop excision of the transformation zone (LLETZ) provide the main types of cervical biopsy specimen for light microscopy. Their collection, fixation and labelling is in the hands of the gynaecologist; therefore it is extremely important for the clinician to have a clear understanding of the requirements of the pathologist in order to provide a specimen that is of maximum diagnostic value. Carefully excised, rapidly fixed, correctly orientated and clearly labelled specimens offer the best opportunity for accurate diagnosis.

The fixatives used for cervical biopsies vary in different laboratories; the fixative most frequently used is buffered 10% formol saline. Bouin's fluid (picric acid, formalin and acetic acid) or its alcohol variant is to be recommended as the fixative of choice at least for small biopsies. The tendency for Bouin's fluid to render the tissue rather brittle for sectioning is outweighed by the excellent preservation of nuclear detail achieved. It allows the pathologist to make valid comparisons between the appearance of the epithelial cells in the histological sections and the exfoliated cells in the cervical smears. Care must be taken to ensure that the specimen is fully immersed in plenty of fixative. Small specimens (punch biopsy, endocervical curettage and wedge biopsy specimens) require fixation for a minimum of 3 hours. Cone biopsies, being larger, must be fixed for 24 hours.

4.2 IDENTIFICATION OF SPECIMEN

It is essential that each specimen is correctly labelled at the time of operation, not only with the patient's details but with the type of specimen clearly indicated. A description of the results of the Schiller test and the cytologic and colposcopic findings should also be included. Endocervical curettage specimens and punch biopsy specimens must be collected into separate containers. Similarly, if endometrial curettage is carried out in the course of the procedure this must also be kept separate from the cervical tissue. If a quadrant biopsy is taken, reference should be made to the cervical segment from which the biopsy comes (right anterior, left posterior lip, etc.). Ideally, every punch biopsy sample should be numbered and submitted in a separate container with the site of origin clearly indicated on an accompanying diagram. This will allow the gynaecologist to identify the area of maximum pathological change when assessing the case for treatment.

4.3 PUNCH BIOPSY

The practice of taking a punch biopsy in the course of colposcopic examination is an integral part of the investigation of the patient with an abnormal cervical smear. Indeed, one of the cardinal rules of colposcopy is that treatment must not be started until a biopsy has been taken. Provided the limits of the lesion are clearly visible and a biopsy is taken from the area of maximum colposcopic change, accurate diagnosis of the underlying cervical pathology can be made in about 75% of cases (Cinel *et al.*, 1990).

Punch biopsy specimens are usually small, difficult to orientate and easily crushed. Careful choice of biopsy forceps will keep these problems to a minimum. The use of Kevorkian biopsy forceps minimizes crush artefacts. These provide a fragment of tissue 3–5 mm across which is large enough to be correctly orientated. A biopsy of this size should include subepithelial connective tissue 3–5 mm in depth so that the possibility of microinvasion can be assessed. It is essential for all the tissue received to be processed no matter how small the fragments.

Both overdiagnosis and underdiagnosis can occur unless care is taken in the assessment of this material. To reduce this risk, sections should be taken from each specimen at several levels as a routine. Overdiagnosis may result from oblique or tangential sectioning causing a false impression that invasive cancer is present (Figure 11.14). This can be minimized if the mucosal surface of the biopsy is identified before embedding and the biopsy is aligned correctly.

To facilitate orientation the punch biopsy specimen should be fixed in Bouin's fluid for about 2 hours (prolonged fixation causes the tissue to harden excessively). After removal from the fixative the specimen will be yellow in colour and show some curling of the epithelium on the smooth convex surface. It should be bisected at right angles to the epithelium and the two halves embedded in wax, the cut surface facing downwards; this is the face from which the first microtome sections are taken. It is usual to take section at three levels from each block.

The assistance of the gynaecologist can be sought if there are persistent problems in the laboratory in identifying the epithelial surface of the specimen. The gynaecologist can be asked to orientate the biopsy at the time of collection by placing the specimen on a fragment of paper towel, connective tissue surface downwards. The specimen will adhere to the towel which can be placed upside down in the fixative.

4.3.1 The punch biopsy report

This should include a note of the size (diameter) and the number of fragments received. In addition, the following histological information should be provided:

- A description of the specimen, i.e. whether it contains stroma, surface epithelium and gland crypts. If the specimen is considered to be unsatisfactory (e.g. crushed or devoid of epithelium or stroma) this must be stated.
- A description of the pathological findings. If CIN is diagnosed, the presence of crypt involvement should be recorded.
- If the nature of the specimen makes it impossible to exclude invasive cancer, a comment to this effect should be included. Evidence of stromal invasion must be reported if present (a punch biopsy is usually too small for a diagnosis of microinvasion to be made). A record should be kept of the number of sections examined.

4.4 WEDGE BIOPSY (SEGMENTAL BIOPSY)

Punch biopsies have largely replaced the segmental or wedge biopsy of the precolposcopic era. However, there is still a place for this time-honoured method of biopsy. Although biopsy is to be avoided in pregnancy as far as possible, it is essential in cases where microinvasive or invasive cancer is suspected. In this situation, wedge biopsy is preferred to punch biopsy as it is easier to control bleeding by suturing the raw edges of the biopsied area. The squamocolumnar junction should be included in the biopsy. A good wedge biopsy specimen should present no problems of orientation. It should be cut into thin slices and all embedded. The principles of reporting are essentially similar to those of punch biopsy (Section 4.3.1).

4.5 ENDOCERVICAL CURETTAGE

Endocervical curettage is a common method of biopsy in North America and is included as part of the routine colposcopic investigation of older women (Dinh *et al.*, 1989). In these patients, the squamocolumnar junction tends to recede into the endocervical canal with age and neoplastic changes at the site may be missed by routine cytology or colposcopy. It is also used when glandular abnormality is suspected cytologically, or when cone biopsy suggests incomplete removal, especially if hysterectomy is contraindicated. Endocervical curettings are characteristically scanty, being composed of thin tissue fragments, blood and mucus. If a bulky specimen is received, adenocarcinoma or endocervical polyp should be suspected. Alternatively, endometrial curettings may be included in the specimen.

4.6 CONE BIOPSY

Cold knife conization of the cervix under general anaesthesia was introduced into gynaecological practice in the 1950s as a practical alternative to random punch biopsy and hysterectomy for the diagnosis and treatment of the patient with an abnormal smear. In recent years colposcopy and colposcopically directed punch biopsy have largely replaced conization for the diagnosis of CIN as the procedures are safer for the patient and the risk of complications is minimal. The trend toward conservative management of patients with CIN has meant that local destruction of the lesion by laser, cryosurgery, diathermy or cautery has partially replaced conization as the treatment mode for these lesions, particularly for low-grade lesions of limited extent. In consequence, the number of cone biopsies submitted for histological examination in many laboratories has fallen considerably. Large loop excision of the transformation zone (LLETZ) is currently the preferred alternative to cold knife cone biopsy for the diagnosis and treatment of CIN.

The indications for cone biopsy or large loop excision (LLETZ), procedures that play an important part in the management of CIN 3, are as follows:

1. Cases where colposcopy reveals that the lesion involves a large area of the ectocervix.
2. Lesions extending into the endocervical canal beyond the limits of colposcopy.
3. Cases where microinvasive cancer is suspected on cytology, colposcopy or histological study of the colposcopic biopsy.
4. Cone biopsy is mandatory when examination of the cervical smear reveals evidence of CIN 3 and colposcopy is negative.

Pregnancy is usually considered to be a contraindication to cone biopsy because of the risk of miscarriage or premature delivery. However this may not always be the case. Klein *et al.* (1991), in a study of 28 women who were pregnant at the time of cone biopsy, found that in the 13 women who had conization in the first 16 weeks of pregnancy, delivery occurred after the 39th week whereas in the 15 women who had conization in the 17th week or later, the mean date of delivery was 34 weeks. They concluded that conization after the 16th week was more likely to result in preterm delivery.

The biopsy specimen is a cone-shaped portion of tissue, the base formed by the

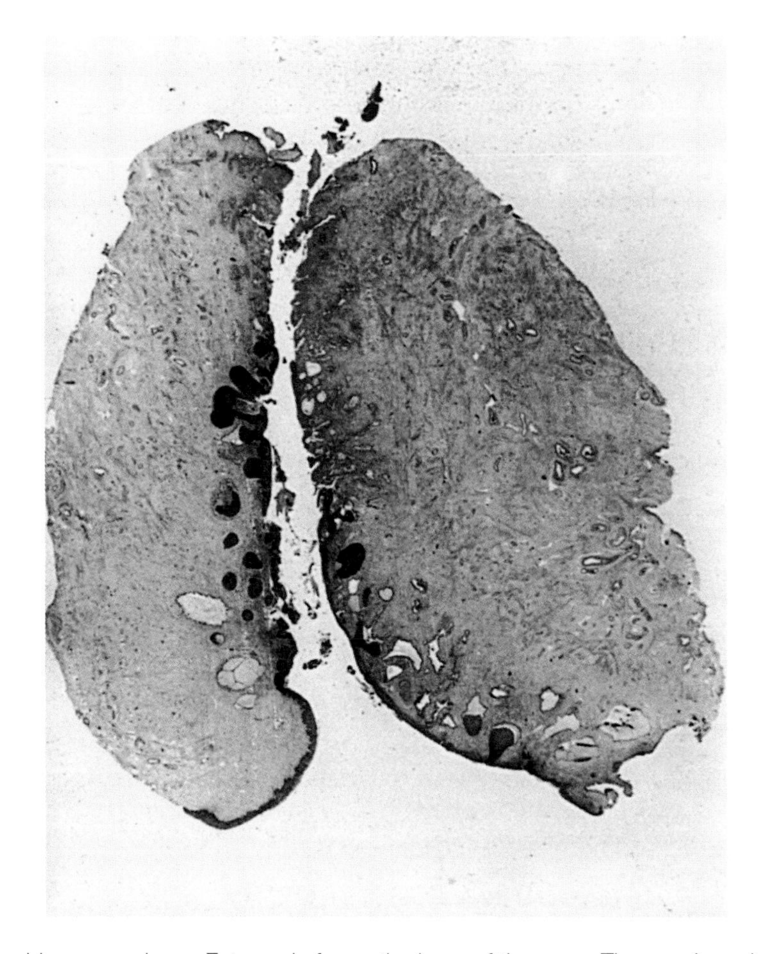

Figure 4.1 Cone biopsy specimen. Ectocervix forms the base of the cone. The canal can be seen extending to the apex with CIN 3 involving its lining and crypts. The proximal but not the distal margins appear free from neoplastic change (H & E, ×2.5).

ectocervix and the apex by the tissues adjoining the endocervical canal (Figure 4.1). The dimensions of the cone will vary from case to case. If the extent and location of the lesion is not known, the cone should include a large area of ectocervix so that the base of the cone may measure 2.5 cm in diameter and the depth may be as much as 3 cm. The dimensions required can be gauged more accurately if the extent of the lesion is defined by colposcopy before the operation. A shallow cone is quite acceptable providing the surgeon is sure that all the abnormal epithelium has been excised. Indeed, the shallower the cone the less the risk of postoperative complication for the patient (Jordan, 1980).

The surgeon should not attempt to open the cone or to remove mucus or blood from the surface as abnormal epithelium is very delicate and may be detached. A suture should be inserted to mark the mid point of the anterior lip in every case, so that the colposcopic and histological findings can be correlated. The specimen should then be placed, intact, in sufficient volume of fixative to cover it completely. No attempt should be made to cut the specimen until it is fixed, as the epithelium is very fragile. Overnight fixation is usually

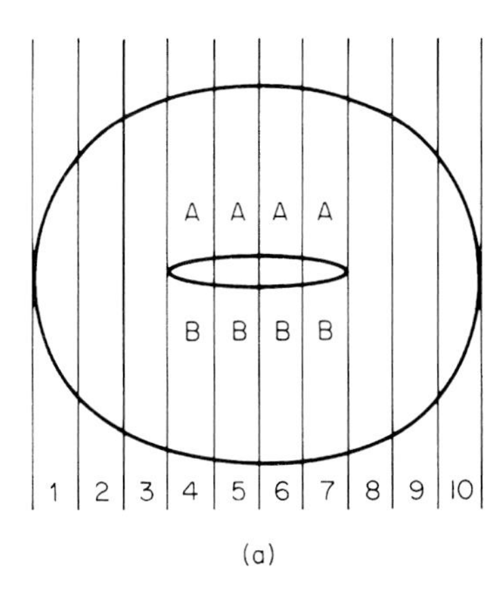

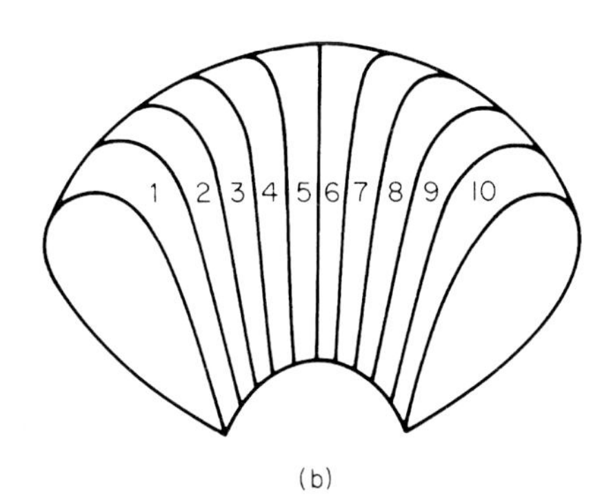

Figure 4.2 Methods of cutting cones: (**a**) step sectioning; (**b**) radial slicing.

required for larger cones, 6 hours of fixation are adequate for smaller ones. Before cutting the cone, the pathologist should record the diameter and depth of the specimen and note the appearance of the surface.

The aim of the histopathologist is to sample the entire area of squamous and columnar epithelium in order to determine the nature and extent of the lesion and whether it has been completely removed. There are several ways of doing this. In our laboratories, the procedure is to step-section all the tissue in sequence (Figure 4.2(a)). The cone is placed with its posterior cut surface downwards on the cutting up board and its ectocervical surface facing the pathologist. No pins are needed but a sharp scalpel is essential. The anterior lip is identified from the position of the suture and parallel blocks 4 mm thick are cut, working from the right hand side of the specimen to the left. The position of each block is recorded on a diagram or preferably a Polaroid photograph of the cervix. The latter is the preferred method because of the variation in size and shape of cone biopsies. As each specimen is cut, it is assigned a number and placed face down in a separate numbered cassette, taking care that in every case the same cut edge is facing downwards in the cassette. Marking one side of the block with Indian ink ensures that the microtomist always knows which surface to cut. Some sections especially those near the cervical os may be too large for the cassette and need to be bisected. Each half should be separately identified, e.g. 3A and 3B. At the site of the external os, the cervix divides naturally so that two fragments of the tissue are obtained at each step, one from the anterior lip and one from the posterior lip of the cervix. Again, each fragment should be separately identified.

The technique described above is but one method of preparing a cone biopsy for examination. Alternative techniques, such as opening up the cone biopsy specimen and slicing radially have also been described (Figure 4.2(b)). This approach may be useful if the cone has already been opened by the gynaecologists. In some centres, the excision margin of the cone is identified by painting it with a 10% solution of silver nitrate; care is needed to avoid discolouring the epithelial surface.

4.6.1 The cone biopsy report

This should include:

1. Dimensions of cone and number of blocks.
2. A description of the pathological findings. If there is CIN or glandular neoplasia the extent of the lesion and its severity should be described.

3. A comment on whether excision is complete or incomplete, i.e. whether the neoplastic lesion extends to the margin of excision.
4. A statement of the presence or absence of invasive or microinvasive cancer. If the latter is present, its depth and lateral extent should be measured (see Chapter 11).
5. A note as to whether the histological findings correlate with the colposcopic or cytological findings.
6. A report on whether there is any lymphatic permeation by cancerous tissue should be included.

Histological assessment of the cone biopsy needs special care and attention. On average there are 7–14 blocks to examine. Several levels from each block, taken at 4 µm intervals, should be examined. Each section is taken in such a way that an area of squamous epithelium of the ectocervix and a portion of the excision margin is present in every one of them. If pieces are missing from a section, deeper sections should be cut. Sections taken from the area of external os will also contain endocervical epithelium. The habit of mentally visualising a three-dimensional map of the histological findings should be cultivated when reporting on cone biopsy specimens.

The aim of cone biopsy is complete removal of the lesion. This is more likely to be achieved if the limits of the lesion have been defined by colposcopy. Recurrent CIN after incomplete removal occurs in 40% of cases. It must be remembered that if a lesion is present in either of the end blocks (Nos 1 or 10) it is necessary to cut sections from the side of the block corresponding to the excision margin of the specimen (i.e. reverse the block) before reporting on whether removal is complete.

Since the majority of cervical biopsies are taken in response to abnormal cytology reports, the pathologist should make a practice of comparing the Papanicolaou smear with the biopsy whenever possible. If there is a discrepancy between the cytological, colposcopic and histological findings the reason for this must be determined. In some instances the smear will have been processed in another laboratory. In these cases, every effort should be made to obtain the pertinent slide or slides for review and at the very least, the cytology report should be traced.

4.7 LASER BIOPSY

Cylindrical biopsy of the cervix using the carbon dioxide laser has been advocated as an alternative to surgical conization (Dorsey and Diggs, 1979; Wright *et al.*, 1984). As with cold knife conization, the procedure is both therapeutic and diagnostic. A distinct advantage of laser biopsy is that it can be carried out as an out-patient procedure and the healing process leaves the squamocolumnar junction intact. Difficulty in the histopathological interpretation has been described due to thermal change (Figure 4.3), especially in smaller cones. When Howell *et al.* (1991) reviewed laser excisional cone biopsy of 77 women with cytological evidence of CIN, they found that 39% were negative for CIN, 36% showed gross epithelial denudation, 13% were unsuitable for diagnosis due to coagulation artefact, and in 14% the same artefact greatly impaired interpretation of the extent of the lesion at the margins of the biopsy. They concluded that such cones were unsuitable for histological assessment.

However, an alternative view was expressed by Tabor and Berget (1990) in a 5-year study of 425 women who had laser or cold knife cone biopsy of the cervix. They concluded that cold knife and laser conization had similar success rates in the management of CIN. Although several other groups including Vergote *et al.* (1992) have similarly demonstrated the value of laser conization, the expense of the equipment and the stringent safety precautions required have effectively limited its popularity as a diagnostic technique and laser excisional biopsy has been superseded by large loop excision of the transformation zone in many centres.

(**a**)

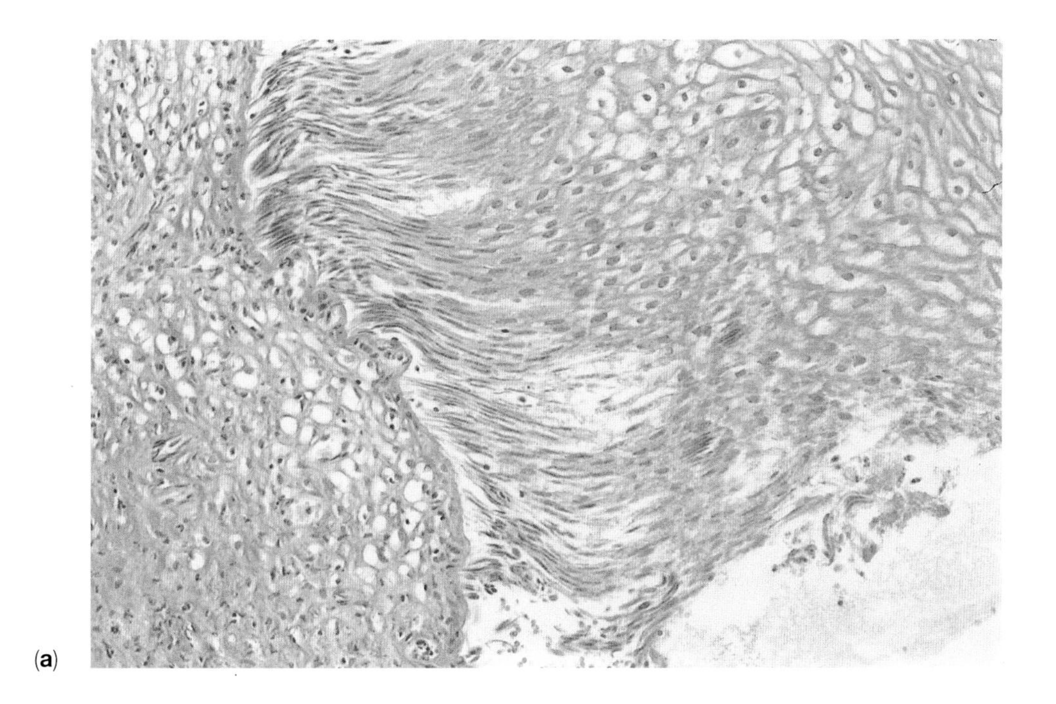

(**b**)

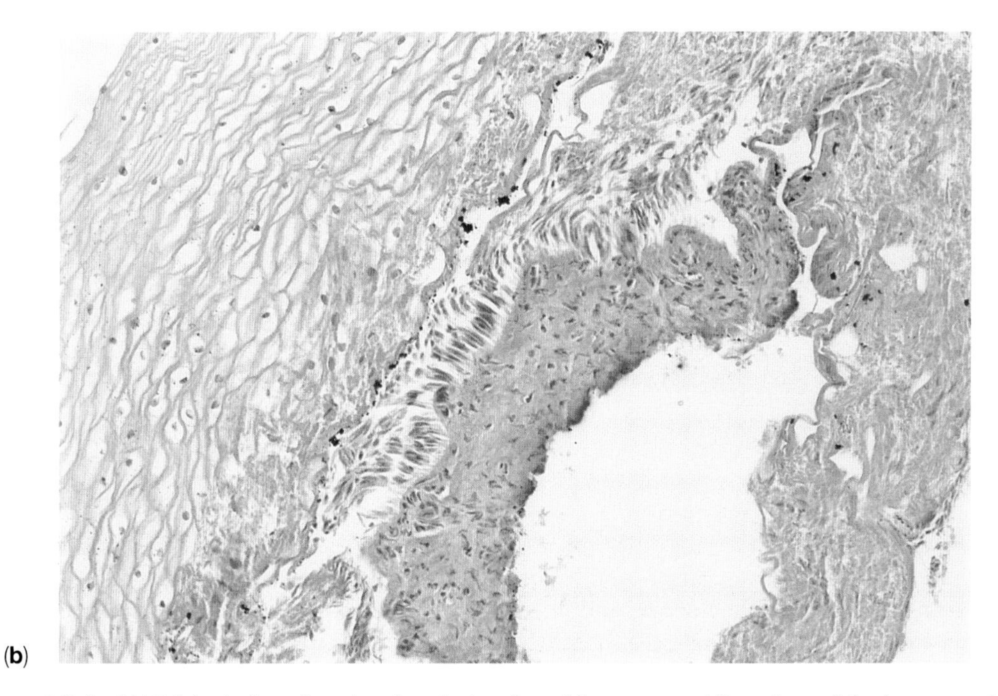

Figure 4.3 (**a, b**) Histological section showing destruction of tissue around the edges of the laser cone biopsy (H&E, ×2.5).

(c)
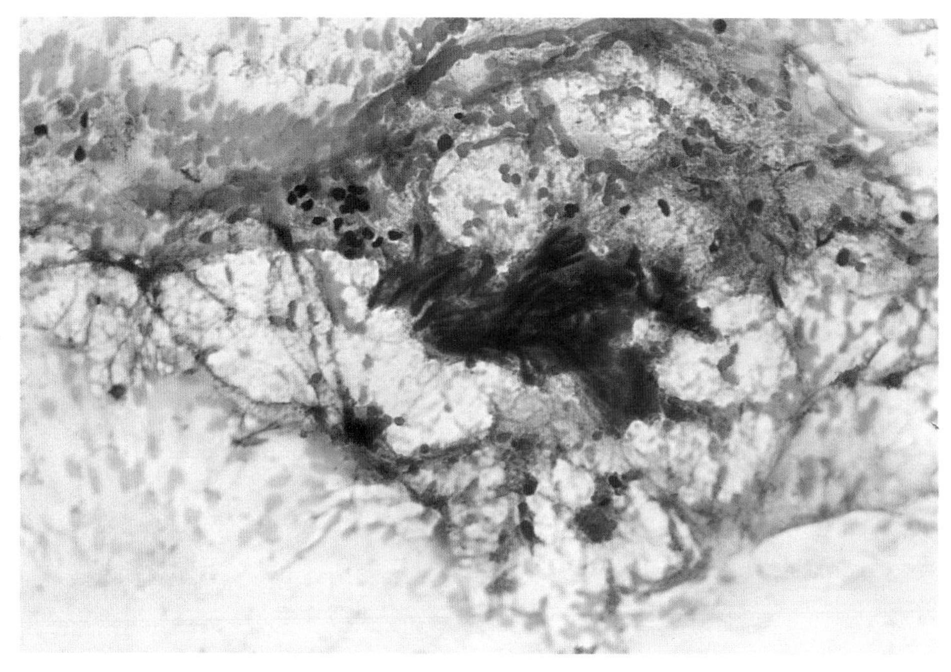

Figure 4.3 (**c**) Cervical smear taken within 48 hours of large loop excision biopsy. Smear shows extreme destruction of epithelial cells.

4.8 LARGE LOOP EXCISION OF THE TRANSFORMATION ZONE (LLETZ)

Large loop excision of the transformation zone (LLETZ) is a low-voltage diathermy technique which was introduced into the United Kingdom by Prendiville *et al.* (1986). It involves the use of fine wire loops energized by a high-frequency current to resect abnormal cervical tissue under direct colposcopic vison. The size of the loop used is determined by the dimensions of the cervical lesion. Ideally, the lesion should be removed by a single sweep of the loop and the specimen sent for histological examination. The specimen should be processed in the same way as cold knife cone biopsy with special attention given to the margins to ensure that clearance of the lesion is complete.

Prendiville *et al.* (1989) showed that the technique was superior to colposcopic punch biopsy because it produces larger biopsies which were more likely to reveal microinvasive disease Moreover, damage to the cut surface of the biopsy by the diathermy loop is minimal so that histological diagnosis is facilitated.

Gunasekera *et al.* (1990) compared LLETZ with carbon dioxide laser conization and found that there was no significant difference in the CIN recurrence rate between the two techniques. An added advantage of LLETZ was that it was well tolerated by the patient, caused less bleeding and was associated with fewer complications. LLETZ was also less expensive and less of a hazard to the surgeon's eyesight. Boulanger *et al.* (1989) compared LLETZ with cold knife conization and laser conization and reached similar conclusions.

Further support for LLETZ has been reported by Minucci *et al.* (1991) and Howe and Vincent (1991) who confirmed that the histological diagnosis based on LLETZ was more accurate than that obtained by colposcopically directed punch biospy. This view is expressed even more forcefully by Luesley (1992) who found that diagnoses based on small directed biopsies are frequently misleading.

However, LLETZ is not totally without its limitations. Cervical stenosis was recorded as a complication by Boulanger *et al.* (1989) although this seemed to be related to the height of the cone rather than the method of conization. Moreover, Gunasekera *et al.* (1990) preferred laser for patients with widespread vaginal involvment. Montz *et al.* (1993) consider that the high rate of surgical margin thermal destruction with related limitation of interpretability may represent a serious diagnostic and therapeutic limitation of the LLETZ procedure when considered as an alternative to cold knife conization. Murdoch *et al.* (1992), in a study of 721 women with CIN, concluded that a histological report of incomplete excision after LLETZ did not always indicate residual disease and close cytological follow-up is required after this form of treatment.

REFERENCES

Bonardi, R., Cecchini, S., Grazzini, G. and Ciatto, S. (1992) Loop electrosurgical excision procedure of the transformation zone and colposcopically directed punch biopsy in the diagnosis of cervical lesions. *Obstet. Gynecol.*, **80**, 1020–1022.

Boulanger, J. C. (1989) Electroconisation of the cervix uteri (French). *Rev. Franc. Gynecol. Obstet.*, **84**, 663–672.

Cinel, A. (1990) The accuracy of colposcopically directed biospy in the diagnosis of cervical intraepithelial neoplasia. *Eur. J. Gynaecol. Oncol.*, **11**, 433–437.

Dinh, T. A., Dinh, T. V., Hannigan, E. V., Yandell, R. B. and Dillard, E. A. (1989) Necessity for endocervical curettage in elderly women undergoing colposcopy *J. Reprod. Med.*, **34**, 621–624.

Dorsey, J. M. and Diggs, E. S. (1979) Microsurgical conisation of the cervix by carbon dioxide laser. *Obstet. Gynecol.*, **54**, 565–570.

Gunasekera, P. C., Phipps, J. H. and Lewis, B. V. (1990) Large loop excision of the transformation zone (LLETZ) compared to carbon dioxide laser in the treatment of CIN: a superior mode of treatment. *Br. J. Obstet. Gynaecol.*, **97**, 995–998.

Howe, D. T. and Vincenti, A. C. (1991) Is large loop excision of the transformation zone (LLETZ) more accurate than colposcopically directed punch biopsy in the diagnosis of cervical intraepithelial neoplasia? *Br. J. Obstet. Gynaecol.*, **98**, 588–591.

Howell, R., Hammond, R. and Pryse Davies, J. (1991) The histologic reliability of laser cone biopsy of the cervix. *Obstet. Gynecol.*, **77**, 905–911.

Jordan, J. A. (1980) The modern treatment of premalignant disease of the cervix. In *Controversies in Gynaecologic Oncology* pp. 25–37. (eds J. A. Jordan and A. Singer), Proceedings of a Scientific Meeting of the Royal College of Obstetricians and Gynaecologists, London, England, 22 February 1980.

Klein, M., Rosen, A., Vavra, N., Gitsch, G. and Beck, A. (1991) Conisation in pregnancy and its significance for the further course of the pregnancy (German). *Geburts Frauen*, **51**, 990–992.

Luesley, D. (1992) Advances in colposcopy and managment of cervical intrepithelial neoplasia. *Curr. Opin. Obstet. Gynecol.*, **4**, 102–108.

Minucci, D., Cinel, A. and Insacco, E. (1991) Diathermic loop treatment for CIN and HPV lesions. A follow-up of 130 cases. *Eur. J. Gynaecol. Oncol.*, **12**, 385–393.

Montz, F. J., Holschneider, C. H. and Thompson, L. D. (1993) Large loop excision of the transformation zone: effect on the pathologic interpretation of resection margins. *Obstet. Gynecol.*, **81**, 976–982.

Murdoch, J. B., Morgan, P. R., Lopes, A. and Monaghan, J. M. (1992) Histological incomplete excision of CIN after large loop excision of the transformation zone (LLETZ) merits careful follow-up not retreatment. *Br. J. Obstet. Gynaecol.*, **99**, 990–993.

Prendiville, W., Davies, R. and Berry, D. J. (1986) A low voltage diathermy loop for taking cervical biopsies: a qualitative comparison with punch biopsy forceps. *Br. J. Obstet. Gynaecol.*, **93**, 773–776.

Prendiville, W., Cullimore, J. and Norman, S. (1989) Large loop excision of the transformation zone (LLETZ). A new method of management for women with cervical intraepithelial neoplasia. *Br. J. Obstet. Gynaecol.*, **96**, 1054–1060.

Tabor, A. and Berget, A. (1990) Cold knife and laser conisation for cervical intraepithelial neoplasia. *Obstet. Gynecol.*, **76**, 633–635.

Vergote, I. B., Makar, A. P. and Kjorstad, K. E. (1992) Laser excision of the transformation zone as treatment of cervical intraepithelial neoplasia with satisfactory colposcopy. *Gynecol. Oncol.*, **44**, 235–239.

Wright, V. C., Davies, E. and Riopelle, M. A. (1984) Laser cylindrical excision to replace conisation. *Am. J. Obstet. Gynecol.*, **150**, 704–709.

5

THE NORMAL CERVIX

5.1 STRUCTURE AND DEVELOPMENT

The cervix is the narrow, inferior segment of the uterus which projects into the vaginal vault. It is a fibromuscular organ lined by mucous membrane, measuring approximately 3 cm in length and 2.5 cm in diameter. In the nulliparous female it is barrel-shaped, but it changes shape and size in pregnancy and at the menopause. Any injuries sustained during parturition may alter its configuration. About half its length (the portia vaginalis or ectocervix) protrudes into the vagina. It is continuous above with the body of the uterus at the 'isthmus', which is an ill-defined region where the cervical mucous membrane changes into endometrial-type epithelium. The cervix is traversed by the endocervical canal which is continuous with the uterine cavity above and opens into the vagina at the external os. The external os is usually circular in nulliparous or slit like in multiparous women (Figure 5.1(a–c)). The endocervical canal is fusiform in shape and approximately 3 cm in length. It is flattened from front to back and measures 8 mm at its greatest width. The measurements vary with the menstrual cycle. It is lined by a single layer of columnar epithelium which is thrown into folds and villus-like projections by stromal clefts and ridges which extend the length of the canal. The clefts extend deep into the stroma so that the canal is pitted with about 100 large branched, tubular crypts to a depth of about 4 mm, normally less than half the thickness of the fibromuscular wall (Hafez, 1982). These, too, are lined by a columnar epithelium. In histological section, the crypts resemble discrete, glandular structures and are often referred to as endocervical glands. However, Fluhmann (Fluhmann, 1957; Fluhmann and Dickmann, 1958) demonstrated, by serial section, that the endocervical crypts were not true glands, but deep infoldings of surface columnar epithelium into the underlying stroma.

The blood vessels and lymphatics ramify near the surface of the ectocervix and endocervical canal, forming a capillary network (Hawkins and Hudson, 1983) which has been studied extensively with the aid of the colposcope. In the normal cervix, the network is flat and regular, whereas in patients with CIN, the vessels proliferate, become markedly tortuous and extend close to the surface producing a variety of irregular colposcopic patterns.

Anteriorly, the supravaginal portion of the cervix is separated from the bladder by a distinct layer of connective tissue (the parametrium) which also extends to the sides of the

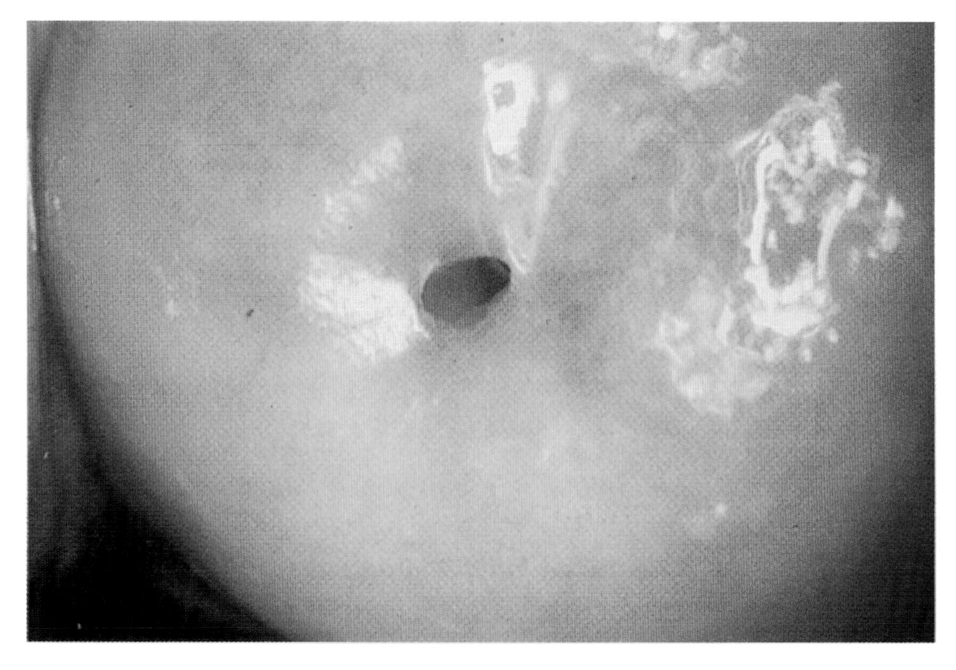

(**a**)

Figure 5.1 The normal cervix as seen through a colposcope. (**a**) A nulliparous cervix. Note the small external os.

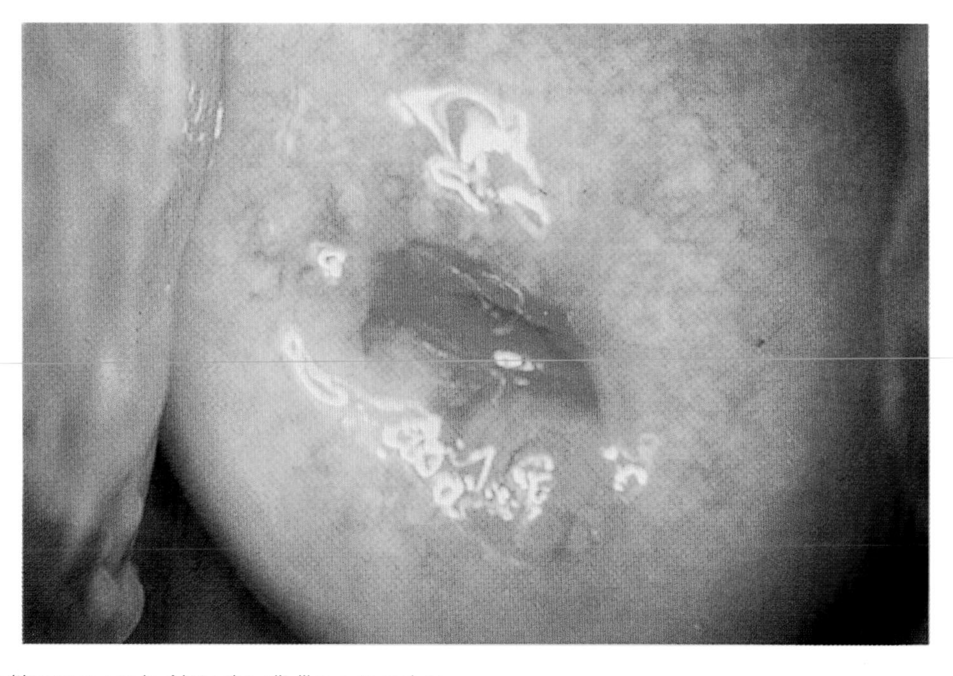

(**b**)

(**b**) A multiparous cervix. Note the slit-like external os.

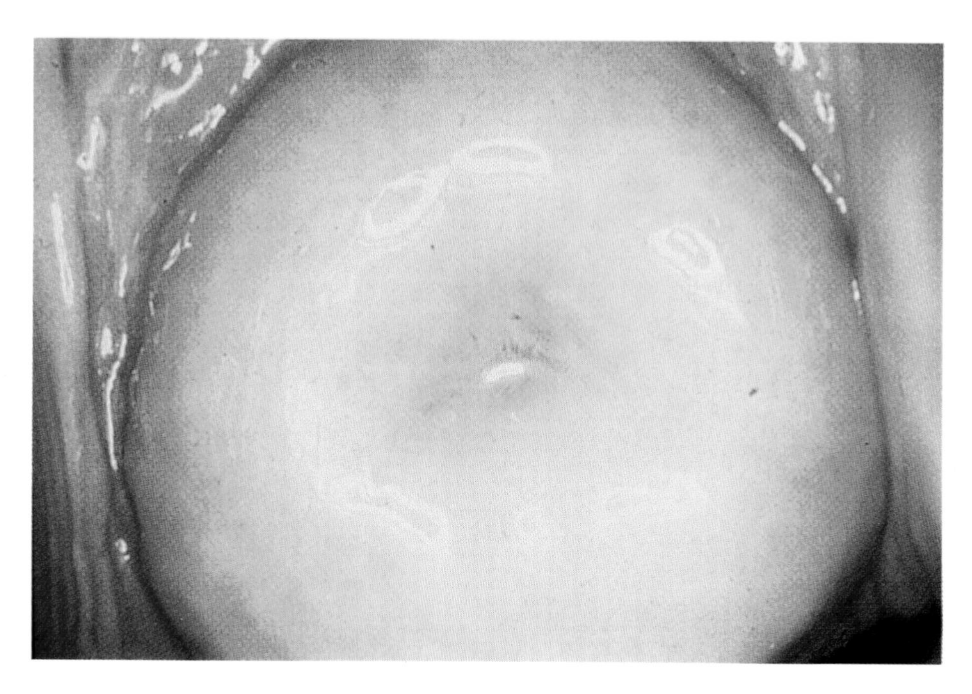

(**c**) An atrophic cervix, where the external os is greatly narrowed.

cervix. The uterine arteries are contained in this tissue. The cervix has a close relationship with the ureters which run downward and forward on either side in the parametrium. Posteriorly, the supravaginal cervix is covered with peritoneum which continues down over the posterior vaginal wall to be reflected off the rectum to form the rectouterine pouch (Pouch of Douglas). The cervix is held in position by the uterosacral and lateral ligaments. These consist mainly of fibrous tissue and some smooth muscle and hold the cervix in its normal anteverted position.

Most of the female genital tract is derived from the paramesonephric (müllerian) ducts. The fallopian tubes are derived from the cranial longitudinal part of the ducts, whereas the intermediate and caudal parts of the ducts fuse to form the body of the uterus and the uterine cervix respectively. The fallopian tubes, endometrial cavity, endocervical canal, ectocervix and upper third of the vagina are lined with epithelium of müllerian duct origin.

Fusion may be incomplete, resulting in a number of variants of partial or complete double genital tract. One variant results in the persistence of the septum dividing the cervical canal. It consists of fibrous or fibromuscular tissue with normal endocervical glands on each side (Figure 5.2).

The mesonephric (wolffian) ducts may be traced alongside the genital tract from epoophoron to hymen. They degenerate slowly, but functionless vestiges may persist as *Gärtner's ducts* (Figures 5.3 and 5.4).

5.2 EPITHELIAL LINING

Three types of epithelium can be recognized in the cervix: (i) columnar epithelium; (ii) original (native) squamous epithelium; and (iii)

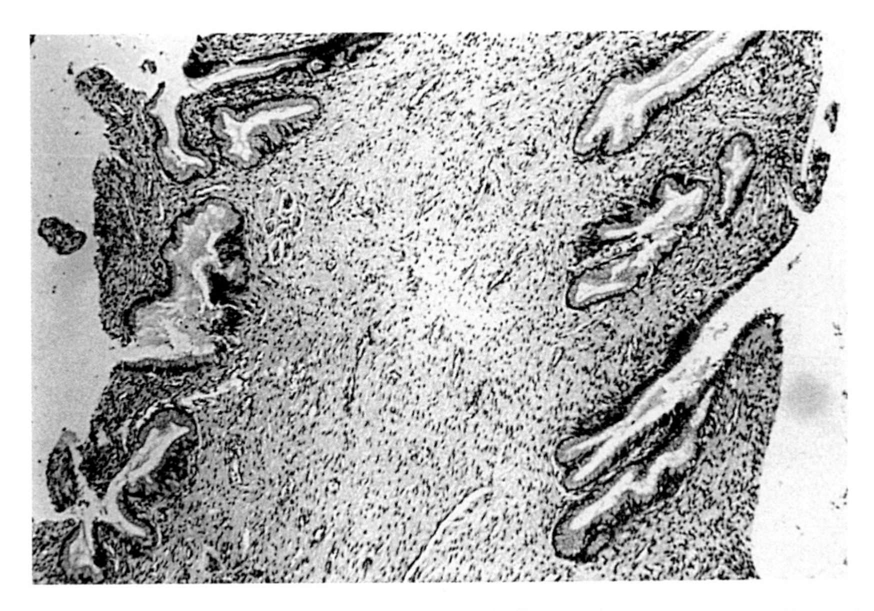

Figure 5.2 Septum dividing the cervical canal composed of a fibrous tissue core covered by endocervical epithelium (H & E, ×100).

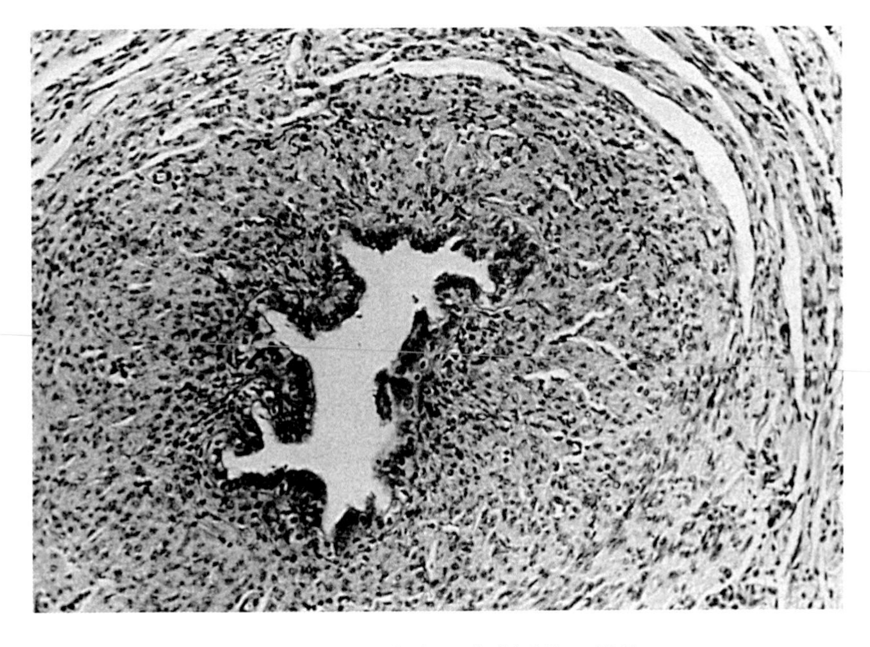

Figure 5.3 Remnants of mesonephric duct in wall of cervix (H & E, ×100).

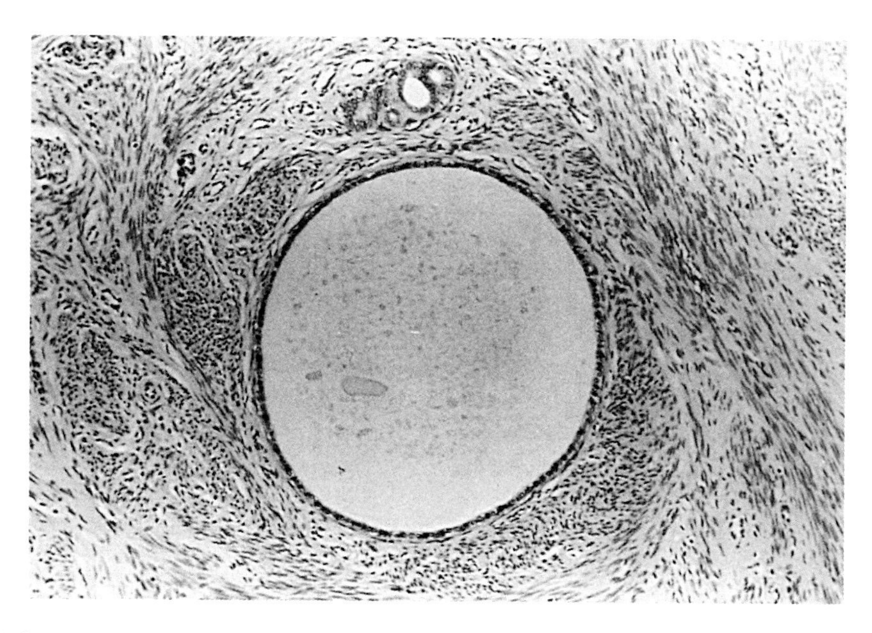

Figure 5.4 Cystic remains of mesonephric duct (Gärtner's duct) in wall of cervix (H & E, ×100).

metaplastic squamous epithelium. Their distribution varies throughout life.

In late foetal life (30 weeks), the cervical canal is lined by *columnar epithelium* of müllerian duct origin and the ectocervix is covered by non-keratinized stratified squamous epithelium (original or native squamous epithelium) derived from the vaginal plate. The interface between the two is termed the *original squamocolumnar junction* (Figures 5.5(a) and 5.6). At 30 weeks' gestation the original squamocolumnar junction is situated at or just cephalad to the external cervical os. However, its position changes as term approaches and in 70% of fetuses at 40 weeks, the original squamocolumnar junction appears caudal to the external os. Studies of the cervical epithelium at this stage (Pixley, 1976) reveal that the columnar epithelium at the squamocolumnar junction frequently undergoes *metaplastic change* to a squamous type of epithelium forming a new squamocolumnar junction. This area of squamous metaplasia between the original (native) and new squamocolumnar junctions is termed the *transformation zone* and persists into adult life.

The extent of metaplastic change in the columnar epithelium and hence the position of the squamocolumnar junction varies throughout a woman's reproductive life in response to hormonal changes and other influences. After puberty and at first pregnancy the cervix increases in bulk and changes in shape. This is accompanied by eversion of the distal endocervical epithelium on to the portio vaginalis of the cervix (ectropion), exposing it to the acid environment of the vagina (Figure 5.5(b)). This provides a stimulus for further metaplastic activity of the columnar epithelium, increasing the extent of the tranformation zone and resulting in the repositioning of the squamocolumnar junction (Figure 5.5(c)). Thus, during adolescence and reproductive life, the squamocolumnar junction is usually located distal to the external os; after the menopause it recedes within the endocervical canal (Figure 5.5(d)).

5.2.1 Columnar epithelium

A single layer of columnar cells lines the luminal surface of the endocervical canal and

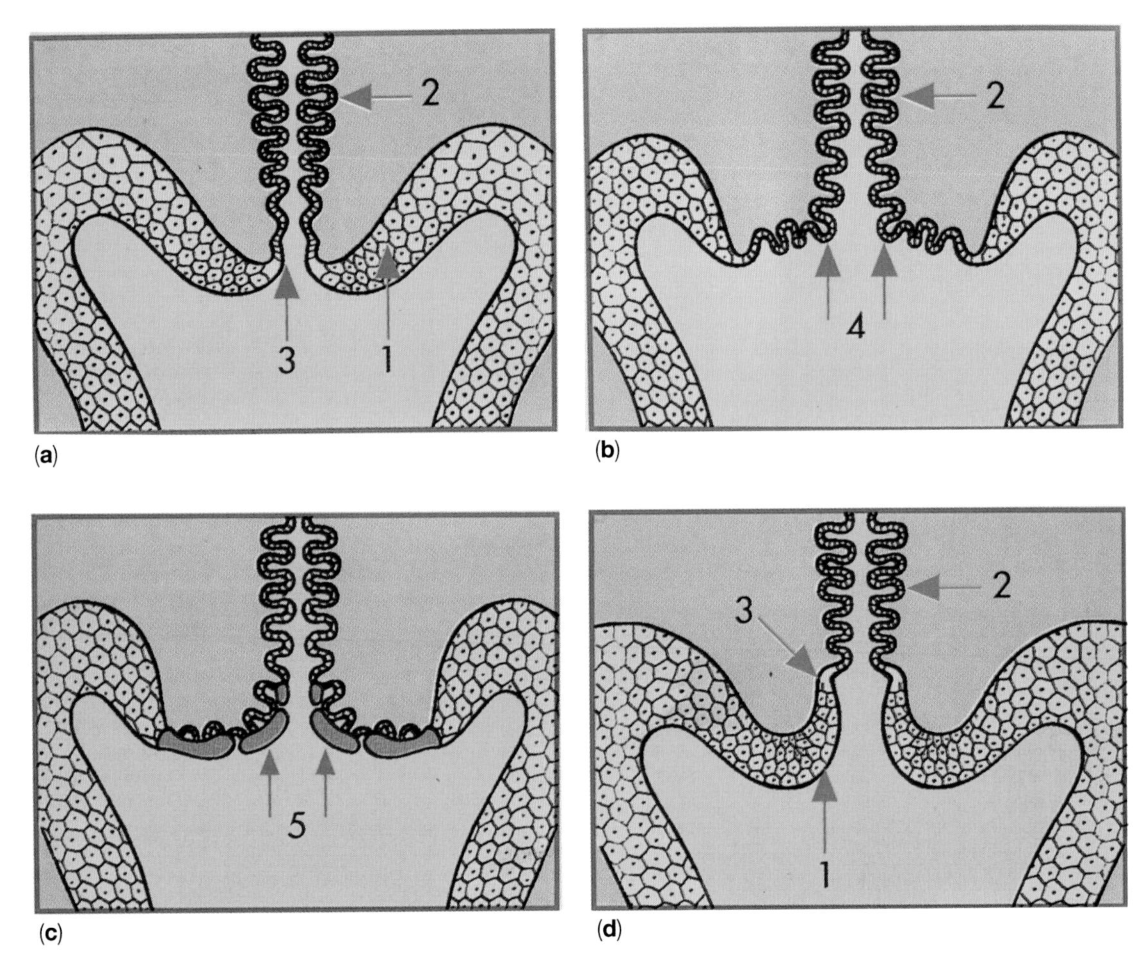

Figure 5.5 Diagram showing the stages in the development of the transformation zone and the changing location of the squamocolumnar junction (**a**) before puberty, (**b**) after puberty and pregnancy, (**c**) during reproductive life, (**d**) after the menopause. 1 = squamous epithelium; 2 = columnar epithelium; 3 = squamocolumnar junction; 4 = eversion of columnar epithelium onto ectocervix; 5 = metaplastic epithelium.

the crypts (Figure 5.7(a,b)). It is made up of tall, cylindrical cells arranged in 'picket' formation. The nucleus of the columnar cells are usually located in the basal part of the cell but when the cells are actively secreting mucus the nucleus may lie suprabasally or in the middle of the cell. Two types of cells can be distinguished: (i) non-ciliated secretory cells; and (ii) ciliated cells. The *secretory cells* produce both acid and neutral mucin, although the relative amounts secreted vary with the menstrual cycle (Wakefield and Wells, 1985). The cells employ both merocrine and apocrine methods of secretion. The cytoplasm of the secretory cells appears pale and vacuolated when stained with haematoxylin; the mucin content can be demonstrated with Alcian blue or periodic acid–Schiff (PAS), both before and after diastase digestion. *Ciliated columnar cells* are interspersed among the secretory cells (Figure 5.7(b)), their main function being to carry the mucus along the membrane. They are found more frequently at the endometrial/endocervical junction than elsewhere in the endocervical canal.

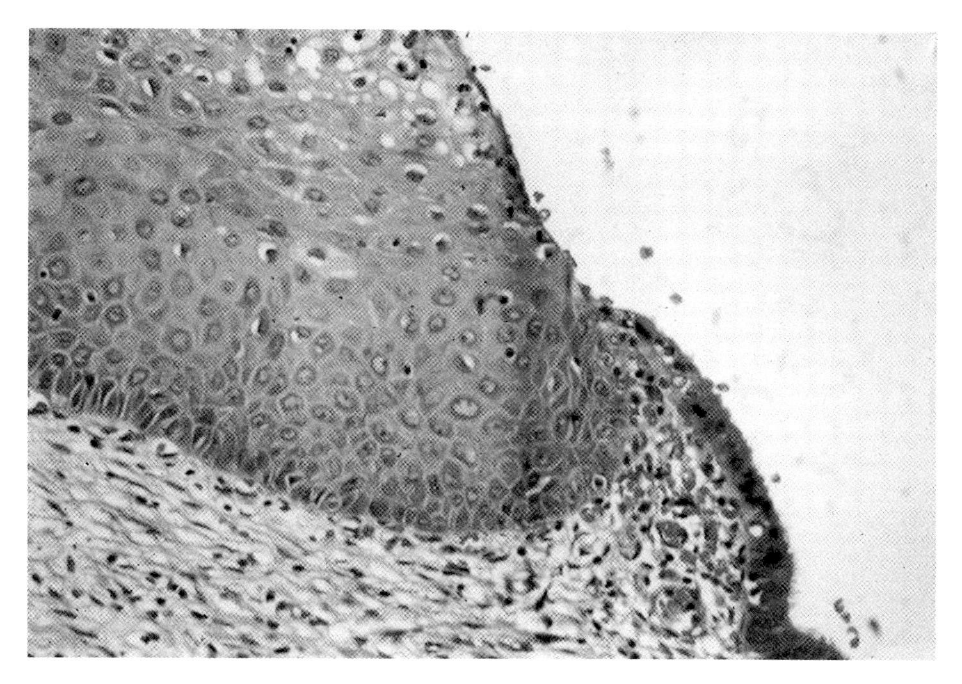

Figure 5.6 Squamocolumnar junction seen at low magnification.

On colposcopy, the glandular epithelium lining the endocervical canal and endocervical crypts has a cobblestone appearance and appears to be thrown into small folds or bunches of grapes. This reflects its villus like structure, each villus being composed of a stromal core lined by columnar cells (Figure 5.8). The villi in the endocervical crypts are frequently the focus of metaplastic change which may occlude the opening of the crypt with the result that colposcopically and histologically they appear as blind tunnels or tubes. The occluded crypts are also foci for the collection of mucus and the development of localized cystic structures known clinically and histologically as Nabothian follicles (see also Chapter 8).

The *cervical mucus* is a watery fluid whose composition varies with the stage of the menstrual cycle. It contains electrolytes, especially sodium chloride and simple sugars such as glucose or fructose, urea and proteins in colloidal solution (Elstein, 1976). The proteins include albumins, immunoglobulins, enzymes and a variety of macromolecules, the amounts of which vary and affect the viscosity of the mucus. Leucocytes, together with desquamated columnar and squamous cells, are suspended in the mucus. The cervical mucus plays an important role in sperm transport.

Two main types of cervical mucus have been described 'oestrogenic' and 'gestogenic', the former being produced in the proliferative phase of the cycle and the latter in the secretory stage (Elstein and Dauntner, 1976). In the follicular phase the mucin is composed of a fine miscellar network of filaments through which there is easy passage of spermatozoa. At this stage 'ferning' of the mucus can be demonstrated when it is spread on a glass slide (Figure 6.14(b)).

In the secretory or luteal phase the viscosity of the mucus increases and the pores of the macromolecular miscellar network are smaller, impeding the progress of spermatazoa. More leucocytes and cell debris are trapped in this thicker mucus and the amounts of protease

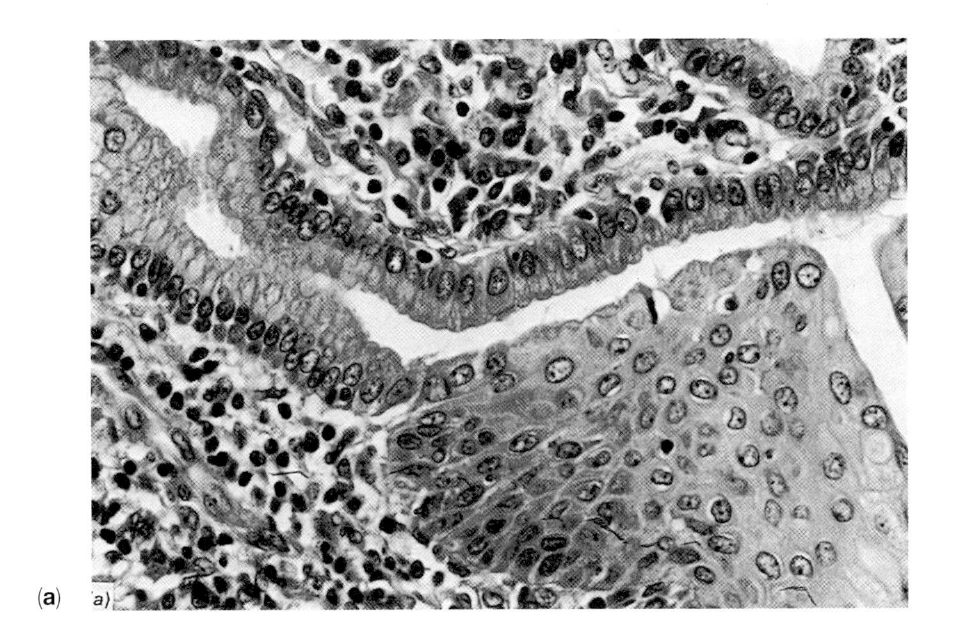

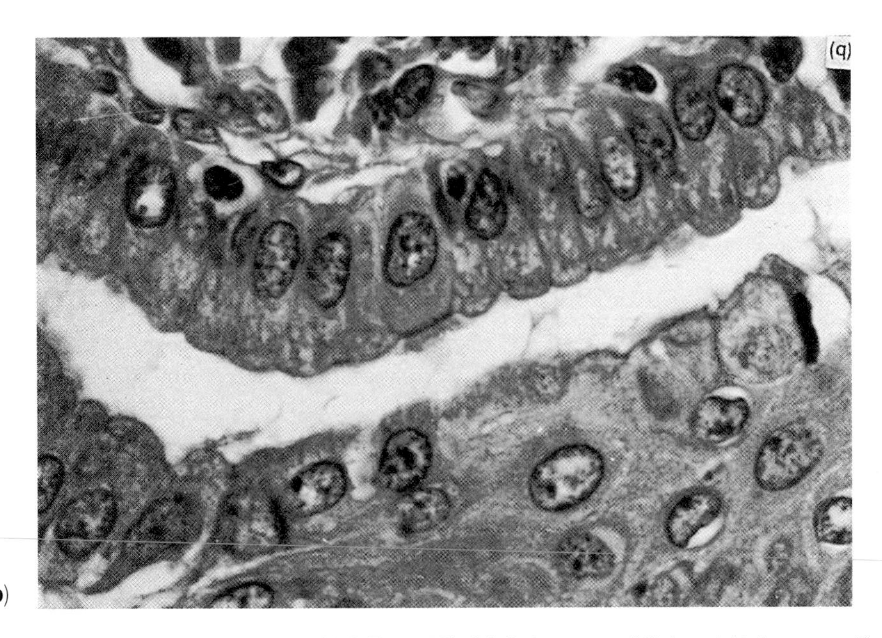

Figure 5.7 (**a**) Squamocolumnar junction (H & E, ×770). (**b**) Columnar cell lining at higher magnification. Note occasional ciliated cell, and interstitial cell (H & E, ×1500).

inhibitor enzymes such as alpha-1-antitrypsin increase, counteracting the potential hydrolytic effect of sperm proteases.

Odeblad (1976) studied the composition of cervical mucus in disease. He noted that prolonged oestrogen secretion resulted in very thin watery fluid which impeded sperm migration while with oral contraceptives the

Figure 5.8 Focus of metaplastic change in the endocervical canal showing fusion of the villi (H & E, ×120).

cervical mucus resembles that found in the luteal phase. Cervicitis also results in a change in the consistency and composition of cervical mucus. In chronic cervicitis the endocervical cells appear to become resistant to hormonal influences.

5.2.2 Original stratified squamous epithelium

The original squamous epithelium lining the portio vaginalis of the cervix is 0.5 mm thick. It is continuous with the squamous epithelium of the vagina but is less sensitive to hormonal influence and lacks the rete pegs seen in the vagina. Its appearance varies with age and with the menstrual cycle, becoming fully mature under the influence of oestrogen (Figure 5.9) but semi-mature with progesterone (Figure 5.10). At this stage it is composed of 5 to 10 layers of flat, polygonal cells. In the absence of these hormones, e.g. after the menopause, the thickness of the epithelium is greatly reduced (Figure 5.11) (see also Section 5.5).

The mature squamous epithelium of the adult menopausal woman at mid-cycle can be conveniently divided into three zones (Figure 5.9): (i) the basal zone (germinal layer); (ii) the mid- or parabasal zone; and (iii) the superficial zone, which is made up of the most mature cells.

The epithelium is separated from the stroma by a delicate network of fibres which constitute the basal lamina. The deep surface of the epithelium is indented in places by stromal papillae which project towards the surface of the epithelium for about one-third of its thickness.

(a) The basal zone

This is composed of a single layer of cylindrical cells 12 μm in diameter whose major function is epithelial regeneration. Their oval nuclei display evidence of active cell division such as nucleoli, numerous chromocentres and very occasional mitotic figures (Figure 5.12). Under normal conditions, the entire process of epithelial regeneration is confined to the basal layers, the remaining zones merely reflecting

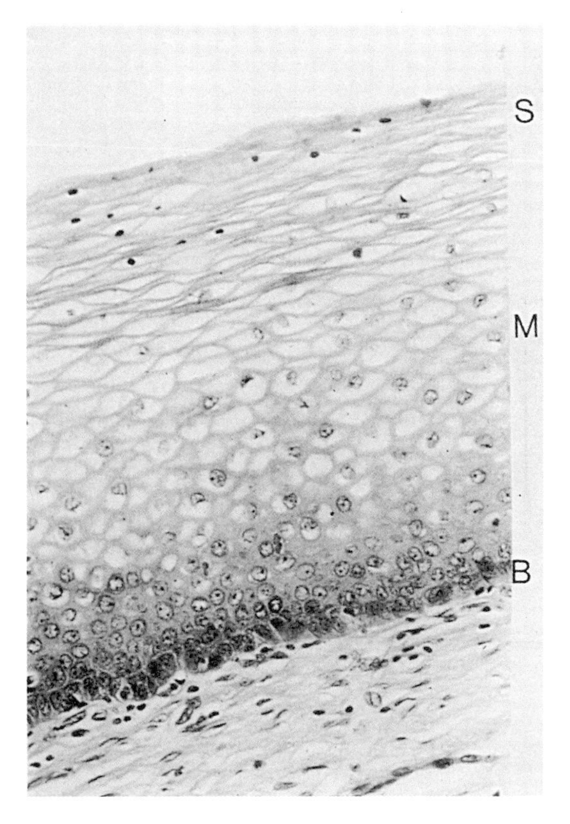

Figure 5.9 Mature squamous epithelium at higher magnification to show basal (B), mid (M) and superficial (S) zones (H & E, ×310).

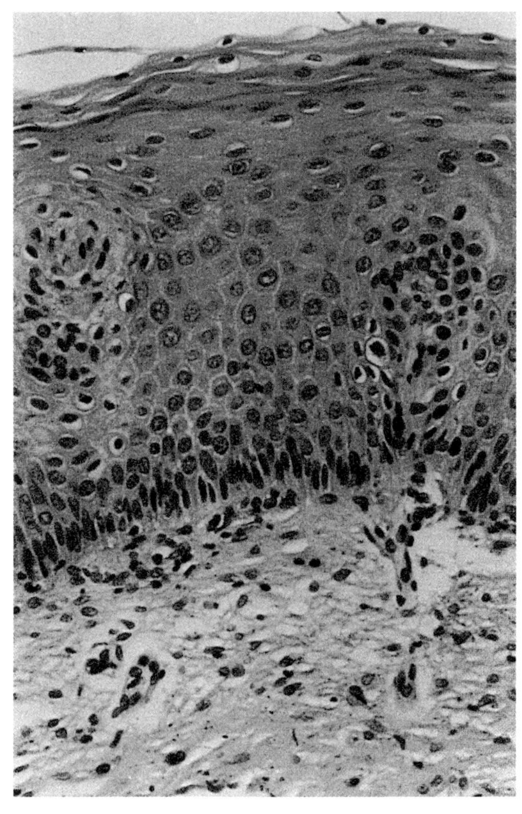

Figure 5.10 Semi-mature squamous epithelium (H & E, ×310).

different stages of cell maturation. In the presence of chronic cervical infection or local trauma the basal zone may be multilayered (Figures 10.20 and 10.21) giving rise to the condition of *basal hyperplasia*. Two or more layers of basal cells are present, some of which may be in mitosis. The configuration of the metaphases is normal. Above the basal zone, normal epithelial maturation will be found.

(b) The mid- or parabasal zone

This is composed of maturing squamous cells which are slightly larger than those of the basal zone with more abundant cytoplasm. The cells in the upper layers of this zone have more cytoplasm than the deeper layers, although the nuclear size remains fairly constant. The parabasal cells of this zone are PAS-positive and electron microscopy shows that the cytoplasm contains abundant tonofilaments and numerous glycogen granules. The integrity of the epithelium is maintained by desmosomal or hemidesmosomal attachments which appear in the light microscope as intercellular bridges (Figure 5.13). In consequence this layer is sometimes referred to as the stratum spinosum. A granular layer of polyhedral cells containing keratohyaline material is sometimes seen in the uppermost layers of the mid zone.

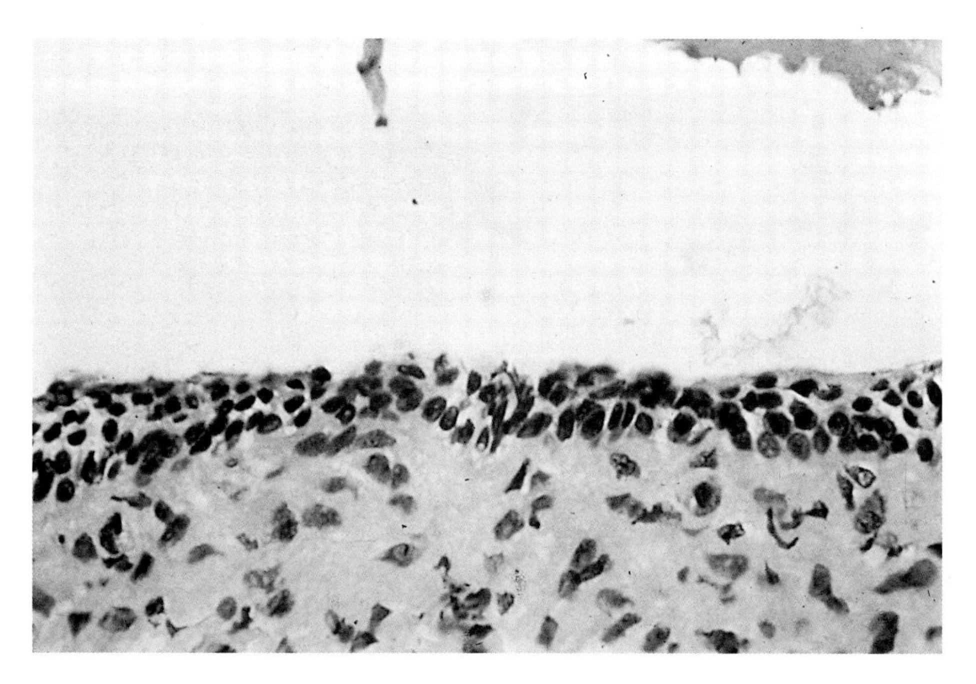

Figure 5.11 Atrophic epithelium (H & E, ×310).

(c) The superficial layers of the epithelium

These are composed of several layers of loosely attached cells that are broader and thinner than the cells of the mid-zone producing a basket weave pattern on histological section. They have small pyknotic nuclei 2–3 μm in diameter. Transmission electron microscopy reveals the presence of occasional membrane-bound keratinosomes which are the source of protein-bound disulphide keratin precursors. Full development of keratin does not occur in normal cervical epithelium. However, under certain conditions, e.g. in the presence of vaginal prolapse, the cells of the superficial layers may synthesize excessive amounts of keratin resulting in hyperkeratosis (Section 9.4). Scanning electron microscopy reveals a network of large, squamous cells which form a mosaic pattern. The surface of each cell is covered by a network of microridges (Figure 5.14) which are believed to increase surface adhesiveness, thereby protecting the deeper layers from trauma and infection (Hafez, 1982).

5.2.3 Squamous metaplasia

Squamous metaplasia in the transformation zone is a patchy process with foci of metaplastic change occurring initially at the tips of the endocervical villae (Figures 5.8 and 5.15) which gradually fuse. The metaplastic process subsequently extends into the crypts (Figure 5.16) so that eventually the whole of the everted endocervical epithelium may be replaced by metaplastic squamous epithelium (Figure 5.17). Metaplastic change is very commonly seen in histological sections of cervices from women of childbearing age and must be considered a normal physiological process. However, it may also occur in cervices which have been traumatized, e.g. by inflammation or cautery. Thus, it is a frequent finding in both

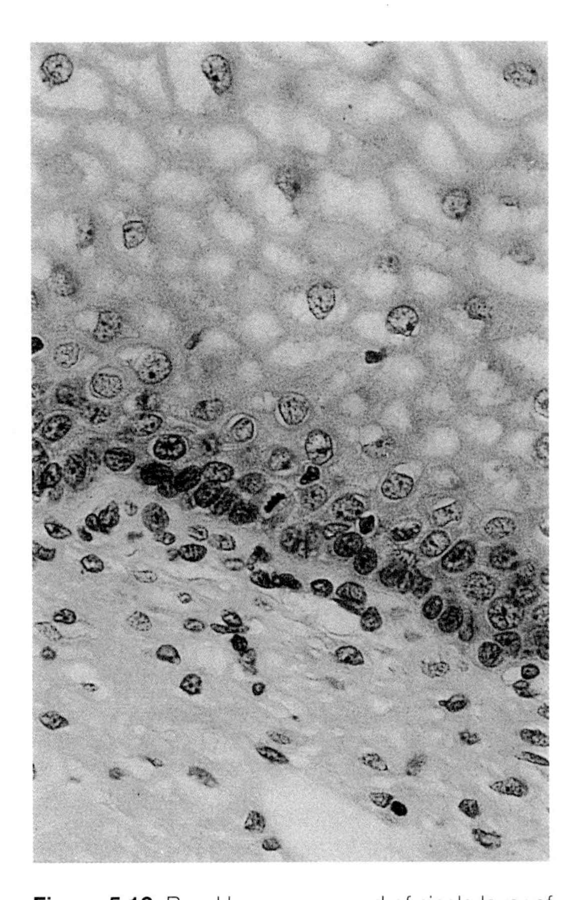

Figure 5.12 Basal layer composed of single layer of cells which have potential for regeneration. Note mitotic figure (H & E, ×770).

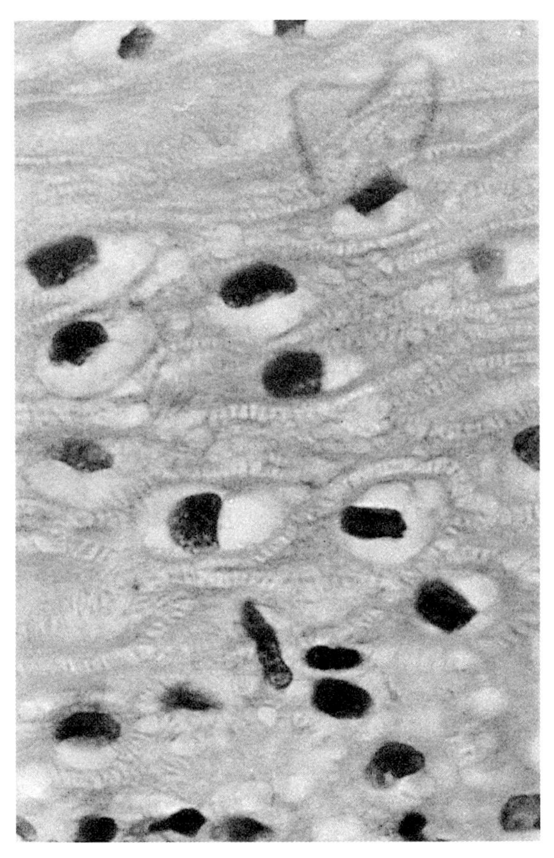

Figure 5.13 Intercellular bridges between cells of the mid zone (H & E, ×1500).

the normal and abnormal cervices and may be the result of physiological change or pathological processes.

Three stages in the development of metaplastic squamous epithelium have been described: (i) reserve cell hyperplasia; (ii) immature squamous metaplasia; and (iii) mature squamous metaplasia. It is an irreversible process which occur over a period of weeks. The histological and cytological changes which characterize each of these stages are presented in detail in Section 9.1.

5.3 CERVICAL STROMA

The connective tissue of the cervix is composed mainly of collagen fibres which are densest in the region of the ectocervix and only loosely surround the endocervical glands. Smooth muscle and elastic fibres are scarce, except in the region of the isthmus. Infiltrates of inflammatory cells may be seen deep to the epithelium in normal cervices particularly in the region of the transformation zone, as in Figure 5.16. Plasma cells are not infrequently seen. It has been suggested that the inflammatory infiltrate is an immunological response to the cell necrosis and regeneration associated with metaplastic change.

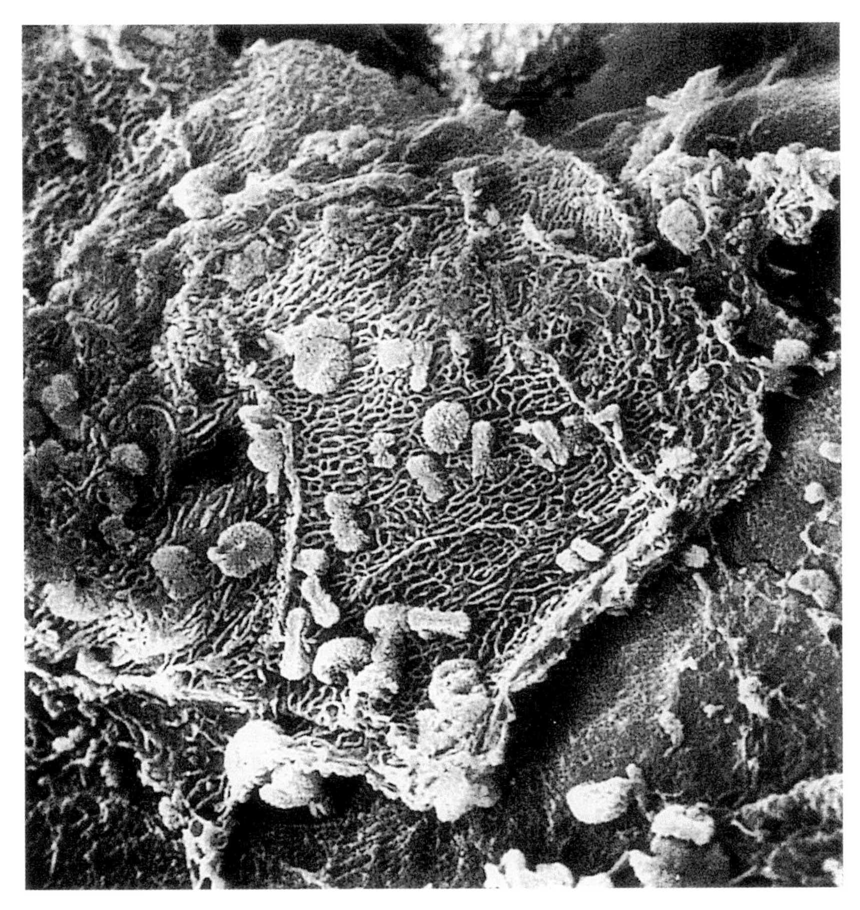

Figure 5.14 Scanning electron micrograph of superficial squamous cells. The surface of each cell is covered by a network of micro-ridges which have a protective function. *Monilia* spores are seen trapped on the surface of the cell (courtesy of Professor C. A. Rubio, Karolinska Institute, Stockholm, Sweden).

5.4 THE CERVIX DURING PREGNANCY

Morphological changes that occur in the pregnant cervix arise in response to the stimulatory effect of the gestational hormones. One of the main features is increase in bulk of the cervix due to increased vascularity and oedema of the stroma. Decidualization of the stroma is common, the stromal fibroblasts becoming enlarged and pale as their cytoplasm fills with glycogen and lipid droplets under the influence of human chorionic gonadotrophin. Decidual changes of this nature may be seen in one-third of biopsies taken in pregnancy (Figures 8.9 and 8.10). Another important feature is the development of ectropion or eversion of the endocervical epithelium on to the portio vaginalis of the cervix (Section 5.2). This has the effect of exposing the columnar epithelium to the vaginal environment where it undergoes metaplastic change. The process of ectropion and metaplasia is most marked in the primiparous woman in the first trimester, but it may also occur in multipara, particularly as the pregnancy advances.

To the naked eye the everted endocervical mucosa often appears as a glistening red area

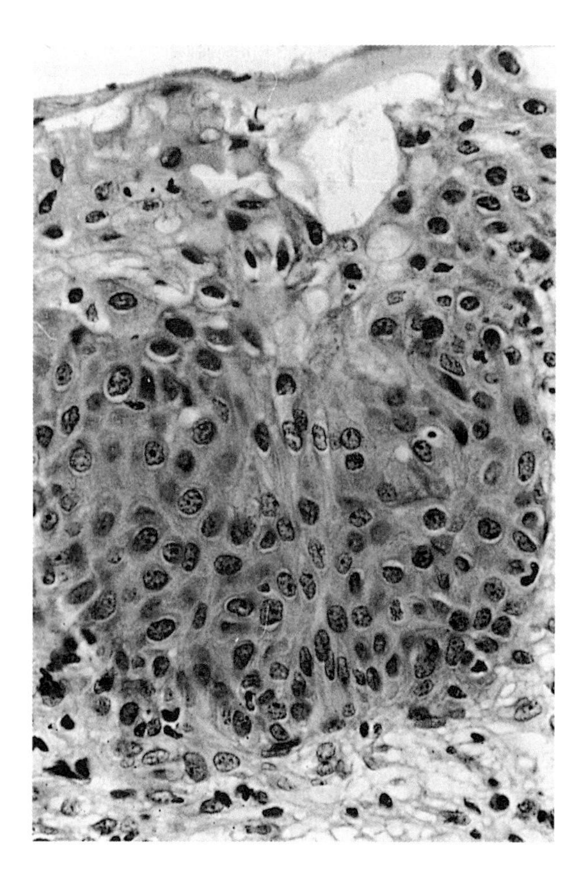

Figure 5.15 Immature metaplastic epithelium at tip of villus shown in Fig. 5.8 (H & E, ×320).

surrounding the external os on speculum examination. For many years this change was erroneously thought to reflect a denuded area of epithelium on the cervix and was described as an 'erosion' by the clinician. It is now more correctly designated *ectropion* (Figure 3.2(a,b)).

As the cervix is very vascular in pregnancy, biopsy should be deferred to the post-partum period in women with an abnormal smear and colposcopic changes suggestive of CIN. However, if invasive squamous cancer is suspected on any grounds, biopsy is essential. Wedge biopsy is preferred as bleeding is easier to control. Foci of decidual change may be seen in the biopsy.

Delivery of the foetus through the cervix causes epithelial injury such as laceration, bruising and ulceration most commonly involving the anterior lip, but repair is swift. The denuded areas are re-epithelialized by ingrowing layers of native squamous

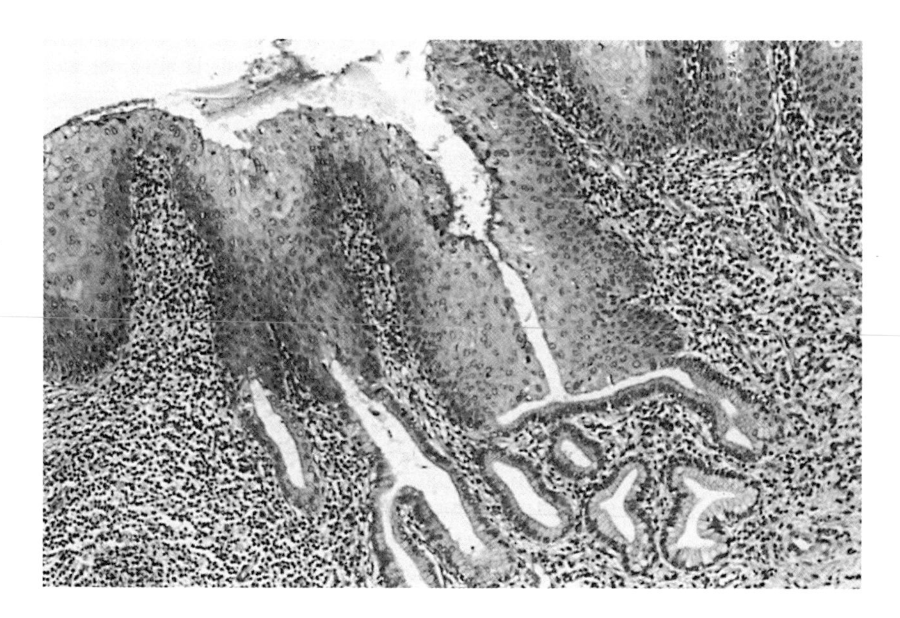

Figure 5.16 Metaplastic change in endocervical crypt (H & E, ×100).

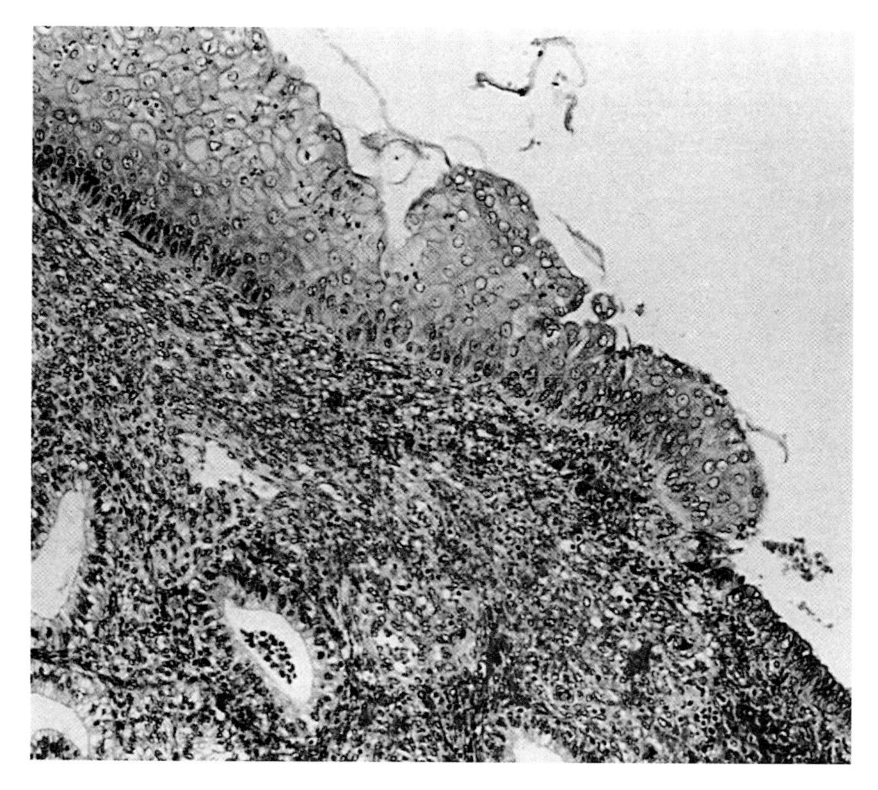

Figure 5.17 Foci of mature and immature metaplasia in endocervical canal. Note endocervical glands in the connective tissue deep to the metaplastic epithelium (H & E, ×100).

epithelium of the portio vaginalis, and the integrity of the epithelium is completely restored within a month (Section 9.6.1).

Although the cervix is continuous with the body of the uterus the amount of smooth muscle and elastic tissue in the cervix is small and the main constituent of the cervix wall is collagenous connective tissue. Elastic fibres, except around blood vessels, are scarce. Despite this the cervix is able to undergo dilation and contraction during parturition. The main factor which enables it to dilate in response to the mechnical force of the contracting uterus is softening of its intercellular substance. A group of peptides collectively known as prostaglandins are important in this respect. Constriction of the cervix after delivery has been attributed to reorganization of fibrous tissue.

5.5 POST-MENOPAUSAL EPITHELIUM

Just as increased volume of the cervix in pregnancy appears to lead to eversion of the endocervical mucosa, so reduction in volume of the cervix occurring after the menopause leads to inversion of the glandular epithelium which recedes into the endocervical canal (Figure 5.5(d)). Typically, the squamous epithelium, becomes atrophic after the menopause (Figure 5.11). It becomes thinner and stratification and glycogenation which are a feature of normal pre-menopausal epithelium are gradually lost so that Schiller's test may be positive (iodine-negative), although no epithelial abnormality may be present in the biopsy. Because the epithelium is so thin, infection is common.

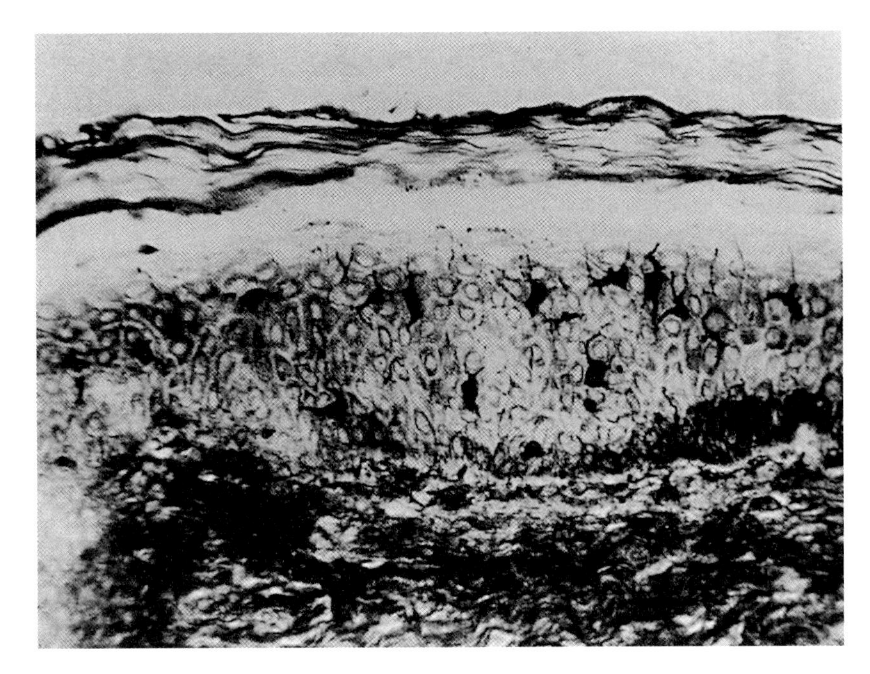

Figure 5.18 Dendritic cells (Langerhans' cell) in cervical epithelium (Gairn's gold chloride, reduced from ×11 000).

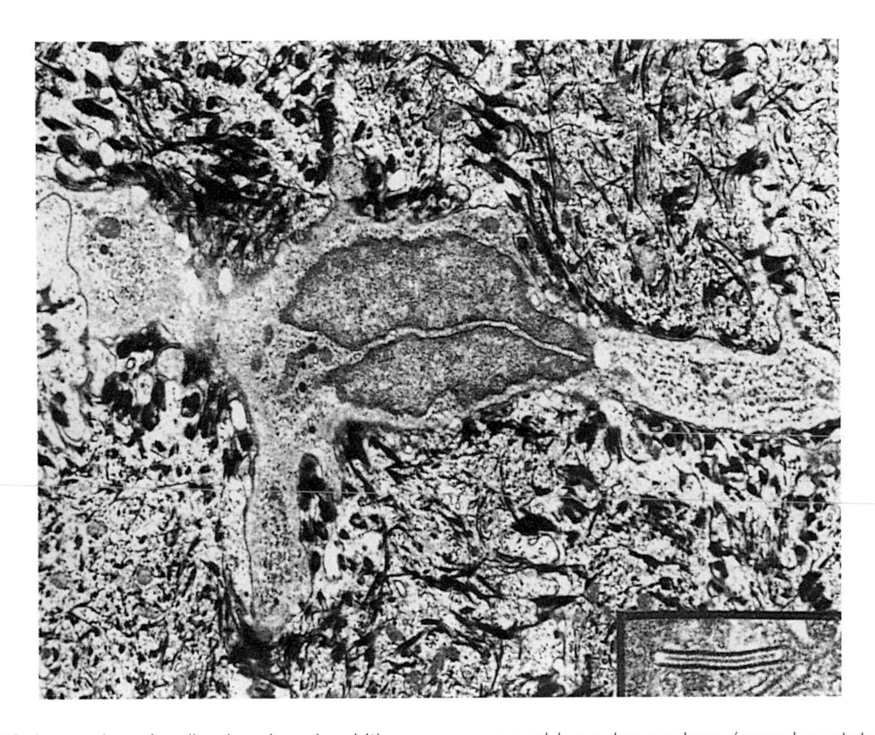

Figure 5.19 Langerhans' cells showing dendritic processes and irregular nucleus (uranyl acetate ×17 000). **Inset**: typical Langerhans' granule (×120 000) (Figs 5.18 and 5.19 courtesy of Professor A. S. Breathnach, Department of Anatomy, St Mary's Hospital Medical School, London).

5.6 LANGERHANS' CELLS IN CERVICAL EPITHELIUM

These intraepithelial dendritic cells (Figures 5.18 and 5.19) which were first described in skin over 100 years ago, are present in the normal cervical epithelium of the transformation zone and ectocervix and are believed to be involved in the recognition and processing of exogenous antigens (Morris *et al.*, 1983a,b; Maclean, 1984). They can be recognized by their typical cytoplasmic granules (Figure 5.19, inset) seen on electron microscopy (Birkbeck *et al.*, 1961) and their ATPase activity. They can also be detected by specific monoclonal antibodies (Murphy *et al.*, 1981; Morris *et al.*, 1983a,b; Puts *et al.*, 1986). Their numbers increase in inflammatory and neoplastic states. The magnitude of the increase and the severity of the neoplastic lesion appear to be closely related phenomena (Morris *et al.*, 1983b; Caorsi and Figueroa, 1986).

REFERENCES

Birkbeck, M. S., Breathnach, A. S. and Everall, J. D. (1961) An electron microscope study of basal melanocytes and high level clear cells (Langerhans' cells) in vitiligo. *J. Invest. Dermatol.*, **37**, 51–64.

Caorsi, I. and Figueroa, C. D. (1986) Langerhans' cell density in the normal exocervical epithelium and in cervical intraepithelial neoplasia. *Br. J. Obstet. Gynaecol.*, **93**, 993–998.

Cinel, A. (1990) The accuracy of colposcopically directed biopsy in the diagnosis of cervical intraepithelial neoplasia. *Eur. J. Gynaecol. Oncol.*, **11**, 433–437.

Elstein, M. (1976) The biochemistry of cervical mucus, in *The Cervix* (eds J. Jordan and A. Singer), W. B. Saunders, London, Chapter 12.

Elstein, M. and Daunter, B. (1976) The structure of cervical mucus, in *The Cervix* (eds J. Jordan and A. Singer), W. B. Saunders, London, Chapter 11.

Fluhmann, C.F. (1957) The nature and development of the so-called glands of the cervix uteri. *Am. J. Obstet. Gynecol.*, **74**, 753–768.

Fluhmann, C. F. and Dickmann, Z. (1958) The basic pattern of the glandular structures of the cervix uteri. *Obstet. Gynecol.*, **11**, 543–555.

Hafez, E. S. E. (1982) Structural and ultrastructural parameters of the cervix. *Obstet. Gynecol. Surv.*, **37**, 507–516.

Hawkins, J. and Hudson, C. N. (1983) *Shaw's Textbook of Operative Gynaecology*, 5th edn., Churchill Livingstone, London, pp. 13–17.

Maclean, A. B. (1984), Cervical healing and Langerhans' cells. *Br. J. Obstet. Gynaecol.*, **91**, 1145–1148.

Morris, H. H. B., Gatter, K. C., Stein, H. and Mason, D. Y. (1983a) Langerhans' cells in human cervical epithelium: an immunohistological study. *Br. J. Obstet. Gynaecol.*, **90**, 400–411.

Morris, H. H. B., Gatter, K. C., Sykes, G., Casemore, V. and Mason, D. Y. (1983b) Langerhans' cells in human cervical epithelium: effects of wart virus infection and intraepithelial neoplasia. *Br. J. Obstet. Gynaecol.*, **90**, 412–420.

Murphy, G. F., Bhan, A. K., Sato, S., Mihm, M. C. and Harris, T. J. (1981) A new immunological marker for human Langerhans' cells. *N. Engl. J. Med.*, **304**, 791–792.

Odeblad, E. (1976) The biophysical aspects of cervical mucus, in *The Cervix* (eds J. Jordan and A. Singer), W. B. Saunders, London, Chapter 13.

Pixley, E. (1976) Morphology of the fetal and prepubertal cervicovaginal epithelium, in *The Cervix* (eds J. Jordan and A. Singer), W. B. Saunders, London, Chapter 7.

Puts, J. J. G., Moesker, O., de Waal, R. M. W., Kenemans, P., Vooijs, G. P. and Ramaekers, F. C. S. (1986) Immunohistochemical identification of Langerhans' cells in normal epithelium and in epithelial lesions of the uterine cervix. *Int. J. Gynecol. Pathol.*, **5**, 151–162.

Wakefield, E. A. and Wells, M. (1985) Histochemical study of endocervical glycoproteins throughout the normal menstrual cycle and adjacent to cervical intraepithelial neoplasia. *Int. J. Gynecol. Pathol.*, **4**, 230–239.

6

Cervical cytology: the normal smear

A cervical smear properly taken with an Ayre spatula contains a variety of cells, some of which have been exfoliated locally and trapped in the cervical mucus, and others which have been detached forcibly by the spatula from the region of the external os. A smear from a normal cervix may therefore contain:

- cells from original squamous epithelium of the cervix and vagina;
- cells from the columnar epithelium of the endocervical canal;
- cells from the metaplastic epithelium of the transformation zone;
- cells from other parts of the genital tract, e.g. endometrial cells;
- leucocytes and red blood cells;
- commensal organisms;
- contaminants, e.g. spermatozoa, talc granules, etc.; and
- cervical mucus strands.

The cytological nomenclature used in this book to describe non-malignant cells seen in cervical smears is based on that recommended by the International Academy of Cytology (1958) as a result of an opinion poll on non-malignant cytological terminology. It has the advantage in that it is widely used and represents the consensus view of cytologists from many nations. It has been largely endorsed by the British Society for Clinical Cytology (Spriggs *et al.*, 1978; Evans *et al.*, 1986) and is used in the WHO atlas on the cytology of the female genital tract (Riotton and Christopherson, 1973).

6.1 CELLS SHED FROM THE CERVICAL EPITHELIUM

Four types of epithelial cells can be recognized in smears; by convention they are designated superficial, intermediate, parabasal and columnar cells. It is important to remember that, despite their names, all the epithelial cells in the smear are derived from the surface layers of the epithelium.

The following descriptions relate to the appearance of these cells in Papanicolaou smears.

6.1.1 Superficial cells (Figure 6.1)

These are shed from the surface of fully mature squamous epithelium which has developed to its full thickness under the influence of oestrogen. They appear in the smear as large polygonal squames with transparent cytoplasm and sharp cell borders, measuring 45–50 µm in

Figure 6.1 Superficial cells in cervical smears. Note transparent cytoplasm, angular shape and pyknotic nuclei (Pap, ×400).

diameter (Figure 6.1). The staining reaction of the cells depends on their maturity: the cytoplasm of the most mature cells is stained a delicate pink colour by the Orange G in the stain; those cells which are less mature react with the light green component in the Papanicolaou stain and stain greenish blue. The nuclei of the superficial cells also reflect the maturity of the epithelium and are shrunken and pyknotic. Occasionally, a small clear zone surrounds the shrunken nucleus serving as a reminder of its former size. Karyorrhexis is sometimes seen which is an indication of cell death. Keratinization of the epithelium is not a feature of the normal cervix.

6.1.2 Intermediate cells (Figure 6.2)

These are shed from the surface of semimature squamous epithelium which is showing a diminished response to oestrogen or the effect of progesterone. These flat squames are only slightly smaller than the superficial cells (35–40 μm diameter), but can be distinguished from them by the structure and size of the nucleus and the less angular appearance of the cell outline (Figure 6.2). The nucleus is 8 μm in diameter, round or oval with a clearly defined chromatin network and prominent chromocentres and sex chromatin. It is frequently described as being 'vesicular' in appearance. The cytoplasm of the intermediate cell is normally more cyanophilic and denser than that seen in superficial cells. The morphology of the intermediate cells alters in pregnancy at which time they appear elongated and boat-shaped and are termed *navicular* cells (Figure 6.3). The cytoplasmic borders are thickened and the vesicular nucleus is usually eccentrically placed. Electron microscopy studies reveal a heavy deposition of glycogen granules in the cytoplasm which may account for the change in shape. The glycogen may be apparent in the light microscope as a brownish deposit in the cells.

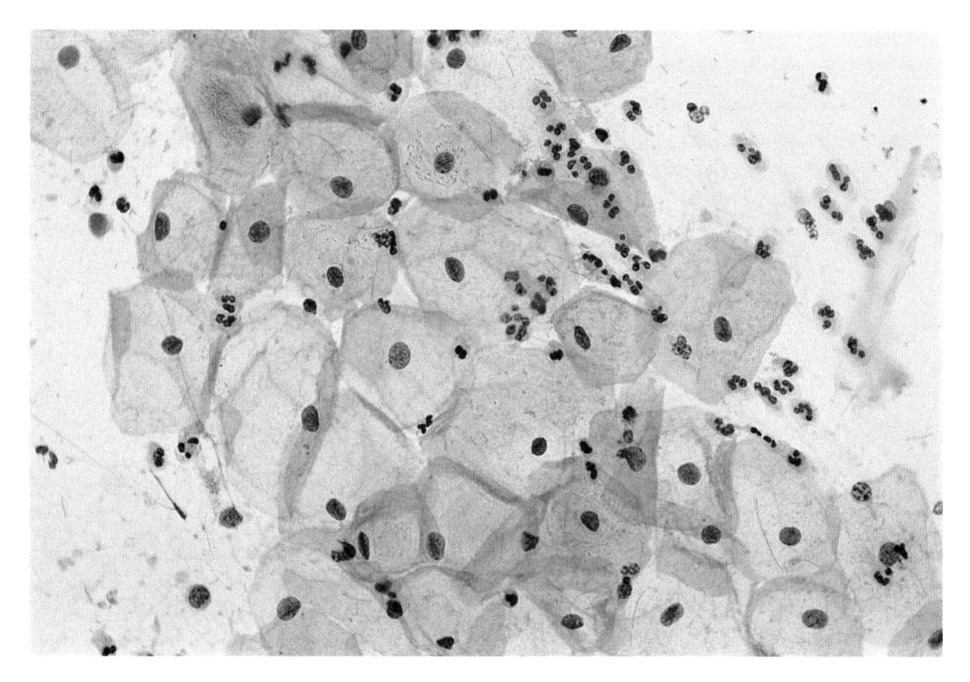

Figure 6.2 Intermediate cells in cervical smear. Note vesicular nucleus, transparent cytoplasm, and slightly rounded borders (Pap, ×400).

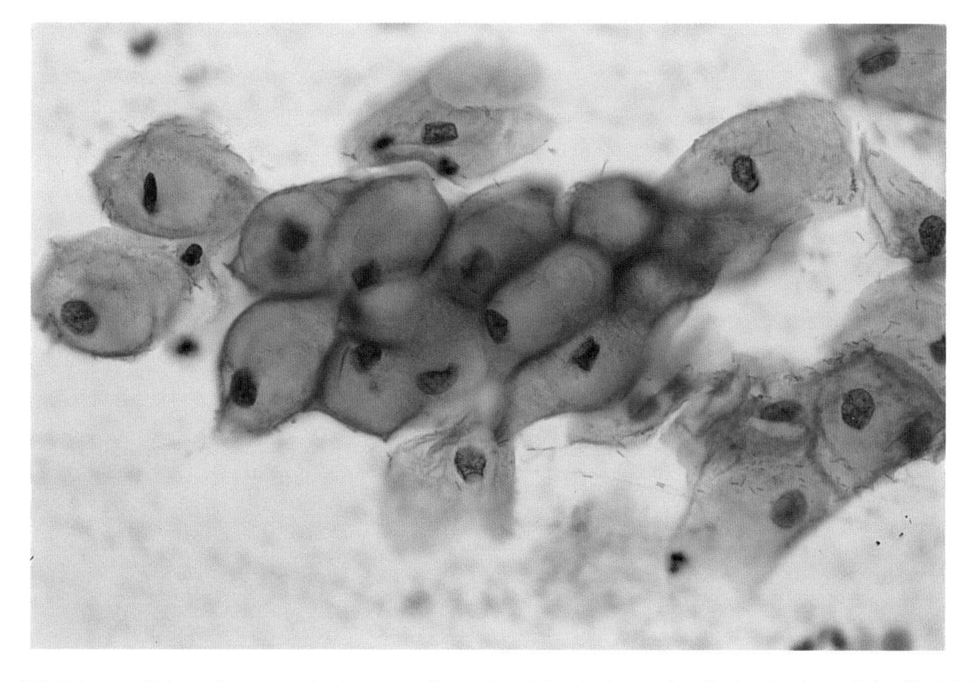

Figure 6.3 Intermediate cells in cervical smear. Occasional boat-shaped cells (navicular cells) with thickened cell borders and eccentric nuclei may be seen. These cells are commonly observed in pregnancy (Pap, ×630).

6.1.3 Parabasal cells (Figures 6.4–6.8)

These are found in atrophic smears such as those obtained post-partum or after the menopause. In such smears, they constitute the predominant epithelial cell type. Parabasal-like cells are also found intermixed with superficial cells in smears from cervices undergoing metaplastic change or epithelial regeneration.

The cells are small (15–30 µm in diameter) with a dense cytoplasm and granular nucleus containing an occasional chromocentre. The nucleus appears large because it occupies a much larger proportion of the cell than the nucleus of superficial or intermediate cells. The cytoplasm is generally cyanophilic, although in an air-dried smear the staining pattern may alter. When the cells are exfoliated spontaneously they appear rounded (Figures 6.4 and 6.5). Cells which have been forcibly detached with an Ayre spatula may appear as cell clusters (Figure 6. 6) or sheets of elongated cells (Figure 6.7). The parabasal cells in atrophic smears are particularly fragile. The cytoplasm frequently disintegrates during smear preparation (Figure 6.8).

6.1.4 Columnar cells (Figures 6.9–6.14)

These cells, which are derived from the epithelium lining the endocervical canal and crypts, may appear in smears as single cells (Figure 6.9) or as a sheet of cells (Figures 6.10 and 6.11). Single cells can be readily recognized when they are seen sideways on by their characteristic columnar shape, tall, delicate, often vacuolated cytoplasm and basal nucleus (Figure 6.9). Occasionally, as in Figure 6.9, the terminal plate supporting a tuft of cilia can be seen. The cell sheets often form a honeycomb pattern or a palisade of cells (Figures 6.10 and 6.11). The endocervical epithelium is fragile and cell morphology is often poorly preserved

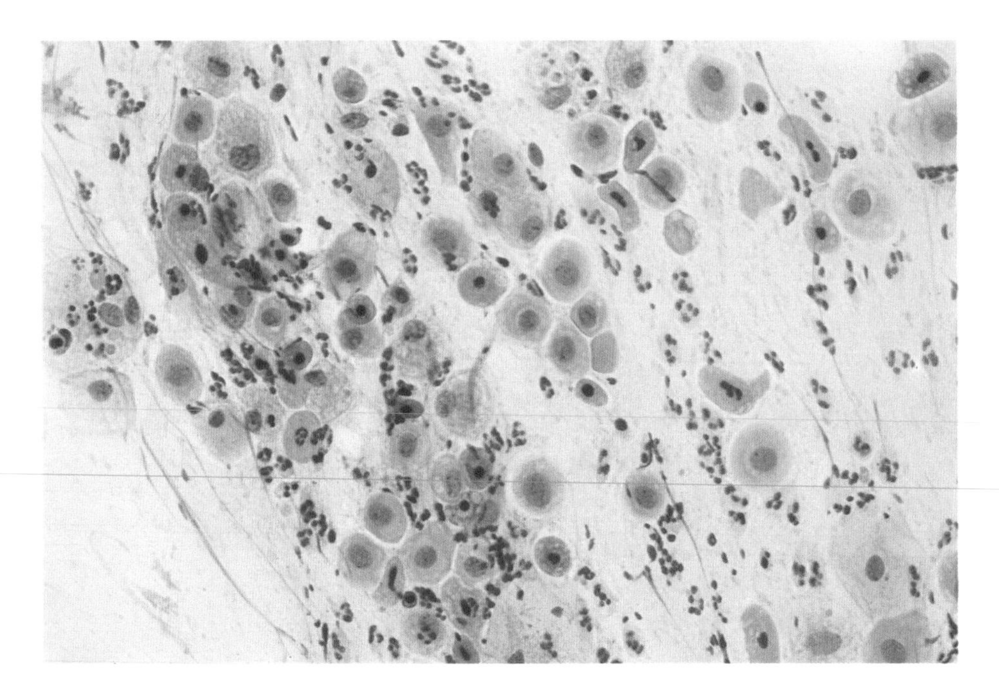

Figure 6.4 Parabasal cells in atrophic post-menopausal smear. Note rounded shape and dense cytoplasm. Many cells have a rounded nucleus occupying one-third of the cytoplasm. In occasional cells, the nucleus is pyknotic or undergoing karyorrhexis. The polymorphs in the background reflect the susceptibility of this fragile epithelium to infection (Pap, ×630).

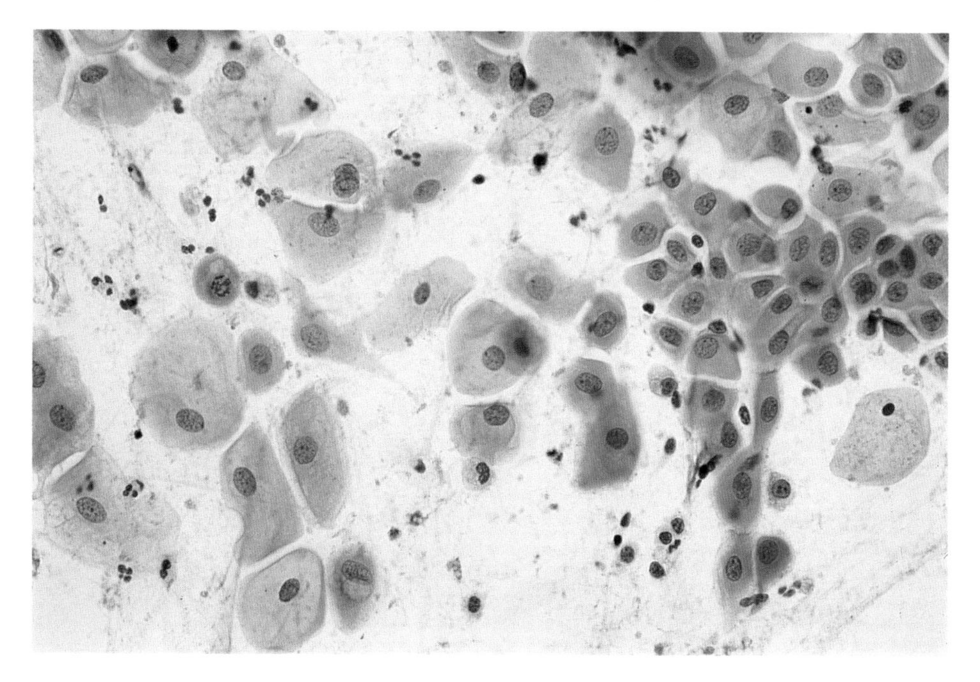

Figure 6.5 Post-partum smear showing a mixture of intermediate and parabasal cells reflecting incomplete maturation of the cervical epithelium during lactation (Pap, ×630).

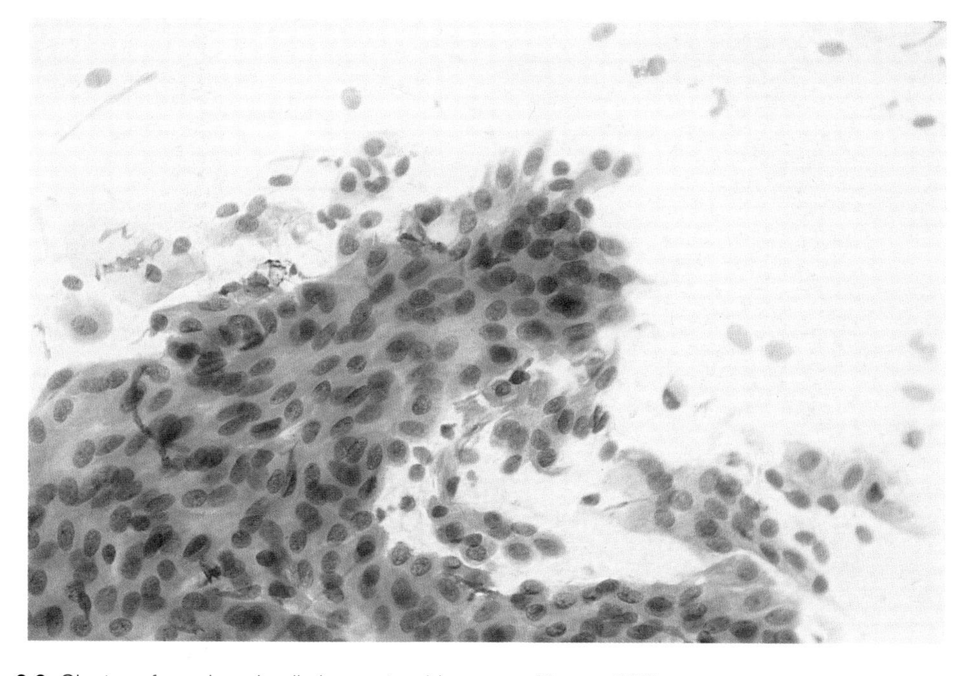

Figure 6.6 Cluster of parabasal cells in an atrophic smear. (Pap, ×630).

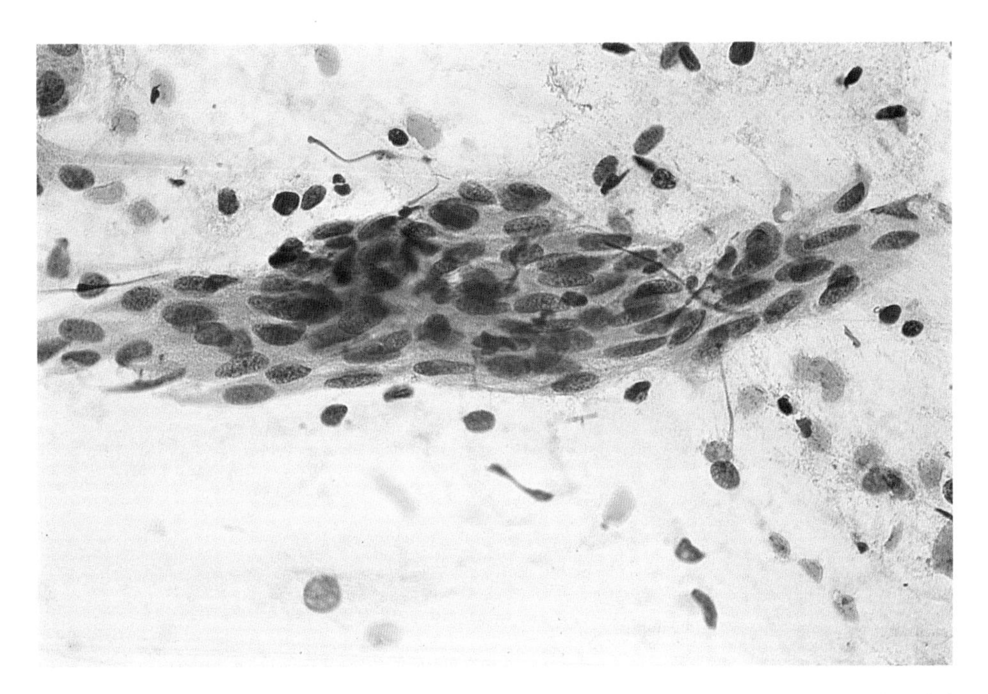

Figure 6.7 Parabasal cells are often detached in syncytial sheets and may appear elongated and drawn out (Pap, ×630).

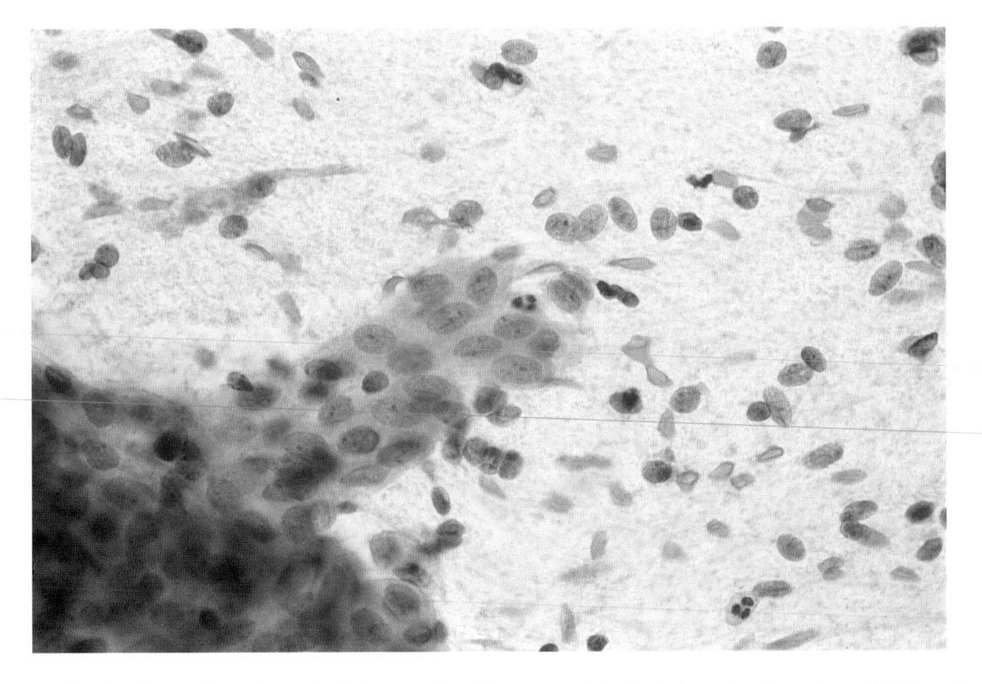

Figure 6.8 A thin sheet of parabasal cells in an atrophic smear. Most of the cytoplasm has disintegrated and only free nuclei are seen (Pap, ×630).

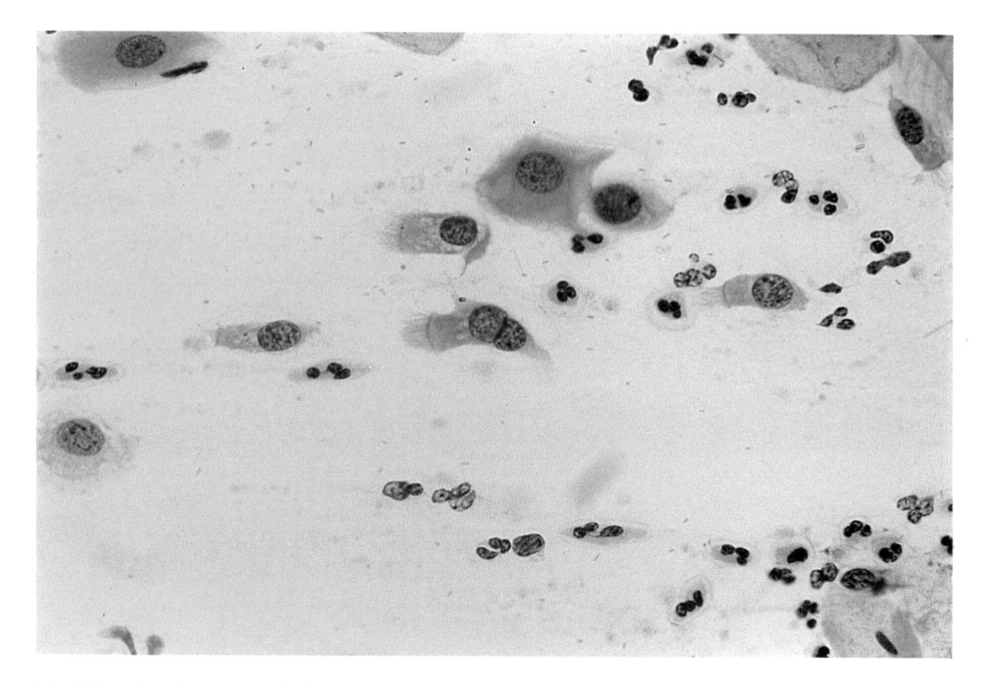

Figure 6.9 Ciliated columnar cells in a cervical smear (Pap, ×630).

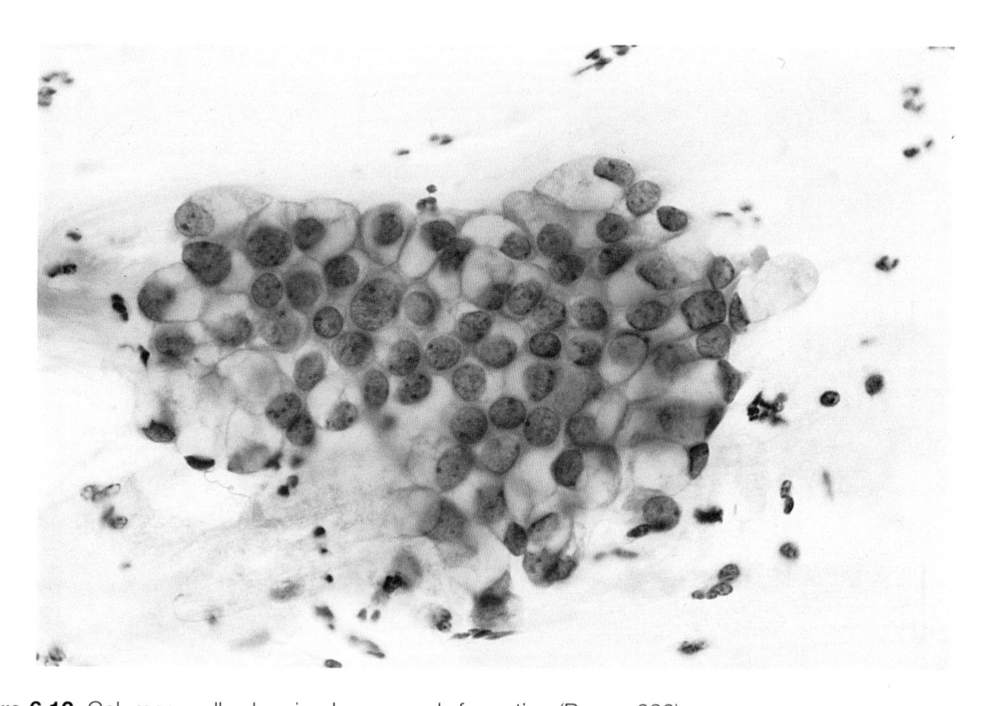

Figure 6.10 Columnar cells showing honeycomb formation (Pap, ×630).

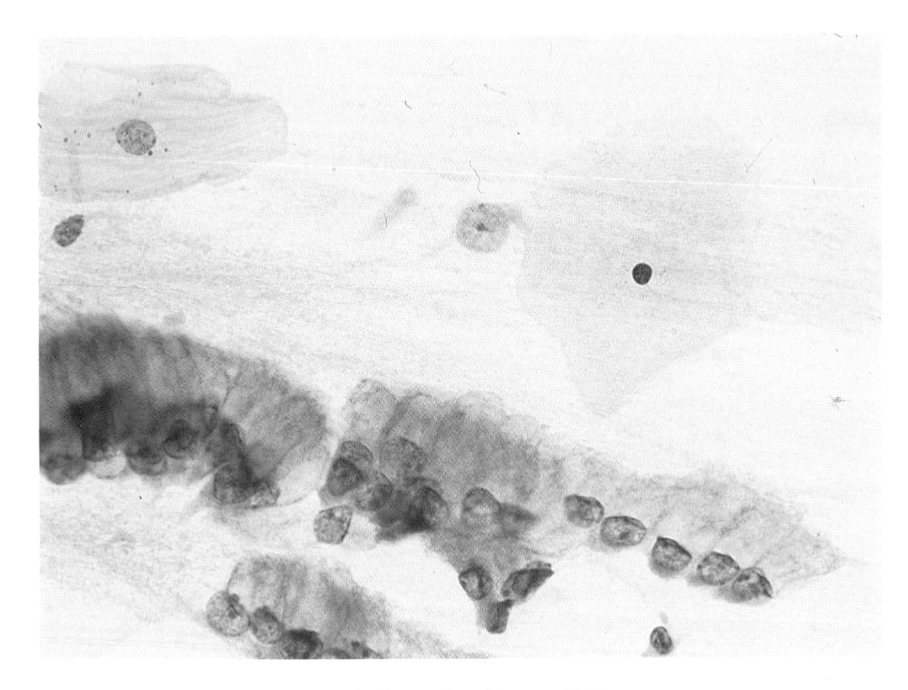

Figure 6.11 Columnar cells showing palisade formation (Pap, ×630).

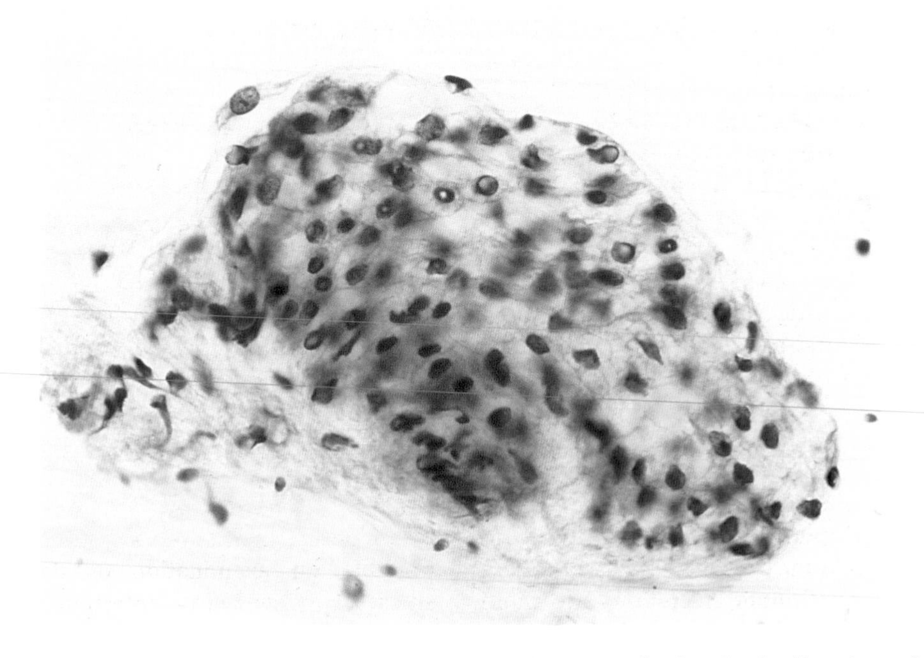

Figure 6.12 Poorly preserved endocervical cells in a cervical smear, reflecting the fragile nature of the cells (Pap, ×630).

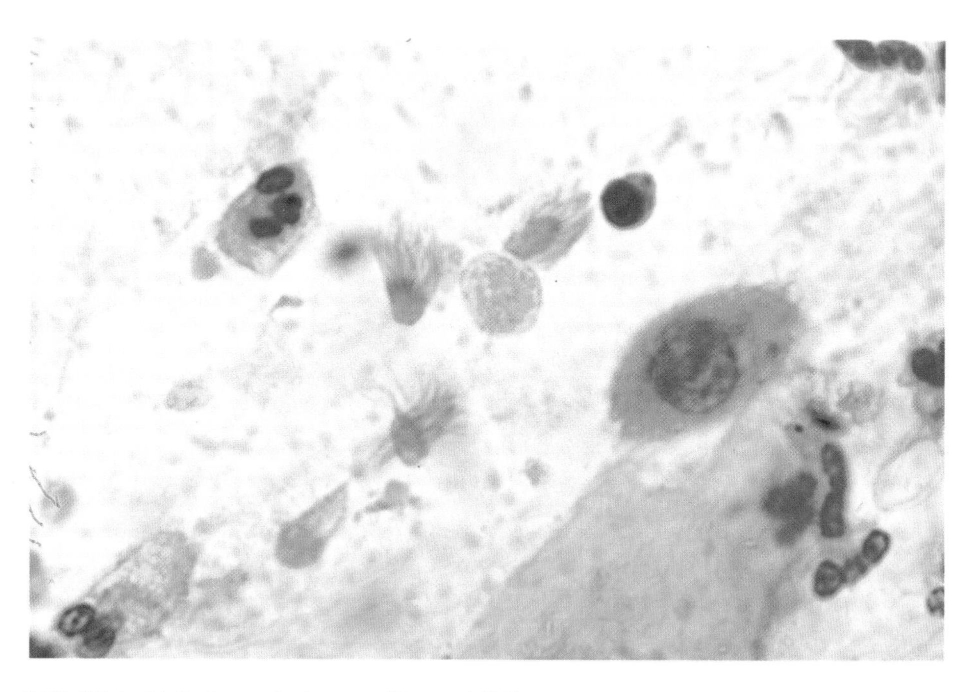

Figure 6.13 Ciliated tufts in cervical smear (Pap, ×1500).

(Figure 6.12). Endocervical cells can usually be distinguished from parabasal cells by their eccentric nuclei and delicate cytoplasm. Occasionally, vacuolated endocervical cells can be seen (Figure 9.14).

Very rarely, ciliated tufts of cytoplasm may be seen and may be mistaken for flagellate protozoa (Figure 6.13). These fragments are found in approximately 1% of smears, although the reason for their appearance in smears is unknown (Rivasi and Ghirardini, 1983). In pregnancy, the nuclei of the columnar cells develop nipple-like protrusions (Figure 6.14(a)).

The mucus produced by the columnar epithelium is a micellar network of complex glycoprotein which varies in its structure throughout the cycle. In the post-ovulatory phase of the cycle, the network is dense and thick so that sperm penetration is difficult. Under oestrogenic influences the mucin becomes much thinner, secretions are more profuse, watery and alkaline so that at ovulation, sperm penetration is easier. Smears made of the cervical mucus taken during different stages of the cycle show different patterns. At mid-cycle, a pattern of ferning (Figure 6.14(b,c)) can be seen.

6.2 NORMAL VAGINAL FLORA

A variety of organisms has been isolated from the vagina of women not suffering from inflammatory disease. The flora varies according to the physiological conditions present in the vagina. The most common saprophytes are *Gardnerella vaginalis (Corynebacterium vaginale)* and Döderlein bacilli, but streptococci (groups B and D), *Staphylococcus aureus*, micrococci, actinomyces, non-pathogenic *Neisseria*, *Pseudomonas*, mycoplasmas, *Leptothrix* and *Candida* species are also found. The precise identification of the majority of cocci and bacilli is not possible on examination of Papanicolaou smears. However, heavy infection with Döderlein bacilli, *Gardnerella vaginalis*, *Candida*, *Actinomyces* and *Leptothrix* can be recognized from the smear pattern.

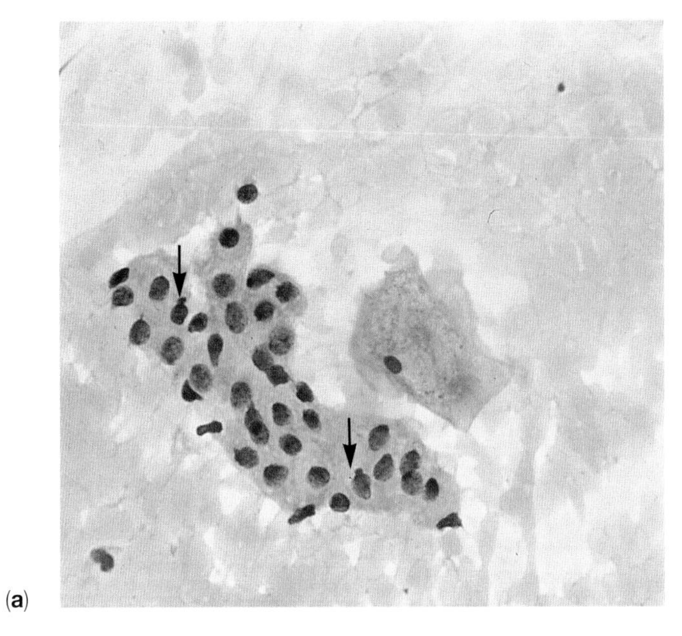

(**a**)

Figure 6.14 (**a**) Nipple-like processes (arrowed) on nuclei of endocervical cells. These may be artefacts of the preparation of the smear, or may reflect degenerative changes (Pap, ×630).

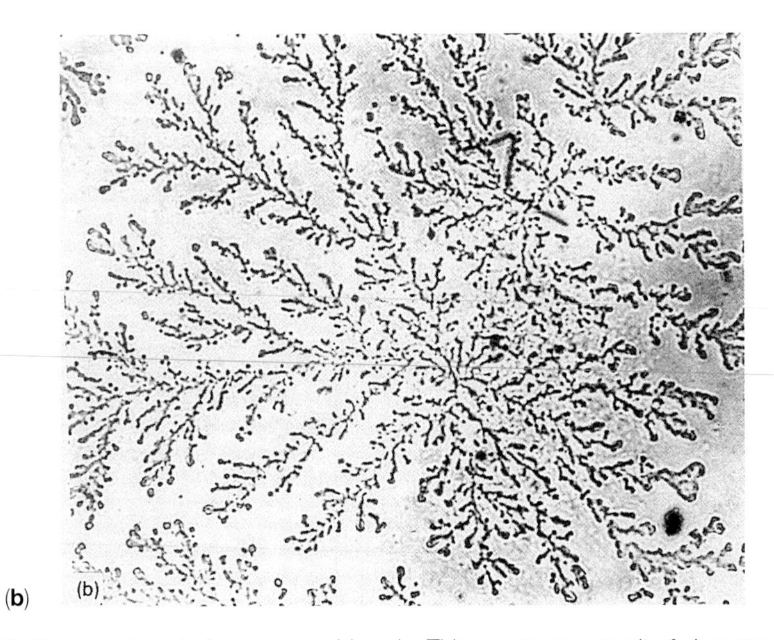

(**b**)

Figure 6. 14 (**b**) Ferning of cervical mucus at mid-cycle. This occurs as a result of changes in the consistency of cervical mucus at ovulation (phase contrast, ×330).

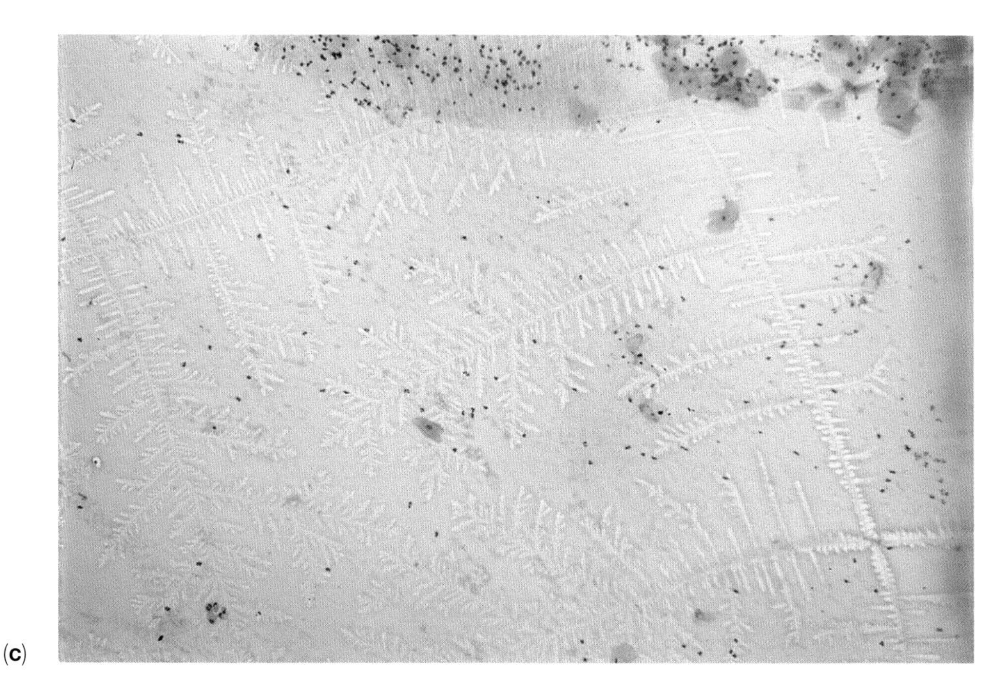

(**c**)

Figure 6.14 (**c**) Ferning mucus as it appears in a Papanicolaou-stained smear (Pap, ×100).

6.2.1 Döderlein bacilli (Figures 6.3 and 6.15)

These bacilli appear as a pale blue-staining rod 1–2 μm in length. They are abundant in smears from pregnant women, metabolizing the glycogen in the intermediate cells, causing cytolysis by destroying the cytoplasm and giving a ragged appearance to the cells. Naked nuclei are commonly seen.

6.2.2 *Gardnerella vaginalis* (Greenwood and Pickett, 1980)

This small Gram-negative, club-shaped organism stains dark blue with the Papanicolaou stain giving the smear a 'dirty' appearance. They tend to accumulate on the surface of large squamous cells which are often called 'clue cells' (Figure 6.16). The organism can be isolated from the genital tract of 20% of asymptomatic women. The number of polymorphs in smears containing clue cells are often rather scanty.

Gardnerella vaginalis is also linked with the pathological condition known as bacterial vaginosis (previously referred to as non-specific vaginitis). This condition is characterized by the presence of a thin, homogeneous vaginal discharge with a characteristic rotten fish smell. This becomes more pronounced on alkalization and can be evoked by placing a drop of potassium hydroxide solution on the fresh exudate on a slide or speculum used for the vaginal examination. The combination of raised vaginal pH (>4.5), the characteristic smell, and the presence of clue cells in the cervical smear is indicative of bacterial vaginosis.

The microbiology of bacterial vaginosis is complex. A variety of organisms including *Gardnerella vaginalis*, *Mobiluncus* species and

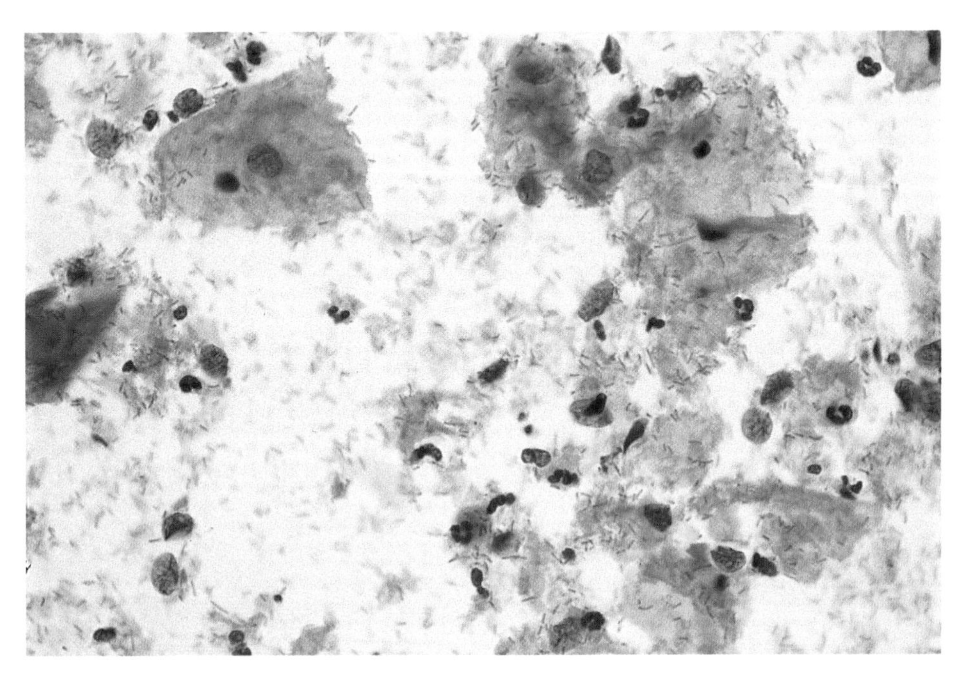

Figure 6.15 Döderlein bacilli in cervical smear. Note cytolysis of epithelial cells and numerous free nuclei. Such a smear would be unsuitable for analysis. These changes may be found in pregnancy (Pap, ×630).

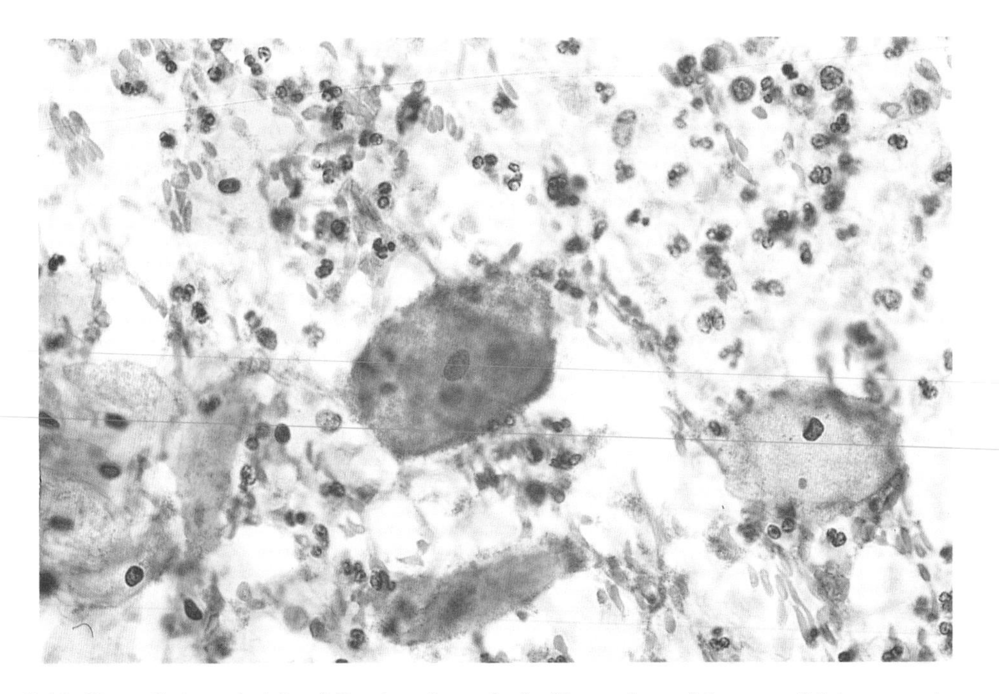

Figure 6.16 Clue cell characteristic of *Gardnerella vaginalis*. The surface of the superficial squame is covered with bacteria. The overall smear pattern is consistent with bacterial vaginosis (Pap, ×630).

Mycoplasma hominis, and anaerobic cocci are frequently isolated in this condition and overgrow the normal flora of the vagina. It is thought that the interaction of *Gardnerella vaginalis* with the above-named organisms is responsible for the development of symptoms.

6.2.3 *Candida albicans* (Figure 6.17)

This yeast-like fungus can be isolated from the genital tract of 20% of pregnant women. Its presence is associated with little damage to the epithelial lining of the cervix or vagina but it may be associated with an irritating white discharge. More often the patient is symptomless. The yeast and pseudomycelial forms form a mesh over the surface of the squamous cells, the filaments of the fungus staining faintly with eosin. The squamous cells themselves may show some degenerative changes such as perinuclear halos and slight nuclear enlargement. *Candida albicans* is a common finding in smears from diabetics and immunosuppressed patients but other *Candida* species and *Torulopsis glabrata* have occasionally been isolated from the vagina. They cannot be distinguished from *Candida albicans* by their appearance in the smear.

6.2.4 *Actinomyces* (Figure 7.29)

This group of bacteria is frequently found in smears from women fitted with an intrauterine contraceptive device (IUCD) (Bhagavan and Gupta, 1978; Traynor *et al.*, 1981). Morphologically the organism appears as a fluffy ball of bacteria 20–100 μm in diameter with slender branching threads streaming from it. Local inflammatory reaction is minimal; however, the real but rare risk of ascending infection should be borne in mind. It has recently been shown that non-pathogenic amoebae resembling *Entamoeba gingivalis* may be found in association with actinomycosis in patients fitted with an IUCD (Ruehsen *et al.*, 1980) (Figure 7.26(a)).

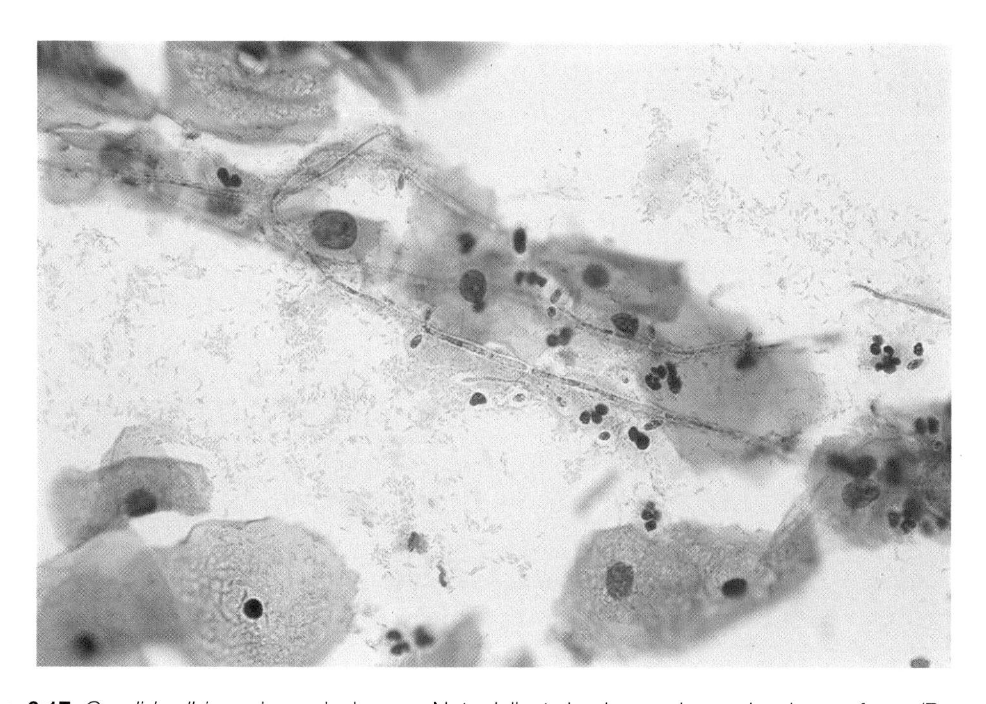

Figure 6.17 *Candida albicans* in cervical smear. Note delicate hyphae and occasional spore forms (Pap, ×630).

6.2.5 *Leptothrix vaginalis* (Figure 7.23)

These long filamentous non-pathogenic actinomyces-like organisms are easily recognized in smears. They vary from 0.4 to 40 μm in length. They are frequently found in association with vaginal trichomoniasis.

6.3 BASIC SMEAR PATTERNS

The types of epithelial cell seen in a normal cervical smear taken from the region of the external os are determined by:

- the degree of maturation of the cervical epithelium;
- the position of the squamocolumnar junction;
- the presence of metaplastic change in the cervix; and
- the stage in the menstrual cycle when the smear is taken.

These factors therefore will influence the general pattern of the smear.

6.3.1 Maturation of the epithelium

As mentioned in Section 5.2.2, the degree of maturation of the squamous epithelium of the female genital tract is hormone-dependent, although the squamous epithelium of the cervix is considerably less sensitive than that of the vagina.

Under the influence of unopposed oestrogen the epithelium thickens and the cells mature rapidly. Smears taken at this time contain an abundance of eosinophilic superficial cells with small pyknotic nuclei (Figure 6.1). Conversely, in the absence of oestrogen, the epithelium atrophies and the smear contains parabasal cells (Figure 6.4). In practice, the majority of cervical smears taken from women of childbearing age reflect an intermediate degree of epithelial maturation probably due to low levels of circulating oestrogen or the presence of endogenous progestogens or exogenous hormones administered as oral contraceptives or other therapy.

The hormonal status of the patient is not the only factor to influence maturation of the cervical epithelium. Superficial cells are frequently seen in the presence of acute infection, particularly *Trichomonas vaginalis*. Similarly, chronic irritation such as that induced by a pessary may result in thickening of an otherwise normal epithelium. Thus, the cervical smear is not always a reliable guide to the hormonal status of a patient and if this information is needed, smears from the upper third of the vaginal wall should be taken.

6.3.2 Position of the squamocolumnar junction

As explained in Section 5.2, the position of the squamocolumnar junction varies and, particularly after the menopause, may recede into the endocervical canal beyond the reach of the Ayre spatula. Smears taken from postmenopausal women rarely contain endocervical cells. Conversely, smears taken in the presence of ectropion contain abundant endocervical cells.

6.3.3 Presence of metaplastic changes in the cervix

Immature metaplasia or reserve cell hyperplasia of the cervix is reflected in smears by the presence of clusters of parabasal type cells among squamous cells of the superficial or intermediate type derived from more mature areas of squamous epithelium (see Section 9.1.3.(a)). This pattern contrasts with that found in an atrophic smear which is composed almost exclusively of parabasal cells (Section 6.5).

6.3.4 Stage in menstrual cycle at which smear is taken

Daily smears taken throughout the menstrual cycle will show some variation in the maturity

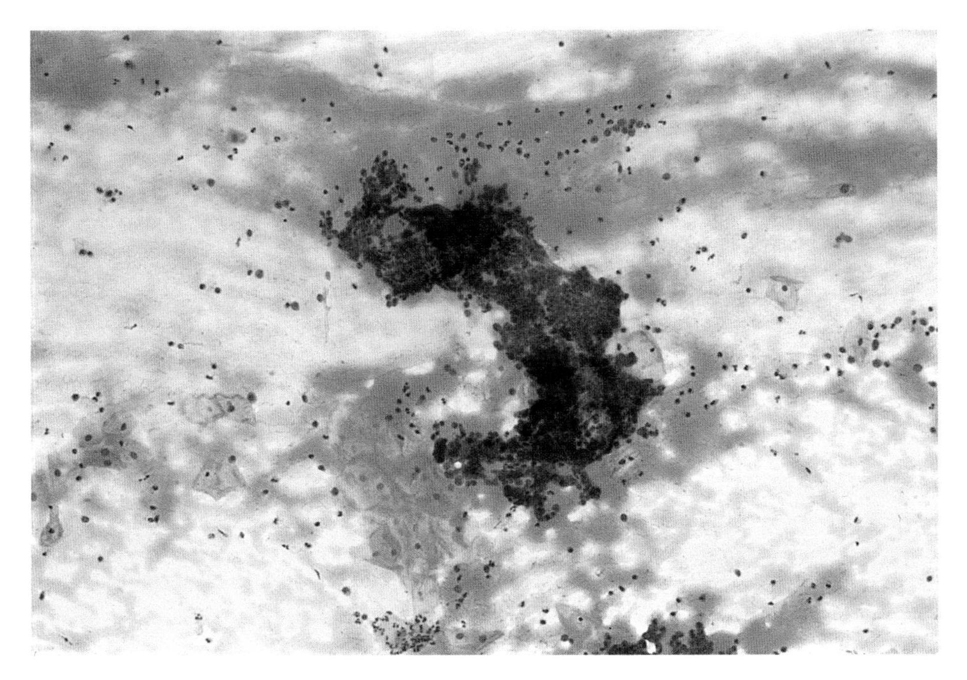

Figure 6.18 Blood-stained menstrual smear containing dense clusters of endometrial cells. Such a smear would probably be unsuitable for analysis (Pap, ×400).

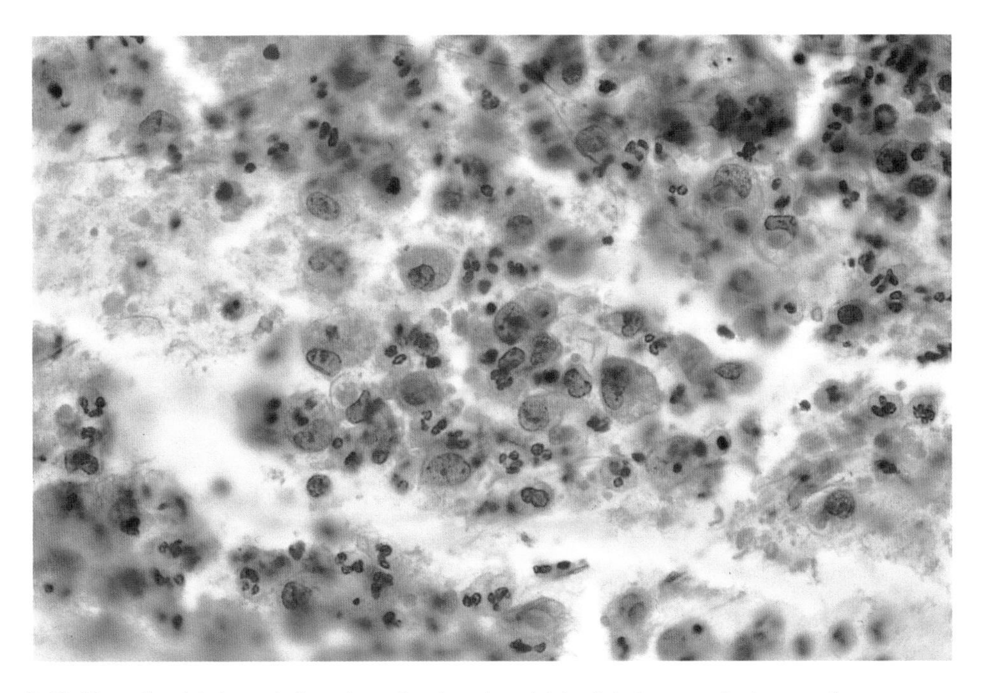

Figure 6.19 Necrotic debris and discrete cells of endometrial origin in a cervical smear from a woman with a history of post-menopausal bleeding, suggesting an endometrial rather than a cervical cause for her symptoms (Pap, ×630).

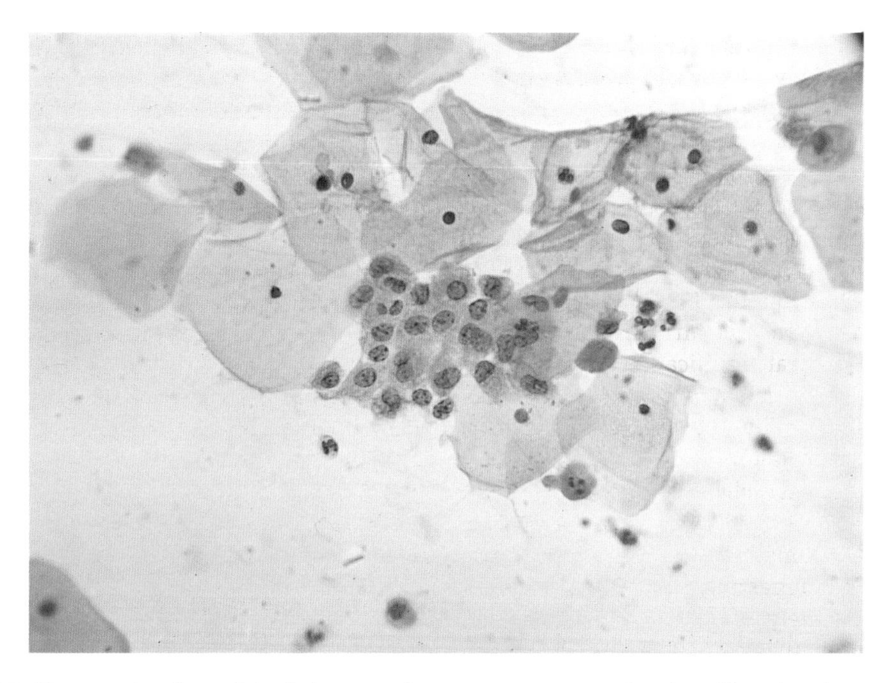

Figure 6.20 Clusters of endometrial cells in smear from post-menopausal patient. Note that the smear is well-oestrogenized. This is an abnormal finding and suggests ovarian and endometrial hyperplasia (Pap, ×400).

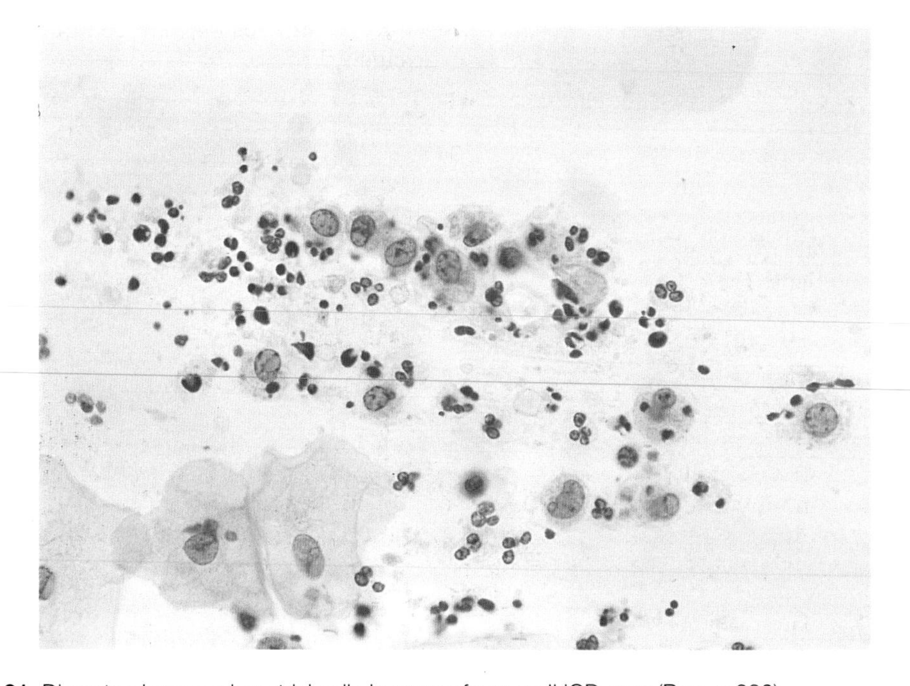

Figure 6.21 Discrete plump endometrial cells in smear from an IUCD user (Pap, ×630).

of the squamous epithelium, so that a smear taken mid-cycle will contain a majority of superficial cells whereas a smear taken just before the onset of menstruation will show a progesterone effect and will contain intermediate cells and possibly some navicular cells. Endometrial cells (Figures 6.18–6.23) may be found in smears taken during menstruation or during the first 10 days of the cycle. Their presence at any other time may have a pathological significance (Figure 6.19). Endometrial cells can be distinguished from endocervical cells by their small size (5–20 μm), delicate cyanophilic cytoplasm and rounded nucleus which has a coarse chromatin structure (Figures 6.20 and 6.21). Berry-like clusters of endometrial cells may be found (Figure 6.22), suggesting the presence of an IUCD; alternatively, endometrial cell clusters consisting of a dense core of stromal cells surrounded by a paler fringe of glandular cells (Figure 6.23) can be seen in menstrual smears.

Towards the end of menstrual flow, it is not uncommon to find numerous histiocytes in the smear (Figure 6.24), a finding that has been designated the 'exodus'. Streaks of delicate cells 15 μm in diameter, with round, oval or bean-shaped nuclei and vacuolated cytoplasm may be seen. The endometrial cells may show evidence of apoptotic change with fragmentation of the cell, chromatin clumping, margination of the chromatin and nuclear pyknosis. Histiocytes and endometrial cells may be found at any stage of the cycle in smears from patients fitted with an intrauterine contraceptive device (Figures 6.21 and 6.22) (see also Section 9.6.5.(b)). Benign endometrial cells may also be found in cervical smears from patients with menorrhagia or other menstrual disorders or hyperplastic changes in the endometrium. Occasionally, especially in women with an intrauterine device, the endometrial cells appear enlarged and atypical and care needs to be taken to distinguish them from adenocarcinoma.

Endometrial cells in a well-oestrogenized smear from a post-menopausal patient suggest endometrial pathology and further investigation of the patient may be indicated (Figure 6.20).

It is important to remember that the appearance of endometrial cells in cervical smears differs from their appearance in smears prepared as a result of endometrial aspiration or endometrial brushing. The endometrial cells in cervical smears have been shed spontaneously, are usually present as single cells or in small dense, three-dimensional clusters and appear degenerate and poorly preserved. In contrast, endometrial cells found in smears prepared from endometrial aspirates have been forcibly detached and are present in large epithelial cell sheets and appear well preserved. (For further information, see Morse, 1981.)

6.4 THE SMEAR IN PREGNANCY

The cytological pattern of pregnancy is that of an intensified progesterone effect, with clustering and folding of the intermediate cells which, in response to the gestational hormones, become packed with glycogen and assume a rounded shape. Such cells have been described as navicular cells and are commonly but not exclusively found in smears taken in the third trimester (Figure 6.3). Döderlein bacilli may cause a marked cytolysis (Figure 6.15) so that cytological interpretation is unreliable and a repeat smear is required post-partum. As term approaches and oestrogen levels rise the number of superficial cells in the smear increases. Smears taken post-partum are often atrophic with large glycogen-packed parabasal cells (Figure 6.5).

6.5 POST-MENOPAUSAL SMEARS

The menopause is associated with diminishing oestrogen activity which is reflected in the smear pattern. This can only be recognized when oestrogen activity is greatly reduced and the pre-menopausal multilayered epithelium is replaced by the thin dry epithelium of the post-menopausal woman (Figure 7.13(a)). The

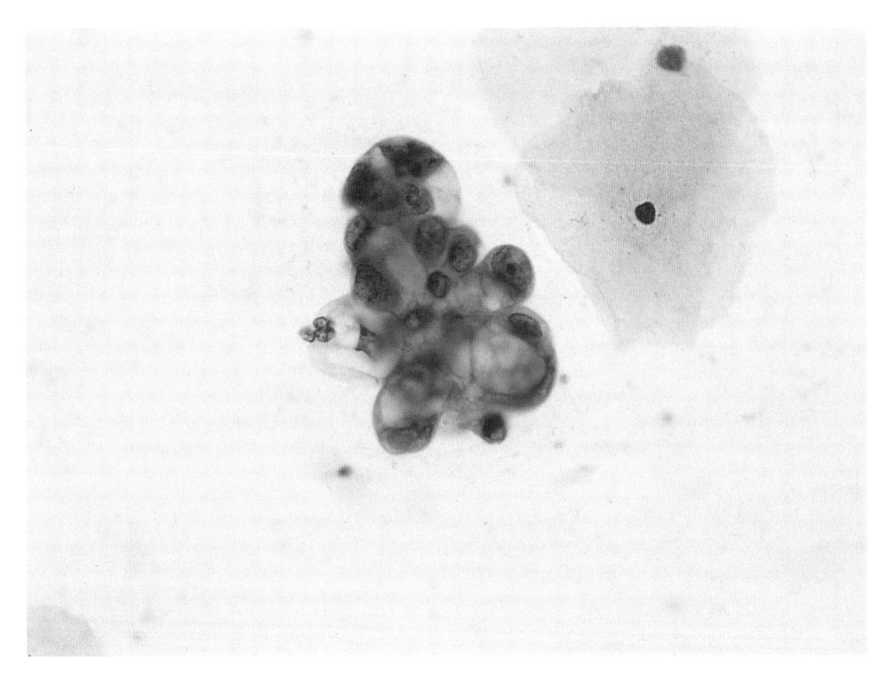

Figure 6.22 Berry-like clusters of endometrial cells associated with an IUCD (Pap, ×630).

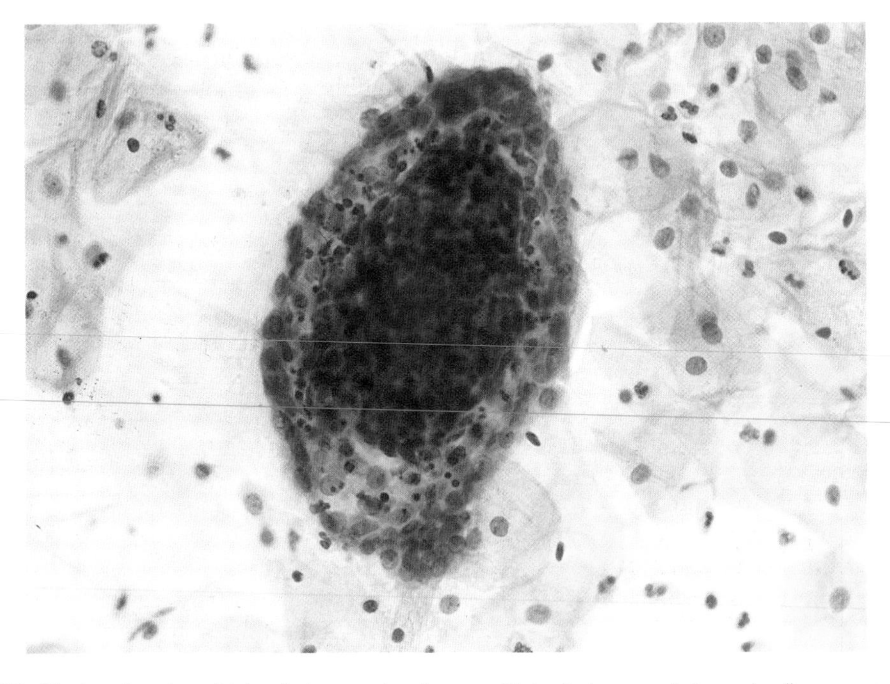

Figure 6.23 Cluster of endometrial cells in menstrual smear. Note dark core of stromal cells surrounded by ring of glandular endometrial cells (Pap, ×630).

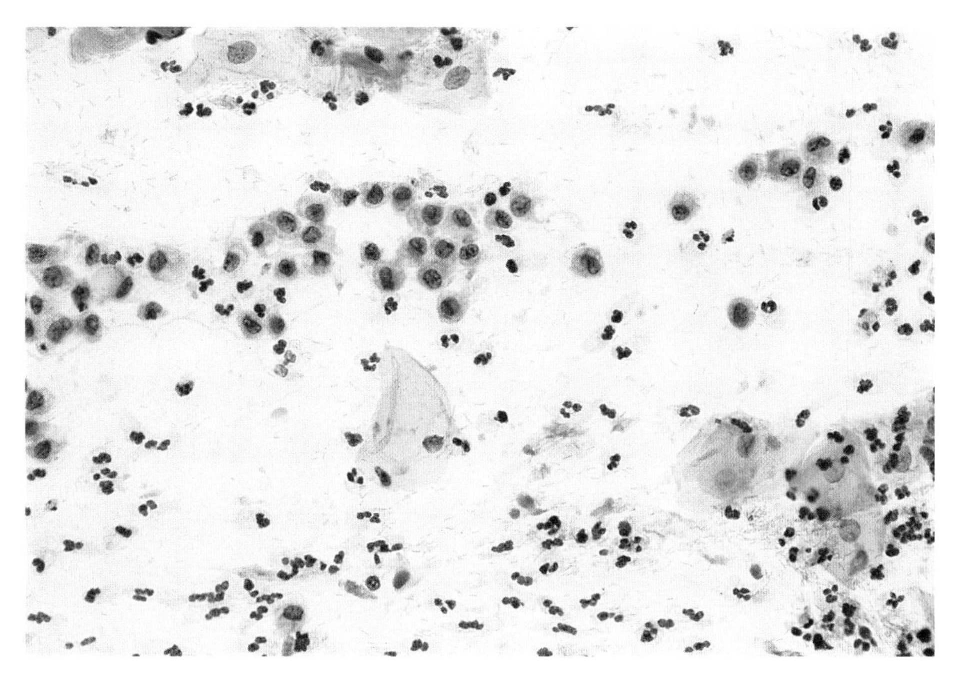

Figure 6.24 Histiocytes in cervical smear. These very delicate cells with vacuolated cytoplasm and bean-shaped, oval or round nuclei are frequently found in menstrual smears (Pap, ×630).

smear is composed of scanty parabasal cells, either as discrete rounded cells (Figure 6.4), in thin sheets (Figure 6.7), or as free nuclei (Figure 6.8). As the smear is thin, the cells may dry before they are fixed. They stain poorly and morphological detail is often indistinct (Figure 6.25). Waddell (unpublished data, 1993) found that women with atrophic cervices are three times more likely to have inadequate smears than women with non-atrophic cervices. She concluded that narrowing of the cervical os and bleeding are the main causes of loss of smear quality in post-menopausal women and special training in smear taking is needed to overcome this problem. Infection is common (Figure 7.13(a,b)). It can be difficult to distinguish between endocervical cells and parabasal cells in atrophic smears, especially as the number of endocervical cells may be small. As the squamocolumnar junction recedes into the canal the endocervical epithelium may remain out of reach of the spatula.

Dense basophilic mucous plugs in the smear resembling free hyperchromatic nuclei ('blue blobs') may be confused with malignant cells (Figure 6.26). When the diagnosis is in doubt it is useful to administer a short course of oestrogen therapy to the patient and repeat the smear in a few days. A normal low proliferative pattern will be seen.

6.6 WHAT CONSTITUTES AN ADEQUATE SMEAR?

An adequate smear should contain abundant squamous cells which are evenly spread and well displayed. It should also contain evidence of transformation zone sampling in the form of endocervical or immature metaplastic cells and strands of mucus. The criteria for judging the adequacy of a cervical smear are set out below. They reflect the opinion of experts in the field but are by no means definitive.

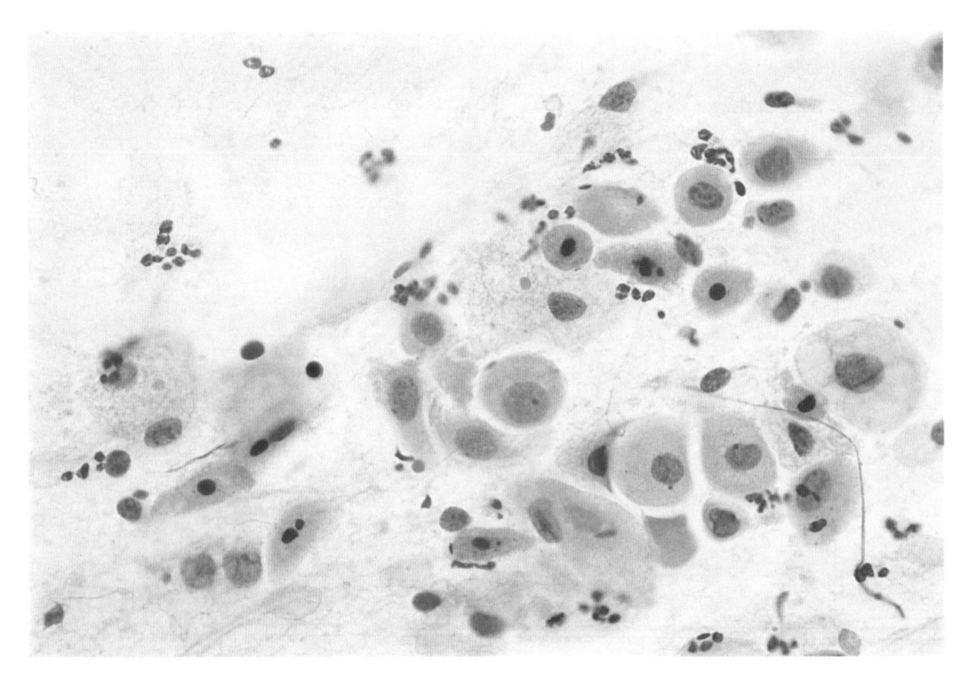

Figure 6.25 Atrophic smear showing typical post-menopausal pattern. The smear is composed of rounded parabasal cells and polymorphs. The smears contain little mucus and frequently dry before fixation so that the cytoplasmic borders and nuclear structure appear faded and indistinct (Pap, ×630).

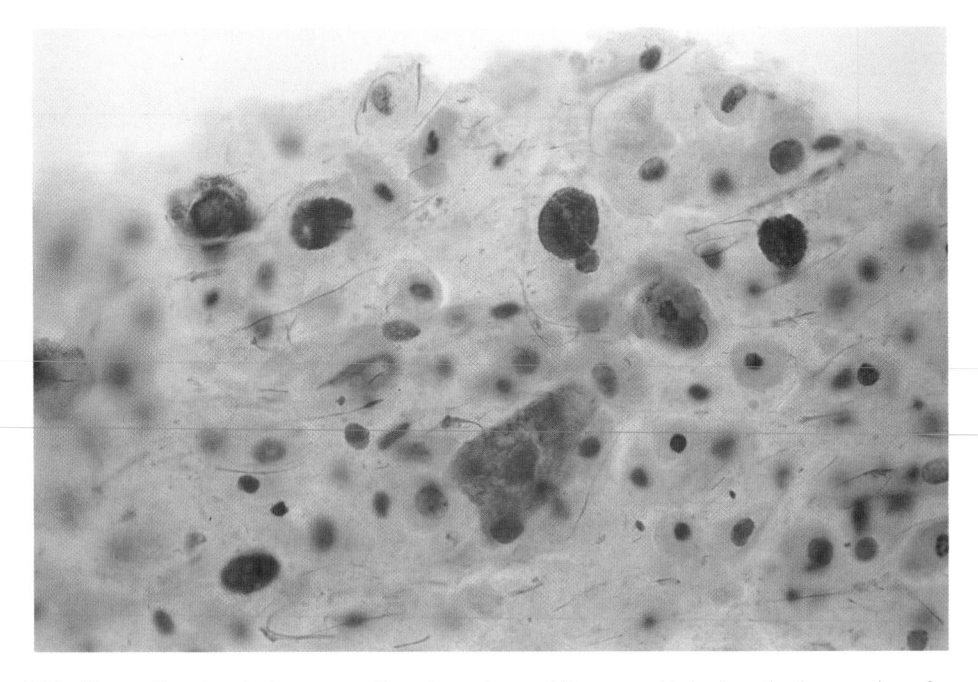

Figure 6.26 Plugs of inspissated mucus, with a dense basophilic core which gives the impression of a nucleus, may be mistaken for tumour cells. The variation in size and lack of cellular structure are useful distinguishing features. These plugs are frequently found in atrophic smears (Pap, ×630).

According to the Bethesda system for reporting cervicovaginal diagnoses (1993, pp. 14, 15) four elements need to be considered when assessing the adequacy of a smear. These are:

1. **Patient and specimen identification**: this must be accurate to permit tracing of prior records and prevent errors in assigning reports.
2. **Clinical information**: this can assist with the interpretation of the smear and the management of the patient and should include, at the very least, date of birth, menstrual status, pregnancy status, information about current contraceptive use, hormonal or drug therapy or radiotherapy therapy, history of previous abnormal smears or cancer, and notification of symptoms which might indicate neoplastic disease in the genital tract. A description of the appearance of the cervix is helpful.
3. **Technical interpretability**: factors which render cervical smears inadequate for light microscopy include air-drying and poor fixation. Under these conditions, the epithelial cells in the smears do not take up the Papanicolaou stain and cannot be visualized clearly in the light microscope (techniques are available for rehydrating air-dried smears for subsequent Papanicolaou staining although the authors have no personal experience of this approach). Similarly, smears that are poorly fixed, improperly dehydrated, or are broken are unsuitable for reliable cytological diagnosis. Cytolytic smears where the cells are poorly preserved are inadequate for cytology. Smears that contain large numbers of leucocytes, red blood cells, spermatazoa (Figure 6.27) or other contaminants which obscure the epithelial cells and prevents their interpretation in the light microscope must also be considered inadequate. A scanty smear in which less than one-third of the slide (322 mm^2) is covered with epithelial cells should not be accepted as evidence of satisfactory sampling. Atrophic smears are an exception to this rule. *It is important to remember that smears that contain abnormal cells should never be categorized as inadequate.*
4. **Cellular composition and evidence of transformation zone sampling**: one of the key factors which determines the effectiveness of the Papanicolaou smear test as a method of preventing cervical cancer is the quality of the smear. The chances of the test detecting precancerous changes in the cervix are greatest when the smear is taken from the transformation zone as this is the site where cancerous changes are most likely to arise. Thus, if a smear has been properly taken it should contain evidence of transformation zone sampling.

At the first Bethesda workshop (1988) it was suggested that presence of endocervical cells, metaplastic cells or strands of cervical mucus in the smear should be regarded as acceptable indicators of transformation zone sampling. Clearly this is not necessarily the case as endocervical cells in a smear may have originated high in the endocervical canal rather than at the squamocolumnar junction and mature metaplastic cells are hard to identify. Nevertheless, according to the Bethesda system of reporting (1993), the presence of two or more endocervical or metaplastic cell clusters each containing a minimum of five cells is regarded as acceptable evidence of transformation zone sampling

The significance of endocervical cells in cervical smears has long been the subject of debate among cytologists. It has been shown that the number of epithelial abnormalities that can be diagnosed is significantly related to the presence or absence of endocervical cells in the smear. Vooijs *et al.* (1985) claim that CIN is more likely to be detected in smears that contain endocervical cells than in smears that do not and therefore consider that endocervical cells should be identified before a smear is pronounced 'adequate'. Conversely, short-term studies have shown that women whose smears do not contain endocervical cells, do not have an increased risk of CIN being missed (Kivlahan and Ingram, 1986). Mitchell and

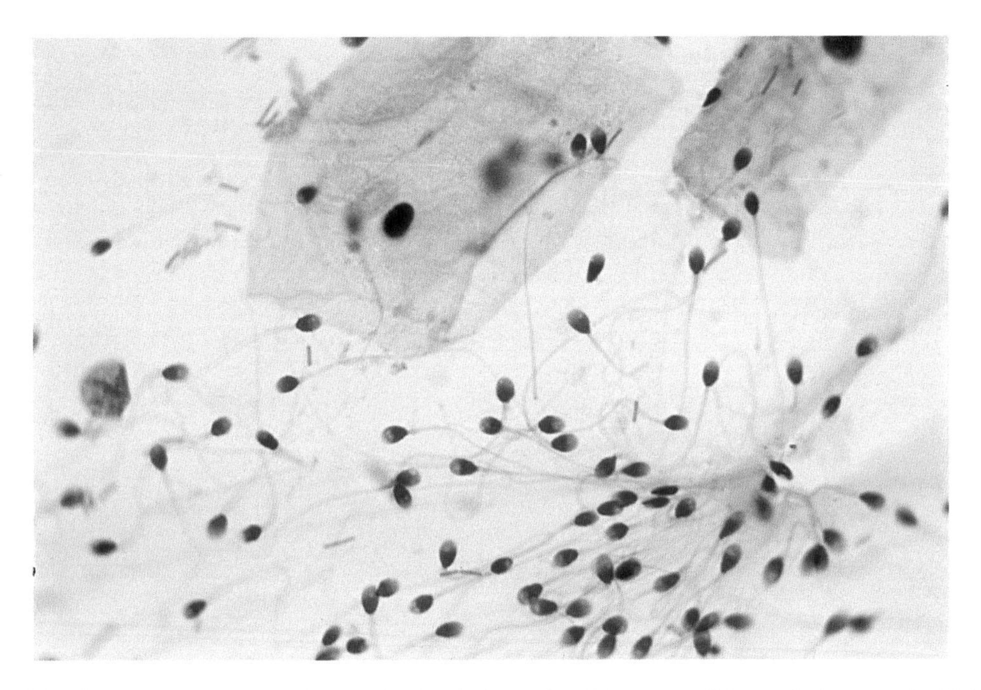

Figure 6.27 Spermatozoa are found as contaminants in post-coital smears. Other cells from seminal fluid may also be present (Pap, ×1000).

Medley (1991, 1992) found that CIN was just as likely to be detected in smears which do not contain endocervical cells as in smears which do. They conclude that the absence of endocervical cells in smears should not be construed as evidence of unsatisfactory sampling or, *per se*, as an indicator of the need for a repeat smear. However, as it can sometimes be very difficult to distinguish endocervical cells from metaplastic cells, endometrial cells or even histiocytes in cervical smears, the validity of these observations can be challenged.

In an attempt to resolve the problem relating to the significance of endocervical cells in cervical smears, the classification proposed by the Bethesda system includes a category of reporting which permits smears to be classified as *'adequate but limited by –'*. This phrase should be applied to smears which are adequate in all respects except that they do not contain endocervical cells. This terminology is not recommended in the UK. In the last analysis, the cytologist is highly dependent on the skill of the smear taker for the quality of the smears he/she receives and a good smear-taking technique will minimize the risk of diagnostic error due to failure to sample the transformation zone.

REFERENCES

Bhagavan, B. S. and Gupta, P. K. (1978) Genital actinomycosis and intrauterine contraceptive devices. *Hum. Pathol.*, **9**, 567–578.

Evans, D. M. D., Hudson, E. A., Brown, C., Boddington, M. M., Hughes, H., Mackenzie, E. F. D. and Marshall, T. (1986) Terminology in gynaecological cytopathology: report of the working party of the British Society for Clinical Cytology. *J. Clin. Pathol.*, **39**, 933–944.

Greenwood, J. C. and Pickett, M. J. (1980) Transfer of *Haemophilus vaginalis* (Gardner and Dukes) to a new genus *Gardnerella*. *Int. J. Syst. Bacteriol.*, **30**, 170.

International Academy of Cytology (1958) Opinion poll on non-malignant cytological terminology. Symposium on terminology. *Acta Cytol.*, **2**, 2S139.

Kivlahan, C. and Ingram, E. (1986) Papanicolaou smears without endocervical cells. Are they adequate? *Act Cytol.*, **30**, 258–260.

Mitchell, H. and Medley, G. (1991) Longitudinal study of women with negative smears according to endocervical cell status. *Lancet*, **337**, 265–267.

Mitchell, H. and Medley, G. (1992) Influence of endocervical status on the cytological prediction of cervical intraepithelial neoplasia. *Acta Cytol.*, **36**, 875–880.

Morse, A. R. M. (1981) The value of endometrial aspiration in gynaecological practice, in *Advances in Clinical Cytology* (eds L. G. Koss and D. V. Coleman). Butterworths, London, pp. 44–59.

Riotton, G. and Christopherson, W. M. (1973) *Cytology of The Female Genital Tract* (International Histological Classification of Tumours, No. 8). World Health Organization, Geneva.

Rivasi, F. and Ghirardini, C. (1983) Ciliated bodies in cervical cytology. *da Pathologica*, **75**, 375–381.

Ruehsen, M. de M., McNeill, R. E., Frost, J. K., Gupta, P. K., Diamond, L. S. and Aoniberg, B. M. (1980) Amoeba resembling *Entamoeba gingivalis* in the genital tract of IUD users. *Acta Cytol.*, **24**, 413–420.

Spriggs, A. I., Butler, E. B., Evans, D. M. D., Grubb, G., Husain, O. A. N. and Wachtel, E. E. (1978) Problems of cell nomenclature in cervical cytology smears. *J. Clin. Pathol.*, **31**, 1226–1227.

Traynor, R. M., Parratt, D., Duguid, H. L. D. and Duncan, I. D. (1981) Isolation of actinomycetes from cervical specimens. *J. Clin. Pathol.*, **34**, 914–916.

Vooijs, G. P., Elias, A. G., Van der Graaf, Y. and Veling, S. (1985) Relationship between the diagnosis of epithelial abnormalities and the composition of cervical smears. *Acta Cytol.*, **29**, 323–328.

7

Infection, inflammation and repair

Inflammatory changes in the cervix may develop as a result of physical or chemical damage to the cervical mucosa, e.g. during parturition or local douching, or may be the consequence of infection with a variety of microorganisms (Table 7.1). Infection may be acquired from an external source as the result of sexual intercourse with an infected partner or by contact with infected fomites. Alternatively the infection may be endogenous as a result of direct spread from the uterus or vagina. A common cause of cervicovaginitis is overgrowth of organisms that are normally commensal in the vagina (Section 6.2) due to altered immune response, change in the pH of the vagina or hormonal activity. In contrast, blood-borne organisms rarely cause cervicitis.

Normally, the cervix is protected from infection by the multilayered stratified squamous epithelium that covers it. Invasion by pathogens is unlikely to be successful as long as this epithelium is intact. However, any breach in the squamous epithelium (e.g. in pregnancy), or any reduction in thickness (e.g. after the menopause) makes the cervix very vulnerable to infection. For the same reason, the presence of ectropion or large areas of immature metaplastic change predispose to infection. As the vaginal and the cervical mucosae are continuous, co-infection is common.

The histological and cytological changes associated with an acute or chronic inflammatory process in the cervix are generally of a non-specific nature and identification of the pathogens causing the cervicitis depends on culture and Gram stain. A few infectious agents, e.g. herpes simplex, papillomaviruses, *M. tuberculosis* and *Trichomonas vaginalis*, induce changes which permit a preliminary diagnosis to be made on the strength of the histological and cytological findings. It should be remembered that in routine Papanicolaou-stained smears it is only possible to recognize the shape and size of bacteria and their distribution in the smear. The appearance may suggest a particular infection but precise identification depends on Gram stain and microbiological culture.

The histological and cytological findings due to some commoner infections is given in Table 7.2.

7.1 ACUTE NON-SPECIFIC CERVICITIS

This may reflect local response to injury or be part of a general cervicovaginitis induced by

Table 7.1 Common cervical pathogens

Viruses	Herpes simplex Cytomegalovirus Human papillomavirus
Bacteria	*Streptococcus* spp. *Staphylococcus aureus* *Neisseria gonorrhoeae* *Escherichia coli* *Gardnerella vaginalis* (syn. *Haemophilus vaginalis corynebacterium vaginalis*) *Clostridium welchii* *Listeria monocytogenes* *Mycobacterium tuberculosis* *Actinomyces israelii* *Proteus* spp. *Treponema pallidum* *Bacteroides fragilis* *Haemophilus ducreyi* – (chancroid) *Calymmatobacterium granulomatis* (granuloma inguinale)
Bedsonia	*Chlamydia trachomatis* (L1, L2, L3)
Fungi	*Candida* spp.
Parasites	*Trichomonas vaginalis* *Entamoeba histolytica* *Schistosoma haematobium* *Schistosoma mansoni* *Filaria bancrofti*

microorganisms that more frequently occur as commensals in the vagina. Common associated factors are obstruction by pessary or tampon, post-menopausal atrophy, parturition and radiotherapy (Figure 7.1). The organism most frequently isolated is *Gardnerella vaginalis* (Pheifer *et al.*, 1978; Ansel *et al.*, 1983) but it is doubtful how often it produces inflammatory change (Section 6.2.2 and Figure 6.16). Important causes of acute cervicitis are *Neisseria gonorrhoeae*, which usually occurs as a result of intercourse with an infected individual and puerperal infections by streptococci or staphylococci.

7.1.1 Histology

The acute inflammatory response to *N. gonorrhoeae* or streptococci is initially located in the stroma and is characterized by hyperaemia of the papillary vessels followed by exudation of fluid into the surrounding tissue and migration of polymorphs from the blood stream to the site of infection. The epithelial cells may show considerable degenerative change (Figures 7.2 and 7.3), the endocervical

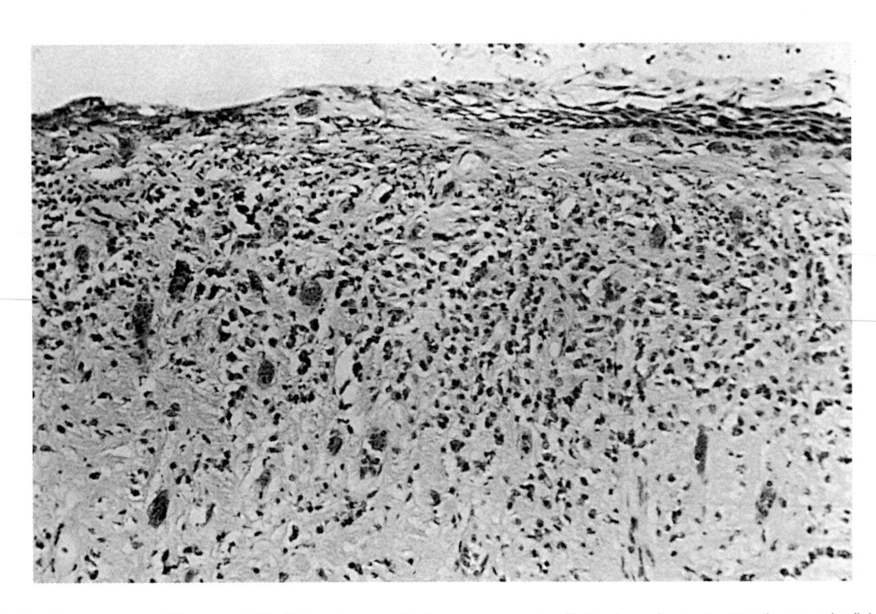

Figure 7.1 Acute non-specific cervicitis following radiotherapy; note dilated and congested vessels (H & E, ×190).

Table 7.2 Infection of the uterine cervix

Disease findings	*Causitive organism*	*Typical histological findings*	*Characteristic cytological findings*	*Confirmatory test*
1. Acute bacterial cervicitis	Streptococci Staphylococci	Acute inflammatory changes	Inflammatory exudate Epithelial cells show degenerative changes	Culture, Gram stain
2. Gonorrhoea	*Niesseria gonorrhoeae*	Necrotizing cervicitis	As above	Gram stain, culture
3. Tuberculosis	*Mycoplasma tuberculosis*	Granulomatous change with caseation	Langhans' cells, endothelioid cells	Ziehl–Neelsen stain
4. Syphilis	*Treponema pallidum*	Ulceration, granulomatous reaction without caseation	As above	Dark-field microscopy, immunofluorescence
5. Lymphogranuloma	*Chlamydia trachomatis (L1, 2, 3 serotypes)*	Granulomatous reaction without caseation	As above	Culture, immunofluorescence
6. Chlamydia (TRIC) infection	*Chlamydia trachomatis*	Follicular cervicitis	Streaks of lymphoid cells	Culture, immunofluorescence
7. Granuloma inguinale	*Calymmatobacterium*	Granulomatous reaction without caseation	Donovan bodies	Giemsa stain, silver stain
8. Genital herpes	Herpes simplex virus	Necrotizing cervicitis Multinucleate giant cells Intranuclear inclusions	Multinucleate giant cells Intranuclear inclusions	Virus isolation, electron microscopy, immunofluorescence
9. Cytomegalic inclusion disease	Cytomegalovirus	Discrete amphophilic intranuclear inclusions in glandular epithelium	Discrete intranuclear inclusion-bearing cells	Virus isolation
10. Condyloma acuminatum	Human papillomavirus (6, 11)	Papillomatosis acanthosis, parakeratosis, koilocytosis	Koilocytosis, dyskeratosis dyskaryosis	Electron microscopy, immunocytochemical staining, DNA hybridization
11. Non-condylomatous wart virus infection	Human papillomavirus (6, 11, 16 and 18)	Koilocytotic atypia (associated with CIN)	Koilocytosis, dyskeratosis dyskaryosis, multinucleation	Electron microcopy, immunocytochemical staining, DNA hybridization
12. Trichomoniasis	*Trichomonas vaginalis*		Trichomonads, lilac hue to smear, perinuclear halos	Wet film, culture
13. Actinomycosis	*Actinomyces israelii*	Granules in colonies	Mycelial-like organisms *Entamoeba gingivalis*-like organisms	Culture under anaerobic conditions
14. Schistosomiasis	*Schistosoma mansoni* *S. haematobium*	Granulomatous lesions Ova present often calcified Pseudoepitheliomatous hyperplasia	Ova present	

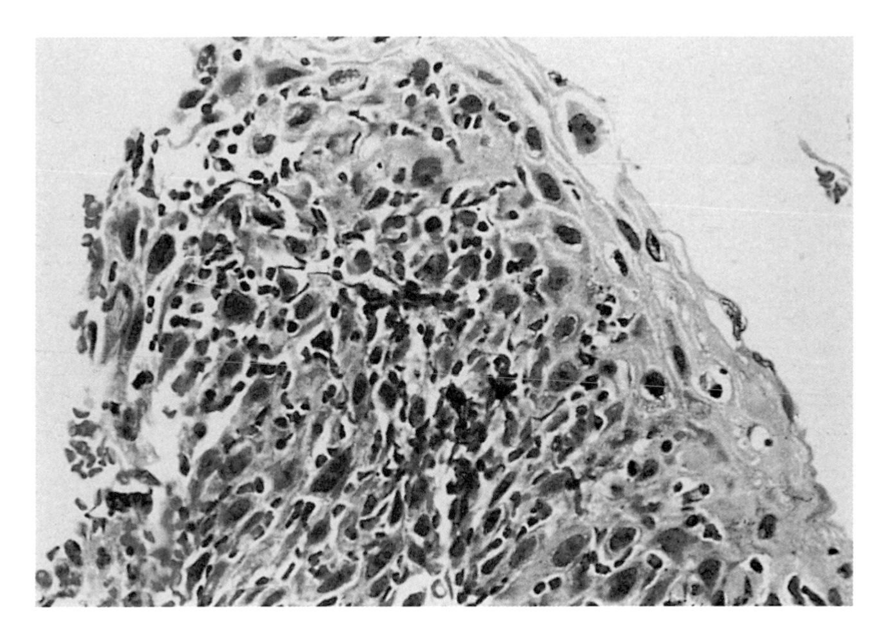

Figure 7.2 Acute cervicitis due to herpes infection, showing polymorph infiltration and degenerative changes of the squamous epithelium (H & E, ×310).

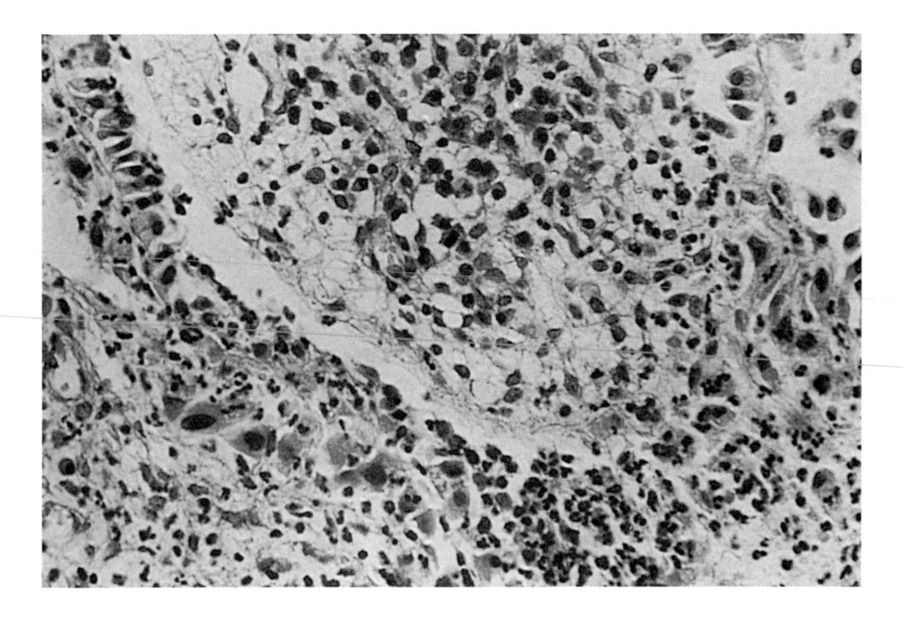

Figure 7.3 Acute cervicitis showing marked degenerative changes in the endocervical cells (H & E, ×480).

Figure 7.4 Acute cervicitis with marked ulceration and surface exudate (H & E, ×190).

crypts often being the worst affected. Necrosis may lead to ulceration with a purulent exudate coating the surface (Figure 7.4). In gonococcal infection, necrosis may be sufficiently marked to mimic carcinoma clinically. On the other hand, acute cervicitis is a common finding in association with advanced cancer. Occasionally, a punch biopsy from a malignant lesion contains only necrotic infected debris and a segmental biopsy may be necessary to ascertain the diagnosis.

7.1.2 Cytology

The outstanding feature of a smear from an infected cervix is the large number of leucocytes in every low-power microscope field. Polymorphs predominate (Figure 7.5(a)) but macrophages and lymphocytes are also seen. Polymorphs may be so numerous that they obscure the epithelial cells in the sample, rendering it useless for analysis. It is not uncommon to find that the smear has a 'dirty' appearance due to bacteria and lysis of the epithelial cells. Phagocytosis of leucocytes and cell debris may be seen (Figure 7.5(b)).

Other features of the smear are the degenerative changes which can be seen in the cytoplasm and nucleus of the epithelial cells. These are most obvious in the parabasal cells and endocervical cells but can also be seen in superficial epithelial cells. There is not normally any need to describe these changes in the smear report.

The *cytoplasmic changes* due to inflammation are characterized by frayed cell borders which confer a ragged appearance to the cells. The characteristic staining pattern of the cytoplasm may be lost so that the epithelial cells appear polychromatic or assume an eosinophilic hue. Vacuolation may be prominent and may be fine or coarse.

The *nuclear changes* are equally varied. At the earliest stages the nucleus undergoes coagulation necrosis with clumping of the chromatin which is manifest in the smear by apparent nuclear hyperchromasia and irregularity of the nuclear membrane. When complete cellular lysis supervenes, the nuclei become swollen and hypochromatic and the nuclear structure and nuclear membrane become pale and indistinct. Eventually the nuclei shrink and become pyknotic and karyorrhexis supervenes. A small perinuclear halo may be apparent at this

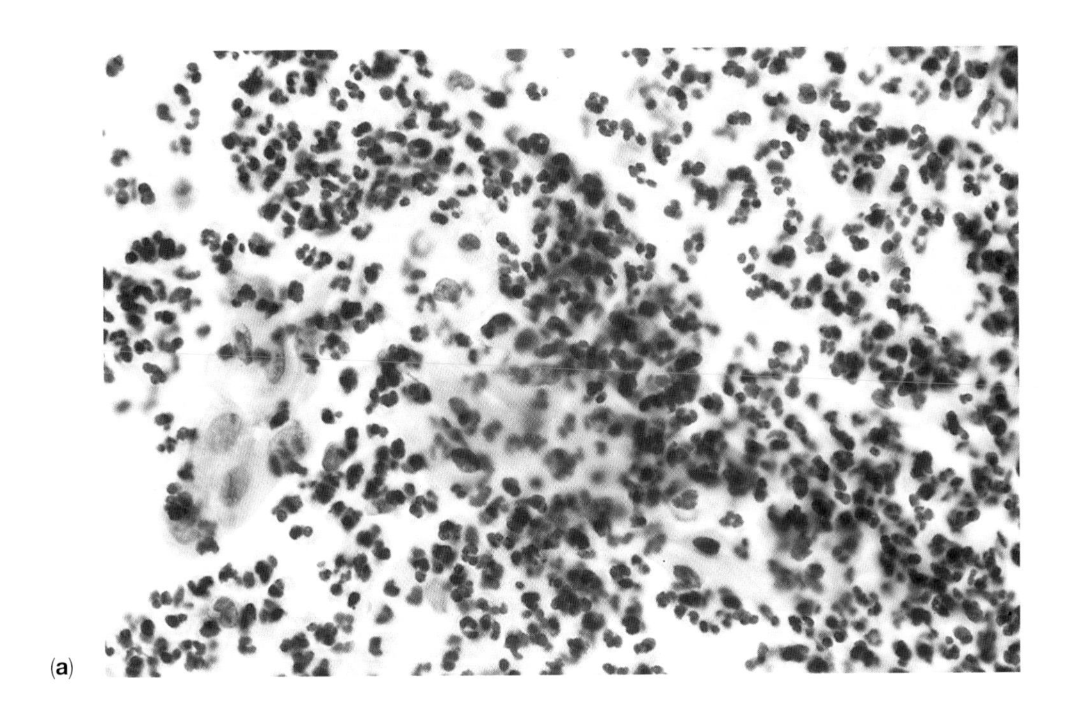

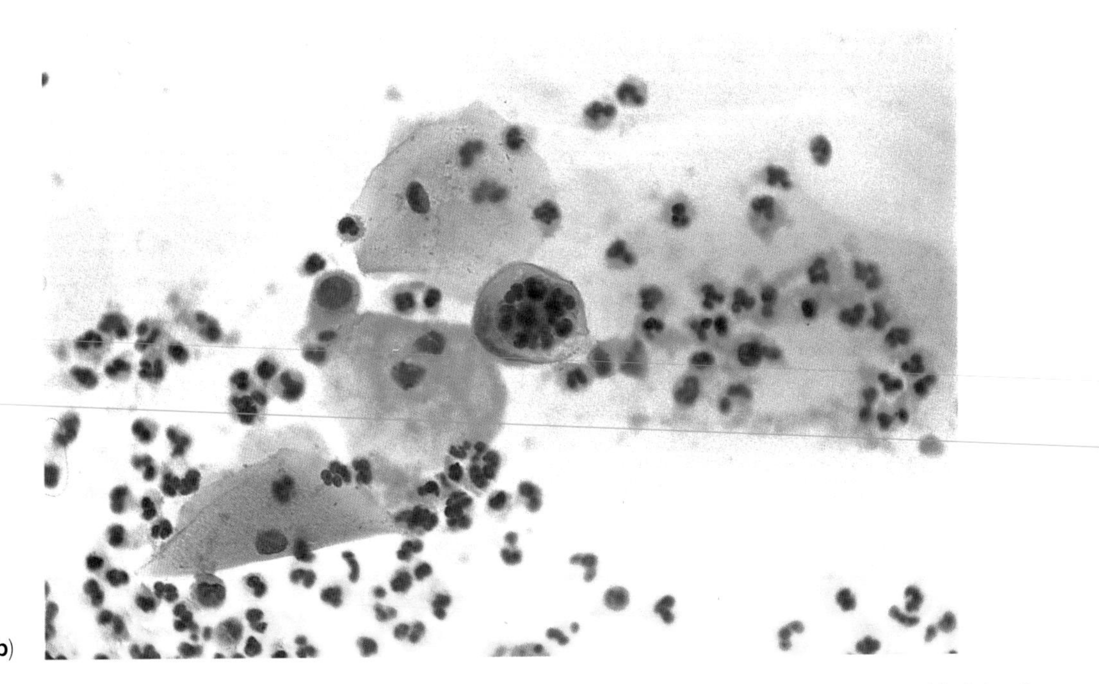

Figure 7.5 (**a**) Cervical smear from a case of acute cervicitis. Polymorphs obscure many of the epithelial cells. Such a smear may be considered unsatisfactory (Pap, ×630). (**b**) Phagocytosis of polymorphs (Pap, ×630).

stage (see Figure 7.6(a–k) for the range of cytoplasmic and nuclear changes associated with inflammation).

Chains of coccoid bacteria may be seen and the paired organisms (diplococci) associated with *N. Gonorrhoeae* have been described in Papanicolaou-stained cervical smears (Arsenault *et al.*, 1976). However, Gram staining and microbiological culture is essential for definitive diagnosis of the organisms.

Distinction between neoplastic and inflammatory change is largely one of degree. Anisonucleosis, irregularity of nuclear outline and abnormal distribution of chromatin is usually much more striking in neoplasia than in inflammatory change. Moreover, the cellular pleomorphism which reflects the disordered growth of neoplastic epithelium is rarely seen in degenerative inflammatory change. However, in some cases, it may be difficult to exclude neoplastic change and a report of 'borderline changes' should be given (Section 2.4).

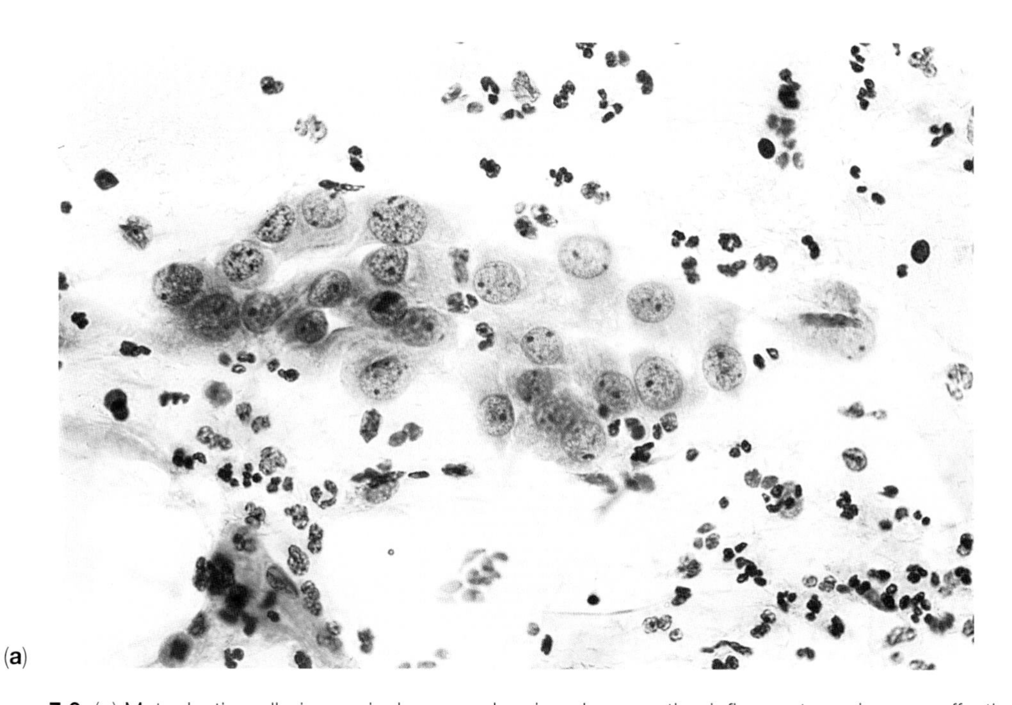

(**a**)

Figure 7.6 (**a**) Metaplastic cells in cervical smear showing degenerative inflammatory changes affecting the nucleus and cytoplasm. Note frayed cytoplasm. The nuclei show little abnormality beyond a slight hypochromasia, some anisonucleosis and prominence of the chromocentres (Pap, ×630).

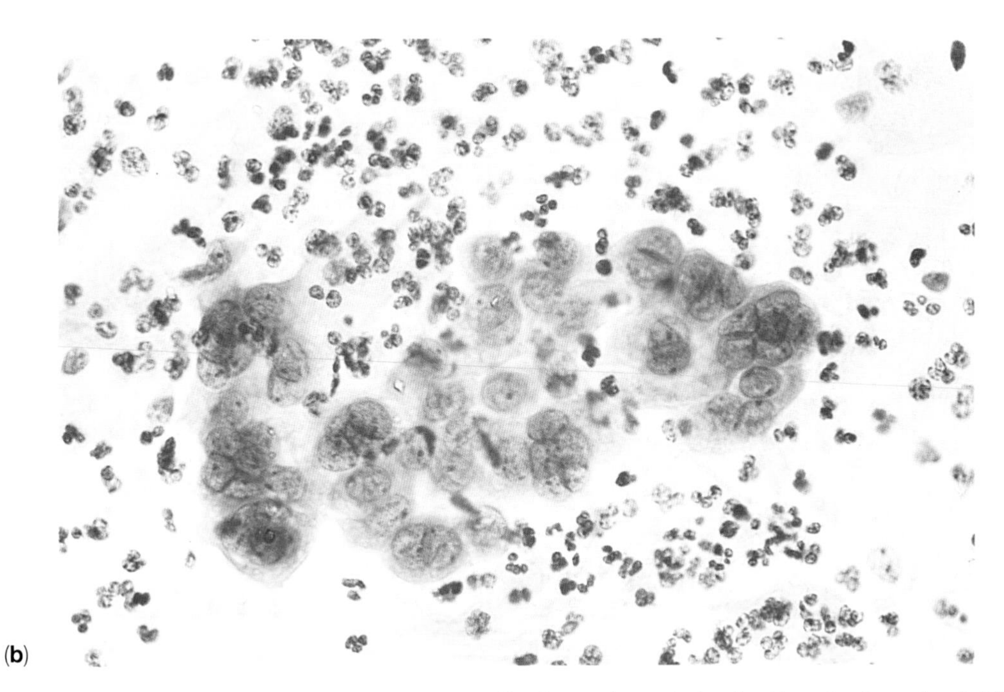

Figure 7.6 (**b**) Degenerative inflammatory changes affecting endocervical cells. Note the irregular distribution of nuclei and loss of honeycomb formation, and the frayed, polychromatic cytoplasm. The cells lie in a background of polymorphs (Pap, ×630).

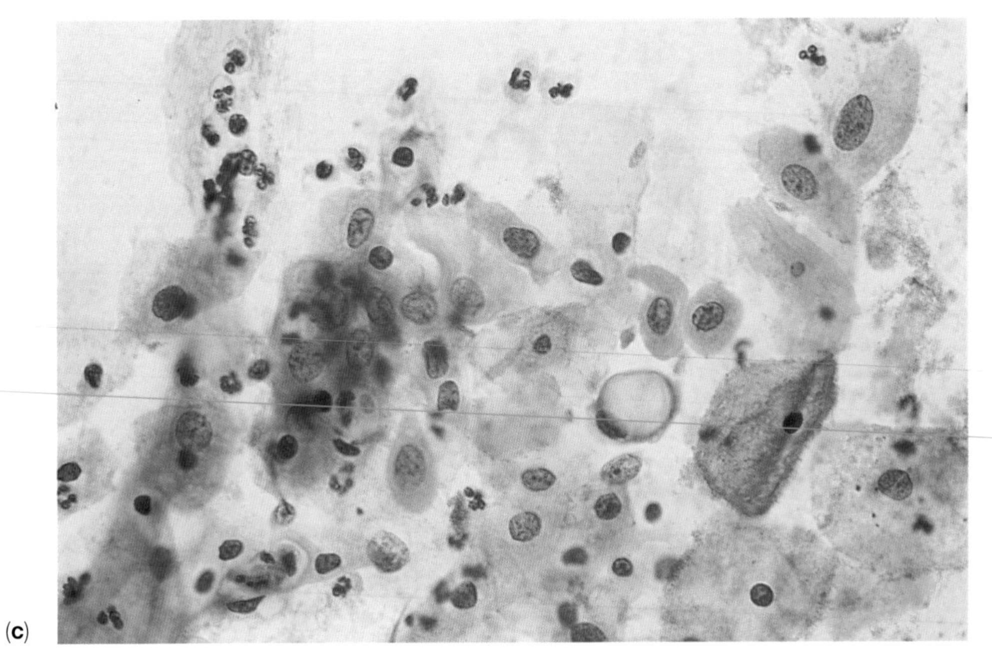

Figure 7.6 (**c**) Degenerative inflammatory change, showing coarse cytoplasmic vacuolation and some nuclear shrinkage, resulting in some anisonucleosis. Note the bacterial background (Pap, ×630).

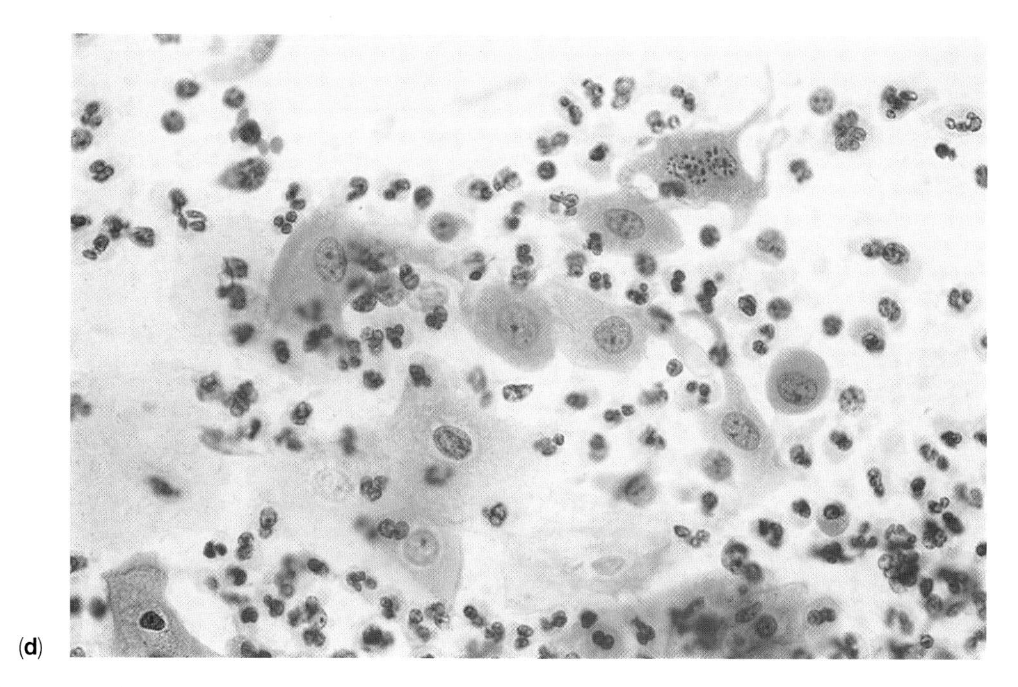
(d)

Figure 7.6 (**d**) Degenerative inflammatory changes in squamous metaplastic cells and superficial cells, undergoing karyolysis. They also have frayed cytoplasm and perinuclear halos (Pap, ×630).

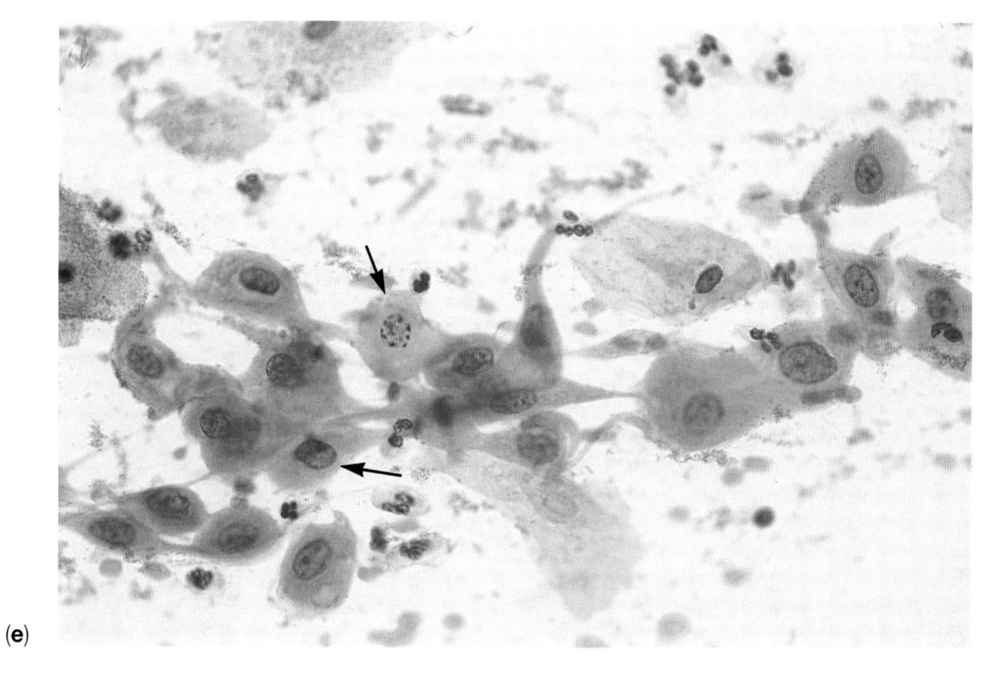
(e)

Figure 7.6 (**e**) Degenerative inflammatory changes in squamous metaplastic cells undergoing karyorrhexis (arrows). They also exhibit anisonucleosis and nuclear shrinkage (arrowed), giving the impression of nuclear hyperchromasia (Pap, ×630).

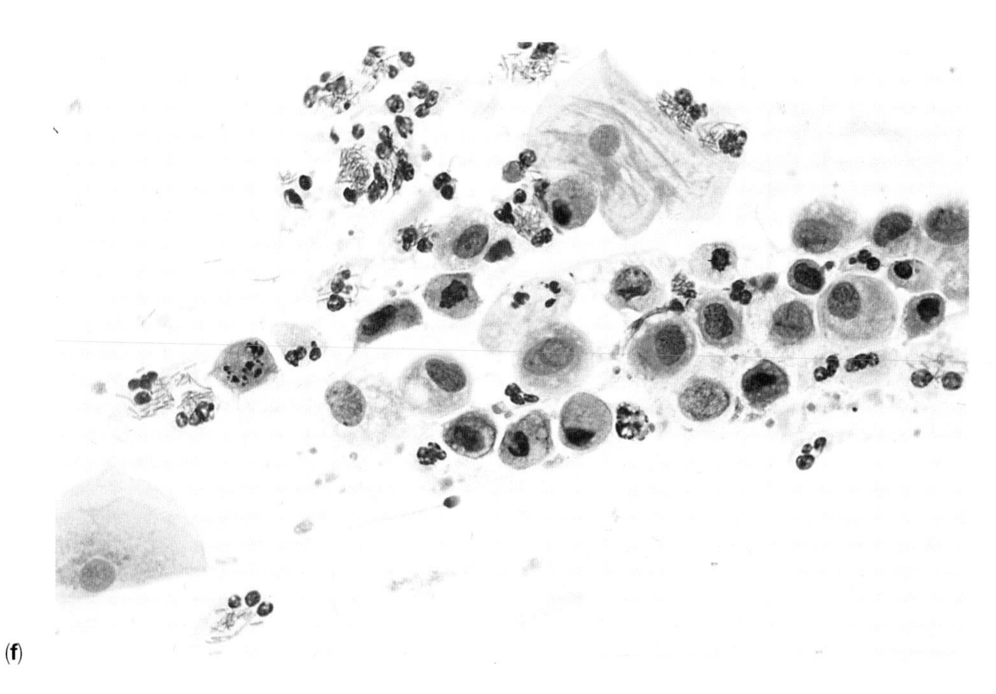

(f)

Figure 7.6 (**f**) Degenerative inflammatory changes in immature metaplastic cells. The cells are undergoing nuclear shrinkage and karyopyknosis, giving an appearance of hyperchromasia. Irregularity of the nuclear membrane and anisonucleosis give a false impression of neoplastic change (Pap, ×630).

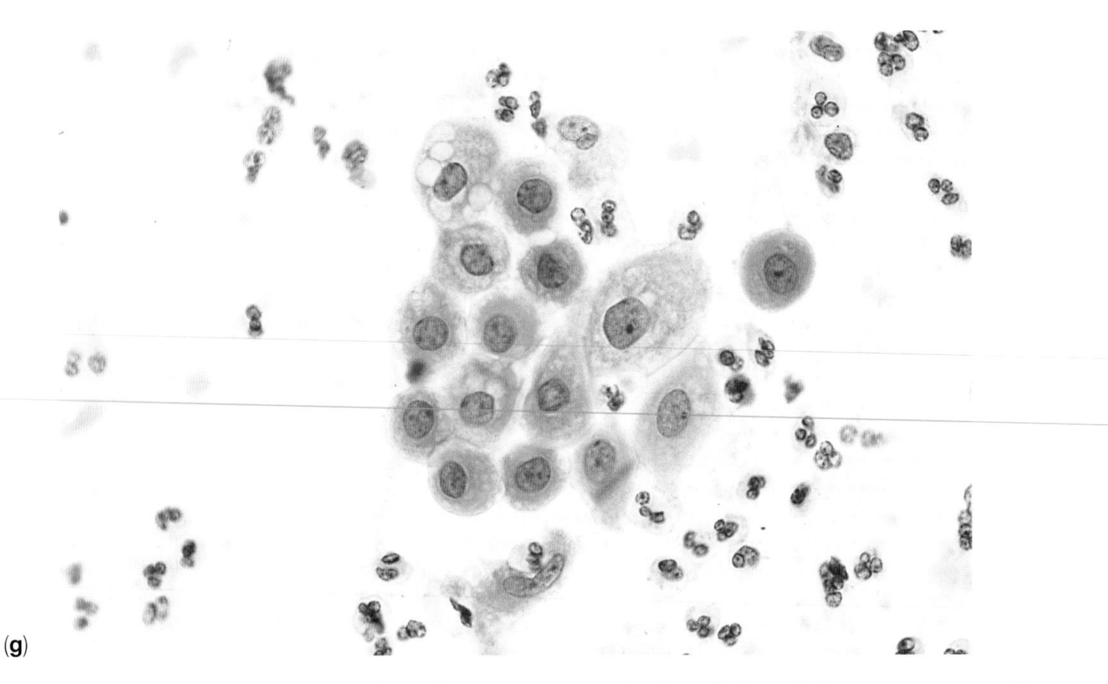

(g)

Figure 7.6 (**g**) Coarse and fine vacuolation of squamous metaplastic cells in an inflammatory smear (Pap, ×630).

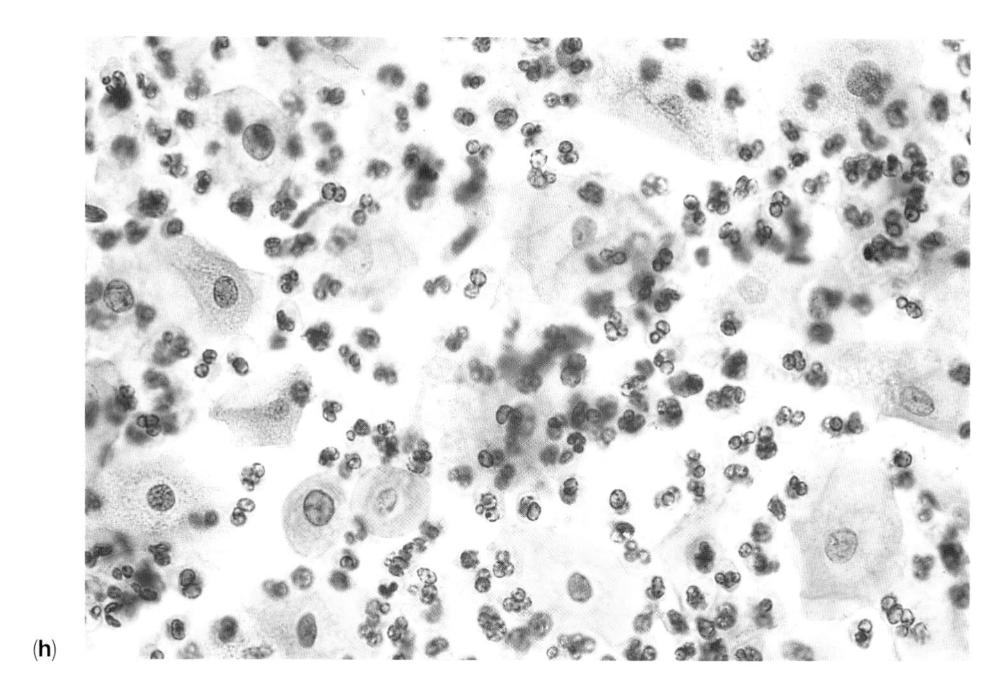

(**h**)

Figure 7.6 (**h**) Degenerative inflammatory changes in superficial squamous cells (Pap, ×630).

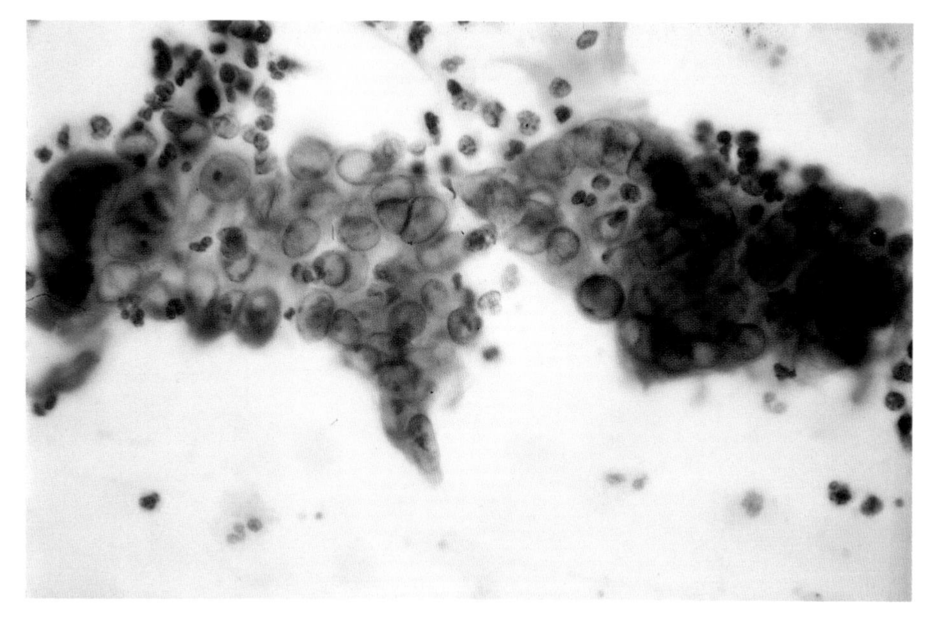

(**i**)

Figure 7.6 (**i**) Degenerative inflammatory changes in endocervical cells. Note swelling and loss of structure (Pap, ×630).

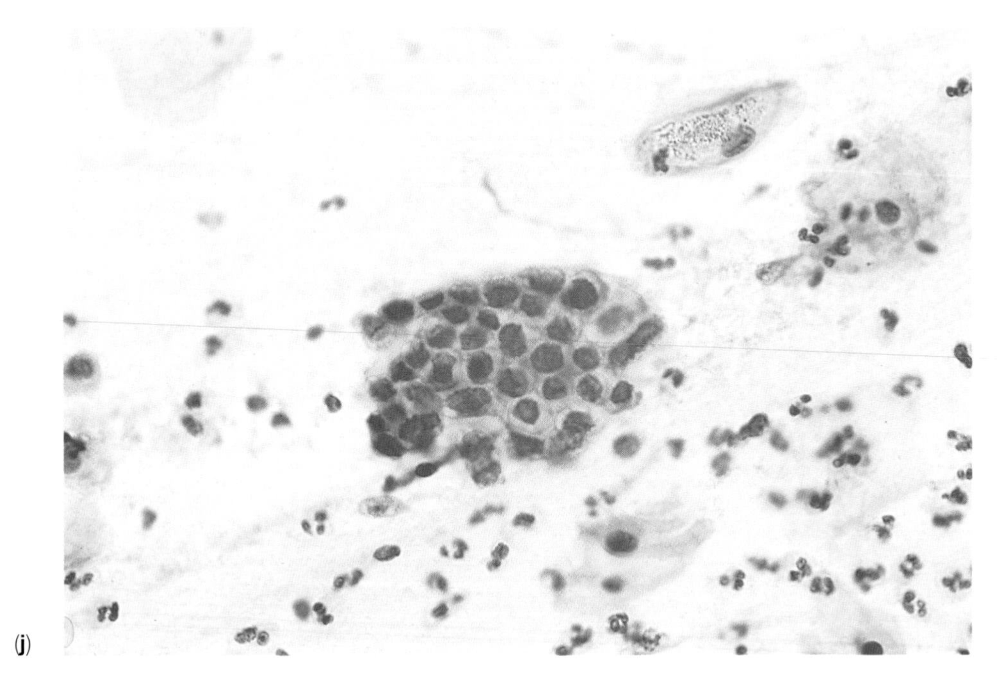

(j)

Figure 7.6 (**j**) Degenerative inflammatory changes in endocervical cells. Note nuclear shrinkage and apparent hyperchromasia (Pap ×630).

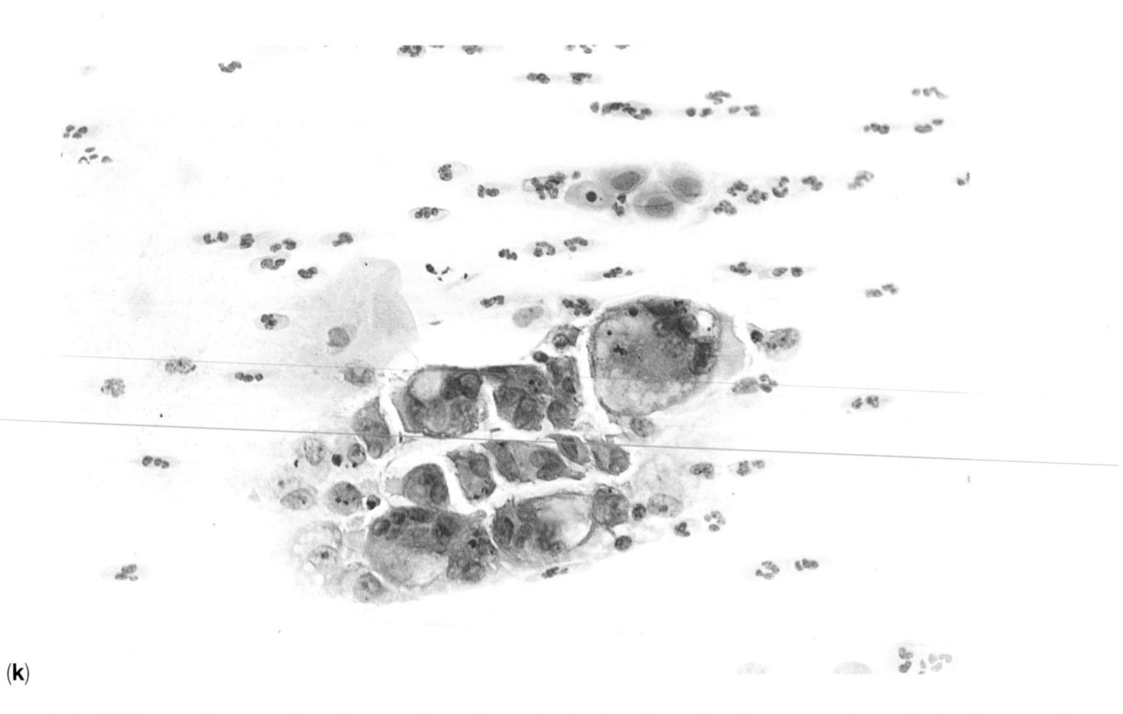

(**k**)

Figure 7.6 (**k**) Severe degenerative changes in endocervical cells with extensive vacuolation of cytoplasm and loss of nuclear structure (Pap, ×400).

7.1.3 Healing and regeneration

Healing of the ulcerated or eroded epithelium is effected by proliferation of adjacent epithelium and extension from local gland crypts. Initially the denuded area is covered by a layer of columnar cells or immature metaplastic cells which is later replaced by mature squamous epithelium. Evidence of regeneration of the cervical epithelium is often more apparent in the smear than in a histological section (Figure 7.7(a–d)). Sheets of cells of columnar or immature metaplastic origin with large nuclei and prominent nucleoli may be confused with adenocarcinoma. A useful way to distinguish these proliferating cells from malignant cells in a smear is to compare the nuclear features of one cell with another of the same type. Actively dividing cells engaged in repair show a uniformity not characteristic of malignant cells, i.e. the same number of chromocentres and nucleoli and an even distribution of the chromatin. The degenerative changes associated with infection and the regenerative changes associated with repair are frequently observed in the same smear (Figures 7.6(g) and 7.7(d) are from the same smear).

7.2 CHRONIC CERVICITIS

7.2.1 Histology

Discrete lymphocytes and plasma cells are commonly found in the submucosa of the transformation zone in women of childbearing age. Only when the infiltrate is associated with the development of large subepithelial collections of lymphocytes of varying maturity and mixed with phagocytic cells and macrophages should a diagnosis of *chronic lymphocytic cervicitis* be made (Figure 7.8). If the infiltrate is accompanied by the formation of subepithelial lymphoid follicles with active germinal centres, the lesion may then be referred to as *chronic follicular cervicitis* (Figures 7.50 and 7.51).

The presence of endocervical papillary processes, each showing lymphocytic infiltration, is characteristic of *chronic papillary cervicitis* (Figure 7.9). When situated distal to the external os it has been incorrectly called 'papillary erosion'.

In a review of 450 consecutive hysterectomy specimens, Roberts and Ng (1975) encountered chronic lymphocytic cervicitis in 30 cases (6.7%). In 29 of the 30 cases the uterus was removed for reasons unrelated to cervical pathology. Some 70% of the women were postmenopausal but the ages ranged from 26 to 69 years. In cases of long-standing cervicitis, basal cell hyperplasia, squamous metaplasia, parakeratosis, hyperkeratosis and mucus retention cysts occur (Chapters 8 and 9). An association between chronic follicular cervicitis and chlamydial infection has been reported (Swanson *et al.*, 1975; Hare *et al.*, 1981) (see Section 7.15).

7.2.2 Cytology

Smears from women with chronic lymphocytic cervicitis contain streaks or swathes of small and large lymphocytes (Figure 7.10). Macrophages, polymorphs and plasma cells are also found (Figure 7.11). Roberts and Ng (1975) pointed out that the lymphoid cells could be confused with endometrial stromal cells or small cell carcinoma of the cervix. The epithelial cells will reflect any associated reaction of the cervical epithelium such as parakeratosis or metaplasia. Parabasal cells are quite commonly shed from ulcerative, reparative or

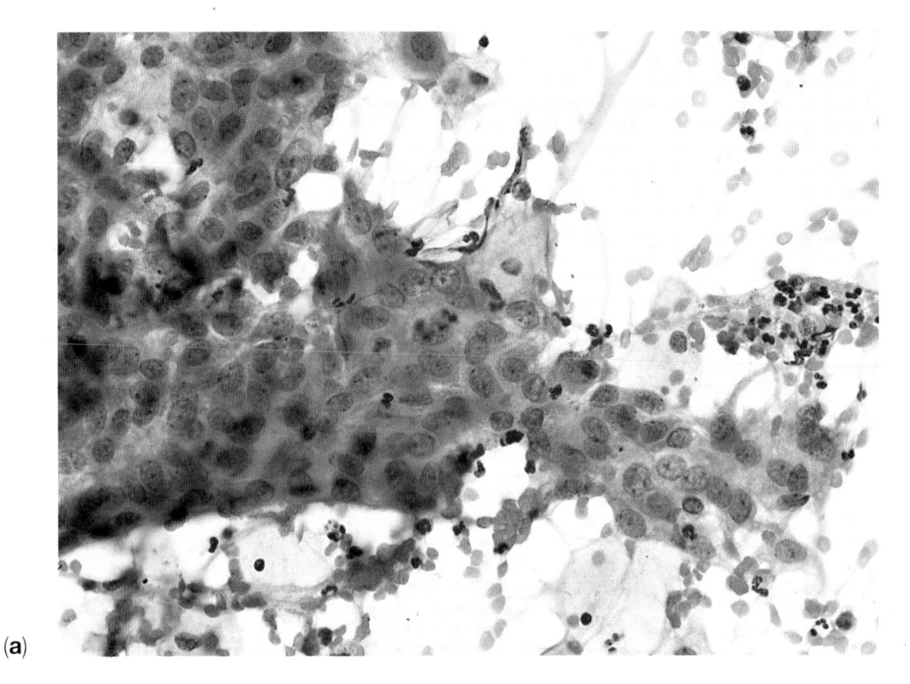

Figure 7.7 (**a**) Sheets of parabasal cells in an inflammatory smear showing regenerative changes. Note nuclear crowding and prominent nucleoli (Pap, ×630).

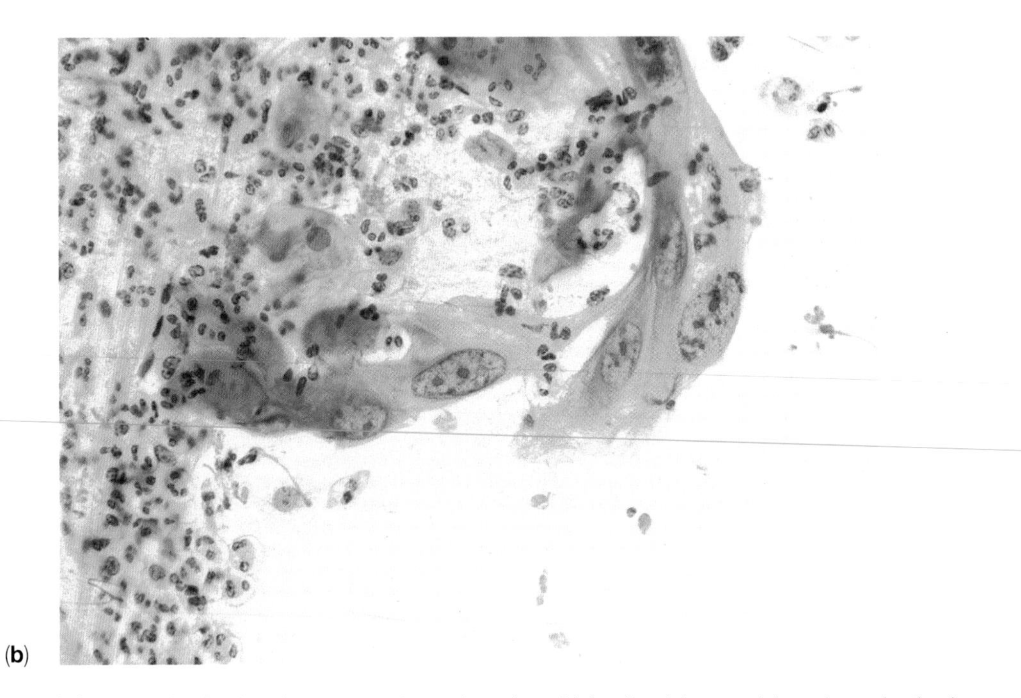

Figure 7.7 (**b**) Clusters of cells showing variation in nuclear size which raised the suspicion of neoplastic change but in this instance was considered to reflect mitotic division of regenerating epithelial cells (Pap, ×630).

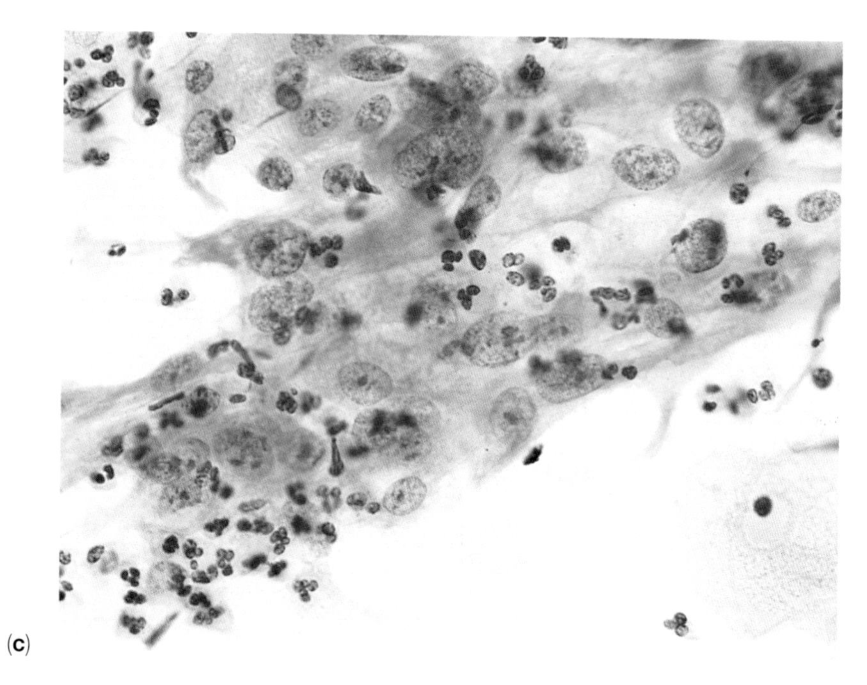

Figure 7.7 (**c**) Actively dividing cells engaged in repair showing a uniformity not characteristic of malignant cells (Pap, ×630).

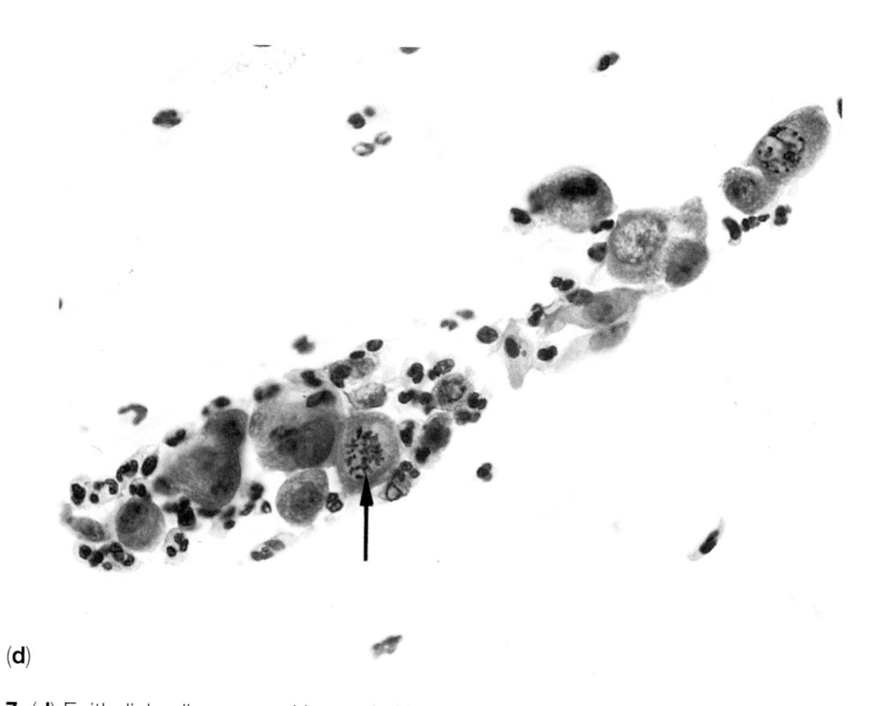

Figure 7.7 (**d**) Epithelial cells engaged in repair. Note mitotic figure (arrowed) (Pap, ×630).

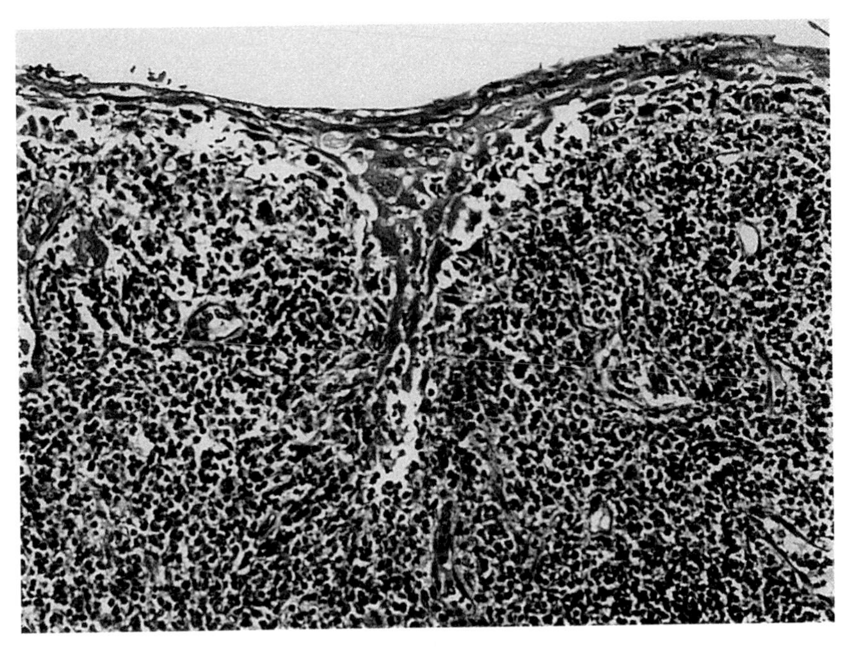

Figure 7.8 Chronic lymphocytic cervicitis showing subepithelial infiltration by lymphocytes and histiocytes (H & E, ×180).

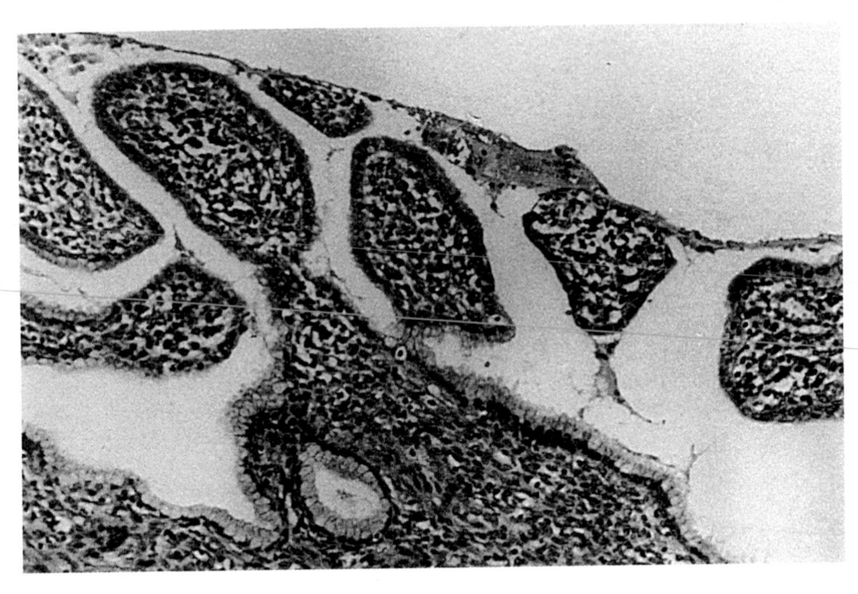

Figure 7.9 Chronic papillary cervicitis, incorrectly designated 'papillary erosion' (H & E, ×190).

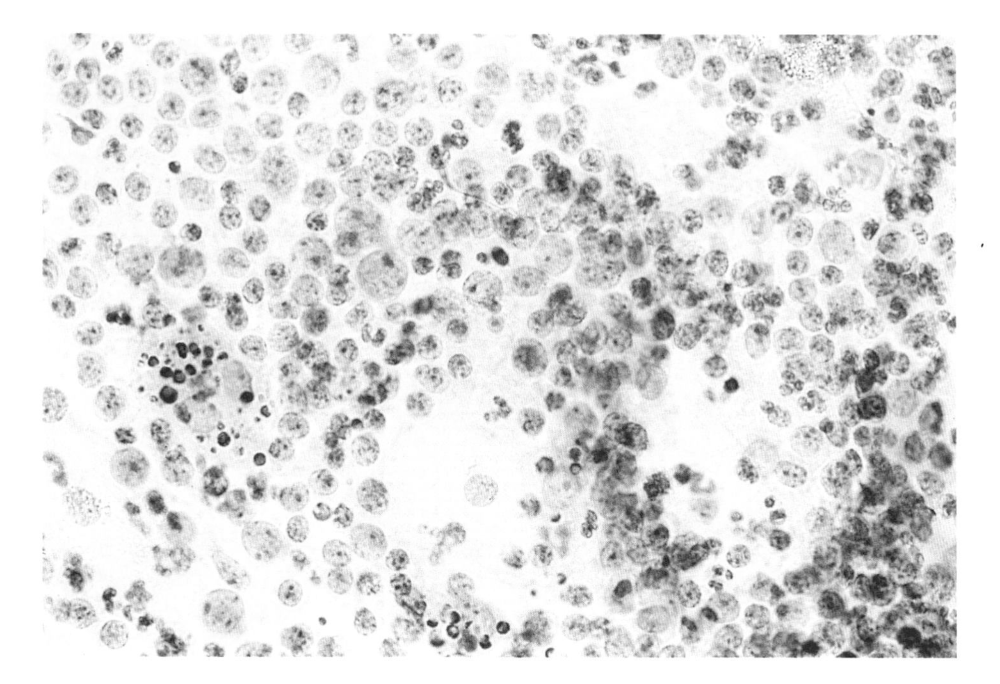

Figure 7.10 Chronic lymphocytic cervicitis in a cervical smear. Note swathes of lymphocytes (Pap, ×630).

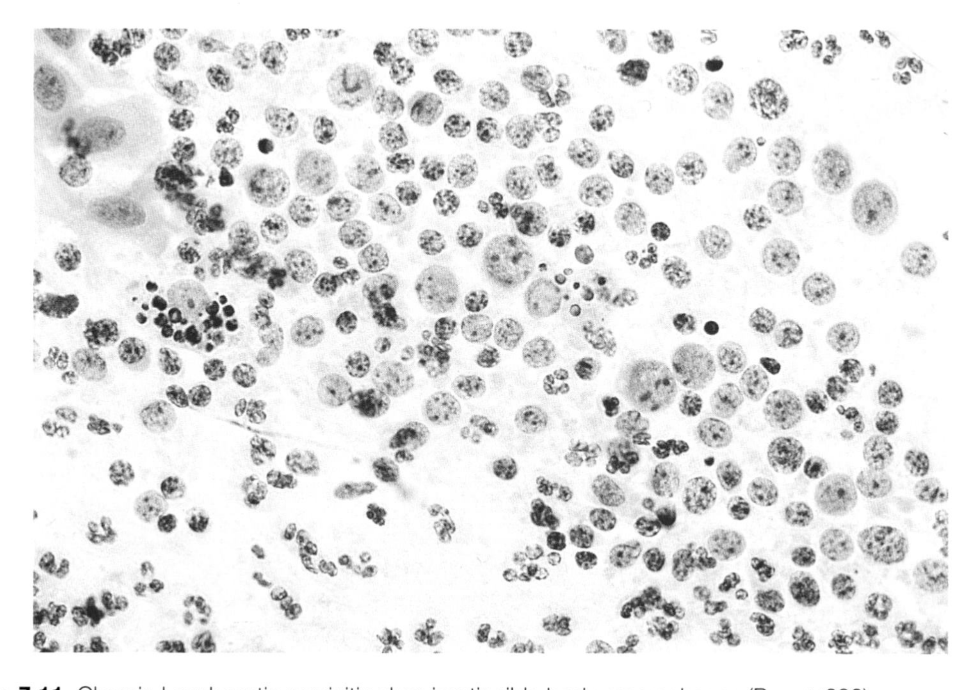

Figure 7.11 Chronic lymphocytic cervicitis showing tingible body macrophages (Pap, ×630).

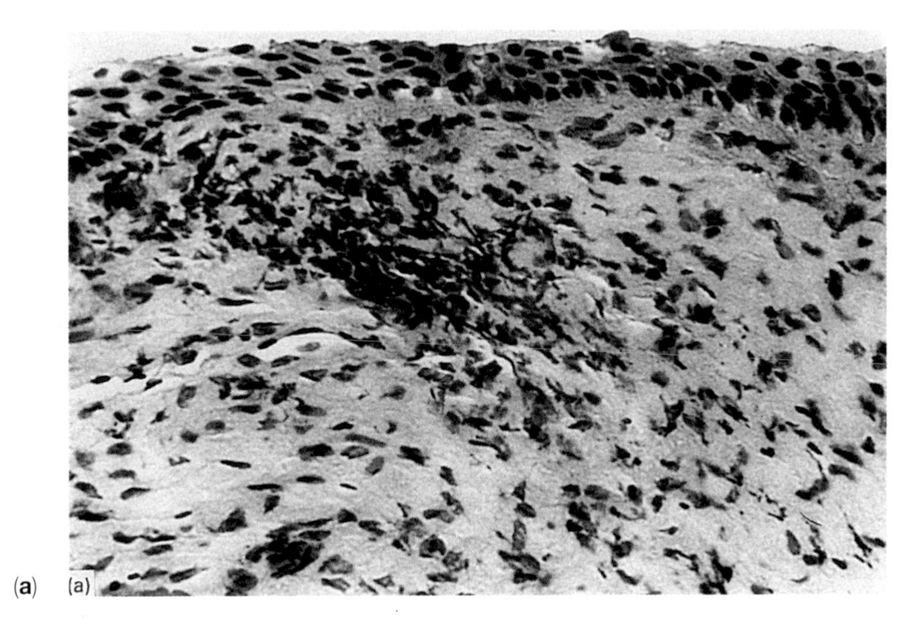

Figure 7.12 (a) Chronic atrophic cervicitis showing post-menopausal atrophy of ectocervical epithelium and underlying inflammatory change (H & E, ×190).

Figure 7.12 (b) Cervical smear from an atrophic cervicitis. Note sparsity of material (Pap, ×80).

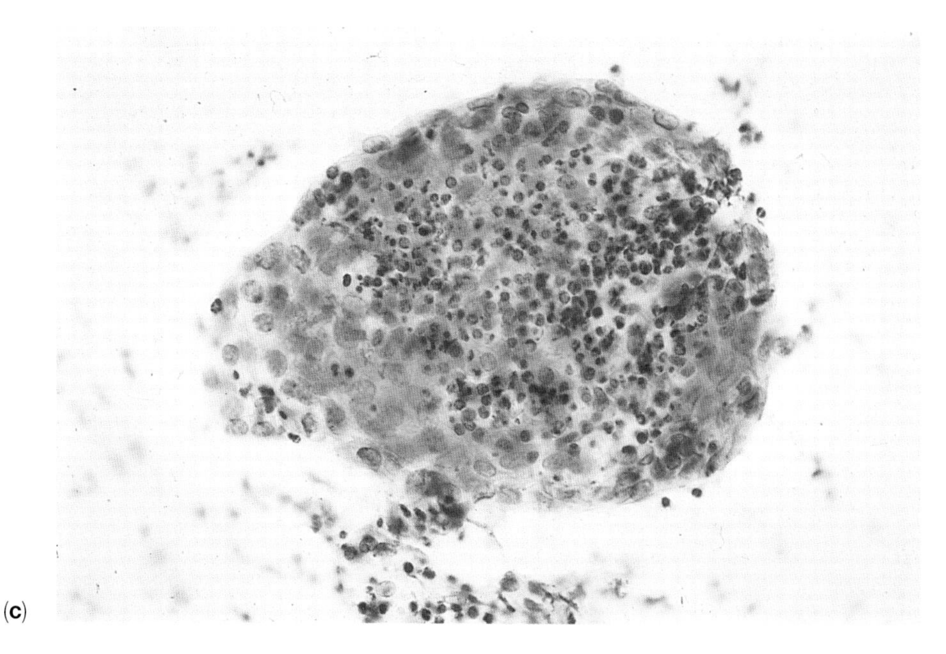

(c)

Figure 7.12 (**c**) Higher magnification reveals polymorph exudate and degenerative changes in parabasal cells (Pap, ×400).

metaplastic areas in the cervix. Eosinophilic anucleate squames (Figure 9.20(b)) may be prominent if keratinization of the epithelium occurs.

7.2.3 Chronic atrophic cervicitis

In post-menopausal women the ectocervical epithelium undergoes atrophy and becomes more susceptible to inflammation (Figure 7.12). Corresponding smears contain parabasal cells showing marked cellular pleomorphism reflecting degenerative changes in the epithelium (Figure 7.13(a)). Some parabasal cells are small and round with eosinophilic cytoplasm and shrunken nuclei; others are swollen and irregular in shape with cyanophilic cytoplasm and large, pale nuclei. Many cells are so fragile that the cytoplasm disintegrates during the preparation of the smear and the smears are composed of sheets of bare nuclei.

Cell death is manifest by nuclear pyknosis, margination of chromatin, karyolysis and karyorrhexis. Nuclear shrinkage may result in the formation of a small perinuclear halo which is similar to that seen in *Trichomonas vaginalis* infection (Figure 7.22) but is also seen in non-specific cervicitis (Figure 7.6(h)). The smears are heavily infiltrated with leucocytes and red blood cells. Multinucleate giant cells are frequently seen (Figure 7.13(b)) and polymorphs and red blood cells abound.

7.3 GRANULATION TISSUE

Repair and replacement of the cervical epithelium destroyed by the inflammatory process may be delayed by the formation of granulation tissue which is composed of proliferating fibroblasts set in a network of capillaries (Figure 7.14(a)). There is usually a heavy infiltrate of leucocytes, including many lymphocytes, macrophages (Figure 7.14(b)) and occasional multinucleate giant cells. Granulation tissue is frequently found at the vaginal vault after hysterectomy and may be mistaken

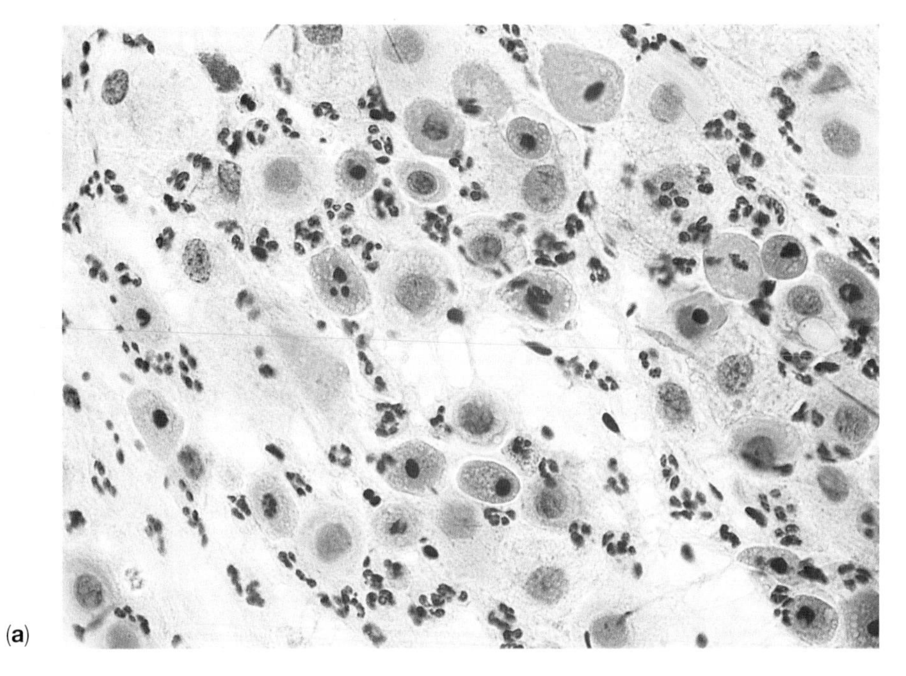

(**a**)

Figure 7.13 (**a**) Cervical smear from an atrophic cervicitis. Note polymorph exudate and marked degenerative changes in parabasal cells, e.g. pyknosis, karyorrhexis and vacuolation (Pap, ×630).

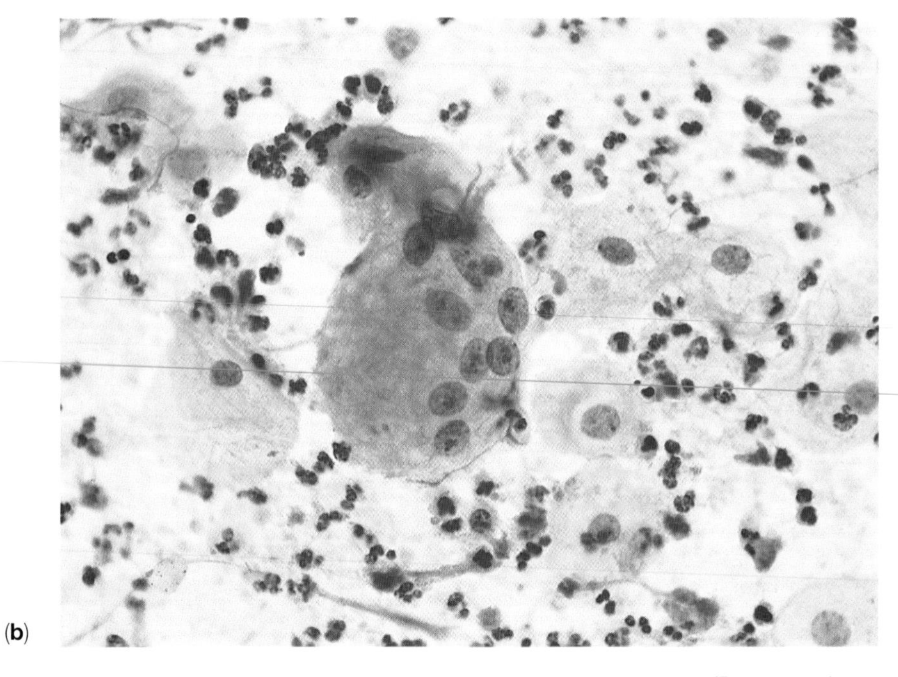

(**b**)

Figure 7.13 (**b**) Foreign body giant cell in cervical smear from atrophic cervicitis (Pap, ×630).

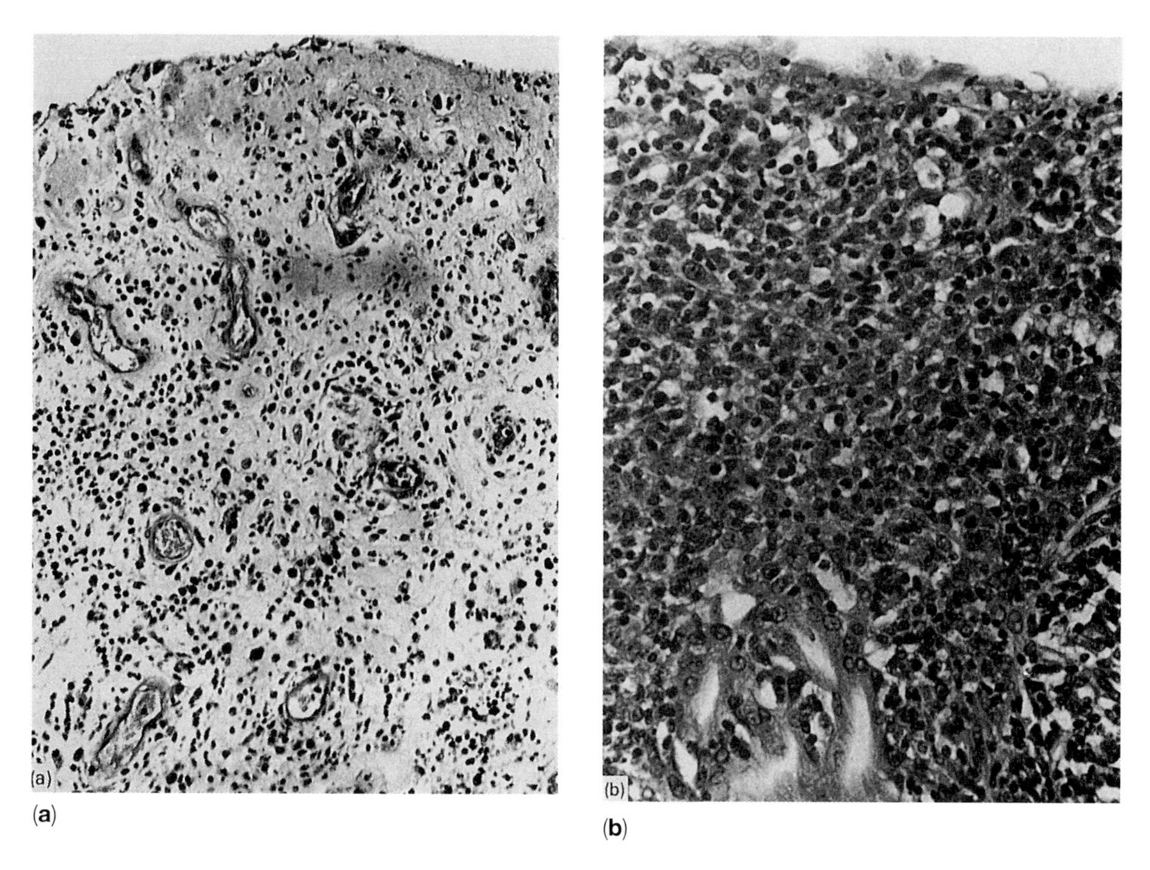

Figure 7.14 (**a**) Granulation tissue showing vascularity and leucocytic infiltration (H & E, ×190). (**b**) Densely cellular granulation tissues in which there are many histiocytes and lymphocytes (H & E, ×310).

for tumour as it is very friable and bleeds easily. Even histologically a tumour may be suspected particularly if, as occasionally happens, Fallopian tube tissue becomes trapped in the vault after hysterectomy (Figure 7.15).

7.4 TRUE EROSION (*EROSIO VERA*)

This term is correctly applied to reddened, glistening lesions of the cervix where there is total or partial loss of a significant area of squamous epithelium revealing the underlying stroma which may be congested and hyperaemic and bleed readily to the touch. Histologically there is ulceration or necrosis (Figure 7.4), which may be due to trauma or treatment.

7.5 TUBERCULOUS CERVICITIS

This is rare except in less well-developed countries. It is secondary to tuberculous infection elsewhere in the genital tract (usually Fallopian tubes) which itself may be associated with primary infection of the lung. Cervical involvement occurs in 3% of cases of genital tuberculosis (Misch *et al.*, 1976). In the advanced stages the cervix may appear ulcerated and necrotic, and mimic carcinoma (Zummo *et al.*, 1955). The characteristic histological findings are foci of caseation in the substance of the cervical connective tissue which is infiltrated with epithelioid cells, lymphocytes and granulomata (Figure 7.16). Langhans'-type giant cells are also present (Figure 7.17). In the absence of caseation, the differential diagnosis rests

(a)

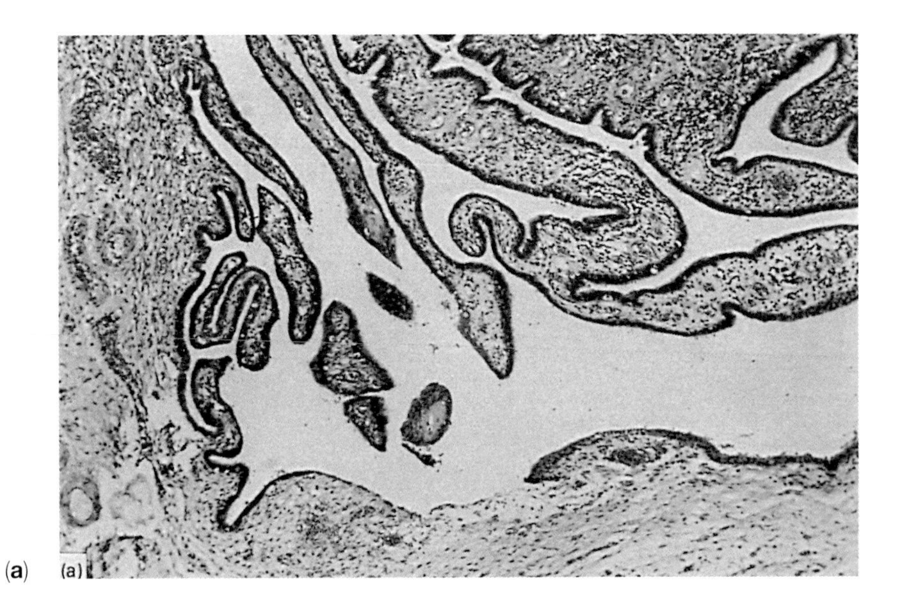

(b)

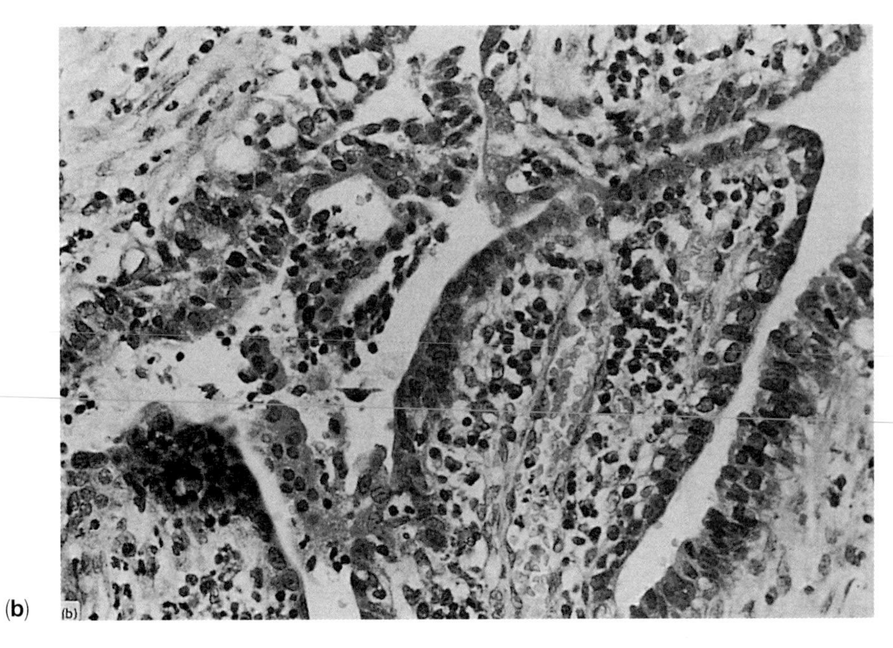

Figure 7.15 Tubal prolapse. (**a**) Fallopian tube tissue entrapped in granulation tissue at vaginal vault following hysterectomy. Note stratified squamous cervical epithelium at bottom right (H & E, ×80). (**b**) Higher-power view to show tubal epithelium (H & E, ×500).

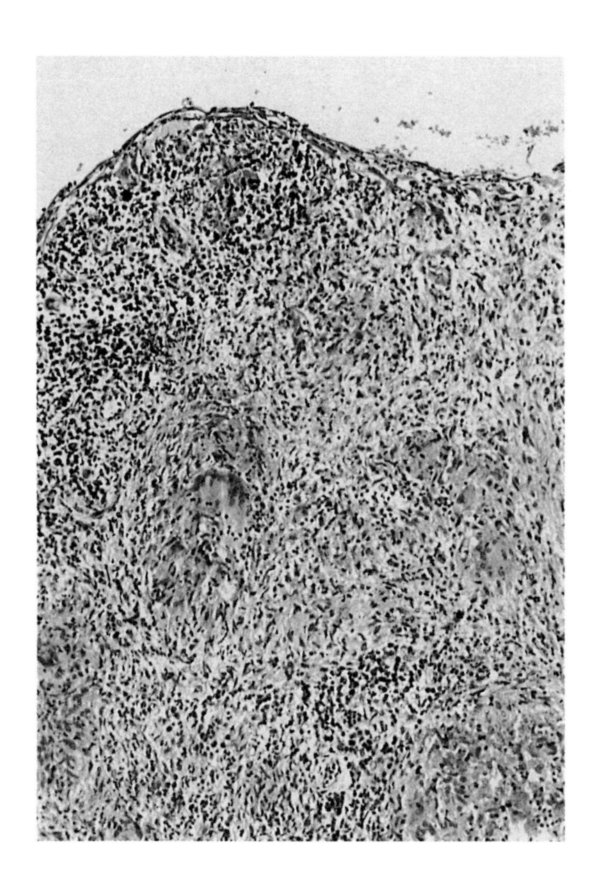

Figure 7.16 Tuberculous cervicitis showing epithelioid cells and some necrosis (H & E, ×180).

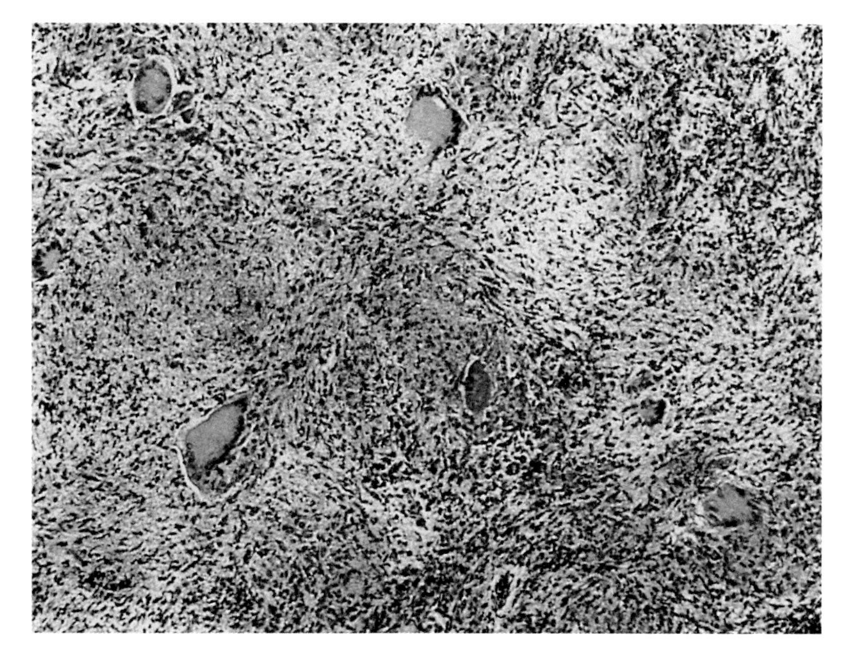

Figure 7.17 Tuberculous cervicitis showing Langhans' giant cells (H & E, ×170).

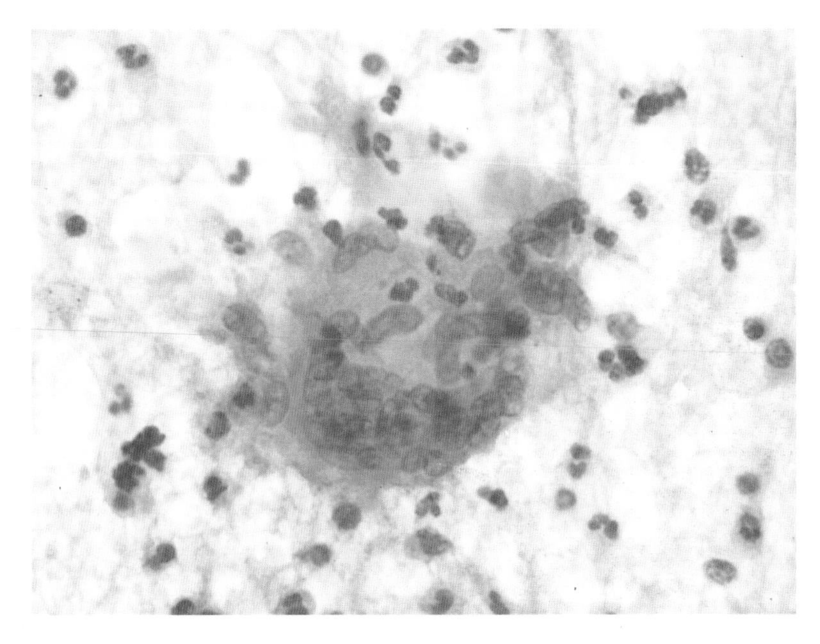

Figure 7.18 Langhans' giant cell in cervical smear from a case of tuberculous cervicitis (Pap, ×630).

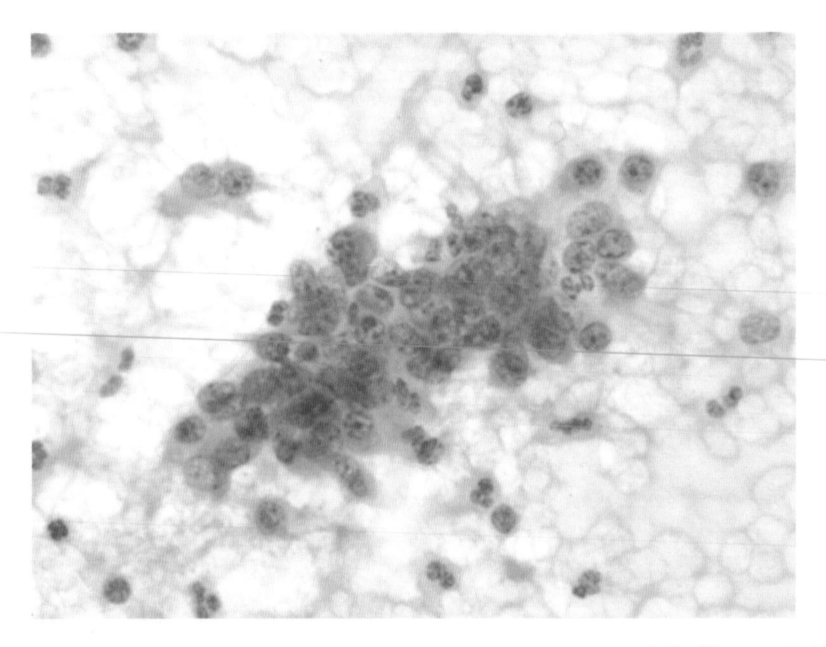

Figure 7.19 Epithelioid cells in cervical smear from a case of tuberculous cervicitis (Pap, ×630).

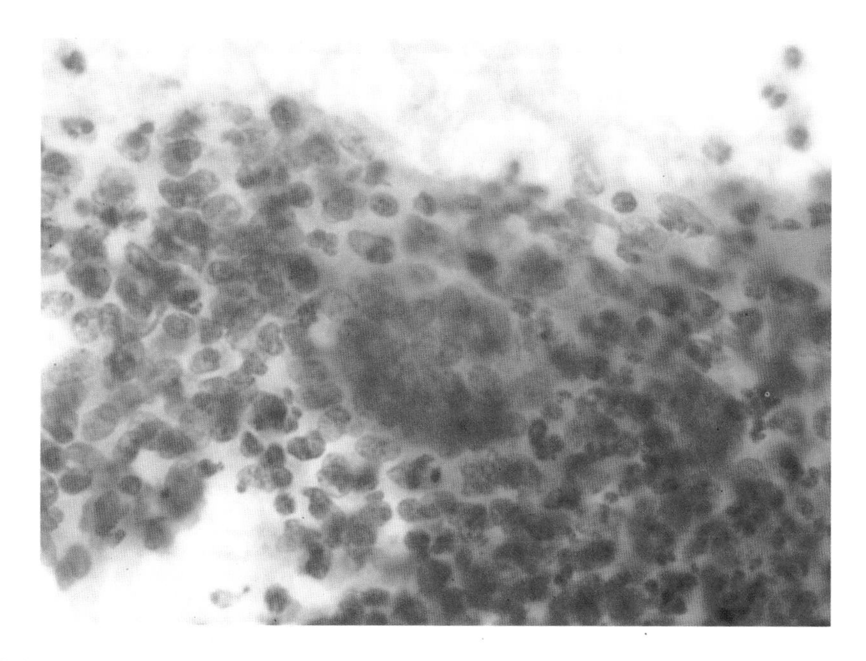

Figure 7.20 Langhans' giant cell in background of epithelioid cells (Pap, ×630).

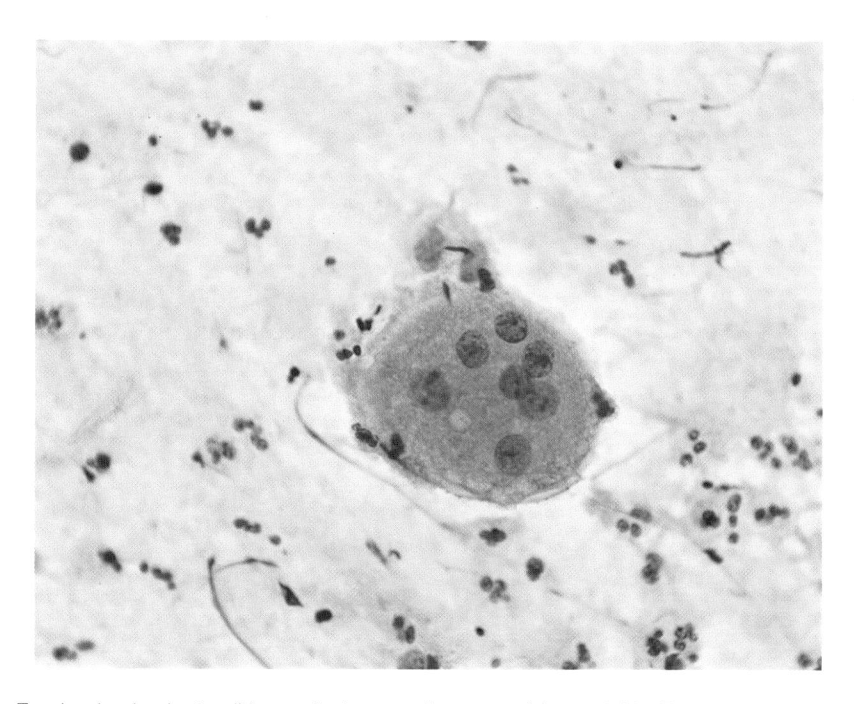

Figure 7.21 Foreign body giant cell in cervical smear from atrophic cervicitis (Pap, ×630).

between tuberculosis and other granulomatous lesions (Section 7.6). The diagnosis of tuberculosis, suspected on the basis of Langhans' giant cells and the epithelioid cells can be established by demonstration of the acid-fast *Mycobacterium tuberculosis*. As culture and animal inoculation are more sensitive methods of detecting infection, fresh biopsy material should be sent for microbiological studies whenever tuberculosis is suspected.

7.5.1 Cytology

The smear is characterized by the presence of Langhans' giant cells (Figure 7.18) and epithelioid cells (Figures 7.19 and 7.20) (Coleman, 1969; Misch *et al.*, 1976; Angrish and Verna, 1981). These are only seen in the smear if there is extensive destruction of the cervical mucosa associated with the infection. The epithelioid cells are characterized by their delicate 'sausage-shaped' nuclei, fine chromatin and ill-defined cytoplasm. They are only slightly larger than polymorphs (12–15 µm) and may lie singly or in dense clumps. Langhans' cells in the smear can be distinguished from other giant cells by virtue of the fact that their nuclei are elongated, of equal size and often distributed unevenly, mainly at the periphery of the cell. The nuclei of foreign body giant cells are rounded, of equal size and distributed evenly throughout the cytoplasm or centrally placed (Figure 7.21). In contrast, herpes simplex giant cells are characterized by nuclear moulding and the presence of intranuclear inclusions (Figure 7.31(c)).

Caseous material may appear as granular amorphous eosinophilic strands but is difficult to recognize against a background of red blood cells and necrotic debris.

7.6 OTHER GRANULOMATOUS LESIONS OF THE CERVIX

The most common cause of a non-caseating granulomatous reaction in the cervix is the presence of a foreign body such as an insoluble suture after operation, often demonstrable by polarized light. Large necrotizing granulomas have been recorded postoperatively (H. Fox, personal communication, 1986). Rare causes of non-caseating granuloma of the cervix include syphilis, schistosomiasis (Section 7.9), granuloma inguinale, lymphogranuloma venereum and non-caseating tuberculosis. Sarcoidosis of the cervix does not appear to have been described (Jones Williams, personal communication, 1986). Syphilis may be suspected by the marked vascular changes and periarteritis. Although special silver stains may reveal *Treponema pallidum* in the primary chancre, it is rarely seen in the gummas and the diagnosis may depend on serological studies. The diagnosis of granuloma inguinale depends on the demonstration of Donovan bodies in macrophages which infiltrate the tissue (de Boer *et al.*, 1984).

7.7 *TRICHOMONAS VAGINALIS*

Trichomonas vaginalis is commonly found in the female genital tract. Combined cytological and microbiological studies revealed its presence in 9% of the adult female population of a large German city (Schnell *et al.*, 1972). In 10% of cases where the organism was identified in the vaginal secretions there was no evidence of concomitant disease. Infection is dependent on the pH of the vagina and the organisms flourish in an alkaline environment. Infection is usually accompanied by an offensive green discharge although some patients are asymptomatic and the organism appears to flourish as a commensal.

7.7.1 Histology

The diagnosis of trichomoniasis is rarely, if ever, made on biopsy material. However, a distinct colposcopic pattern ('the strawberry cervix') has been noted in some patients (Kolstad, 1964). This reflects the underlying engorgement of the superficial vessels and focal haemorrhages described by Koss and Wolinska (1959) in 68% of cervical biopsies from women with trichomoniasis. These authors also described oedema of the superficial layer of the epithelium as a common finding and very occasionally full thickness necrosis of the epithelium.

7.7.2 Cytology

A definitive diagnosis of *Trichomonas vaginalis* can be made in 50% of cases by examination of a cervical smear (Figure 7.22(a,b)). The unicellular organisms can be identified in a Papanicolaou-stained smear by their slatey grey colour and pear-shaped appearance. Usually slightly larger than leucocytes, their sizes range from 8 to 30 μm in length. The larger sizes prevail *in vitro* when culture conditions are unfavourable but whether this is true *in vivo* has yet to be established. Identification of the fusiform nucleus and the axostyle (a dark rod running through the centre of the body) will help distinguish the trichomonads from inspissated mucus and fragments of cytoplasm or free nuclei. The organisms are usually present in large numbers and cluster around the edge of the superficial squames. Although the organisms have four flagellae and an undulating membrane which has its origin at the forward pole of the cell, these features are rarely recognizable in smears. Confirmation by wet film or culture is desirable whenever there is diagnostic doubt; the motile organisms can be readily recognized by the whip-like movements of the flagellae.

Associated cytological features should alert the mind to the diagnosis. The smear frequently assumes a 'lilac' hue due to the effect of the alkaline vaginal secretions on the Papanicolaou stain. The squamous epithelium may show marked changes in response to the infection, particularly nuclear reduplication, enlargement and irregularity. The cells are well-oestrogenized, even in post-menopausal smears. Small perinuclear haloes are prominent (Figure 7.22). Filamentous organisms of the genus *Leptothrix* (Figure 7.23) are commonly found in the smear.

Trichomonas infection is frequently found in patients with cervical cancer but there is no evidence that it predisposes to the development of cervical cancer (Koss and Wolinska, 1959). Multiple infections are not uncommon. We have detected trichomoniasis, herpes genitalis and cervical neoplasia in the same smear.

7.8 CERVICOVAGINITIS EMPHYSEMATOSA

This rare condition is characterized by multiple blue-grey air-containing subepithelial cysts (Figure 7.24(a)) affecting the portio vaginalis and the vaginal wall (Wilbanks and Carter, 1963; Gardner and Fernet, 1964). It is often associated with trichomoniasis, disappearing after eradication of the infection. The cysts lack an epithelial lining but often have foreign body giant cells in the walls which may be flattened at the surface of the lumen (Figure 7.24(b)) (Evans and Hughes, 1961). The demonstration of gas production by *Trichomonas vaginalis* in guinea pig tissue indicates that these protozoa could be the causative agent (Newton *et al.*, 1960).

7.9 SCHISTOSOMIASIS

The majority of cases are due to infection with *Schistosoma haematobium,* but occasionally *Schistosoma mansoni* is involved. Infection is common in Egypt, South America and South-East Asia. Cervical infection is usually coincidental with infection at other sites and microscopically the lesions appear as non-caseating granulomas. Identification depends on recognition of the ova and larval stages in smear and sections (Figure 7.25). The ova of both species have a chitinous shell and vary in length from 80 to 186 μm and breadth from 30 to 93 μm (Berry, 1966). The ova may be calcified or surrounded by leucocytes or multinucleate

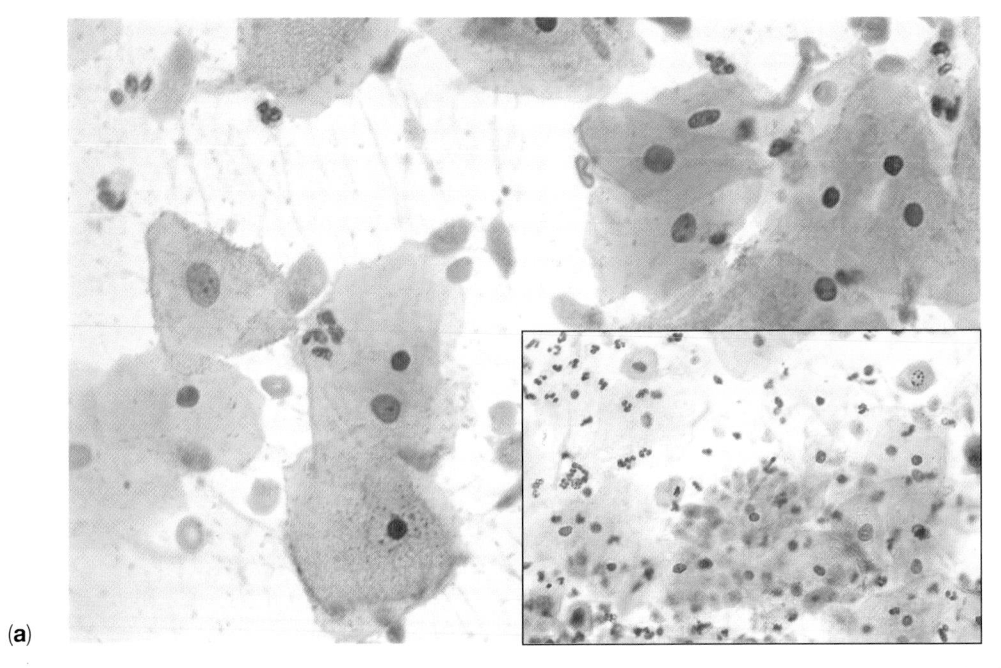

Figure 7.22 (**a**) *Trichomonas vaginalis* in cervical smear. Note perinuclear halos in the epithelial cells, the lilac background, and the brassy redness of the superficial cells (Pap, ×630) **Inset** An unusually heavy infestation of *Trichomonas vaginalis*. Trichomonads are attached in a cluster to the margin of a squamous cell (Pap, ×400)

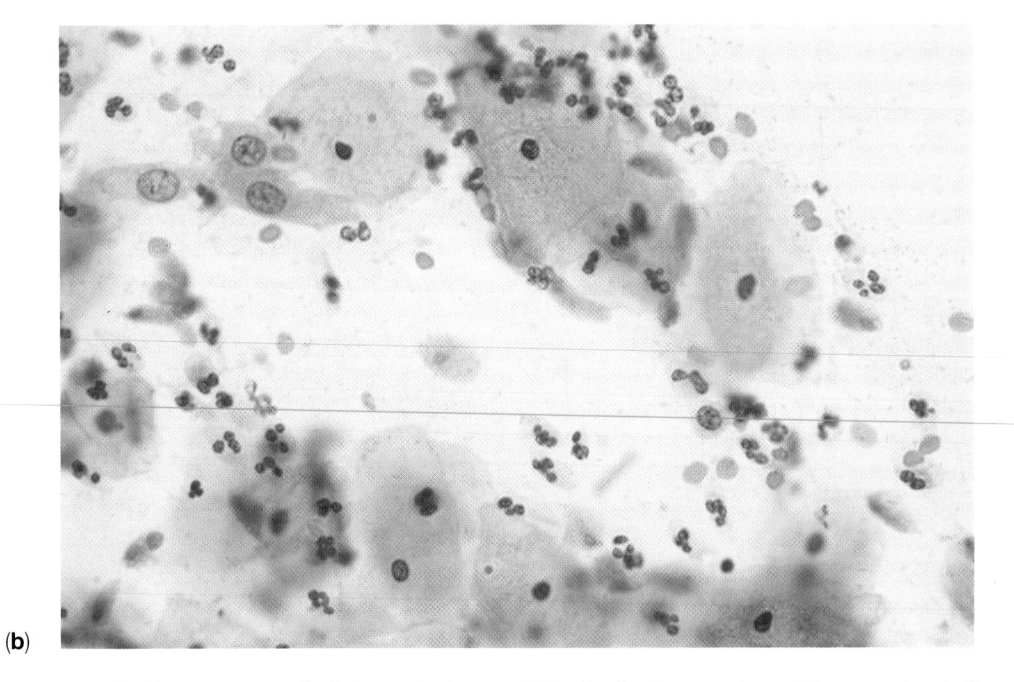

Figure 7.22 (**b**) *Trichomonas vaginalis* in cervical smear. Note the fusiform nucleus of the organism in the centre (Pap, ×630).

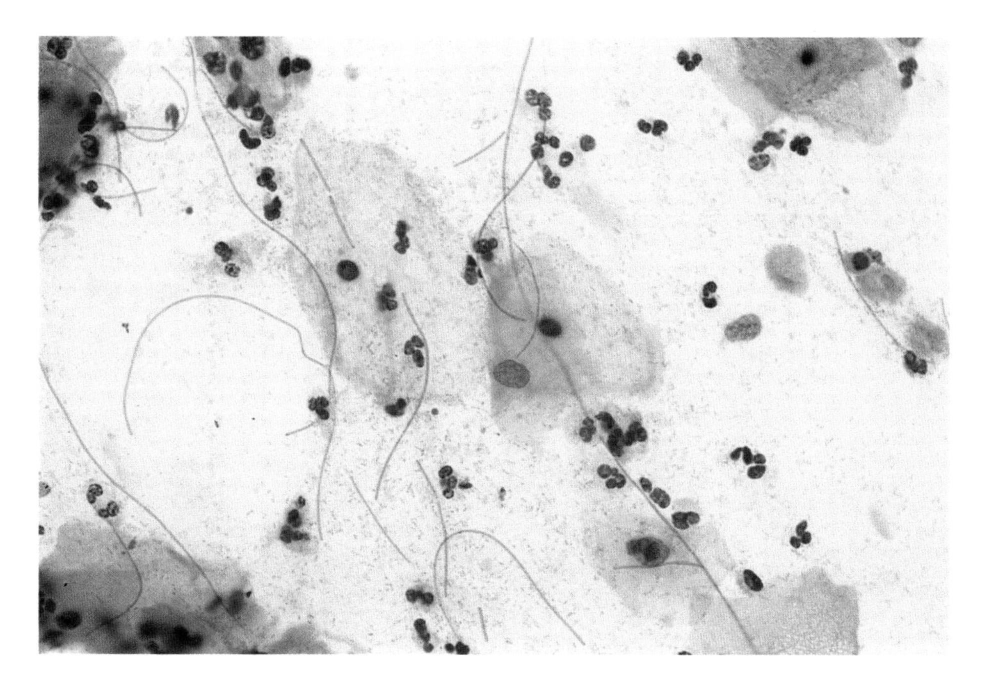

Figure 7.23 *Leptothrix* associated with trichomonas in a cervical smear (Pap, ×630).

giant cells. The ova of *Schistosoma mansoni* are distinguished by their long lateral spine from those of *Schistosoma haematobium* with their short terminal spine. Infection may be accompanied by pseudoepitheliomatous hyperplasia which must be distinguished from epithelial neoplasia.

The importance of genital schistosomiasis relates to its potential as a carcinogen and its effect on fertility. El Tabbah and Hamza (1989) have reported an association with cervical cancer. Sterility, abortion and ectopic pregnancy have all been associated with this infection, although cause and effect have yet to be proved (Notelovitz, 1982).

7.10 AMOEBIASIS, FILARIASIS AND OTHER PARASITIC INFECTIONS

Entamoebae, microfilariae and various worms have been described in cervical smears (Chaterjee, 1975; Wikeley, 1988). Cervical infection with *Entamoeba histolytica* has been recorded in endemic areas (Cohen, 1973). Most female patients with this infection present with pain, tenderness and a profuse, sanguineous vaginal discharge, sloughing ulcers or recto-vaginal fistula. Biopsy shows acutely inflamed necrotic tissue with numerous amoebae. Papanicolaou smears contain numerous trophozoites about 30 µm in diameter, with single eccentric round nuclei and a deeply stained central karyosome. Ingested red blood cells are frequently found in the cytoplasm in the pathogenic form of infection. Confirmation of the diagnosis can be made by hanging-drop preparations and immunocytochemical staining of the parasites (Bhaduri, 1957; Braga and Teoh, 1964; Cohen, 1973).

Reports by McNeill and Ruehsen (1978) and Ruehsen *et al.* (1980) indicate that amoebae resembling *Entamoeba gingivalis* (Figure 7.26(a)) may be found in 1% of cervical smears from women fitted with an IUCD. Distinction between *E. histolytica* and *E. gingivalis*-like organisms can be made on morphological grounds. The commensals are small with

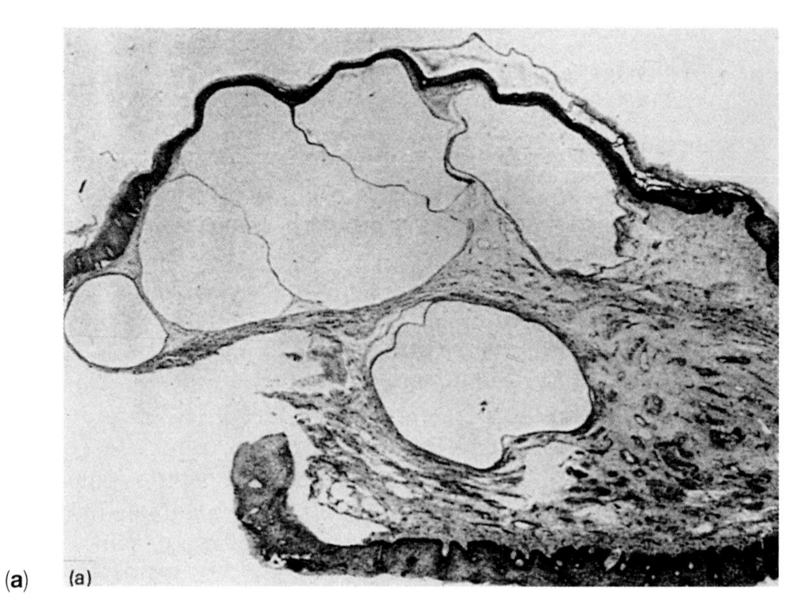

(**a**)

Figure 7.24 (**a**) Cervicovaginitis emphysematosa showing multiple subepithelial cysts (H & E, ×10).

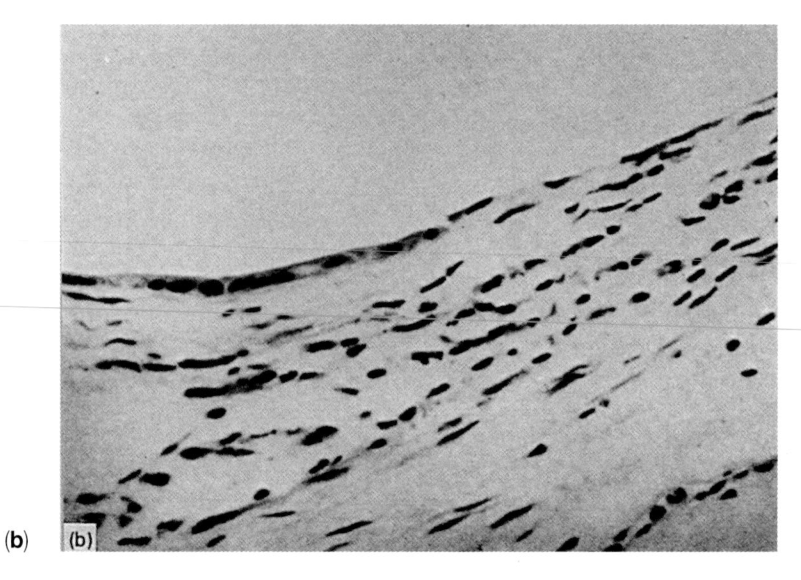

(**b**)

Figure 7.24 (**b**) Flattened multinucleate giant cell lining the cyst. Same lesion as (**a**) (H & E, ×190). (Courtesy of *J. Obstet. Gynaecol. Br. Comm.*).

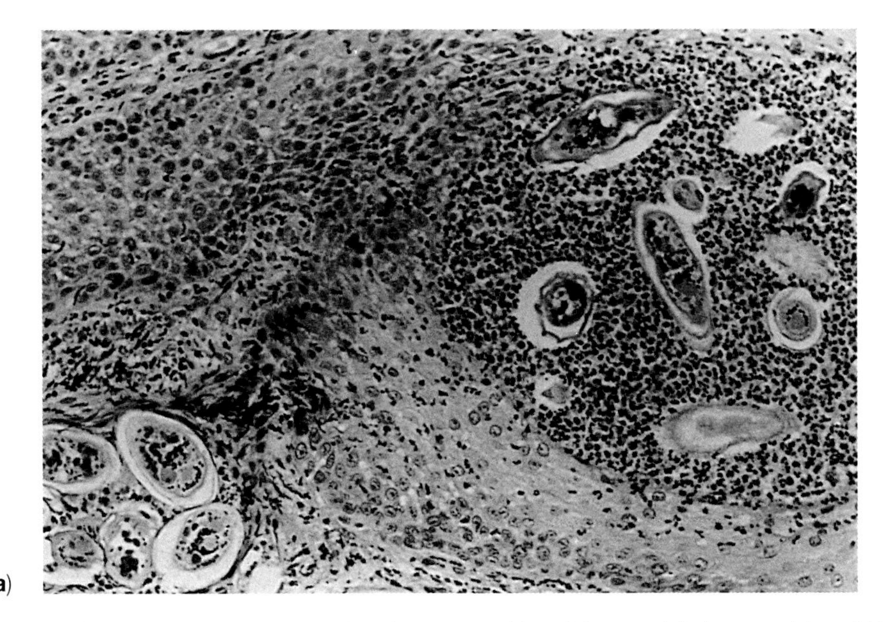

(**a**)

Figure 7.25 (**a**) Schistosomiasis of cervix showing ova and larval forms of *S. haematobium* (H & E, ×120).

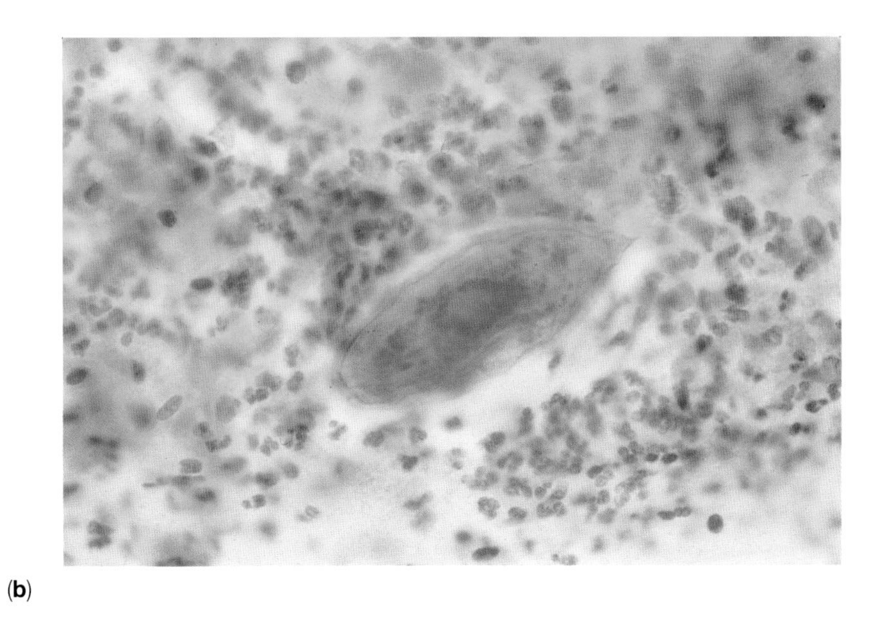

(**b**)

Figure 7.25 (**b**) Cervical smear containing ovum of *S. haematobium* (Pap, ×630).

numerous blunt pseudopodia. They never ingest red blood corpuscles although they may contain small dark granules.

Various species of microfilaria (Figure 7.26(b)) have been described (De Borges, 1971; Sharma *et al.*, 1971; Acharya and Das, 1981). Ova of *Enterobius vermicularis* (pin worm) (Figure 7.27(a)), *Trichiuris trichiuria* (thread worm) and *Ascaris lumbricoides* (Figure 7.27(b)) as well as the ova of *Taenia solium* (Figure 7.28) may be found as faecal contaminants of cervical smears (Berry, 1966; Bhambhani, 1984; Bhambhani *et al.*, 1985).

7.11 *ACTINOMYCES* INFECTION

Actinomyces-like organisms are commonly found in cervical smears from women fitted with an IUCD (Gupta *et al.*, 1976), particularly of the plastic variety (Duguid *et al.*, 1980). *Actinomyces* is usually present as a saprophytic growth on the string of the IUCD. It can also occur following abortion and surgical instrumentation or in smears from women fitted with vaginal pessaries. Diagnosis depends on the demonstration of a radial arrangement of branching filamentous threads or 'sulphur granules' in smears (Figure 7.29) or biopsy material (Schiffer *et al.*, 1975). The colonies resemble bundles of cotton wool with thin, radiating filaments protruding from the mass. Lactobacilli are usually absent having been replaced by a coccoid flora.

Most patients with actinomyces infection are symptomless. The presence of the organism in the smear is only rarely associated with an inflammatory response and as such the presence of actinomyces in a cervical smear is not an indication for treatment. Should there appear to be a risk of pelvic inflammatory disease, removal of the IUCD is sufficient to clear the infection. Penicillin treatment is an alternative.

It should be remembered that the presence of sulphur granules in a smear are not *per se* pathognomonic of actinomyces. Both *Actinomyces israelii* and *Nocardia asteroides* may produce them. Differential diagnosis depends on histochemical staining; *N. asteroides* is acid-fast whereas *A. israelii* is not. Moreover, the sulphur granules of *A. israelii* may be confused with clusters of other bacteria growing in an anaerobic environment. The greater affinity of *A. israelii* for silver stains provides one method of distinguishing between the two conditions in histological sections. However, culture provides the most reliable method of diagnosing this infection.

7.12 HERPES CERVICITIS

The herpes viruses are large, enveloped DNA viruses 120 nm in diameter, which replicate in the nucleus of the host cell with the formation of Cowdray type A intranuclear inclusions. The majority of cases of genital herpes are due to infection with herpes simplex virus, type 2, although type 1 has been isolated from 12% of women with genital herpes. Genital infection is venereally transmitted and is present in the cervices of 8% of women attending VD clinics compared with less than 1% of patients attending gynaecological clinics or antenatal clinics (Morse *et al.*, 1974; Coleman *et al.*, 1983). Primary infection with type 2 virus occurs after puberty when the patient becomes sexually active (Figure 7.32(a)). Recurrence due to reactivation of latent virus which is thought to reside in the sacral ganglia often coincides with *T. vaginalis* or *N. gonorrhoeae* infection. Herpes infection may produce a necrotizing cervicitis (Figure 7.32(b)). The changes are occasionally mistaken for malignancy, especially in pregnancy when the herpetic changes resemble a fungating carcinoma. In 40% of patients with cytologic or virologic evidence of genital infection with herpes simplex, the disease is clinically inapparent.

7.12.1 Histology

Biopsy of the early herpetic lesions on the ectocervix may reveal the presence of multinucleate

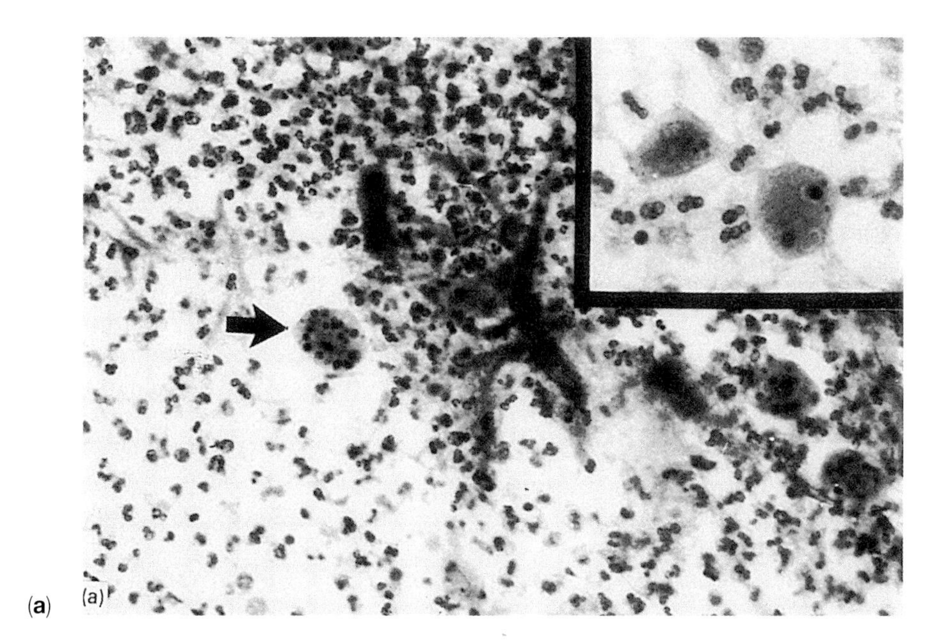

(**a**)

Figure 7.26 (**a**) Amoebae in cervical smear associated with IUCD. Note multiple nuclei and absence of ingested red blood cells distinguishing them from *Entamoeba histolytica* (Pap, ×250; inset, ×770).

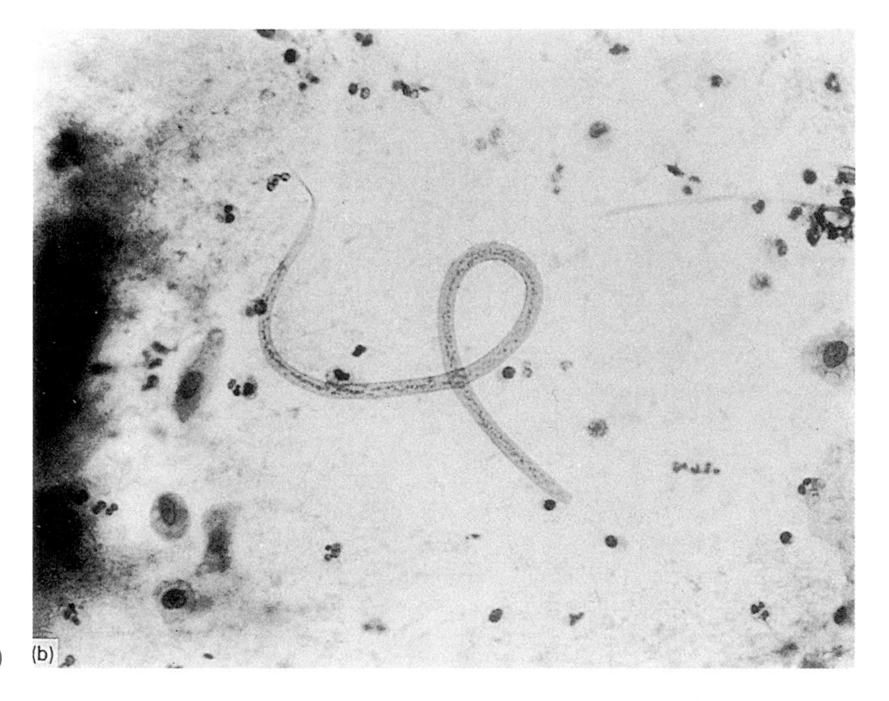

(**b**)

Figure 7.26 (**b**) Microfilaria in cervical scrape (Pap, ×550). (Courtesy of Dr S. Bhambani, New Delhi).

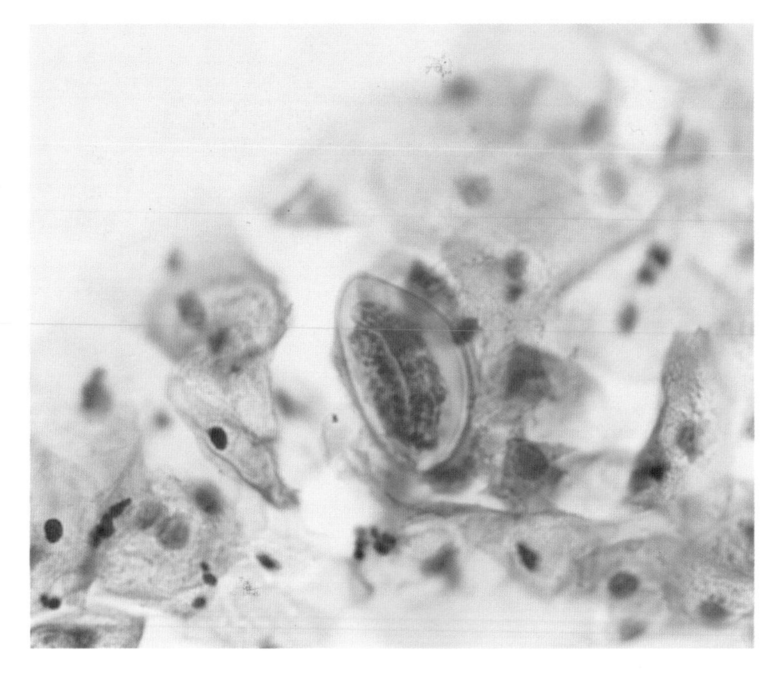

Figure 7.27 (**a**) Ovum of *Enterobius vermicularis* in a cervical smear.

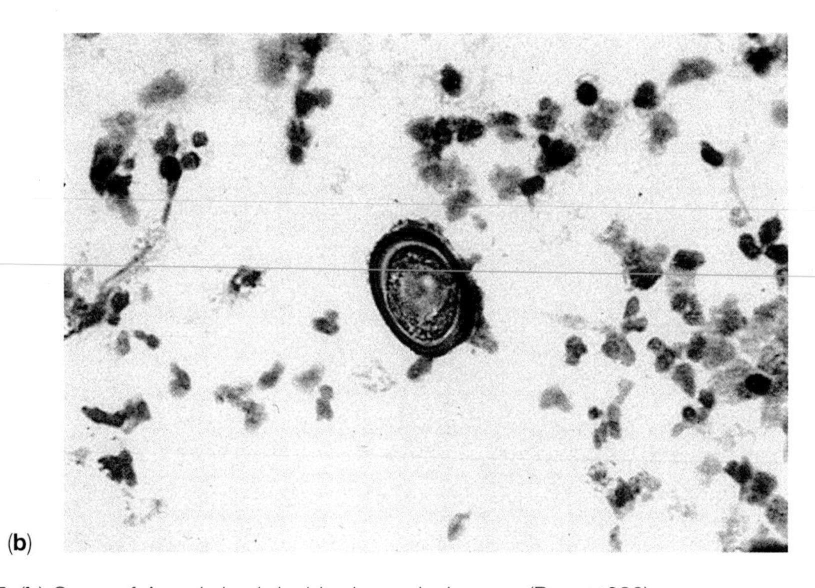

Figure 7.27 (**b**) Ovum of *Ascaris lumbricoides* in cervical smear (Pap, ×630).

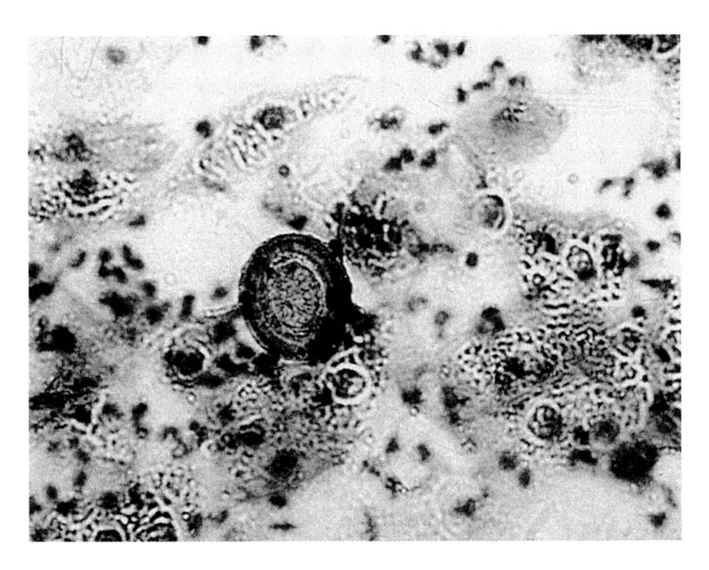

Figure 7.28 Ovum of *Taenia solium* in cervical smear showing slightly striated cortex, which is thicker than that in Figure 7.27(**b**) (Pap, ×630).

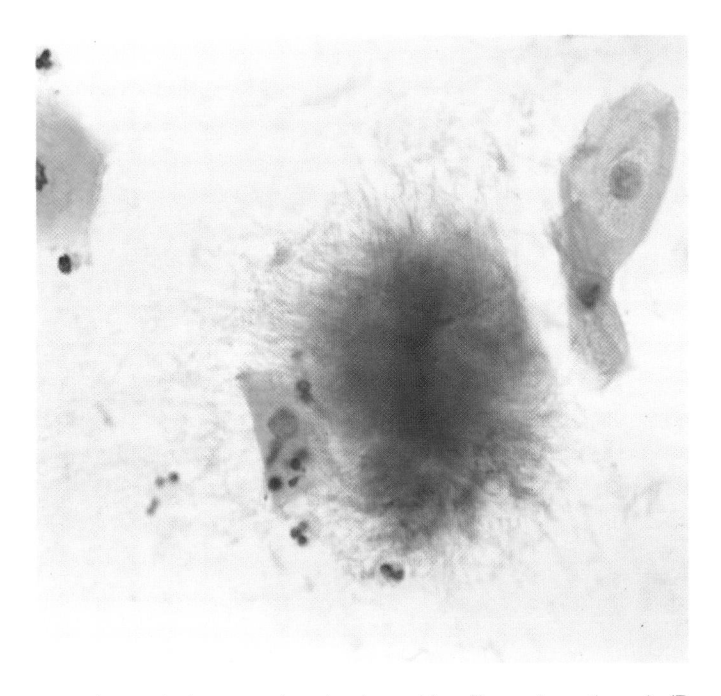

Figure 7.29 Actinomyces in cervical smear showing branching filamentous threads (Pap, ×630).

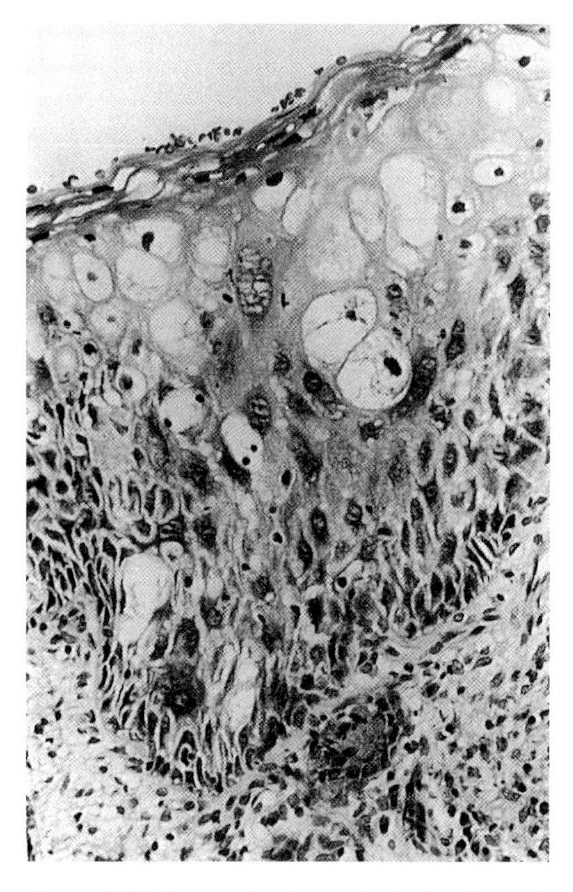

Figure 7.30 Herpes simplex cervicitis showing epithelial vacuolation and multinucleation (H & E, ×340).

giant epithelial cells in the deeper layers of the cervical epithelium with some vacuolation of the cells in the surface layers (Figure 7.30). Occasionally, intranuclear inclusions can be seen in these cells.

7.12.2 Cytology

Herpes simplex virus generally infects squamous cells and only occasionally endocervical cells (Coleman, 1979). It is characterized by cell fusion and the development of intranuclear inclusions. The multinucleate giant cells are readily recognizable by virtue of their large size (50–100 μm). They contain up to 50 nuclei which may appear structureless and glassy or may contain a few angular fragments of chromatin according to the stage in the infection at which the diagnosis is made (Figure 7.31). Amphophilic inclusion bodies and nuclear moulding are a feature of late infection (Figure 7.31(b,c)). The cytoplasm is usually abundant and often vacuolated. Single cells with glassy nuclei may also be present in the smear but these are not of diagnostic value as similar cells can be found in many inflammatory smears. The herpes-infected smear generally contains numerous polymorphs and red blood cells. It is not possible to distinguish between primary infection and recurrence from the smear pattern or histological findings. Some 50% of cervical herpes infections can be detected by cytology (Morse *et al.*, 1974).

The herpes-infected cell must be distinguished from: (i) foreign body giant cells; (ii) malignant cells; (iii) Langhans' giant cells; and (iv) syncytiotrophoblast.

Foreign body giant cells contain nuclei that are all of the same size dispersed evenly throughout the cytoplasm (Figure 7.21). No inclusions or moulding are seen. In contrast, malignant cells with multiple nuclei exhibit a marked anisonucleosis and nuclear hyperchromasia (Figure 12.12(b)). Langhans' cells are characterized by the peripheral distribution of their nuclei (Figure 7.18) and the presence of epithelioid cells in the smear (Figures 7.19 and 7.20). Syncytiotrophoblast is very rarely found in smears and then only when a pregnancy miscarries. In cases of doubt, such as that shown in Figure 7.33(a,b) virus isolation or electron microscopy (Figure 7.33(c)) is of value.

The clinical significance of a diagnosis of genital herpes must not be overlooked even if the patient is asymptomatic. Genital herpes frequently occurs in patients with gonorrhoea and tests for the gonococcus should be performed. In pregnancy, the infection may be transmitted to the foetus *in utero* or during parturition. The risk to the foetus is variously estimated at between 3 and 10%. The greatest hazard to the foetus occurs when primary infection develops in the mother within

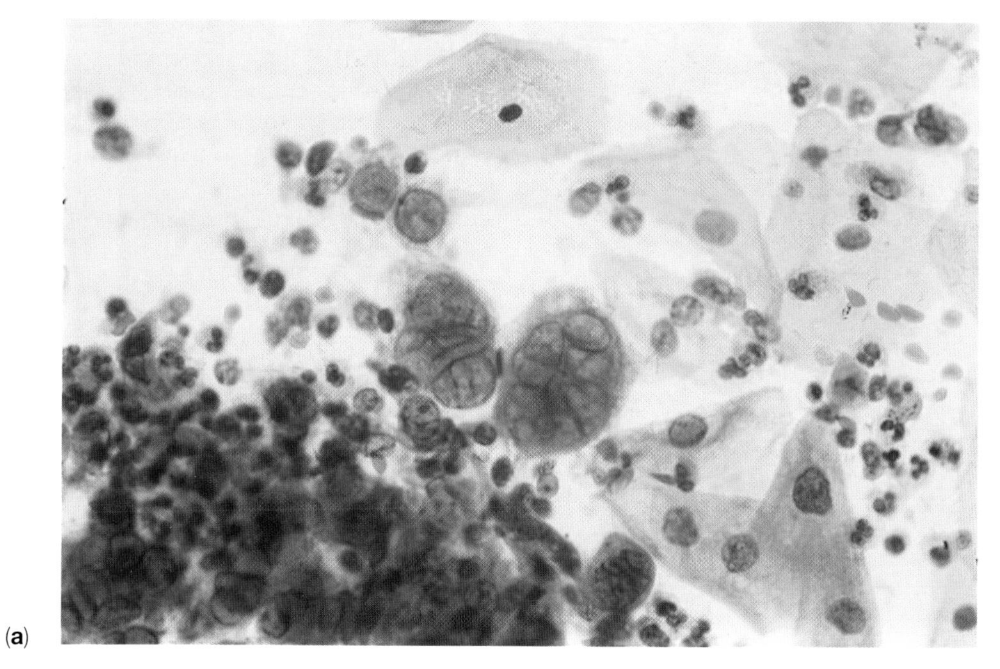

(**a**)

Figure 7.31 (**a**) Multinucleate giant cells showing changes characteristic of Herpes simplex in cervical smear. Note ground-glass appearance of nuclei, margination of chromatin and nuclear moulding (Pap, ×400).

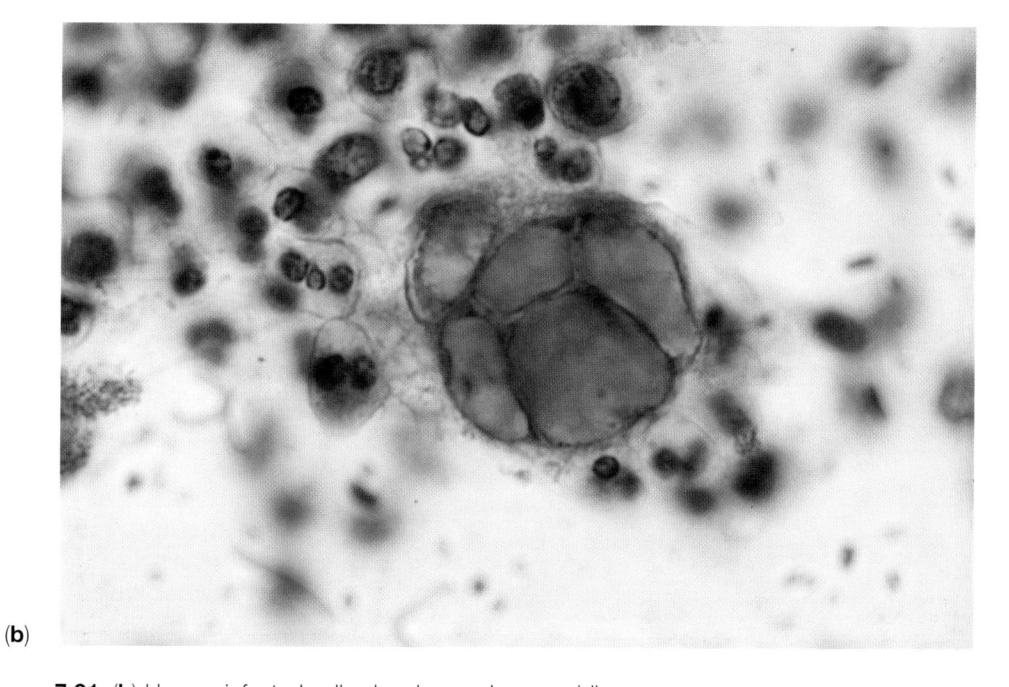

(**b**)

Figure 7.31 (**b**) Herpes-infected cells showing nuclear moulding.

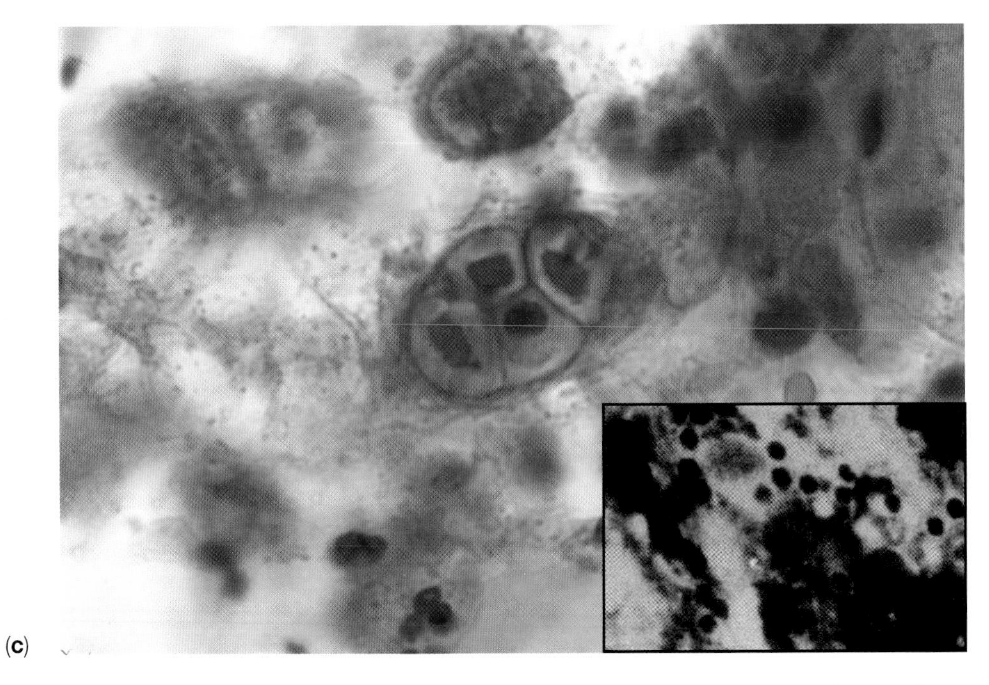

(**c**)

Figure 7.31 (**c**) Herpes-infected cells showing intranuclear inclusion bodies (Pap, ×630). **Inset** Electron microscopy of cell to show herpes virus particles (uranyl acetate, ×30 000).

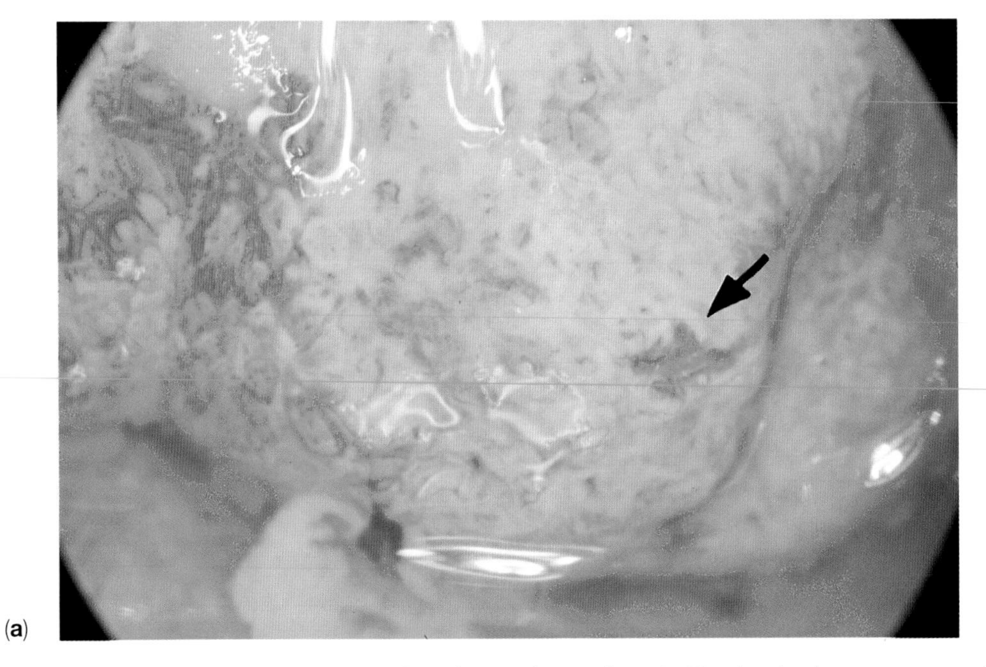

(**a**)

Figure 7.32 (**a**) Colposcopic appearance of an ulcerated area of cervix. Herpes simplex was seen on the cervical smear.

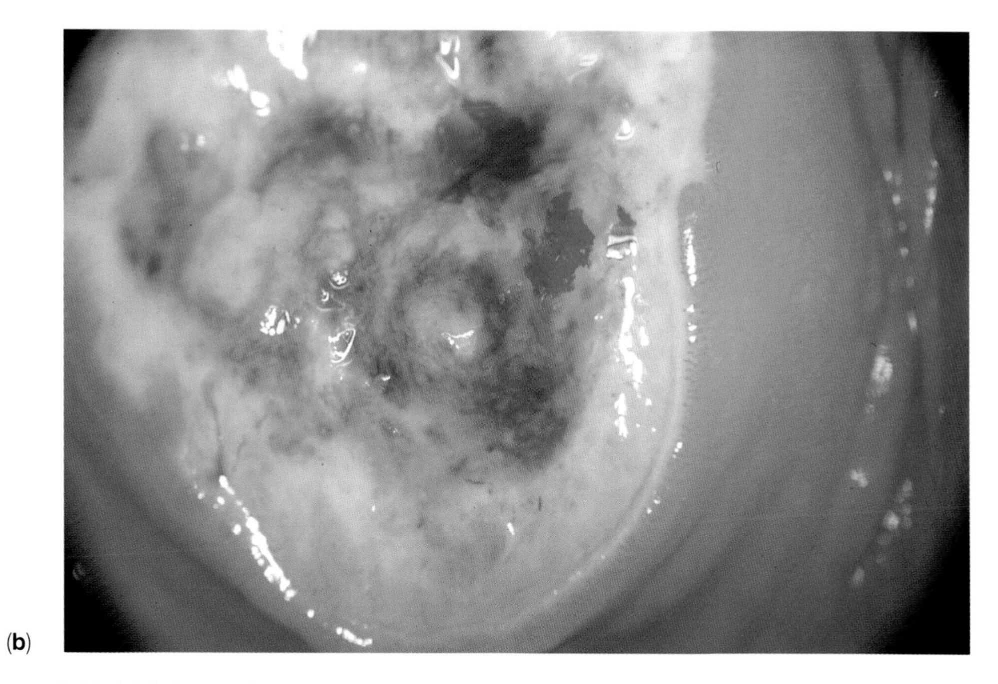

(**b**)

Figure 7.32 (**b**) Colposcopic appearance of necrotising Herpes cervicitis.

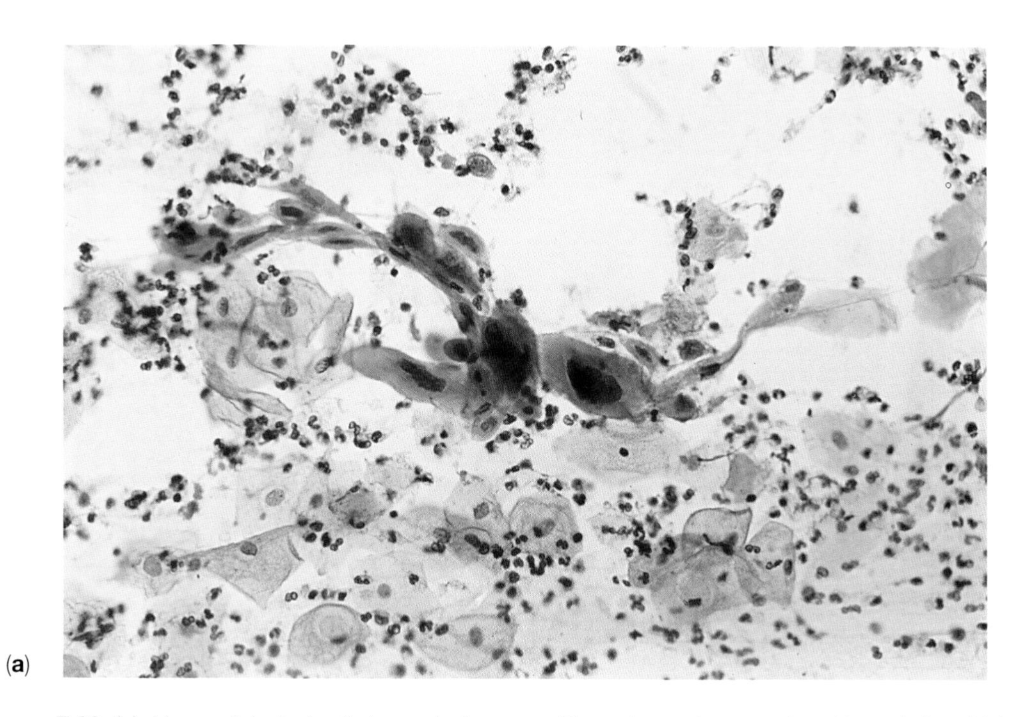

(**a**)

Figure 7.33 (**a**). Herpes-infected cells in cervical smear with nuclear enlargement and irregularity which could be misdiagnosed as malignant cells (Pap, ×630).

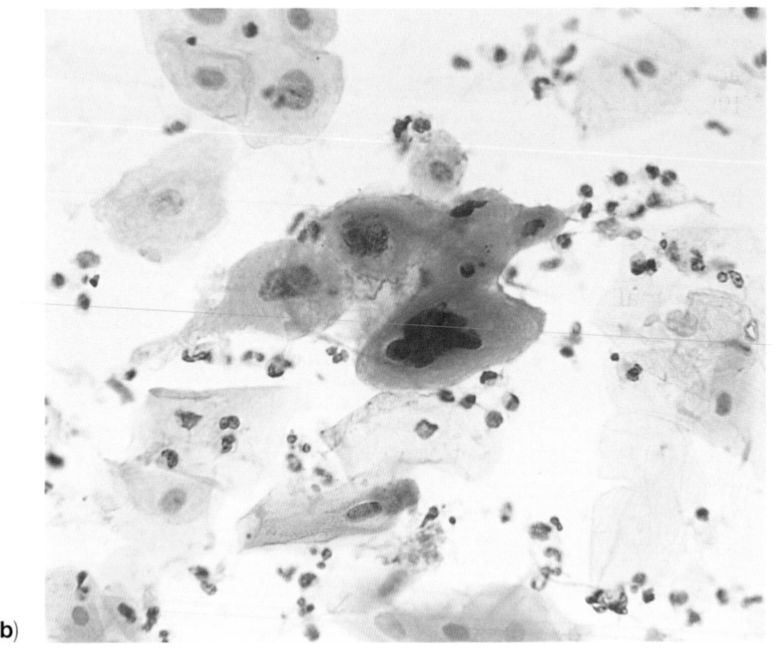

(**b**)

Figure 7.33 (**b**) Herpes-infected cells in cervical smear with nuclear enlargement and irregularity which could be misdiagnosed as malignant cells (Pap, ×400).

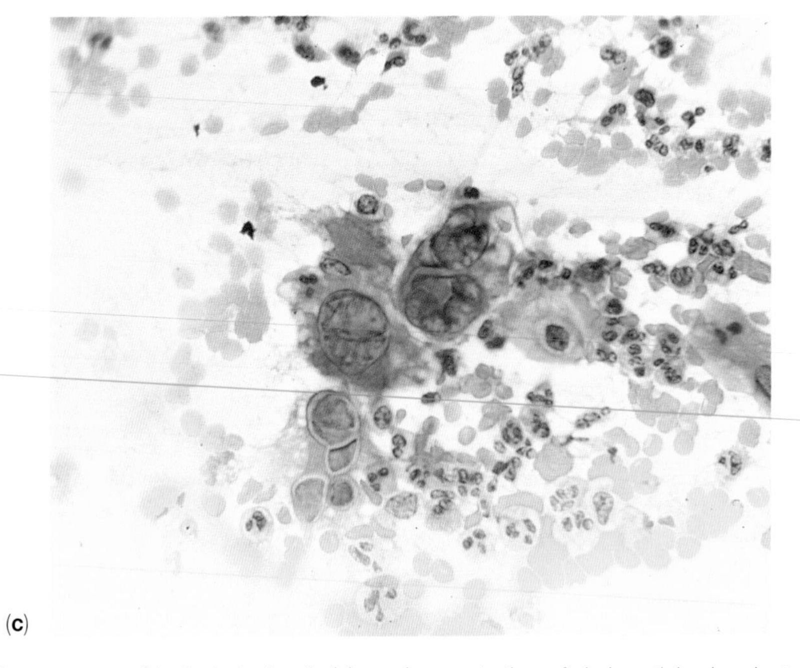

(**c**)

Figure 7.33 The presence of typical giant cells (**c**), or demonstration of viral particles by electron microscopy (as shown in 7.31 inset) in the smear would suggest the correct diagnosis (Pap, ×630).

1 month of delivery. For a review of the evidence linking herpes simplex type 2 and cervical cancer, the reader is referred to Galloway and McDougall (1983).

7.13 HUMAN PAPILLOMAVIRUS (HPV) INFECTION

Papillomaviruses are small, non-enveloped, double stranded DNA viruses of 45–54 nm diameter which are members of the papovavirus group. The application of molecular hybridization techniques to the study of the viruses has greatly increased our knowledge of the biological properties and epidemiology of this virus group. They have a circular genome approximately 8 kilobases in length. The genome comprises early (E) and late (L) genes, the E genes being involved in viral replication and the L genes coding for capsid proteins. The E1 gene plays a part in viral replication and the E2 gene codes for a transcriptional regulator. The E6 and E7 genes, and to a lesser extent the E5 gene, are involved in cellular transformation.

Although papillomaviruses are found in a wide variety of mammals and birds they are, like most viruses, species specific. They replicate exclusively in the nucleus of the host cell (Figure 7.43) and HPV are the causative agents of a variety of warts and papillomatous lesions that are found on the skin and mucous membranes in humans. There are more than 70 distinct genotypes of HPV, each genotype being associated with a particular type of wart.

Until recently, papillomaviruses have been numbered in order of their discovery according to the degree of homology assessed by cross-hybridization in solution (Herrington, 1995). Subtypes and variants were initially defined by differences in restriction digestion pattern and small variations in nucleotide sequence of the viral genome. More recently, classification has been based on sequencing of only a short segment of the E6 region. Classification using this system is easier and permits the establishment of a central database. There is considerable heterogeneity between HPV isolates both between and within populations. Although the biological relevance of these minor variations in HPV sequence is not known, they may have a clinical significance which has yet to be determined (Herrington, 1995).

Genital infection is acquired by sexual contact although the exact mechanism of virus entry is not known. *In situ* hybridization studies indicate that the viruses infect the basal layers of the squamous epithelium (Beckmann *et al.*, 1985); however, viral particles and viral antigens are found mainly in the surface layer, indicating that virus replication may be restricted to keratinized cells in the epithelium. Replication of HPV in squamous epithelium is frequently associated with morphological changes in the cells, the most striking of which is *koilocytotic atypia* (Koss and Durfee, 1956). Cells showing this change are characterized by the presence of a large clear zone or halo which surrounds the nucleus of the cell and displaces the cytoplasm to the periphery of the cell (Figures 7.34–7.42). The halo has a sharp border bounded by the thickened rim of cytoplasm. The nucleus of the koilocyte is usually enlarged, hyperchromatic and irregular in outline. The degree of nuclear abnormality may vary from slight enlargement and hyperchromasia to a marked nuclear atypia indistinguishable from neoplasia (Section 10.2.3). Multiple nuclei are sometimes seen. The presence of koilocytes in cervical biopsies and smears is considered to be pathognomonic of HPV infection.

Three patterns of HPV infection have been identified in the genital tract: (i) clinical infection which is visible to the naked eye; (ii) subclinical infection which is not visible to the naked eye but is visible at the microscopic or colposcopic level; and (iii) latent infection in which HPV DNA is detectable by molecular means but without clinical, histological, cytological or colposcopic evidence of infection. Thus, in most cases, diagnosis of HPV infection depends on the demonstration of virus particles at the electron microscopic level or its antigens or its nucleic acid either by DNA or RNA hybridization techniques.

Clinical infection is manifest by two distinct lesions, namely exophytic warts (condylomata acuminata) and the rare endophytic wart. Subclinical HPV infection presents in the form of non-condylomatous wart virus infection (synonyms flat warts) in the genital tract. This latter form of infection can be recognized colposcopically and microscopically and is commonly associated with cervical intraepithelial neoplasia; it is described in the appropriate chapter (Section 10.6).

7.13.1 Condyloma acuminatum (exophytic wart) and inverted (endophytic) wart

Cervical condylomata are found in less than 6% of women with genital warts (Oriel, 1971; Walker *et al.*, 1983). On speculum examination, the cervical condylomata appear as knob-like lesions on the ectocervix, not infrequently situated outside the transformation zone. They are usually multiple and may measure 1 cm or more across, although single lesions (syn: squamous papillomas) do occur. These single lesions may be 1 mm in diameter and are just visible to the naked eye. The crops of warts often extend into the vagina and vulva and are deemed to reflect field change. This is reflected by the fact that large crops of cervical warts have been recorded after laser therapy and in patients whose immune response has been impaired by therapy or disease. They have also been recorded during pregnancy when they may produce huge masses which obstruct delivery; in the puerperium they regress.

Colposcopically, exophytic warts (cervical condylomata) have a vascular, papilliferous or frond-like surface (Anderson *et al.*, 1992). After the application of acetic acid the epithelium may appear pearly white or dense white if the wart has a hyperkeratotic surface (Figure 7.34(a)). Biopsy of cervical condylomata shows that, histologically, the squamous epithelium is thrown into folds to produce numerous papillary processes each with its central core of connective tissue (Figures 7.34(b) and 7.35) which may be infiltrated with inflammatory cells. The epithelium shows marked acanthosis and one or two of the surface layers of cells may be keratinized. Parakeratosis and hyperkeratosis are not as prominent in genital warts as in warts at other body sites. Koilocytotic atypia (Figures 7.35, 7.36 and 7.38) is a prominent feature. Mitotic figures (Figure 7.36) and multinucleate cells may be seen. Individual cells may show nuclear irregularity with marked variation in size and shape (Figure 7.36). In some cases, the nuclear irregularity may be sufficient to suggest CIN (Figures 7.37–7.39). Occasionally, wart formation may be inverted and simulate a carcinoma (Figure 7.40). Differential diagnosis is from verrucous carcinoma (Section 12.6.1).

7.13.2 Cytology

Cervical infection with HPV may be suspected from the examination of the cervical smear. Koilocytes with their slightly enlarged, hyperchromatic, irregular nucleus, large perinuclear halo and thickened cytoplasm (Figure 7.41) are readily recognized at low magnification. Care must be taken not to confuse the large perinuclear halo of the koilocyte with the small area of perinuclear clearing seen in trichomonas and other infections. The sharp outline and large size of the perinuclear halo and the thickened rim of cytoplasm are useful markers of koilocytosis.

Keratinized plaques studded with pyknotic nuclei may be seen reflecting the abnormal keratinization associated with HPV infection (Figure 7.44); multinucleate cells (Figure 7.45) and unusual keratinized squames (dyskeratocytes) with bizarre shapes (Figure 7.46) may also be seen.

7.13.3 Diagnosis of human papillomavirus infection

Although koilocytotic atypia is a frequent finding in HPV-infected epithelium, a definitive

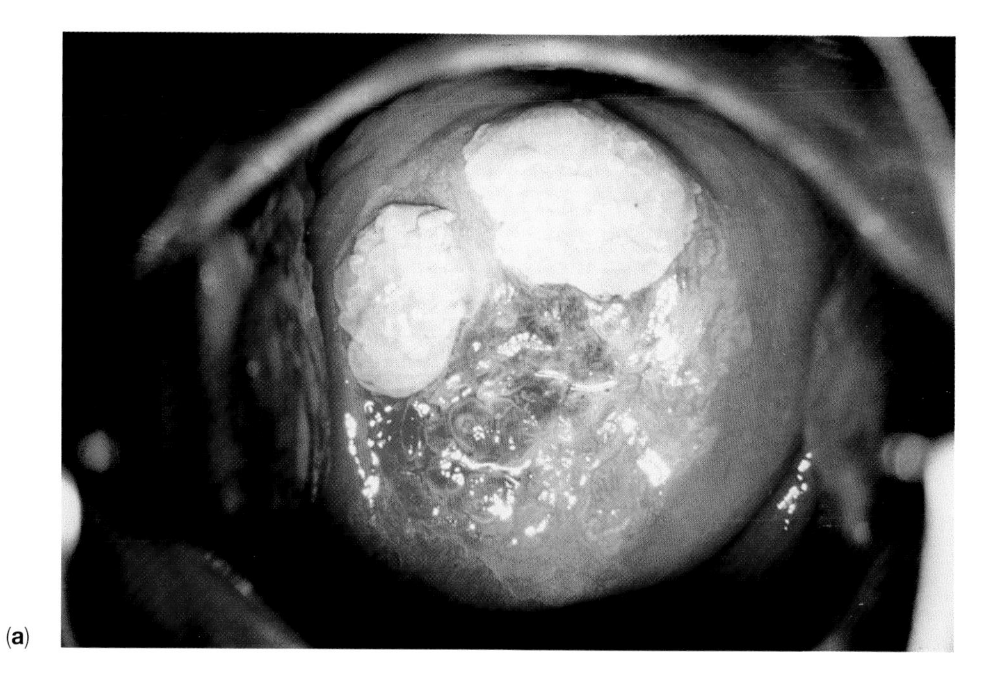

(**a**)

Figure 7.34 (**a**) Colposcopic appearance of cervical condylomata.

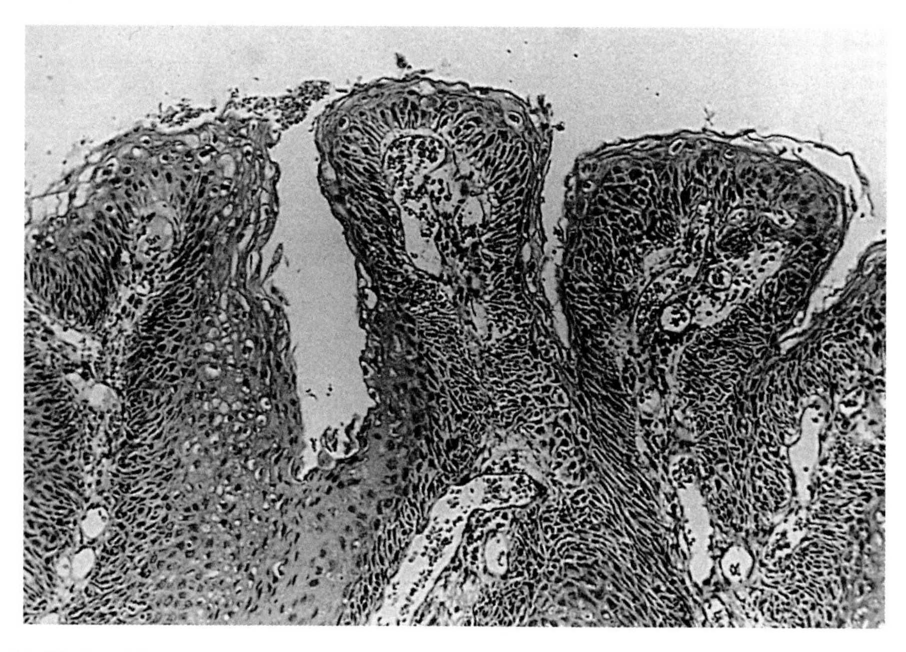

(**b**)

Figure 7.34 (**b**) Condyloma acuminatum showing papillary process with squamous epithelial lining and central core of vascular connective tissue (H & E, ×130).

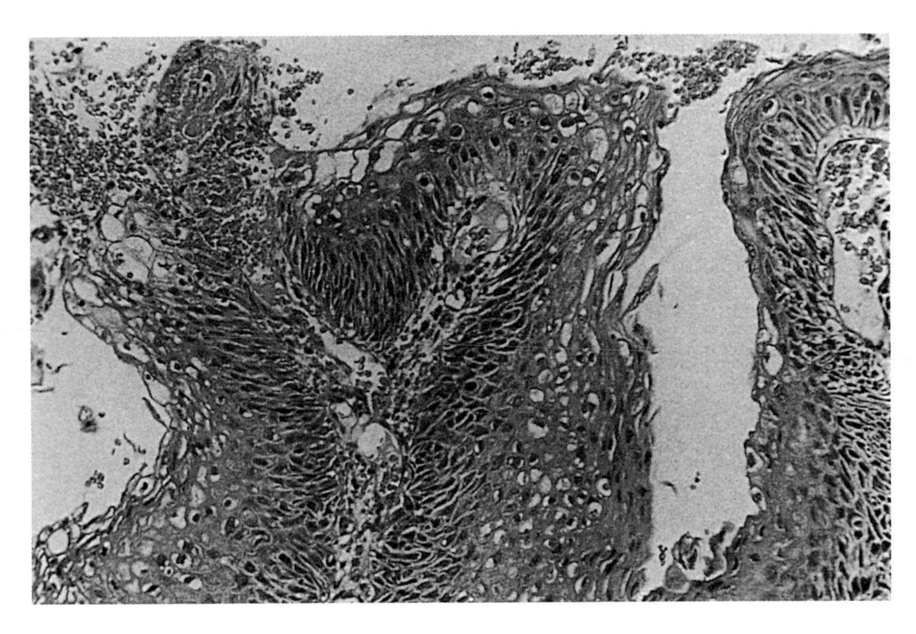

Figure 7.35 One of the papillary processes shown in Figure 7.34(**b**) demonstrating koilocytosis (H & E, ×210).

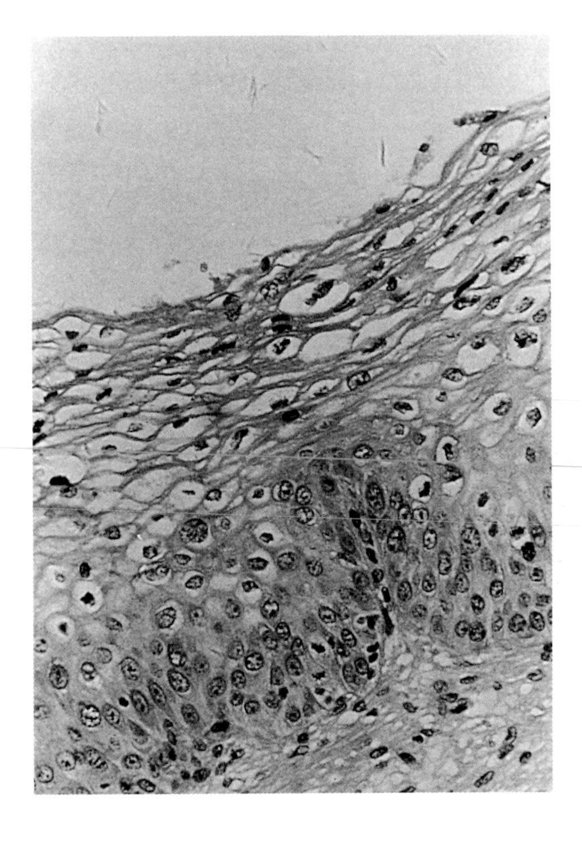

Figure 7.36 Condyloma acuminatum at higher magnification. Note nuclear irregularity, koilocytotic atypia and occasional mitotic figures (H & E, ×480).

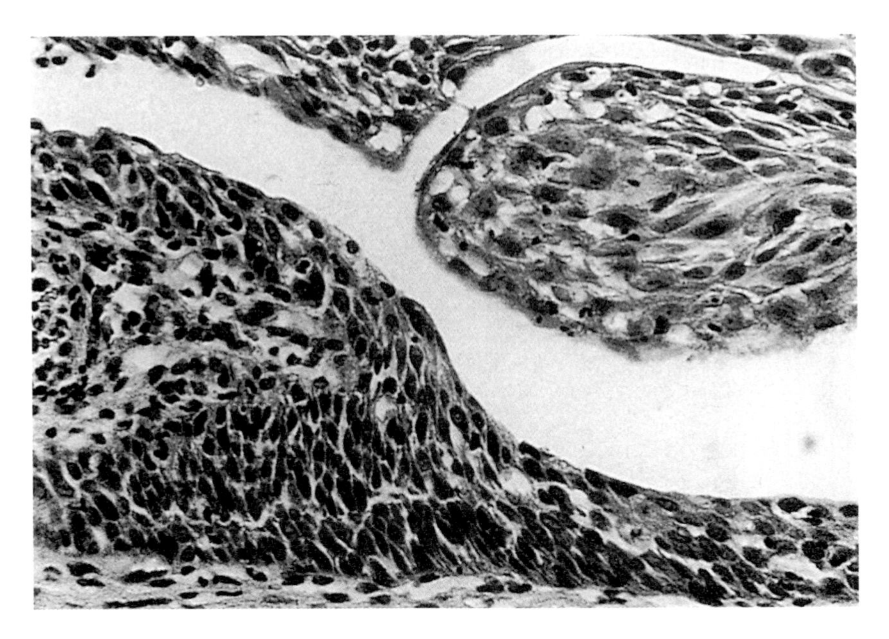

Figure 7.37 Nuclear atypia in condyloma shown in Figure 7.34(**b**) (H & E, ×480).

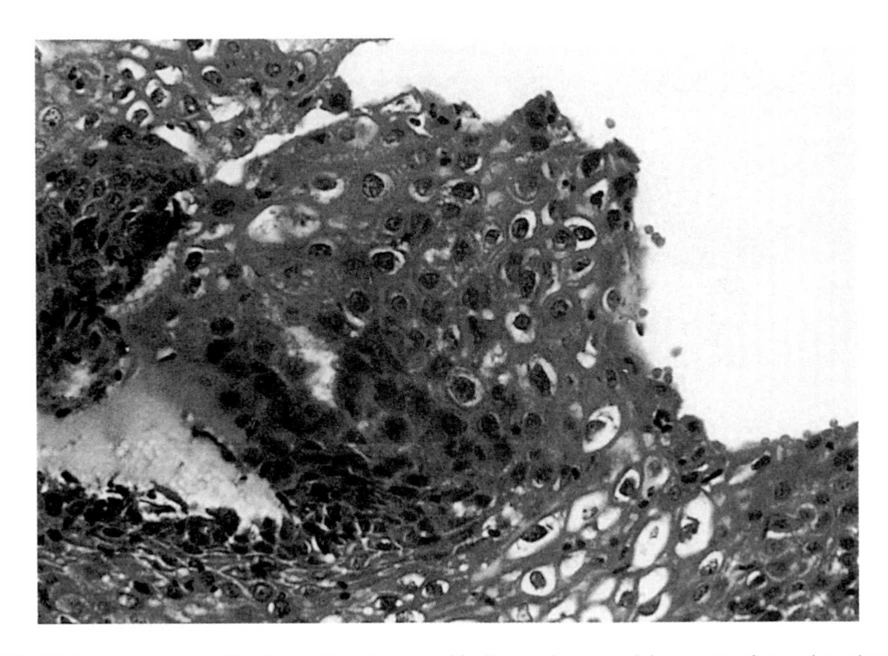

Figure 7.38 High-power magnification of surface epithelium of a condyloma to show the abnormal cells (H & E, ×600).

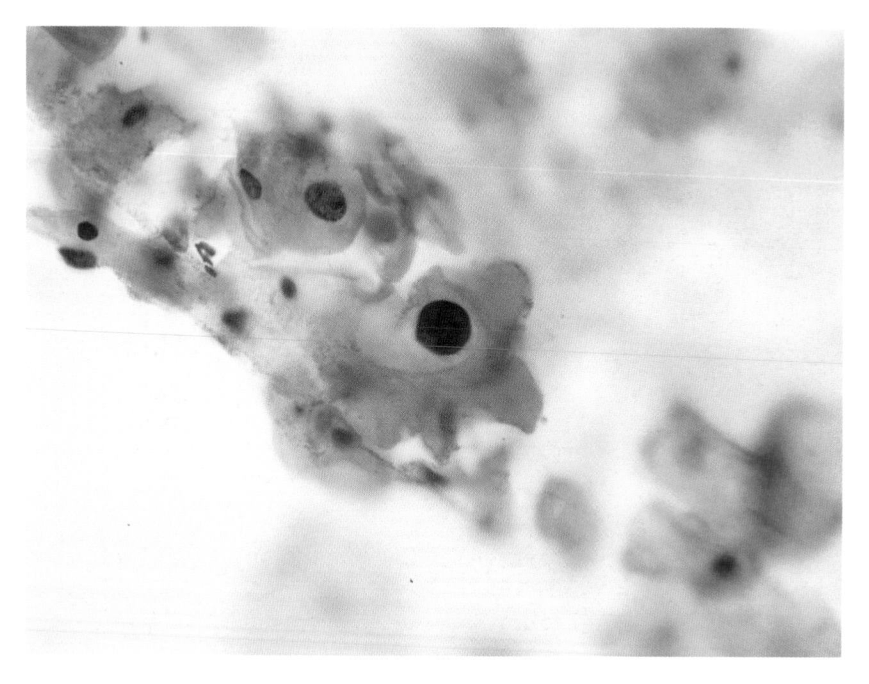

Figure 7.39 Cervical smear from case shown in Figure 7.38. The smear was reported as showing borderline cases (Pap, ×630).

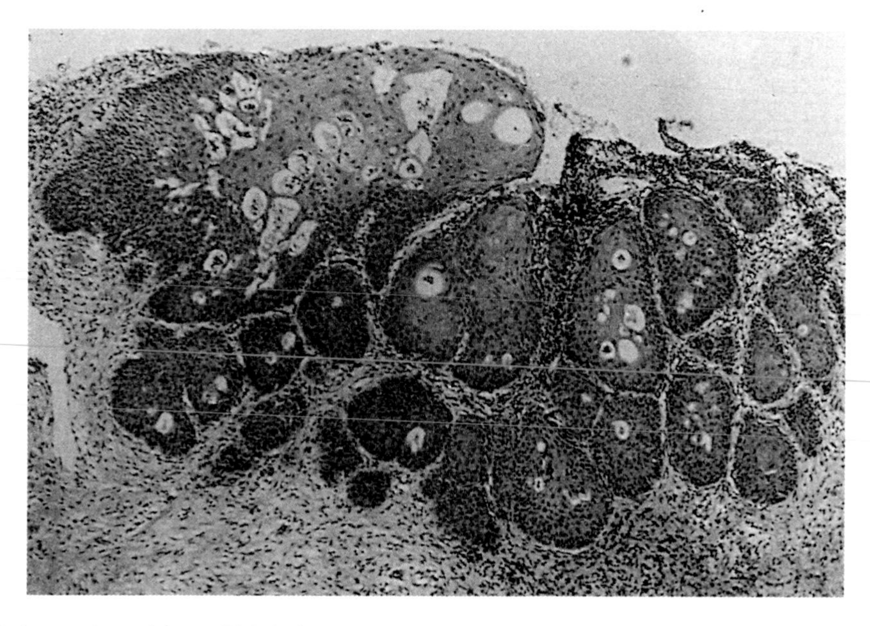

Figure 7.40 Inverted condyloma. This lesion was mistaken for a carcinoma from which it can be distinguished by the regularity of its cells (H & E, ×80).

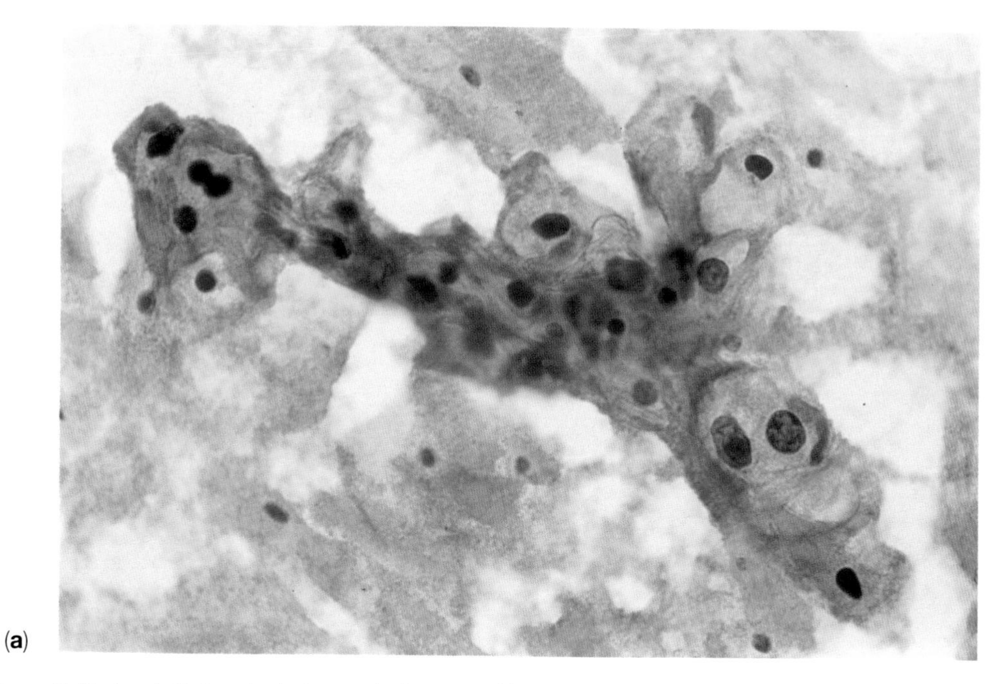

(**a**)

Figure 7.41 (**a–c**) Koilocytosis in cervical smear. Note slight anisonucleosis and nuclear hyperchromasia (Pap, ×630).

(**b**)

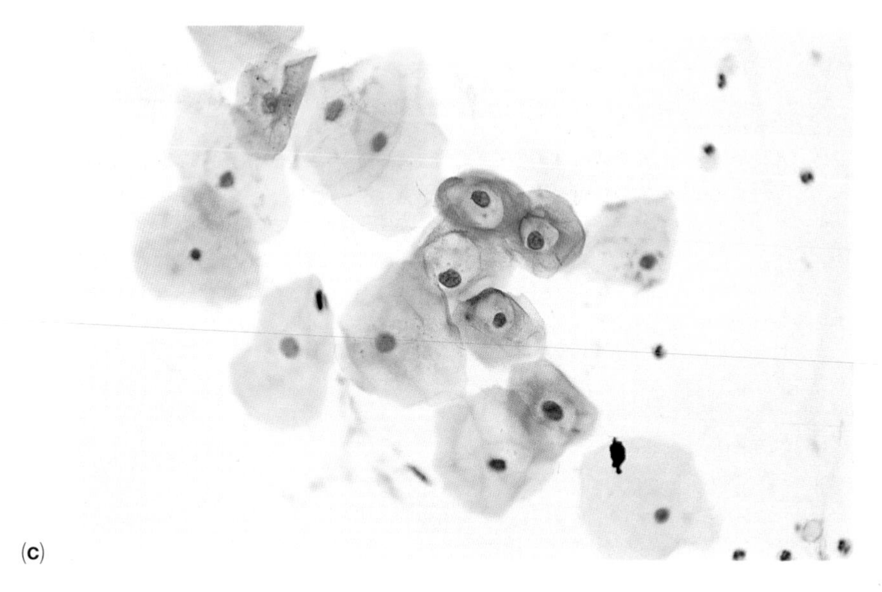
(c)

Figure 7.41 (**a–c**) Koilocytosis in cervical smear. Note slight anisonucleosis and nuclear hyperchromasia (Pap, ×630).

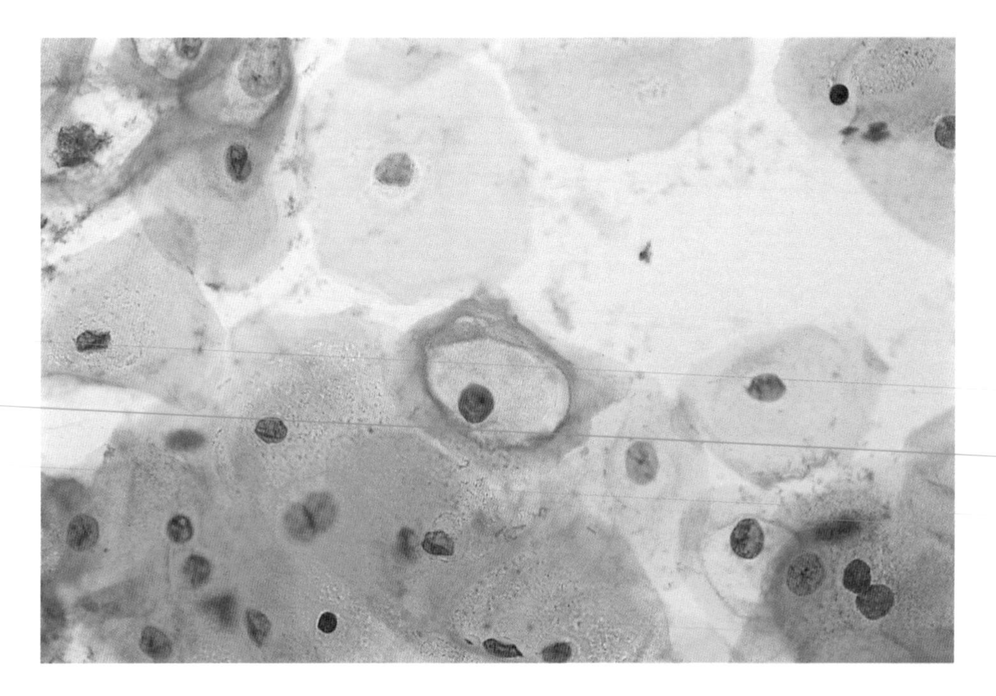

Figure 7.42 Large perinuclear halo and peripheral condensation of the cytoplasm in a koilocyte, which is pathognomonic for HPV infection (Pap, ×630).

diagnosis of HPV infection depends on the demonstration of virus particles, viral antigen or viral DNA. When no fresh material is available, diagnosis can be made retrospectively by electron microscopy of alcohol-fixed smears (Figure 7.43) or paraffin-embedded tissue sections using the techniques described in Section 2.5.1. An alternative method of establishing the diagnosis is by immunocytochemical staining of paraffin-embedded sections using an antibody raised against an internal capsid antigen common to all papillomaviruses (Figure 2.7). DNA hybridization of unfixed biopsy specimens or cervical scrapings using radioactive or biotin-labelled probes (dot blot method) is used to detect viral genome. This approach has the added advantage over electron microscopy and immunocytochemistry in that it is a more sensitive method of detecting the virus and can be used to identify the virus type (Wickenden *et al.*, 1985; Byrne *et al.*, 1986). *In situ* hybridization of paraffin-embedded histological sections and Papanicolaou smears using non-radioactive probes has also been used to study distribution of HPV DNA in cervical epithelium (Beckmann *et al.*, 1985) but is less sensitive than the dot blot technique. Sensitivity can be improved using the PCR but specificity is easily lost by this approach. The hybrid capture technique which is based on the ELISA method has been developed for quantitative analysis of the amount of viral genome in the sample.

The viruses as yet cannot be propagated *in vitro*, although interestingly, koilocytotic atypia can be induced in heterografted cervical tissue infected with HPV 11 (Kreider *et al.*, 1985).

7.14 CYTOMEGALOVIRUS

Cytomegalovirus has been isolated from the cervices of 10% of women attending a venereal disease clinic (Morse *et al.*, 1974). A similarly high incidence of cervical infection with these viruses has been recorded in pregnant women. The majority of cases represent reactivation of latent infection. Primary infection of the mother in pregnancy is rare but significant because of the small but real risk of transplacental transmission to the foetus (Goldman *et al.*, 1969). The bird's eye inclusions characteristic of this infection in kidney and lung in the neonate are only occasionally seen in the cervical biopsies (Figure 7.47) or smears (Figure 7.49) (Goldman *et al.*, 1969; Morse *et al.*, 1974). The diagnosis can be confirmed by immunofluorescence (Figure 7.48).

7.15 *CHLAMYDIA TRACHOMATIS*

Chlamydia trachomatis is a common venereal infection. The organism has been isolated from the cervices of nearly 50% of consorts of males with non-gonococcal urethritis and one-third of all women attending venereal disease clinics. Most women with cervical infection are asymptomatic. Nevertheless, direct evidence from clinical and experimental studies indicates that ascending infection can cause tubal damage (Westrom and Mardh, 1983). There is also indirect evidence that pelvic inflammatory disease, infertility and ectopic pregnancy may result from ascending infection (Taylor-Robinson and Thomas, 1980). *Chlamydia* infection has been associated with radiotherapy (Maeda *et al.*, 1990).

7.15.1 Histology

Swanson *et al.* (1975) and Hare *et al.* (1981) have shown that infection with *Chlamydia trachomatis* is frequently associated with follicular cervicitis (Figures 7.50 and 7.51). Colposcopic biopsy (Hare *et al.*, 1981) revealed the presence of lymphoid follicles in the submucosa of the cervix deep to endocervical epithelium or immature metaplastic epithelium in seven of 15 women who were contacts of males with non-specific urethritis. In a further five women, a lymphocytic cervicitis was noted. *Chlamydia trachomatis* was isolated from the cervices of five of 11 women investigated (45%). In 1974, Schachter *et al.* reported an

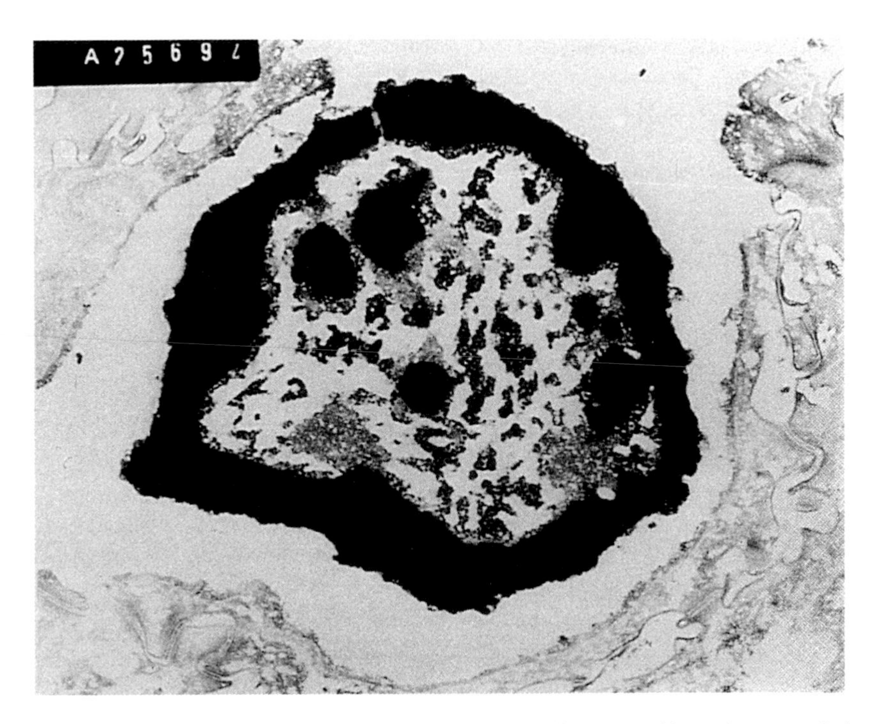

Figure 7.43 Intranuclear papillomavirus particles in pseudocrystalline array. Note absence of virus particles in clear perinuclear zone (uranyl acetate, ×30 750).

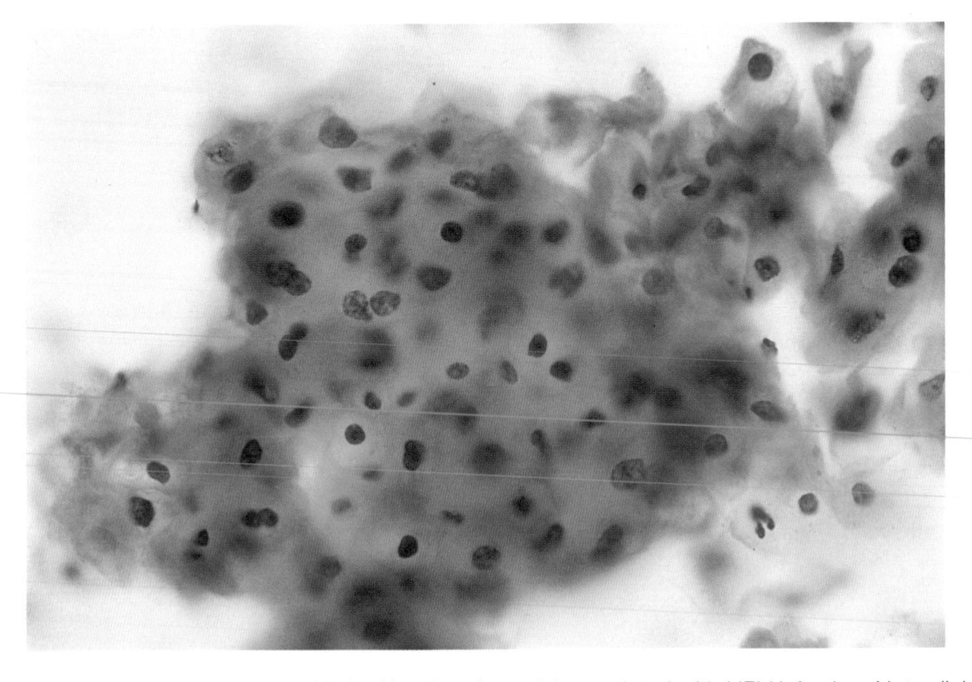

Figure 7.44 Keratinized plaques studded with pyknotic nuclei associated with HPV infection. Note slight variation in nuclear size and multilayering of the epithelium (Pap, ×630).

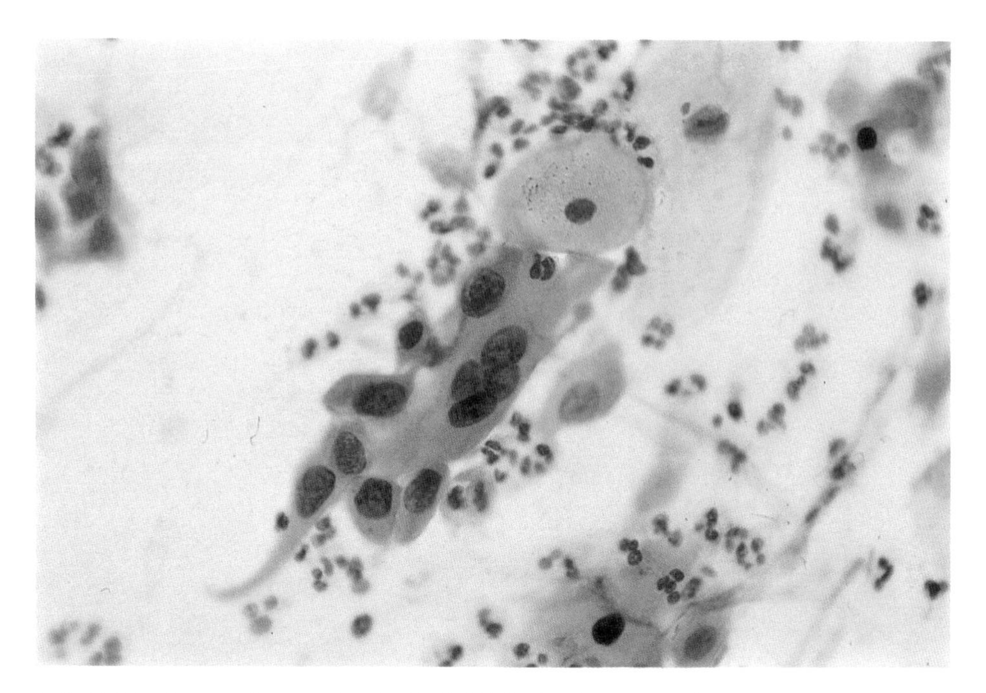

Figure 7.45 Multinucleate cells associated with HPV infection (Pap, ×630).

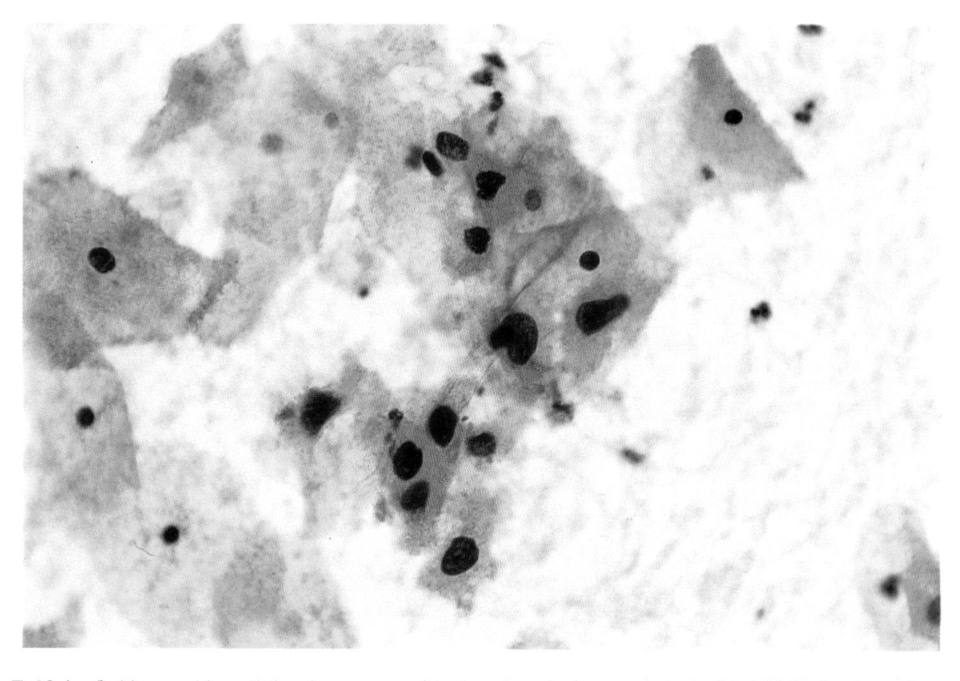

(**a**)

Figure 7.46 (**a–f**) Unusual keratinized squames (dyskeratocytes) associated with HPV infection. Note tadpole cells and fibre cells similar to those seen in squamous carcinoma but lacking the nuclear atypia (Pap, ×400).

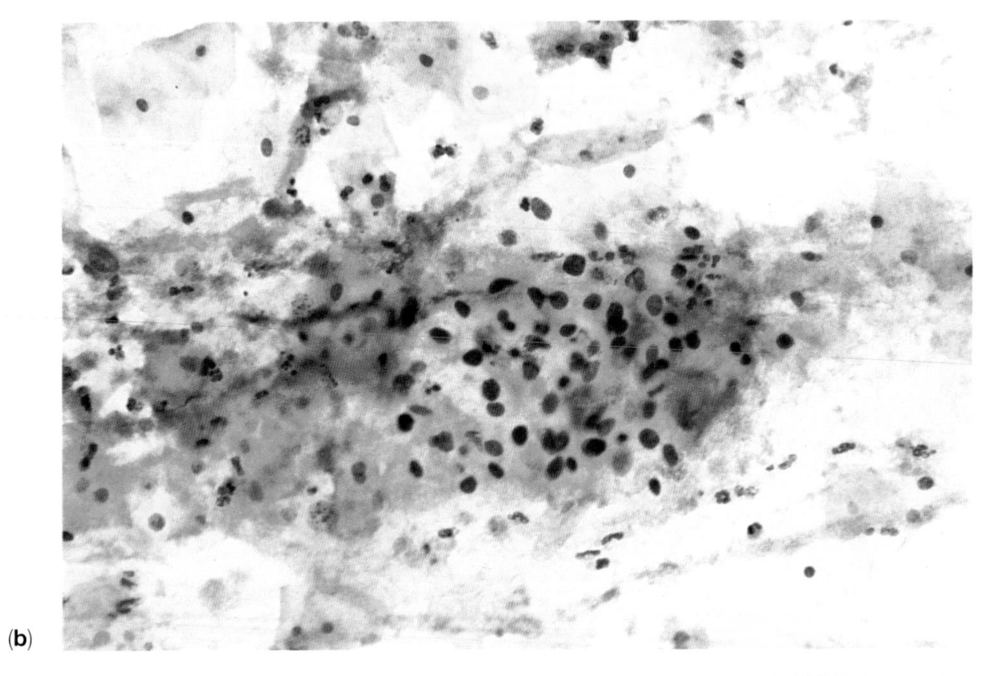

(**b**)

Figure 7.46 (**a–f**) Unusual keratinized squames (dyskeratocytes) associated with HPV infection. Note tadpole cells and fibre cells similar to those seen in squamous carcinoma but lacking the nuclear atypia (Pap, ×400).

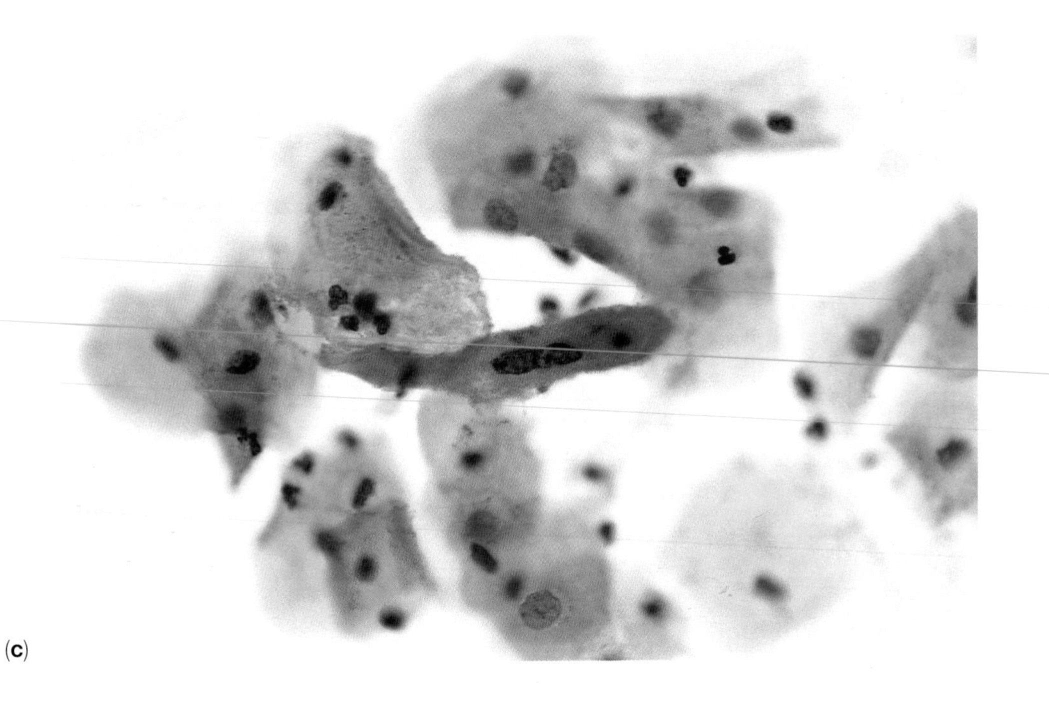

(**c**)

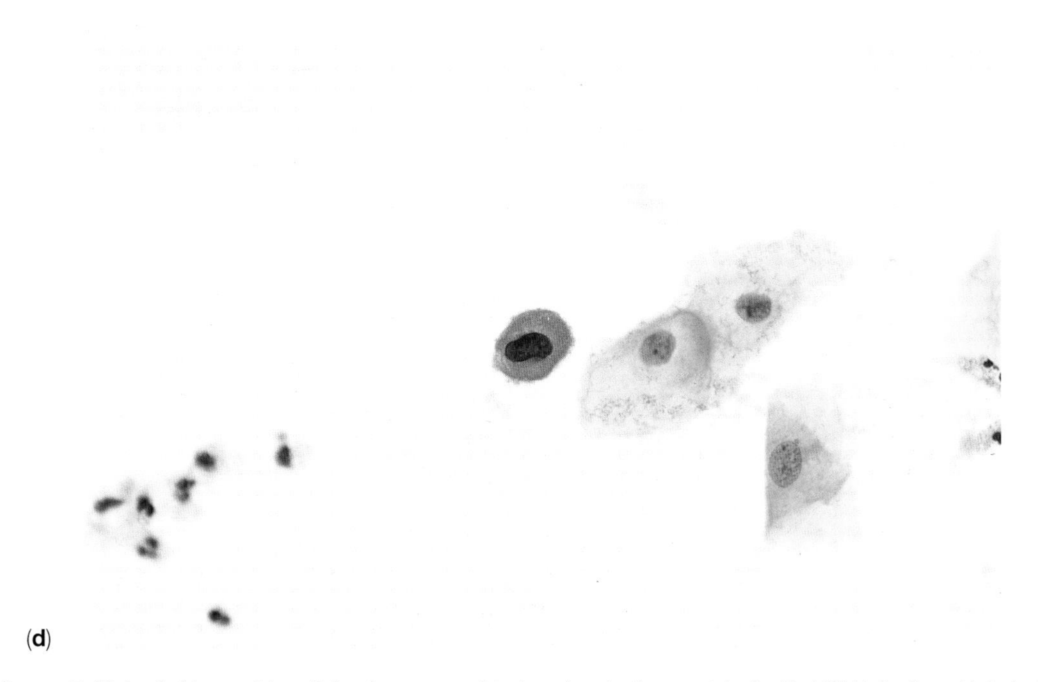

(**d**)

Figure 7.46 (**a–f**) Unusual keratinized squames (dyskeratocytes) associated with HPV infection. Note tadpole cells and fibre cells similar to those seen in squamous carcinoma but lacking the nuclear atypia (Pap, ×400).

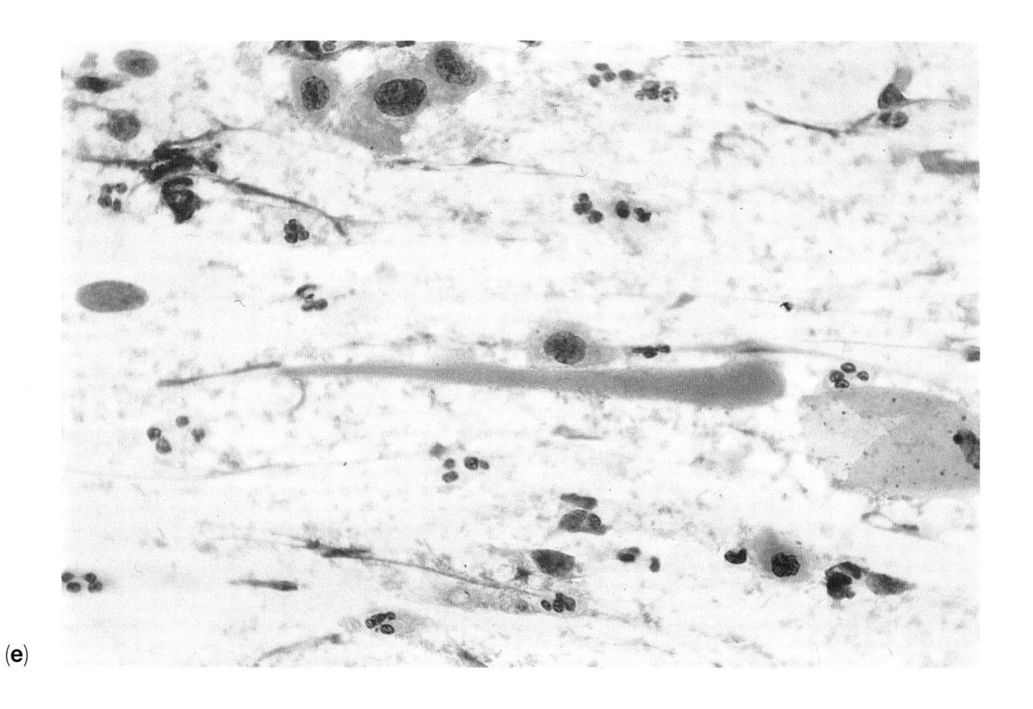

(**e**)

(f)

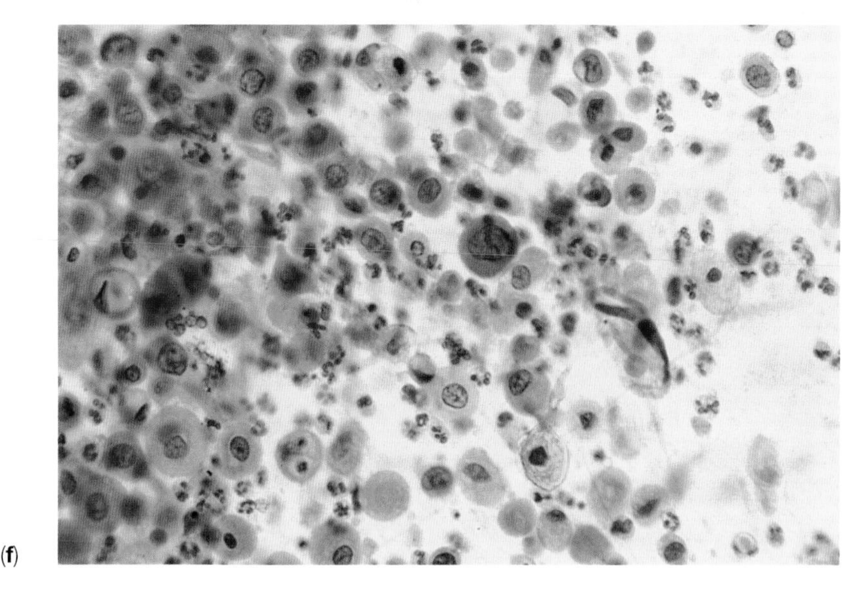

Figure 7.46 (**a–f**) Unusual keratinized squames (dyskeratocytes) associated with HPV infection. Note tadpole cells and fibre cells similar to those seen in squamous carcinoma but lacking the nuclear atypia (Pap, ×400).

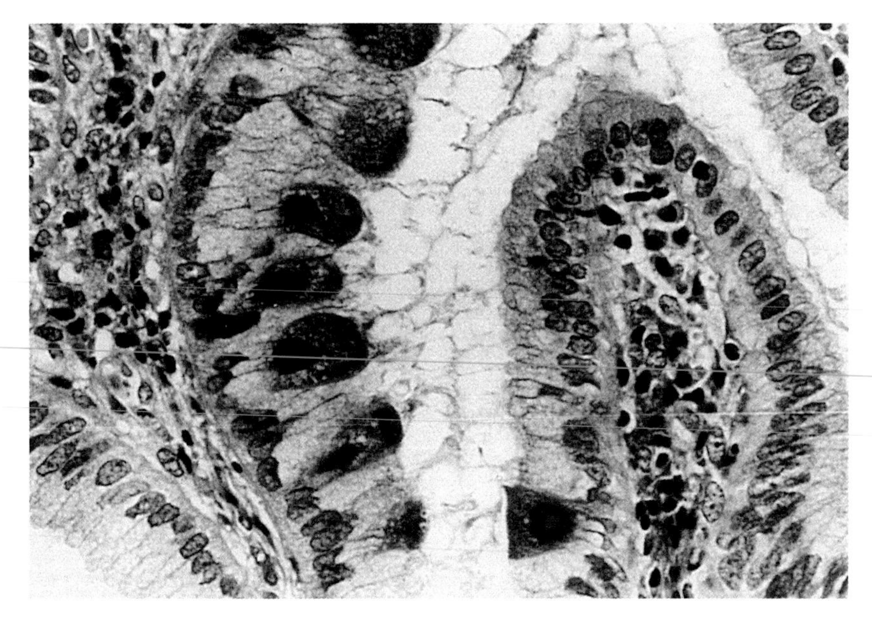

Figure 7.47 Cytomegalic inclusions in cervical biopsy (H & E, ×750) (courtesy of Mr R. Francis, London Hospital).

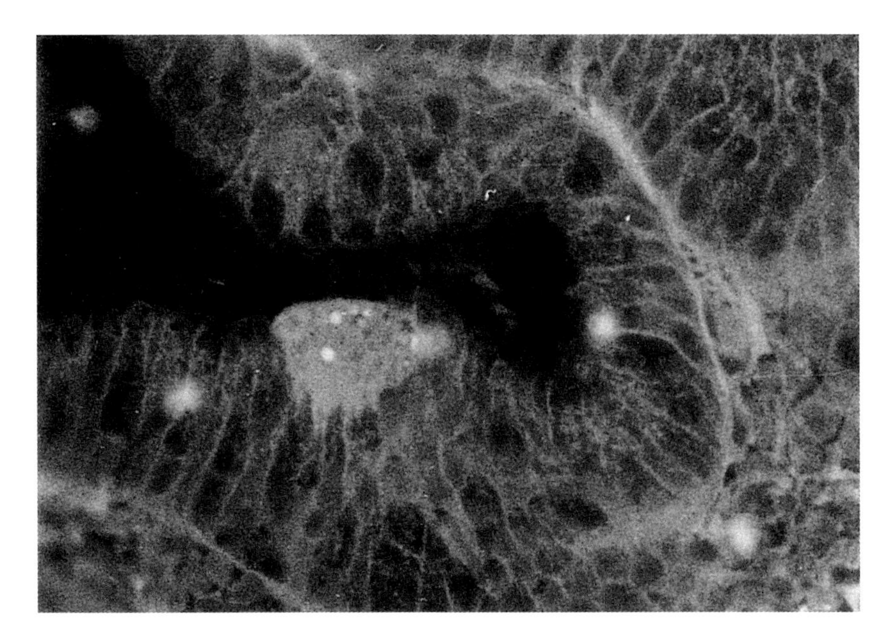

Figure 7.48 Cytomegalovirus infection of cervix shown in Figure 7.47 confirmed by immunofluorescence using virus-specific monoclonal antibodies (FITC ×750) (courtesy of Dr S. Tyms, St Mary's Hospital, London).

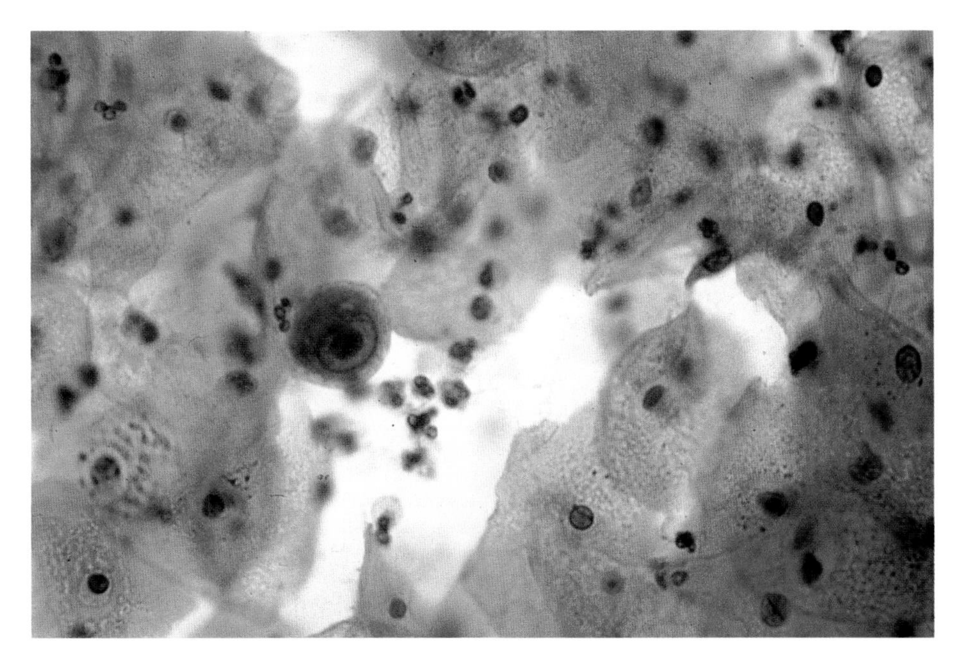

Figure 7.49 Cervical smear containing a cytomegalic inclusion-bearing cell. The typical 'owl's eye' appearance is visible (Pap, ×630).

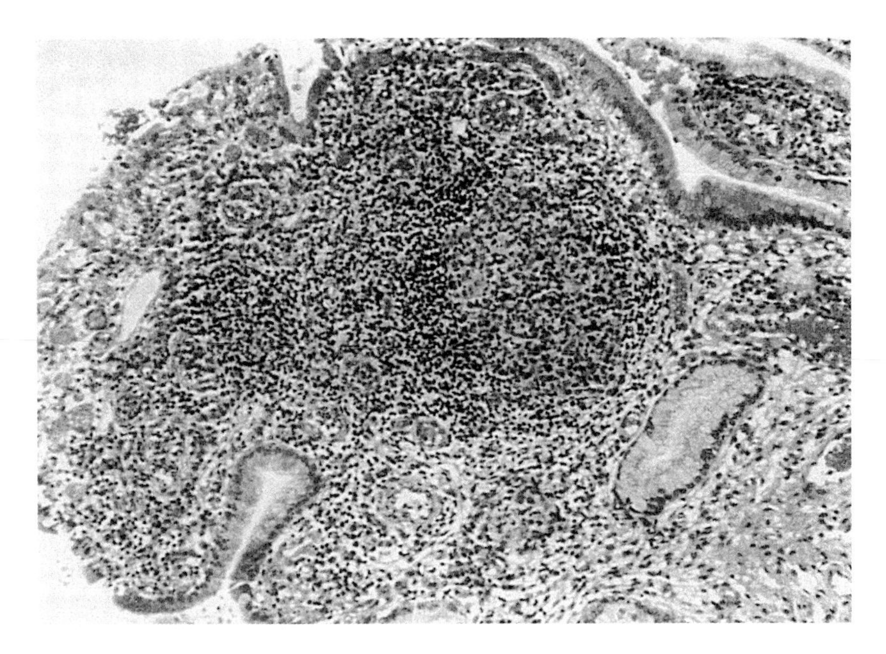

Figure 7.50 Chronic follicular cervicitis with formation of subepithelial lymphoid follicles (H & E, 180) (courtesy of Dr P. Cooper, Addenbrooke's Hospital, Cambridge).

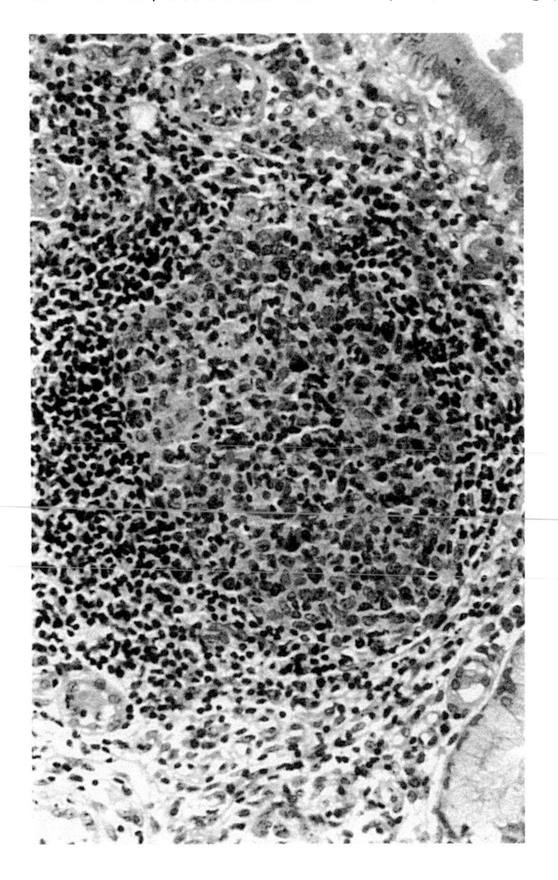

Figure 7.51 Chronic follicular cervicitis. Same case as Figure 7.50 to show germinal centre (H & E, ×350).

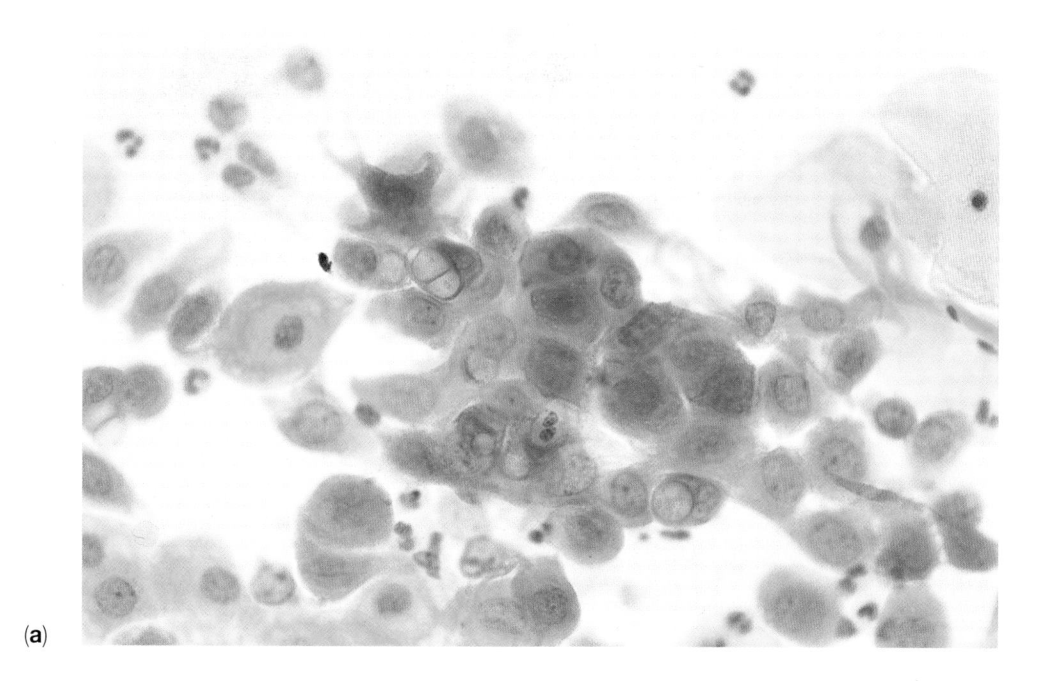

(**a**)

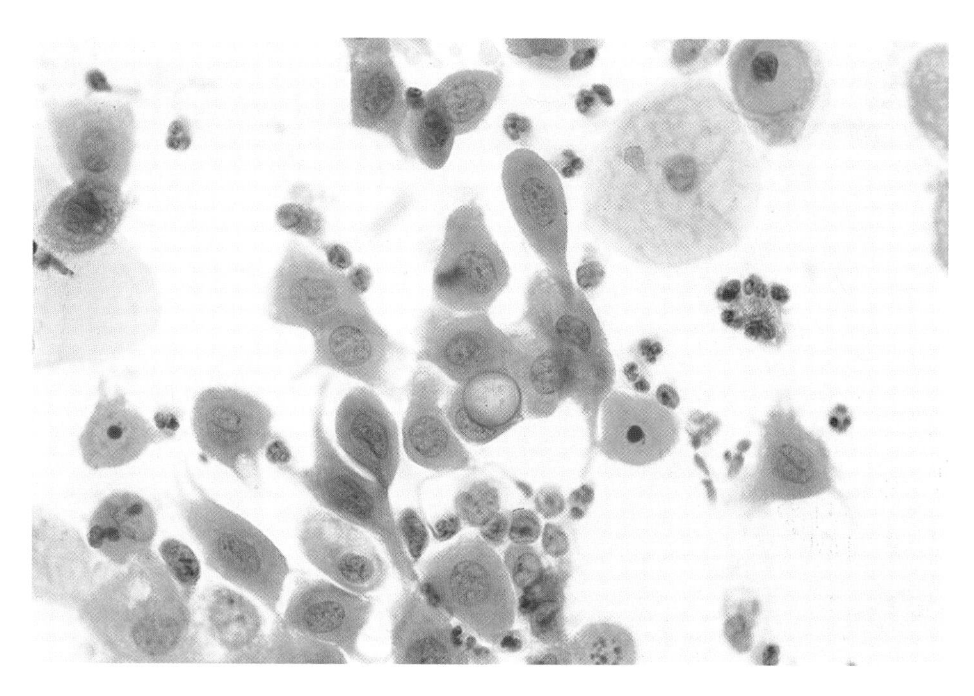

(**b**)

Figure 7.52 (**a–b**) Cells with vacuolated cytoplasm and intracytoplasmic inclusion purported to be due to *Chlamydia* infection (Pap, ×630).

association between *Chlamydia trachomatis* and CIN but a causal relationship has not been established.

7.15.2 Cytology

There have been several reports claiming that a diagnosis of chlamydial infection of the cervix can be made from the appearances of exfoliated epithelial cells in Papanicolaou-stained smears (Naib, 1970; Gupta *et al.*, 1979). The smears contained cells with vacuolated cytoplasm or cytoplasmic inclusion bodies (Figure 7.52) reminiscent of those seen in conjunctival scrapes. However, Dorman *et al.* (1983) compared the cytological findings with cell culture of cervical scrapings from 487 women attending a venereal disease clinic and found a poor correlation. A similar study by Forster *et al.* (1985) confirmed this finding. Moreover, the diagnosis of chlamydia in cervical smears using monoclonal antibodies shows that the organisms are present as extracellular elementary bodies. No intracellular inclusions of the type found in conjunctival scrapes were seen. Forster concluded that the intracytoplasmic inclusions in Papanicolaou-stained cervical smears are probably bacterial aggregates, cell debris or other artefacts.

REFERENCES

Acharya, G. S. and Das, S. R. (1981) Microfilariae in cervical smears. *J. Obstet. Gynaecol. India*, **31**, 438–439.

Angrish, K. and Verna, K. (1981) Cytological detection of tuberculosis of the uterine cervix. *Acta Cytol.*, **25**, 160–162.

Ansel, R., Totten, P. A., Spiegel, C. A., Chenk, C. S., Esenbach, D. and Holmes, K. K. (1983) Non-specific vaginitis, diagnostic criteria and microbial and epidemiologic association. *Am. J. Med.*, **74**, 14–22.

Arsenault, G. M., Kalman, C. F. and Sorensen, K. W. (1976) The Papanicolaou smear as a technique for gonorrhoea detection: a feasibility study. *J. Am. Ven. Dis. Ass.*, **2**, 35–38.

Beckmann, A. M., Myerson, D., Daling, J., Kiviat, N. B., Fenoglio, C. M. and McDougall, J. K. (1985) Detection and localization in human genital condylomas by *in-situ* hybridization with biotinylated probes. *J. Med. Virol.*, **16**, 265–273.

Berry, A. (1966) A cytopathological and histopathological study of bilharziasis of the female genital tract. *J. Pathol. Bacteriol.*, **91**, 325.

Bhaduri, K. P. (1957) *Entamoeba histolytica* in leukorrhoea and salpingitis. *Am. J. Obstet. Gynecol.*, **74**, 434.

Bhambhani, S. (1984) Eggs of *Ascaris lumbricoides* in cervical vaginal smear. *Acta Cytol.*, **28**, 92.

Bhambhani, S., Milner, A., Pant, J. and Luthra, U. (1985) Ova of *Taenia* and *Enterobius vermicularis* in cervicovaginal smears. *Acta Cytol.*, **29**, 913–914.

Braga, C. A. and Teoh, T. B. (1964) Amoebiasis of the cervix and vagina: a report of two cases. *J. Obstet. Gynaecol. Br. Commonwlth*, **71**, 299–301.

Byrne, M. A., Møller, B. R., Taylor-Robinson, D., Harris, J. R. W., Wickenden, C., Malcolm, A. D. B., Anderson, M. C. and Coleman, D. V. (1986) The effect of interferon on human papillomaviruses associated with CIN monitored by cytology, colposcopy and DNA hybridisation. *Br. J. Obstet. Gynaecol.*, **93**, 1136–1144.

Chaterjee, K. D. (1975) *Parasitology in Relation to Clinical Medicine*. Sree Saraswathy Press Ltd, Calcutta, pp. 700–709.

Cohen, C. (1973) Three cases of amoebiasis of the cervix uteri. *J. Obstet. Gynaecol. Br. Commonwlth.*, **80**, 476–479.

Coleman, D. V. (1969) A case of tuberculosis of the cervix. *Acta Cytol.*, **13**, 104–107.

Coleman, D. V. (1979) Cytological diagnosis of virus infected cells in Papanicolaou smears and its application in clinical practice. *J. Clin. Pathol.*, **32**, 1075–1089.

Coleman, D. V., Morse, A. R., Beckwith, P., Anderson, M. C., Knowles, W. A. and Skinner, G. R. B. (1983) Prognostic significance of herpes simplex antibody status in women with cervical intraepithelial neoplasia (CIN). *Br. J. Obstet. Gynaecol.*, **90**, 421–427.

De Boer, A. L., de Boer, F. and van der Merwe, J. V. (1984) Cytologic detection of Donovan bodies in granuloma inguinale. *Acta. Cytol.*, **28**, 126–128.

De Borges, R. (1971) Findings of microfilarial larval stages in gynaecologic smears. *Acta Cytol.*, **15**, 476–478.

Dorman, S. A., Danos, L. M., Wilson, D. J. *et al.* (1983) Detection of chlamydial cervicitis by Papanicolaou stained smears and culture. *Am. J. Clin. Pathol.*, **79**, 421–425.

Duguid, H. L. D., Parratt, D. and Traynor, R. (1980) Actinomyces-like organism in cervical smears from women using intrauterine contraceptive devices. *Br. Med. J.*, **281**, 534.

El Tabbakh, G. and Hamza, M. A. (1989) Carcinoma of the uterine cervix and schistosomiasis. *Int. J. Gynaecol. Obstet.*, **29**, 263–268.

Evans, D. M. D. and Hughes, H. (1961) Cysts of the vaginal wall. *J. Obstet. Gynaecol. Br. Commonwlth.*, **68**, 247–253.

Forster, G. E., Cookey, I., Mandray, P. E., Richman, P. I., Jha, R., Coleman, D. V., Thomas, B. J., Hawkins, D. A., Evans, R. T. and Taylor-Robinson, D. (1985) Investigation into the value of Papanicolaou stained cervical smears for the diagnosis of chlamydial cervical infection. *J. Clin. Pathol.*, **38**, 399–402.

Galloway, D. A. and McDougall, J. K. (1983) The oncogenic potential of HSV: evidence for a hit and run mechanism. *Nature*, **302**, 21–24.

Gardner, H. L. and Fernet, P. (1964) Aetiology of vaginitis ephysematosa: report of ten cases and review of literature. *Am. J. Obstet. Gynecol.*, **88**, 680–694.

Goldman, R. I., Bank, R. W. and Warver, D. E. (1969) Cytomegalovirus infection of the cervix. An incidental finding of possible clinical significance. *J. Obstet. Gynaecol.*, **34**, 326–329.

Gupta, P. K., Hollander, D. H. and Frost, J. K. (1976) Actinomyces in cervico-vaginal smears; an association with IUCD usage. *Acta Cytol.*, **20**, 295–297.

Gupta, P. K., Lee, E. F., Erozan, Y. S., Frost, J. K., Geddes, S. T. and Donovan, P. A. (1979) Cytologic investigation of chlamydia infection. *Acta Cytol.*, **23**, 315–320.

Hare, M. J., Toone, E., Taylor-Robinson, D., Evans, R. T., Furr, P. M., Cooper, P. and Oates, J. K. (1981) Follicular cervicitis – colposcopic appearances and association with *Chlamydia trachomatis*. *Br. J. Obstet. Gynaecol.*, **88**, 174–180.

Kolstad, P. (1964) The colposcopical picture of trichomonas vaginitis. *Acta Obstet. Gynaecol. Scand.*, **43**, 388–398.

Koss, L. G. and Durfee, G. R. (1956) Unusual pattern of squamous epithelium of uterine cervix; cytologic and pathologic study of koilocytotic atypia. *Ann. N. Y. Acad. Sci.*, **63**, 1245–1261.

Koss, L. G. and Wolinska, W. H. (1959) *Trichomonas vaginalis* cervicitis and its relationship to cervical cancer. A histocytological study. *Cancer*, **12**, 1171–1193.

Kreider, J. W., Howett, M. K., Wolfe, S. A. *et al.* (1989) Morphological transformation *in vivo* of human uterine cervix with papillomavirus from condylomata acuminata. *Nature*, **317**, 639–641.

Maeda, M. Y., Longatto-Filho, A., Shih, L. W., Cavaliere, M. J., Oyafuso, M. S., Marziona, F., Santos, D. R. and Carvalho, M. I. (1990) *Chlamydia trachomatis* in cervical uterine irradiated patients. *Diagnostic. Cytopathol.*, **6**, 86–88.

McNeill, R. E. and Ruehsen, M. de M. (1978) Amoeba trophozoites in cervico-vaginal smear of a patient using an intrauterine device. A case report. *Acta Cytol.*, **22**, 91–92.

Misch, K. A., Smithies, A., Twomey, D., O'Sullivan, J. C. and Onuigobo, W. (1976) Tuberculosis of the cervix: cytology as an aid to diagnosis. *J. Clin. Pathol.*, **29**, 313–316.

Morse, A. R., Coleman, D. V. and Gardner, S. D. (1974) An evaluation of cytology in the diagnosis of herpes simplex virus infection and cytomegalovirus infection of the cervix uteri. *J. Obstet. Gynaecol. Br. Commonwlth.*, **81**, 393–398.

Naib, Z. M. (1970) Cytology of TRIC agent infection of the age of new born infants and their mother's genital tracts. *Acta Cytol.*, **14**, 390–395.

Newton, W. L., Reardon, L. V. and Deleva, A. M. (1960) A comparative study of the subcutaneous inoculation of germ free and conventional guinea pigs with two strains of *Trichomonas vaginalis*. *Am. J. Trop. Med.*, **9**, 56–61.

Notelovitz, M. (1982) Schistosomiasis, in *Infectious Diseases in Obstetrics and Gynaecology* (ed. G. R. G. Monio), 2nd edn., Harper and Row, pp. 357–362.

Oriel, J. D. (1971) Natural history of genital warts. *Br. J. Ven. Disease.*, **47**, 1–13.

Pheifer, T. A., Forsyth, P. S., Durfee, M. A., Pollack, H. M. and Holmes, K. K. (1978), Non-specific vaginitis. Role of haemophilus vaginalis and treatment with metronidazole. *N. Engl. J. Med.*, **298**, 1429–1434.

Roberts, T. H. and Ng, A. B. P. (1975) Chronic lymphocytic cervicitis. Cytological and histopathologic manifestations. *Acta Cytol.*, **19**, 235–243.

Ruehsen, M. de M., McNeill, R. E., Frost, J. K., Gupta, P. K., Diamond, L. S. and Honigberg, B. M. (1980) Amoeba resembling *Entamoeba gingivalis* in the genital tract of IUD users. *Acta Cytol.*, **24**, 413–420.

Schacter, J., Hill, E. C., King, E. B., Coleman, V. R., Jones, P. and Meyer, K. F. (1974) Chlamydial infection in women with cervical dysplasia. *Am. J. Obstet. Gynecol.*, **123**, 753–757.

Schiffer, M. D., Elguezebal, A., Sultana, M. and Allen, A. C. (1975) Actinomycosis infections associated with intrauterine contraceptive devices. *Obstet. Gynecol.*, **45**, 67–72.

Schnell, J. D., Andrews, P. and Plempel, M. (1972) Die Vaginale Kontamination der weiblichen

Bevölkerung einer Grossstadt mit Trichomonaden und Hefen. *Geburtshilfe Frauenheilkd.*, **32**, 1007–1014.

Sharma, S. D., Zeigler, O. and Trussell, R. R. (1971) A cytological study of *Dipetolenema perstans*. *Acta Cytol.*, **15**, 479–481.

Swanson, J., Eschenbach, D. A., Alexander, R. E. and Holmes, K. K. (1975) Light and electronmicroscopic study of *Chlamydia trachomatis* infection of the uterine cervix. *J. Infect. Dis.*, **131**, 678–687.

Taylor-Robinson, D. and Thomas, B. J. (1980) The role of *Chlamydia trachomatis* in the genital tract and associated diseases. *J. Clin. Pathol.*, **33**, 205–233.

Walker, P. G., Singer, A., Dyson, J. L., Shah, D. V., To, A. and Coleman, D. V. (1983) The prevalence of human papillomavirus antigen in patients with cervical intra-epithelial neoplasia. *Br. J. Cancer*, **48**, 99–101.

Westrom, L. and Mardh, P.A. (1983) Chlamydial salpingitis. *Br. Med. Bull.*, **39**, 145–50.

Wickenden, C., Steele, A., Malcolm, A. D. B. and Coleman, D. V. (1985) Screening for wart virus infection in normal and abnormal cervices by DNA hybridisation of cervical scrapes. *Lancet*, **i**, 65–7.

Wikeley, G. W. (1988) Parasitic infections, in *Clinical Cytotechnology* (eds. D. V. Coleman and P. Chapman), Butterworths, London.

Wilbanks, G. D. and Carter, B. (1963) Vaginitis emphysematosa. *Obstet. Gynecol.*, **22**, 301.

Zummo, B. P. H., Sered, H. and Falls, F. H. (1955) The diagnosis and prognosis of female genital tuberculosis. *Am. J. Obstet. Gynecol.*, **70**, 34–1.

8

POLYPS, CYSTS AND OTHER BENIGN LESIONS

In this chapter the following cervical lesions are considered: polyps, pregnancy changes, endometriosis, squamous cell papilloma, cysts (mucus retention, epidermoid and mesonephric), mesonephric duct hyperplasia, mesonephric papilloma and villous adenoma.

The frequency with which lesions in this heterogeneous group occur in the cervix varies considerably. Cervical polyps and mucus retention cysts (Nabothian follicles) are common lesions whereas mesonephric papilloma and villous adenoma are rare. Decidual change, endometriosis or mesonephric remnants are often unexpected findings in biopsy material. Cytology is rarely of value in the diagnosis of these benign lesions.

8.1 CERVICAL POLYP

This is defined as a localized, pedunculated or sessile outgrowth of the endocervical mucosa (Poulsen *et al.*, 1975). This very common benign lesion usually arises from the lining of the endocervical canal from enlargement of a papillary process, often prolapsing through the external os. Occasionally it may arise from the ectocervix, due to enlargement of a papillary process in the transformation zone. Its surface is lined by cervical columnar epithelium and it contains glands similarly lined (Figure 8.1). Inflammation (Figure 8.2), ulceration and squamous metaplasia (Figure 8.3) frequently occur. Its stroma is usually composed of loose, oedematous fibrous tissue often densely infiltrated by inflammatory cells. Several histological types of polyps are recognized. A *fibrous polyp* is composed almost entirely of fibrous connective tissue with few glands. In a *vascular polyp*, large numbers of blood vessels form the major component (Figure 8.4). *Mixed polyps* arise from the region of the isthmus and contain both endocervical and endometrial glands (Figure 8.5). A *decidual polyp* can occur during pregnancy, the stroma of the polyp undergoing decidual change (Section 8.2). Microglandular endocervical hyperplasia (MEH, see Section 9.2.1) also occurs in a cervical polyp (Figure 8.6). The smear from the latter lesion may contain atypical glandular cells (Figure 8.7(a)).

Carcinoma of either glandular or squamous type arising in a polyp is very rare, less than 0.3% (14 in 5015), being limited to the polyp in four of the 14 cases described by Aaro *et al.* (1963). Cytology is of value in detecting such changes (Figure 8.7(b,c)). Adenocarcinoma limited to the excised polyp has a very good prognosis (Abell and Gosling, 1962). If either *in situ* or invasive neoplasia is discovered it is

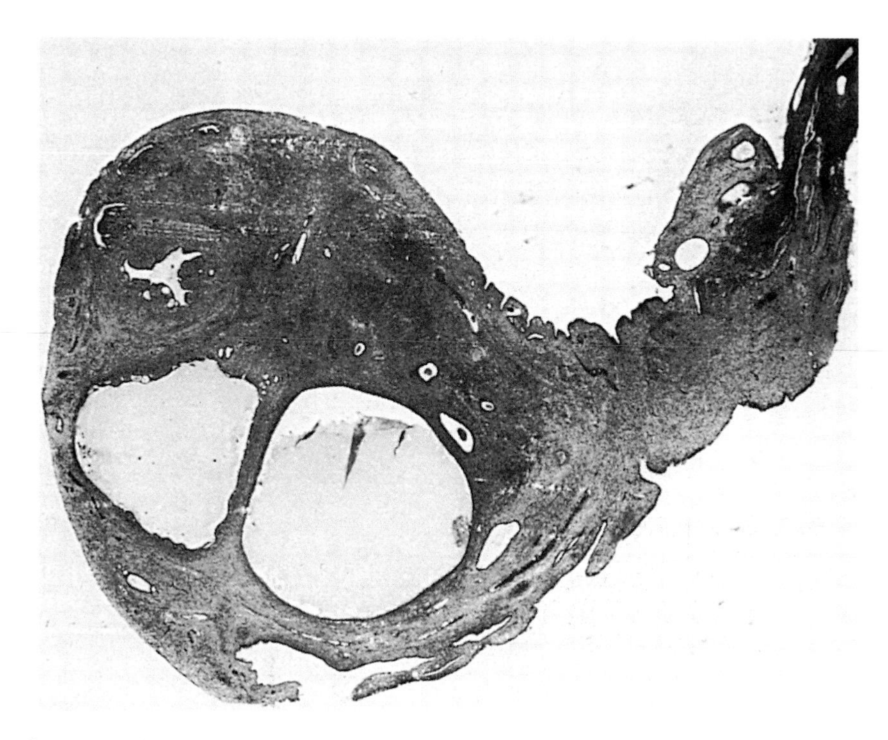

Figure 8.1 Cervical polyp, showing characteristic shape with an elongated peduncle. It contains mucus-secreting glands, some of which are dilated (H & E, ×7).

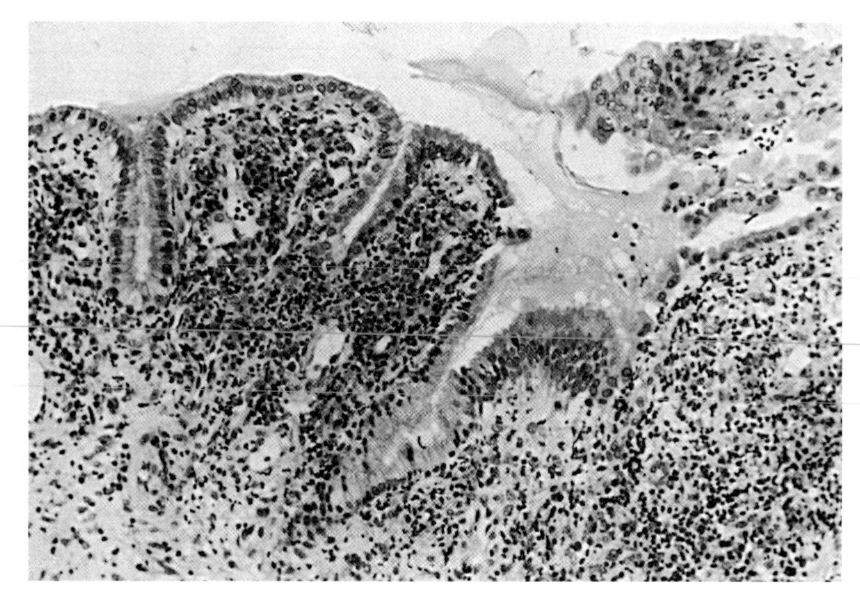

Figure 8.2 Inflammatory change in a cervical polyp, associated with minimum surface erosion and slight cellular atypia (H & E, ×190).

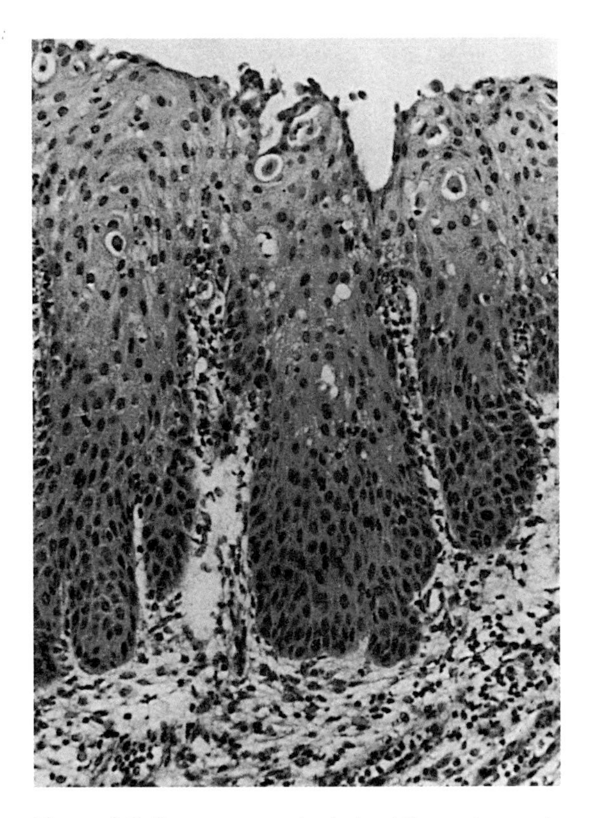

Figure 8.3 Squamous metaplasia at the surface epithelium of a cervical polyp (H & E, ×190).

important to scrutinize the base of the stalk to determine whether complete removal has been achieved. Even if removal appears complete there is still a slight possibility that another zone of neoplasia may be present in the residual cervical tissue. Very occasionally, both benign polypoid change and papillary adenocarcinoma may be present in the same cervix and close examination may be required if the malignant lesion is not to be overlooked (Figure 8.8).

8.1.1 Glial polyp

Occasionally, endocervical polyps and, more rarely, ectocervical polyps composed mainly of neuroglial tissue have been described (Grönroos *et al.*, 1983). Neurones and ganglia were usually absent but sometimes cartilage, bone and in one case, stratified squamous epithelium were also demonstrated. Implantation of foetal tissues from a preceding abortion was considered to be the most probable source of this lesion. Young *et al.* (1981) reported a glial polypoid tumour replacing the endometrium and infiltrating

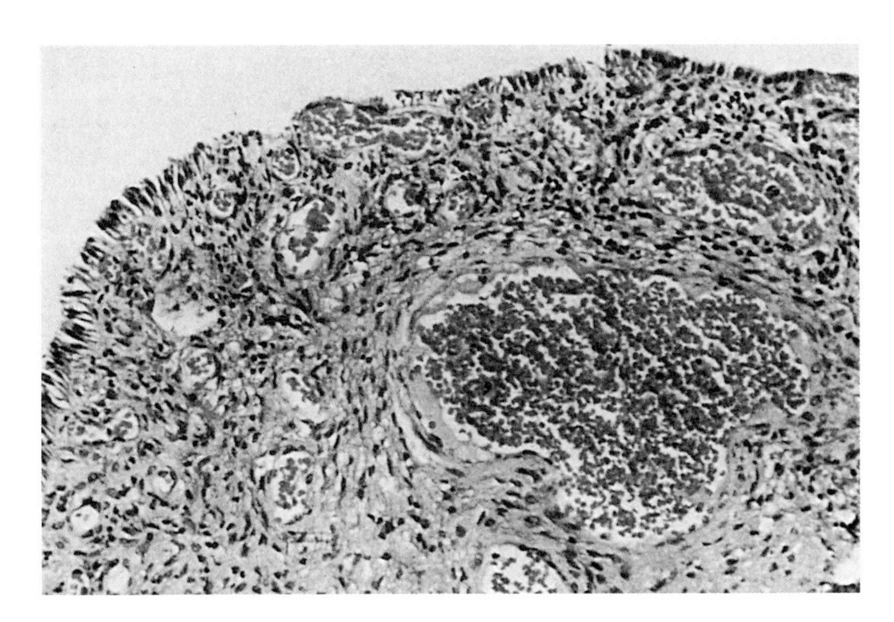

Figure 8.4 Vascular polyp containing a large number of blood vessels of varying calibre (H & E, ×190).

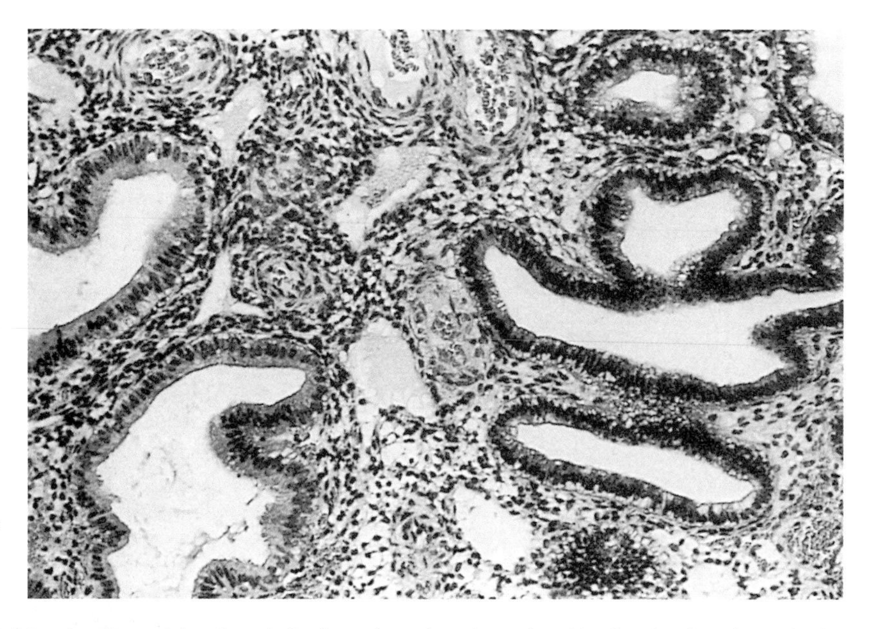

Figure 8.5 Mixed polyp, arising from isthmic region of endocervix with glands of endocervical type on left and endometrial type on right (H & E, ×190).

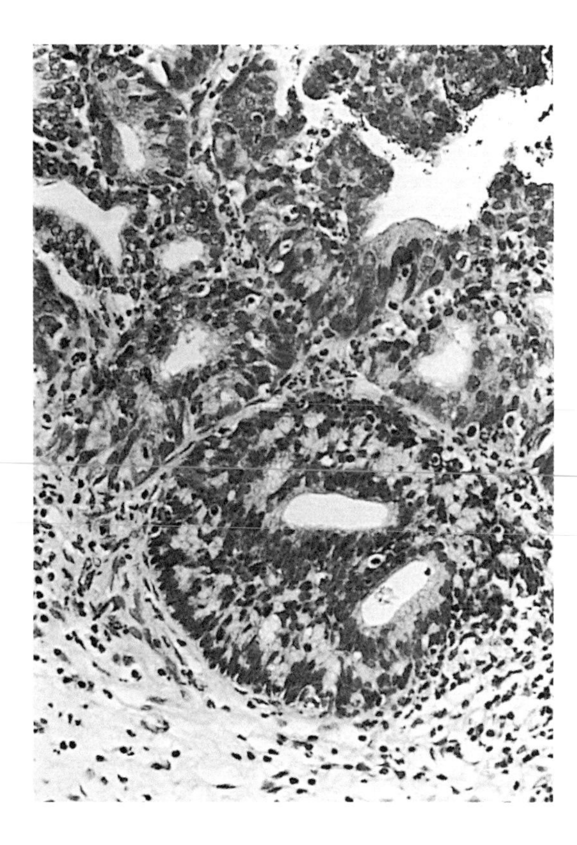

Figure 8.6 Microglandular endocervical hyperplasia (MEH) in a cervical polyp (H & E, ×310).

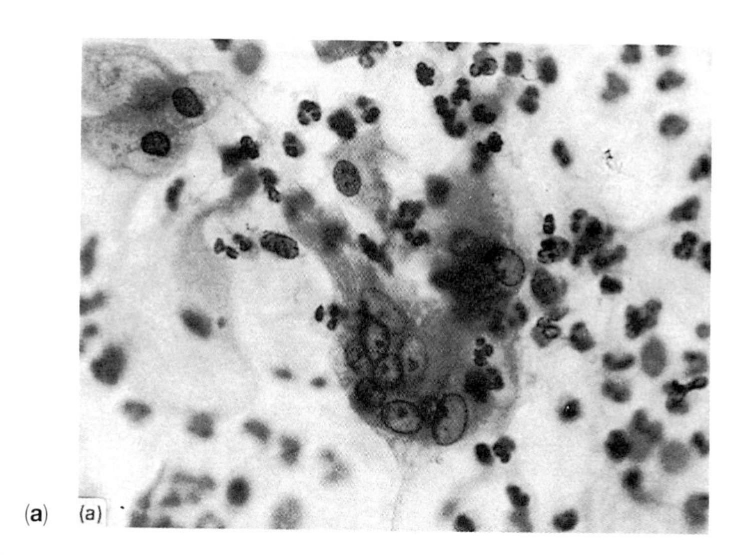

Figure 8.7 (**a**) Glandular cells showing reactive change in smear from cervical polyp showing MEH (Pap, ×770).

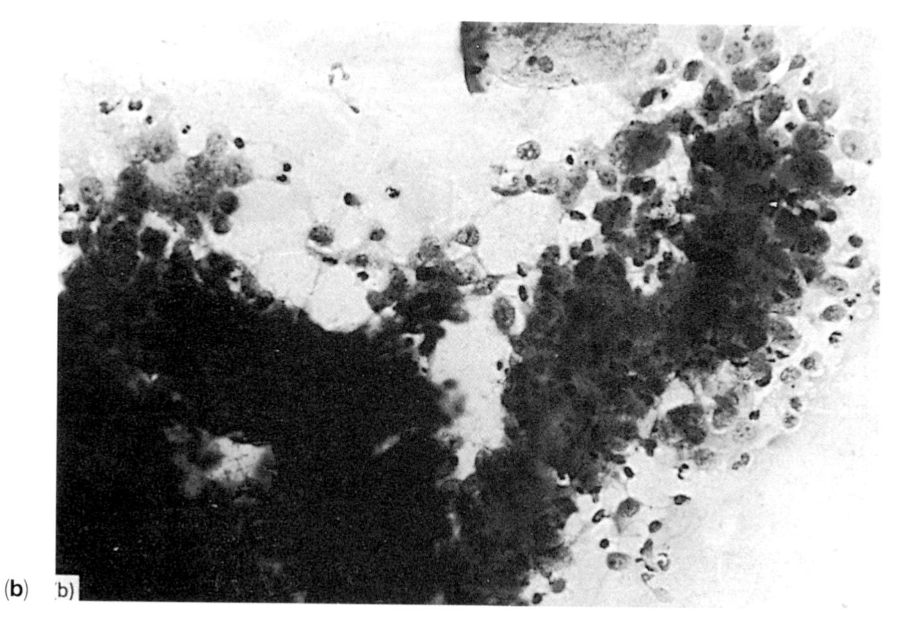

Figure 8.7 (**b**) Malignant change in a polyp detected by cytology. Tissue fragments with a papillary structure were present in the smear (Pap, ×310).

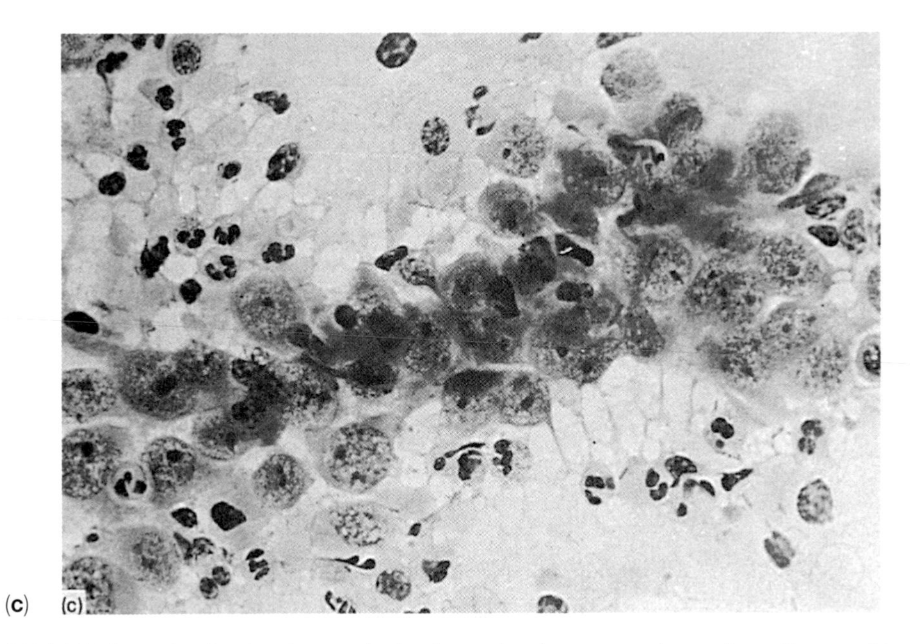

(**c**)

Figure 8.7 (**c**) Malignant cells showing marked anisonucleosis, coarse chromatin pattern and large nucleoli. Same case as shown in (**b**). Biopsy confirmed adenocarcinoma arising in a polyp (Pap, ×770).

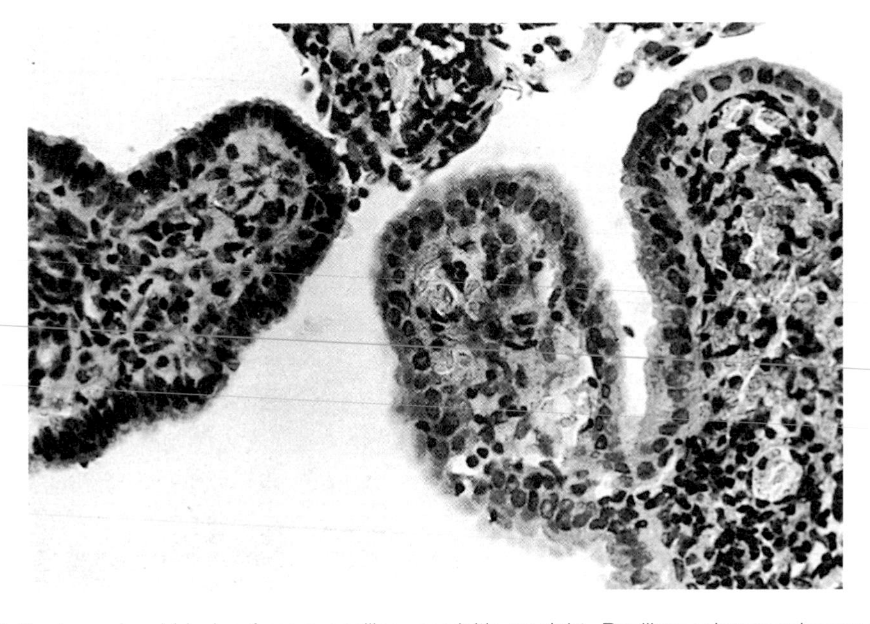

Figure 8.8 Benign polypoid lesion from a papillary cervicitis on right. Papillary adenocarcinoma on left from same tumour as in Figures 14.13 and 14.14 (H & E, ×480).

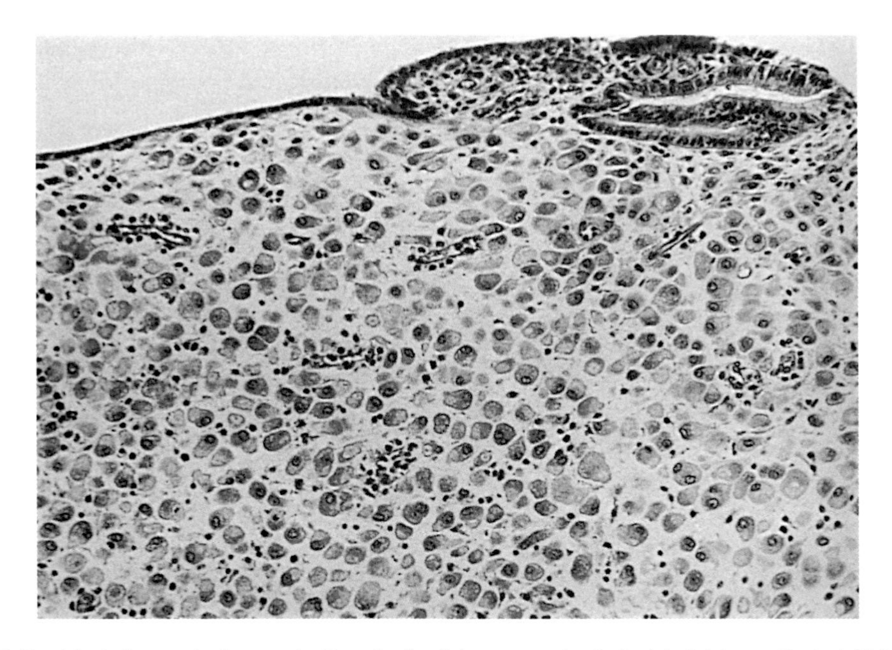

Figure 8.9 Decidual change in the cervix. Practically all the stromal cells in this field are affected (H & E, ×210).

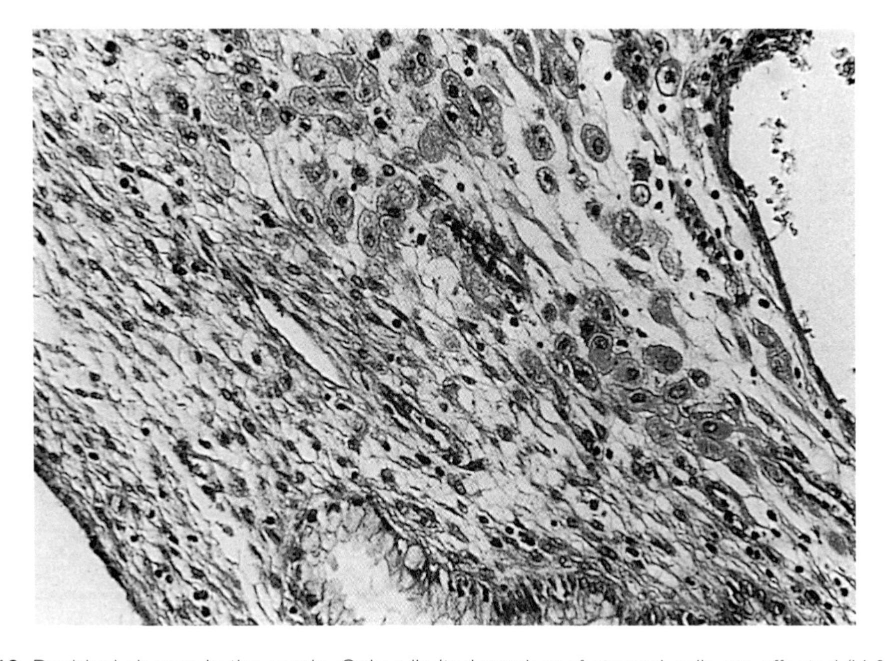

Figure 8.10 Decidual change in the cervix. Only a limited number of stromal cells are affected (H & E, ×240).

the myometrium of a 15-year-old girl who had not previously been pregnant. They considered it to be a tumour of congenitally ectopic glial tissue or a one-sided development of a teratoma or a tumour of deviant metaplastic müllerian tissue.

8.1.2 Mesodermal stromal polyp

This lesion is a benign fibroepithelial polyp, occurring in women of reproductive age. It has a hypocellular fibrous stroma in which there are bland spindle cells and occasionally foci of multinucleated stromal giant cells (Clement, 1985). Mitotic figures are rare. During pregnancy, changes resembling sarcoma botryoides may occur (pseudosarcoma botryoides; see Section 8.2, Pregnancy changes).

8.1.3 Other types of benign cervical polyp

These include:

- Leiomyoma and fibromyoma (see p. 393)
- Adenomyoma and fibroadenoma (see p. 399)
- Papillary adenofibroma and adenosarcoma (see p. 408)
- Prolapsed decidua and decidual pseudopolyp (see p. 161)
- Placental site trophoblastic nodule (Young *et al.*, 1988): this lesion very occasionally involves the cervix; it is not usually visible to the naked eye but at low magnification appears well-circumscribed, composed of single or multiple nodules, oval or rounded with lobulated margins. Histologically, it is usually densely eosinophilic with a central zone of hyalinized material which is relatively acellular, surrounded by a cellular zone containing degenerate trophoblastic cells. Mitotic figures are few or absent.
- Postoperative spindle cells nodule (Scully *et al.*, 1994): this is a localized benign lesion composed of closely packed proliferating spindle cells and capillaries, occurring several weeks to several months postoperatively in the region of an incision. Mitotic figures may be numerous and it may closely resemble a leiomyosarcoma, from which the history should assist in the differentiation.

8.2 PREGNANCY CHANGES

During pregnancy, the stroma of the cervix may undergo decidual change (Figures 8.9 and 8.10). It occurs in about one-third of cervices in pregnant women, disappearing about 2 months post-partum (Johnson, 1973). Cervical polyps, if removed during pregnancy, also frequently show decidual change. Because of the epithelioid appearance of the decidua, the condition may be misinterpreted as carcinoma (Poulsen *et al.*, 1975).

Lesions resembling sarcoma botryoides (Section 16.4) can occur in the cervix during pregnancy. The condition is termed pseudosarcoma botryoides (Elliott *et al.*, 1967) and is usually polypoid. The decidual stroma resembles sarcomatous spindle cell mesenchyme; there may also be giant cells with irregular nuclei and occasional mitotic figures (up to one per high-power field). It differs from the true sarcoma by its slow growth, and the absence of a densely packed subepithelial 'cambial' layer of mesenchymal tumour cells (Figure 16.15) which in the neoplastic lesion frequently migrate through the surface epithelium. Pseudosarcoma is also much more likely to be associated with pregnancy than is sarcoma botryoides.

Another type of pregnancy change, the Arias–Stella reaction, affects the glands which can develop a hobnail appearance (Cove, 1979) not unlike a mesonephroid adenocarcinoma (Figures 14.39 and 14.43) but lacking mitotic figures (Figure 8.11(a)); similar pregnancy changes can occur in foci of endometriosis in the cervix.

The papillary fronds shown in Figure 8.11(b,c) were found in a cervical smear and erroneously reported as 'suggestive of adenocarcinoma'. Biopsy of the cervix revealed an

(**a**)

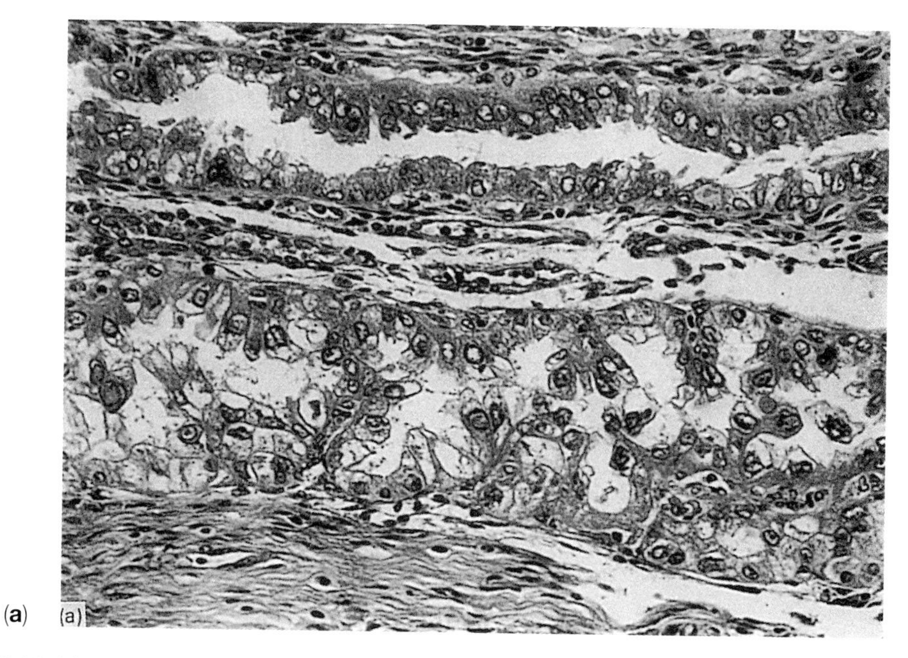

Figure 8.11 (**a**) Arias–Stella reaction involving endocervical gland cells. The cells are enlarged and show a hob-nailed pattern. Less obvious changes can be seen in the upper gland (H & E, ×310).

(**b**)

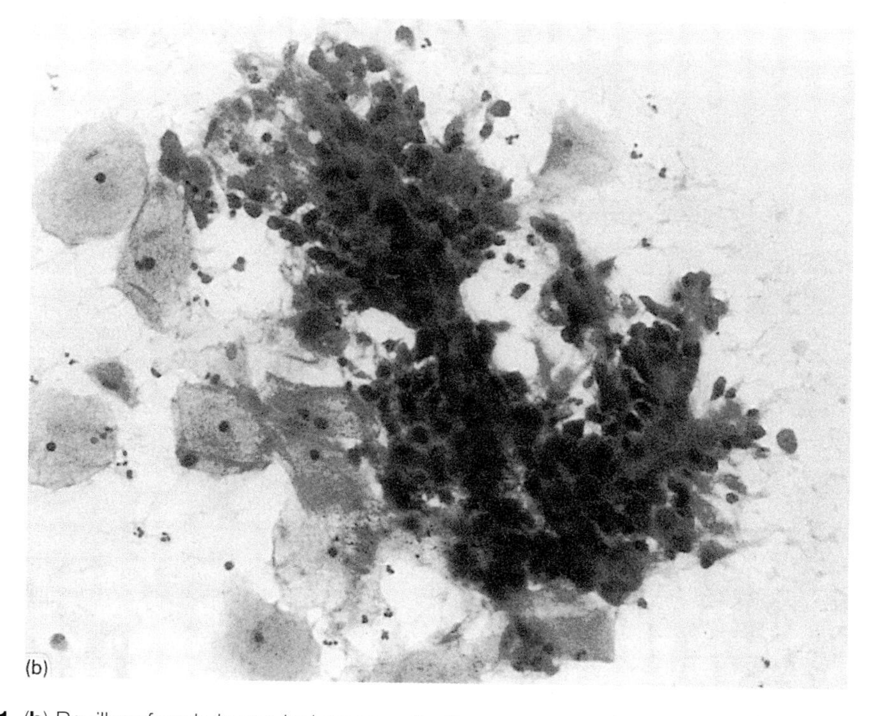

Figure 8.11 (**b**) Papillary fronds in cervical smear reflecting pregnancy changes in the cervix. The smear was erroneously reported as 'suggestive of adenocarcinoma' (Pap, ×320).

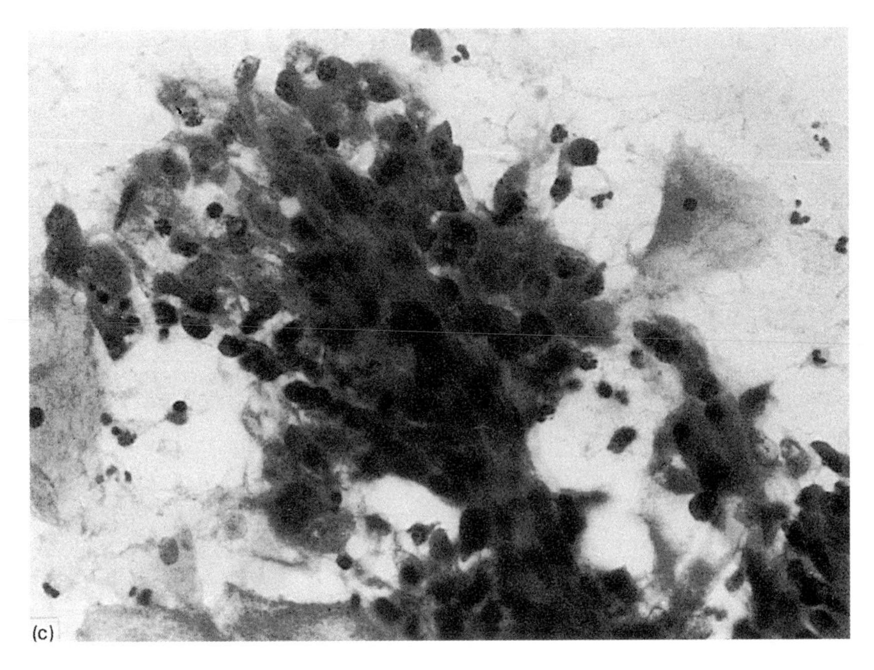

(**c**)

Figure 8.11 (**c**) Papillary frond shown in (**b**) at higher magnification (Pap, ×720).

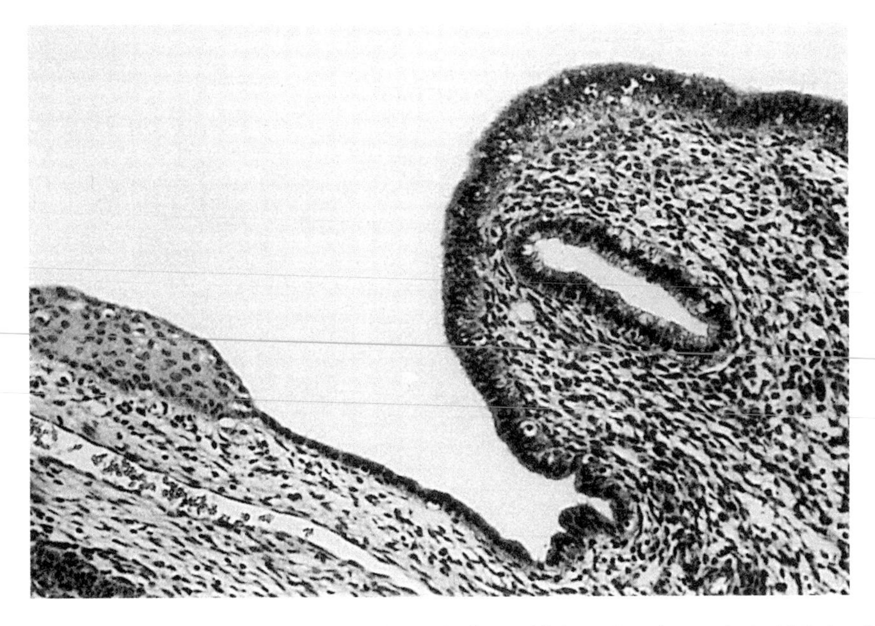

Figure 8.12 Endometriosis of cervix involving endocervical canal lining, showing endometrial gland and stroma (H & E, ×190).

Arias–Stella reaction. The pathologist had not been informed that the patient was pregnant. Very rarely, intracervical pregnancy can occur.

8.2.1 Decidual pseudopolyp

Occasionally, the decidual changes occurring in the cervix already described may produce a raised zone or pseudopolyp which can be mistaken for a carcinoma, both colposcopically and microscopically. Distinguishing features are the lack of nuclear pleomorphism and the sparsity of mitotic figures (Wright and Ferenczy, 1994).

8.3 ENDOMETRIOSIS

Endometriosis is the presence of displaced endometrial glands and stroma (Figure 8.12). It occurs in the cervix fairly frequently, either on the portio or adjoining the endocervical canal (Ridley, 1968). It sometimes occurs at the squamocolumnar junction following cone biopsy. It may be accompanied by extravasated red cells (Figure 8.13). True endometriosis must be distinguished from endometrial metaplasia of the endocervical epithelium (Section 9.3.2) which lacks surrounding endometrial stroma (Figure 8.14). Very occasionally, adenocarcinoma *in situ* (Figure 8.15) or invasive adenocarcinoma (Chang and Maddox, 1971) may be seen arising from endometriotic glands (Figure 8.16). Features which distinguish a neoplastic gland from a non-neoplastic gland of endometrial type are discussed in the last two paragraphs of Section 13.1.1.

8.4 SQUAMOUS CELL PAPILLOMA (SYNONYMOUS WITH EPIDERMOID PAPILLOMA)

This is a benign neoplasm composed of papillary processes with a core of fibrous tissue covered by stratified squamous epithelium (Poulsen *et al.*, 1975). The latter often shows atypicality. The papillary processes are frequently fused together and the central connective tissue core contains blood vessels (Figures 8.17 and 8.18). It usually occurs as a focal outgrowth on the vaginal aspect of the cervix and is seldom larger than 1–2 cm in diameter (Gilbert and Palladino, 1966); most large growths are probably examples of verrucous carcinoma (Section 12.6.1).

Squamous cell papilloma is typically a solitary lesion often showing hyperkeratosis with a well-developed granular layer, features considered to distinguish it from condylomata acuminata (Section 7.13.1) which are usually multiple and show parakeratosis with little hyperkeratosis (Woodruff and Peterson, 1958). However, the histological structure of a papilloma is so similar to that of a condyloma that it has become a distinction without a difference (Figure 8.19), particularly as both lesions have a tendency to recur after removal. In fact Ferenczy (1982) no longer recognizes squamous cell papilloma as a separate entity, considering it to be a papillomatous condyloma acuminatum.

8.5 CYSTS

8.5.1 Mucus retention cyst

This is an extremely common benign lesion which can be recognized colposcopically as a small swelling in the transformation zone; it is known by the odd designation 'Nabothian follicle'. Rarely it reaches sufficient size to present at the vaginal outlet (Novak and Woodruff, 1974). It results from the occlusion of a cervical gland orifice by inflammation or inspissated mucus. The gland secretion accumulates and the gland dilates to form a cyst (Figure 8.20). The mucus-secreting lining cells often become flattened and the mucus appears laminated and both cells and secretions give a positive Periodic acid-Schiff (PAS) reaction. Squamous metaplasia also occurs. Occasionally, papillary processes may be seen protruding from the

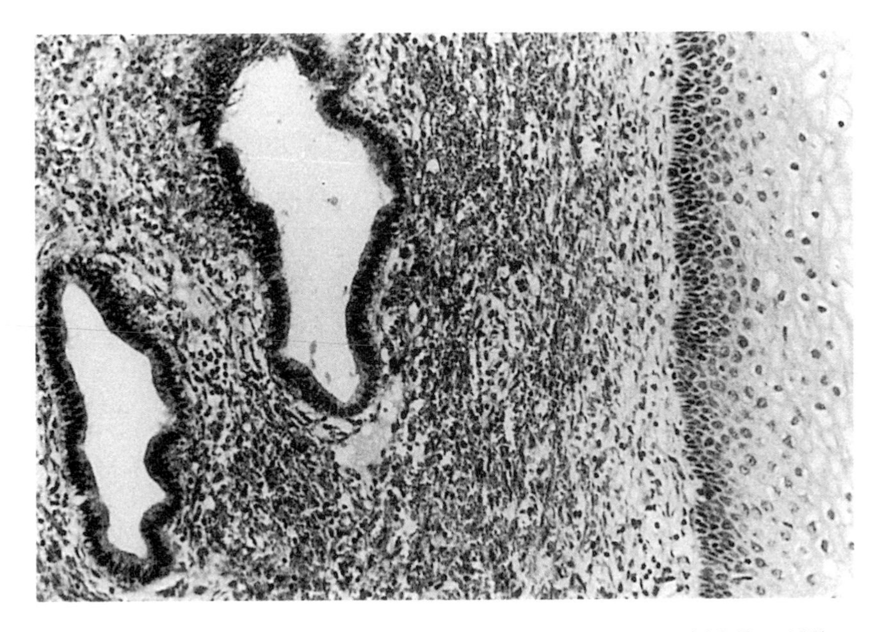

Figure 8.13 Cervical endometriosis with extravasation of red cells into stroma (H & E, ×190).

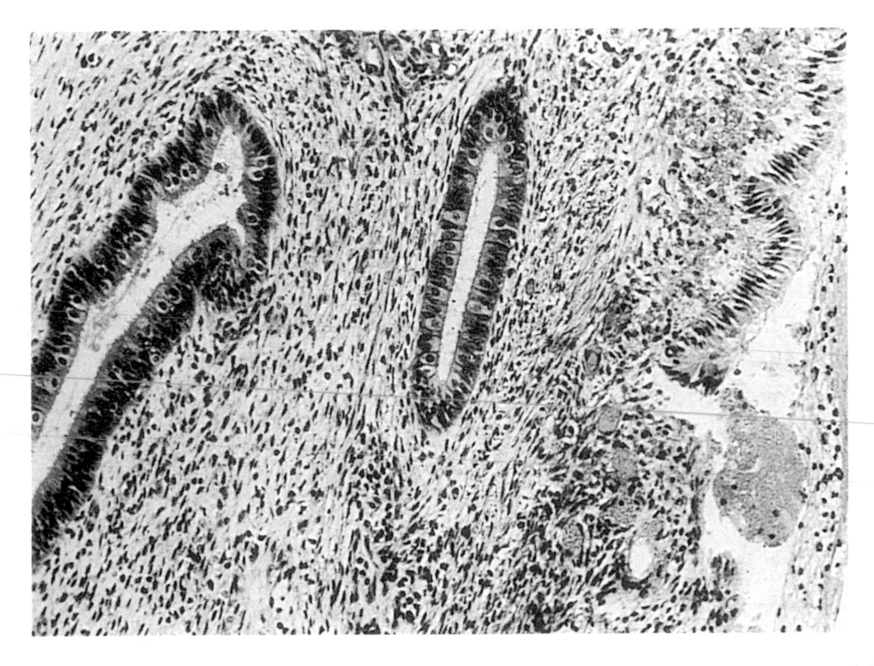

Figure 8.14 Endometrial metaplasia: glands of endometrial type without endometrial stroma lying deep to endocervical canal lining (H & E, ×190).

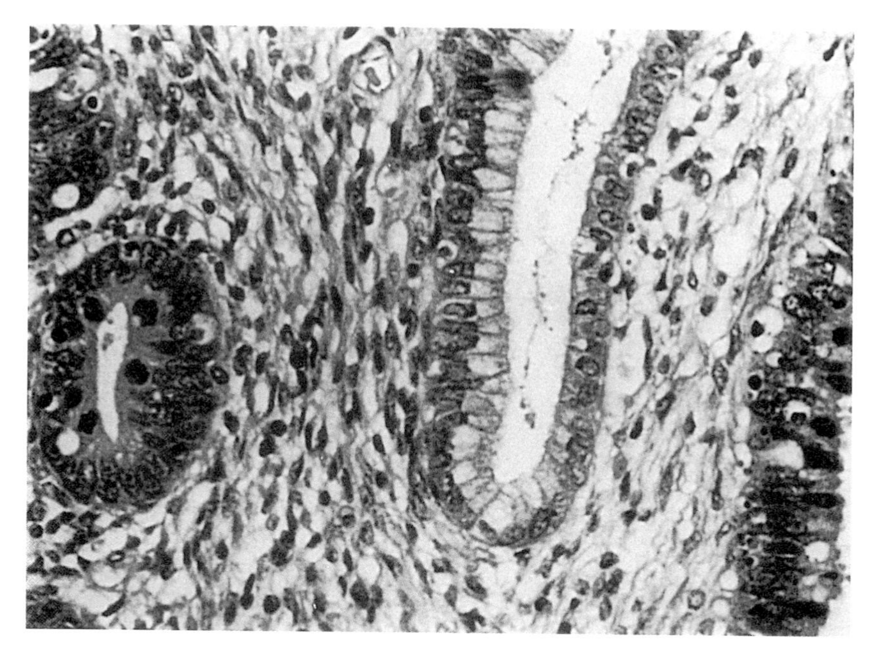

Figure 8.15 Adenocarcinoma *in situ* arising in a focus of cervical endometriosis. Glands on right and left show neoplastic change but centre gland is unaffected (H & E, ×480).

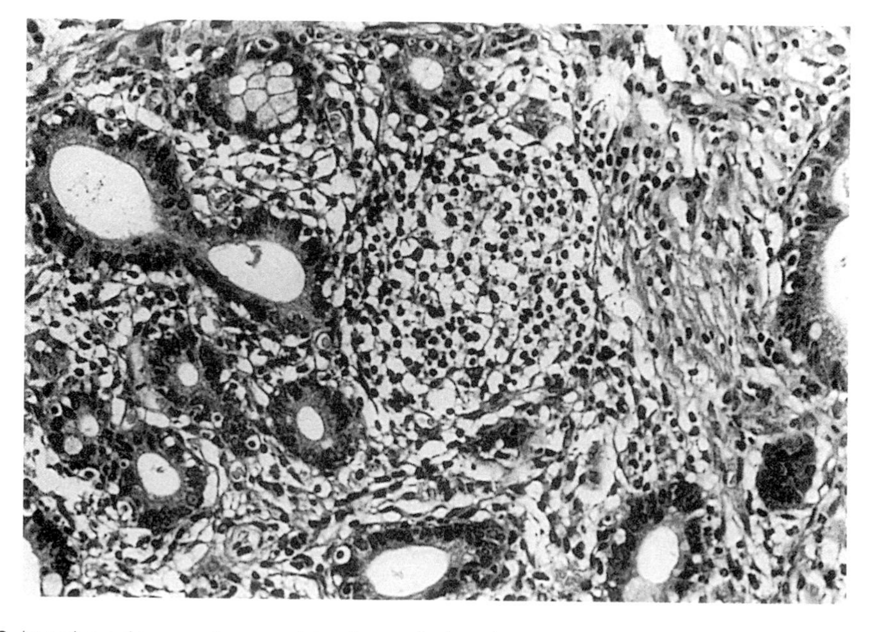

Figure 8.16 Invasive adenocarcinoma arising in cervical endometriosis (H & E, ×310).

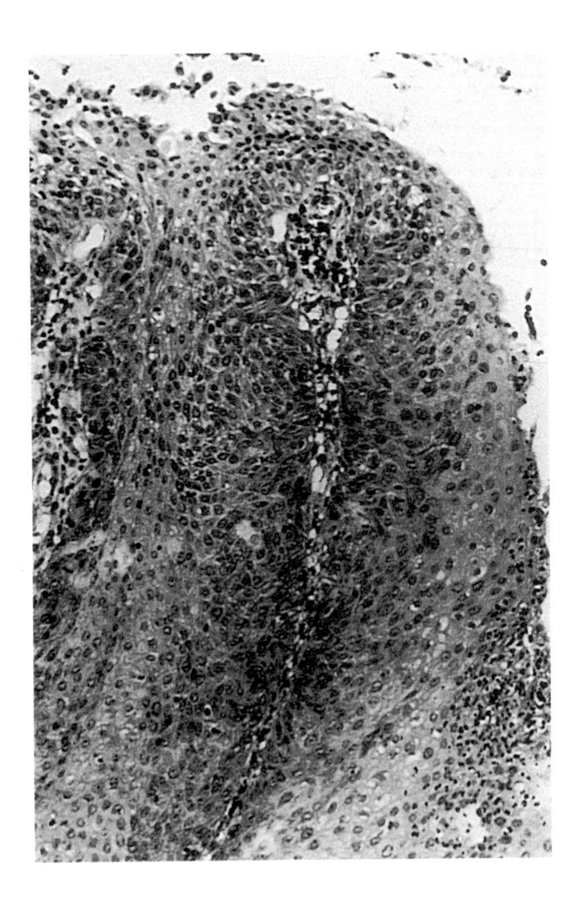

Figure 8.17 Squamous cell papilloma with fused papillary processes, each having a central connective tissue core containing blood vessels (H & E, ×190). See also Figure 8.18.

wall of the cyst into the lumen (Figure 8.21). The cells lining the papillary processes may show nuclear atypicality (Figure 8.22), probably of little significance, so that the processes bear a morphological resemblance to the intraluminal papillary processes seen in adenocarcinoma *in situ* (AIS; see Figure 13.7). However, malignant change in a mucus retention cyst appears to be exceedingly rare.

(a) Endocervical tunnel clusters

Fluhmann (1961) described clusters of closely packed endocervical glands which may be non-cystic and lined by tall columnar epithelium, or cystic and lined by cuboidal or flattened epithelium. We believe that both forms

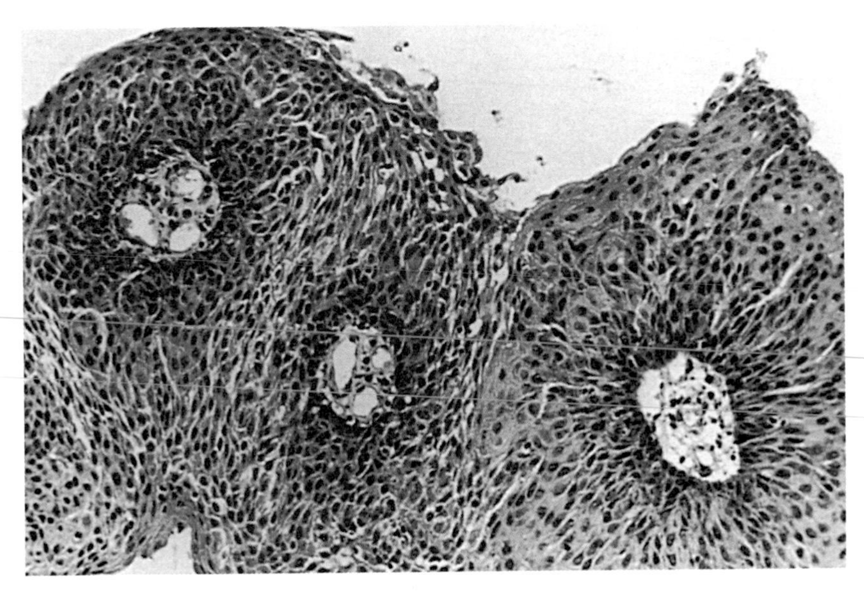

Figure 8.18 Transverse section across papillary processes of squamous cell papilloma, each having a central fibrovascular core (H & E, ×230).

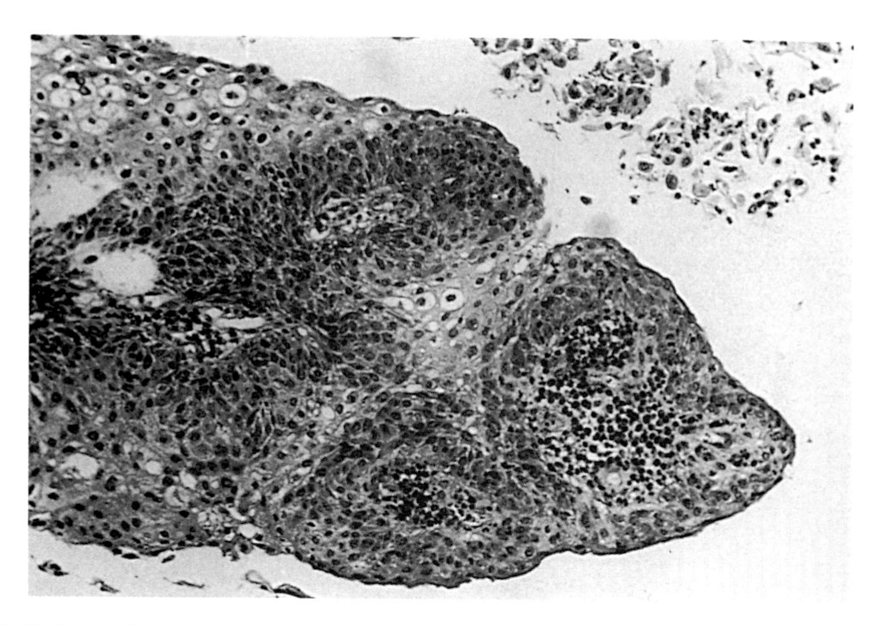

Figure 8.19 Koilocytosis is present in the squamous cells of this papilloma, indicating that the lesion is a papillomatous condyloma acuminatum (H & E, ×190). See also Figure 7.34.

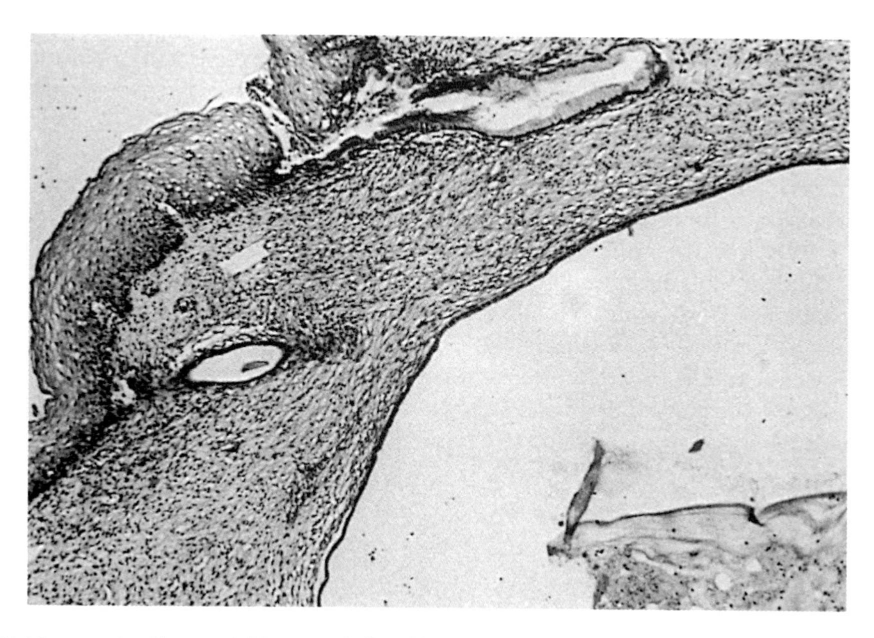

Figure 8.20 Mucus retention cyst. The cyst is lined by endocervical cells which tend to become flattened as the cyst enlarges (H & E, ×80).

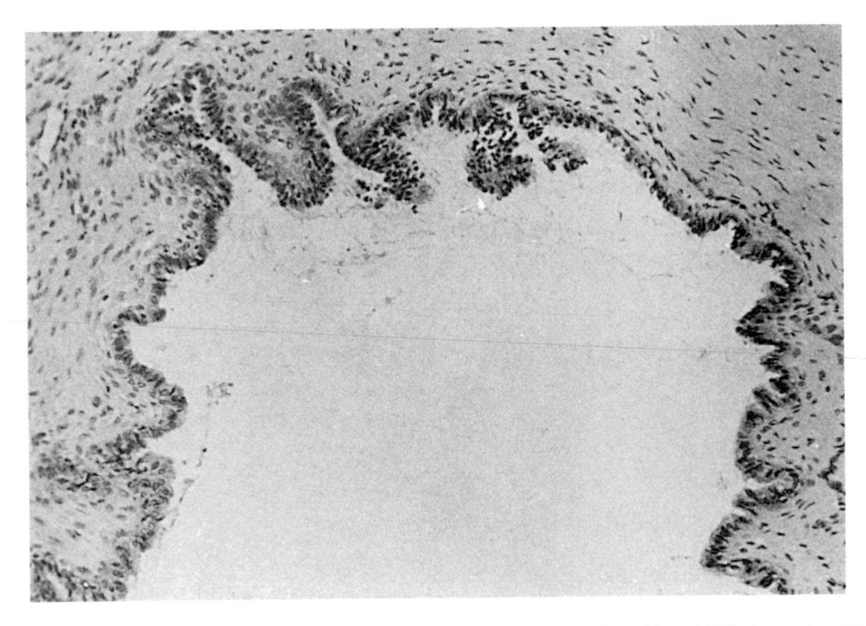

Figure 8.21 Papillary processes in the wall of a mucus retention cyst (H & E, ×190). See also Figure 8.22.

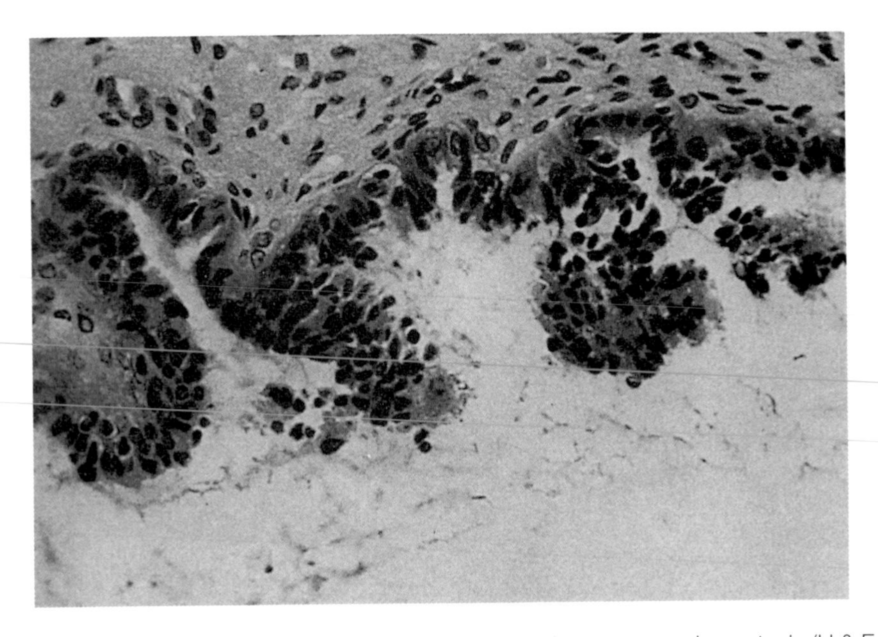

Figure 8.22 Cells lining the papillary processes in a mucus retention cyst may show atypia (H & E, ×480). Cf. papillary processes in AIS (Figure 13.7).

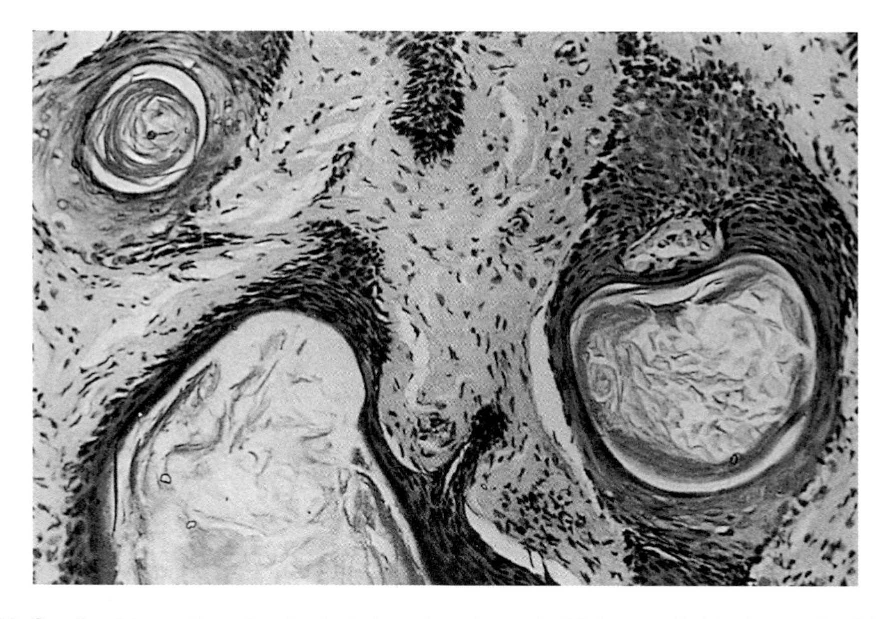

Figure 8.23 Small epidermoid cysts situated deep to ectocervical lining, probably the result of implantation during delivery (H & E, ×190).

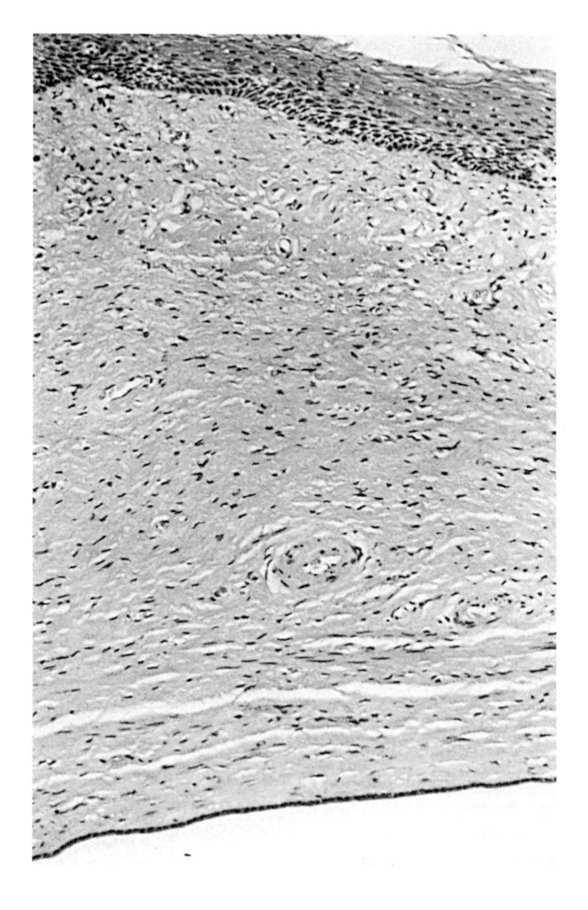

of this lesion and simple adenomatous hyperplasia (Section 9.2.2 and Figure 9.17) are one and the same entity.

8.5.2 Epidermoid cyst

Small epidermoid cysts (Figure 8.23) occur fairly frequently deep to native squamous epithelium. It is probable that they are the result of implantation due to trauma during delivery. They rarely reach macroscopic proportions and are of little clinical significance.

8.5.3 Mesonephric (Gärtner's) duct cyst

A mesonephric duct cyst can arise in the outer lateral fibromuscular tissue of the cervix but is more commonly seen in the lateral vaginal wall where it may occasionally reach a

Figure 8.24 Mesonephric duct cyst lined by a single layer of cubical epithelium (H & E, ×190).

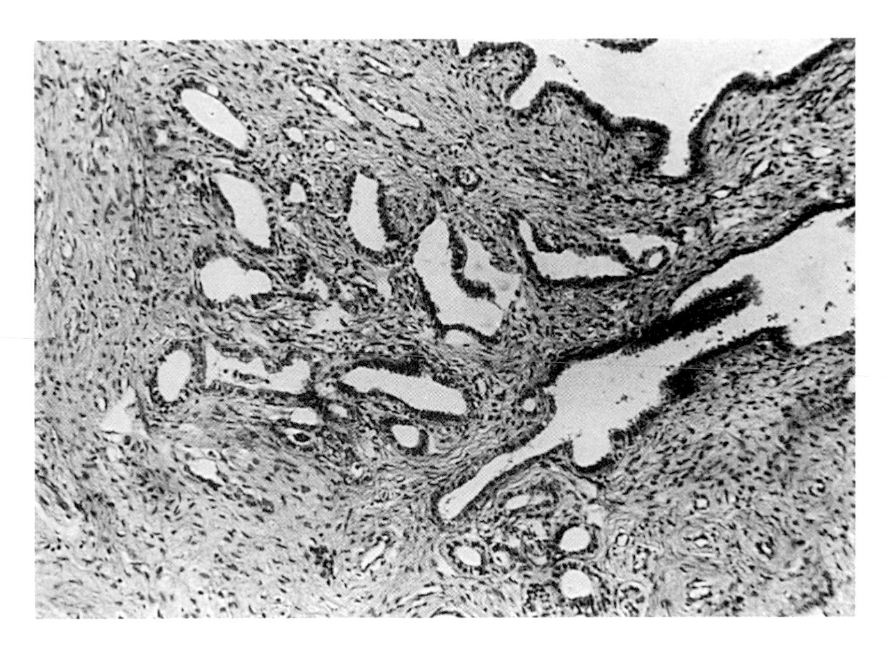

Figure 8.25 Mesonephric (Gärtner's) duct hyperplasia. The lesion was situated in the outer fibromuscular tissue of the cervix (H & E, ×310).

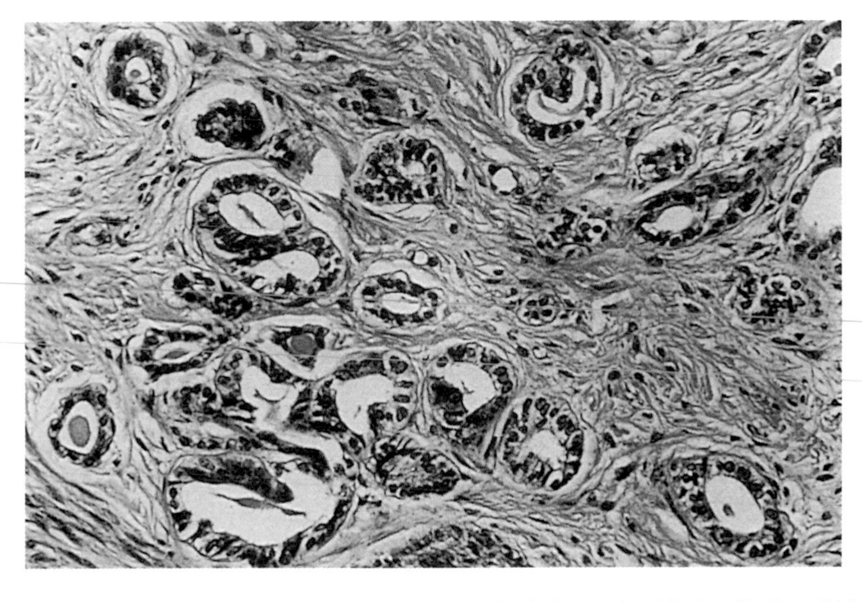

Figure 8.26 Proliferation of mesonephric tubules lined by a single layer of cubical epithelium (H & E, ×310).

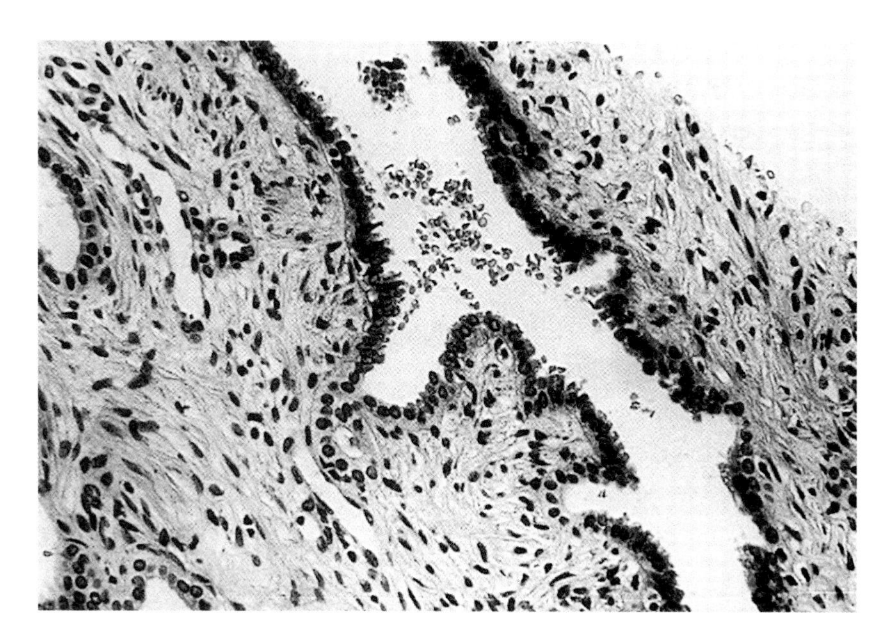

Figure 8.27 Mesonephric tubules lined by cubical epithelium showing a hobnailed pattern (H & E, ×310).

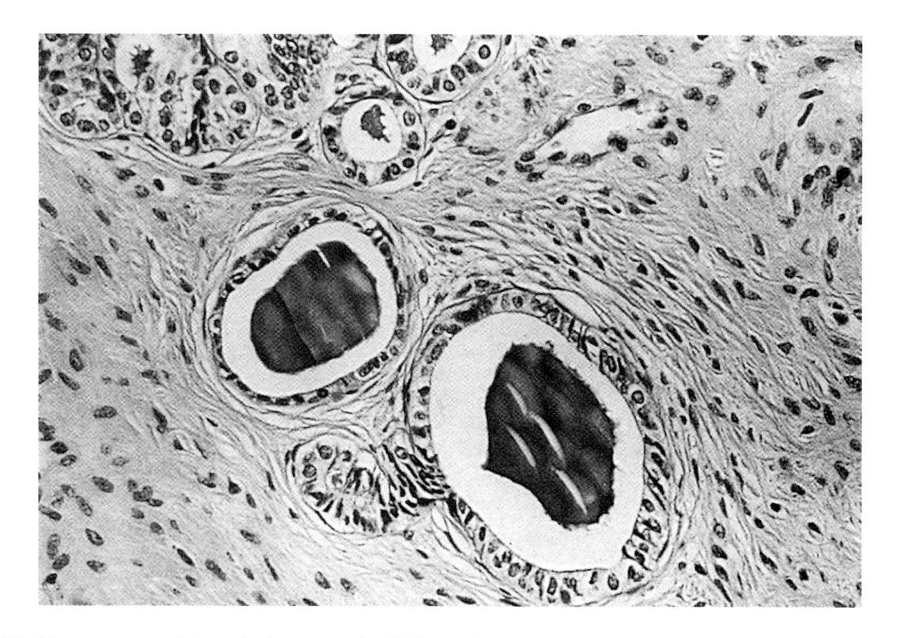

Figure 8.28 The mesonephric tubules contain PAS-positive material in the lumen but the cytoplasm of the cubical epithelial cells gives a negative reaction (PAS, ×310).

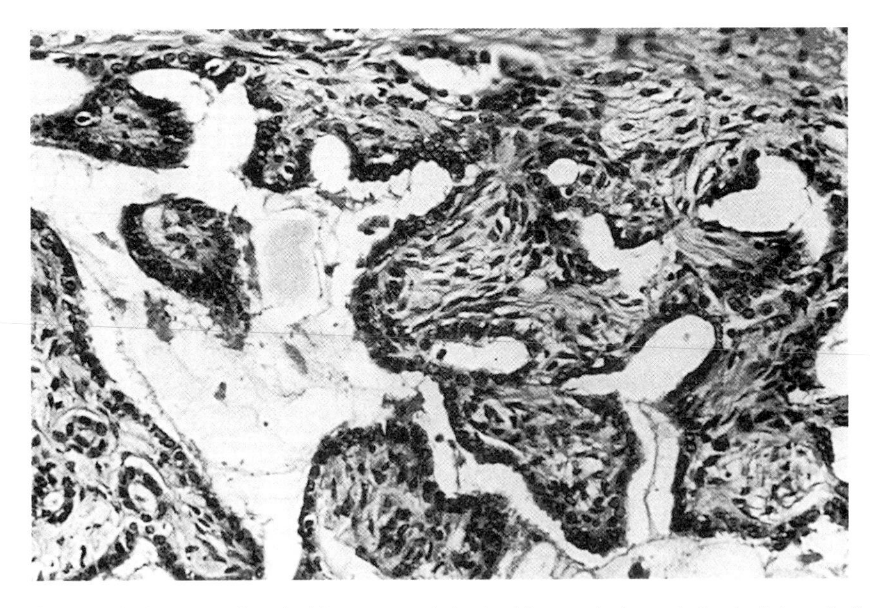

Figure 8.29 Papillary lesion associated with mesonephric duct hyperplasia and situated deep in the lateral wall of the cervix (H & E, ×310).

sufficient size to obstruct labour. It is lined by a single layer of cubical epithelium (Figure 8.24), usually quite regular, which may be ciliated but does not show evidence of mucin secretion (Evans and Hughes, 1961). The lining cells may be distinguished from those of endocervical origin which line mucus retention cysts by the negative PAS reaction of the cytoplasm, although PAS-positive material may be present in the lumen. The cells are set on a basement membrane which can give a weakly positive PAS reaction. Surrounding smooth muscle may be demonstrable, e.g. by Van Gieson's stain.

8.6 MESONEPHRIC (GÄRTNER'S) DUCT HYPERPLASIA

Occasionally remnants of the mesonephric duct (Figures 5.3 and 5.4) proliferate deep in the substance of the cervix. The gland-like and tubular structures are regular and lined by a single layer of cubical or low columnar epithelium (Figures 8.25 and 8.26) some of which is hobnail in type (Figure 8.27), contrasting with the tall mucus-secreting epithelium of the endocervix (Poulsen *et al.*, 1975). Like the mesonephric duct itself the cubical cells lining the tubules are non-ciliated and contain no glycogen or mucin, consequently being PAS-negative. Their lumen however may contain eosinophilic material which is PAS-positive (Figure 8.28). These illustrations are from a lesion which was an incidental finding in a cone biopsy specimen removed for CIN 3. The lesion was situated in the outer fibromuscular tissue of the cervix well away from the endocervical glands. The patient was aged 53 years but there is no reason why the lesion should not be found at any age.

8.7 MESONEPHRIC PAPILLOMA

The so-called mesonephric papilloma of the cervix or upper vagina is a rare lesion of infants or children in which epithelium of the hobnail

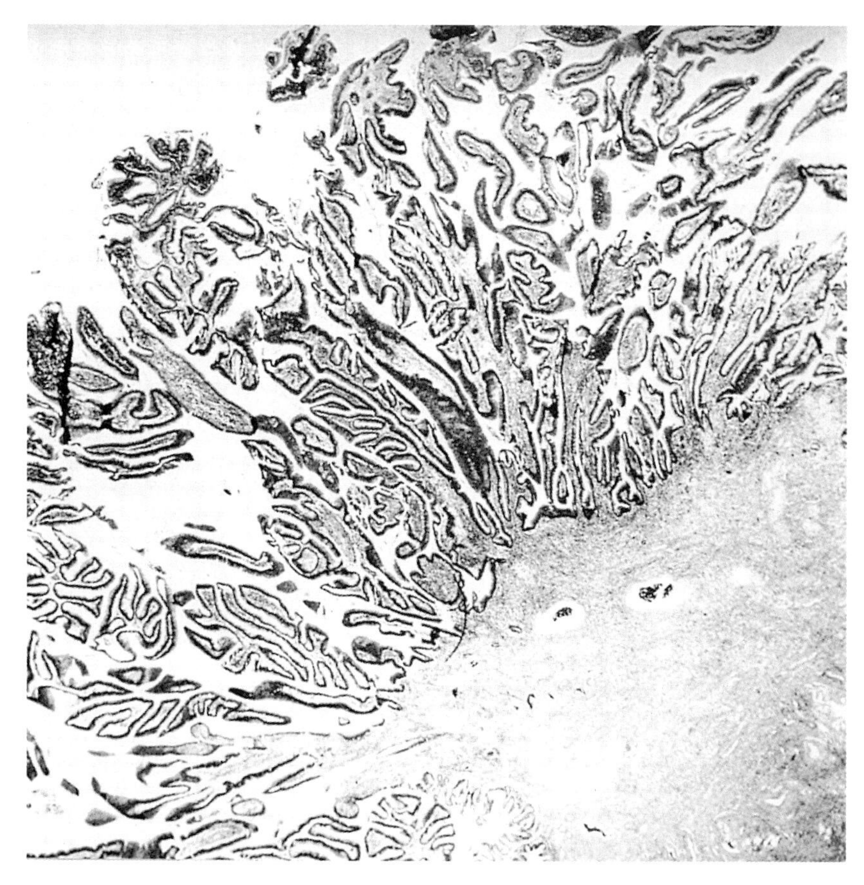

Figure 8.30 Villous adenoma showing long finger-like papillary processes (H & E, ×290) (courtesy of Dr H. Michael, Indiana University Hospital, Indianapolis, USA).

type covers loose oedematous stroma (Poulsen *et al.*, 1975). This exceedingly rare lesion has been described by several authorities (Novak *et al.*, 1954; Selzer and Nelson, 1962; Janovski and Kasdon, 1963). There is some doubt about the mesonephric origin of some of the lesions which have been described. Superficially situated papillomas probably have an endocervical origin. If situated more deeply in the lateral wall of the cervix and if associated with mesonephric duct remnants (Section 5.1), the likelihood of a mesonephric duct origin must be accepted. A papillary lesion associated with mesonephric duct hyperplasia is shown in Figure 8.29.

8.8 MÜLLERIAN PAPILLOMA

This is a unifocal or multifocal papillary lesion, characteristically arising at the squamocolumnar junction, composed of thin fibrovascular processes lined by müllerian cells, e.g. endocervical epithelium, which may undergo squamous metaplasia (Scully *et al.*, 1994).

8.9 VILLOUS ADENOMA

This unusual neoplasm was reported by Michael *et al.* (1986) in a pregnant multiparous woman aged 32 years with a 3-month history of profuse vaginal mucorrhoea and a cervical mass. Histologically, it was a broad-based lesion with long villous processes (Figure 8.30) resembling a villous adenoma of colon and showing the same malignant potential. Deep to the adenoma was an adenocarcinoma 3 mm in depth. In both lesions the cytoplasm gave a strongly positive reaction for carcino embryonic antigen (CEA). Electron microscopy of the villous adenoma revealed evidence of intestinal differentiation reflecting the potential of the cervix for intestinal metaplasia (Section 9.3.4). The related lesion, an intestinal type of mucinous adenocarcinoma, is described in Section 14.1.2.

8.10 HETEROTOPIC CARTILAGE

Roth and Taylor (1966) report the occasional occurrence of heterotopic cartilage in cervical stroma (more commonly found in the endometrium). They quote a case where hard fragments were palpated in the cervix during labour and were removed manually after delivery. The importance of recognizing the lesion as benign heterotopic cartilage lies in not misinterpreting it as a component of mixed mesodermal tumour.

REFERENCES

Aaro, L. A., Jacobson, L. J. and Soule, E. H. (1963) Endocervical polyps. *Obstet. Gynecol.*, **21**, 659–665.

Abell, M. R. and Gosling, J. R. G. (1962) Gland cell carcinoma (adenocarcinoma) of the uterine cervix. *Am. J. Obstet. Gynecol.*, **83**, 729–755.

Chang, S. H. and Maddox, W. A. (1971) Adenocarcinoma arising within cervical endometriosis and invading the adjacent vagina. *Am. J. Obstet. Gynecol.*, **110**, 1015–1017.

Clement, P. B. (1985) Multinucleated stromal giant cells of the uterine cervix. *Arch. Pathol. Lab.*, **109**, 200–203.

Cove, H. (1979) Arias–Stella reaction occurring in the endocervix in pregnancy. *Am. J. Surg. Pathol.*, **3**, 567–568.

Elliott, G. B., Reynolds, H. A. and Fidler, H. K. (1967) Pseudosarcoma botryoides of cervix and vagina in pregnancy. *J. Obstet. Gynaecol. Br. Commonwlth.*, **74**, 728–733.

Evans, D. M. D. and Hughes, H. (1961) Cysts of the vaginal wall. *J. Obstet. Gynaecol. Br. Commonwlth.*, **68**, 247–253.

Ferenczy, A. (1982) Benign lesions of the cervix, in *Pathology of the Female Genital Tract* (ed. A. Blaustein), Springer-Verlag, New York, pp. 136–155.

Fluhmann, C. F. (1962) Focal hyperplasia (tunnel clusters) of the cervix uteri. *Obstet. Gynecol.* **17**, 206–214.

Gilbert, E. F. and Palladino, A. (1966) Squamous papillomas of the uterine cervix: review of the literature and report of a giant papillary carcinoma. *Am. J. Clin. Pathol.*, **46**, 115–121.

Grönroos, M., Meurman, L. and Kahra, K. (1983) Proliferating glia and other heterotopic tissues in the uterus: foetal homografts? *Obstet. Gynecol.*, **61**, 261–266.

Janovski, N. A. and Kasdon, E. J. (1963) Benign mesonephric papillary and polypoid tumours of the cervix in childhood. *J. Paediatr.*, **63**, 211–216.

Johnson, L. D. (1973), Dysplasia and carcinoma *in situ* in pregnancy, in *The Uterus* (eds H. J. Norris, A. T. Hertig and M. R. Abell), Williams and Wilkins, Baltimore, pp. 382–412.

Michael, H., Sutton, G., Hull, M. T. and Roth, L. M. (1986) Villous adenoma of the uterine cervix associated with invasive adenocarcinoma: a histologic, ultrastructural and immunohistochemical study. *Int. J. Gynecol. Pathol.*, **5**, 163–169.

Novak, E., Woodruff, J. D. and Novak, E. R. (1954) Probable mesonephric origin of certain female genital tumours. *Am. J. Obstet. Gynecol.*, **68**, 1222–1239.

Novak, E. R. and Woodruff, J. D. (1974) *Novak's Gynecologic and Obstetric Pathology*, 7th edn., W. B. Saunders, Philadelphia, p. 81.

Poulsen, H. E., Taylor, C. W. and Sobin, L. H. (1975) *Histological Typing of Female Genital Tract Tumours*, WHO, Geneva, pp. 61–62.

Ridley, J. H. (1968) The histogenesis of endometriosis. A review of facts and fancies. *Obstet. Gynecol. Surv.*, **23**, 1–35.

Roth, E. and Taylor, H. B. (1966) Heterotopic cartilage in the uterus. *Obstet. Gynecol.*, **27**, 838–844.

Scully, R. E., Bonfiglio, T. A., Kurman, R. J. *et al.* (1944) *WHO International histological classification of tumours. Histological typing of female genital tract tumours*. Springer-Verlag, Berlin, p. 54.

Selzer, I. and Nelson, H. M. (1962) Benign papilloma

(polypoid tumour) of the cervix uteri in children: report of two cases. *Am. J. Obstet. Gynecol.*, **84**, 165–169.

Woodruff, J. D. and Peterson, W. F. (1958) Condyloma acuminata of the cervix. *Am. J. Obstet. Gynecol.*, **75**, 1354–1362.

Wright, T. C. and Ferenczy, A. (1944) Benign diseases of the cervix, in *Blaustein's Pathology of the Female Genital Tract*, 4th edn (ed. R. J. Kurman), Springer-Verlag, New York, pp. 203–227.

Young, R. H., Kleinman, G. M. and Scully, R. E. (1981) Glioma of the uterus: report of a case with comments on histogenesis. *Am. J. Surg. Pathol.*, **15**, 695–699.

Young, R. H., Kurman, R. J., and Scully, R. E. (1988) Proliferations and tumours of intermediate trophoblast of the placental site. *Semin. Diagn. Pathol.*, **5**(2), 223–237.

9

METAPLASTIC, HYPERPLASTIC AND REACTIVE CHANGES IN THE CERVICAL EPITHELIUM

9.1 SQUAMOUS METAPLASIA

It is important for the pathologist to have a clear understanding of the full range of histological changes which occur in the transformation zone as it is from this area that the majority of cervical biopsies will be taken.

The normal physiological changes occurring in this region at puberty, first pregnancy and at intervals thereafter are described in Section 5.2. At these key periods, columnar epithelium is exposed to the vaginal environment, undergoes metaplastic change and is transformed into squamous epithelium. Three histological patterns can be recognized, viz. reserve cell hyperplasia, immature squamous metaplasia, and mature squamous metaplasia. Squamous metaplasia of the cervical epithelium also occurs as a pathological change in response to persistent infection or chronic irritation of the cervix. The various stages in the metaplastic process are often clearly seen in endocervical mucous polyps and the lesions of microglandular endocervical hyperplasia. The histological changes characteristic of the three stages of squamous metaplasia are as follows.

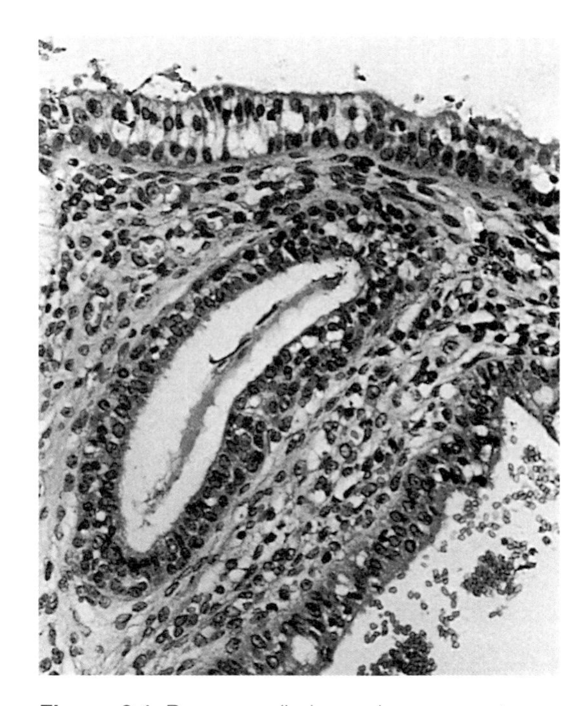

Figure 9.1 Reserve cells becoming apparent as a layer of cubical cells deep to the endocervical cells, in places giving the impression that the columnar epithelium possesses two rows of nuclei and suggesting that the reserve cells are derived from columnar cells (H & E, ×310).

9.1.1 Reserve cell hyperplasia

The first stage in the metaplastic process is heralded by the appearance of a layer of small cuboidal cells, one cell in depth, lying immediately beneath the normal endocervical cells

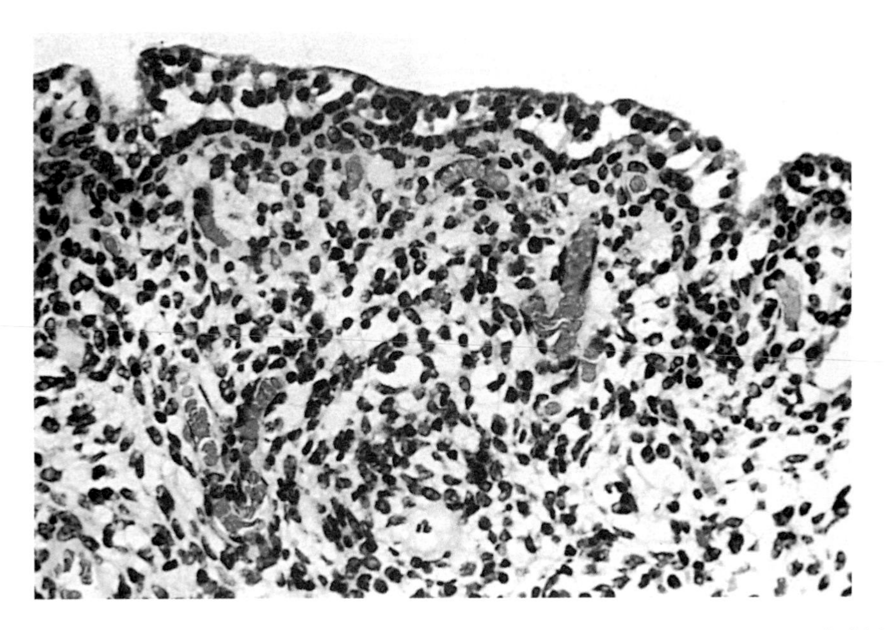

Figure 9.2 Reserve cell hyperplasia, the appearance suggesting a stromal origin for reserve cells (H & E, ×230).

giving the impression that the columnar epithelium is composed of two rows of nuclei (Figure 9.1). These subcolumnar cells or 'reserve' cells are normally inconspicuous in the columnar epithelium of the cervix. They become obvious the moment when the metaplastic process becomes manifest (Burghardt, 1973). There is still some uncertainty about their origin. There are those who believe that new cells arise from primitive epithelial cells

(a)

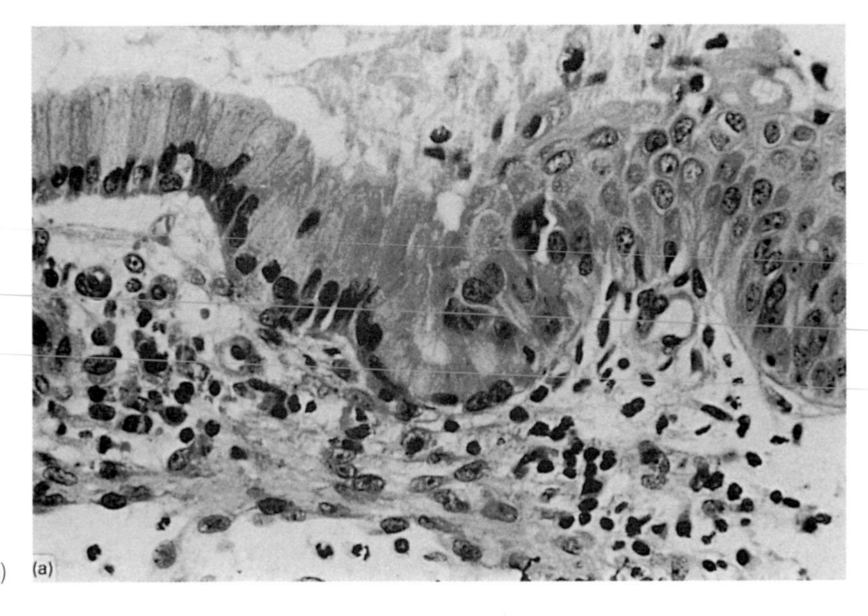

Figure 9.3 (**a**) Immature squamous metaplasia (H & E, ×480).

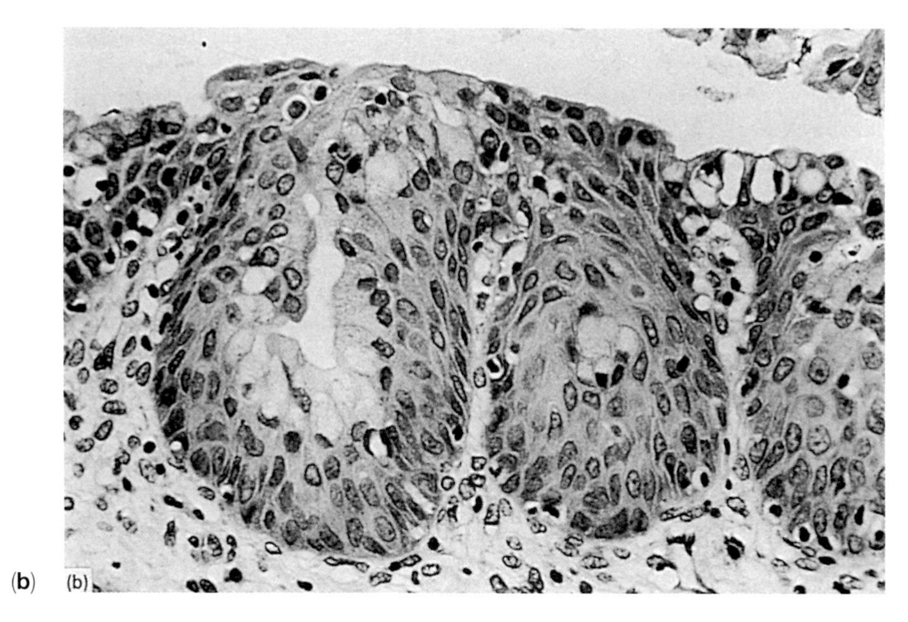

Figure 9.3 (**b**) Immature squamous metaplasia with columnar cells on the surface and trapped within the metaplastic epithelium (H & E, ×480).

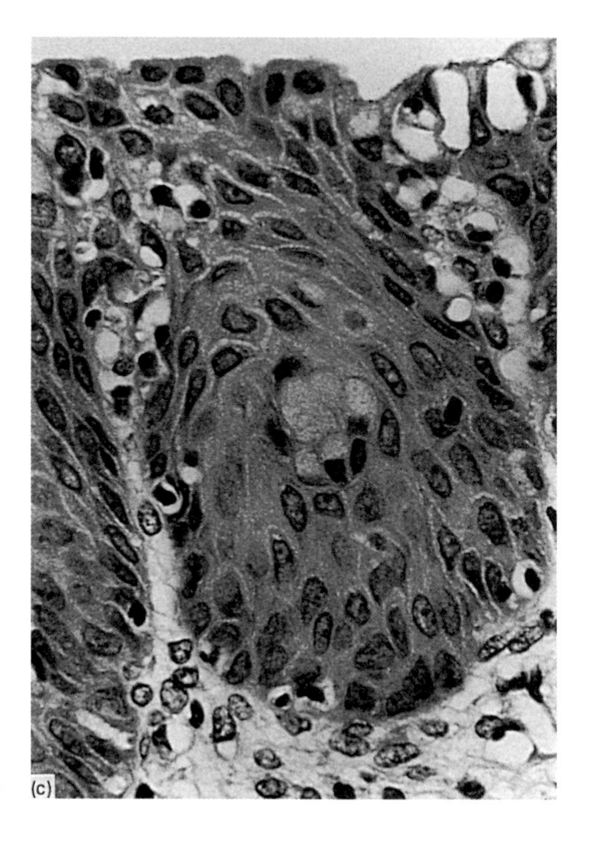

located between columnar cells and the basement membrane (Fluhmann, 1961), while others (Coppleson and Reid, 1967) have presented evidence that they may have a stromal origin (Figure 9.2).

Studies of the regrowth of epithelium after laser destruction (Maclean, 1984) indicate that metaplastic cells originate from mononuclear cells which migrate to the area from a stromal source. It is possible that metaplastic cells originate from both sources.

9.1.2 Immature squamous metaplasia

Progressive growth and stratification of reserve cells leads to the development of a multilayered epithelium showing some of the characteristics of a squamous epithelium (Figure 9.3(a,b)). Intercellular bridges are apparent and in its earliest stages (Figure 9.3(b,c)), columnar epithelial

Figure 9.3 (**c**) Intercellular bridge formation in the immature squamous metaplasia shown in (**b**) (H & E, ×770).

cells may be present on its surface or trapped in the layers of the epithelium. As the metaplastic process continues, the columnar epithelium is replaced by immature squamous epithelium. Immature metaplastic epithelium differs from its mature counterpart by lack of surface maturation and inconspicuous intracytoplasmic glycogen. It gives a negative reaction with Best's carmine which stains for glycogen. Scanning electron microscopy has shown that the surfaces of the superficial cells of the young transformation zone are covered with microvilli. In this respect, they contrast with the microridges seen on mature squamous epithelial cells (Figure 5.14). The immature metaplastic epithelium is often demarcated from the mature squamous epithelium by a sharp line (Figure 9.4(a)) and the contrast may be so marked that immature metaplastic epithelium may be mistaken for CIN. As it lacks glycogen, it appears pale when stained with Schiller's iodine compared with the surrounding native squamous epithelium leading to further confusion with CIN.

9.1.3 Mature metaplastic squamous epithelium

As the metaplastic epithelium matures, the appearance becomes identical with that of native squamous epithelium. It can be differentiated from it, however, by the presence of endocervical glands in the connective tissue deep to the metaplastic epithelium (Figure 9.4(a,b)). Glands are not normally found deep to native squamous epithelium. Original and metaplastic squamous epithelium can also be distinguished by immunocytochemical staining of cervical epithelium (Fray *et al.*, 1984; Jha *et al.*, 1984). It has been shown that epithelial cell markers such as epithelial membrane antigen, CEA and Ca-1 are expressed strongly by metaplastic epithelium (Figure 9.4(b)) and only weakly by the original squamous epithelium. The term *atypical squamous metaplasia* has been used to describe metaplastic epithelium associated with an overall increase in nuclear size and nucleolar prominence. Such changes may be seen in inflammation (Figure 9.5(a)) but it is a term best avoided.

(a) Cytology of squamous metaplasia

Cells shed from mature metaplastic squamous epithelium cannot be distinguished on morphological grounds from those shed from oestrogenized original squamous epithelium. Cells shed from immature metaplastic epithelium resemble parabasal cells seen in

(a)

Figure 9.4 (**a**) Sharp demarcation between immature and mature squamous epithelium. Note endocervical gland deep to latter, indicating its origin from metaplastic epithelium (H & E, ×130).

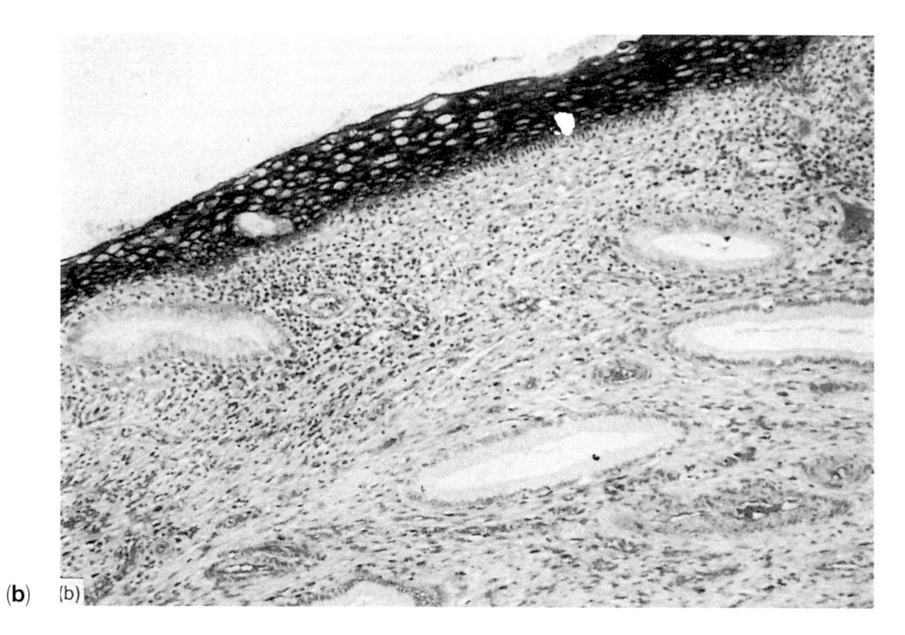

(b)

Figure 9.4 (**b**) Immature metaplastic epithelium staining positively for epithelial membrane antigen. Endocervical glands are negative (immunoperoxidase, ×310).

atrophic smears (Figure 6.4). They can be distinguished from them however, by the company they keep. They are usually intermingled with superficial and intermediate cells derived from other areas of the cervix where the epithelium is mature (Figure 9.5(b)). In addition, immature metaplastic cells show less tendency to shed spontaneously than parabasal cells from atrophic epithelium and often appear in the smear as a sheet of cells with dense cytoplasm (Figure 9.5(c)) indicating they have been forcibly detached by the action of the spatula. Some of the cells in the cluster have a vacuolated cytoplasm (Figure 9.5(d)) consistent with a glandular origin. Some have long cytoplasmic processes as a result of disruption of the intracellular bridges (Cf. Figures 6.7 and 9.5(e)).

Immature metaplastic cells are frequently seen in smears from patients with chronic cervicitis and show associated inflammatory changes (Section 7.2.2). The degree of nuclear atypia may be quite marked and this group of smears is often categorized as showing 'borderline' changes. Very occasionally, cells from an area of reserve cell hyperplasia can be identified in the smear (Figure 9.5(f)). They resemble glandular cells with an eccentric nucleus but their cytoplasm is more solid.

9.1.4 Transitional metaplasia

Transformation of native squamous or glandular epithelium of the cervix into transitional epithelium resembling that of the urinary tract has been described by Scully *et al.* (1994). It is characterized by cells with a relatively clear cytoplasm and nuclei that usually contain grooves. Occasionally, if the cells are tightly packed and the cytoplasm sparse, this entity may resemble CIN 3 but the absence of hyperchromasia, cellular and nuclear pleomorphism or increased mitotic activity should permit a correct diagnosis.

9.2 GLANDULAR HYPERPLASIA

This was defined by Poulsen *et al.* (1975) as a lesion in which the gland-like structures of the cervix proliferate. These are differentiated from adenocarcinoma by their orderly arrangement and bland nuclear features (Scully *et al.*, 1994). Oral contraceptives have greatly increased the prevalence of this condition, but it can also occur during or following pregnancy (Nichols and Fidler, 1971). Once it has occurred, it may persist for a long period without further stimulation by oral progestogen or pregnancy. Two histological patterns have been described: (i) microglandular endocervical hyperplasia; and (ii) simple adenomatous hyperplasia (adenoid proliferation).

9.2.1 Microglandular endocervical hyperplasia (MEH)

This is characterized by the proliferation of endocervical epithelium to form numerous small glands, varying in size and shape, lined by essentially normal endocervical cells and lacking in mitotic activity (Figure 9.6). It is associated with reserve cell hyperplasia (Figure 9.7) and immature squamous metaplasia (Figure 9.3), sometimes progressing to mature squamous metaplasia. MEH can occur in one or more small foci which may protrude into the endocervical canal (Figure 9.8) and is sometimes associated with glandular dilation (Figure 9.9). It can also take the form of polypoid enlargement of one or more endocervical papillary processes (Figures 9.10 and 9.11). The lesion may be so extensive that it involves not only much of the endocervical canal but also quite a large portion of the ectocervical surface where an ectropion may have been present (Figure 9.12). A striking feature of all these lesions is the prominent vacuolation. This occasionally involves the intercellular spaces but is, to a very large extent, intracellular and may result in a marked degree of ballooning of the cells (Figure 9.13). Clusters of vacuolated endocervical cells may be seen in a cervical scrape (Figure 9.14). It is of note that we have seen the condition constantly associated with koilocytotic atypia of the ectocervix, in which cell vacuolation is also a prominent feature (Figure 9.15) suggesting the possibility of

(a)

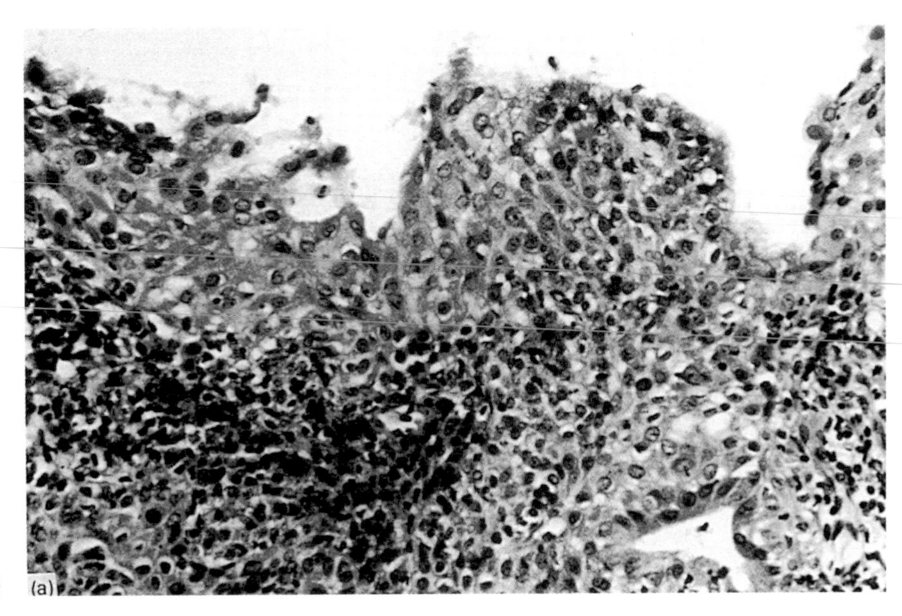

Figure 9.5 (**a**) Inflammatory change in a zone of immature squamous metaplasia (H & E, ×310).

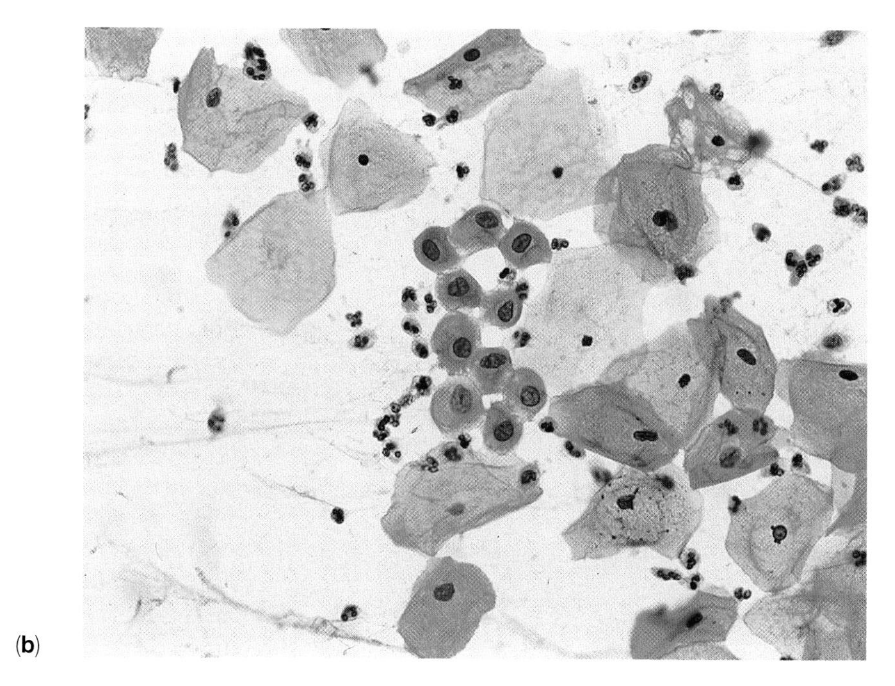

(**b**)

Figure 9.5 (**b**) Epithelial cells from an area of immature squamous metaplasia intermixed with mature superficial squamous cells in a cervical smear (Pap, ×400).

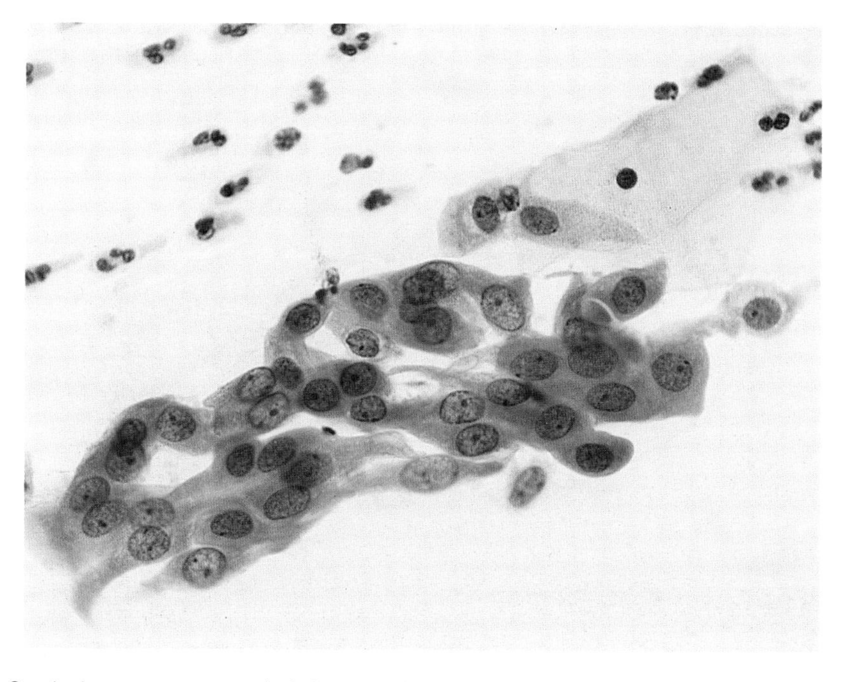

(**c**)

Figure 9.5 (**c**) Cervical smear composed of sheets of immature metaplastic cells and mature squames (Pap, ×630).

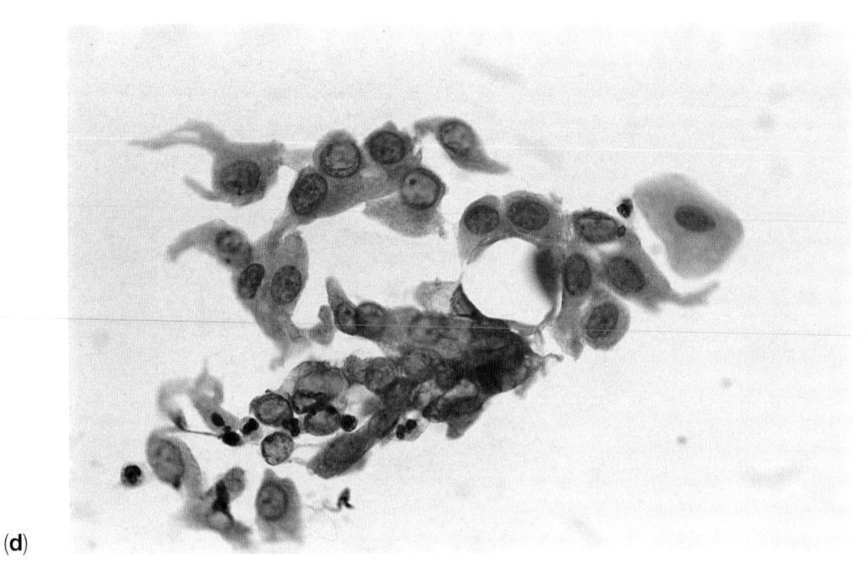

(d)

Figure 9.5 (**d**) Clusters of cells from immature metaplastic epithelium. Note vacuolated cytoplasm reflecting the presence of columnar cells on the surface of the epithelium (Pap, ×630).

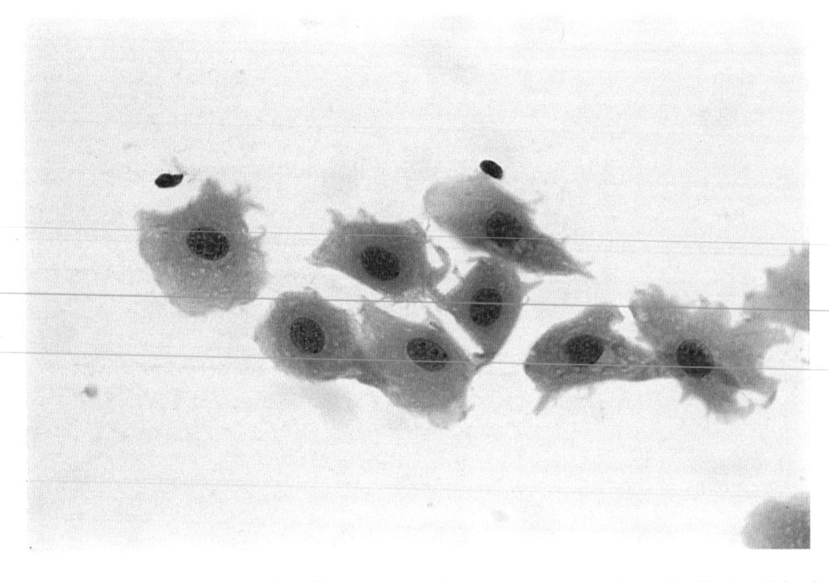

(e)

Figure 9.5 (**e**) Immature metaplastic cells with cytoplasmic processes consistent with forcible detachment from the epithelium. Intercellular bridges can be seen (Pap, ×630).

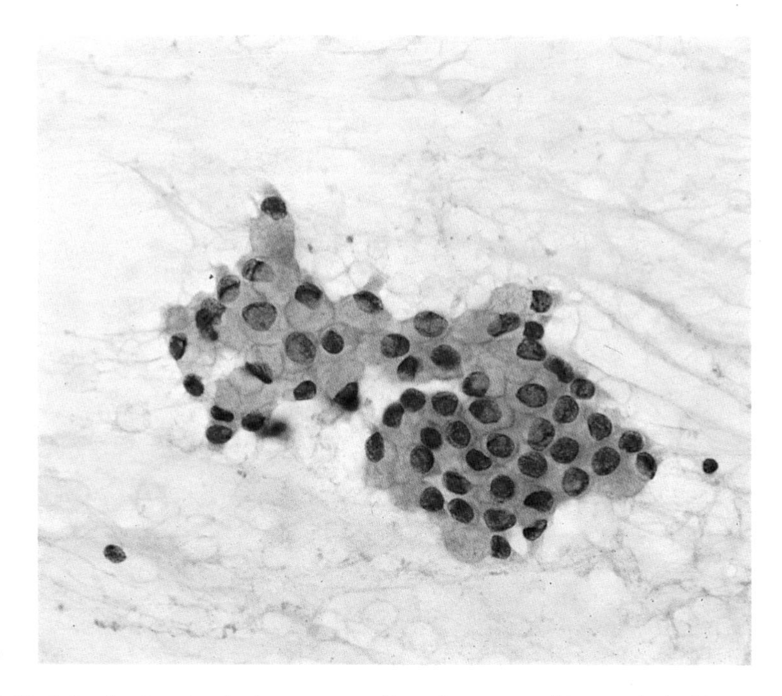

(f)

Figure 9.5 (**f**) Epithelial cells showing features suggestive of reserve cell hyperplasia. They resemble glandular cells in shape, but have the dense cytoplasm of squamous cells (Pap, ×630).

involvement of the endocervical epithelium in wart virus infection.

Sometimes the number of cells with clear cytoplasm may be so numerous that the MEH lesion could be mistaken for clear cell adenocarcinoma (Robboy and Welch, 1977). No statistical relationship between MEH and endocervical adenocarcinoma has been demonstrated (Jones and Silverberg, 1989). However, we have seen a cervix with MEH (Figure 9.16(a)) in which adenocarcinoma *in situ* (AIS, Section 13.1.1) was present (Figure 9.16(b)), the latter also having a microglandular structure suggesting that it could have been developed from a focus of MEH. We have seen a similar close association with invasive adenocarcinoma having extensive lymphatic spread (Figure 14.15(a,b)). Microglandular hyperplasia may also be encountered in foci of vaginal adenosis (Robboy and Welch, 1977), the latter condition often being associated with DES exposure (Section 9.6.3).

(a) Atypical microglandular endocervical hyperplasia

Young and Scully (1989) have described several variants of microglandular endocervical hyperplasia which could be misinterpreted as adenocarcinoma, particularly if the atypical pattern is the predominant feature. One such variant, resembling a clear cell carcinoma, is the solid proliferation of vacuolated endocervical cells, sometimes with an occasional 'signet ring' cell. A pseudoinfiltrative pattern can be produced

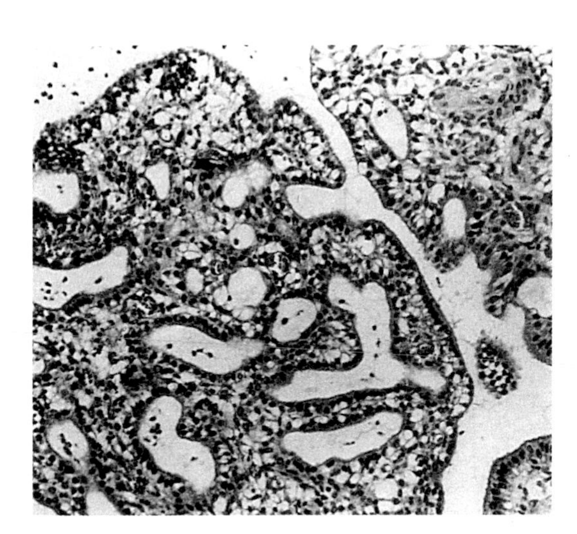

Figure 9.6 Microglandular endocervical hyperplasia (MEH) (H & E, ×190).

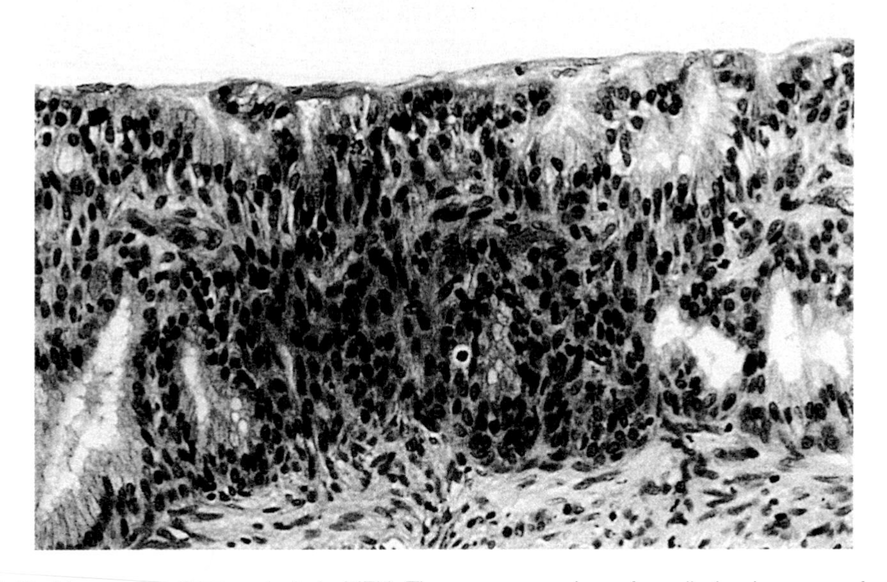

Figure 9.7 Florid reserve cell hyperplasia in MEH. There are a number of small glands, some of which have been cut tangentially to give a solid appearance (H & E, ×310).

by nest and cords of endocervical cells irregularly distributed in a myxoid stroma. The impression of malignancy is enhanced if a hobnail pattern prevails. The benign nature of these lesions is supported by 12 years of uneventful follow-up under conservative management. The glycogen-rich cytoplasm of clear cell carcinomas, sometimes with papillary pattern, does not occur in MEH. Atypia and mitotic figures are also significantly less likely to occur in MEH, which in the latter rarely exceed one per high-power field. Occasionally

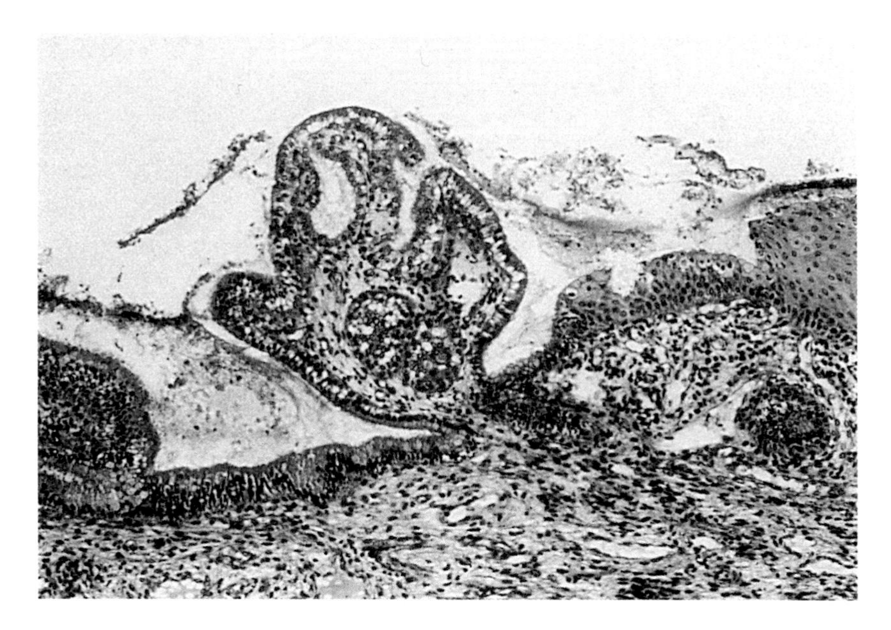

Figure 9.8 Small focus of protuberant MEH (H & E, ×60).

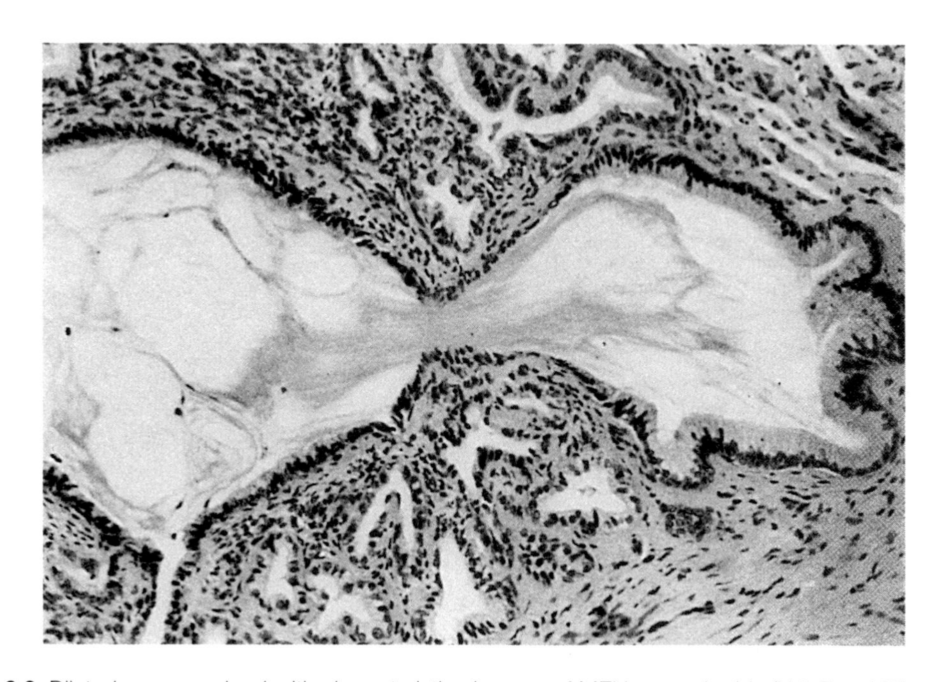

Figure 9.9 Dilated mucous gland with characteristic changes of MEH on each side (H & E, ×190).

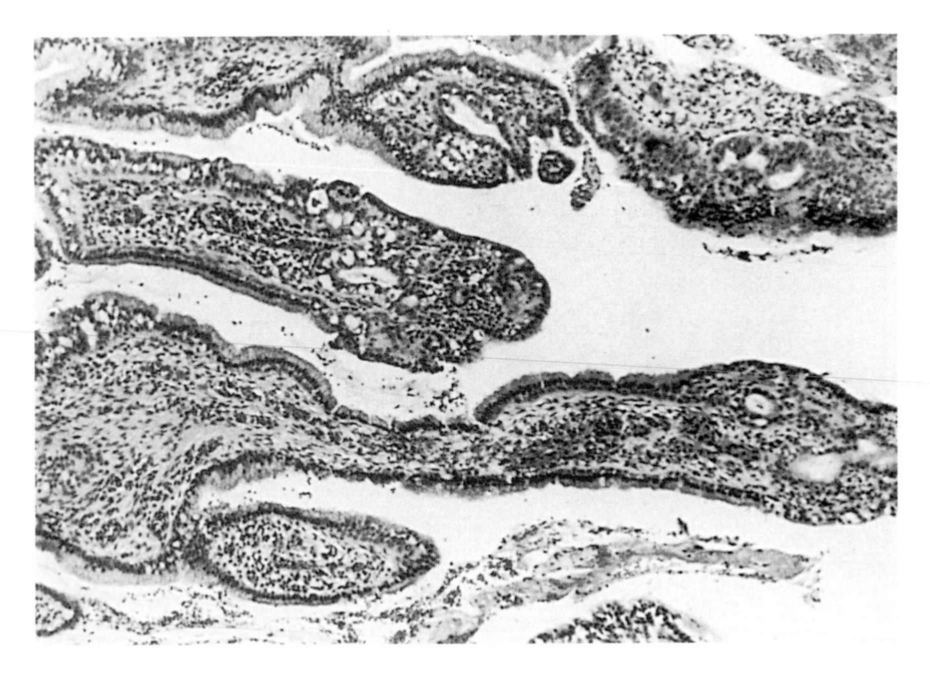

Figure 9.10 Florid MEH associated with enlargement of papillary processes (H & E, ×30).

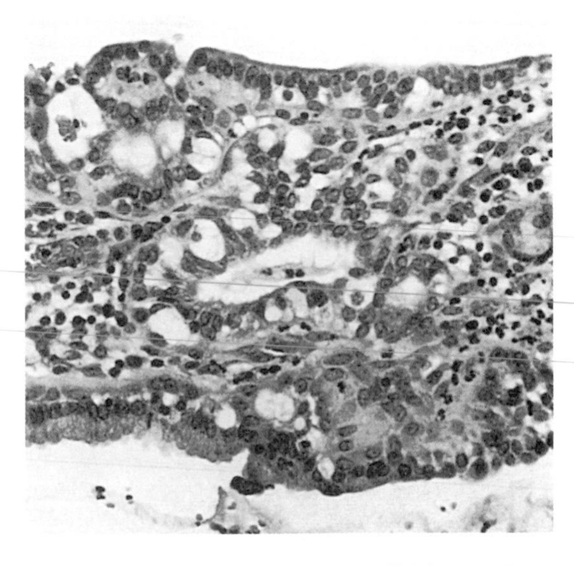

Figure 9.11 Typical pattern of MEH in papillary process of lesion shown in Figure 9.10 (H & E, ×310).

MEH may have a reticular pattern which mimics a yolk sac tumour. The older age of the patients with MEH and the absence of primitive cells, together with the great rarity of yolk sac tumours of the cervix, all serve to assist in making the correct diagnosis.

9.2.2 *Simple adenomatous hyperplasia (tunnel clusters, diffuse laminar endocervical glandular hyperplasia)* (Figure 9.17)

This form of glandular hyperplasia differs from MEH by the greater regularity of the glands and the sparsity of reserve cells. Another important difference is the absence of either intracellular or intercellular vacuolation and the absence of any accompanying koilocytotic

Figure 9.12 Florid MEH. Very low-power view to show extent of lesion shown in Figures 9.10 and 9.11 (H & E, ×7).

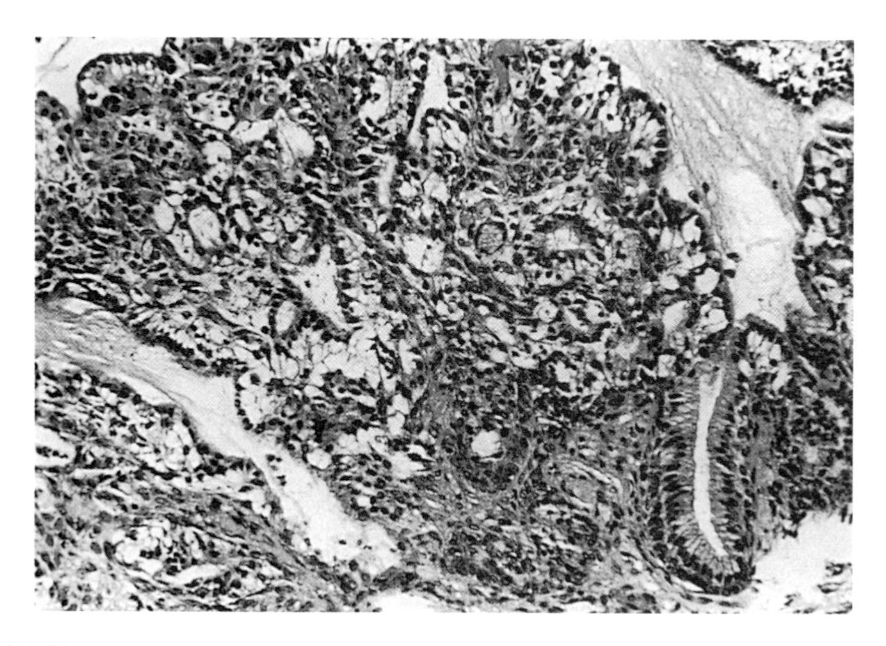

Figure 9.13 MEH showing ballooning of cells (H & E, ×190).

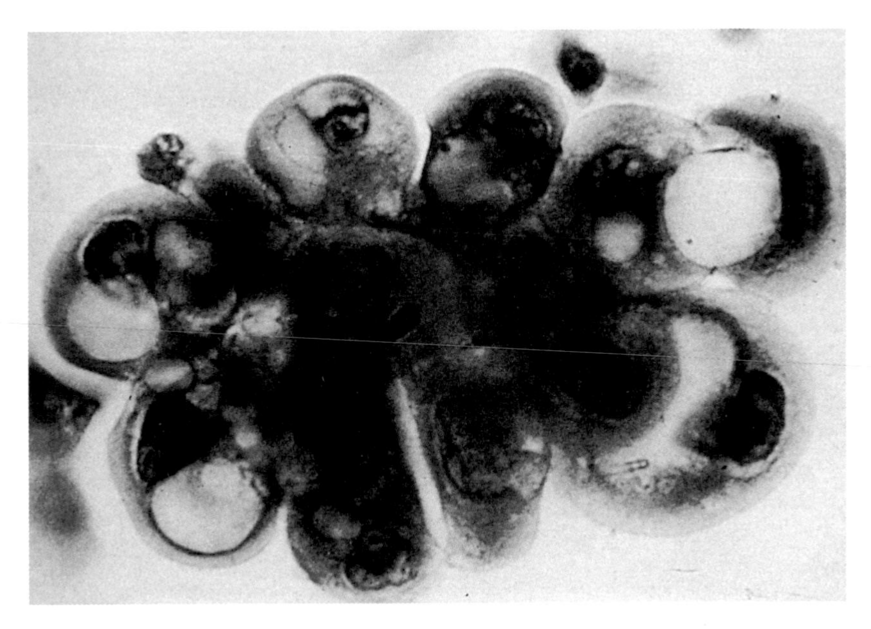

Figure 9.14 Cluster of vacuolated endocervical cells in cervical scrape from MEH (Pap, ×1500) (courtesy of Dr Costa, Madrid, Spain).

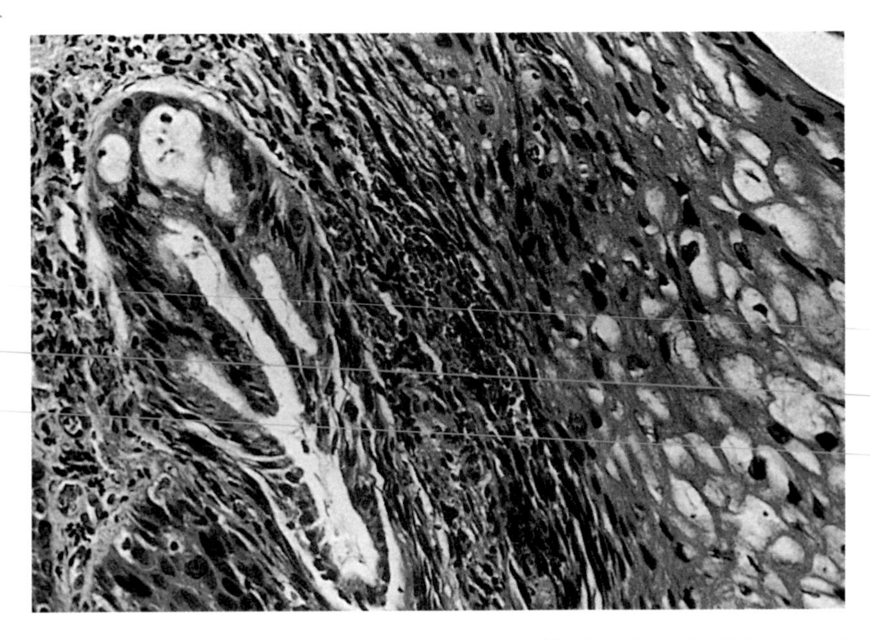

Figure 9.15 Koilocytotic atypia of the ectocervix associated with the microglandular hyperplasia shown in Figures 9.10–9.12. There is vacuolation of both ectocervical and endocervical cells (H & E, ×310).

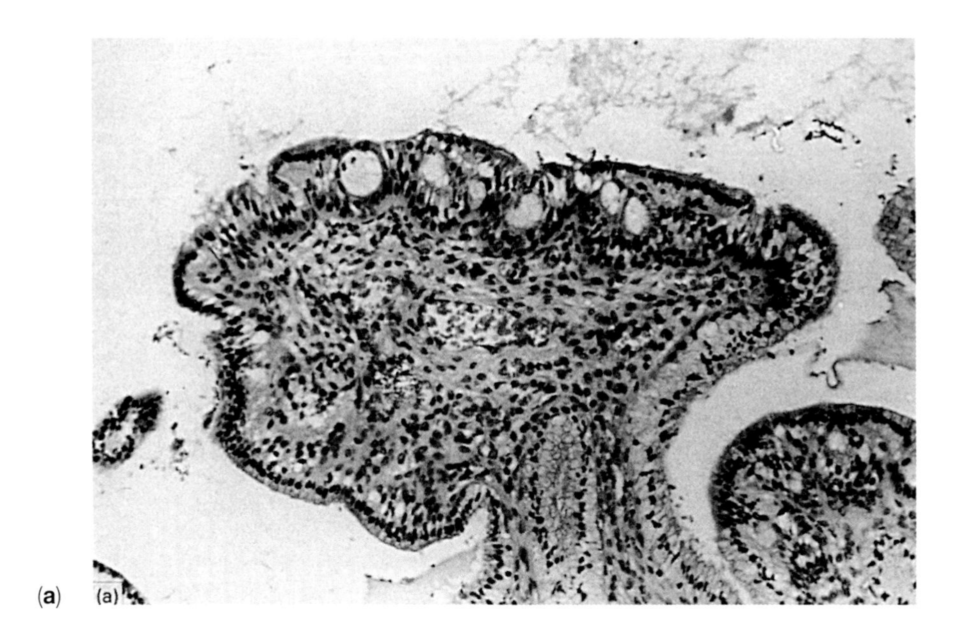

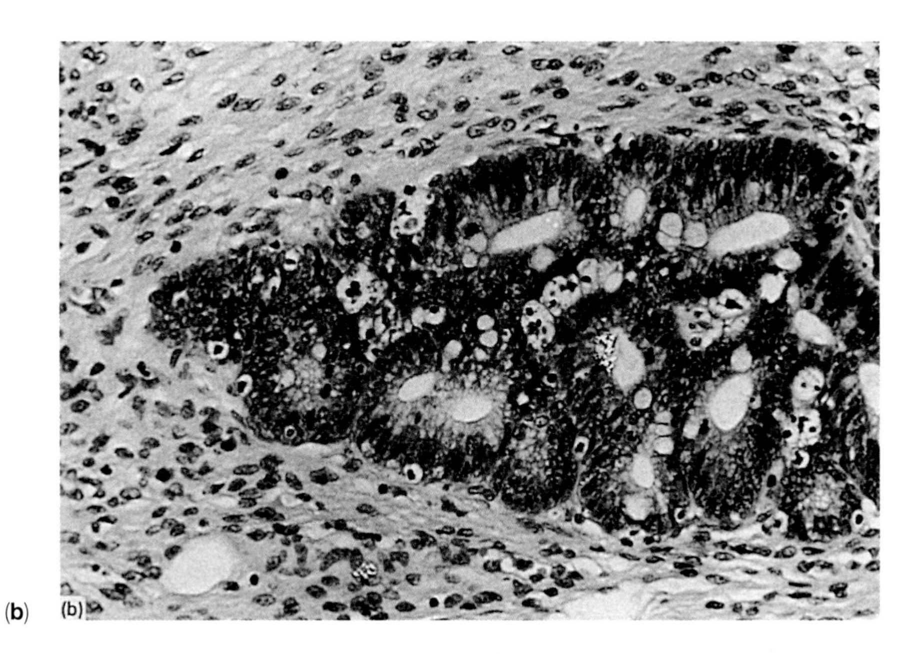

Figure 9.16 (**a**) MEH occurring in close relationship to adenocarcinoma *in situ*, shown in (**b**) (H & E, ×190). (**b**) Adenocarcinoma *in situ* occurring in close relationship to MEH shown in (**a**) (H & E, ×310).

atypia of the ectocervix. It is situated within the inner third of the cervical wall with clearly demarcated moderate-sized glands as in Figure 9.17. Jones *et al.* (1991) reported seven examples of this lesion, emphasizing its benign nature and the need to differentiate it from minimal deviation adenocarcinoma (Section 14.1.1a). In the latter, occasional mitotic figures in the endocervical gland cells, usually situated near the luminal margin of the lining epithelium (Figure

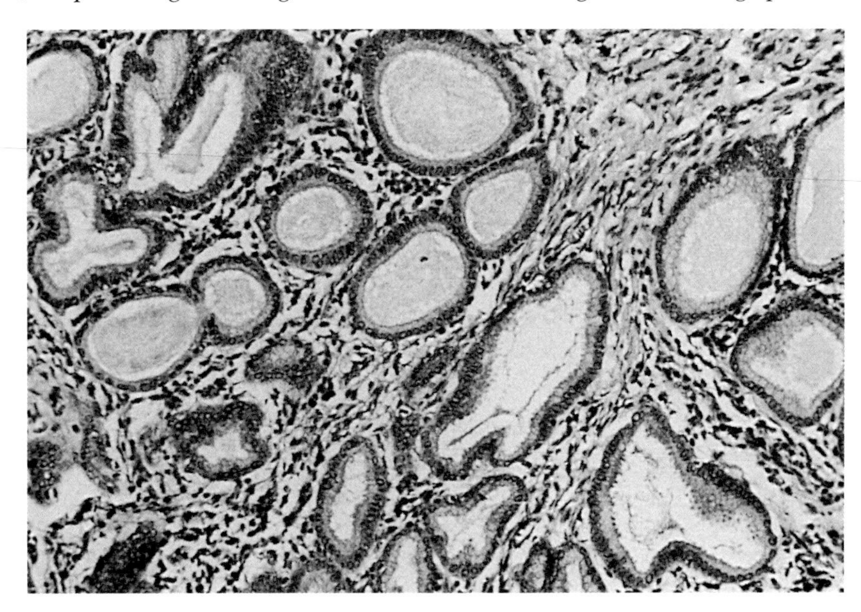

Figure 9.17 Adenomatous hyperplasia of endocervical glands. It differs from MEH by the absence of cell ballooning and the sparsity of reserve cell hyperplasia, none being present in this field (H & E, ×190).

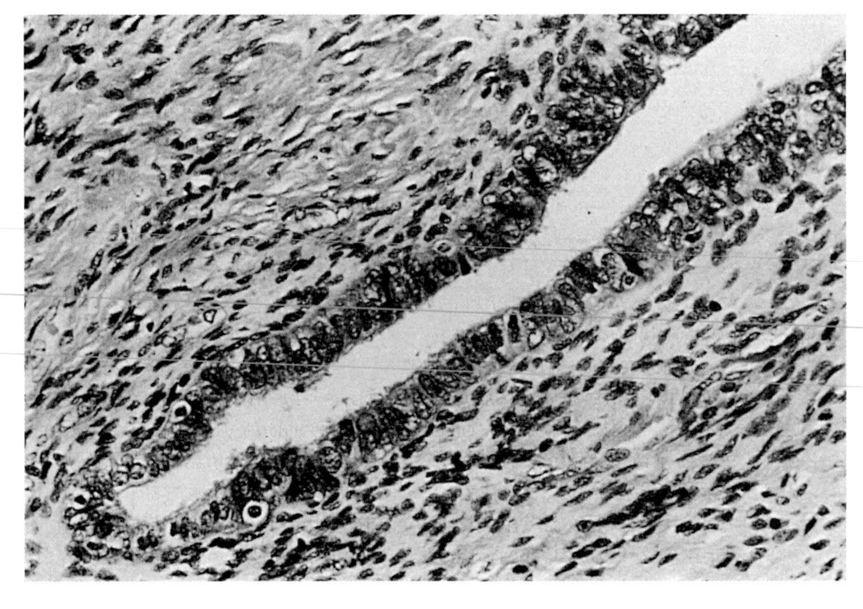

(a)

Figure 9.18 (**a**) Tubal metaplasia in an endocervical gland. Columnar ciliated and intercalated cells are more readily identifiable than secretory cells (H & E, ×350) (courtesy of Dr A.S. Hill, Jessop Hospital for Women, Sheffield).

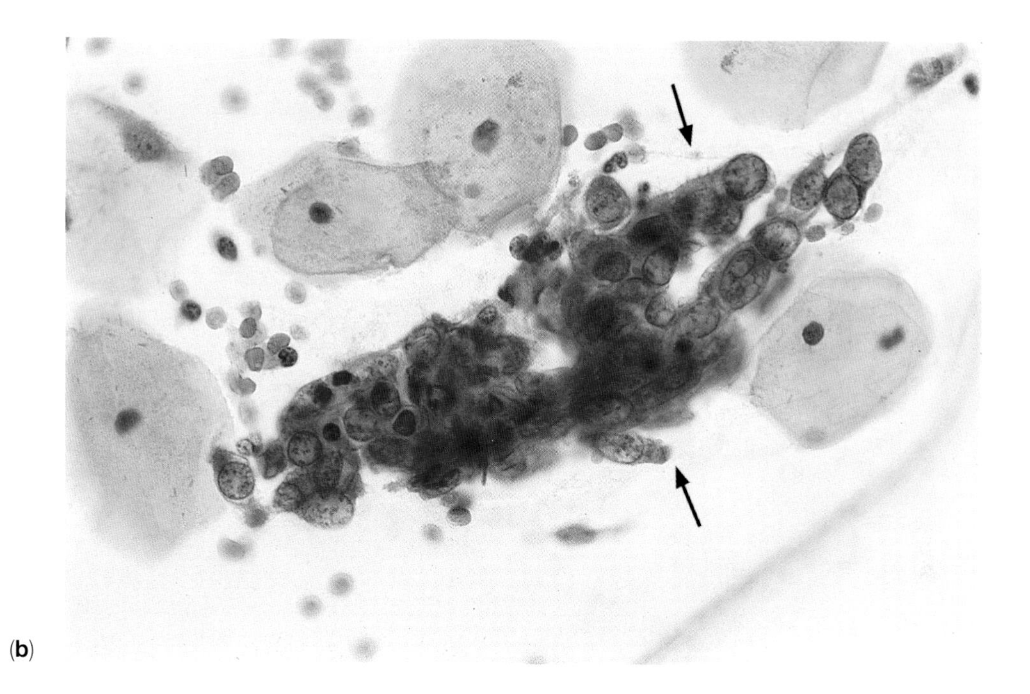

(**b**)

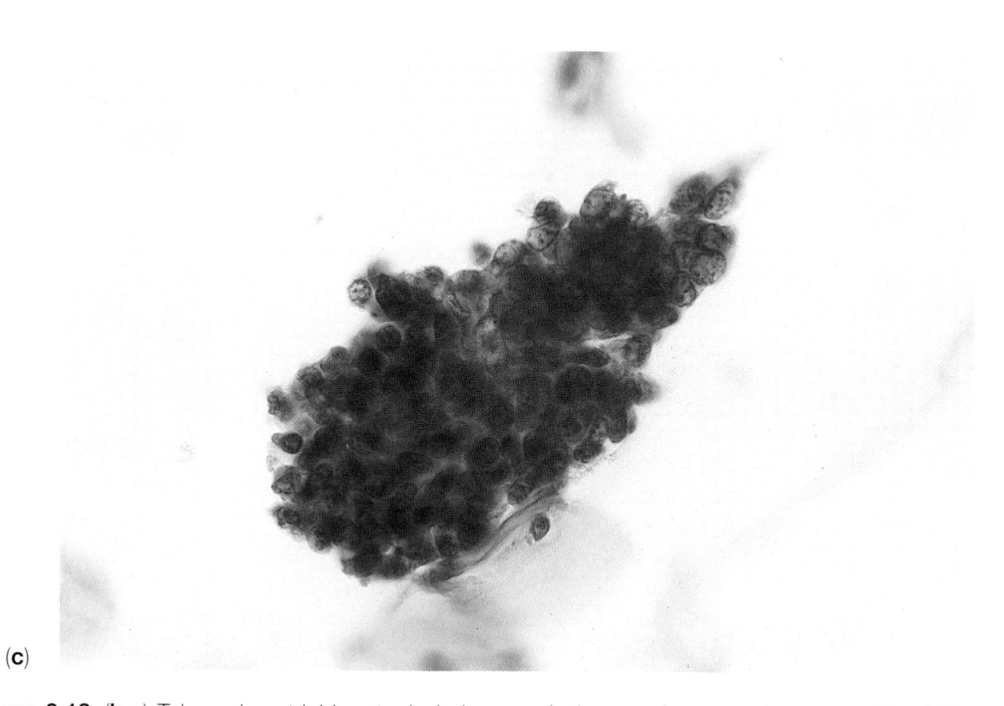

(**c**)

Figure 9.18 (**b,c**) Tuboendometrioid metaplasia in a cervical smear from a patient treated for CIN with laser therapy. Note endometrioid appearance of cell clusters (**b**) and ciliated cell borders in (**c**). (Courtesy of Dr Winifred Gray, John Radcliffe Hospital, Oxford).

14.9) provide an important distinguishing feature. Practically no mitotic figures are seen in adenomatous hyperplasia which is characterized by its limited extent and the regularity of its cells.

9.3 GLANDULAR METAPLASIAS

9.3.1 Tubal (or ciliated) metaplasia

Brown and Wells (1986) found cervical glands lined by epithelium of Fallopian tube-type (Figure 9.18(a)) in 4% of 105 cervices in which CIN 3 had been diagnosed. The glands were lined by columnar ciliated epithelium, and small dark intercalated cells. On review of the cervical biopsies of 50 women with cytological evidence of glandular dysplasia, Novotny *et al.* (1992) found that 38 women (76%) had histological evidence of tubal metaplasia. The initial cytological diagnoses of glandular dysplasia, based largely on Cytobrush material, were consequently revised. Ducatman *et al.*, (1993) using Cytobrush material taken shortly before biopsy of 20 cases diagnosed histologically as tubal metaplasia found evidence of tubal metaplasia in only two (10%) of the cervical smears. This was consistent with their histological findings that tubal metaplasia of the endocervix typically involved the intraglandular epithelium of the upper endocervix rather than the surface epithelium of the lower half of the endocervix. Suh and Silverborg (1990) however, found that eight out of 11 cases of tubal metaplasia involved endocervical glands near the squamocolumnar junction. Although tubal metaplasia of the uterus is thought to be related to the administration of oestrogen (Fruin and Tighe, 1967) no known aetiological factors for tubal metaplasia of the cervix have been identified (Hirschowitz, 1994).

9.3.2 Endometrial metaplasia

Endometrial type glands without stroma (Figure 8.14), probably due to aberrant müllerian differentiation, are not uncommon in the cervix. They are to be distinguished from true endometriosis (Section 8.3) and adenocarcinoma *in situ* (Section 13.1.1).

9.3.3 Tuboendometrioid metaplasia (Figure 9.18 (b,c))

Hirschowitz *et al.* (1994) described the condition of tuboendometrioid metaplasia of the cervix in three women who had been treated two years previously for CIN. A follow-up smear was found to contain abnormal cells which were erroneously thought to represent recurrence of CIN. The women were referred for surgical treatment and the surgical specimens showed extensive tuboendometrial metaplasia of the cervix. Histological examination showed tuboendometrioid glands in the endocervix well away from the uterine isthmus with no associated endometrial stroma. The glands showed endometrial features including pseudostratification of the nuclei, nuclear hyperchromasia and secretory apical snouting in addition to luminal ciliation. No CIN was seen in the surgical material.

These cases show that cells shed from an area of tuboendometrioid metaplasia of the cervix may be misinterpreted as dysplastic in cervical smears and clinicians should be alert to this possibility. The metaplastic cells in the smears differ from those found in glandular intraepithelial neoplasia or CIN in that they are small, uniform, arranged in three-dimensional glandular structures and lack the feathering typical of glandular neoplasia (Figure 9.18(b,c)). Most but not all cases of tuboendometrioid metaplasia have been associated with previous surgery and there is some evidence that this condition is increasing due to the widespread use of the diathermy loop for cervical biopsy (Burnett, 1992; Ismail, 1992).

9.3.4 Intestinal metaplasia

Intestinal metaplasia in the cervix (Figure 9.19(a,b)) has been described by several

(**a**)

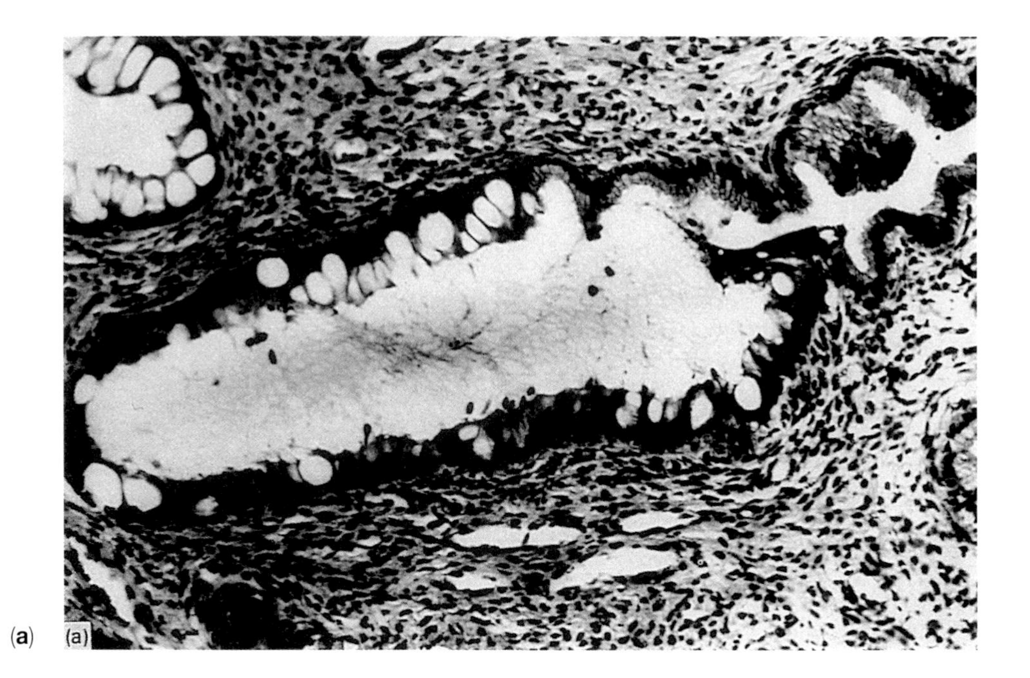

(**b**)

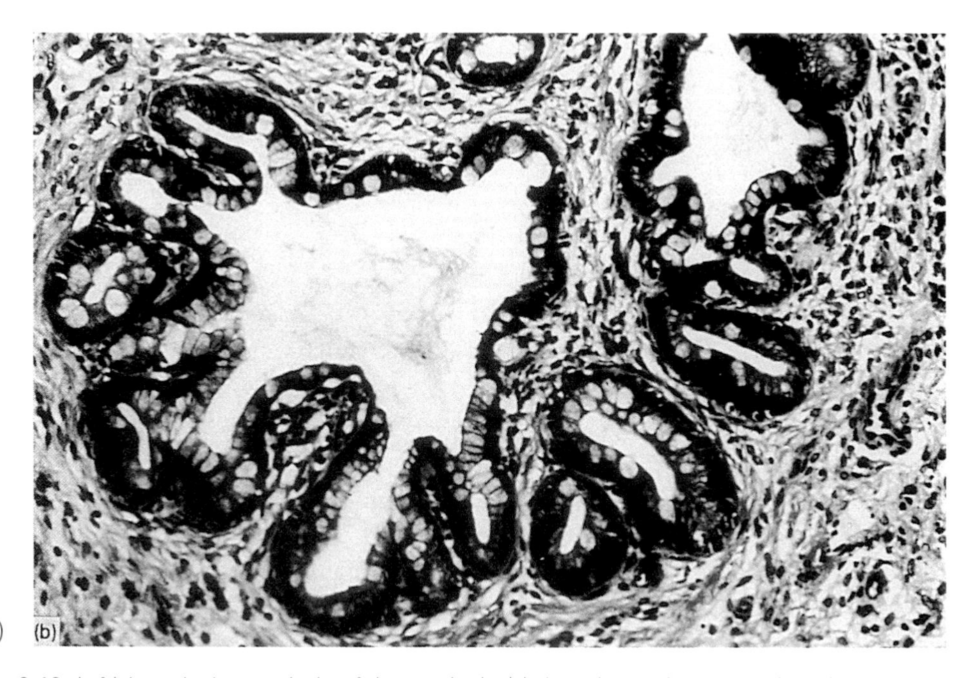

Figure 9.19 (**a,b**) Intestinal metaplasia of the cervix. In (**a**) there is an abrupt transition from a normal endocervical epithelium to that of intestinal type. In (**b**) numerous goblet cells can be seen (H & E, ×310) (courtesy of Dr J.E. Trowell, Ipswich Hospital).

authors. It is characterized by the presence of goblet cells which may be accompanied by argentaffin and Paneth cells (Scully *et al.*, 1994). This unusual phenomenon is often associated with malignant transformation. Azzopardi and Hou (1965) reported intestinal metaplasia with Paneth and argentaffin cells in a cervical adenocarcinoma; Michael *et al.* (1986) found it in a cervical villous adenoma with an adjacent adenocarcinoma (Section 8.9) and Trowell (1985) recorded two cases associated with CIN 3. Trowell observed that the intestinal epithelium itself may show varying degrees of glandular atypia sometimes amounting to adenocarcinoma. In most of the cases reported so far, argentaffin cells could be demonstrated in glands containing the goblet cells.

9.3.5 Epidermal metaplasia

See Section 9.5, Epidermidization.

9.4 HYPERKERATOSIS AND PARAKERATOSIS

Hyperkeratosis of the ectocervix is frequently seen when the uterus has undergone prolapse. The ectocervical epithelium develops a thick outer layer of keratin (Figure 9.20(a,b)) which may be apparent macroscopically as a whitish plaque (leukoplakia) on the cervix. A characteristic feature of hyperkeratosis is the absence of nuclei in the layer of keratin distinguishing it from *parakeratosis* in which the nuclei persist (Figure 9.20(c)). There is often hypertrophy of the whole cervix. Deep to the keratin layer, there is usually a well-defined granular layer. Other features of the epithelium are acanthosis with prominent intercellular bridges and elongation of the rete pegs (Ferenczy, 1982).

In the absence of cellular atypia, both hyperkeratosis and parakeratosis are essentially benign, particularly where there is an identifiable cause such as prolapse, and the changes are clearly reactive. It should, however, be remembered that hyperkeratosis and parakeratosis sometimes feature quite prominently in cervical intraepithelial neoplasia (Figure 10.12(a)) and invasive squamous cell carcinoma, and may obscure the underlying neoplastic changes.

9.4.1 Cytology

The histological findings in hyperkeratosis are reflected in the cervical smear which contains

(a)

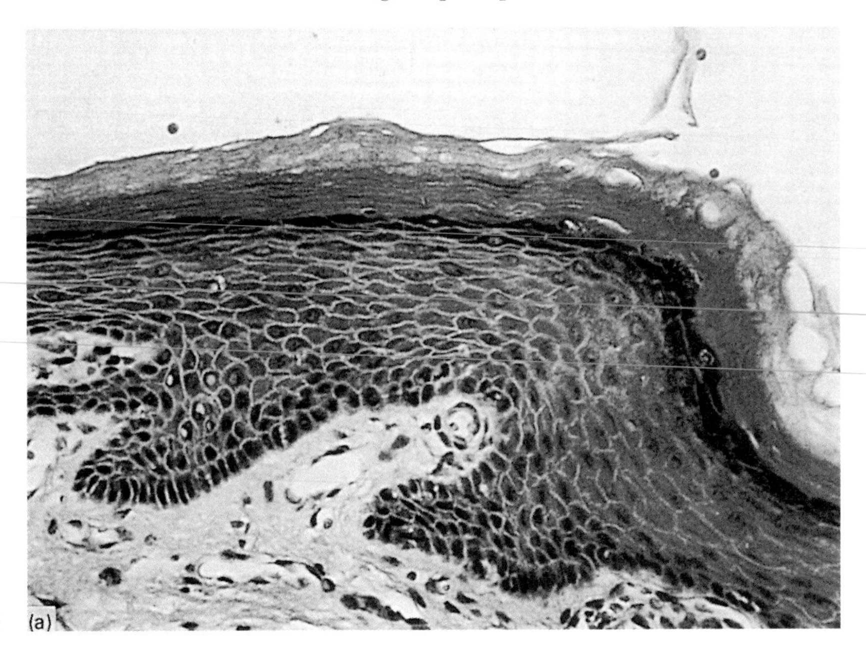

Figure 9.20 (**a**) Hyperkeratosis of ectocervix associated with prolapse. Deep to the thick keratin layer is a prominent granular layer beneath which intercellular bridges can be distinguished (H & E, ×310).

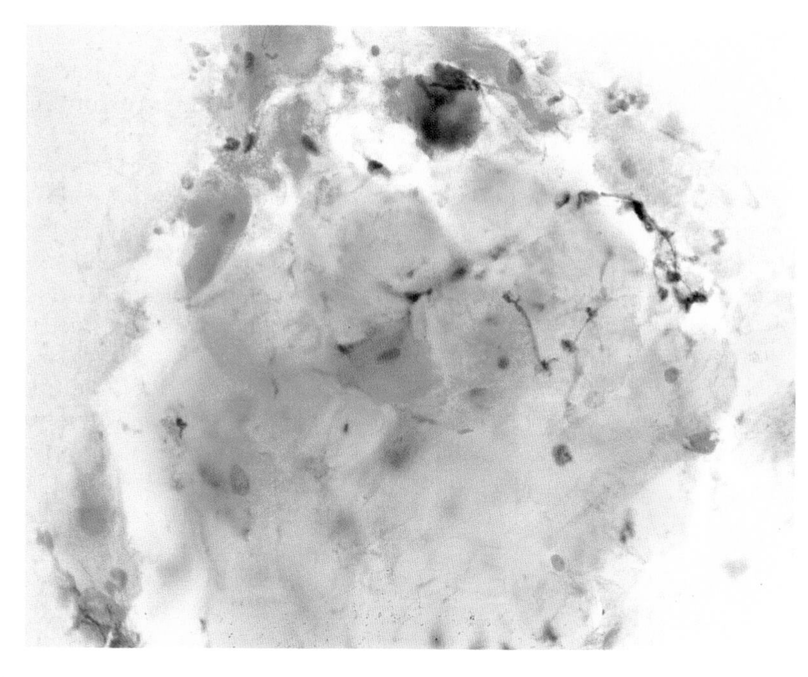

(**b**)

Figure 9.20 (**b**) Non-nucleated squames from hyperkeratosis. These appear as yellow-staining plaques in cervical smears. They are commonly found in papillomavirus infection and squamous carcinoma (Pap, ×400).

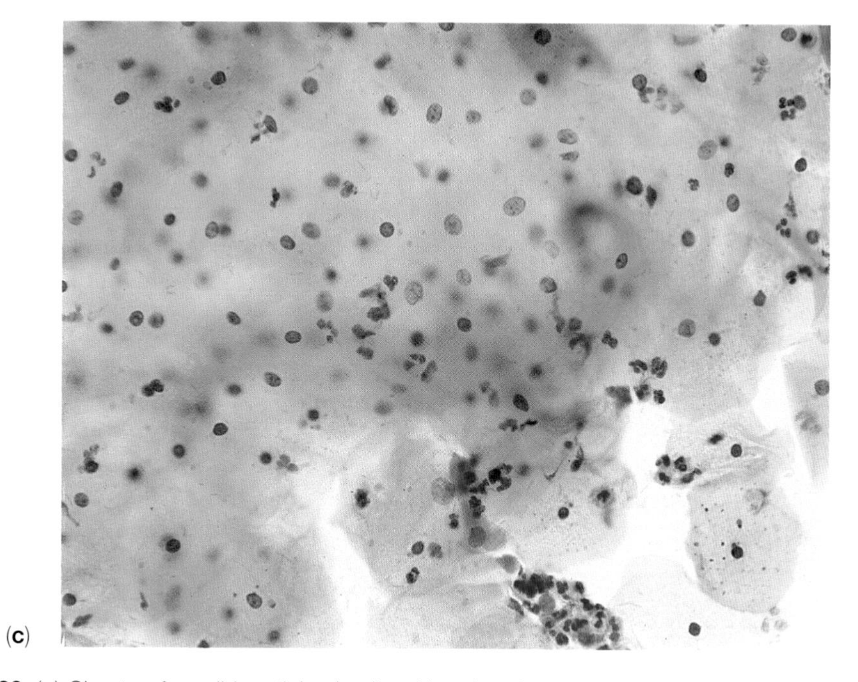

(**c**)

Figure 9.20 (**c**) Sheets of small keratinized cells with pyknotic nuclei reflecting parakeratosis of cervical epithelium (Pap, ×400).

sheets of anucleate squames staining orange or yellow with the Papanicolaou stain (Figure 9.20(b)). Parakeratosis in the cervix may be reflected in the smear by the presence of sheets of small keratinized cells with slightly irregular pyknotic nuclei (Figures 9.20(c) and 7.44). The presence of anucleate, keratinized squames in a smear may mask underlying neoplastic change and colposcopic biopsy is advisable in such cases (Section 10.6.2).

9.4.2 Dyskeratosis

This term describes abnormal keratinization. It is frequently applied to describe collectively the changes seen in papillomavirus infection, namely hyperkeratosis (Figure 9.20(b)), parakeratosis (Figures 7.44 and 9.20(b)) and individual cell keratinization (Figure 10.2(a)). It is also used to describe abnormal keratinization of neoplastic epithelium.

Examples of keratinizing CIN are shown in Figures 9.21(a) and 10.12(a). In cervical smears, highly keratinized anucleate squames or keratinized plaques studded with abnormal pyknotic nuclei may be found (Figures 9.21(b) and 10.12(b)). Biopsy is essential in such cases as the dyskeratotic cells may not reflect the severity of the underlying lesion.

9.5 EPIDERMIDIZATION

Very occasionally, the ectocervix is lined by stratified squamous epithelium closely resembling true epidermis (H. Fox, personal communication, 1986). Sebaceous glands are present but no hair follicles, the appearance being similar to that of the epithelial lining of the labium minor. It may possibly reflect a more extensive embryological contribution from the urogenital sinus than the usual lower third of the vagina.

9.6 IATROGENIC CHANGES

Included under this heading are changes induced by heat coagulation, cryosurgery and laser, ionizing radiation, cytotoxic therapy and exposure to diethylstilboestrol.

9.6.1 Electrodiathermy, cryotherapy and laser therapy

These modalities are frequently used for the treatment of CIN. Biopsy is inappropriate immediately after any of these forms of therapy as there is considerable tissue damage and destruction which makes interpretation difficult. Damage may be manifest as coagulation necrosis of the tissue and elongation of the cells due to the action of shearing forces. Two stages have been recognized in the repair process. Initially, a single layer of regenerating epithelial cells is formed which subsequently becomes multilayered and stratified. In most women, inspection of the cervix 4 months after laser therapy will reveal that the ectocervix is lined with mature squamous epithelium and a new squamocolumnar junction develops at the external os. In a small proportion of cases however, a new transformation zone develops lined by immature metaplastic epithelium which may persist unchanged for many months or years (Byrne *et al.*, 1988).

Several studies have shown that the integrity of the epithelium is restored within a month and repair is generally complete after 3 months. Biopsies taken after 4 months have elapsed should accurately reflect the state of the underlying tissue (Holmquist *et al.*, 1976; MacLean, 1984).

(a) Cytology

A smear taken within a week of ablative therapy for CIN will contain cells showing degenerative changes, e.g. vacuolation, pyknosis and karyorrhexis and peculiar elongated cells similar to those shown in Figures 9.28 and 9.33(b). There may be necrotic debris in the background. As the cervix regenerates, there may be difficulty distinguishing regenerating cells from residual neoplastic cells in the

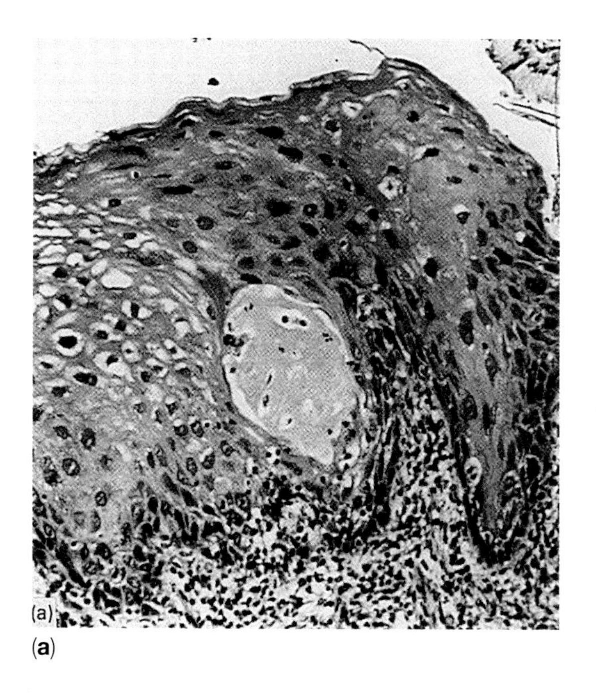

smear. Therefore, it is unwise to evaluate the cervix cytologically within 16 weeks of laser therapy.

9.6.2 Ionizing radiation

In contrast to the transient effect of ablative therapies described in the preceding section, the effect of ionising radiation may persist for many years.

(a) Histology

The non-specific inflammatory changes following irradiation have been mentioned (Section 7.1). Multinucleate histiocytes are often present in large numbers (Figure 9.22). Epithelial cells may show bizarre changes

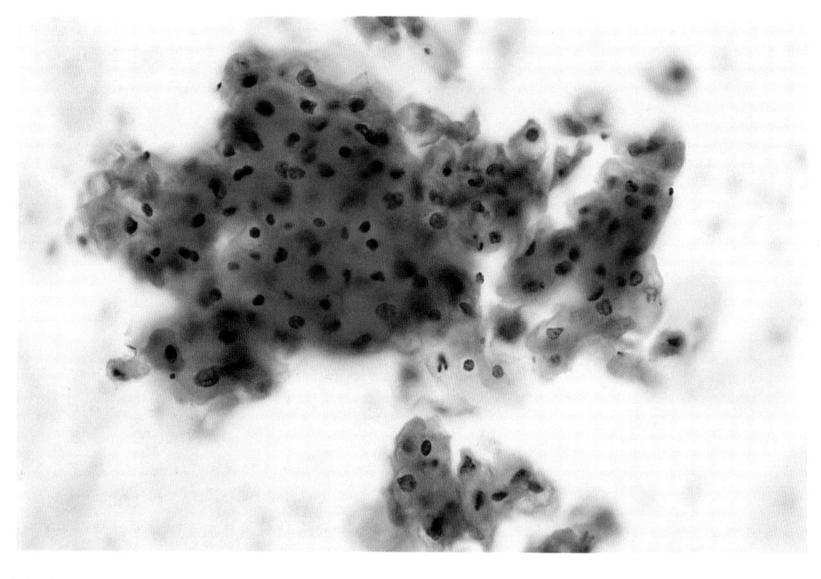

Figure 9.21 (**a**) Dyskeratosis in an area of CIN 3 (H & E, ×190). (**b**) Dyskeratosis in a cervical smear. Note marked variation in nuclear size. Biopsy revealed underlying CIN 3 (Pap, ×630).

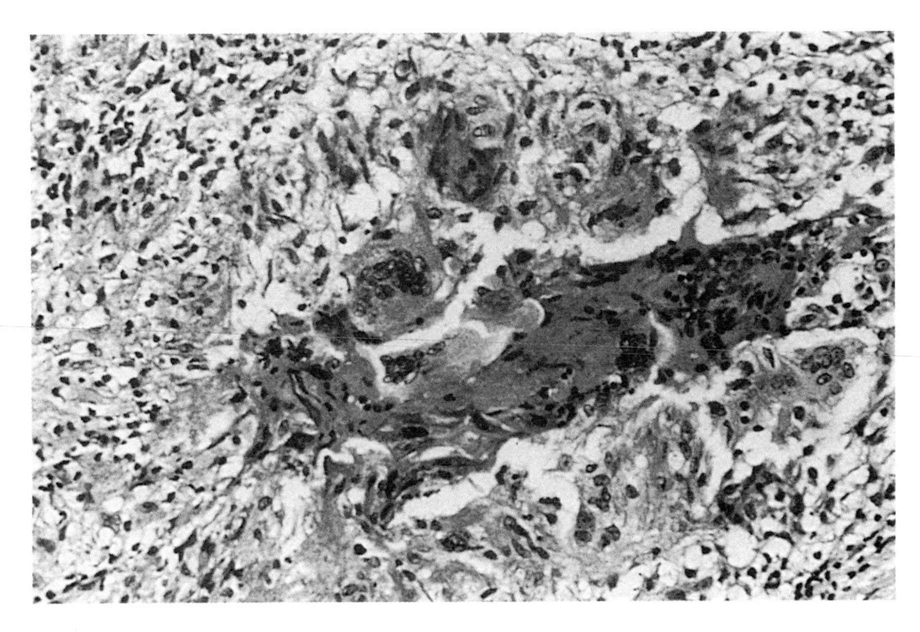

Figure 9.22 Collection of multinucleated giant cell histiocytes in a cervix which had been irradiated for carcinoma (H & E, ×310).

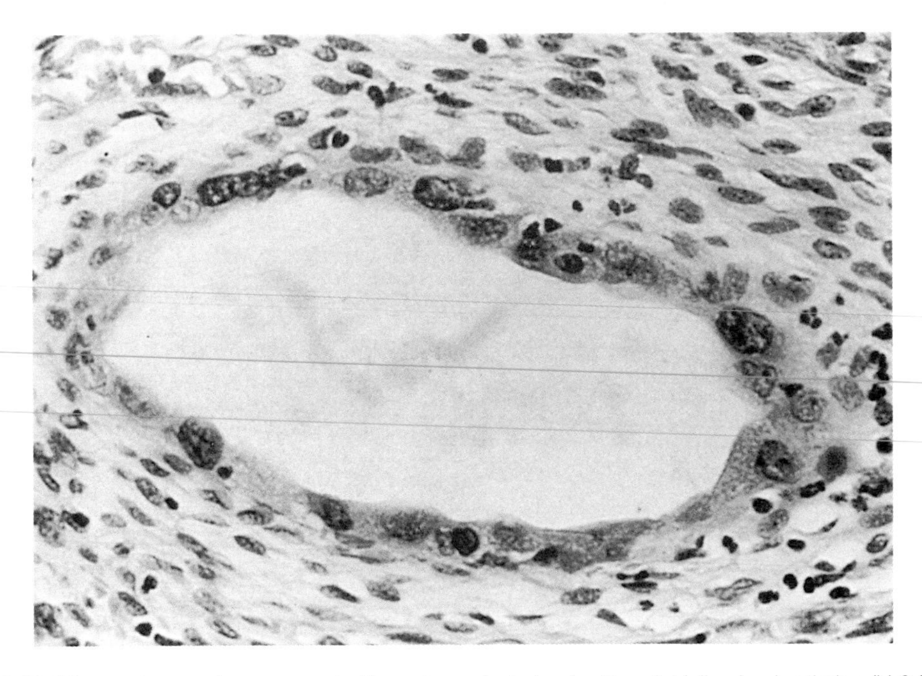

Figure 9.23 Bizarre changes in non-neoplastic endocervical gland cell nuclei following irratiation (H & E, ×480).

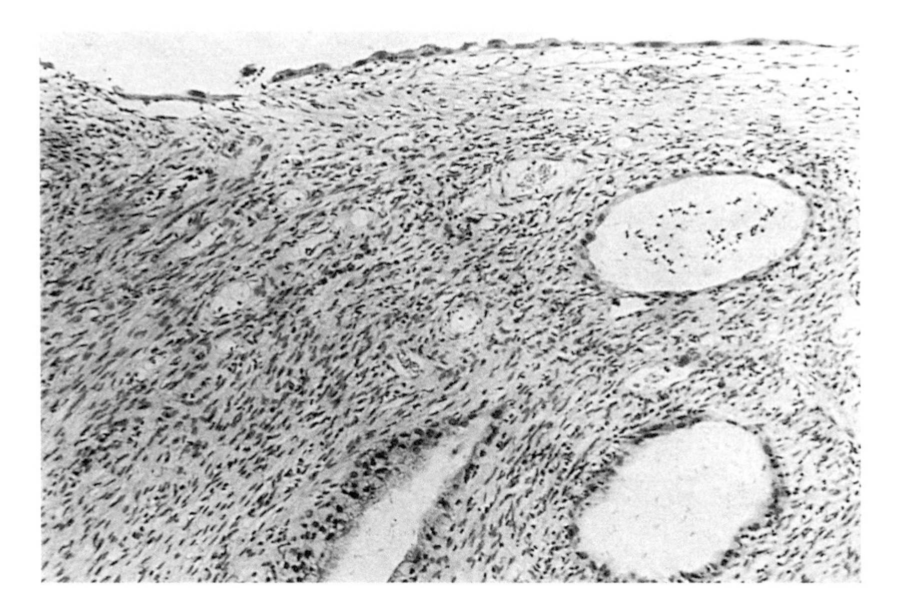

Figure 9.24 Fibrosis following irratidation occurring in the same cervix as shown in Figure 9.23 (H & E, ×120).

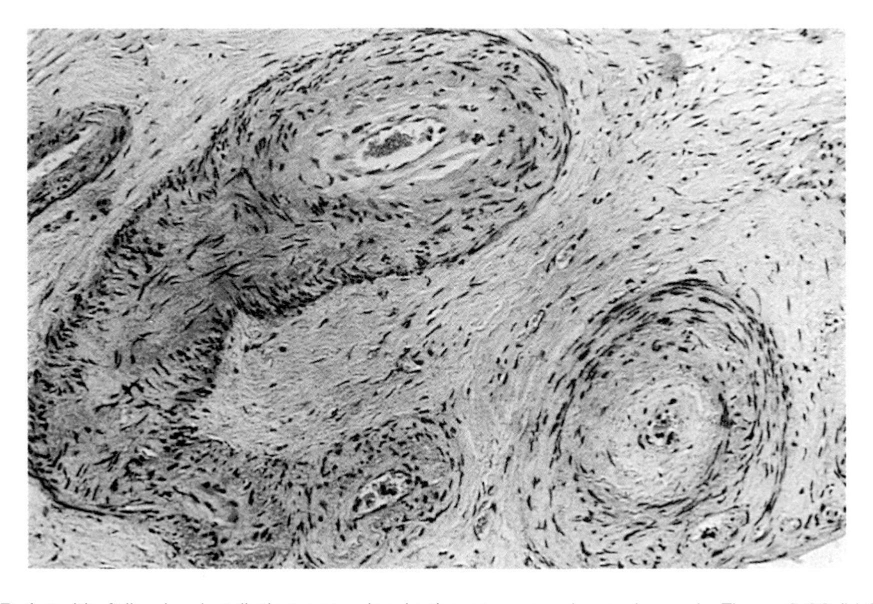

Figure 9.25 Arteritis following irradiation occurring in the same cervix as shown in Figure 9.23 (H & E, ×190).

(Figure 9.23). Fibrosis occurs (Figure 9.24) and also arteritis (Figure 9.25).

(b) Cytology

The non-specific inflammatory changes associated with atrophic cervicitis are a common finding after radiotherapy (Section 7.1). Coarse vacuolation (Figure 9.26(a)), multinucleation (Figure 9.26(b)) and phagocytosis (Figure 9.26(c)) are often a prominent feature of the smear. Enlarged squames with swollen, structureless nuclei similar to those seen in histological sections (Figure 9.26(d–g)) may cause diagnostic difficulty. The presence of abnormal cells in mitosis (Figure 9.27(a,b)) is a reliable guide to recurrence or persistence of malignant disease.

9.6.3 Changes associated with diethylstilboestrol exposure

An association between prenatal exposure to diethylstilboestrol (DES) and related drugs with the development of clear cell adenocarcinoma of the vagina and cervix (Section 14.3) was first reported 25 years ago by Herbst *et al.* (1971) and later by Horowitz *et al.* (1988). Since that time, a number of non-neoplastic changes have been noted in the genital tract in DES-exposed females. These include cervical ectropion, various types of cervicovaginal ridges and the development of ectopic foci of glandular epithelium (adenosis) in the vagina and ectocervix. The frequency with which these benign changes occur depends upon the duration of drug exposure and the stage in pregnancy at which the drug was given. Regression has been observed with age and, in the absence of symptoms, treatment is rarely necessary (Ng *et al.*, 1977).

The most common lesion found in DES-exposed females is ectropion with extension of the glandular epithelium beyond the external os on to the portia vaginalis of the cervix. Metaplastic change is frequently seen and may involve both the surface epithelium and gland crypts so that in histological section the basement membrane of the metaplastic epithelium appears undulating and the dermal papillae

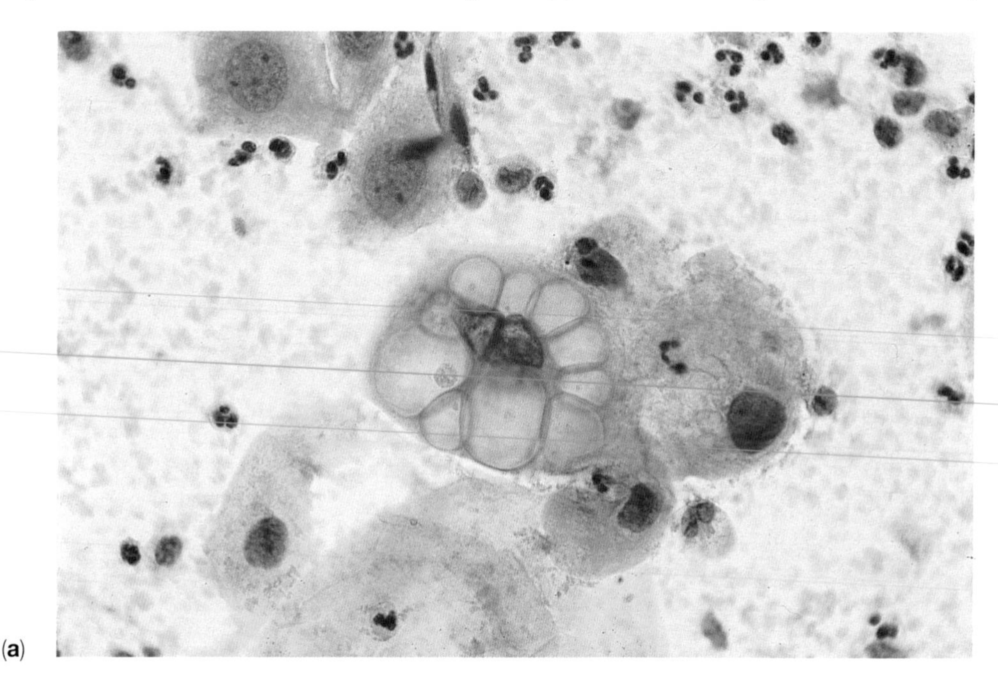

(**a**)

Figure 9.26 Irradiation atypia. (**a**) Coarse vacuolation of parabasal cells (Pap, ×630).

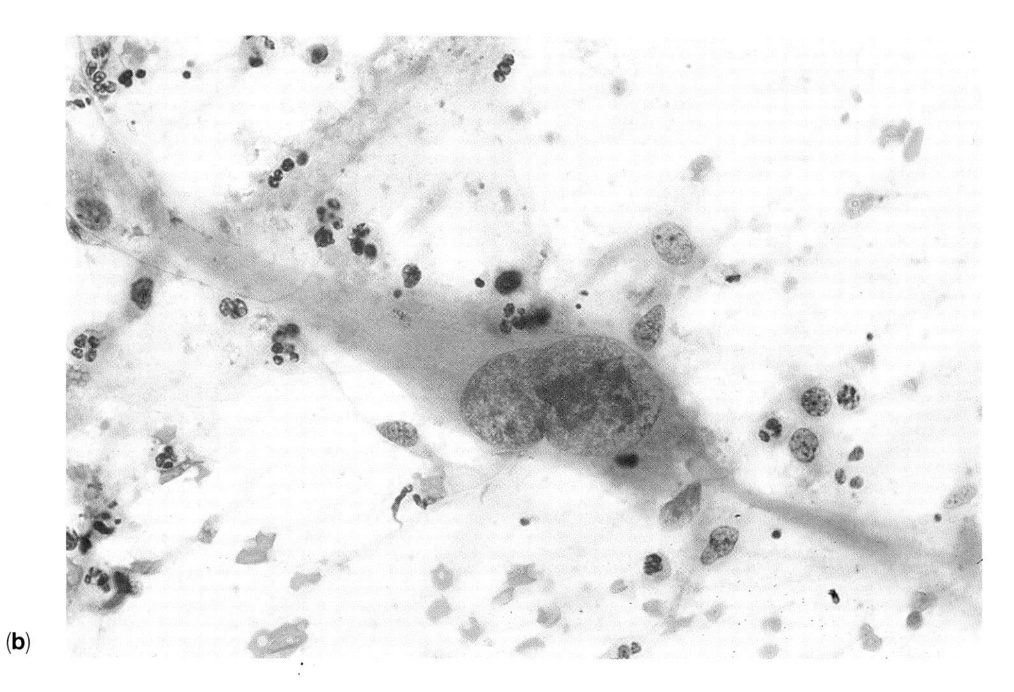

(**b**)

Figure 9.26 Irradiation atypia. (**b**) Bizarre multinucleate cell (Pap, ×630).

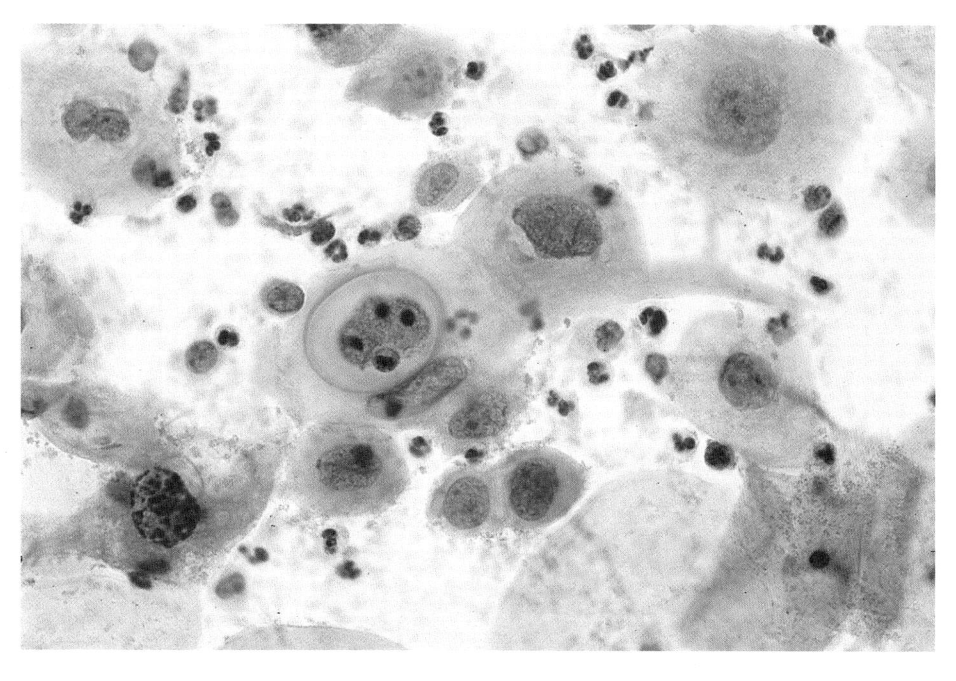

(**c**)

Figure 9.26 Irradiation atypia. (**c**) Radiation changes in parabasal cells. Note nuclear wrinkling, smudged nuclei and phagocytosis (Pap, ×630).

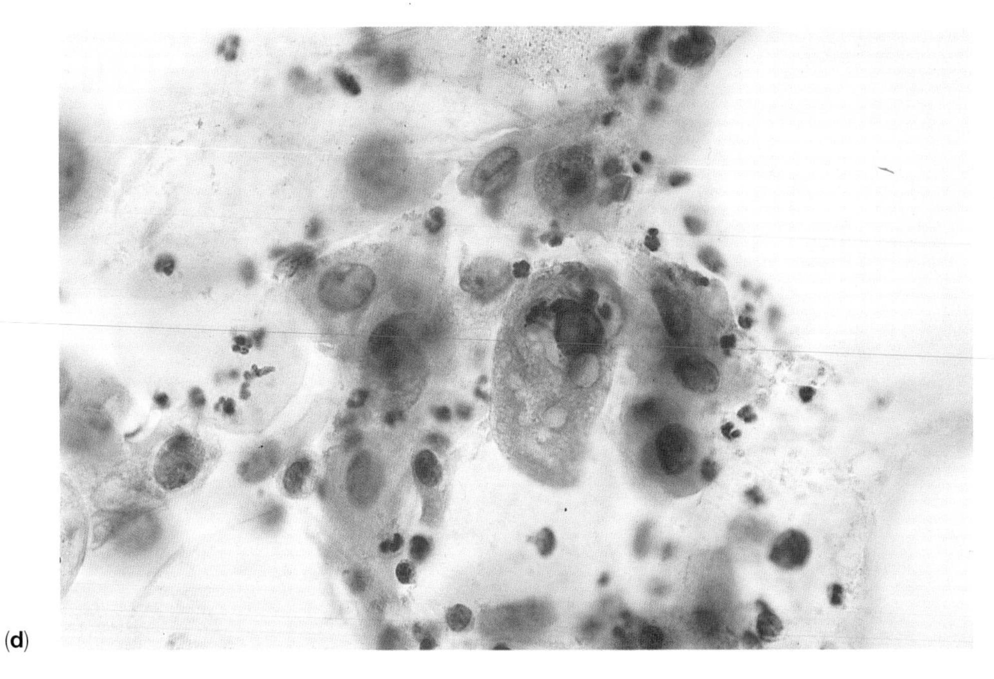

(**d**)

Figure 9.26 Irradiation atypia. (**d**) Abnormal cells with large nuclei and smudged chromatin (Pap, ×630).

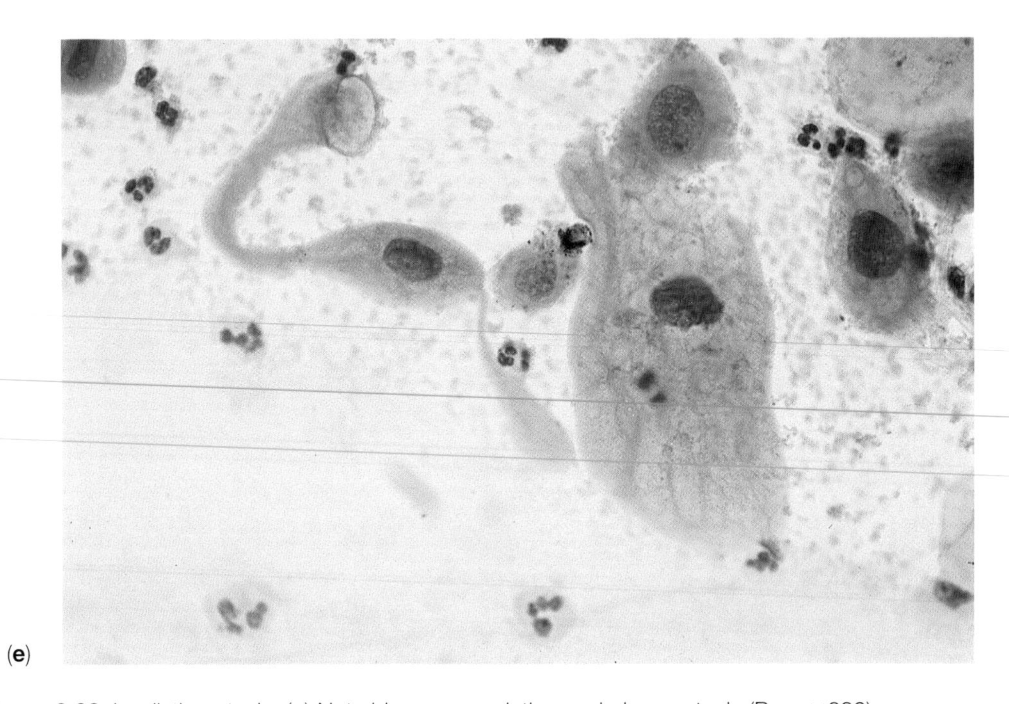

(**e**)

Figure 9.26 Irradiation atypia. (**e**) Note bizarre vacuolation and phagocytosis (Pap, ×630).

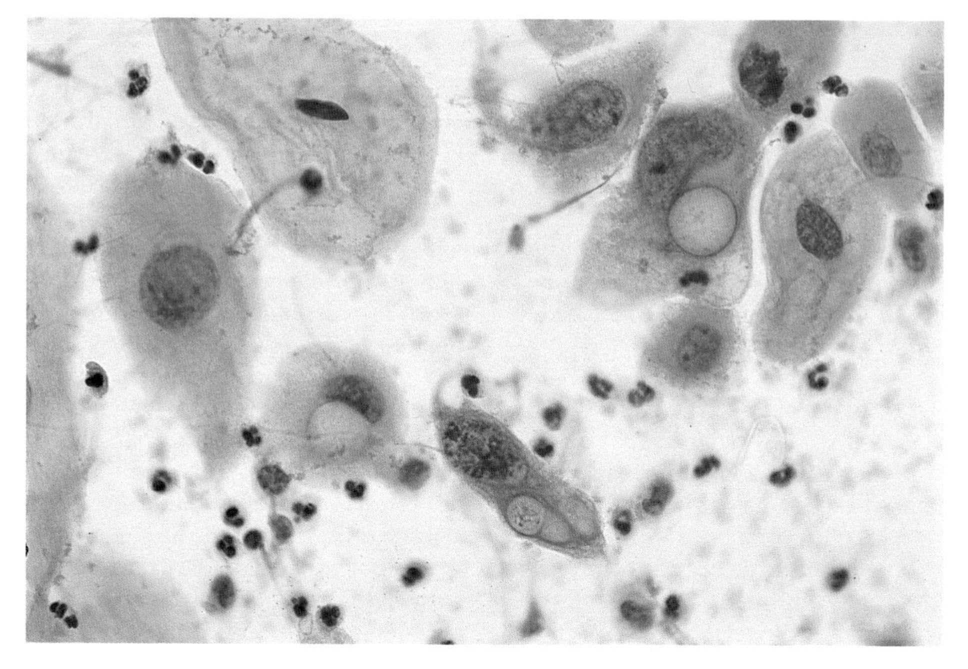

(**f**)

Figure 9.26 Irradiation atypia. (**f**) Note discrete punched out vacuoles and anisonucleosis (Pap, ×630).

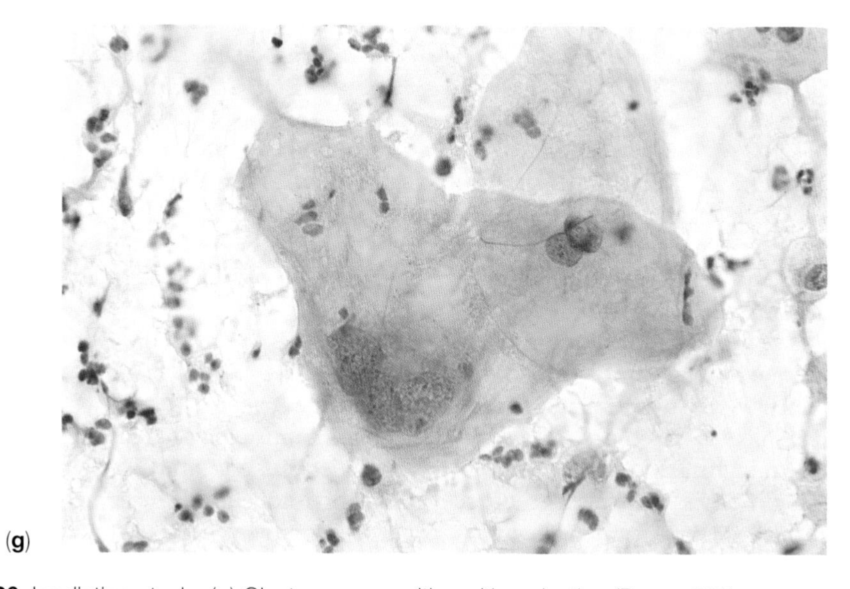

(**g**)

Figure 9.26 Irradiation atypia. (**g**) Giant squames with multi-nucleation (Pap, ×630).

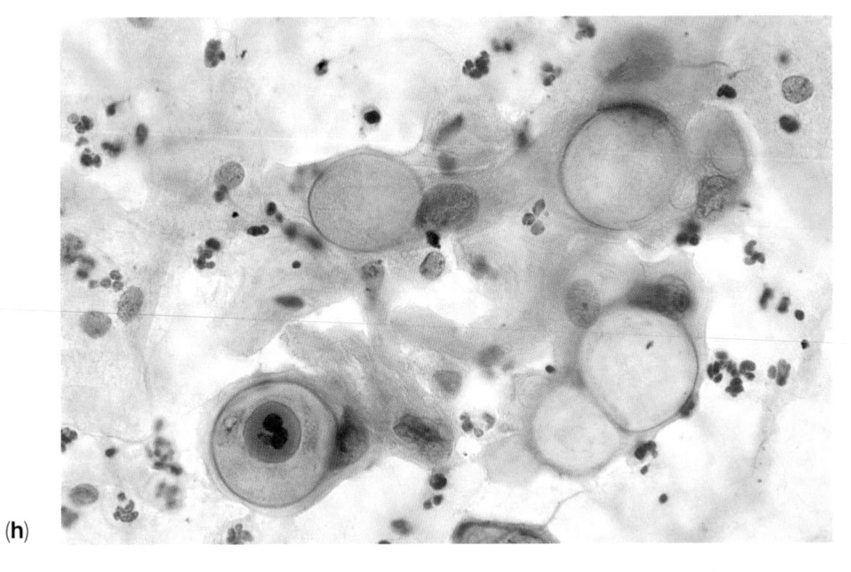

(**h**)

Figure 9.26 Irradiation atypia. (**h**) Note vacuolation and phagocytosis (Pap, ×630).

are very pronounced. The area involved in this type of metaplastic change is sometimes termed the 'congenital transformation zone' and this pattern is not confined to DES-exposed females. Dysplastic changes may supervene. Although Robboy *et al.* (1979) and McDonnell *et al.* (1984) considered that there was no evidence to support the suggestion of Stafl and Mattingly (1974) that these patients have an increased risk of squamous intraepithelial neoplasia, most authors disagree. Ben Baruch *et al.* (1991) found a consistently higher rate of abnormal colposcopic, cytological and histological findings in 89 DES-exposed Israeli women than in 318 women in a control group although the difference only reached statistical significance in women aged between 25 and 34. However, Vessey (1989) noted that the risk of vaginal and cervical neoplasia was significantly different in DES- and non-DES-exposed women. He found that DES-exposed women had twice the risk of neoplasia of non-DES-exposed women. The figures reported by Piver *et al.* (1988) imply that DES-exposed women have an even higher risk of neoplasia than that quoted by Vessey.

Structural abnormalities of the cervix such as the presence of cervical peaks, collars or hoods have been noted in about one-fifth of DES-exposed women. These deformities are composed of fibrous connective tissue covered by columnar, metaplastic or squamous epithelium. Cervicovaginal adenosis is the next most common lesion occurring in about one-third of DES-exposed women. The ectopic glands are usually endocervical in type, although they may resemble Fallopian tube or endometrial mucosa. Clinically, adenosis may be suspected when red spots or patches are noted in the upper vagina. The ectopic glandular epithelium is subject to metaplastic change. Thus, in time, the ectopic glands will be replaced by metaplastic squamous epithelium which is continuous with and virtually indistinguishable from the normal squamous epithelium of the vagina. Cytological studies of

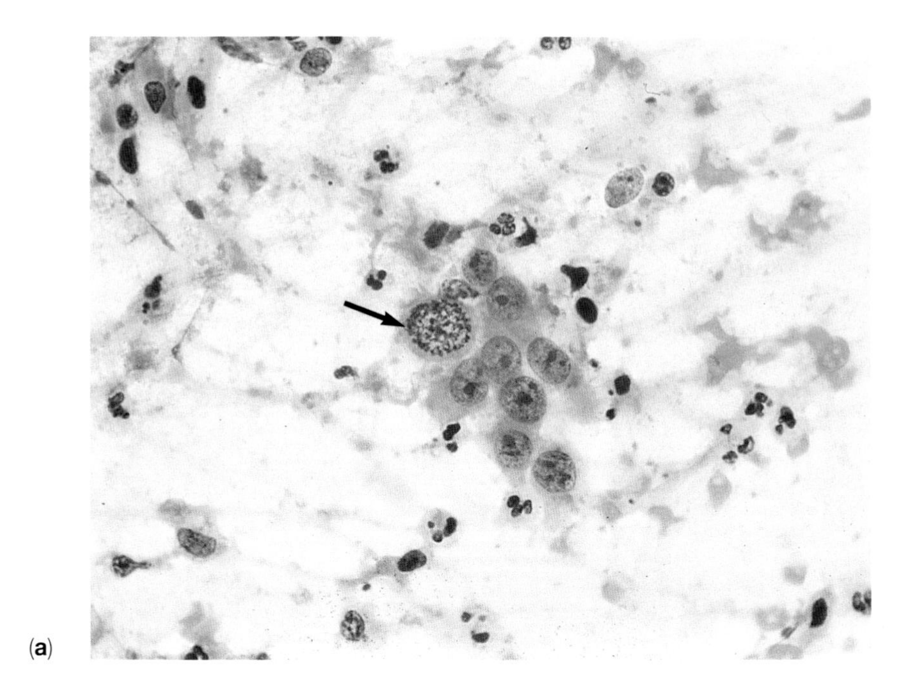

(a)

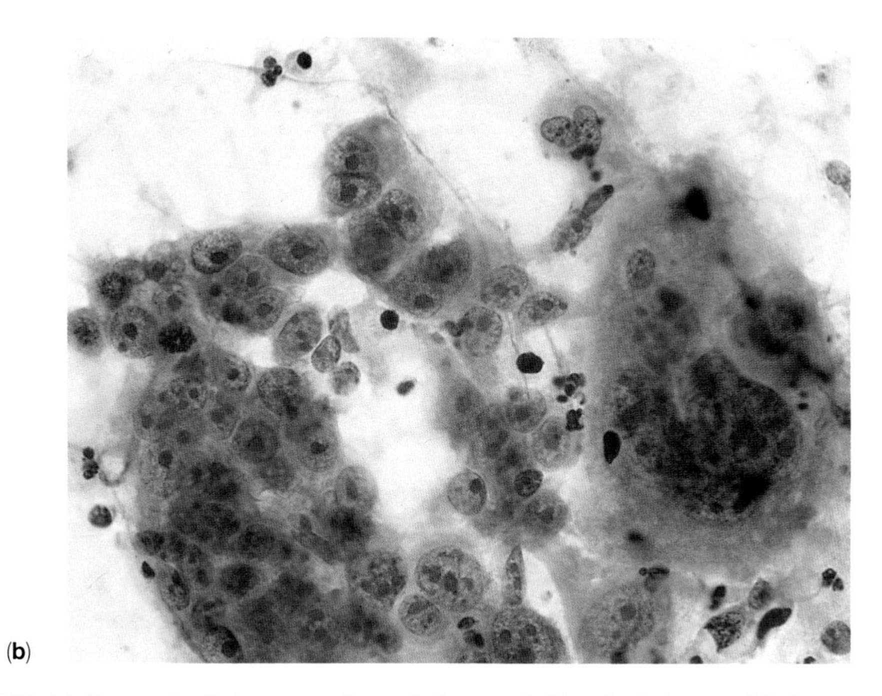

(b)

Figure 9.27 (**a**) Abnormal cells in smear after radiotherapy. Cell in mitosis (arrowed) indicates cells are viable (Pap, ×630). (**b**) Another area from same case showing viable malignant cells.

the cervix or vagina can be helpful in the detection of metaplastic or neoplastic change in the epithelium.

9.6.4 Oral contraceptives and hormone replacement therapy

The use of steroid contraceptives has been associated with the development of microglandular hyperplasia (Section 9.2.1). Very occasionally, atypical cells may be seen in cervical smears (Figure 9.28). An increased risk of cervical neoplasia in women who are long-term users of the contraceptive pill has been reported (Harris *et al.*, 1980; Vessey *et al.*, 1983), although this has not been confirmed by other groups (Worth and Boyes, 1972; Boyce *et al.*, 1977). These conflicting results probably reflect the difficulty in controlling the studies for variables such as sexual activity and until this can be achieved, it is unwise to ascribe a carcinogenic role to the contraceptive pill.

Hormone replacement therapy may result in an unexpectedly well-oestrogenized smear in a post-menopausal woman. The smear may also contain atypical endometrial cells reflecting the effect of hormone replacement therapy on the endometrium.

9.6.5 Intrauterine contraceptive device (IUCD)

(a) Histology

The use of the IUCD has been associated with the development of microglandular hyperplasia, basal hyperplasia (Figure 9.29) and with chronic cervicitis.

(b) Cytology

Endometrial cells are frequently found in smears from IUCD users, reflecting chronic endometritis. The cells may be discrete (Figure 9.30) or in clusters and atypical forms may be seen with enlarged nuclei (Figures 9.31 and 9.32). Large fragments of endocervical epithelium may also be found in smears of IUCD users possibly reflecting chronic cervicitis

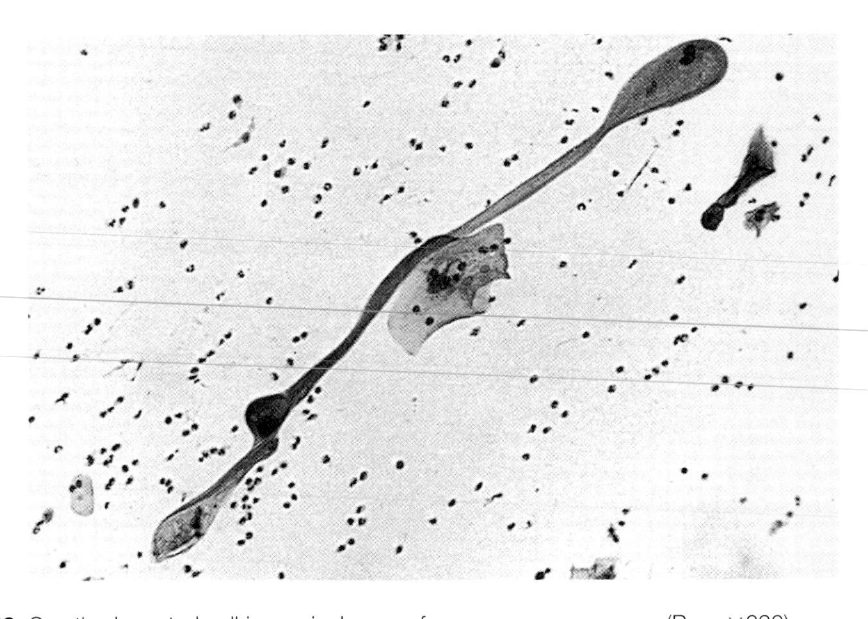

Figure 9.28 Greatly elongated cell in cervical smear from woman on provera (Pap, ×630).

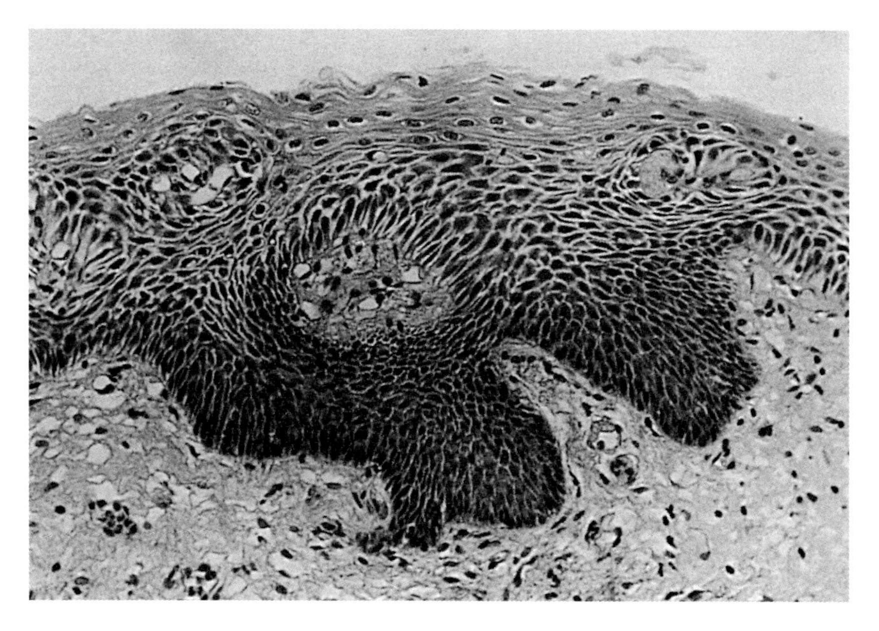

Figure 9.29 Basal cell hyperplasia occurring in association with an IUCD (H & E, ×310).

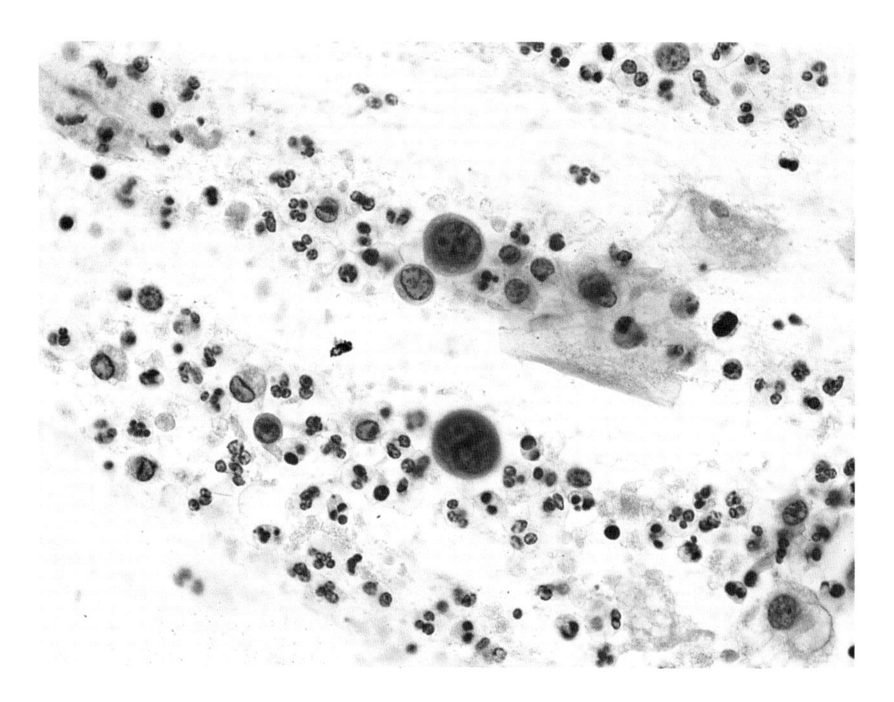

Figure 9.30 Discrete endometrial cells in cervical smear from IUCD user (Pap, ×630).

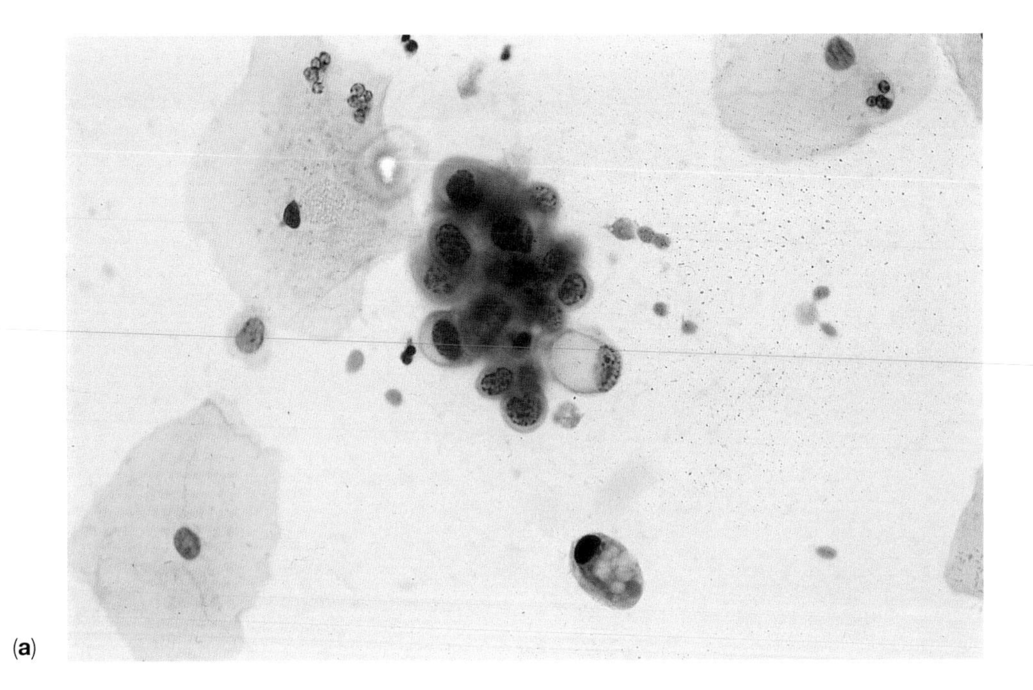

(**a**)

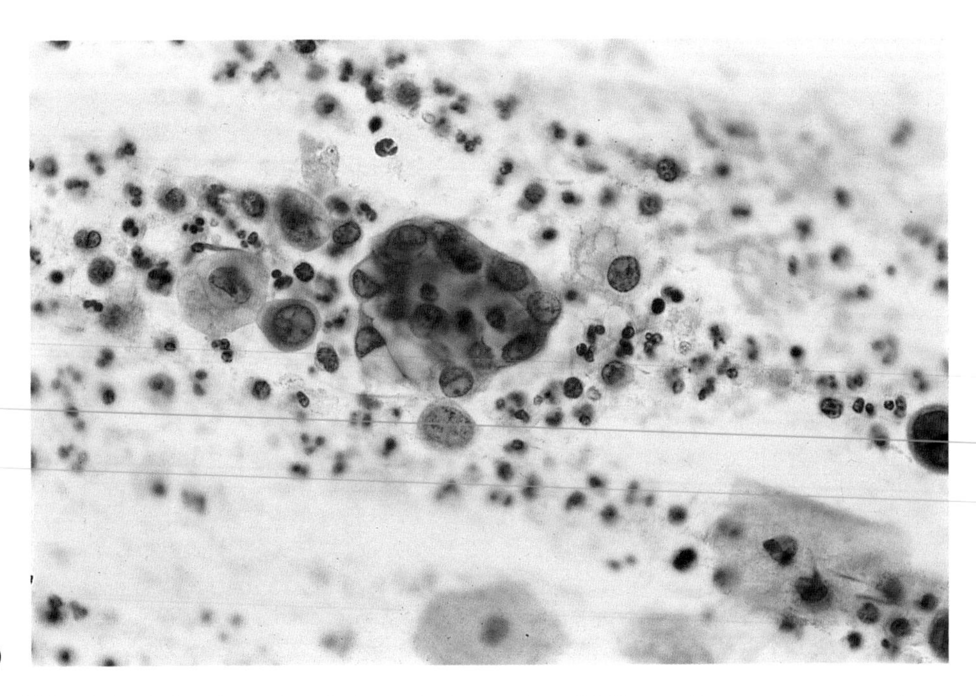

(**b**)

Figure 9.31 (**a**) Atypical endometrial cells in cervical smear from IUCD user. Note coarse chromatin and slightly enlarged nuclei. (**b**) Smear from IUCD user. Note cohesive cluster of atypical endometrial cells (all Pap, ×630).

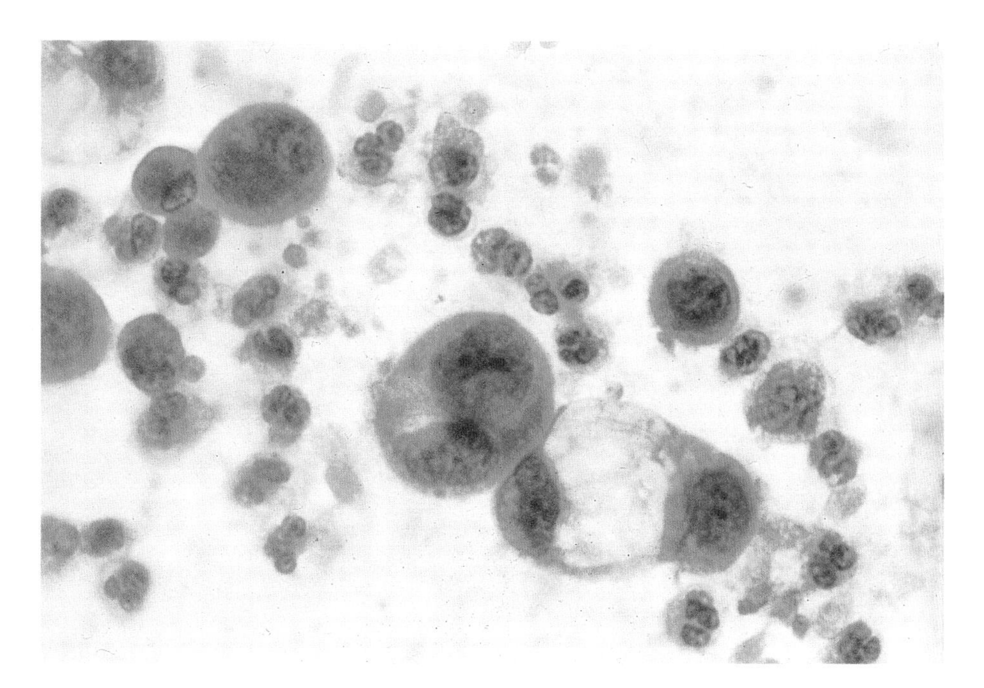

Figure 9.32 Atypical endometrial cells in smear of IUCD user. Note enlarged nucleoli (Pap, ×630).

(Figure 9.33(a)). The filaments of *Actinomyces israelii* grow on the string of the device and may be found in the smears of 1% of users (Figure 7.29). Non-pathogenic amoebae have also been found (Section 6.2.4).

9.6.6 Pessaries

Cervical smears from women wearing a ring pessary may occasionally contain large keratinized squames with a very bizarre shape (Figure 9. 33(b)).

9.6.7 Cytotoxic therapy

The topical application of chemotherapeutic agents such as fluorouracil to the genital mucous membrane in cases of multifocal intraepithelial neoplasia and treatment with busulphan and cyclophosphamide is associated with the development of cellular changes in the epithelium which may themselves mimic cancer (Koss *et al.*, 1965; Koss, 1969). The presence of squamous cells with irregular, hyperchromatic but largely structureless nuclei (Figure 9.34) should raise the suspicion that these cells are the result of treatment. However, it must be remembered that cytotoxic therapy may induce herpes simplex virus and cytomegalovirus reactivation. We have seen polyomavirus-infected transitional cells (Figure 9.35) from the urinary tract contaminating the cervical smear from an immunosuppressed patient (Coleman *et al.*, 1977). Similar inclusion-bearing cells were found in smears of urinary sediment and the diagnosis was confirmed by electron microscopy.

Athanassiadou *et al.* (1992) studied the effect of Tamoxifen on the vaginal epithelium of 33 pre-menopausal and 99 post-menopausal women with primary breast cancer. They determined the karyopyknotic indices (KPI) of vaginal smears before therapy and at monthly intervals during therapy. They showed a slight

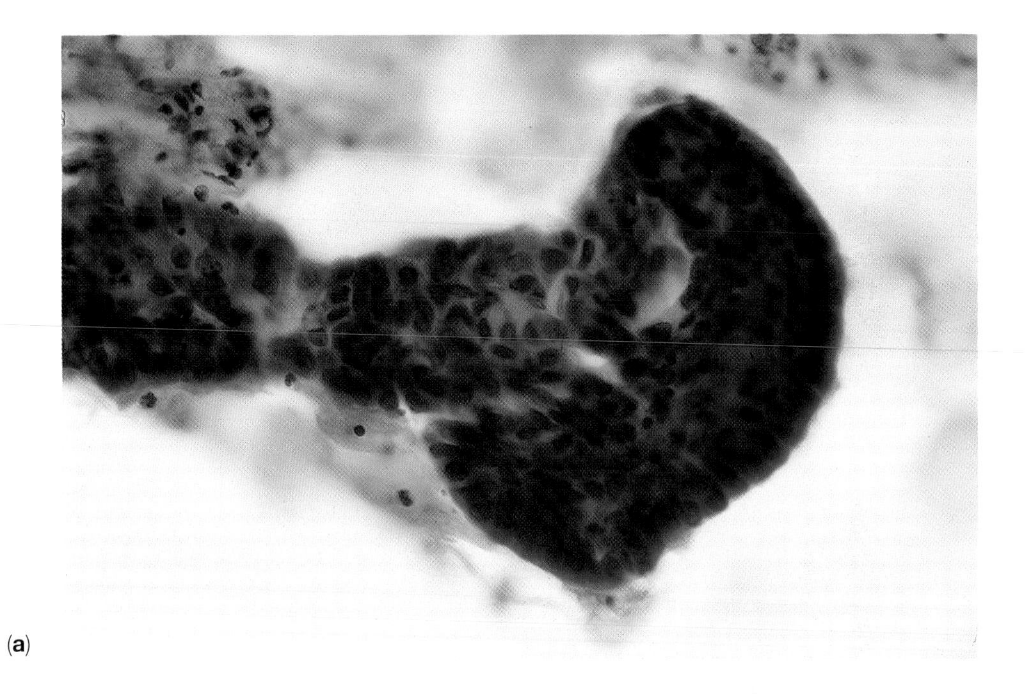

(**a**)

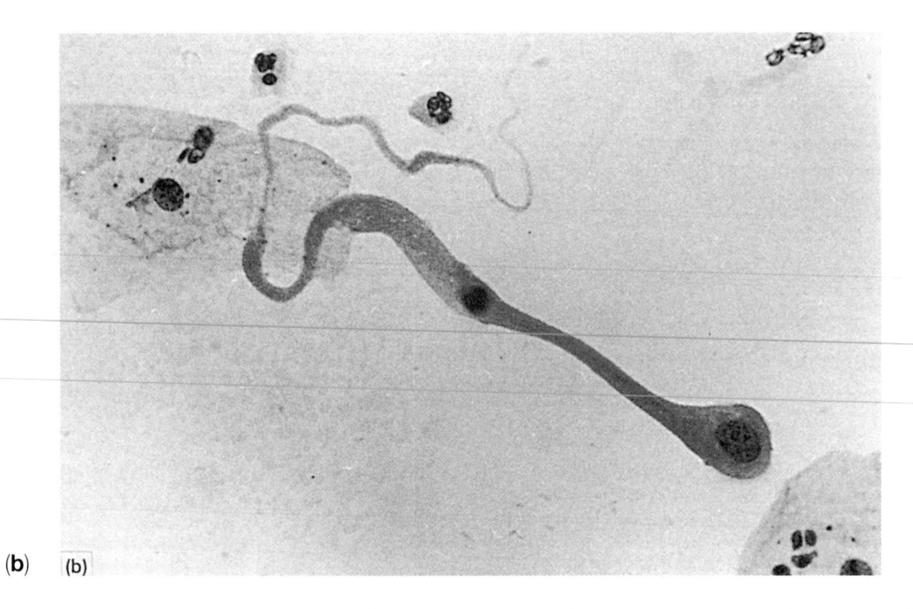

(**b**)

Figure 9.33 (**a**) Fragment of endocervical epithelium in smear of IUCD user (Pap, ×630). (**b**) Odd cell with long tail occurring in association with ring pessary (Pap, ×630).

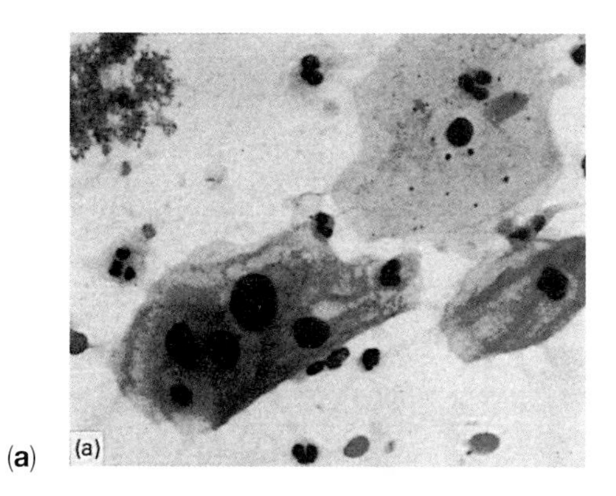
(a)

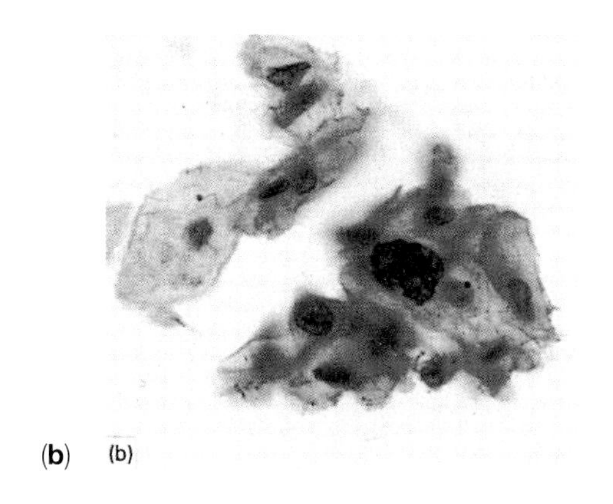
(b)

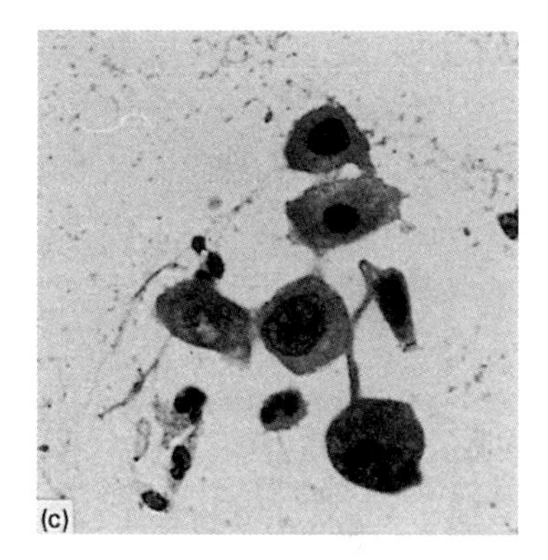
(c)

Figure 9.34 (**a–c**) Squamous cells with structureless, enlarged, hyperchromatic nuclei in smear of patient who had received fluorouracil for multicentric intraepithelial neoplasia of the vagina (Pap, ×630).

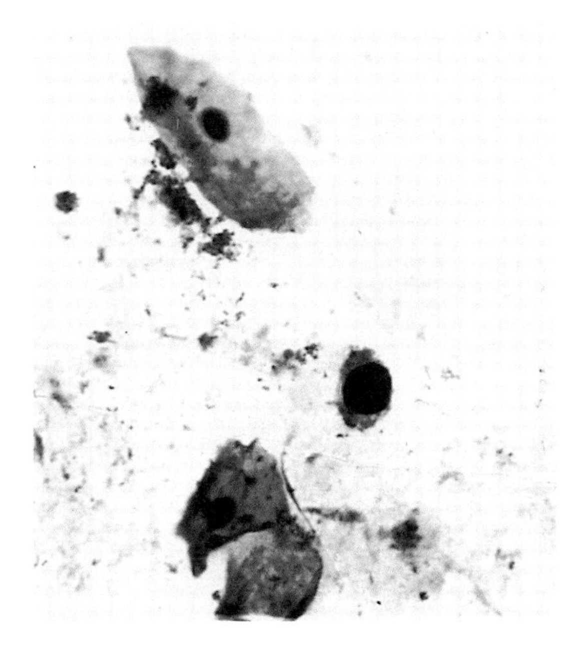

Figure 9.35 Large inclusion-bearing transitional cells contaminating cervical smear from patient receiving cytotoxic therapy for leukaemia. Electron microscopy confirmed the presence of polyomavirus particles in the nucleus (Pap, ×630).

decrease in KPI in the pre-menopausal women and a definite increase in the post-menopausal group. The mechanism by which Tamoxifen achieves this effect is not known. It has been suggested that the drug has an anti-oestrogen effect on the ovaries – hence the reduction in KPI in pre-menopausal women. In post-menopausal women the increase in KPI is thought to reflect stimulation of the pituitary–hypothalamic pathway by the drug. It may result in endometrial proliferation as reflected in the cervical smear shown in Figure 9.36.

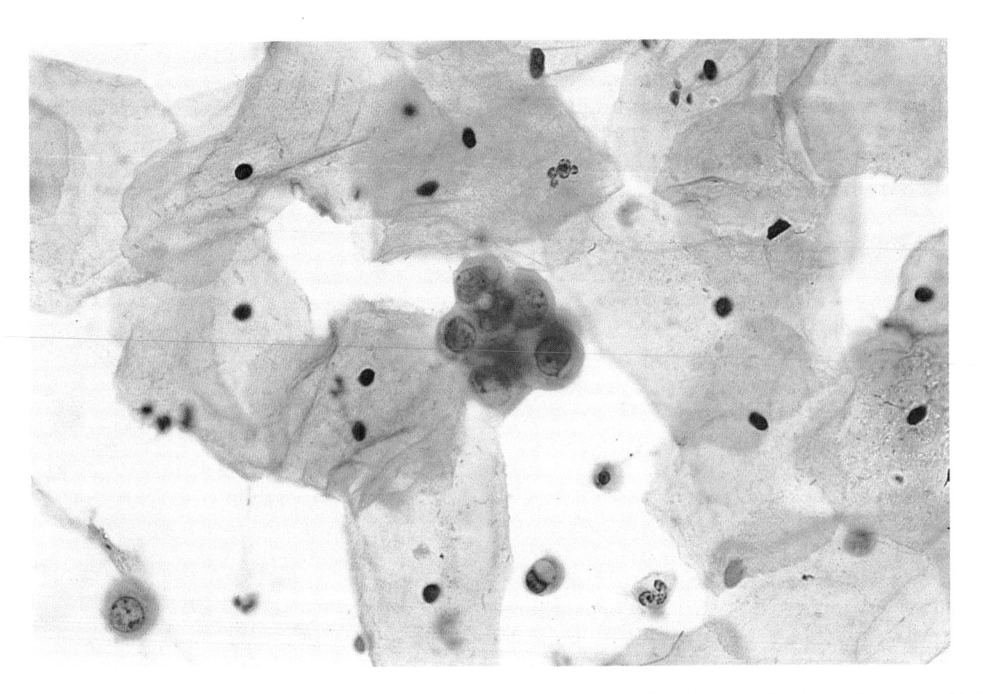

Figure 9.36 Cervical smear from a post-menopausal patient on Tamoxifen therapy for breast cancer. Note the somewhat paradoxical effect of increased maturation of the epithelial cells, and the endometrial cell cluster present. These patients are at increased risk of endometrial pathology, and require careful gynaecological surveillance (Pap, ×630).

REFERENCES

Athanassiadou, P. P., Kyrkou, K. A., Antoniades, L. G. *et al.* (1992) Cytological evaluation of the effect of Tamoxifen in pre-menopausal and post-menopausal women with primary breast cancer by analysis of the karyopyknotic indices of vaginal smears. *Cytopathology*, **3**, 203–208.

Azzopardi, J. G. and Hou, L. T. (1965) Intestinal metaplasia with argentaffin cells in cervical adenocarcinoma. *J. Pathol. Bacteriol.*, **90**, 686–690.

Ben Baruch, G., Rothenberg, D., Modan, M. and Menczer, J. (1991) Abnormal cervical cytologic colposcopic and histologic findings in DES exposed young Israeli Jewish women. *Clin. Exp. Obstet. Gynaecol.* **18**, 71–74.

Boyce, J. G., Lu, T., Nelson, J. H. and Fruchter, R. G. (1977) Oral contraceptives and cervical carcinoma. *Am. J. Obstet. Gynecol.*, **128**, 761–766.

Brown, L. J. R. and Wells, M. (1986) Cervical glandular atypia associated with squamous intraepithelial neoplasia: a premalignant lesion. *J. Clin. Pathol.*, **39**, 22–28.

Burghardt, E. (1973) Early histological diagnosis of cervical cancer, in *Major Problems in Obstetrics and Gynaecology* (ed. E. A. Friedman), W. B. Saunders, Philadelphia, London, Toronto, vol. 6, pp. 61–65.

Burnett, R. A. (1992) Cervical endometriosis and tuboendometrial metaplasia. *Histopathology*, **20**, 279.

Byrne, M., Taylor Robinson, D., Wickenden, C., Malcolm, A. D. B., Anderson M. L. and Coleman, D. V. (1988) Prevalence of HPV subtypes in the cervix before and after laser therapy. *Br. J. Obstet. Gynaecol.*, **95**, 201–202.

Coleman, D. V., Russell, W. J., Hodgson, J., Tun Pe and Mowbray, J. F. (1977) Human papovavirus in Papanicolaou smears of urinary sediment detected by transmission electron microscopy. *J. Clin. Pathol.*, **30**, 1015–1020.

Coppleson, M. and Reid, B. (1967) *Preclinical Carcinoma of the Cervix Uteri*, Pergamon Press, Oxford, pp. 223–253.

Ducatman, B. S., Wang, H. H., Jonasson, J. G., Hogan, C. L. and Antonioli, D. A. (1993) Tubal metaplasia: a cytological study with comparison to other neoplastic and non neoplastic conditions of the endocervix. *Diagn. Cytopathol.*, **9**, 98–103.

Ferenczy, A. (1982) Benign lesions of the cervix, in *Pathology of the Female Genital Tract*, 2nd edn (ed. A. Blaustein), Springer-Verlag, New York, Heidelberg, Berlin, pp. 138–155.

Fluhmann, C. F. (1961) *The Cervix Uteri and its Diseases*, W. B. Saunders, Philadelphia, pp. 58–59.

Fray, R. E., Husain, O. A., To, A. C., Watts, K. C., Lader, S., Rogers, G. T., Taylor-Papadimitriou, J. and Morris, N. F. (1984), The value of immunohistochemical markers in the diagnosis of cervical neoplasia. *Br. J. Obstet. Gynaecol.*, **91**, 1037–1041.

Fruin, A. H. and Tighe, J. R. (1967) Tubal metaplasia of the endometrium. *J. Obstet. Gynaecol. Br. Commonwlth*, **74**, 93–97.

Harris, R. W. C., Brinton, L. A., Cosvdell, R. H., Skegg, D. C. G., Smith, P. G., Vessey, M. P. and Doll, R. (1980) Characteristics of women with dysplasia or carcinoma *in situ* of the cervix uteri. *Br. J. Cancer*, **42**, 359–369.

Herbst, A. L., Ulfelder, H. and Poskanzer, D. C. (1971) Adenocarcinoma of the vagina: association of maternal stilboestrol therapy with tumour appearance in young women. *N. Engl. J. Med.*, **284**, 878–881.

Hirschowitz, L., Eckford, D., Philpotts, B. and Midwinter, A. (1994) Cytological changes associated with tuboendometrioid metaplasia of the uterine cervix. *Cytopathology*, **5**, 1–8.

Holmquist, N. D., Bellina, J. H. and Danos, M. L. (1976) Vaginal and cervical cytologic changes following laser treatment. *Acta Cytol.*, **20**, 290–294.

Horowitz, R. I., Viscoli, C. M., Merino, M., Brennan, T. A., Flannery, J. T. and Robboy, S. J. (1988) Clear cell adenocarcinoma of the vagina and cervix: incidence, undetected disease and diethylstilbestrol. *J. Clin. Epidemiol.*, **41**, 593–597.

Ismail, S. M. (1992) Cervical endometriosis and tuboendometrioid metaplasia. *Histopathology*, **20**, 279–280.

Jha, R. S., Wickenden, C., Anderson, M. C. and Coleman, D. V. (1984) Monoclonal antibodies for the histopathological diagnosis of cervical neoplasia. *Br. J. Obstet. Gynaecol.*, **91**, 483–488.

Jones, M. A., Young, R. H. and Scully, R. E. (1991) Diffuse laminar endocervical glandular hyperplasia. A benign lesion often confused with adenoma malignum (minimal deviation adenocarcinoma). *Am. J. Surg. Pathol.*, **15**, 1123–1129.

Jones, M. W. and Silverberg, S. G. (1989) Cervical adenocarcinoma in young women: possible relationship to microglandular endocervical hyperplasia and use of oral contraceptives. *Obstet. Gynecol.*, **73**, 984–989.

Koss, L. G. (1969) Some effects of alkylating agents on epithelia in man and in an experimental system in the rat. *Ann. N. Y. Acad. Sci.*, **163**, 931–935.

Koss, L. G., Melamed, M. R. and Mayer, K. (1965) The effect of busulphan on human epithelia. *Am. J. Clin. Pathol.*, **44**, 385–397.

Maclean, A. B. (1984) Healing of cervical epithelium changes after laser treatment of cervical intraepithelial neoplasia. *Br. J. Obstet. Gynaecol.*, **91**, 697–706.

Michael, H., Sutton, G., Hull, M. T. and Roth, L. M. (1986) Villous adenoma of the uterine cervix associated with invasive adenocarcinoma; a histologic, ultrastructural and immunohistochemical study. *Int. J. Gynecol. Pathol.*, **5**, 163–169.

McDonnell, J. M., Emens, J. M. and Jordan, J. A. (1984) The congenital cervicovaginal transformation zone in young women exposed to diethylstilboestrol *in utero*. *Br. J. Obstet. Gynaecol.*, **91**, 574–579.

Ng, A. B. P., Reagan, J. W., Nadji, M. and Greening, S. (1977) Natural history of vaginal adenosis in women exposed to diethylstilboestrol *in utero*. *J. Reprod. Med.*, **18**, 1–13.

Nichols, T. M. and Fidler, H. K. (1971) Microglandular hyperplasia in cervical cone biopsies taken for suspicious and positive cytology. *Am. J. Clin. Pathol.*, **56**, 424–429.

Novotny, D. B., Maygarden, S. J., Johnson, D. E. and Frabel, W. J. (1992) Tubal metaplasia. A frequent potential pitfall in cytologic diagnosis of endocervical glandular dysplasia on cervical smears. *Acta. Cytol.*, **36**, 1–10.

Piver, M. S., Lele, S. B., Baker, T. R. and Dandeck, A. (1988) Cervical and vaginal cancer detection at a regional diethylstilbestrol (DES) screening clinic. *Cancer Detect. Prevent*, **11**, 197–202.

Poulsen, H. E., Taylor, C. W. and Sobin, L. H. (1975) *Histological Typing of Female Genital Tract Tumours*. WHO, Geneva, pp. 61–62.

Robboy, S. J. and Welch, W. R. (1977) Microglandular hyperplasia in vaginal adenosis associated with oral contraceptives and prenatal diethylstilboestrol exposure. *Obstet. Gynecol.*, **49**, 430–434.

Robboy, S. J., Kaufman, R. H., Prat, J. *et al.* (1979) Pathological findings in young women enrolled in the National Co-operative Diethylstilboestrol Adenosis (DESAD) project. *Obstet. Gynecol.*, **53**, 309–317.

Scully, R. E., Bonfiglio, R. J. and Kurman, R. J. (1994) *WHO International histological classification of tumours. Histological typing of female genital tract tumours*. Springer-Verlag, Berlin, pp. 39–54.

Stafl, A. and Mattingly, R. F. (1974) Vaginal adenosis: a precancerous lesion? *Am. J. Obstet. Gynecol.*, **120**, 666–673.

Suh, K.-S. and Silverberg, S. G. (1990) Tubal metaplasia of the uterine cervix. *Int. J. Gynecol. Pathol.*, **9**, 122–128.

Trowell, J. E. (1985) Intestinal metaplasia with argentaffin cells in the uterine cervix. *Histopathology*, **9**, 551–559.

Vessey, M. P. (1989) *Epidemiological studies of the effects of diethylstilboestrol.* IARC Scientific Publications, Lyon, **96**, 335–348.

Vessey, M. P., Lawless, M., McPherson, K. and Yeats, D. (1983) Neoplasia of the cervix uteri and contraception: a possible adverse effect of the pill. *Lancet*, **2**, 930–934.

Worth, A. J. and Boyes, D. A. (1972) A case control study into the possible effect of birth control pills on preclinical carcinoma of the cervix. *J. Obstet. Gynecol.*, **79**, 673–679.

Young, R. H. and Scully, R. E. (1989) Atypical forms of microglandular endocervical hyperplasia of the cervix simulating carcinoma. A report of five cases and review of the literature. *Am. J. Surg. Pathol.*, **13**, 50–56.

10

Cervical intraepithelial neoplasia (CIN)

10.1 TERMINOLOGY

The concept of early neoplastic change confined to the cervical epithelium originated at the turn of the century with Rubin's description of incipient cancer (Rubin, 1910). He described histological changes in the cervical epithelium which he believed were precursors of invasive squamous carcinoma and which, if left untreated, would develop into the infiltrating lesion. Subsequently the term 'carcinoma *in situ*' was used to describe these changes (Schottlander and Kermauner, 1912; Broders, 1932), but synonyms such as intraepithelial carcinoma, Bowen's disease of the cervix, incipient cancer and surface cancer were also used for many years.

The precursor lesions described by Rubin were characterized by the complete replacement of the normal cervical epithelium with abnormal cells which bore a striking resemblance to the malignant cells in invasive cancer. With the introduction of cytology and the widespread use of colposcopy and colposcopic biopsy it became clear that the range of abnormality of the cervical epithelium was much wider than was anticipated from the early studies and disordered growth patterns less severe than those of carcinoma *in situ* were common. Reagan *et al.* (1953) introduced the term 'dysplasia' to describe these lesions and reported that the majority of dysplastic lesions, if left untreated, regressed or remained unchanged for many years. Poulsen *et al.* (1975) defined dysplasia for the World Health Organization as a lesion in which part of the thickness of the epithelium is replaced by cells showing varying degrees of atypia. The lesions were further graded as mild, moderate and severe, although there were no nationally agreed criteria for these gradings.

Although the terms dysplasia and carcinoma *in situ* are commonly applied in clinical practice, there are many pathologists and gynaecologists who have reservations about their use (Koss, 1978). The terms lack precision so that reporting is a very subjective exercise. This has been illustrated in several studies in which two pathologists reporting on the same specimen frequently give very different opinions (Siegler, 1961; Kirkland, 1963; Cocker *et al.*, 1968). A particularly contentious point has been the significance of surface differentiation in an otherwise undifferentiated neoplastic epithelium. There was often disagreement on whether the lesion should be classified as severe dysplasia or carcinoma *in situ*.

The dual terminology has led gynaecologists to assume that carcinoma *in situ* and dysplasia are two biologically distinct lesions with different malignant potentials and that dysplastic lesions may not require treatment. Further studies have shown that the behaviour of any individual lesion cannot be accurately predicted from the morphological findings, and that cases of dysplasia if left untreated may progress to carcinoma *in situ* (Kottmeier, 1961; Fidler *et al.*, 1968). Furthermore, it is now accepted that invasive carcinoma can develop directly from dysplasia without prior progression to carcinoma *in situ* (Burghardt, 1976).

One concept of dysplasia and carcinoma *in situ is* that they are a lesional continuum in which one abnormality merges into the next. Support for this concept has been provided by many laboratory and clinical studies: radioautography (Richart, 1963), time lapse cinematography (Richart and Lerch, 1966; Richart *et al.*, 1967), chromosome studies (Kirkland *et al.*, 1967), DNA cytophotometry (Wilbanks *et al.*, 1967), biological behaviour (Fox, 1967; Hulka, 1968; Richart and Barron, 1969), electron microscopy (Shingleton *et al.*, 1968) and light microscopy (Burghardt, 1973; Poulsen *et al.*, 1975). As a result there was pressure from gynaecologists and pathologists for a terminology that would more accurately reflect this view. In 1967, Richart suggested Cervical Intraepithelial Neoplasia (CIN) as a single descriptive term which would embrace all grades of dysplasia as well as carcinoma *in situ* under a single disease heading and accurately convey the morphological unity and malignant potential of these lesions. This proposal was gradually accepted by gynaecologists and pathologists and the CIN terminology has been widely adopted in clinical practice. Three grades of CIN are recognized (Buckley *et al.*, 1982; Ferenczy, 1982):

1. CIN 1 corresponding to mild dysplasia;
2. CIN 2 corresponding to moderate dysplasia; and
3. CIN 3 corresponding to both severe dysplasia and carcinoma *in situ*.

The site at which these changes occur is the metaplastic epithelium of the transformation zone in the region of the squamocolumnar junction. In younger women, this junction is in the vicinity of the external os or on the ectocervical surface. In older women, it may be situated well within the endocervical canal. CIN occurring at the latter site is liable to be missed by a routine cervical scrape using the usual Ayre spatula (Section 2.1.1)

10.1.1 Bethesda and WHO terminology

The Bethesda group (Luff, 1992) which is concerned with the nomenclature used to report cervical smears has recently suggested that the lesions currently categorized as dysplasia/carcinoma *in situ* or cervical intraepithelial neoplasia (CIN) be known collectively as **squamous intraepithelial lesions** (SIL). The term **low-grade squamous intraepithelial lesion** (LSIL) is recommended to describe lesions currently categorized as mild dysplasia or CIN 1 and the term **high-grade intraepithelial lesions** (HSIL) should be used to describe the lesions formerly categorized as moderate or severe dysplasia, carcinoma *in situ* or CIN 2 and CIN 3. Cellular changes characteristic of human papillomavirus infection are included in the LSIL category. It was anticipated that this approach would promote consensus in the classification of these lesions and thereby make it easier to standardize treatment. The Bethesda working group also justify these changes on the grounds of distribution of HPV types and clinical behaviour. Several studies have shown that LSIL harbours HPV 6 and 11 and are likely to regress, whereas HSIL harbours high-risk HPV DNA (HPV 16 and 18) and has a greater risk of progression to invasive cancer (see also pp. 14 and 15).

The Committee on Classification and Nomenclature of the International Society of Gynaecological Pathologists have endorsed the use of the term **'squamous intraepithelial lesions'** to describe the group of lesions comprising cervical intraepithelial neoplasia (CIN) and provides a description as follows:

> Squamous intraepithelial lesions (are) characterised by disordered maturation and nuclear abnormalities such as loss of polarity, pleomorphism, coarsening of nuclear chromatin, irregularities of the nuclear membrane and mitotic figures, including atypical forms, at various levels in the epithelium. The lesions have been subdivided into three or four grades depending on their extent and severity. They typically occur in the transformation zone and usually involve endocervical glands and surface epithelium. A small minority begin in the ectocervical epithelium.
>
> Scully *et al.* (1994)

As the SIL two-tier system is not universally accepted, we have retained the term cervical intraepithelial neoplasia (CIN) to describe this group of lesions.

10.2 DIAGNOSTIC CRITERIA

10.2.1 Histological features of CIN

In all cases of CIN, the normal epithelium of the surface and crypts is replaced by neoplastic cells showing varying degrees of differentiation. Undifferentiated cells of basaloid or parabasaloid type may occupy the whole thickness of the epithelium or may differentiate as they approach the surface. All the neoplastic cells have abnormal nuclei. The nuclei of the undifferentiated neoplastic cells are large, irregular and vary in size and shape. The chromatin content is increased and abnormal in structure and the nucleocytoplasmic ratio is very high. As the neoplastic cells differentiate, the amount of cytoplasm increases; nuclear pleomorphism and irregularity persist, although there is usually a decrease in nuclear size. Disorganized growth results in loss of cell polarity. Mitotic activity is disturbed, being no longer confined to the basal layer and atypical mitoses may be seen at any level.

10.2.2 Histological grading of CIN

This is based on the proportion of the epithelial thickness occupied by basaloid or parabasaloid neoplastic cells, viz.: not more than the lower third in CIN 1; not more than the lower two-thirds in CIN 2; more than two-thirds of the full thickness in CIN 3.

Secondary features which are of value in determining the grade of CIN are the degree of nuclear abnormality of the neoplastic cells, and their potential for normal or abnormal mitotic division. The degree of nuclear abnormality is best regarded as a function of nuclear size, shape and chromatin content. Mitotic activity can be assessed by the frequency with which mitotic figures occur, the height in the epithelium at which mitoses may be found and the presence of abnormal metaphases, e.g. three group and multipolar metaphases.

The degree of epithelial abnormality often varies from one zone of neoplasia to the next and it is not uncommon for CIN 1, 2 and 3 to be present in the same biopsy. The three gradings are described in more detail in Sections 10.3, 10.4 and 10.5.

10.2.3 Cytological diagnosis of CIN

Neoplastic cells from the surface of the CIN lesion are found in the smear. They can be distinguished from the normal cells in the smear by the following features:

1. disproportionate nuclear enlargement usually resulting in abnormal nucleocytoplasmic ratio;
2. variation in nuclear size and shape;
3. irregularity of nuclear outline;
4. hyperchromasia (the chromatin may appear stippled or clumped);
5. multinucleation (only significant if accompanying other changes);
6. large, irregular and sometimes multiple nucleoli.

Papanicolaou invented the word 'dyskaryosis' (literally 'abnormal nucleus') to describe any epithelial cell nucleus showing these changes. We reserve this term to describe cells which we consider have been shed from a neoplastic lesion. The degree of dyskaryosis may be mild, moderate or severe depending on nuclear size, shape, chromatin content and structure and nucleocytoplasmic ratio. Mild dyskaryosis is found in the surface cells shed from an area of CIN 1, moderate dyskaryosis reflects the presence of CIN 2, and severe dyskaryosis reflects the presence of CIN 3. Thus, the cytological picture can often predict the probable histological picture (Section 2.4). Should the cervix contain foci of CIN 1, 2 and 3, the smear may contain cells having a mild, moderate and severe dyskaryosis.

The degree of dyskaryosis is more readily assessed in cells which are discrete or in small clumps. Sheets or large clusters of dyskaryotic cells may be more difficult to define, as cytoplasmic boundaries are less clear and nuclei may be overlapping. Many of the morphological changes which are characteristic of dyskaryotic cells are also found in epithelial cells in inflammatory smears. The difference is mainly one of degree since inflammatory cell nuclei rarely show the pleomorphism of neoplastic cells. If there is an element of doubt as to whether a cell shows inflammatory changes or dyskaryosis, we suggest the term 'borderline changes' be used (see Section 2.4 and also Section 7.1.2).

The reader may, in the older textbooks, come across a different terminology for the neoplastic cells in cervical smears. They are described as showing superficial, intermediate or parabasal dyskaryosis depending upon the degree of cytoplasmic differentiation of the cell (Papanicolaou, 1949). We prefer to use the terms mild, moderate or severe dyskaryosis as they place emphasis on nuclear rather than cytoplasmic features in the assessment of CIN. These terms are also recommended by the British Society for Clinical Cytology (Evans *et al.*, 1986).

10.3 CIN 1 (MILD DYSPLASIA)

Undifferentiated neoplastic cells occupy not more than the lower third of the epithelium (Figures 10.1 and 10.2(a)). Cells in the upper two-thirds show differentiation and orientation to form a stratified pattern. The transition between the lower third and the stratified upper layers is usually well demarcated.

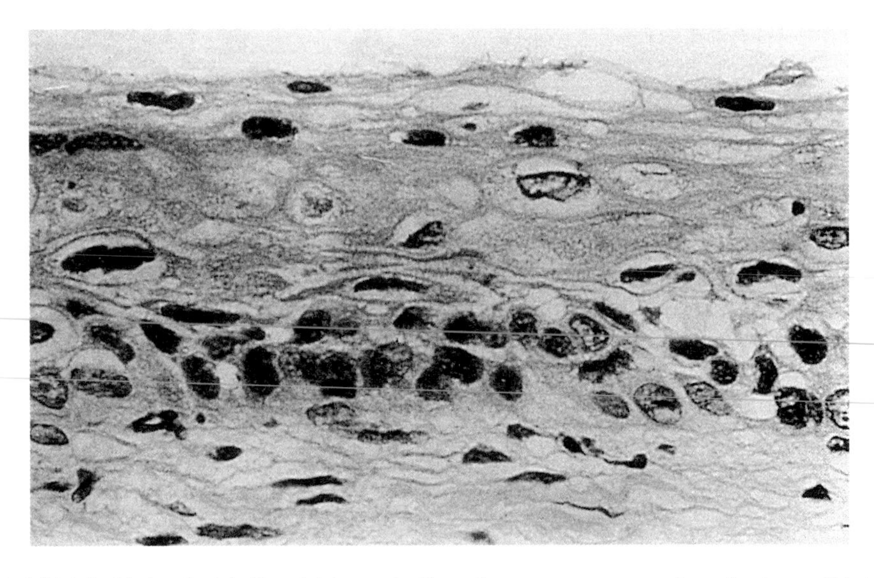

Figure 10.1 CIN 1 (mild dysplasia). Basaloid neoplastic cells occupy no more than the lower third of the epithelium. Koilocytosis is also present, indicative of papillomavirus infection (see Sections 7.13 and 10.6) (H & E, ×770).

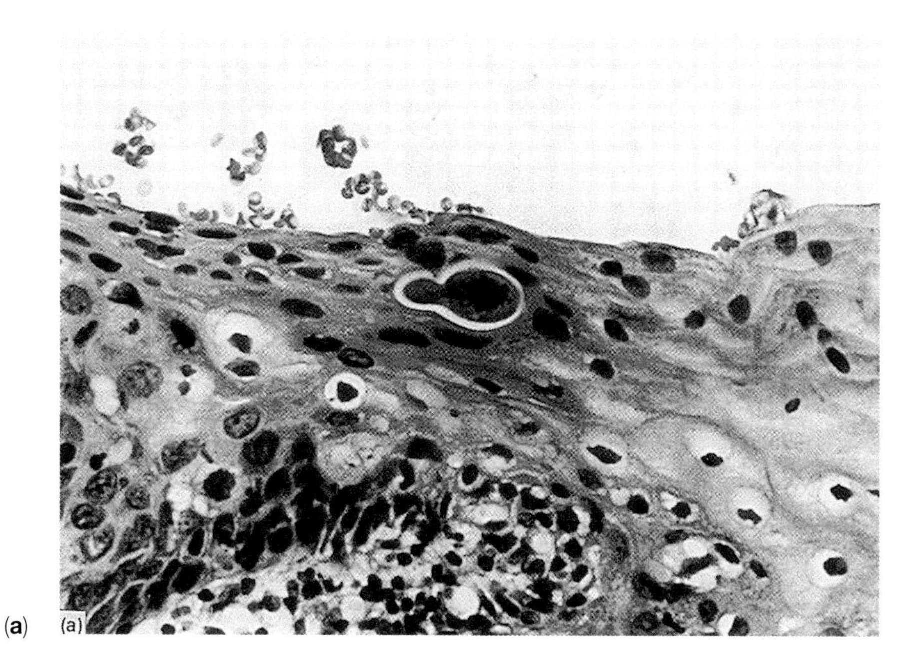

(**a**)

Figure 10.2 (**a**) CIN 1 (mild dysplasia). Note individual cell keratinization (dyskeratosis) in the upper part of the epithelium. There is also koilocytosis (H & E, ×480).

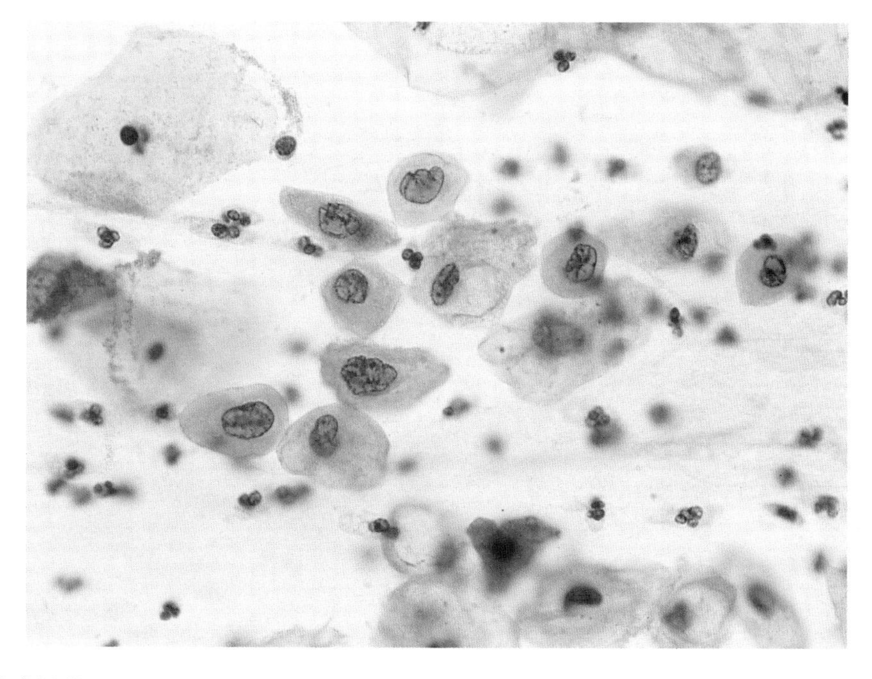

(**b**)

Figure 10.2 (**b**) Mild dyskaryosis suggestive of CIN 1. Note large squamous cells clearly resembling superficial or intermediate squames with enlarged nuclei of varying size. In no case does the nucleus exceed half the diameter of the cell (Pap, ×630).

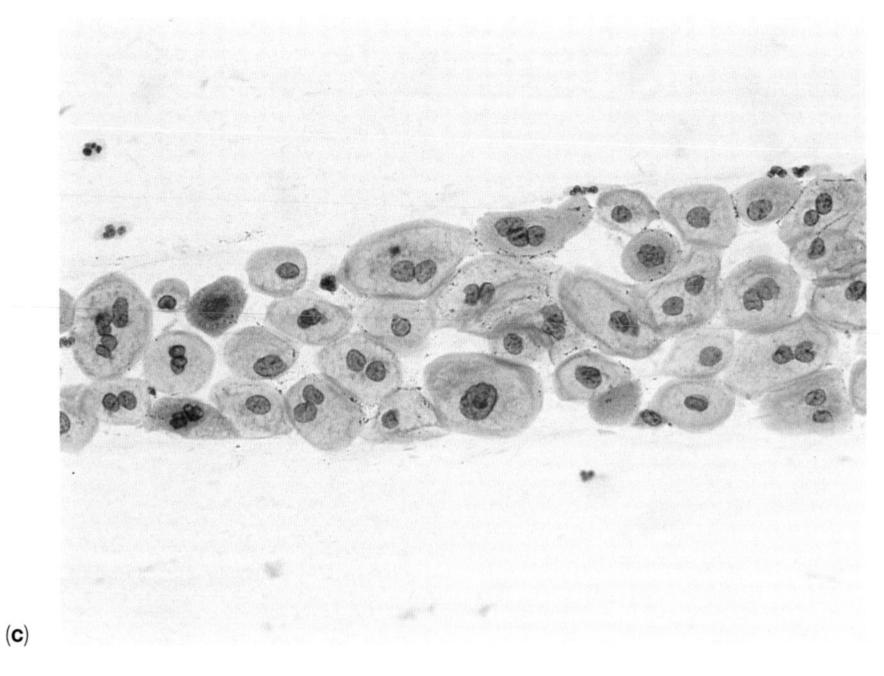

(**c**)

Figure 10.2 (**c**) Mild dyskaryosis suggestive of CIN 1 (Pap, ×630).

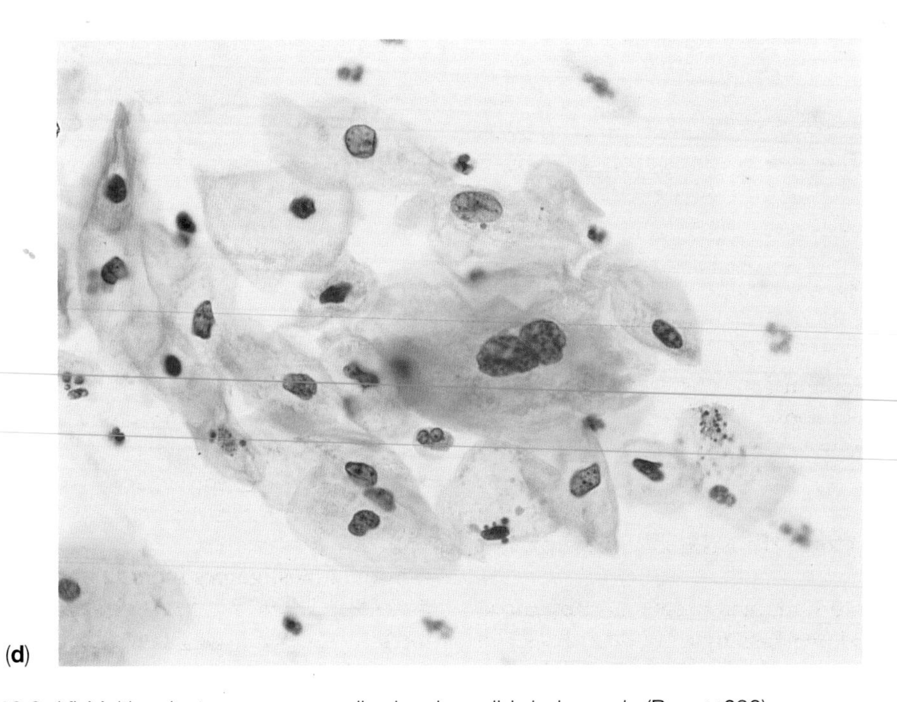

(**d**)

Figure 10.2 (**d**) Multinucleate squamous cells showing mild dyskaryosis (Pap, ×630).

Nuclear abnormality is present at all levels, but is most marked in the cells in the deeper layers. Mitotic figures are infrequent and occur mainly in the lower third of the epithelium. Abnormal forms are rarely seen.

According to Scully *et al.* (1994), cervical biopsies which contain evidence of *koilocytotic atypia* in the upper third of the epithelium without nuclear abnormalities in the lower third should now be classified as mild dysplasia/CIN 1. Koilocytosis must be identified unequivocally to warrant this diagnosis. The presence of multinucleated cells, binucleated cells or individual cell keratinization provides additional evidence to support the diagnosis but are not specific for HPV infection. The koilocytes should contain enlarged irregular hyperchromatic nuclei, a rim of cytoplasm and thick cell membrane (Scully *et al.*, 1994).

10.3.1 Cytology

The smears contain neoplastic cells showing mild dyskaryosis (Figure 10.2(b–d)). These are characterized by the presence of one or more irregular nuclei, 3–4 μm in diameter which occupy less than one-half of the total area of the cytoplasm. The chromatin content of the nucleus is increased and chromatin clumping may be seen. Frequently, degenerative changes supervene and the nucleus is pyknotic. Overall, the cell resembles a superficial or intermediate squamous cell with a slightly enlarged irregular nucleus.

The presence of koilocytes with enlarged irregular nuclei is sufficient to classify the smear as containing cells showing mild dyskaryosis or borderline nuclear changes, depending on the degree of nuclear atypia.

10.4 CIN 2 (MODERATE DYSPLASIA)

Undifferentiated neoplastic cells occupy more than one-third and less than two-thirds of the total thickness of the epithelium (Figures 10.3–10.5(a)). The cells in the upper third of the epithelium are differentiated but stratification is often less clearly defined than in CIN 1 (Figure 10.2(a)). The degree of nuclear abnormality in the cells in the superficial layers is also more marked than in CIN 1. Mitotic figures are more frequent, being found mainly in the lower two-thirds of the epithelium and include some abnormal forms (Figure 10.4).

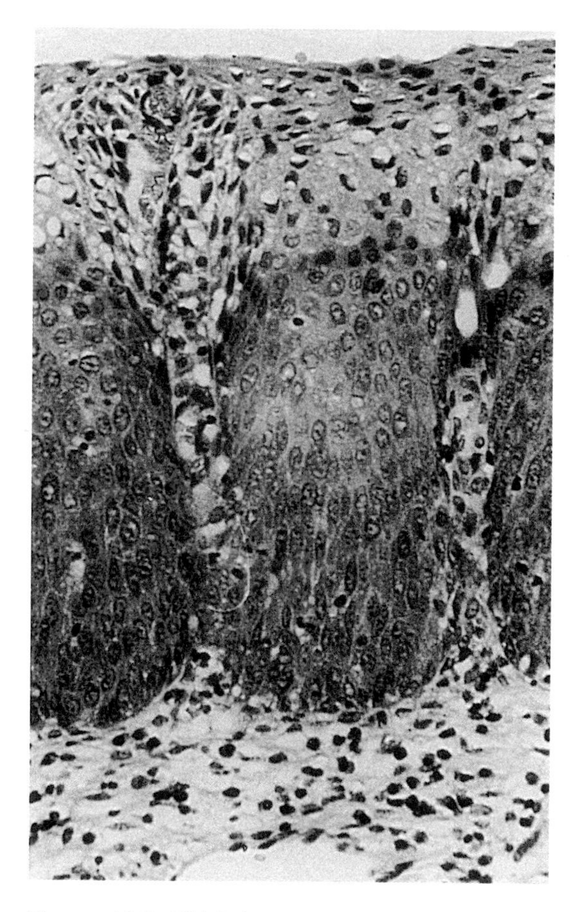

Figure 10.3 CIN 2 (moderate dysplasia). Basaloid neoplastic cells occupy up to two-thirds of the thickness of the epithelium (H & E, ×310).

Some investigators include under moderate dysplasia/CIN 2, those lesions in which nuclear abnormalities confined to the lower third of the epithelium are unusually severe (Scully *et al.*, 1994).

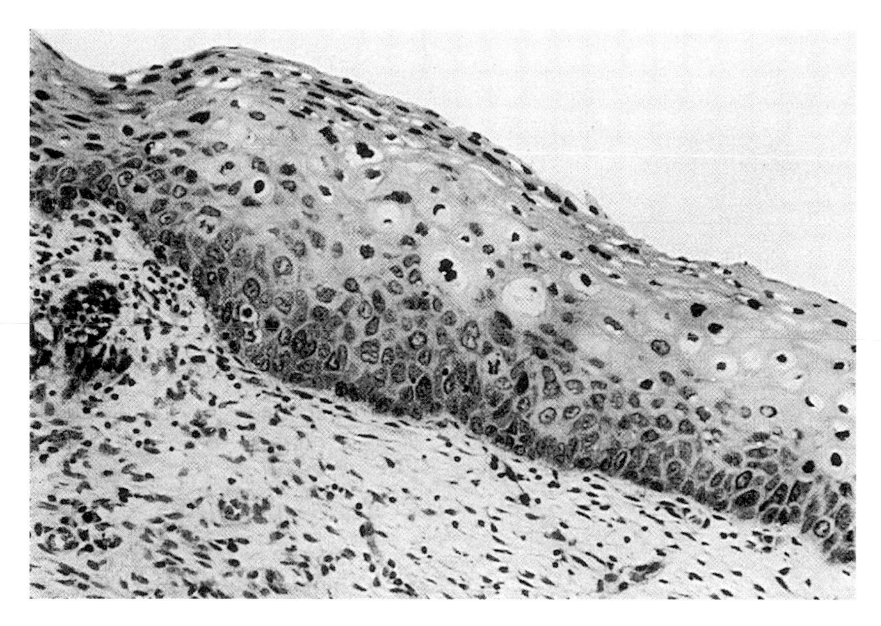

Figure 10.4 CIN 2. Mitotic figures, mainly in the lower two-thirds of the epithelium, are more frequent than in CIN 1. Koilocytosis is also present (H & E, ×310).

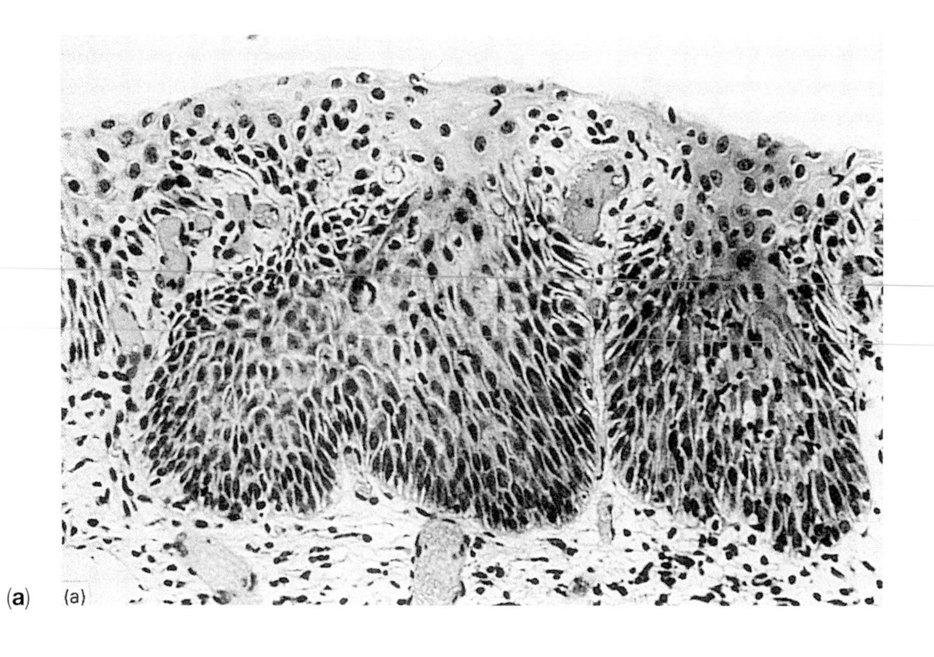

(**a**)

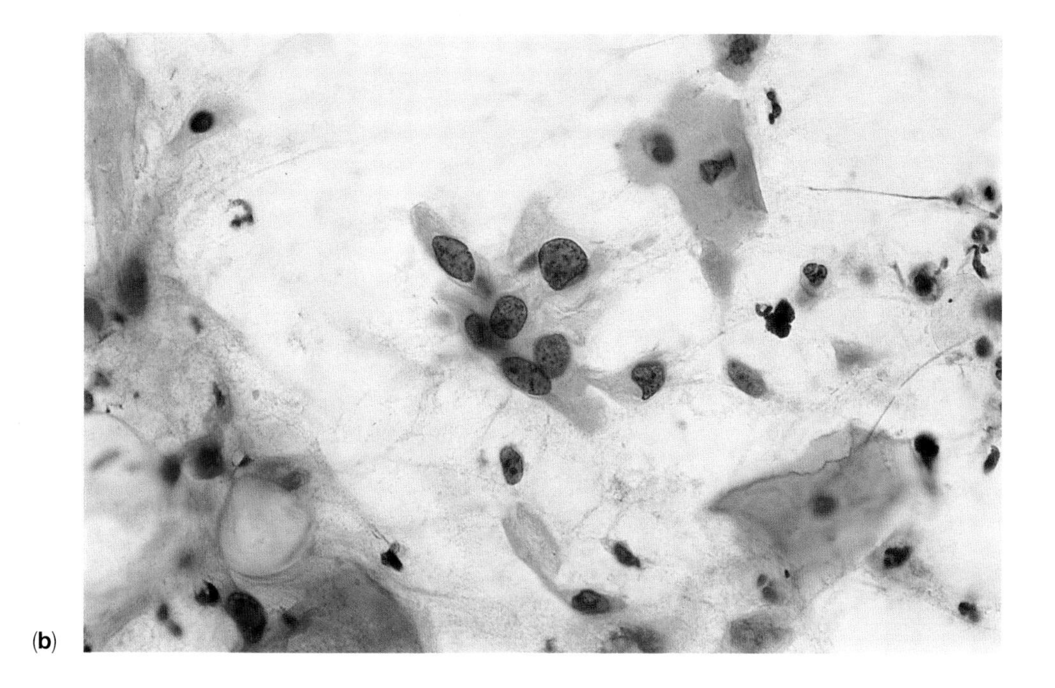
(**b**)

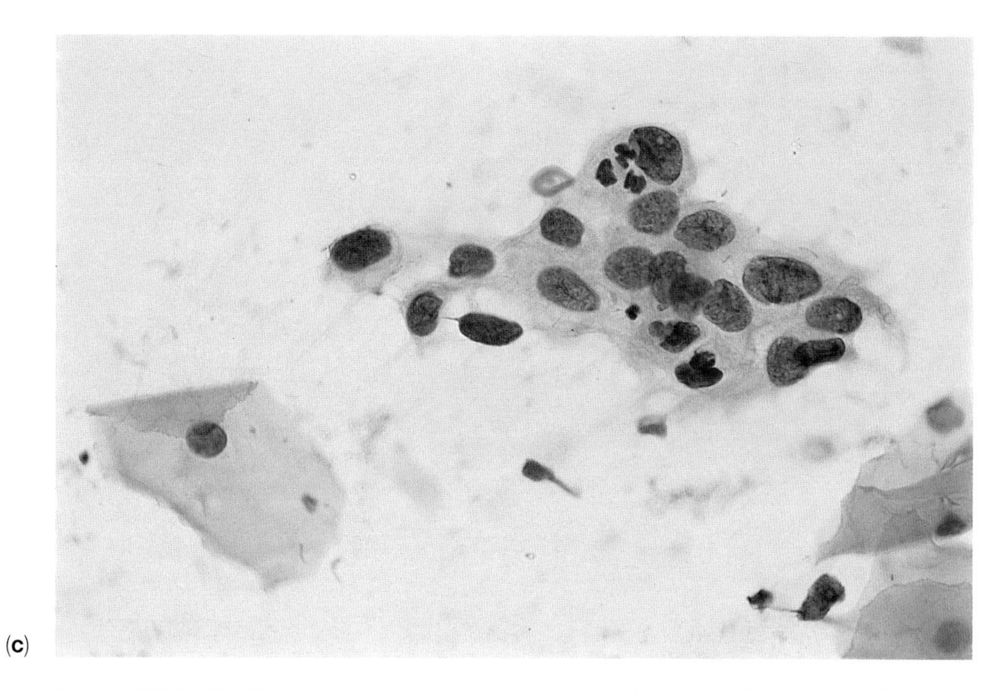
(**c**)

Figure 10.5 (**a**) CIN 2. Stratification of the upper third is less clearly defined than in CIN 1 (H & E, ×310). (**b**) Moderate dyskaryosis suggestive of CIN 2. Note similarity of epithelial cells to those on surface of lesion shown in (**a**) (Pap, ×630). (**c**) Moderate dyskaryosis suggestive of CIN 2 (Pap, ×630).

Figure 10.6 CIN 3. Basaloid and parabasaloid neoplastic cells occupy more than two-thirds of the epithelium. The surface cells are vacuolated (H & E, ×480).

(**a**)

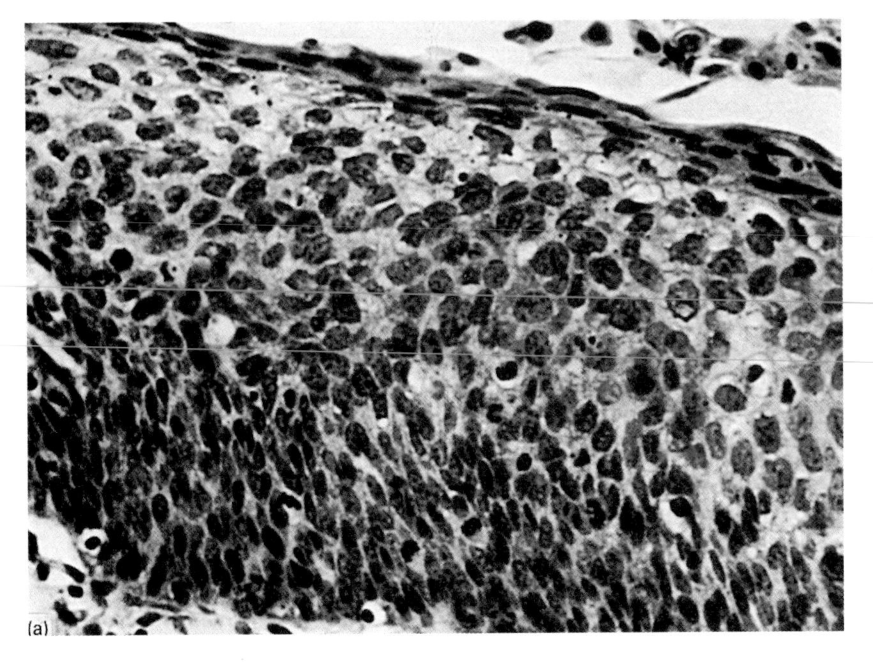

Figure 10.7 (**a**) CIN 3 (severe dysplasia). Basaloid neoplastic cells occupy more than two-thirds of the thickness of the epithelium and show characteristic nuclear crowding. Surface maturation is present. Note similarity of exfoliated cells in right-hand corner of photograph to those shown in (**b**) (H & E, ×480).

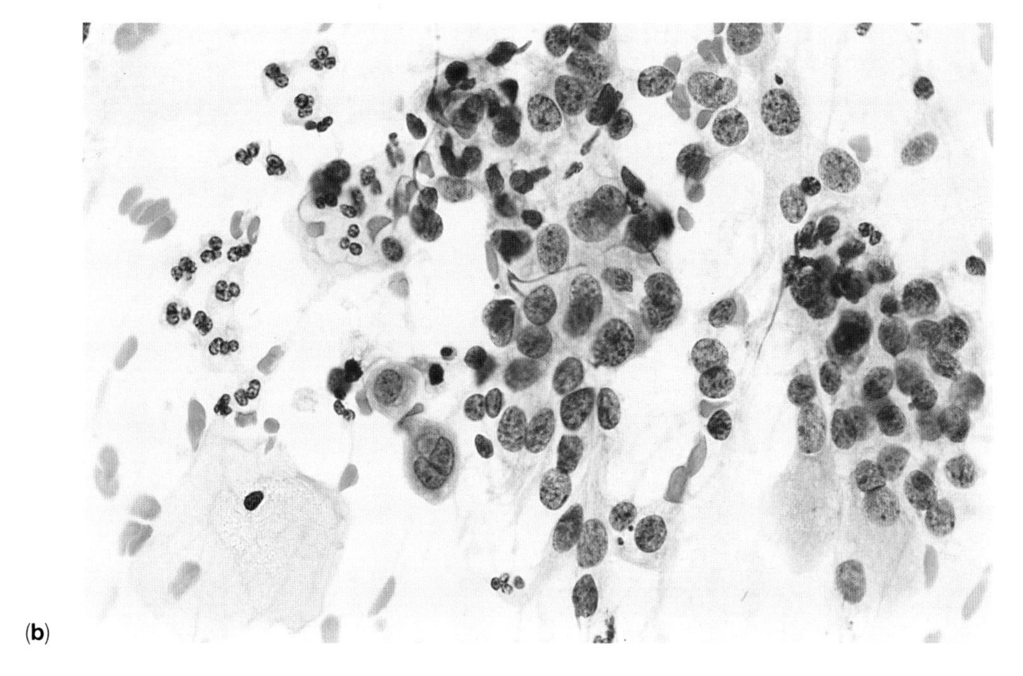

(**b**)

Figure 10.7 (**b**) Severe dyskaryosis in cervical smear (Pap, ×630).

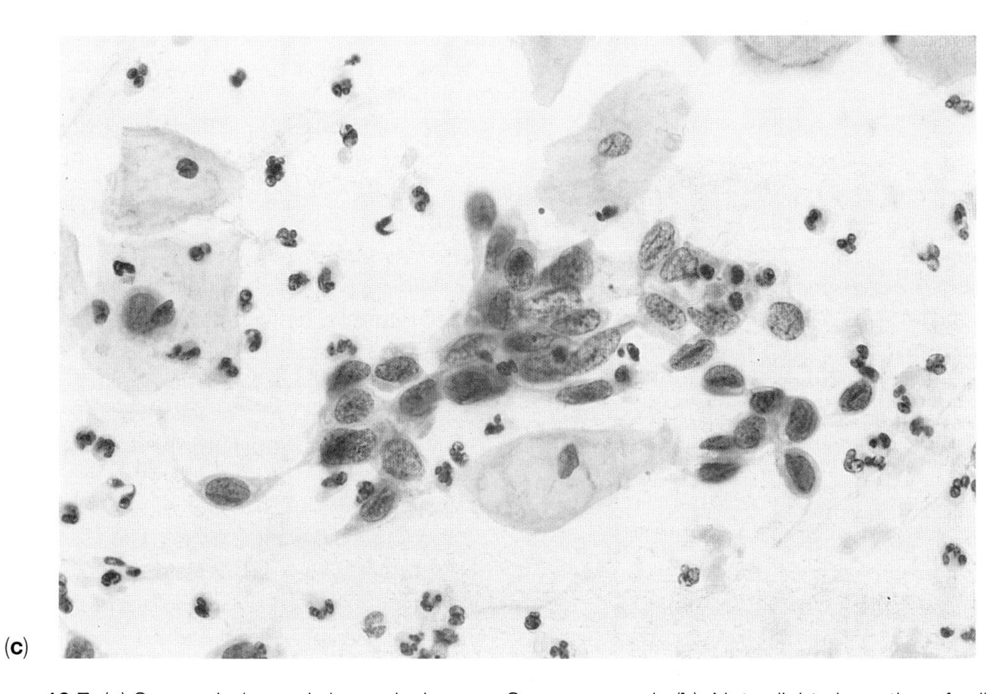

(**c**)

Figure 10.7 (**c**) Severe dyskaryosis in cervical smear. Same case as in (**b**). Note slight elongation of cells which corresponds to flattening of the cells on the surface of the biopsy (Pap, ×630).

10.4.1 Cytology

Neoplastic cells showing moderate dyskaryosis are present in the smear (Figure 10.5(b,c)). They contain comparatively larger nuclei than those showing mild dyskaryosis. The nuclei occupy one-half to two-thirds of the cytoplasmic area. Overall, the cell resembles a superficial or intermediate cell with a considerably enlarged nucleus.

10.5 CIN 3 (SEVERE DYSPLASIA AND CARCINOMA *IN SITU*)

Undifferentiated neoplastic cells with a high nucleocytoplasmic ratio occupy more than two-thirds of the total thickness of the epithelium (Figures 10.6–10.11). The cells have ill-defined boundaries and scanty cytoplasm, so that loss of polarity and nuclear crowding are characteristic of this lesion. The nuclei are large, irregular and hyperchromatic and show clumping of the chromatin, often with clearing of the intervening spaces. Surface maturation with flattening of the surface layers or is a common finding. A layer of keratin or a thin layer of desiccated cells on the surface does not exclude a diagnosis of CIN 3.

Mitotic figures may be found at all levels; abnormal forms such as tripolar mitoses are also seen (Figure 10.11(c)).

Some pathologists include in the category of CIN 3 those lesions in which the nuclear abnormalities in the lower and middle third of the epithelium are unusually severe (Scully *et al.*, 1994).

The transition between CIN 3 and non-neoplastic epithelium may be quite abrupt (Figure 10.11(a)), the epithelial junction then being known as Schiller's line. Often the transition is more gradual (Figure 10.11(b)), and there are large zones of CIN 2 and CIN 1 between the CIN 3 and the normal epithelium. When reporting on such a lesion, emphasis is placed on the most severe histological abnormality found. Nevertheless it is helpful, particularly if an attempt is being made to correlate cytological with histological findings, to indicate the presence of lesser changes.

Glandular involvement by high-grade CIN (CIN 2 or 3) is common. Anderson and Hartley (1980) found it in over 80% of cone biopsies containing CIN 3. Unless it is recognized it may be reported as stromal invasion (Section 11.5).

10.5.1 Keratinizing CIN 3

This is a variant of CIN 3 in which the whole thickness of the epithelium is composed of neoplastic cells with a relatively low nucleocytoplasmic ratio (Figure 10.12(a)). There is a complete disorganization of the growth pattern and loss of stratification. Individually the cells have large, irregular, hyperchromatic nuclei and abundant eosinophilic cytoplasm. The marked nuclear abnormality and pleomorphism, loss of polarity and evidence of mitotic activity at all levels, are consistent with a diagnosis of CIN 3. Anderson (1985) described an unusual example of CIN 3 in which the surface of the neoplastic epithelium was covered by a thick layer of keratin.

Several other variants of CIN 3 have been described (Poulsen *et al.*, 1975; Buckley *et al.*, 1982) including small cell and large cell forms.

10.5.2 Cytology

The smear contains neoplastic cells with a severe dyskaryosis. The nucleocytoplasmic ratio is high and the nuclei occupy more than two-thirds of the cell cytoplasmic area (Figures 10.7(b, c), 10.8(b), 10.9(b), 10.10(b), 10.12(b–d), 10.13 and 10.14). If the epithelium is composed of undifferentiated neoplastic cells throughout its thickness, the cytoplasm of the dyskaryotic cells will be very scanty indeed. If there is some surface maturation or flattening, the amount of cytoplasm surrounding the nucleus will be greater and may be keratinized. Occasionally, dyskaryotic nuclei devoid of any cytoplasm (Figure 10.15(a,b)) may be found in cytolytic smears (Figure 10.16) or in smears from older

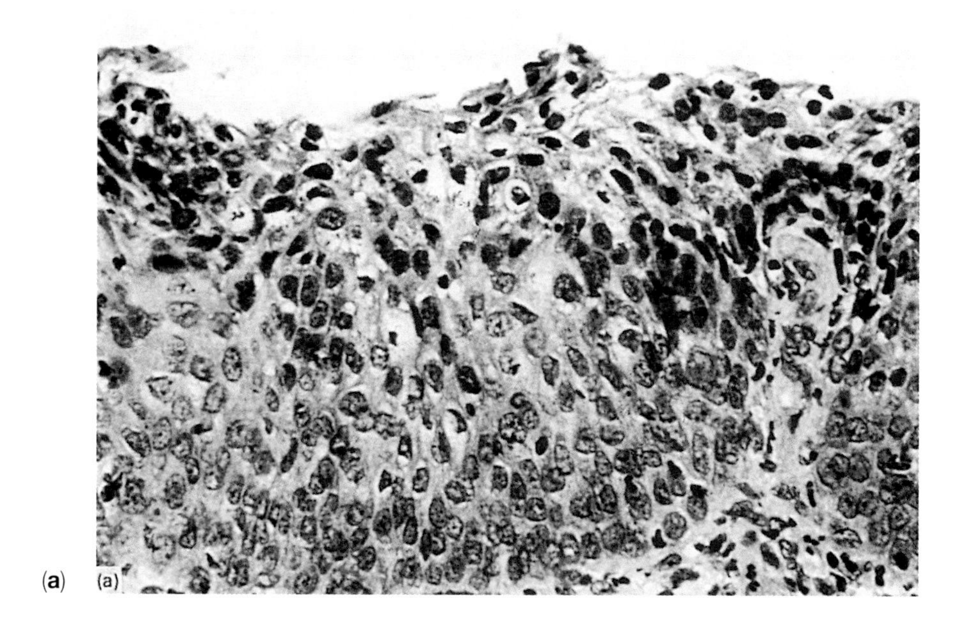

(**a**)

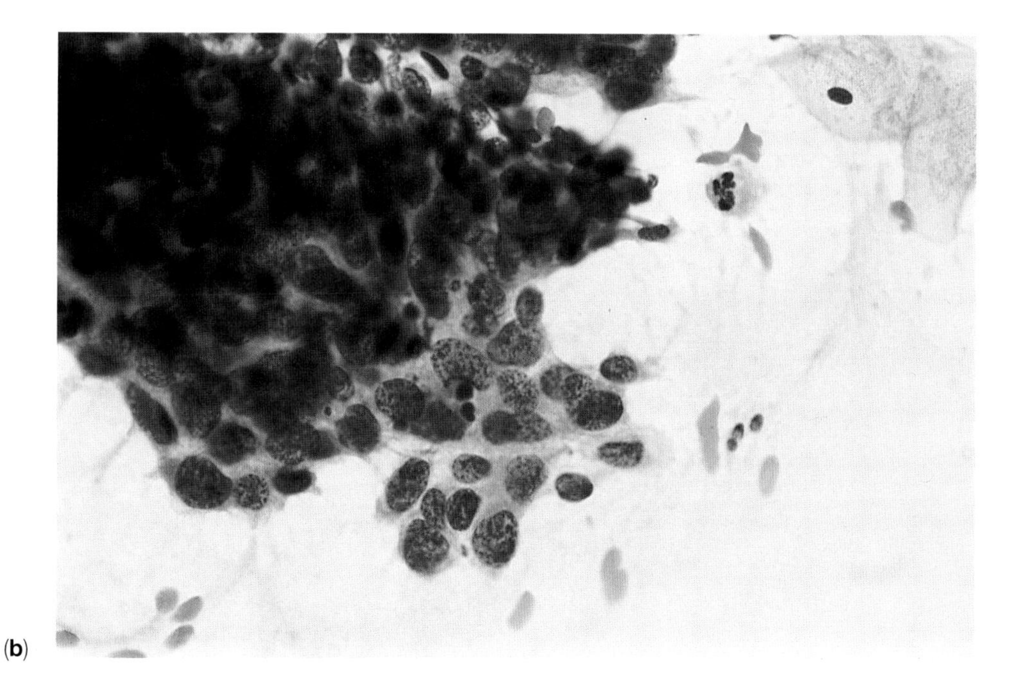

(**b**)

Figure 10.8 (**a**) CIN 3 (carcinoma *in situ*). Basaloid neoplastic cells occupy the full thickness of the epithelium. Surface maturation is lacking (H & E, ×480). (**b**) Severe dyskaryosis. Smear contains fragments of tissue composed of hyperchromatic nuclei. The irregularity of the nuclei and the coarse chromatin structure are evident at higher magnification in cells at the periphery of the cluster (Pap, ×630).

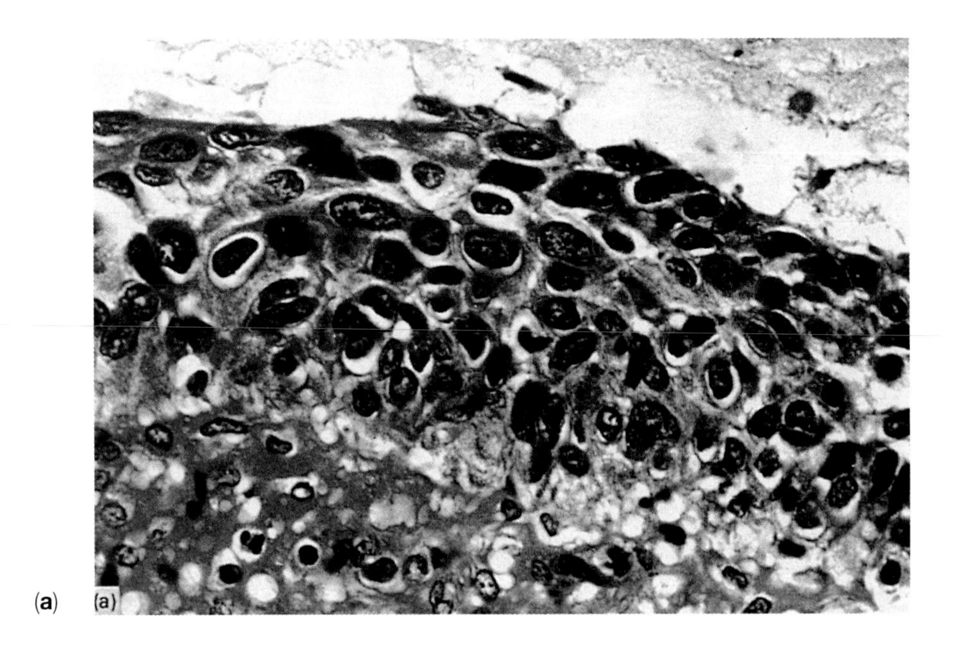

(**a**)

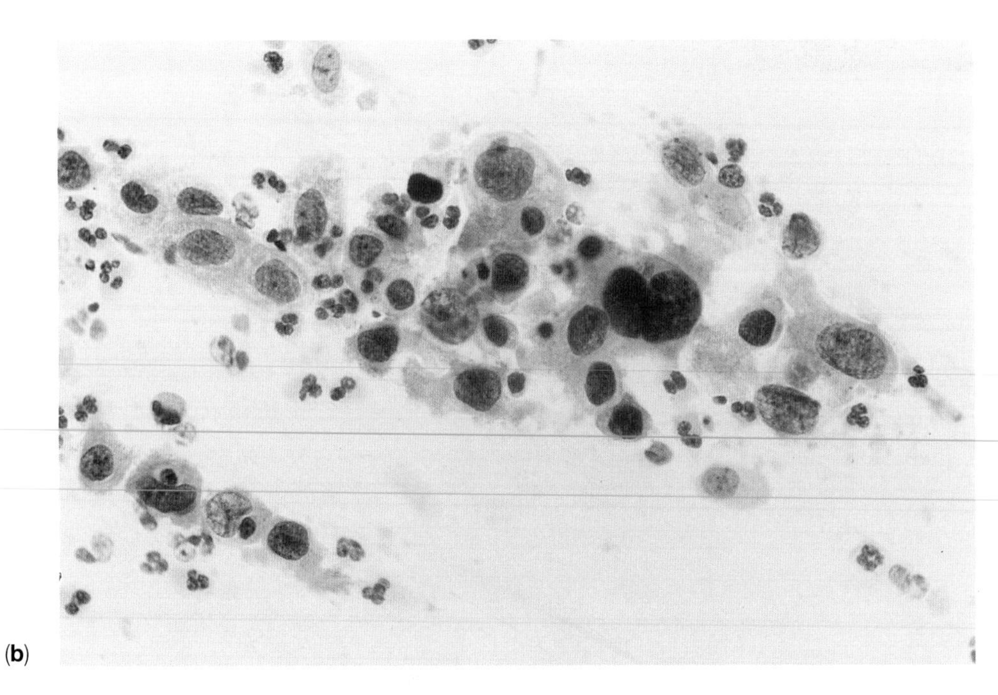

(**b**)

Figure 10.9 (**a**) CIN 3 (carcinoma *in situ*). Basaloid neoplastic cells occupy the full thickness of the epithelium. Surface maturation is absent (H & E, ×480). (**b**) Severe dsykaryosis. Corresponding cervical smear to case shown in (**a**). Note large nuclei, varied size, irregular shape and coarse chromatin (Pap, ×630).

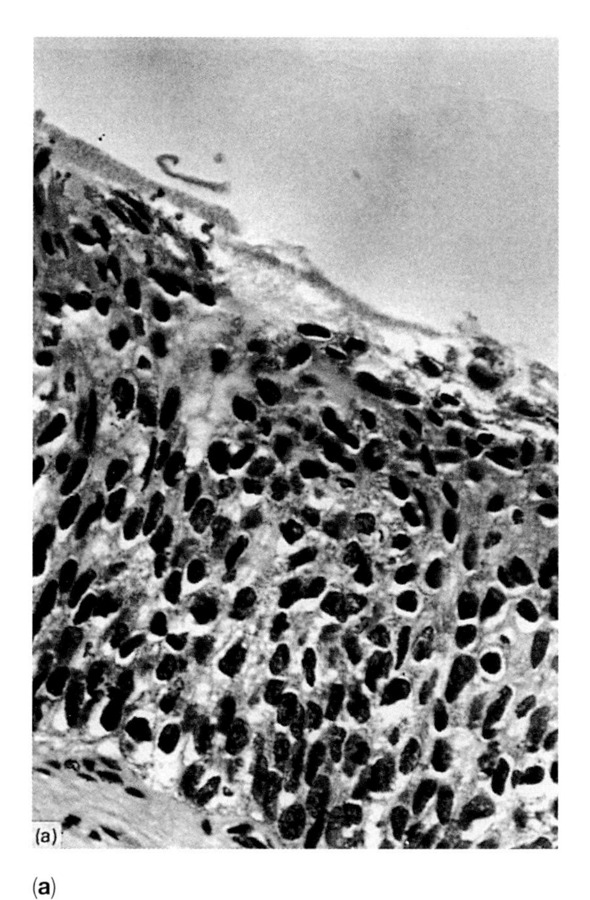

(**a**)

Figure 10.10 (**a**) CIN 3. Surface keratinization is present. Abnormal nuclei are present in the surface layers (H & E, ×230). (**b**) Cervical smear from same cervix as in (**a**) showing streaks of highly keratinized cells with moderate or severe dyskaryosis (Pap, ×630).

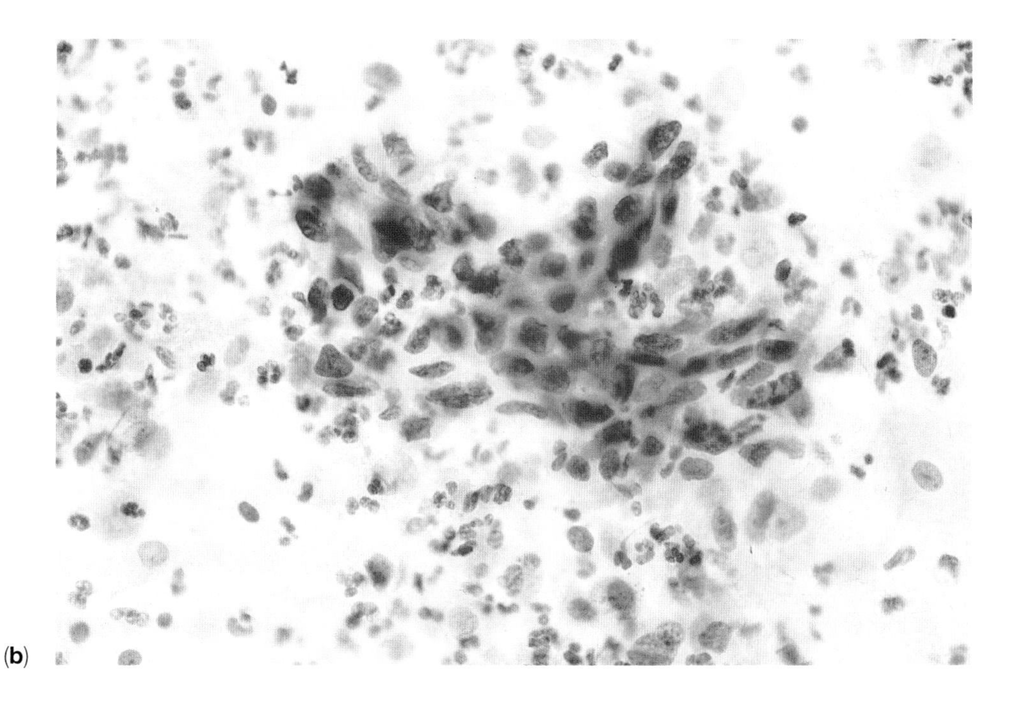

(**b**)

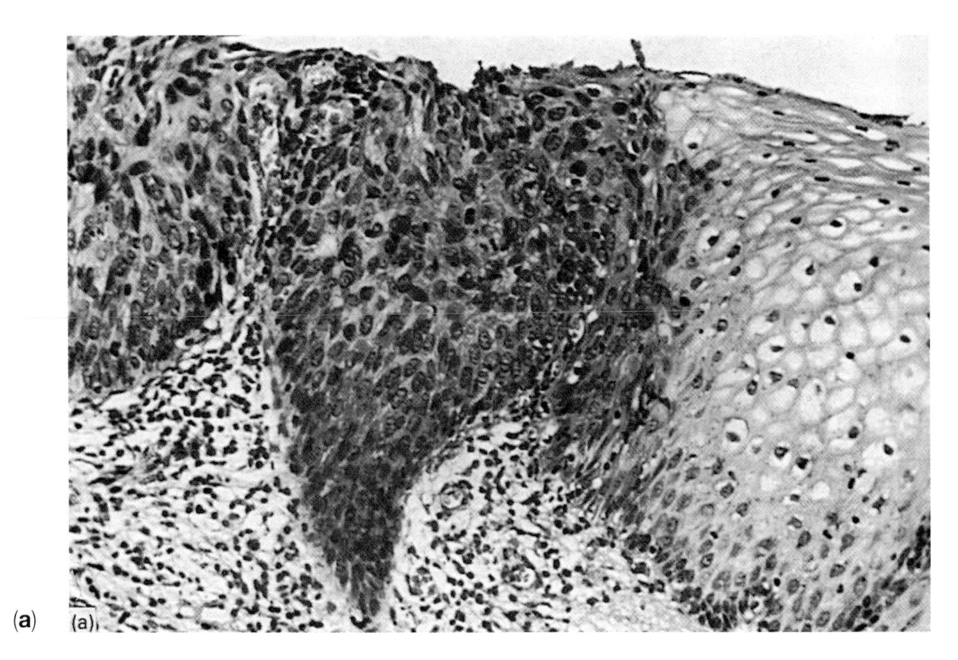

Figure 10.11 (**a**) CIN 3 (carcinoma *in situ*) with abrupt transition (Schiller's line) to normal epithelium (H & E, ×250).

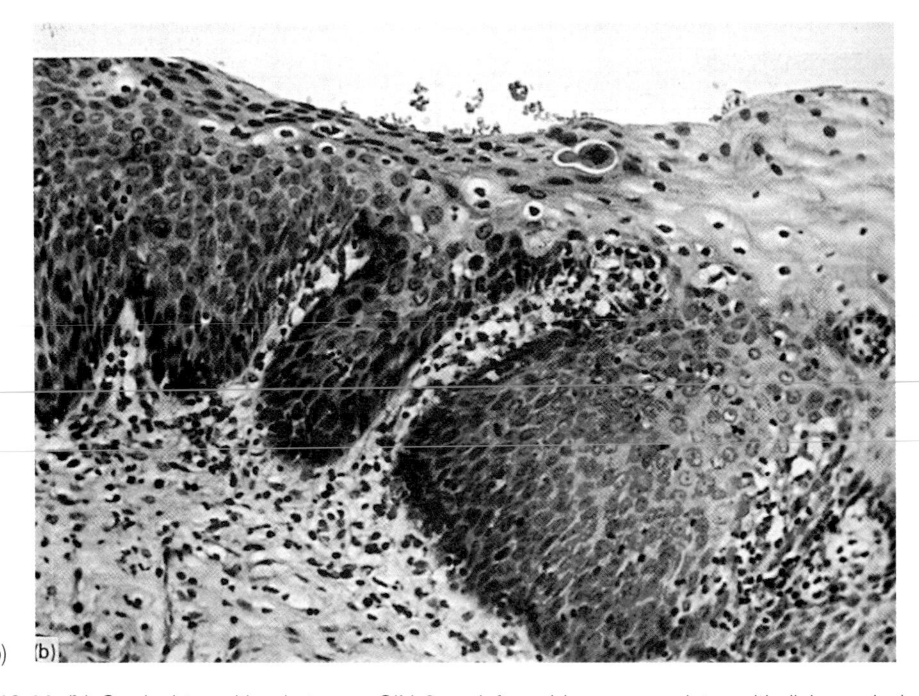

Figure 10.11 (**b**) Gradual transition between CIN 3 on left and less severe intraepithelial neoplasia on right. Same lesion as shown in Figure 10.2. Note individual cell keratinization and koilocytosis (H & E, ×240).

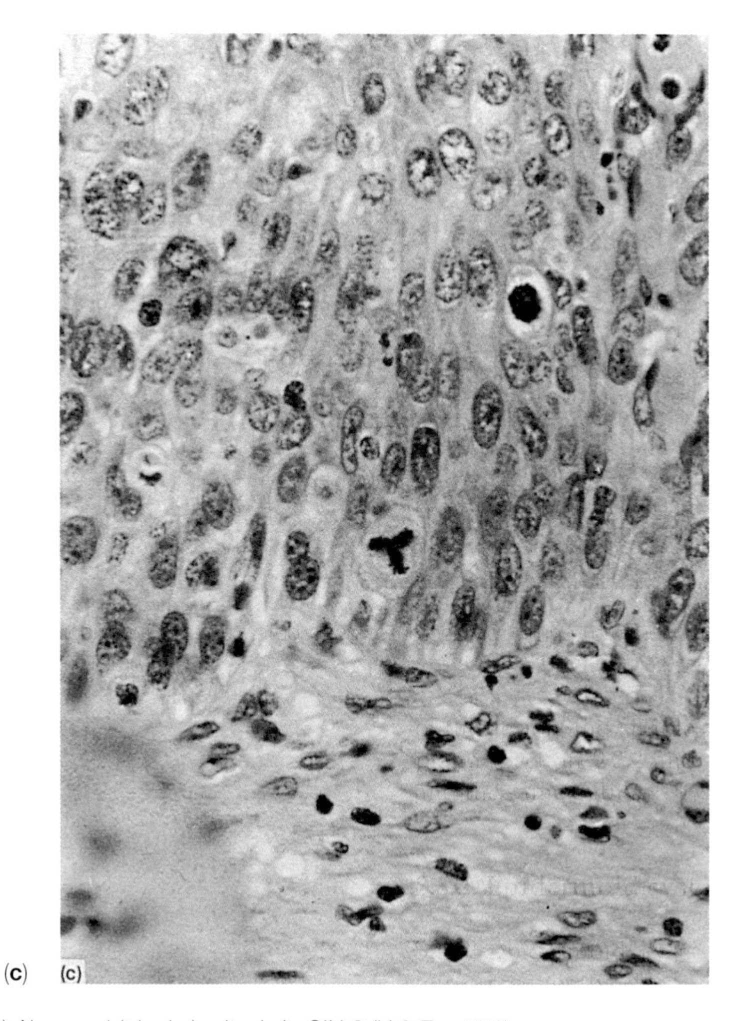

(**c**)

Figure 10.11 (**c**) Abnormal (tripolar) mitosis in CIN 3 (H & E, ×770).

women where the lesion is arising in atrophic epithelium (Figure 10.17). Hypochromasia may rarely be a feature of the dyskaryotic cells, the so-called pale dyskaryosis (Figure 10.15(a), and (b)). Macronucleoli (Figure 10.18) or bizarre-shaped cells more commonly associated with invasive cancer are sometimes seen.

In CIN 3, the neoplastic cells may be shed in sheets or dense clusters sometimes designated microbiopsies (Figures 10.8(b) and 10.19). The large hyperchromatic nuclei often overlap, reflecting the loss of polarity of the epithelium. Nuclear pleomorphism can be recognized if the high-power objective is used. Mitotic figures may be seen. However, the distinction between CIN, invasive squamous carcinoma and glandular neoplasia is not always possible and this should be reflected in the report.

A report by Selvaggi (1994) suggested that certain cytological features of cervical smears

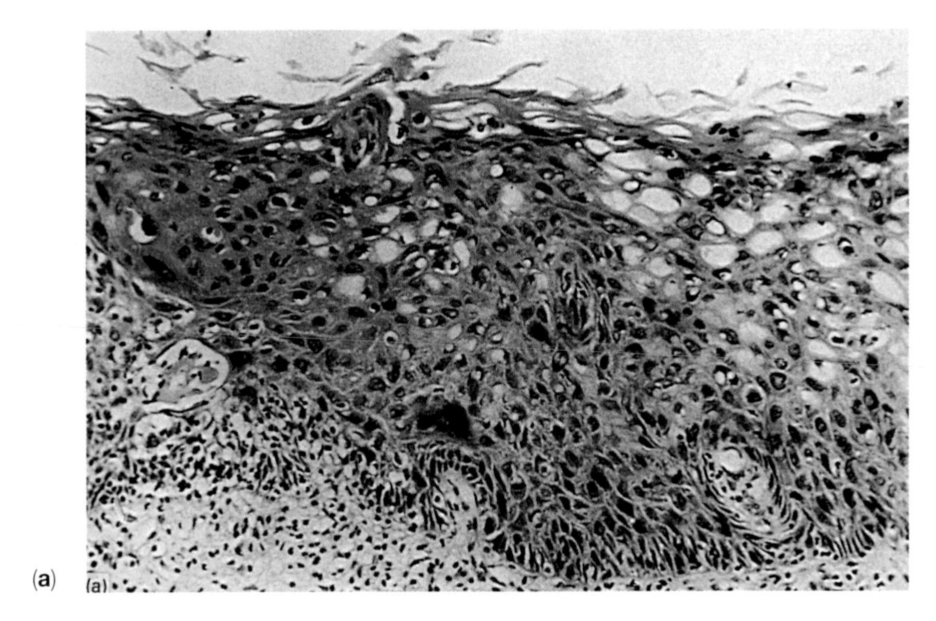

Figure 10.12 (**a**) Keratinizing CIN 3. Surface keratinization is marked. There is hyperkeratosis and parakeratosis. The neoplastic cells are well-differentiated and have a relatively low nucleocytoplasmic ratio but the nuclei show considerable pleomorphism and disorientation (H & E, ×230).

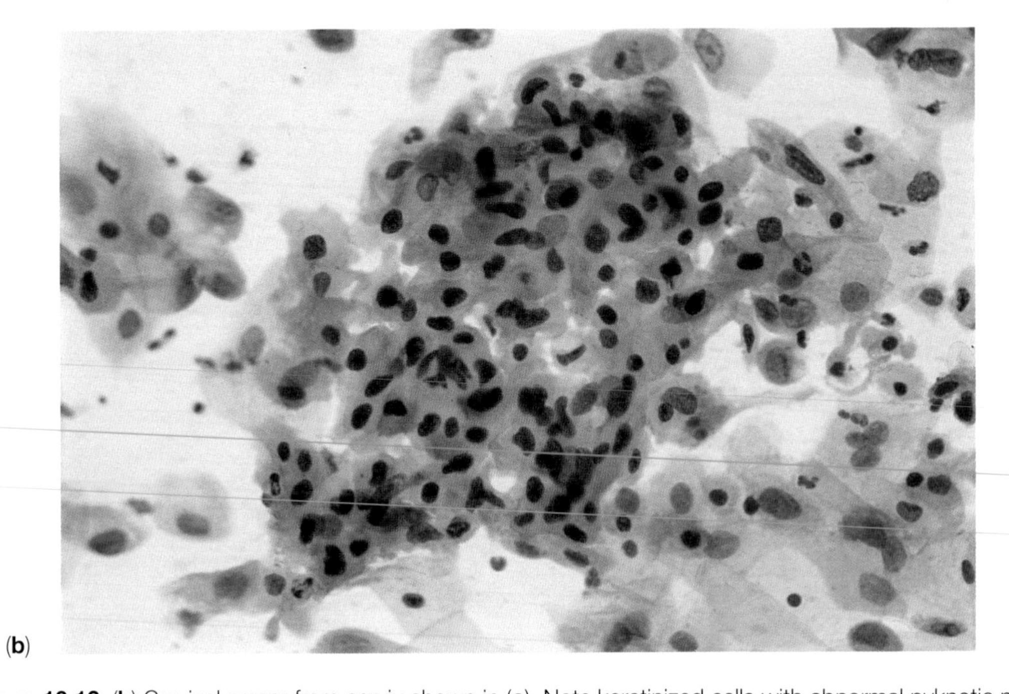

Figure 10.12 (**b**) Cervical smear from cervix shown in (**a**). Note keratinized cells with abnormal pyknotic nuclei and anucleate keratinised material. The slight degree of nuclear abnormality belies the severity of the underlying lesion (Pap, ×630).

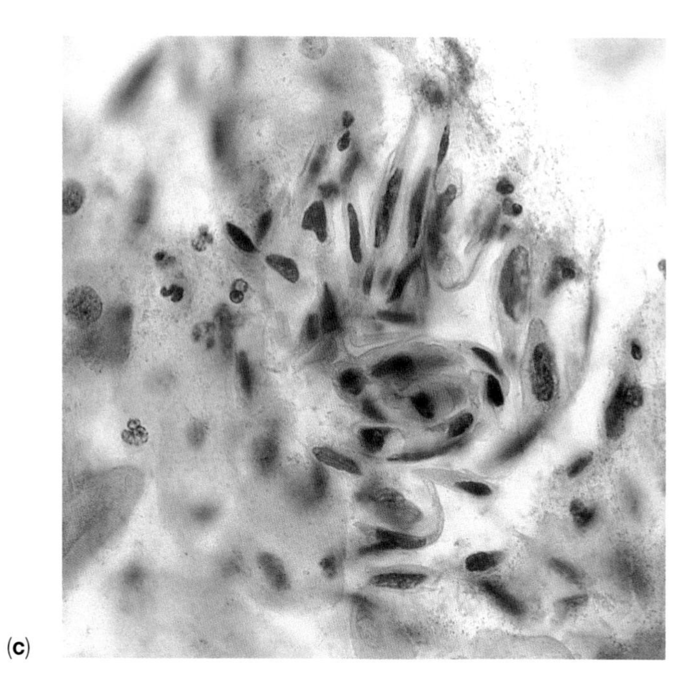

(**c**)

Figure 10.12 (**c**) Fibre cells from keratinising CIN 3 shown in (**a**) (Pap, ×630).

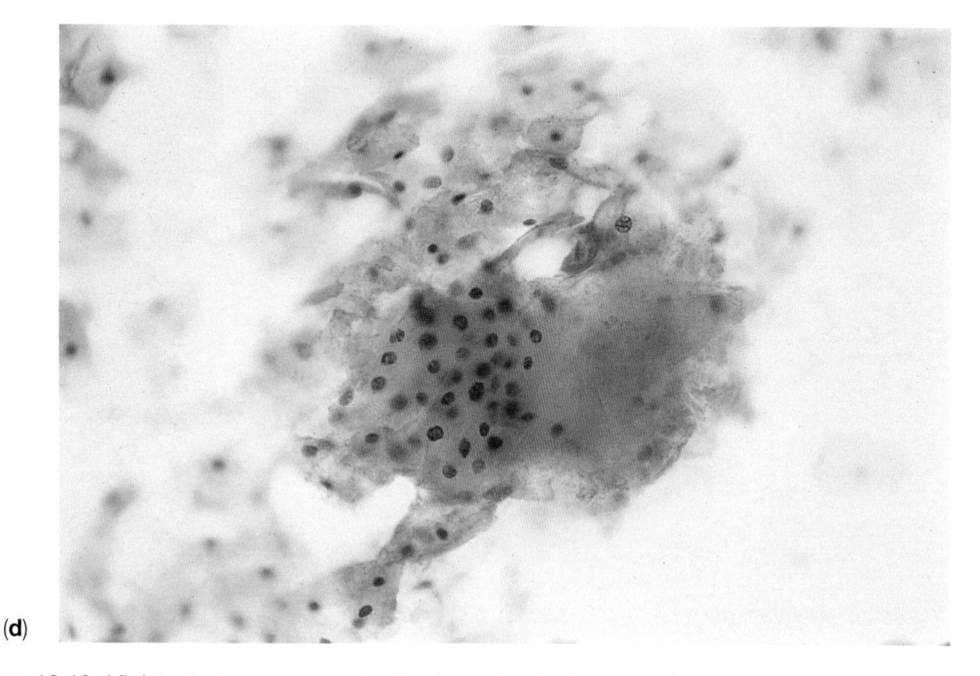

(**d**)

Figure 10.12 (**d**) Anucleate squames reflecting hyperkeratosis seen in (**a**).

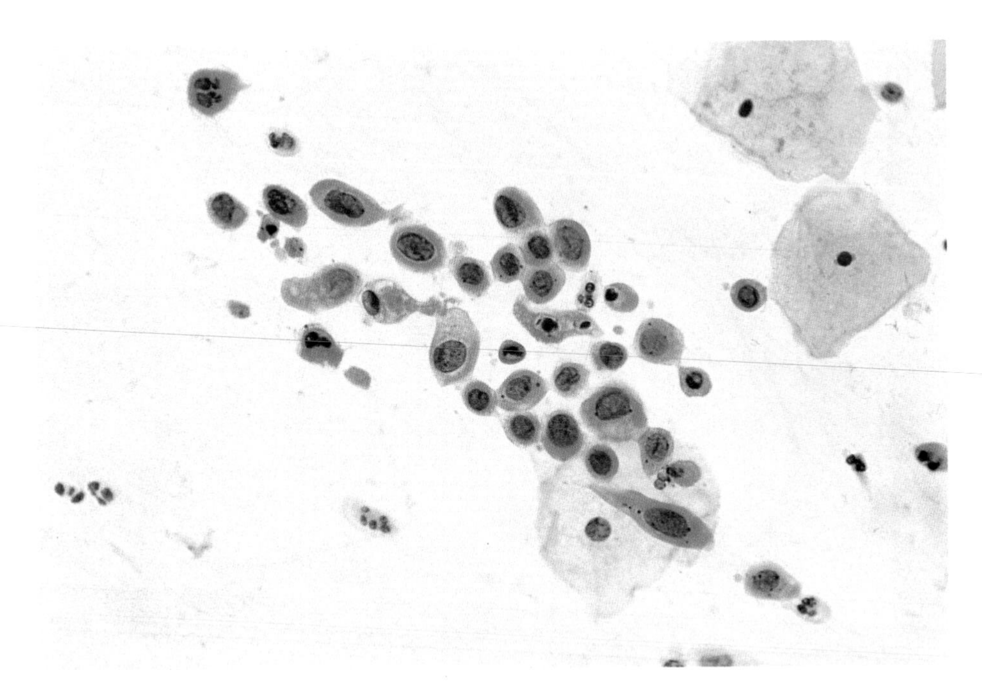

Figure 10.13 Severe dyskaryosis. Note large nuclei with coarse, irregular chromatin and scanty cytoplasm. The cells are often found in streaks reflecting the loss of cohesion which is characteristic of neoplastic cells (Pap, ×400).

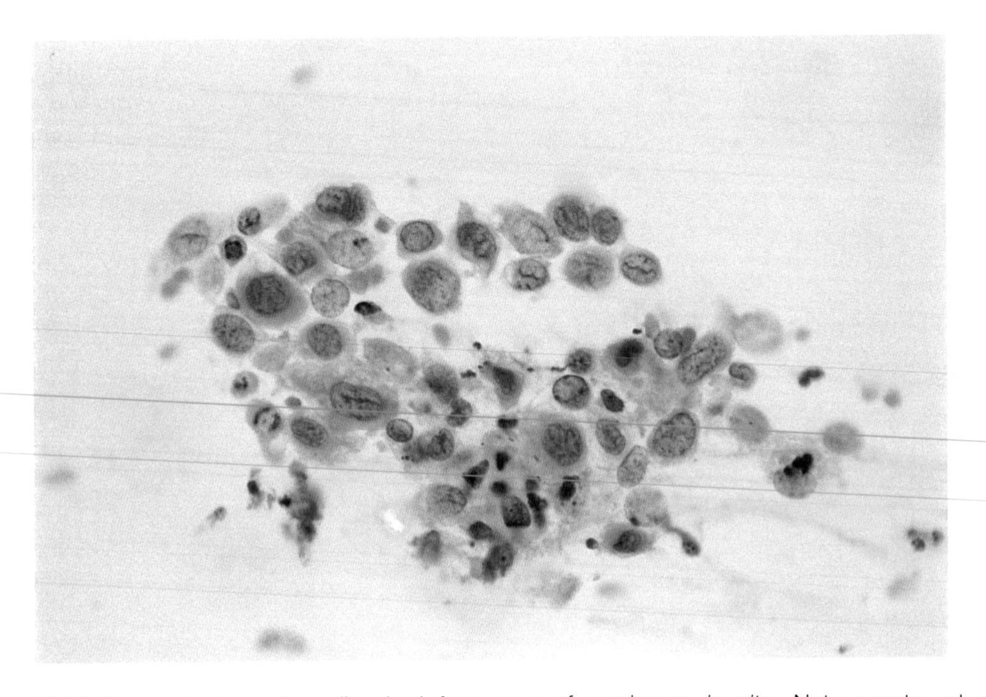

Figure 10.14 Severe dyskaryosis, cells shed from area of carcinoma *in situ*. Note scanty cytoplasm (Pap, ×630).

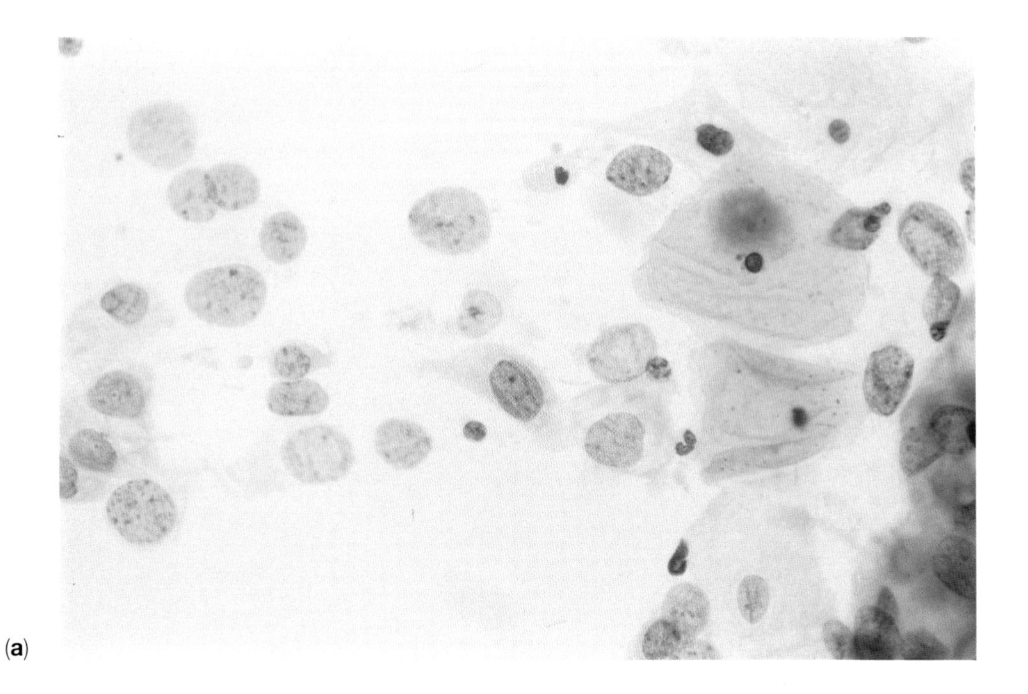
(**a**)

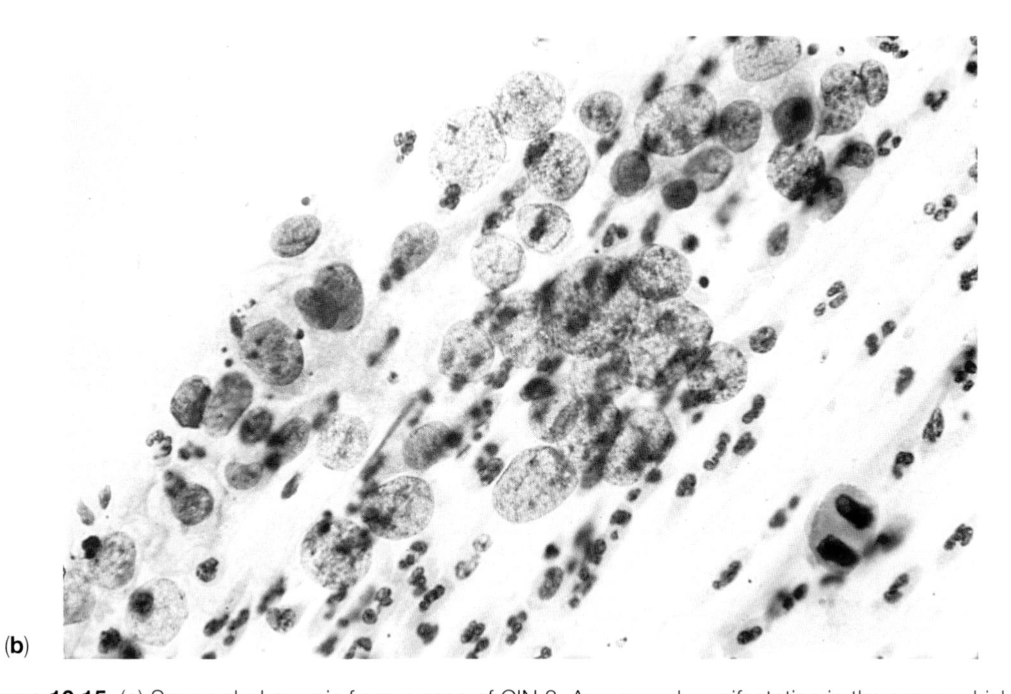
(**b**)

Figure 10.15 (**a**) Severe dyskaryosis from a case of CIN 3. An unusual manifestation in the smear which contained clusters of these hypochromatic nuclei, sometimes termed 'pale dyskaryosis'. They were recognized as dyskaryotic by the variation in size and irregular chromatin (Pap, ×630). (**b**) Severe dyskaryosis. Same case as shown in (**a**). The abnormal nuclei were devoid of cytoplasm (Pap, ×630).

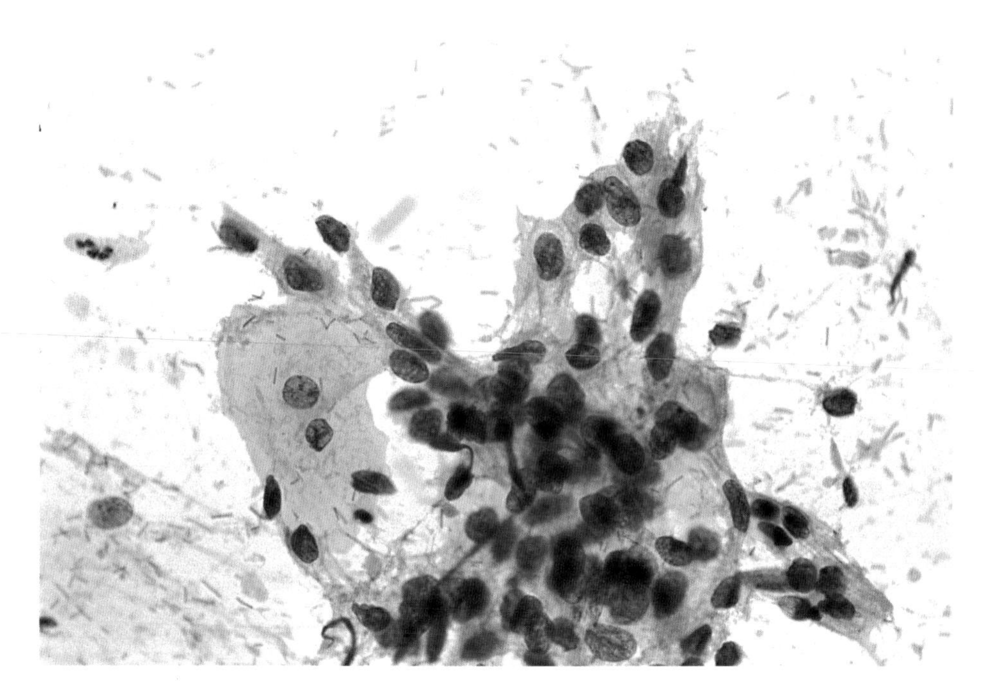

Figure 10.16 Severe dyskaryosis in a cytolytic smear from a pregnant woman. The dyskaryotic cells may be difficult to identify among the cell debris, Döderlein bacilli and free nuclei. Careful screening is needed. CIN was confirmed on biopsy (Pap, ×630).

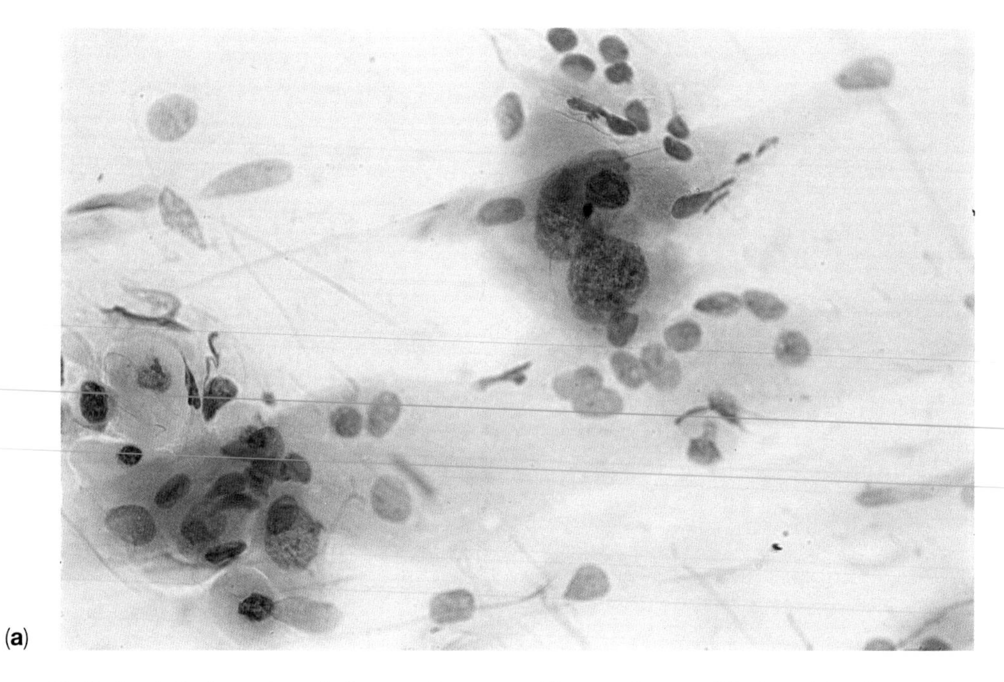

(**a**)

Figure 10.17 (**a–c**) Dyskaryotic cells in atrophic smear. These cells are difficult to identify, and the smears require careful screening. CIN 3 was diagnosed on biopsy (Pap, ×630).

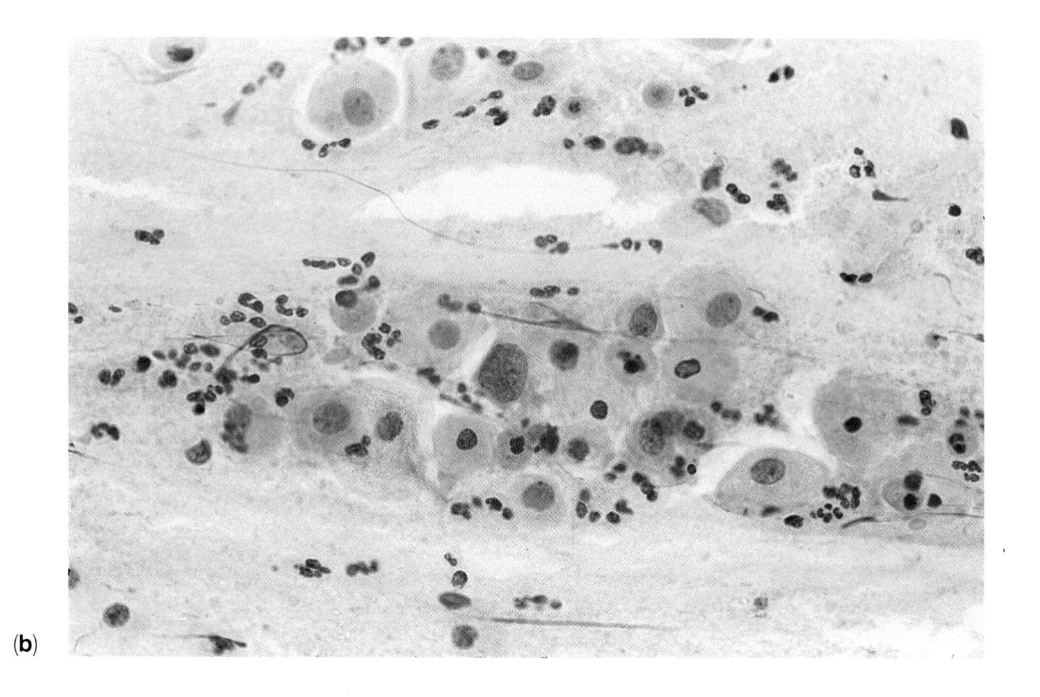

(**b**)

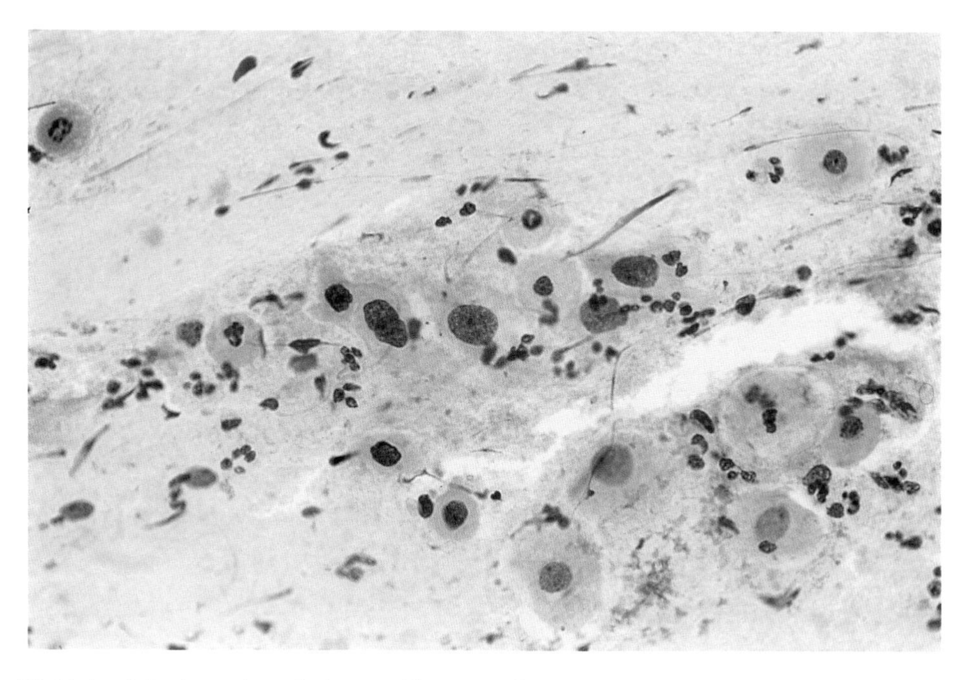

(**c**)

Figure 10.17 (**a–c**) Dyskaryotic cells in atrophic smear. These cells are difficult to identify, and the smears require careful screening. CIN 3 was diagnosed on biopsy (all Pap, ×630).

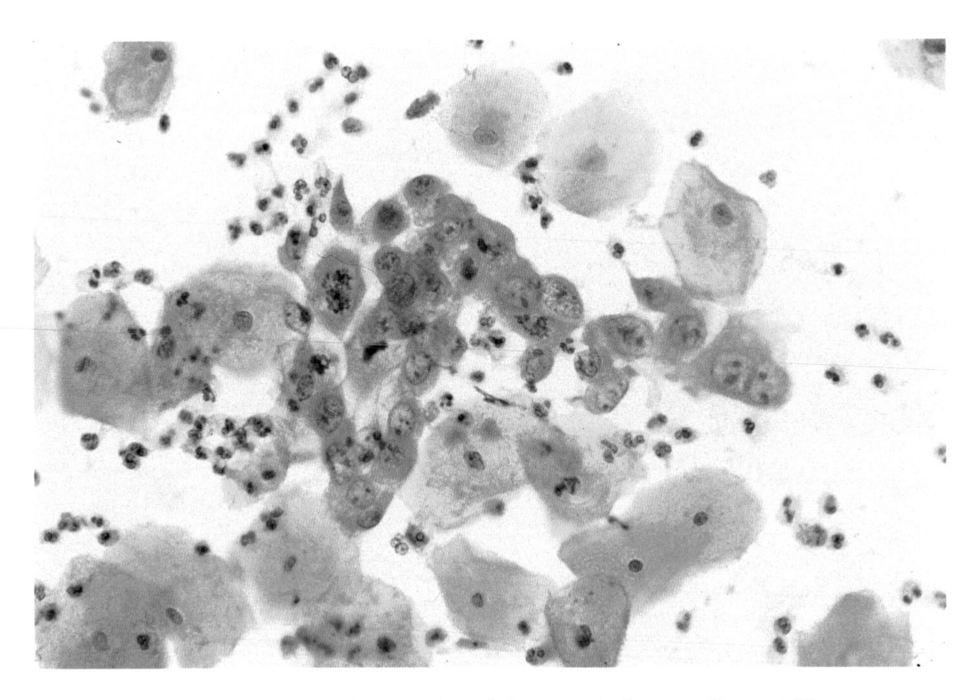

Figure 10.18 Macronucleoli in severely dyskaryotic cells in a cervical smear (Pap ×400).

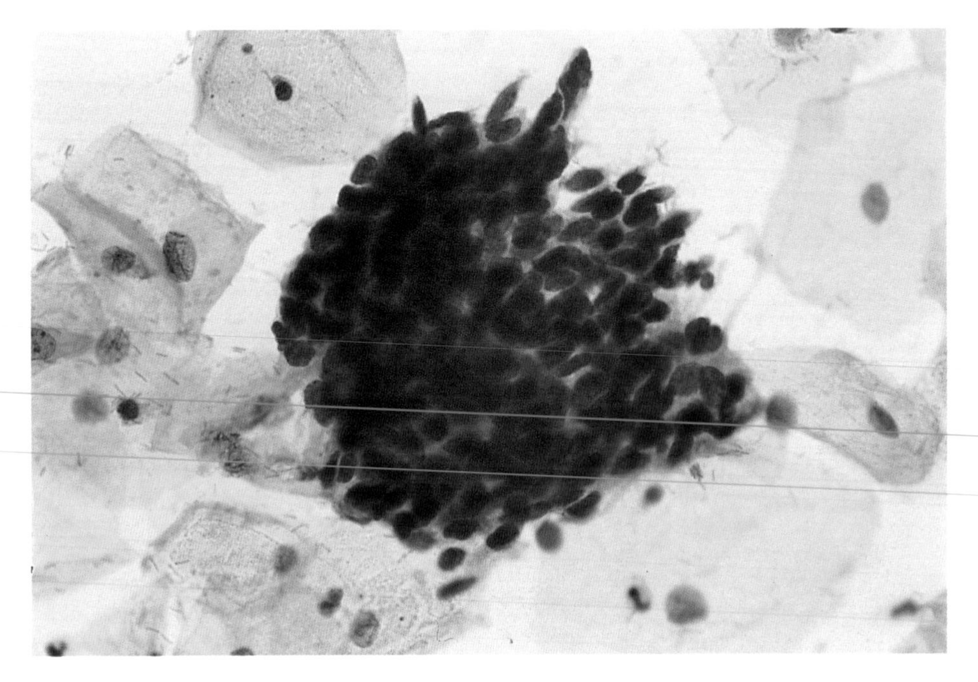

Figure 10.19 Dense clusters of cells with hyperchromatic nuclei in cervical smear. Examination under oil immersion revealed marked nuclear pleomorphism consistent with severe dyskaryosis (Pap, ×630).

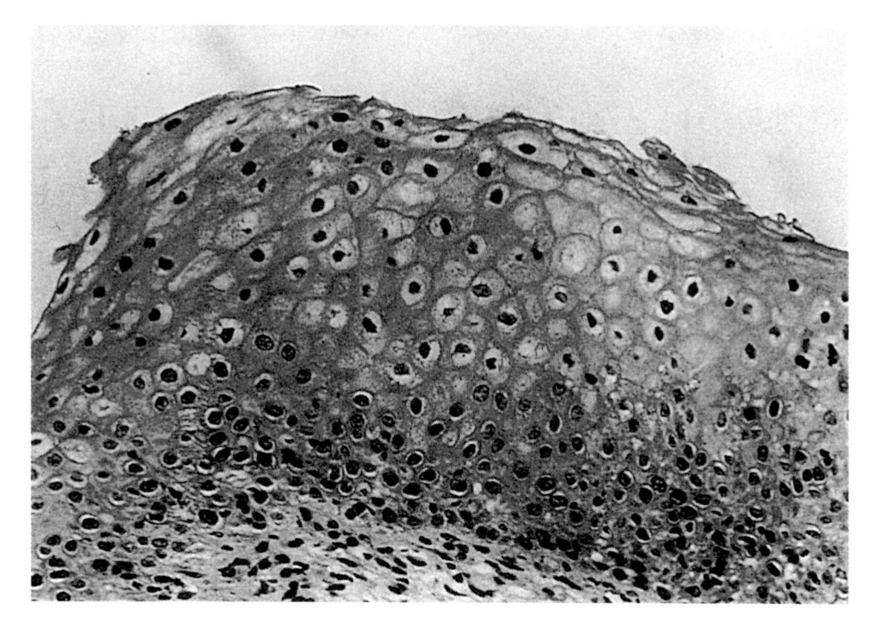

Figure 10.20 A discrete focus of SPI. Note koilocytosis in surface layers with slight loss of stratification. There is some basal hyperplasia (H & E, ×310).

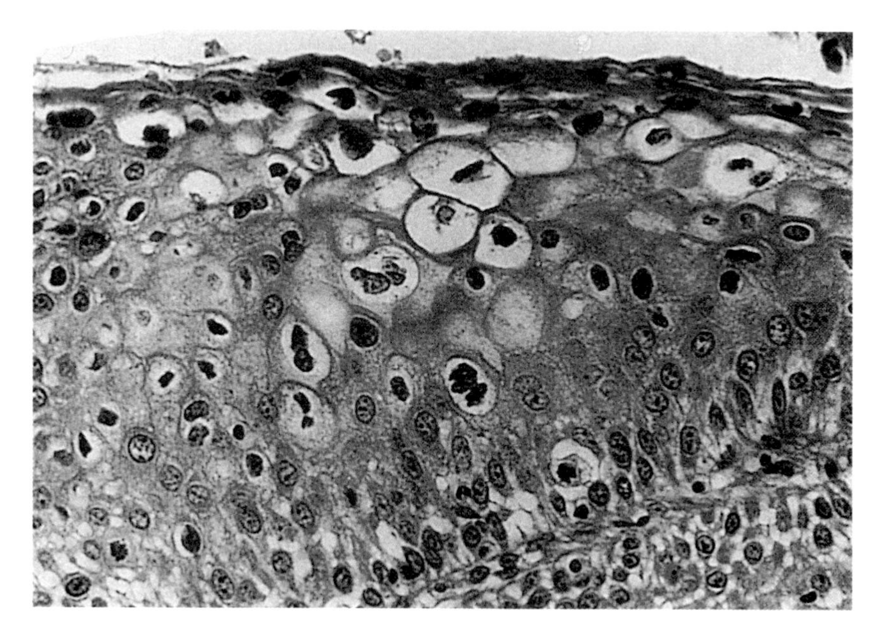

Figure 10.21 SPI. Note koilocytotic atypia, multinucleation and basal hyperplasia. This lesion was found separate from but adjacent to an area of CIN 3 (H & E, ×480).

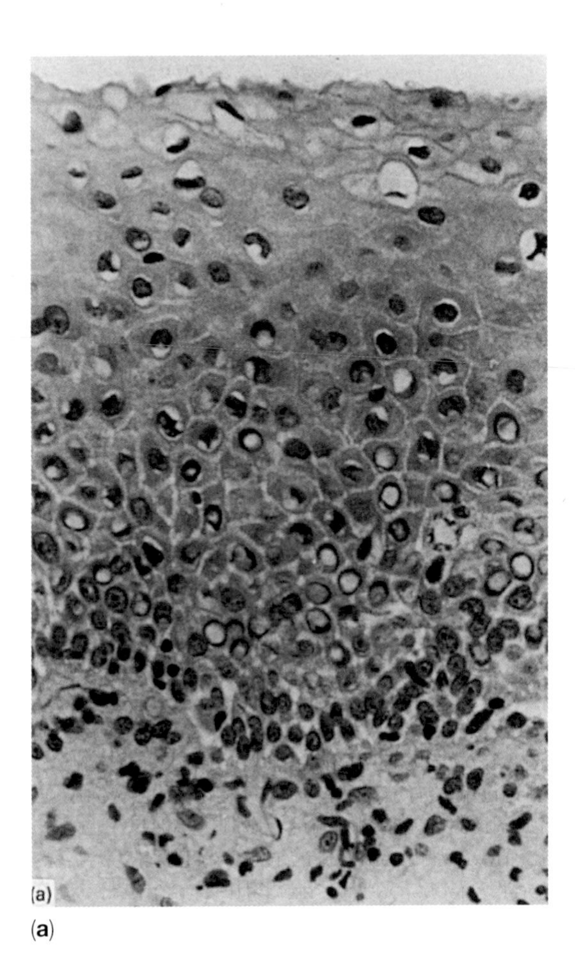

(**a**)

Figure 10.22 (**a**) SPI. Note vacuolation of nuclei in basal layers – an occasional finding (Pap, ×480). (**b**) SPI. Note vacuolated cytoplasm giving foamy appearance (H & E, ×770).

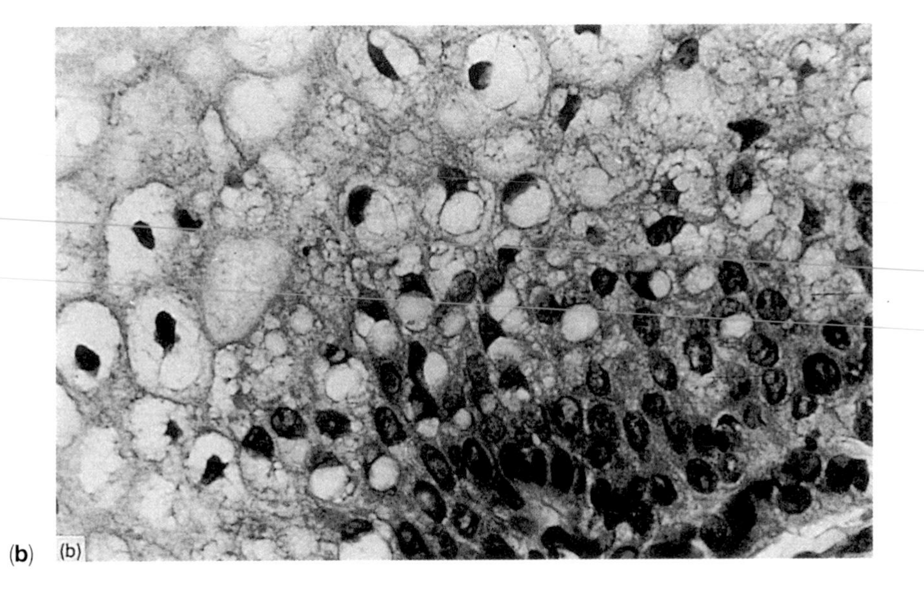

(**b**)

taken with an endocervical brush reflect endocervical crypt involvement by CIN 3. These include: (i) the presence of sheets or clusters of abnormal cells with pseudostratification and peripheral palisading; (ii) the presence of round or oval cells clusters with smooth borders; and (iii) central whorling in clusters of abnormal cells with flattening.

A more recent study of endocervical brush specimens and cone biopsies from patients with CIN 3 by Van Hoeven *et al.* (1996) showed that that there is no statistical association between these abnormal fragments in cervical smears prepared from endocervical brush specimens and endocervical gland involvement by CIN 3 in cone biopsies.

10.6 HUMAN PAPILLOMAVIRUS INFECTION AND CIN

The first evidence of a link between HPV infection and CIN was presented by Meisels and Fortin (1976) and Purola and Savia (1977) who were investigating the prevalence of wart virus infection in the cervix by light microscope examination of normal and abnormal cervical smears and biopsies. Meisels described morphological changes suggestive of a subclinical form of papillomavirus infection in 70% of abnormal smears and biopsies and in 2.4% of normal smears, indicating a strong association between papillomaviruses and neoplasia.

The validity of these observations was quickly confirmed by electron microscopy (Ferenczy *et al.*, 1981; Stanbridge *et al.*, 1981), immunocytochemistry (Ferenczy *et al.*, 1981; Walker *et al.*, 1983a) and DNA hybridization (Gissmann *et al.*, 1983; McCance *et al.*, 1983). Analysis of the viral genomes revealed that several different HPV types were present in the cervix, the most common being HPV 6, 11 and 16.

The cellular changes in the biopsies and smears were very similar to those seen in cervical condylomata (Section 7.13) and Meisels *et al.* (1977) suggested the term 'flat condyloma' to describe these lesions. Laverty *et al.* (1978) proposed the term 'non-condylomatous wart virus infection' (NCWVI) and Reid *et al.* (1982) used the phrase 'subclinical papillomavirus infection' (SPI), a term which is generally preferred.

10.6.1 Histological diagnosis of subclinical papillomavirus infection (SPI) of the cervix (synonyms: non-condylomatous wart virus infection (NCWVI), flat warts, flat condyloma, condyloma planum)

Comparison of the histology of colposcopic biopsy material with immunocytochemical staining of the same tissue for papillomavirus antigen has shown that the morphological changes in the cervical epithelium which are most consistently found in SPI are: koilocytotic atypia (Section 7.13); individual cell keratinization (Figure 10.2); and multinucleation (Figures 10.20 and 10.21).

These three features when found together provide indirect evidence of SPI. Other features such as surface keratinization, vacuolation of the nuclei or cytoplasm of cells in the parabasal zone (Figure 10.22) may occasionally be found but are also frequently found in other conditions and therefore are of little diagnostic value. It is important to remember that definitive evidence of papillomavirus infection depends on demonstration of virus particles, viral antigen or viral genome as described in Section 7.13.3.

Careful observation by Fletcher (1983) has shown that SPI may present histologically in three ways:

1. As a discrete focus of infection in an otherwise normal cervical epithelium. In these cases, koilocytosis, individual cell keratinization and multinucleation are present in the superficial and mid-zones of the epithelium while the basal zone appears normal or may exhibit basal hyperplasia. Although the nuclei of the koilocytes may

be enlarged, hyperchromatic or irregular, the cells in the basal layers are essentially normal. The focus of SPI may be the only pathological finding in the biopsy (Figure 10.20). This pattern is commonly found in women who have been subjected to colposcopic biopsy because their smears have repeatedly shown borderline nuclear changes or mild dyskaryosis.

2. As a discrete focus of infection *adjacent to but distinct from* an area of CIN (Figure 10.21).
3. As an area of cervical squamous epithelium which show changes of both SPI and CIN (Figures 10.1, 10.2(a), 10.4, 10.6 and 10.12). In these biopsies the neoplastic cells exhibit the changes of koilocytosis, individual cell keratinization and multinucleation, reflecting infection of the neoplastic epithelium with HPV. Meisels *et al.* (1977) suggested the term 'atypical condyloma' for these lesions but the term has not been widely accepted. SPI *associated with* CIN is a more acceptable way of describing these lesions.

This form of SPI is found most frequently in the low-grade CIN lesions. Meisels *et al.* (1977) reported finding these changes in 70% of colposcopic biopsies showing the changes of CIN 1 or 2 and 58% of biopsies showing changes of CIN 3. Markowitz *et al.* (1986) report the changes in 100% of cone biopsies. They attribute their success in finding the koilocytes to the greater amount of cervical tissue available for assessment in cone biopsies compared with colposcopic biopsies.

In fact, the increased prevalence of SPI in low-grade CIN reflects the fact that wart virus replication only occurs in keratinocytes; these are much more likely to be found in low-grade CIN lesions which always show surface differentiation than in CIN 3.

10.6.2 Colposcopic diagnosis of SPI

The claims by some gynaecologists (Meisels *et al.*, 1977, 1979) that SPI can be diagnosed with a high degree of accuracy on colposcopy have not been substantiated by others. Walker *et al.* (1983b) compared the histologic and colposcopic findings in 156 patients referred for colposcopy with an abnormal cervical smear. He found that in most cases, SPI was colposcopically indistinguishable from CIN. This is not surprising when the histological presentation of SPI is considered (Section 10.6.1).

Anderson *et al.* (1992) have described some colposcopic patterns which suggest the presence of a discrete focus of SPI infection: the presence of shiny, snow-white lesions with an irregular outline after the application of acetic acid to the cervix; satellite lesions extending well beyond the transformation zone; and capillary patterns which can rarely be distinguished from those of CIN. Lugol's iodine may be helpful in making the distinction between a discrete area of SPI and an area of CIN, as the neoplastic epithelium (in contrast to the HPV-infected epithelium) is unlikely to contain glycogen and will not take up the stain. The final diagnosis will be based on histological assessment of the biopsy.

10.6.3 Cytological diagnosis of SPI

Koilocytosis, multinucleation and dyskeratosis in the smear are indicative of human papillomavirus infection of the cervix (Figures 10.23 and 10.24) and reflect the histological changes described in Section 10.6.1. The koilocytes often show a degree of nuclear irregularity or hyperchromasia amounting to mild or moderate dyskaryosis. In such cases it is not possible to determine whether the virus-infected cells have been shed from a discrete focus of SPI or from SPI superimposed on a focus of CIN.

Occasionally, the nuclei of the koilocytes show only a slight deviation from the norm, appearing only slightly larger and darker-staining than normal intermediate cell nuclei. In these cases a diagnosis of borderline dyskaryosis (ASCUS) may be offered and a discrete focus of SPI may be suspected (Figure 10.25(a–c)). It is reasonable to assume that in

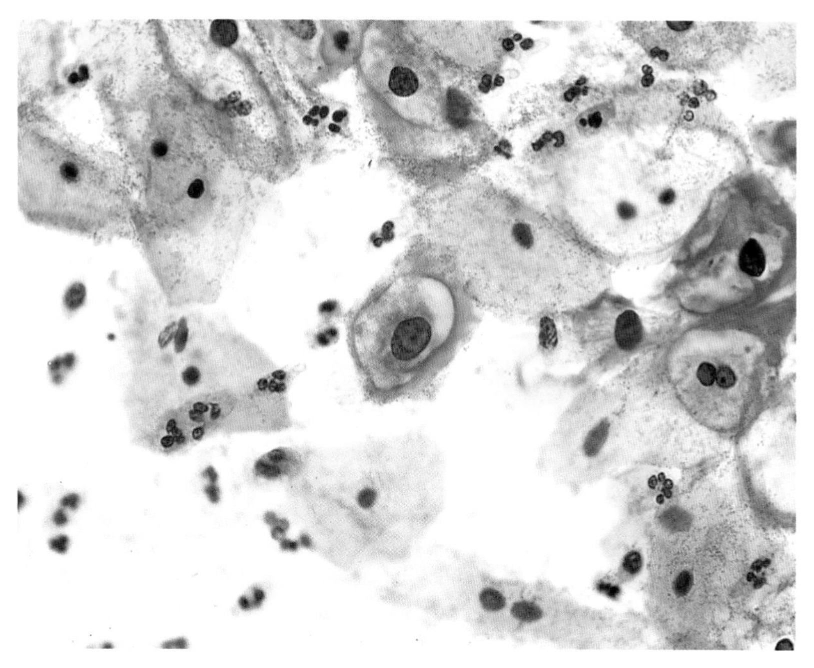

Figure 10.23 Koilocytes in cervical smear. Note clear zone around enlarged, hyperchromatic nuclei (mild dyskaryosis) and thickened rim of cytoplasm (Pap, ×630).

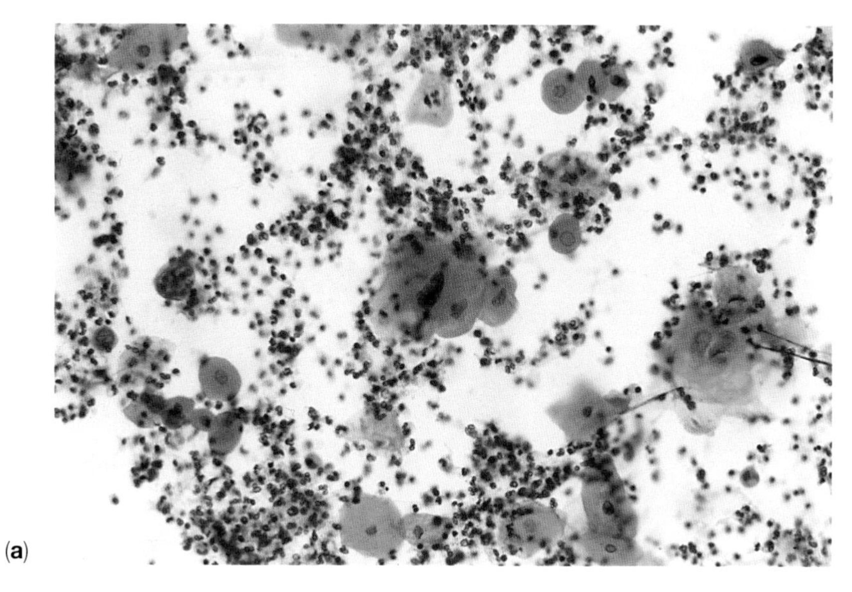

(**a**)

Figure 10.24 (**a–g**) Koilocytosis, multinucleate cells and sheets of highly keratinized anucleate squames and dyskeratotic cells suggestive of papillomavirus infection in a cervical smear: the nuclei of the koilocytes are enlarged and hyperchromatic (Pap, ×630).

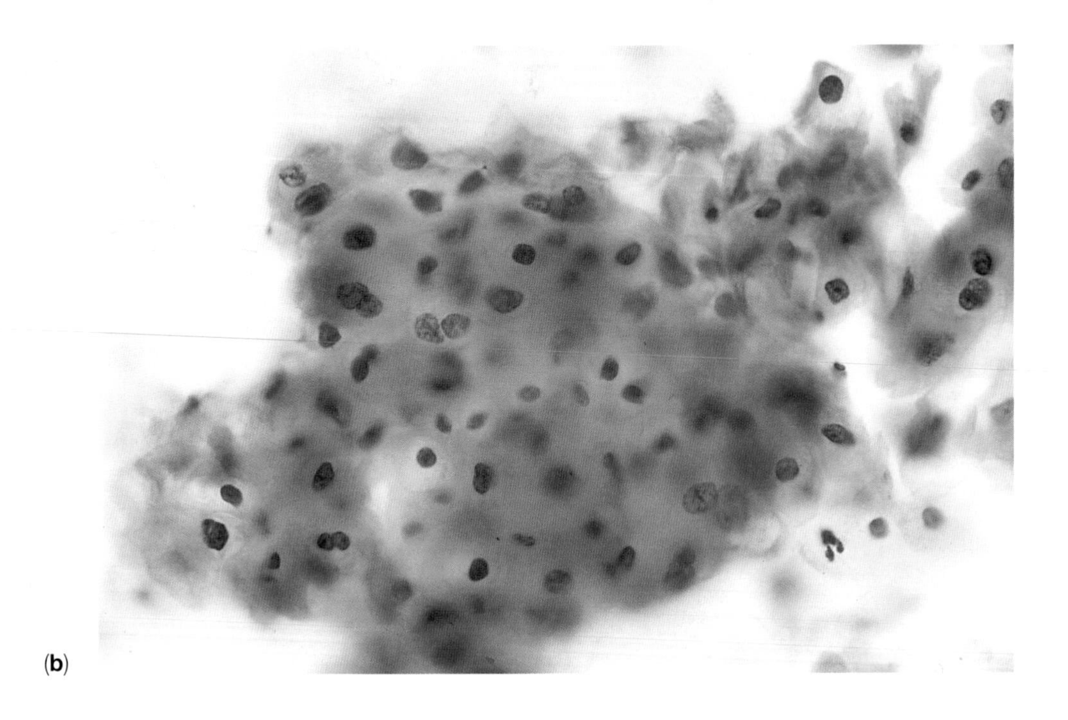

(b)

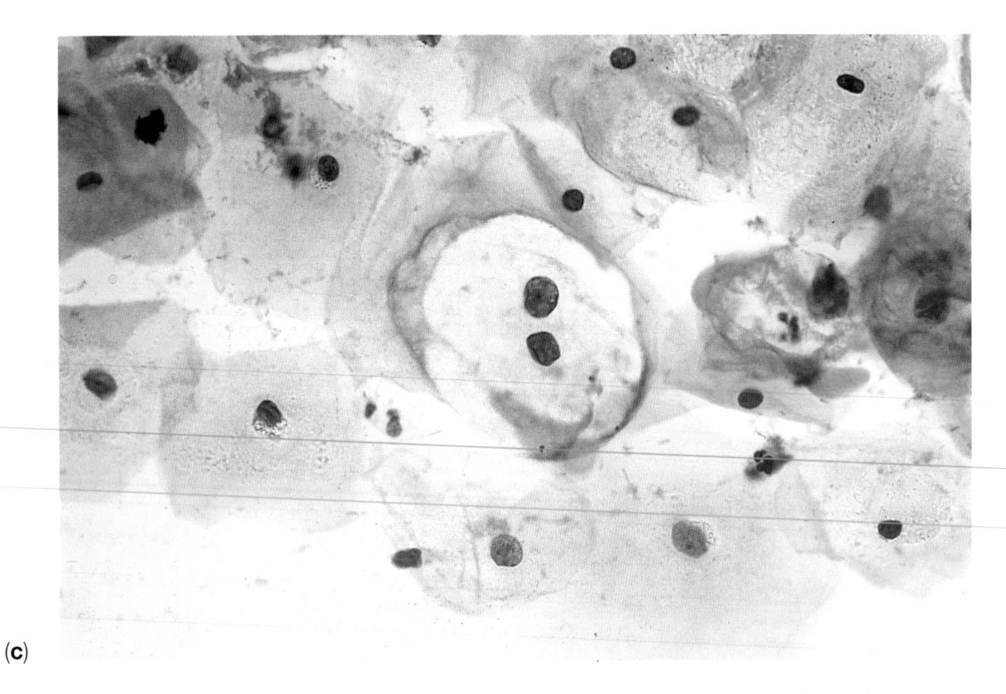

(c)

Figure 10.24 (**a–g**) Koilocytosis, multinucleate cells and sheets of highly keratinized anucleate squames and dyskeratotic cells suggestive of papillomavirus infection in a cervical smear: the nuclei of the koilocytes are enlarged and hyperchromatic (Pap, ×630).

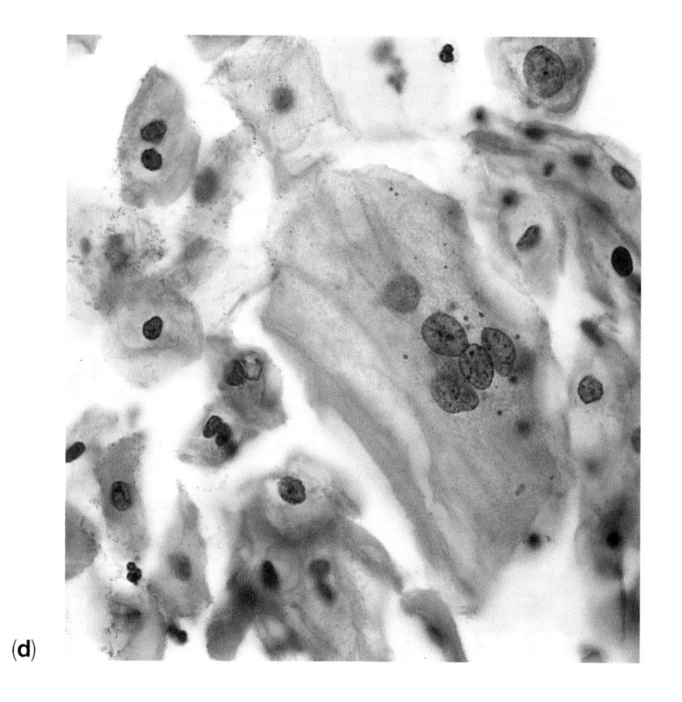

(**d**)

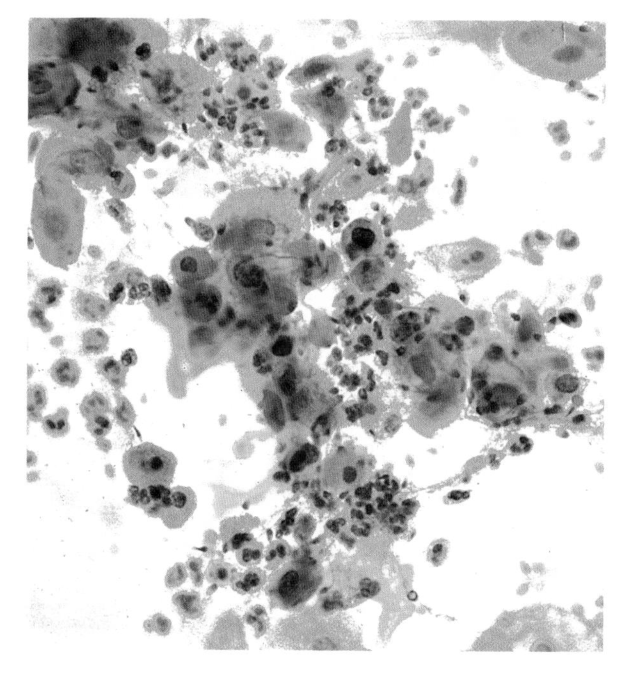

(**e**)

Figure 10.24 (**a–g**) Koilocytosis, multinucleate cells and sheets of highly keratinized anucleate squames and dyskeratotic cells suggestive of papillomavirus infection in a cervical smear: the nuclei of the koilocytes are enlarged and hyperchromatic (Pap, ×630).

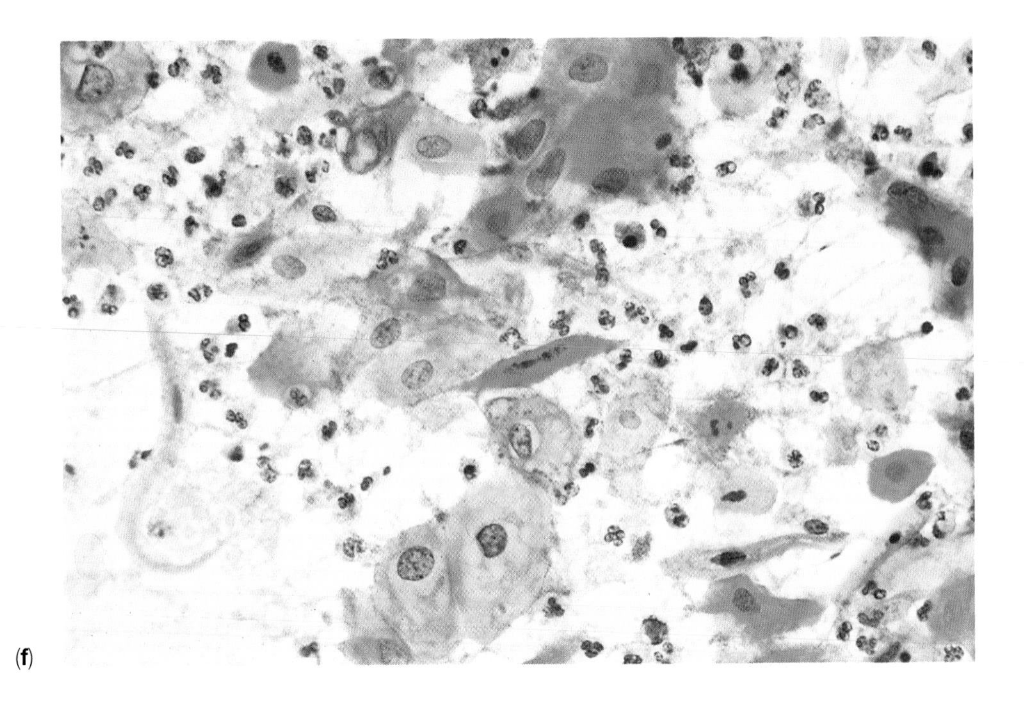

(**f**)

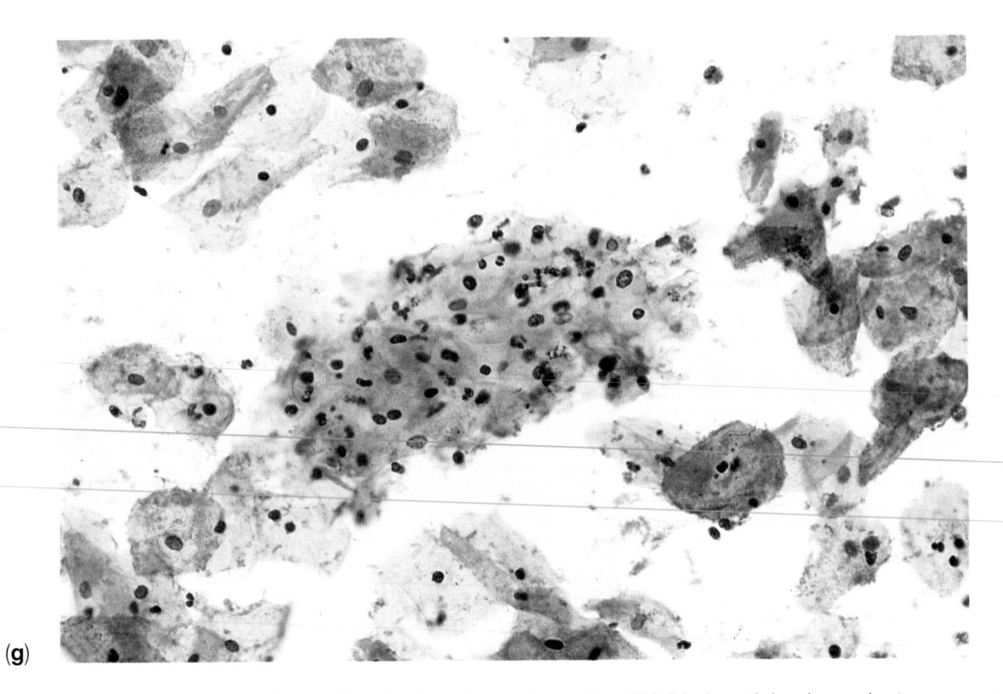

(**g**)

Figure 10.24 (**a–g**) Koilocytosis, multinucleate cells and sheets of highly keratinized anucleate squames and dyskeratotic cells suggestive of papillomavirus infection in a cervical smear: the nuclei of the koilocytes are enlarged and hyperchromatic (Pap, ×630).

the absence of mild, moderate or severe dyskaryosis, the virus infection is unassociated with neoplasia and may be the sole pathological change in the cervix.

10.6.4 Significance of human papillomavirus infection in the cervix

Women most at risk of cervical cancer are those who have multiple sexual partners, intercourse at an early age, and recurrent venereal infection (Rotkin, 1973). This strong association between sexual activity and cervical cancer led to speculation that a sexually transmitted agent may be involved in the initiation or promotion of cervical neoplasia.

Spirochaetes, spermatozoa, *Trichomonas vaginalis*, *Chlamydia trachomatis* and herpes virus type 2 have all been considered as potential carcinogens but proof has been lacking in every case. Recent studies have implicated the human papillomavirus in cervical carcinogenesis and the evidence to support this hypothesis is strong.

DNA studies of cervical biopsy material have shown that HPV DNA can be detected in 70% of cervical biopsies of intraepithelial or invasive cancer. The prevalence of the different HPV types in the biopsies varies with the severity of the cervical lesion. Thus, HPV 6 and 11 were found most commonly in low-grade CIN lesions (CIN 1 or 2) whereas HPV 16 or HPV 18 is found in 100% of CIN 3 and 90% of invasive cancers (Durst *et al.*, 1983). This preference of HPV 16 and HPV 18 for the high-grade CIN lesion and invasive cancers has led to the suggestion that women who harbour HPV 16 or HPV 18 have a high risk of developing invasive cancer. Indeed, there are reports that women with CIN 1 who harbour HPV 16 or 18 have a greater risk of progression to CIN 3 than women with CIN 1 who harbour HPV 6 or 11. A number of prospective studies of women with CIN 1 appear to demonstrate this (Kataja *et al.*, 1989; Koutsky *et al.*, 1992). In addition, the risk of progression increases with the severity of the lesion. Thus, in a clinical study of 532 women with cervical HPV infection, Kataja *et al.* (1989) found that there was clinical progression in 6% of those with no initial CIN compared with 12% for CIN 1, 20% for CIN 2 and 55% for CIN 3.

The hypothesis that HPV types appear to vary in their pathological potential, is strengthened by the fact that HPV 16 and HPV 18 are present in the invasive lesions in an integrated form (whereas in the preinvasive lesions the virus is episomal) and by the demonstration of HPV DNA sequences (E6 and E7 coding regions) in cancer cells. The HPV 16 E6 and E7 coding regions are analogous to similar regions of the bovine papillomavirus which are known to have transforming functions. Normally the E6 and E7 genes are under the control of the E2 gene and their transforming functions are suppressed. The effect of integration of HPV into the genome of the host cells is the loss of expression of the E2 gene and activation of E6 and E7. The E6 and E7 sequences code for proteins which bind to the product of the tumour suppressor genes *p53* and *pRB* respectively. This leads to degradation of these proteins' function with consequent acceleration of cell proliferation and cell cycle time. The *p53* also functions as a promoter of apoptosis (programmed cell death) and this too is interfered with by *p53* binding to the E6 protein

The oncogenic potential of HPV and their gene products have been demonstrated in several ways. Crawford and his colleagues have shown that the HPV E6 and E7 genes can immortalize baby rat kidney cells and, with the cooperation of the activated H-*ras* gene, transform them so that they are tumorigenic in syngeneic animals (Matlashewski *et al.*, 1987) This group also showed that this transforming activity was conferred by HPV 16, 18, 31 and 33 but not by HPV 6 and 11. Greenhalgh *et al.* (1994) demonstrated that transgenic mice which expressed targeted HPV 18 E6 and E7 in their epidermis develop verrucous lesions and spontaneous papillomas. Montgomery *et al.* (1995) have shown that a human keratinocyte cell line transfected with HPV 16 and HPV 18

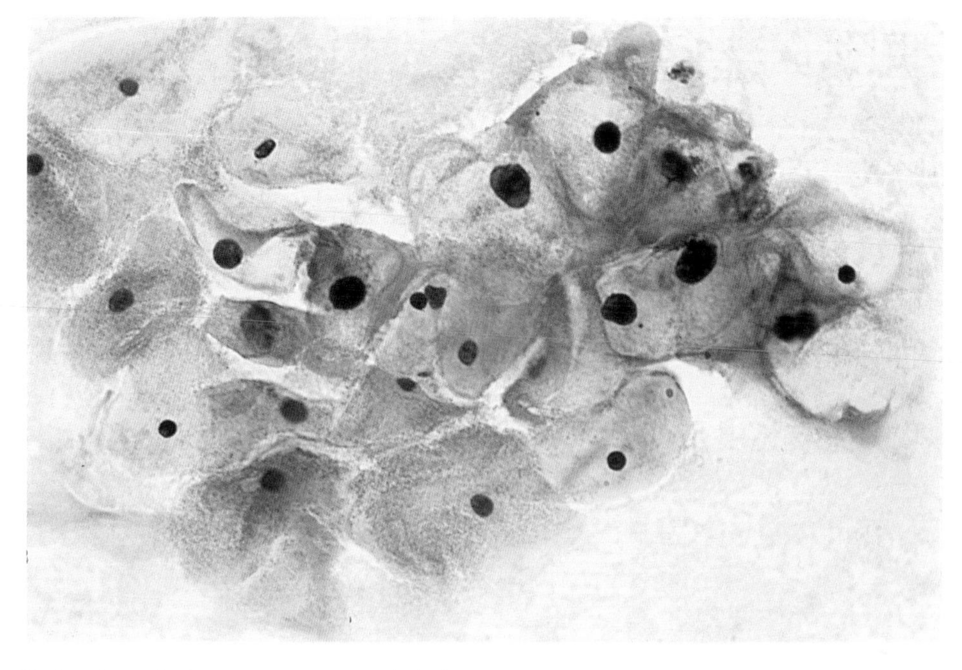

Figure 10.25 (**a**) Borderline nuclear changes in koilocytes.

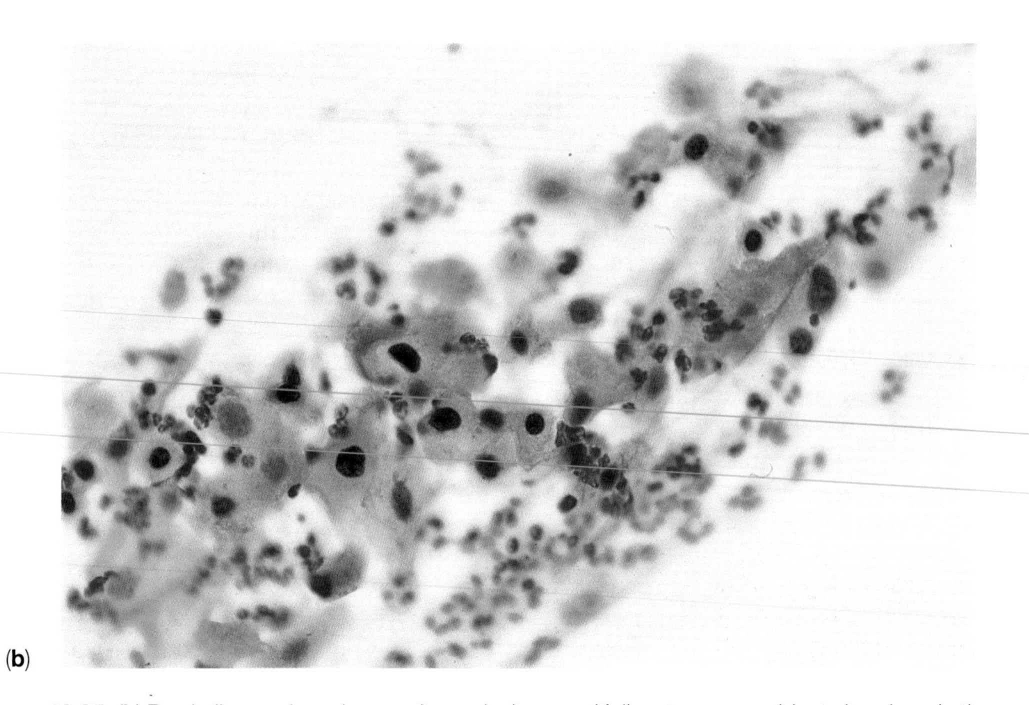

Figure 10.25 (**b**) Borderline nuclear changes in cervical smear. Koilocytes were evident elsewhere in the smear.

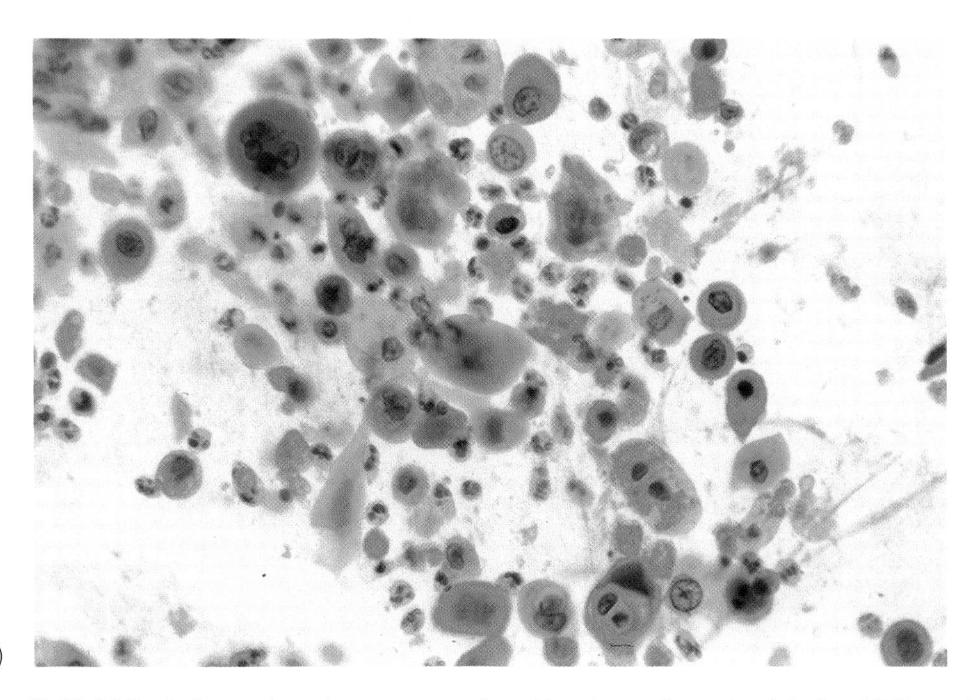

Figure 10.25 (**c**) Borderline nuclear changes suggestive of human papillomavirus infection (all Pap, ×630).

respectively will produce invasive squamous cell carcinomas when injected into nude mice.

Although the evidence linking HPV with cervical cancer is very strong, it cannot be assumed that the mere presence of HPV 16 or 18 in the female genital tract increases cancer risk. It must be remembered that papillomavirus infection of the genital tract is extremely common and HPV DNA has been demonstrated in cervical scrapes from up to 30% of women with cytologically and colposcopically normal cervices (Coleman *et al.*, 1986; Schiffman, 1992; Melkert *et al.*, 1993). At least 35 different HPV types have been isolated from the female genital tract, some of which have been shown *in vitro* to have an oncogenic potential similar to that of HPV 16 and 18, yet none shows the same association with cervical neoplasia. Moreover, epidemiological studies have shown that the incidence of cervical cancer in a population bears little relationship to the incidence of HPV infection in the same population (Van Herckenrode *et al.*, 1992). In addition, the studies which claim to show rapid progression of CIN 1 to CIN 3 in women with HPV 16 infection of the cervix are highly subjective and subject to bias. It is probable that other factors such as integration of the viral genome into the host DNA, hormonal or immunological factors, play an important role in the neoplastic process.

HPV infection of the male partner in cervical carcinogenesis has been extensively studied particularly in relation to the sexual transmission of high-risk HPV types from male to female, but the results of two major studies are conflicting. Bosch *et al.* (1996) studied genital HPV infection in husbands of Spanish women with cervical cancer and a control group of Spanish women who had normal cervixes. They reported that the presence of HPV DNA in the husband's penis conveyed a five-fold risk of cervical cancer in their wives. The risk was particularly associated with HPV type and number of extramarital partners. Munoz *et al.* (1996) carried out a similar study in Columbia but were unable to demonstrate the increased risk of cervical cancer in wives of husbands who were HPV DNA-positive. They drew attention to the difficulty in interpreting such studies, since Spain has a low incidence of cervical cancer and Columbia has one of the highest. These findings confirm the view stated above that other factors have a role to play in the genesis of cervical cancer.

Until the relationship between CIN and human papillomaviruses is clarified, evidence of this infection in a smear or biopsy should not modify the clinical approach. Thus, a woman whose smear shows wart virus changes and dyskaryosis should be managed according to the degree of dyskaryosis (Kaufman *et al.*, 1983).

Because the evidence linking HPV with cervical cancer is so compelling, there are epidemiologists and pathologists who advocate HPV testing of women at risk of cervical cancer. They propose that HPV testing should be carried out either to complement the Papanicolaou cervical smears test or to replace it (Cuzick *et al.*, 1994, 1995). They cite studies which show that HPV DNA analysis of cervical cells and tissue for oncogenic virus types, e.g. HPV 16, 18, 31 and 33 can be used to predict CIN 3 with a high degree of reliability (Cuzick *et al.*, 1992, 1994, 1995; Cox *et al.*, 1995), and suggest that HPV testing could be used to triage the large number of women with borderline (ASCUS) or mildly dyskaryotic smears who are deemed to have a low but real risk of developing CIN 3 or invasive cancer. This group of women have to be monitored closely by repeated smear tests or referred for colposcopy. Both management strategies are costly and provoke anxiety in the patient; thus, a realistic alternative approach would be welcomed.

At the time of writing, most clinicians are reluctant to introduce HPV testing into the cervical screening programme. There are good reasons for this. Firstly, the sensitivity and specificity of current methods of testing for HPV are not known. Secondly, screening for HPV will inevitably miss the few cases of CIN 3 and invasive cancer that are not HPV-associated. Thirdly, screening for oncogenic human papillomaviruses will inevitably identify the 12–40% of women with normal cervices who harbour the virus. No protocol has been proposed with regard to the management of these women who may be wrongly labelled an 'at-risk' group for the rest of their lives.

For a comprehensive review of the relationship between HPV and cervical neoplasia readers are referred to Herrington (1995).

10.7 DIAGNOSTIC PITFALLS IN BIOPSIES AND SMEARS

A number of physiological and pathological changes in the cervix may give rise to an erroneous diagnosis of CIN in smears or biopsies or alternatively cause the presence of CIN to be overlooked. These include reaction to infection, trauma and irradiation, atrophic and inflammatory changes, reserve cell hyperplasia, immature squamous metaplasia and basal hyperplasia. Recognition of these pitfalls is the most important factor in their avoidance.

10.7.1 Diagnostic pitfalls in histology

(a) Papillomavirus infection

The distinction between a discrete focus of SPI in mature squamous epithelium and SPI associated with CIN 1 or 2 may be difficult (Figures 10.26 and 10.27), especially if the koilocytes have enlarged, irregular and hyperchromatic nuclei. The problem can usually be resolved by careful examination of the basal layers. In the absence of CIN, the basal nuclei will appear monomorphic and regular in outline and the mitotic figures will be normal. A diagnosis of SPI associated with CIN should be given if the cells in the basal layers show neoplastic change (nuclear enlargement,

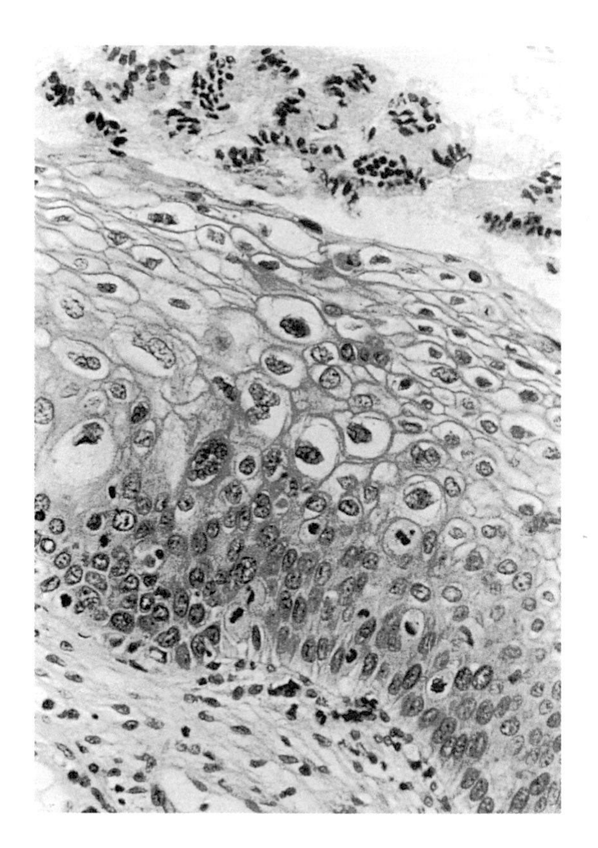

Figure 10.26 The distinction between CIN 1 with superimposed wart virus changes and a discrete focus of SPI may be difficult. In this example, a diagnosis of wart virus changes superimposed on CIN 1 was made because of the presence of abnormal mitoses high in the epithelium, although the basal cells show only a slight nuclear atypia (H & E, ×310).

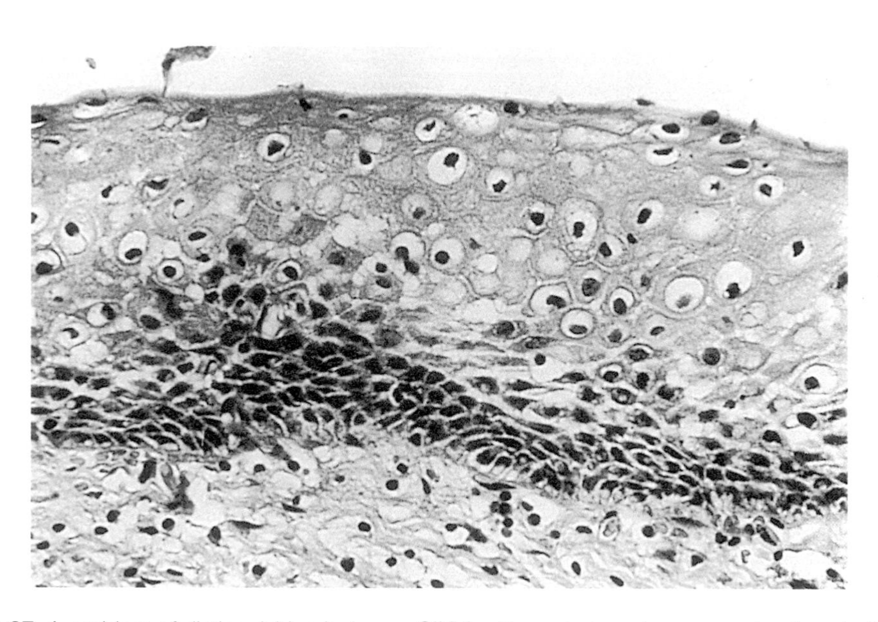

Figure 10.27 A problem of distinguishing between CIN 1 with wart virus changes and a discrete focus of SPI. This section was diagnosed as a focus of SPI. There were very few mitoses and although the basal cells appear hyperchromatic, they were monomorphic. Note intercellular bridges (H & E, ×310).

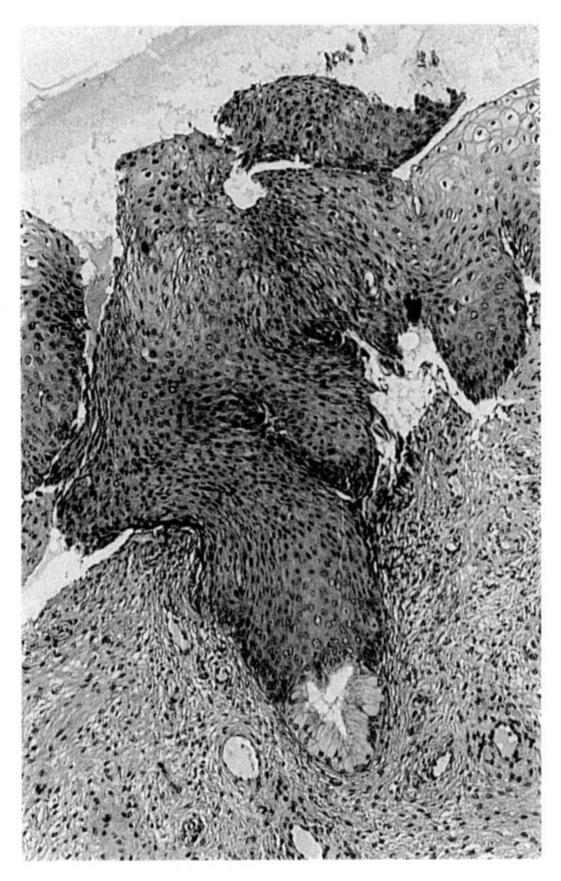

Figure 10.28 Squamous metaplasia in gland (H & E, ×310).

hyperchromasia and pleomorphism) and abnormal mitotic figures are seen.

Koilocytosis in the surface layers of the cervical epithelium must be differentiated from the normal basket-weave pattern of the cervix. The loss of stratification and the presence of abnormal nuclei associated with koilocytotic atypia should not be confused with the regular basketweave pattern which is probably a histological artefact due to the glycogen content of the surface epithelial cells.

The presence of koilocytotic atypia in a focus of immature squamous metaplasia may also be confused with CIN 3 (Crum *et al.*, 1983). Again, careful scrutiny of the nuclei in the deeper layers of the epithelium will ensure that a correct diagnosis is made.

(b) Basal hyperplasia

In basal hyperplasia, two or more layers of basal cells are present. It characteristically affects the ectocervix (Burghardt, 1973) and in simple basal hyperplasia, maturation of the overlying epithelial cells is normal. Basal hyperplasia can occur in response to oral contraceptives and the IUCD (Figure 9.29), during pregnancy (Maqueo *et al.*, 1966) and as part of the healing process, e.g. following ulceration, or in response to infection. It is very commonly seen in association with koilocytotic atypia in a discrete focus of subclinical papillomavirus infection (Figures 10.20 and 10.21). The presence of normal cells in the upper layers of the epithelium and the regularity of the cells in the hyperplastic basal layers should make an accurate diagnosis possible. Burghardt (1973) has described atypical basal hyperplasia in which the proliferating cells in the basal layers of the epithelium are quite abnormal, whereas the superficial layers are normal. This rare lesion may be an unusual manifestation of CIN.

(c) Immature squamous metaplasia, reserve cell hyperplasia and tuboendometrioid metaplasia

Immature squamous metaplasia may present a problem when it extends into glands (Figures 10.28 and 10.29). The absence of surface differentiation and the apparent downgrowth of the epithelium may mislead the pathologist into reporting CIN or even invasive cancer. Careful examination of the morphology of the individual cells will permit a correct diagnosis. Coincident infection may result in a very bizarre pattern (Figure 9.5(a)) and frequent mitotic figures may give cause for concern. The overall pattern, more readily appreciated at low magnification, should enable a distinction to be made. Tuboendometrioid metaplasia may also give rise to errors of diagnosis. It is most likely to be mistaken for glandular neoplasia (Section 9.3.3).

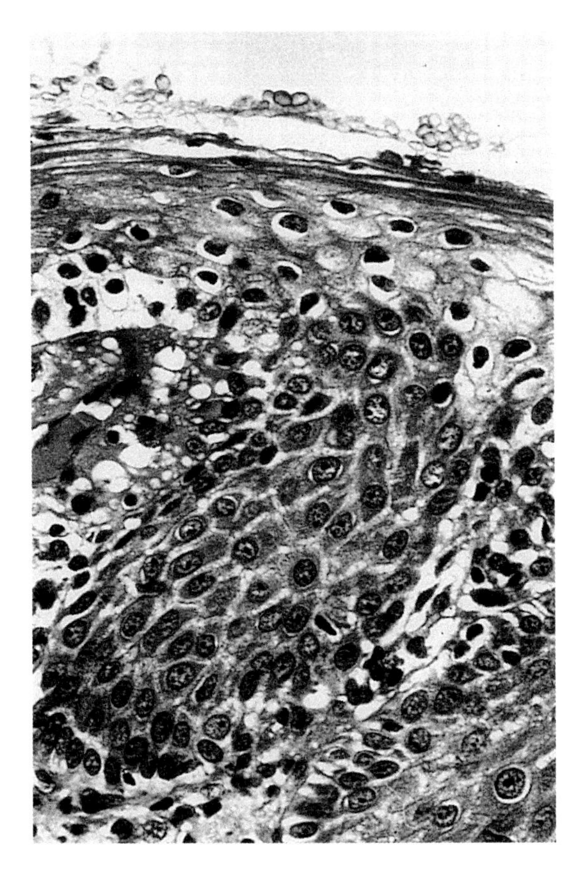

Figure 10.29 Squamous metaplasia in gland. The appearance is misleading due to tangential cutting (H & E, ×770).

(d) Atrophic squamous epithelium

The thin epithelium in post-menopausal women may be composed of cells with little cytoplasm giving an impression of nuclear crowding (Figure 7.12). The nuclei may show degenerative changes and appear irregular. There may be slight hyperchromasia. If there is marked pleomorphism CIN must be considered, and the presence of mitoses in the epithelium make a diagnosis of CIN more likely. Grading is difficult on thin epithelium and should be based on the degree of nuclear abnormality. 'CIN grade not specified' may be offered as a diagnosis (Anderson *et al.*, 1991).

10.7.2 Diagnostic pitfalls in cytology

A number of pathological changes in the cervix may give rise to problems of interpretation of the cervical smear resulting in genuine doubt as to whether the cytological changes seen in the smear are neoplastic or reflect inflammatory or reactive change. The British Society for Clinical Cytology have recommended that this type of cytological pattern be designated 'borderline change'. In the Bethesda System this pattern is described as 'atypical squamous cells of unknown significance (ASCUS) or atypical glandular cells of unknown significance (AGUS)'. The cytological patterns which may fall into this category are itemized below.

(a) Immature squamous metaplasia and reserve cell hyperplasia

These changes may produce clusters or sheets of cells which have enlarged nuclei and prominent chromocentres (Figure 9.5(b–f)). They can usually be recognized by the overall uniformity of size and shape and the absence of angularities in the nuclear outline.

(b) Inflammation and repair

Inflammatory changes not infrequently cause problems in the interpretation of cervical smears, especially if they occur in immature metaplastic cells or endocervical cells. The cells may reflect degenerative change if the smear is taken during the destructive phase of the inflammatory process, or regenerative changes as the epithelium undergoes repair. Degenerating nuclei may appear large but pale-staining in cervical smears or they may have an abnormal chromatin pattern due to intranuclear coagulation necrosis. Regenerating epithelium may result in appearance of sheets of large hyperchromatic nuclei with prominent nucleoli in the smear (Figure 7.7 (a–d)). The distinction between dyskaryosis and inflammatory change depends on careful examination of the nuclei of the cells. Dyskaryotic cells show a degree of pleomorphism rarely seen in inflammatory smears. The uniformity of regenerating cells in cervical epithelium is particularly striking. If doubt remains, a report of borderline changes (ASCUS) should be given and follow-up advised (Section 2.4).

(c) Irradiation and cytotoxic therapy

Both these forms of treatment produce abnormal cellular changes (Sections 9.6.2 and 9.6.7). In a histological section the general pattern (Figures 9.23 and 9.24) is usually sufficiently different from CIN for confusion to be unlikely. For a cytologist basing an opinion solely on individual cells, the distinction is not always so obvious (Figures 9.26(e, f), 9.34 and 9.35). Following irradiation, there are often features such as histiocytic reaction, karyorrhexis or giant epithelial cells (Section 9.6.2) which provide a clue. The provision of an adequate history on the request form is really the key to correct diagnosis in these cases. However, it must be remembered that irradiation can itself induce malignant change, or may have failed to eradicate completely the neoplastic change which it was used to treat.

(d) Atrophic cervicitis

The nuclei of the epithelial cells are often pyknotic and hyperchromatic, and the cytoplasm may be strongly eosinophilic. The lack of nuclear structure becomes apparent on scrutiny so that a false-positive report can be avoided.

The identification of dyskaryotic cells in atrophic smears is particularly difficult (Figure 10.17). They usually have irregularly shaped nuclei that are only slightly enlarged. The chromatin content is increased but the chromatin pattern is indistinct and the cell may have lost its cytoplasm due to degenerative changes. Blood and pus in the smear also affect the staining adversely and careful screening is required to prevent a false-negative report being given.

(e) Herpes genitalis, human polyomavirus, human papillomavirus

The multinucleate giant cells of herpes (Figure 7.33) have in the past been misdiagnosed as dyskaryotic cells. However, an increasing awareness of this infection makes misdiagnosis less likely today. The structureless appearance of the nucleus of the herpes-infected cell and the margination of the chromatin should point to the correct diagnosis. Human polyomavirus-infected cells (Figure 9.35) contain large basophilic intranuclear inclusions and closely resemble dyskaryotic cells. The lack of structure in the nucleus, thickened nuclear membrane and the uniformity of nuclear size are indicative of this virus. Electron microscopy may be needed to confirm the diagnosis.

The cytological changes associated with human papillomavirus infection are discussed in Sections 10.6.3 and 7.13.2 (see also Figures 10.25 and 7.39–7.46).

(f) Cervical smear in pregnancy

Screening of pregnant women for cervical cancer is undertaken at many antenatal clinics in the UK. The smears are often cytolytic and it can be very difficult to distinguish dyskaryotic cells from the cell debris, Döderlein bacilli and free nuclei in the smear (Figure 10.16). The dyskaryotic cells may also be devoid of cytoplasm but the abnormal nuclei can be recognized by virtue of their hyperchromasia and large size.

(g) Hyperkeratosis and dyskeratosis

CIN may occasionally be associated with marked keratinization of the surface layers of the epithelium (Figure 10.12(a)). In these cases, the smear may contain numerous highly keratinized anucleate squames or dyskeratocytes (Figure 10.12(c,d)) and a significant underlying cervical lesion may not be apparent. Colposcopic biopsy is advisable in these cases (Section 9.4.2).

10.7.3 Pale dyskaryosis

The key morphological features of a dyskaryotic squamous cell are nuclear enlargement, increased nuclear/cytoplasmic ratio, irregularity of nuclear outline and abnormal chromatin pattern. Hyperchromasia is a common finding and makes the diagnosis easier for the cytotechnologist as the dyskaryotic nucleus stands out against a background of normal nuclei. However, it tends to be forgotten that not all dyskaryotic nuclei are hyperchromatic and in some cases the chromatin is sparse and the nuclei appear pale. These cells have been described as showing 'pale dyskaryosis' (Smith and Turnbull, 1997). Examples of pale dyskaryosis in a smear from a woman with invasive squamous cancer are shown in Figure 10.15(a,b). There is some controversy regarding the origin of such cells. Dudding (1996) considers them to be a fixation artefact, while others consider them to be examples of cancer cells with a hypodiploid chromosome complement. Whatever the explanation, care should be taken to ensure that these cells are not overlooked and a false-negative report issued.

10.8 OBSERVER VARIATION IN REPORTING CIN

Ismail *et al.* (1989) reported a study in which eight experienced histopathologists (including one of the authors of this book), based at different hospitals examined the same set of 100 consecutive colposcopic biopsies and assigned them to six diagnostic categories, namely normal epithelium, non-neoplastic squamous proliferation, CIN 1, CIN 2, CIN 3 and other. Agreement between observers was excellent for invasive lesions, moderately good for CIN 3 and poor for low-grade lesions. The most serious disagreement was over reactive change and CIN 1. Similar findings were reported by Robertson *et al.* (1989). It was suggested by Ismael *et al.* that histologists should assign such sections to a 'borderline' category (comparable with the category of 'borderline' in cytology), entailing follow-up without treatment. This idea is considered by Anderson *et al.* (1991) who also propose a borderline histological category designated 'basal abnormality of uncertain significance'. The histological features of this entity are either: (i) a minimal degree of nuclear pleomorphism limited to the basal layers in the absence of inflammation; (ii) features of CIN 1 in the presence of severe inflammation; or (iii) a thin epithelium in which a diagnosis of CIN (grade not specified) would have been appropriate but for the presence of severe inflammation. However, the concept of a fourth histological category did not find widespread support and Anderson *et al.* concluded that the three-grade CIN system should be retained for the present.

10.9 RESIDUAL CIN AND RECURRENCE AFTER ABLATIVE THERAPY AND CONE BIOPSY

The colposcopic, cytological and histological findings after ablative therapy of the cervix may be confusing for several weeks after therapy (Section 9.6.1); therefore, the changes in the smear or biopsy taken within 4 months of therapy should be interpreted with caution. Following inadequate treatment, dyskaryotic cells from residual CIN may not be evident in the first smears after treatment as the focus of CIN may be too deep in a gland crypt or too high in the cervical canal to be detected with the Ayre spatula. For this reason a combined endocervical brush and cervical scrape is recommended for the follow-up of women after treatment of CIN. Residual CIN (Figure 10.30) may be suspected if dyskaryotic cells are present in the smear within one year of treatment. Thereafter, recurrence would seem to be more likely (Figures 10.31 and 10.32).

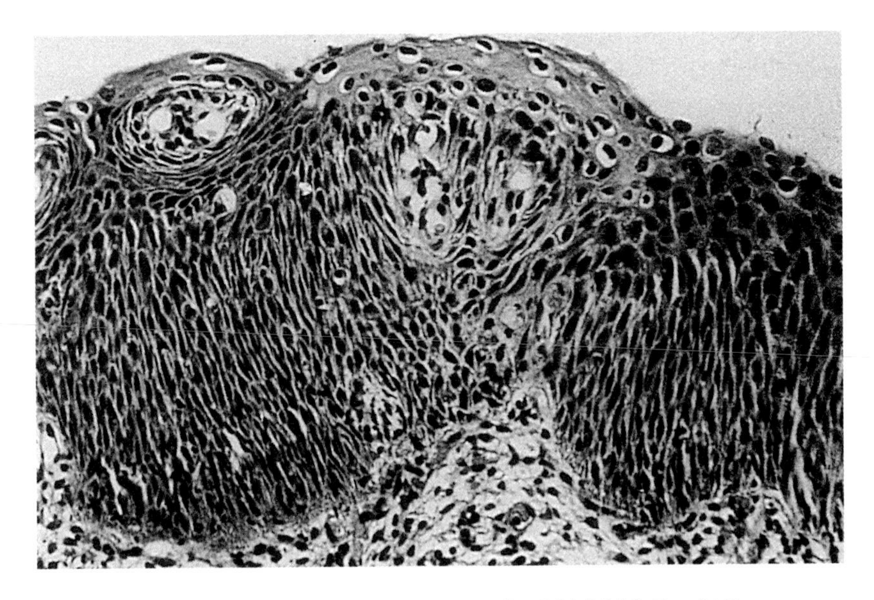

Figure 10.30 Residual CIN 2, 10 months after laser therapy for CIN 3 (H & E, ×310).

Figure 10.31 Recurrence of CIN 3, 18 months after laser therapy (H & E, ×310).

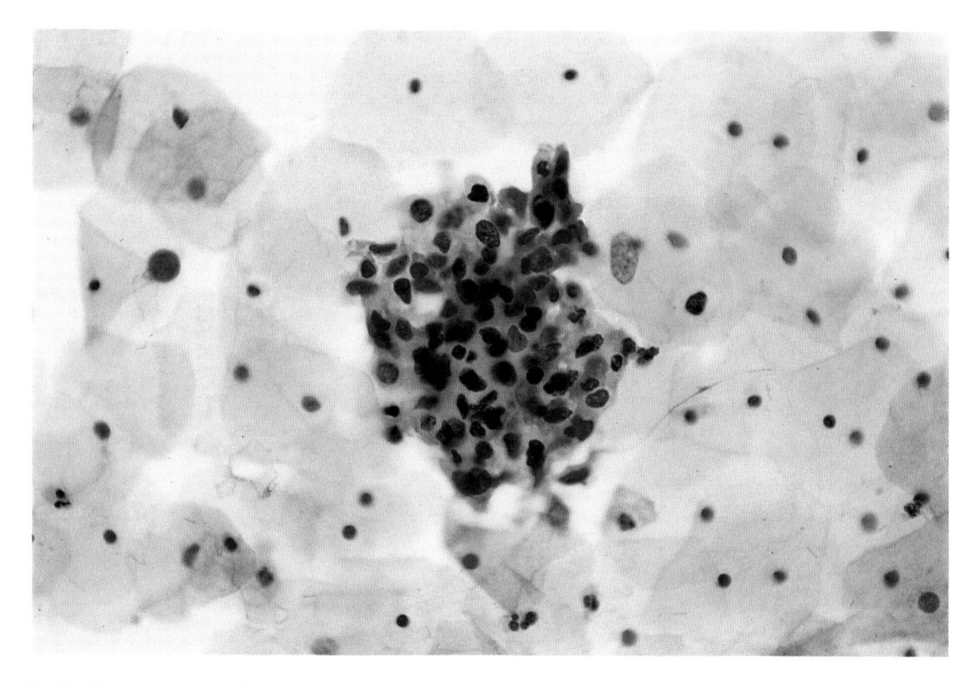

Figure 10.32 Recurrence of CIN detected by follow-up cervical cytology (Pap, ×630).

Paraskevaidis *et al.* (1991) reported that of 2130 women who received laser treatment for CIN, 119 (5.6%) were found to have subsequent CIN lesion; 71% were detected in the first year of follow-up, 24% during the second year, 3.3% in the third year and 1.7% in the sixth year. Most were detected by cytology, but 18% were detected colposcopically in spite of negative cytology.

In a series of 96 women treated for hysterectomy within 8 weeks of cone biopsy for CIN 3, Demopoulus *et al.* (1991) found that the only variable in the cone biopsy that could predict the presence of residual disease at hysterectomy was positive excision margins. In view of this, laser loop excision is preferred to colpobiopsy followed by cryotherapy for the diagnosis and treatment of CIN (Vergote *et al.*, 1992).

REFERENCES

Anderson, M. C. (1985) The pathology of cervical cancer. *Clin. Obstet. Gynecol.*, **12**, 87–119.

Anderson, M. C. and Hartley, R. B. (1980) Cervical crypt involvement by intraepithelial neoplasia. *Obstet. Gynecol.*, **55**, 546–550.

Anderson, M. C., Brown, C. L., Buckley, C. H. *et al.* (1991) Current view on cervical intraepithelial neoplasia. *J. Clin. Pathol.*, **44**, 969–978.

Anderson, M. C., Jordan, J., Morse, A., Sharp, F. (1992) *A Text and Atlas of Integrated Colposcopy.* Chapman & Hall, London, p. 131.

Bosch, F. X., Castellsague, X., Munoz, N., *et al.* (1996) Male sexual behaviour and human papillomavirus DNA: key risk factors for cervical cancer in Spain. *J. Natll. Cancer Inst.*, **88**, 1060–1067.

Broders, A. C. (1932) Carcinoma *in situ* contrasted with benign penetrating epithelium. *JAMA*, **99**, 1670–1674.

Buckley, C. H., Butler, E. B. and Fox, M. (1982) Cervical intraepithelial neoplasia. *J. Clin. Pathol.*, **35**, 1–13.

Burghardt, E. (1973) Early histological diagnosis of cervical cancer, in *Major Problems in Obstetrics and Gynaecology* (ed. E. A. Friedman), W. B. Saunders, Philadelphia, vol. 6, p. 43.

Burghardt, E. (1976) Premalignant conditions of the cervix. *Clin. Obstet. Gynecol.*, **3**, 257–295.

Cocker, J., Fox, H. and Langley, F. A. (1968) Consistency in the histological diagnosis of epithelial abnormalities of the cervix uteri. *J. Clin. Pathol.*, **21**, 67–70.

Coleman, D. V., Wickenden, C. and Malcolm, A. D. B. (1986) Association of human papillomavirus with squamous carcinoma of the uterine cervix. In *Papillomaviruses* (Ciba Foundation Symposium No. 120), Wiley, Chichester, pp. 175–189.

Cox, J., Lorincz, A. T., Schiffman, M. H. *et al.* (1995) Human papillomavirus testing by hybrid capture appears to be useful in triaging women with a cytologic diagnosis of atypical squamous cells of undetermined significance. *Am. J. Obstet. Gynecol.*, **172**, 946–954.

Crum, C. P., Egawa, L. and Fu, Y. S. (1983) Atypical immature metaplasia (AIM) a subset of human papillomavirus infection of the cervix. *Cancer*, **51**, 2214–2219.

Cuzick, J., Terry, G. and Ho, L. (1992) Human papillomavirus type 16 DNA in cervical smears as a predictor of high grade cervical cancer. *Lancet*, **339**, 959–960.

Cuzick, J., Terry, G., Ho, L. *et al.* (1994) Type specific human papillomavirus DNA in abnormal smears as a predictor of high grade cervical intraepithelial neoplasia. *Br. J. Cancer*, **124**, 13–20.

Cuzick, J., Szarewski, A., Terry, G. *et al.* (1995) Human papillomavirus testing in primary cervical screening. *Lancet*, **345**, 1533–1536.

Demopoulos, R. I., Horowitz, L. F. and Vamrakis, E. C. (1991) Endocervical gland involvement by cervical intraepithelial neoplasia grade III. Predictive value for residual and/or recurrent disease *Cancer*, **68**, 1932–1936.

Dudding, N. (1996) Pale dyskaryosis in cervical smears. *Cytopathology*, **7**, 221.

Durst, M. I., Gissmann, L., Ikenberg, H. and Zur Hausen, H. (1983) A papillomavirus DNA from a cervical carcinoma and its prevalence in cancer biopsy samples from different geographic regions. *Proc. Natl Acad. Sci. USA*, **80**, 3812–3815.

Evans, D. M. D., Hudson, E. A., Brown, C. L., Boddington, M. M., Hughes, H. C., Mackenzie, E. F. D. and Marshall, T. (1986) Terminology in gynaecological cytopathology: report of the working party of the British Society for Clinical Cytology. *J. Clin. Pathol.*, **39**, 933–944.

Ferenczy, A., Braun, L. and Shah, K. V. (1981) Human papillomavirus (HPV) in condylomatous lesions of cervix: a comparative ultrastructural and immunohistochemical study. *Am. J. Surg. Pathol.*, **5**, 661–670.

Ferenczy, A. (1982) Cervical intraepithelial neoplasia, in *Pathology of the Female Genital Tract* (ed. A. Blaustein), 2nd edn., Springer-Verlag, New York, Heidelberg, Berlin, pp. 156–177.

Fidler, H. K., Boyes, D. A. and Worth, A. J. (1968) Cervical cancer detection in British Columbia. *J. Obstet. Gynaecol. Br. Commonwlth.*, **75**, 392–404.

Fletcher, S. (1983) Histopathology of papillomavirus infection of the cervix uteri: the history, taxonomy, nomenclature and reporting of koilocytotic dysplasia. *J. Clin. Pathol.*, **36**, 615–624.

Fox, C. H. (1967) Biologic behaviour of dysplasia and carcinoma *in situ*. *Am. J. Obstet. Gynecol.*, **99**, 960–972.

Gissmann, L., Wolnik, L., Ikenberg, H., Koldovsky, U., Schnurch, H. G. and Zur Hausen, H. (1983) Human papillomavirus type 6 and 11 DNA sequences in genital and laryngeal papillomas and some clinical cancers. *Proc. Natl Acad. Sci. USA*, **80**, 560–563.

Greenhalgh, D. A., Wang, X. J., Rothnagel, J. A. *et al.* (1994) Transgenic mice expressing targeted HPV 18 E6 and E7 oncogenes in the epidermis, develop verrucous lesions and spontaneous ras-Ha-activated papillomas *Cell Growth Diff.*, **5**(6), 667–675.

Ismail, S. M., Cocklough, A. B., Dinnen, J. S. *et al.* (1989) Observer variation in histopathological diagnosis and grading of cervical intraepithelial neoplasia. *Br. Med. J.*, **298**, 707–710.

Herrington, C. S. (1995) Human papillomaviruses and cervical neoplasia II. Interaction of HPV with other factors. *J. Clin. Pathol.*, **48**, 1–6.

Hulka, B. S. (1968) Cytologic and histologic outcome following an atypical cervical smear. *Am. J. Obstet. Gynecol.*, **101**, 190–199.

Kataja, V., Syrjanen, K., Mantyjarvi, R. *et al.* (1989) Prospective follow-up of cervical. HPV infection: life table analysis of histopathological, cytological and colposcopic data. *Eur. J. Epidemiol.*, **5**, 1–7.

Kaufman, R., Koss, L. G. and Kurman, R. J. *et al.* (editorial) (1983) Statement of caution in the interpretation of papillomavirus associated lesions of the epithelium of the uterine cervix. *Acta Cytol.*, **27**, 107–108.

Kirkland, J. A. (1963) Carcinoma *in situ*. Diagnosis and prognosis. *J. Obstet. Gynecol. Br. Commonwlth.*, **70**, 232–243.

Kirkland, J. A., Stanley, M. A. and Cellier, K. M. (1967) Comparative study of histologic and chromosomal abnormalities in cervical neoplasia. *Cancer*, **20**, 1934–1952.

Koss, L. G. (1978) Dysplasia: a real concept or a misnomer? *Obstet. Gynecol.*, **51**, 374–379.

Kottmeier, H. L. (1961), Evolution et traitement des epitheliomas. *Rev. Franc. Gynaecol. et Obstet.*, **56**, 821–825.

Koutsky, L. A., Holmes, K. K., Critchlow, C. *et al.* (1992) Cohort study of risk of CIN 2 or 3 associated with human papillomavirus infection. *N. Engl. J. Med.*, **327**, 1272.

Laverty, C. R., Russell, P., Hills, E. and Booth, N. (1978) The significance of non condylomatous wart virus infection of the cervical transformation zone. A review with discussion of two illustrative cases. *Acta Cytol.*, **22**, 195–201.

Luff, R. D. (1992) The Bethesda system for reporting cervical/vaginal cytologic diagnosis: report of the 1991 Bethesda workshop. *Hum. Pathol.*, **23**, 719–721.

McCance, K. J., Walker, P. G., Dyson, J. L., Coleman, D. V. and Singer, A. (1983) Presence of human papillomavirus DNA in cervical intraepithelial neoplasia (CIN). *Br. Med. J.*, **287**, 784–788.

Maqueo, M., Azuela, J. C., Calderon, J. J. and Goldzieher, J. W. (1966) Morphology of the cervix in women treated with synthetic progestins. *Am. J. Obstet. Gynecol.*, **96**, 994–998.

Markowitz, S., Leiman, G. and Margolius, K. A. (1986) Human papillomavirus and cervical intraepithelial neoplasia in an African population. *S. Afr. J. Epidemiol. Infect.*, **1**, 65–69.

Matlashewski, G., Osborn, K., Murray A., Banks, L. and Crawford, L. (1987) Transformation of mouse fibroblasts with HPV type 16 DNA using a heterologous promoter. *Cancer Cells*, **5**, 195–199.

Meisels, A. and Fortin, R. (1976) Condylomatous lesions of the cervix and vagina. I. Cytologic patterns. *Acta Cytol.*, **20**, 505–509.

Meisels, A., Fortin, R. and Roy, M. (1977) Condylomatous lesions of the cervix II. Cytologic, colposcopic and histopathologic study. *Acta Cytol.*, **21**, 379–390.

Meisels, A., Roy, M., Fortier, M. and Morin, C. (1979) Condylomatous lesions of the cervix; morphologic and colposcopic diagnosis. *Am. J. Diagn. Gynecol. Obstet.*, **1**, 109–116.

Melkert, P. W., Hopman, E., van den Brule, A. *et al.* (1993) Prevalence of HPV in cytomorphologically normal cervical smears as determined by the polymerase chain reaction is age dependent. *Int. J. Cancer*, **53**, 919–923.

Montgomery, K. D., Tedford, K. L. and McDougall, J. K. (1995) Genetic instability of chromosome 3 in HPV -immortalised and tumorigenic human keratinocytes. *Genes, Chromosomes and Cancer*, **14**, 97–105.

Munoz, N., Castellsague, X., Bosch, F. X. *et al.* (1996) Difficulty in elucidating the male role in cervical cancer in Columbia a high risk area for the disease. *J. Natl Cancer Inst.*, **88**, 1068–1075.

Papanicolaou, G. N. (1949) A survey of the actualities and potentialities of exfoliative cytology in cancer diagnosis. *Am Int. Med.*, **31**, 661–674.

Paraskevaidis, E., Jandial, L., Mann, E. M. *et al.* (1991) Pattern of treatment failure following laser for cervical intraepithelial neoplasia, implications for follow-up protocol. *Obstet. Gynecol.*, **78**, 80–83.

Poulsen, M. E., Taylor, C. W. and Sobin, L. M. (1975) *Histological Typing of Female Genital Tract Tumours*. World Health Organization, Geneva, pp. 15–18, 55–62.

Purola, E. and Savia, E. (1977) Cytology of gynaecologic condyloma acuminatum. *Acta Cytol.*, **23**, 26–31.

Reagan, J. W., Seidemand, I. L. and Saracusa, Y. (1953) The cellular morphology of carcinoma *in situ* and dysplasia or atypical hyperplasia of the uterine cervix. *Cancer*, **6**, 224–235.

Reid, R., Stanhope, C. R., Herschman, B. R., Booth, E., Phibbs, D. G. and Smith, P. J. (1982) Genital warts and cervical cancer I: evidence of an association between subclinical papillomavirus infection and cervical malignancy. *Cancer*, **50**, 377–387.

Richart, R. M. (1963) A radioautographic narrative of cellular proliferation in dysplasia and carcinoma *in situ* of the uterine cervix. *Am. J. Obstet. Gynecol.*, **86**, 925–930.

Richart, R. M. (1967) Natural history of cervical intraepithelial neoplasia. *Clin. Obstet. Gynecol.*, **10**, 748–784.

Richart, R. M. and Barron, B. A. (1969) A follow-up study of patients with cervical dysplasia. *Am. J. Obstet. Gynecol.*, **105**, 386–393.

Richart, R. M. and Lerch, V. (1966) Time lapse cinematographic observations of normal human cervical epithelium, dysplasia and carcinoma *in situ*. *J. Natl. Cancer Inst.*, **7**, 317–329.

Richart, R. M., Lerch, V. and Barron, B. A. (1967) A time lapse cinematographic study *in vitro* of mitosis in normal human cervical epithelium, dysplasia and carcinoma *in situ*. *J. Natl Cancer Inst.*, **39**, 571–577.

Robertson, A. J., Anderson, J. M., Swanson-Beck, J. *et al.* (1989) Observer variability in histopathological reporting of cervical biopsy specimens. *J. Clin. Pathol.*, *42*, 231–238.

Rotkin, I. D. (1973) A comparison review of key epidemiological studies in cervical.cancer related to current searches for transmissible agents. *Cancer Res.*, **33**, 1353–1357.

Rubin, I. C. (1910) The pathological diagnosis of

incipient carcinoma of the uterus. *Am. J. Obstet.*, **62**, 668–676.
Schiffman, M. H. (1992) Recent progress in defining the epidemiology of human papillomavirus infection and cervical neoplasia. *J. Natl. Cancer Inst.*, **84**, 394–398.
Schottlander, J. and Kermauner, F. (1912) Zur Kenntis des Uterus Karzinoms: Monographische: *Studie über Morphologie, Entwichlung, Waschstum nebst Beitragen zur klinik der Erkrankung*. S. Karger, Berlin.
Scully, R., Bonfiglio T. A., Kurman R. J. *et al.* (1994) *Histology typing of female genital tract tumours*. 2nd edition WHO International Classification of Tumours, Springer-Verlag, Berlin.
Selvaggi, S. M. (1994) Cytologic features of squamous cell carcinoma *in situ* involving endocervical glands in endocervical brush specimens. *Acta. Cytol.*, **38**, 687–692.
Shingleton, H. M., Richart, R. M., Weiner, J. and Spiro, D. (1968) Human cervical intraepithelial neoplasia. Fine structure of dysplasia and carcinoma *in situ*. *Cancer Res.*, **28**, 695–706.
Siegler, E. E. (1961) Histomorphology of carcinoma *in situ*. *Acta Cytol.*, **5**, 275–278.
Smith, P.A. and Turnbull, L. S. (1997) Small cell and 'pale' dyskaryosis cytopathology. *Cytopathology*, **8**, 3–8.
Stanbridge, C. M., Mather, J., Curry, A. and Butler, E. B. (1981) Condylomata of the uterine cervix and koilocytosis of cervical intraepithelial neoplasia. *J. Clin. Pathol.*, **34**, 524–531.
Van Herckenrode, M. C., Barbera G. E., Malcolm, A. D. B., Coleman, D. V. (1992) Prevalence of human papillomavirus (HPV) infection in Basque country women using slot blot hybridisation: a survey of women at low risk of developing cervical cancer. *Int. J. Cancer*, **51**, 581–586.
Van Hoeven, K. H., Hanau C. A and Hudock, J. A. (1996) The detection of endocervical gland involvement by high grade intraepithelial lesions in smears prepared from endocervical brush specimens. *Cytopathology*, **7**, 310–315.
Vergote, I. B., Makar, A. P. and Kjorstad, K. E. (1992) Laser excision of the transformation zone as treatment of cervical intraepithelial neoplasia with satisfactory colposcopy. *Gynecol. Oncol.*, **44**(3), 235–239.
Walker, P. G., Singer, A., Dyson, J. L., Shah, K. V., To, H. and Coleman, D. V. (1983a) The prevalence of human papillomavirus antigen in patients with cervical intraepithelial neoplasia. *Br. J. Cancer*, **48**, 99–101.
Walker, P. G., Singer, A., Dyson, J. L., Shah, K. V., Wilters, J. and Coleman, D. V. (1983b) Colposcopy in the diagnosis of papillomavirus infection of the uterine cervix. *Br. J. Obstet. Gynaecol.*, **90**, 1082–1086.
Wilbanks, G. D., Richart, R. M. and Terner, J. Y. (1967) DNA contents of cervical intraepithelial neoplasia studied by two-wavelength Feulgen cytophotometry. *Am. J. Obstet. Gynecol.*, **98**, 792–799.

11

MICROINVASIVE AND OCCULT INVASIVE SQUAMOUS CARCINOMA

Microinvasive carcinoma and occult invasive carcinoma represent the preclinical stages of invasive cervical carcinoma.

11.1 MICROINVASIVE CARCINOMA

Microinvasive carcinoma is the earliest stage of invasive cancer that can be recognized histologically. The lesion may appear in tissue sections as a tiny bud of cancer cells arising from a focus of CIN and penetrating the underlying stroma (Figure 11.1). It is most commonly found arising from a focus of CIN 3, but may arise from a lower grade of CIN. As might be expected, the larger the size of the CIN lesion the greater the risk of finding a focus of microinvasion. This has recently been confirmed by Tidbury *et al.* (1992) who observed that the mean size of CIN 3 lesions showing microinvasion was seven times greater than that for CIN 3 without invasion.

It is particularly important for the pathologist to recognize and define microinvasive carcinoma because the diagnosis implies little risk of lymph node metastases and a good prognosis for the patient. This, in turn, has a direct bearing on the management of these lesions; a conservative approach to treatment may well be appropriate.

11.1.1 Definition of microinvasive carcinoma

The concept of microinvasive carcinoma was introduced by Mestwerdt in 1947. He applied it to a group of early invasive carcinomas with stromal invasion of 5 mm or less deep to the basement membrane. Unfortunately, other authors have recommended various limiting depths to define microinvasive carcinoma, for example 5 mm (Ng and Reagan, 1969; Rubio *et al.*, 1974), 3–4 mm (Ullery *et al.*, 1965), 3 mm (Seski *et al.*, 1977; Ferenczy, 1982) and 1 mm (Nelson *et al.*, 1975) The most recent working definition proposed by the International Federation for Gynaecology and Obstetrics is cited by Scully *et al.* (1994) as follows: an area of cancer cells which extends 0.5 cm or less below the surface epithelial basement membrane or the basement membrane of an endocervical gland and which has a horizontal extent of 0.7 cm or less. Vascular space invasion is noted if present but does not itself exclude a tumour from being placed in the microinvasive category.

Microinvasive carcinoma is a histological diagnosis which should be made on a large biopsy specimen which removes the whole lesion, usually a cone biopsy. Two groups can be recognized:

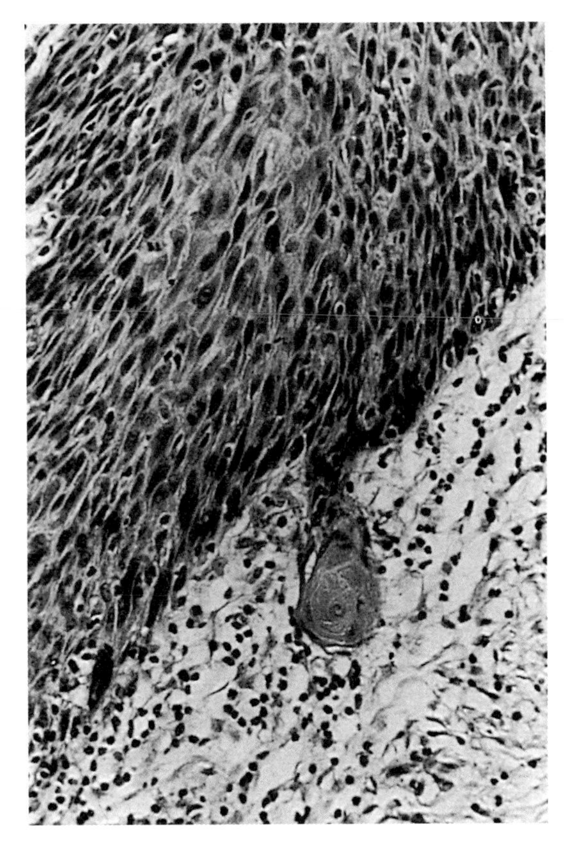

Figure 11.1 Small peg of microinvasive carcinoma showing greater differentiation than the CIN from which it has arisen (H & E, ×310).

1. *Early stromal invasion (FIGO stage 1a, 1)*: this is too small for accurate measurement and is characterized microscopically as invasive buds which are present either in continuity with an *in situ* lesion or as apparently separated cells, not more than 1 mm from the nearest surface or crypt basement membrane.
2. *Measurable lesions (FIGO stage 1a, 2)*: these are measured in two dimensions. The depth of invasion, measured from the base of the epithelium from which it develops, should not exceed 5 mm and the largest diameter (lateral extent) should not exceed 7 mm on the section that shows the greatest extent. Lesions that exceed these dimensions and are not detectable clinically should be assigned to the category of occult invasive cancer (FIGO 1b; see Section 11.3).

From the foregoing definitions, it can be seen that a diagnosis of a microinvasive carcinoma can only be made after the whole lesion has been delineated by a very thorough examination of the entire specimen. Examples of microinvasive carcinoma are shown in Figures 11.1–11.5. They illustrate the three morphological features which enable the pathologist to discriminate between a small focus of microinvasive cancer and artefact due to tangential cutting, etc.

One of the most useful markers of microinvasion is the 'striking differentiation' of the invasive peg. This was first emphasized by Burghardt (1973) and is clearly shown in

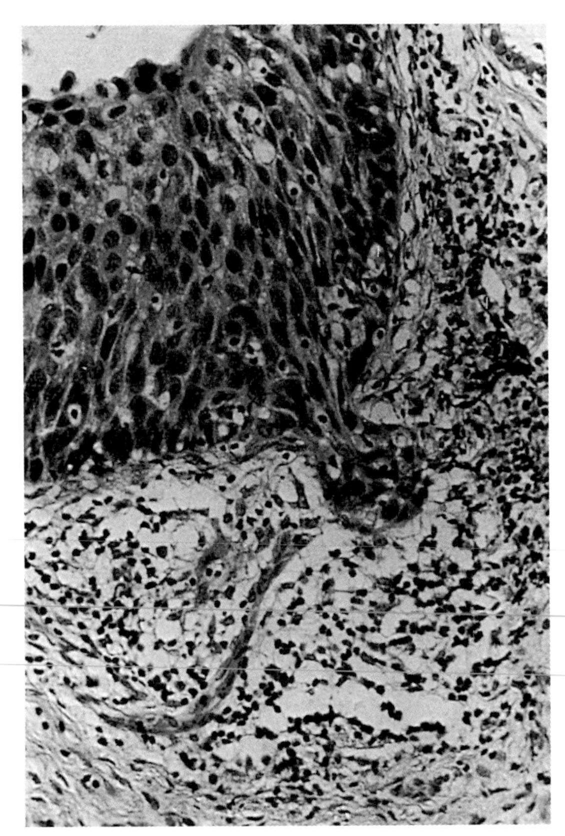

Figure 11.2 Bud of microinvasive carcinoma protruding through the disrupted basement membrane and surrounded by round cells (H & E, ×310).

Figure 11.1; the cells of the invasive peg are better differentiated than the cells in the CIN from which it arises.

Another useful feature which is suggestive of microinvasive carcinoma is disruption of the basement membrane at the site of the invasive peg (Figures 11.2 and 11.3). Although this is most clearly seen on electron microscopy (Ferenczy, 1982), it is usually visible by light microscopy and gives the edge of the microinvasive bud a ragged appearance.

The third feature which should alert the pathologist to the possibility of microinvasion is a prominent stromal reaction in the form of lymphocytes and plasma cells surrounding the invading tongue (Figures 11.1–11.5). This inflammatory infiltrate is believed to reflect the immunological response of the cervix to the invading cells (Sano and Veki, 1987). Crissman *et al.* (1985) reported that the degree of inflammatory reaction was of no prognostic significance. However, Reinthaller *et al.* (1991) considered that with a heavy inflammatory stromal reaction the risk of nodal metastases and tumour recurrence was significantly lower, independent of the histological stage.

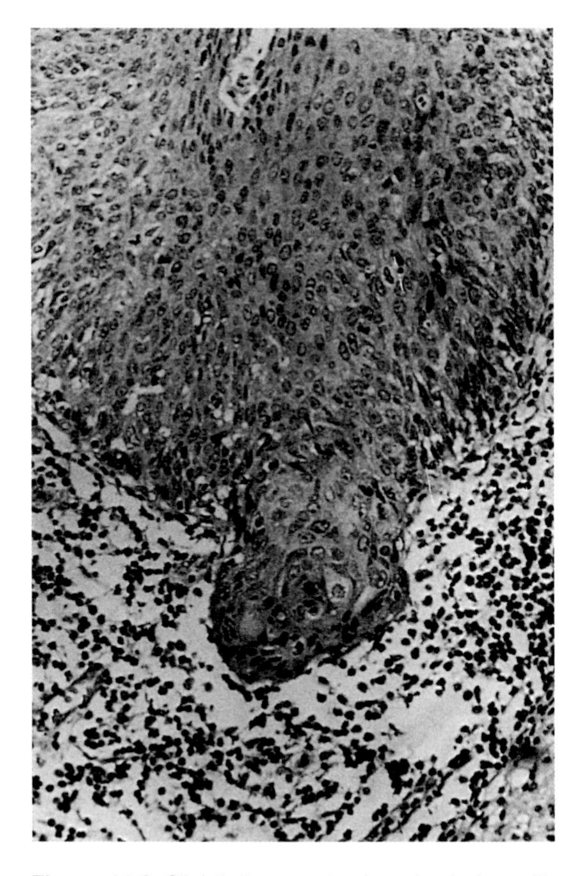

Figure 11.3 Slightly larger microinvasive lesion with ragged edge and stromal round cell reaction (H & E, ×310).

Whenever microinvasion is suspected, a search should be made for the presence of tumour cells in lymphatic channels. Ferenczy (1982) considers the presence of lymphatic channel involvement to be of such critical importance that its demonstration should exclude the lesion from the category of microinvasive cancer whatever the depth of invasion, although not all authorities hold this view (Section 11.1.2(c)). Roche and Norris (1975) pointed out that it is not always possible to be certain whether the tumour is in a lymphatic or vascular channel and suggested the term 'capillary-like' (CL) to describe these spaces. Whatever term is used, it is essential that a diagnosis of lymphatic channel or CL space involvement be made only when endothelial lining cells (Figure 11.7) can be identified (Burghardt, 1973). Misdiagnosis may occur when tumour cells are found in spaces produced by tissue shrinkage. In these cases, no endothelial lining cells are seen.

11.1.2 Risk factors

Several studies have shown that the risk of microinvasive carcinoma becoming a frankly invasive carcinoma depends on a number of factors:

- depth of invasion;
- lateral extent of lesion;
- lymphatic (capillary-like) involvement; and
- completeness of removal.

Less readily assessed factors which could affect prognosis include the growth pattern of the microinvasive lesion and the stromal

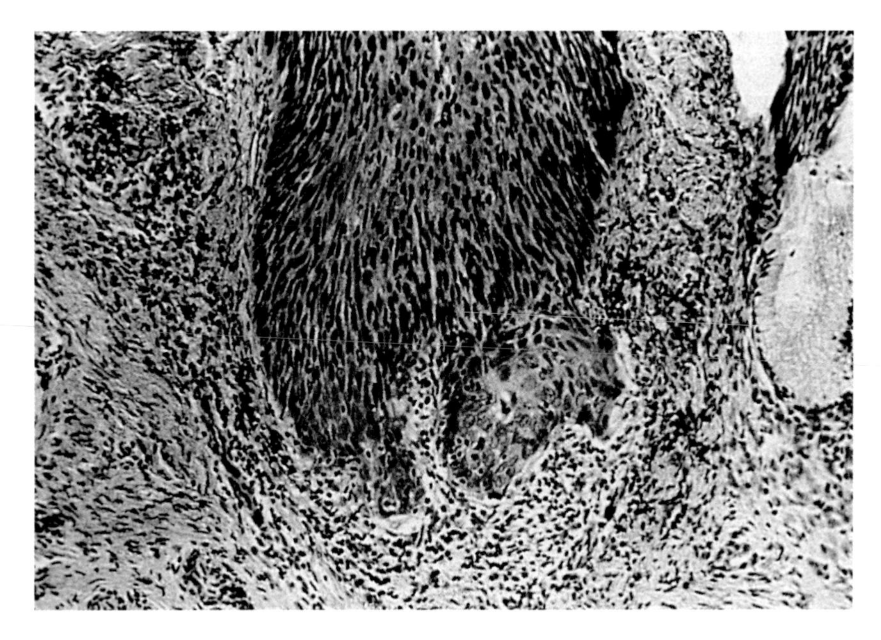

Figure 11.4 Irregular focus of microinvasive carcinoma showing evidence of increased differentiation. Note reactive change in surrounding stroma (H & E, ×190).

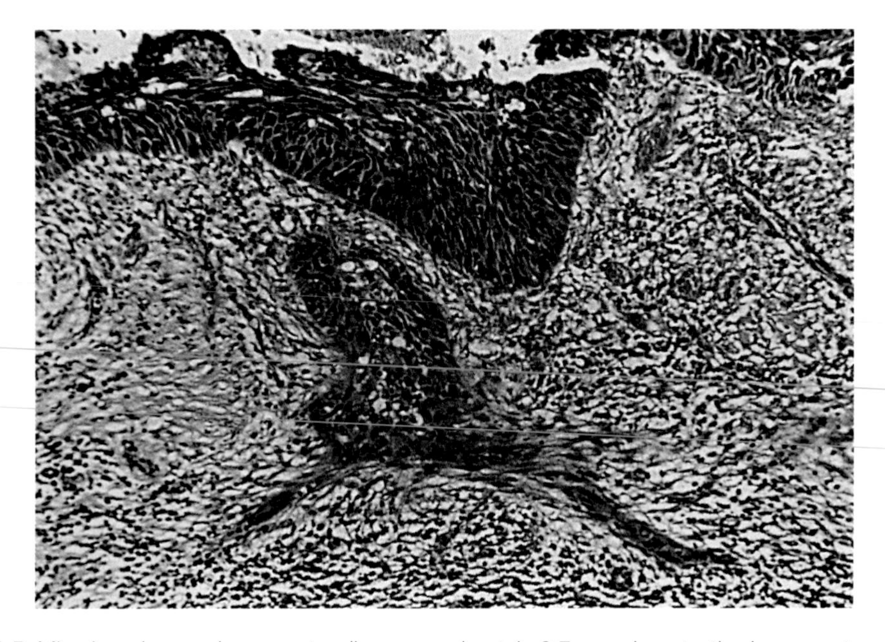

Figure 11.5 Microinvasive carcinoma extending approximately 0.7 mm deep to the basement membrane of the overlying CIN 3 (H & E, ×130).

lymphocytic reaction to the tumour. Confluence, convergence or divergence (Figure 11.5) of invading foci does not appear to affect prognosis except in so far as it contributes to the overall measurable extent of the lesion.

(a) Depth of invasion

This is measured from the basement membrane of the epithelium from which the invasion is occurring. Where the basement membrane is uneven, as in Figure 11.5, it is only possible to give an approximate estimate of the depth of invasion. When the depth of invasion is not more than 1 mm, it should be reported as such and the lesion described as 'early stromal invasion'. When the depth of invasion is 1–5 mm, the measured depth should be stated in the report.

The most accurate method of measuring this distance is by means of an ocular micrometer which is calibrated for each objective by a stage micrometer. Alternatively, the latter can be used for measuring the diameter of the microscope field which is then used as a measure. A simpler method is to mark the coverslip with a felt pen at the level of the basement membrane and at the deepest point of invasion and then measure the distance between the two marks with a ruler. This method is less accurate than a calibrated eyepiece but provides a useful check on the micrometer measurements. By definition (Anderson *et al.*, 1982; Scully *et al.*, 1994), the depth of invasion should not exceed 5 mm for a measurable microinvasive carcinoma or 1 mm for early stromal invasion.

The significance of depth of invasion is illustrated by the findings of Rubio *et al.* (1974) who found lymphatic permeation in 3% (5 in 157) of lesions less than 3 mm in depth and in 14% (4 in 27) of those 3–5 mm in depth. Benson and Norris (1977) reviewed lymphadenectomized women and found no lymph node deposits in 57 women with a depth of invasion of less than 3 mm compared with 2.7% (1 in 37) lymph node deposits with a depth of invasion of 3–5 mm. In a critical review of the literature, Lohe (1978) concluded that lymph node deposits occurred in less than 1% of microinvasive carcinomas of up to 5 mm in depth. Larsson *et al.* (1983) analysed 343 cases of early invasive cancer and found that variations in depth of invasion between 0.2 and 9.0 mm did not affect prognosis in the absence of other risk factors (Section 11.2) However, Maiman *et al.* (1988) studied 115 women with early invasive cancer and reported 2% lymph node deposits if invasion was 3 mm or less and 13 % if invasion was 3–5 mm.

(b) The lateral extent of the microinvasive lesion

This is measured similarly, being the distance between the two lateral extremes of each focus of penetration. If more than one focus of invasion is present, the sum of such distances gives the total lateral extent of microinvasion. According to the FIGO definition, microinvasive carcinoma should not exceed 7 mm in lateral extent. Figure 11.6 shows a lesion which conforms to the definition of microinvasion in depth but has a lateral extent of 15 mm, thus excluding it from the microinvasive category. The significance of measuring the lateral extent of a microinvasive carcinoma is illustrated by the results of Sedlis *et al.* (1979). They found that the proportion of post-cone hysterectomy specimens containing residual cancer was 2% when the lateral extent was 4 mm or less, 27% when 8 mm or less, and 35% when greater than 8 mm. In contrast, Larsson *et al.* (1983) reported that variations in lateral extent between 0.4 and 17.2 mm were of no prognostic importance in the absence of other risk factors (Section 11.2)

Burghardt (1982) has suggested that tumour volume gives a more reliable indication of the prognosis than measurement of lateral extent and depth of invasion alone. This involves examining several step sections of the tumour and is not widely used. However, measurement of tumour size in at least two dimensions is strongly recommended.

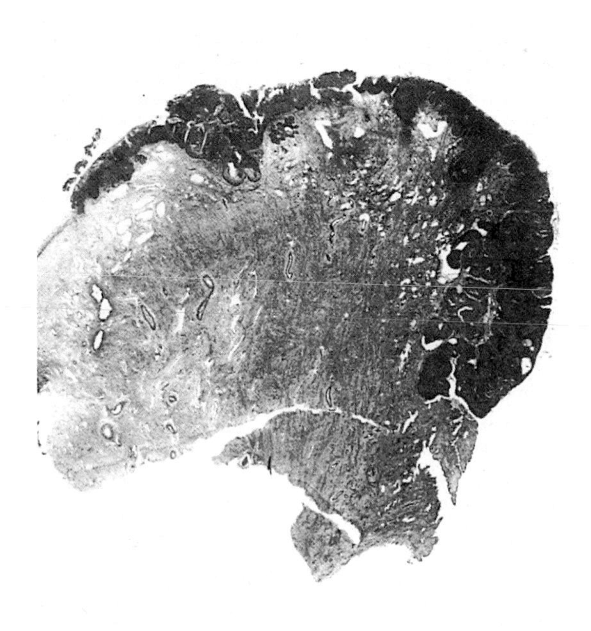

Figure 11.6 Although the greatest depth of invasion in this lesion was only 2.5 mm, the lateral extent was found to be 15 mm, which excludes it from the microinvasive category (H & E, ×7).

(c) Lymphatic channel (capillary-like) involvement (Figure 11.7)

There are two schools of thought regarding the risk associated with lymphatic channel or CL involvement. Roche and Norris (1975) found no lymph node deposits in 30 lymphadenectomized women with microinvasion of 2–5 mm (average 3.2 mm) in depth, 57% of whom had CL space invasion. Burghardt (1973) reported no recurrences after 3–11 years in five cases of early stromal invasion but four with invasion up to 5 mm deep, all with lymphatic channel involvement. On the other hand, Seski *et a1.* (1977) found that one out of a group of four women with up to 3 mm deep invasion with CL space involvement had lymph node metastasis whereas none of the 37 women with a similar degree of invasion but no vascular involvement had lymph node deposits. Furthermore, Sedlis *et al.* (1979), found residual carcinoma in 52% of post-cone hysterectomy specimens in the 23 microinvasive cases with CL space involvement and only 4% of the 77 comparable cases without CL space involvement. Although the results of these studies are conflicting, we consider that lymphatic or vascular channel involvement is a significant risk factor and implies the need for more radical treatment. This view is supported by Boyce *et al.* (1984) who found that in 138 patients with stage 1 carcinoma, vascular invasion was significantly associated with a poor outcome. This point of view has received further endorsement in a recent study by Reinthaller *et al.* (1991). They studied 158 cases of invasive carcinoma ranging from stages 1a to 2b and found that when no vascular space involvement was obvious, lymph nodes were tumour free in 94% of the cases.

(d) Completeness of removal

This is probably the single most important criterion for subsequent management, so it is essential to check very carefully whether there is any involvement of the resection margin by tumour tissue. Steps to be taken to establish that the edge of a section coincides with the margin of the specimen are described in Section 4.6.1. Sedlis *et al.* (1979) found residual invasive carcinoma in 80% of the 15 post-cone hysterectomy specimens where microinvasive tumour had been present in the deep or lateral margins of the cone biopsy specimen. In two-thirds of these cases, the invasion was more extensive in the uterus than in the cone.

11.2 RISK FACTOR ASSESSMENT AND IMPLICATIONS FOR MANAGEMENT

In order to assess the prognostic significance of histological findings, it appears helpful to tabulate the risk of significant invasion and metastatic spread in relation to the four parameters outlined above. Table 11.1 is a guide based on

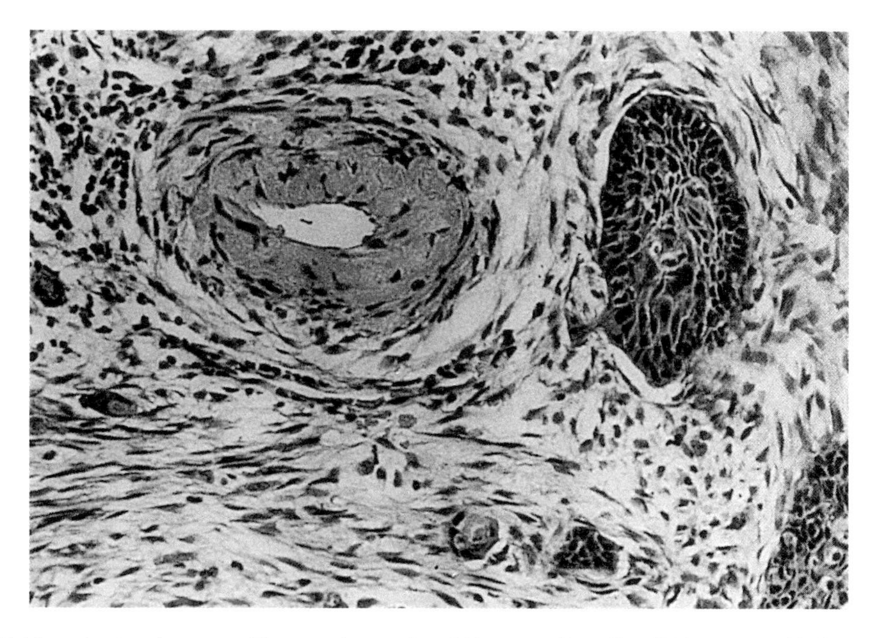

Figure 11.7 Vascular involvement. The carcinoma is within a capillary-like (CL) space with flattened cells of endothelial type, usually interpreted as lymphatic permeation (H & E, ×310).

currently available information, the category into which a case is placed being determined by its worst feature.

Many gynaecologists and oncologists find a graded assessment such as that outlined below to be of great value in deciding on the appropriate management of individual cases. For example, they may consider that cases in the very low-risk category require no further action other than cone biopsy and follow-up cytology. Low-risk cases are now often considered to be appropriately treated by hysterectomy only. Medium-risk cases have a slightly increased chance of lymph node involvement, although the only lymph nodes likely to be involved are those of the pelvic wall (Lohe, 1978). After histologically complete removal of microinvasive carcinoma

Table 11.1 Assessment of risk of significant invasion and metastatic spread for microinvasive and occult carcinoma

		Microinvasive		*Occult*
Degree of risk	Very low (early stromal invasion)	Low	Medium	High
Depth of invasion	< 1 mm	1–3 mm	3–5 mm	> 5 mm
Lateral extent of invasion	< 2 mm	2–5 mm	5–7 mm	> 7 mm
Vascular (CL) involvement	–	–	–	+
Completeness of removal	Complete	Complete	CIN to margin	Invasion to margin

(5 mm or less) by cone biopsy or hysterectomy, Ng and Reagan (1969) found no residual carcinoma in any of 60 cases after 5 years. In a comparable but larger series of 210 cases, Rubio *et al.* (1974) reported seven deaths from disseminated cervical cancer and three cases with local recurrence giving an approximate 97% survival rate over an average of 7 years.

Another possible parameter in risk factor assessment is the degree of inflammatory reaction. Reinthaller *et al.* (1991) found that heavy inflammatory infiltration at the tumour periphery correlated with good prognosis.

11.3 OCCULT INVASIVE CARCINOMA

The term occult invasive carcinoma is currently used to describe invasive lesions which do not correspond to the definition of microinvasive carcinoma given in Section 11.1.1 but which are not yet clinically perceptible. They are generally considered to have a worse prognosis than microinvasive carcinoma and are categorized as FIGO stage 1B tumours for the purposes of clinical statistics and treatment. The mean age of diagnosis for occult

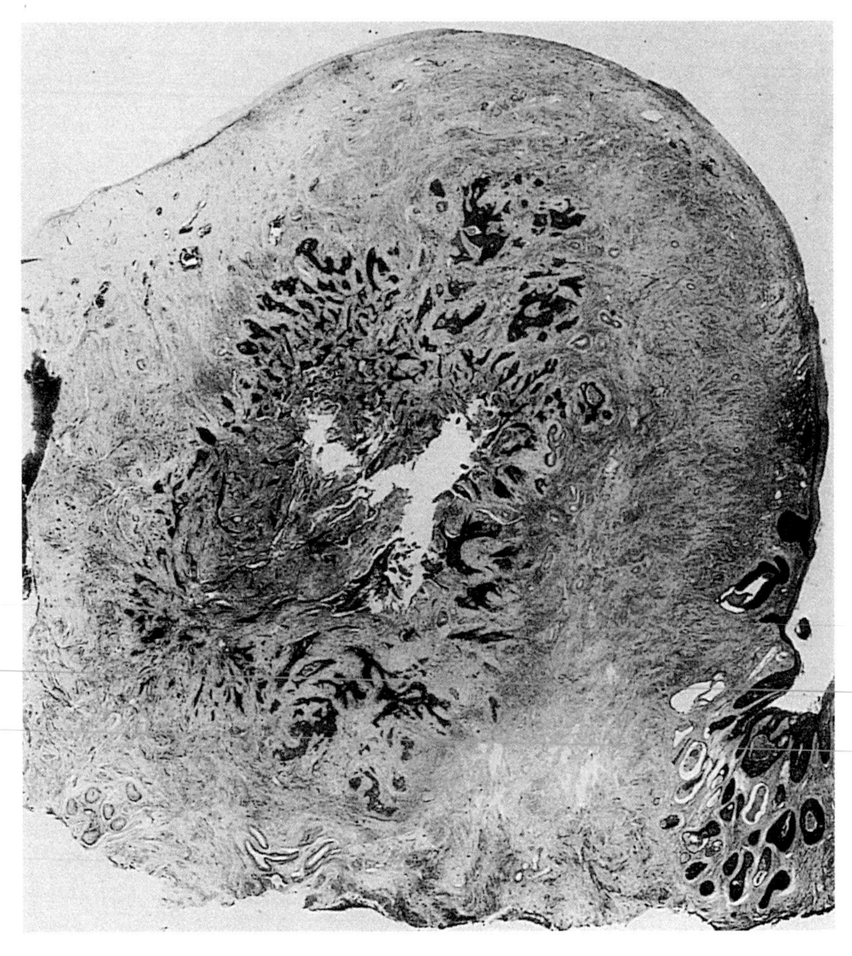

Figure 11.8 Occult invasive carcinoma. The greatest depth of invasion is approximately 10 mm with evidence of lymphatic permeation, shown in Figure 11.7. Lymph node metastases were present (H & E, ×4).

carcinoma is 48.6 years compared with 52 years for clinically invasive carcinoma (Walton Report, 1976).

Boyes *et al.* (1970) found that patients with occult carcinoma who eventually die from carcinomatosis generally have lymphatic infiltration and a tumour measuring more than 5 mm in diameter. In their series, the 5-year survival rate for occult squamous carcinoma is 93.5%. Crissman *et al.* (1992), in evaluating occult invasive carcinomas, reported that the 36 patients with 5 mm or less depth of tumour invasion had a 5-year survival rate of 95% and none died of their cancer, whereas 30 patients with greater than 5 mm invasion in their resected uteri had a 5-year survival rate of only 58%. Vascular (lymphatic or capillary) invasion were the only predictors of poor outcome in this study. Kinney *et al.* (1992) reviewed 27 women (median age 60 years) with occult carcinoma and reported a 5-year survival rate of 82%. These lesions comprise the high-risk group in Table 11.1.

An example of occult invasive carcinoma found unexpectedly in a cone biopsy specimen is shown in Figure 11.8. The tumour measured 10 mm in greatest depth with evidence of lymphatic permeation (Figure 11.7). Lymph node metastases were found.

11.4 CYTOLOGY

The cytological features of smears from microinvasive and occult invasive squamous carcinomas may be virtually indistinguishable from those found in smears from patients with CIN 3 and it is not possible to predict the limited extent of invasion implied by the terms microinvasion or occult invasion from the cytological findings. Most cases of microinvasive squamous carcinoma are characterized by the presence of squamous epithelial cells showing a severe degree of dyskaryosis (Figures 11.9 and 11.10). Occasionally, abnormal cells showing features characteristic of

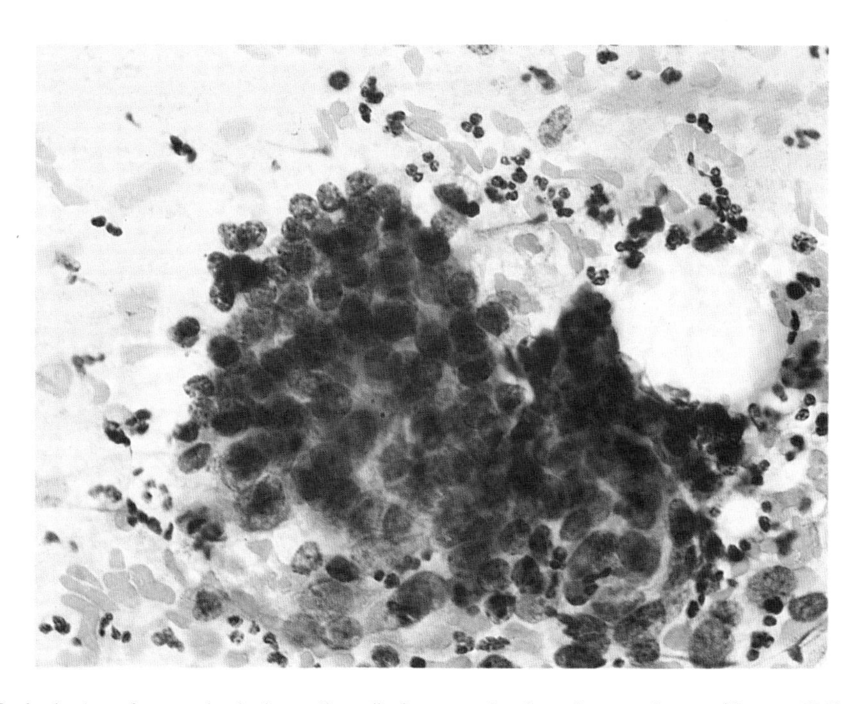

Figure 11.9 A cluster of severely dyskaryotic cells from a microinvasive carcinoma (Pap, ×630).

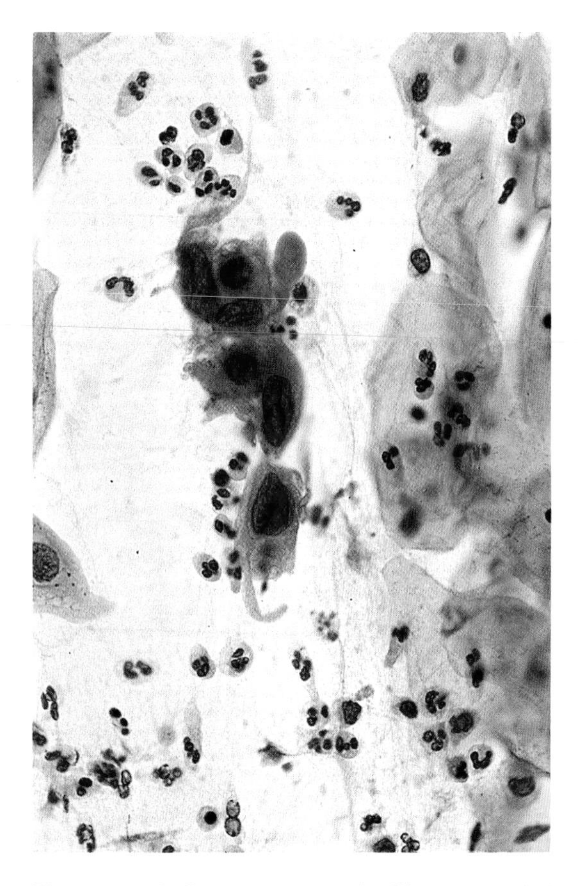

Figure 11.10 Severe dyskaryosis. Biopsy revealed microinvasive carcinoma (Pap, ×630).

invasive squamous carcinoma of the cervix may also be seen, but the degeneration and necrosis which are features of deeply invasive cancer are rarely present. When fibre cells (Figure 11.11), dense clusters of undifferentiated malignant cells (Figure 11.12) or abnormal keratinization (Figure 11.13) are present in a smear containing severely dyskaryotic cells, a cytodiagnosis of probable invasive cancer can be made.

11.5 QUESTIONABLE STROMAL INVASION

In the first edition of the handbook published by the WHO Collaborating Centre for the histological classification of female genital tract tumour (Poulsen *et al.*, 1975), the term 'questionable stromal invasion' is used for lesions having the features of CIN with focal areas suspected of penetrating the underlying stroma but lacking definite evidence of invasion. This term is not included in the most recent edition (Scully *et al.*, 1994) although the problem of recognizing stromal invasion is a real one.

There are several possible reasons why overdiagnosis of stromal invasion may occur. Tangential cutting is probably the most common (Figure 11.14). Sometimes suspicion of stromal invasion can be allayed by cutting multiple step sections from the relevant tissue block. This may reveal, for example, that a downward protuberance of the lower margin of the epithelium is merely the extension of CIN into an endocervical gland crypt (Figures 11.15–11.17). Anderson and Hartley (1980) studied 343 cervical conization specimens containing CIN 3 and found that 88.6% showed some involvement of the crypts by CIN. They showed that the mean depth of involvement was 1.24 mm and the maximum 5.55 mm and pointed out that this amount of crypt involvement must be taken into account when CIN is being treated; otherwise there is a risk that a focus of neoplasia may be left in the depths of the crypt after ablative therapy.

Alternative, multiple step sections may reveal bizarre pleomorphism (Figure 11.18) or an uneven outline with a suggestion of subsidiary protuberances apparently lacking a basement membrane (Figure 11.19). In such cases the lesion may be interpreted as incipient stromal invasion. Suspicion of microinvasion becomes stronger when there is not only absence of the basement membrane but also an alteration in the pattern of the basal cells together with an inflammatory reaction in the underlying stroma (Figure 11.20). Even in this example, there is still sufficient uncertainty to warrant the designation 'questionable stromal invasion'. When a downward extension of the suspect epithelium has secondary protuberances as prominent as those shown in

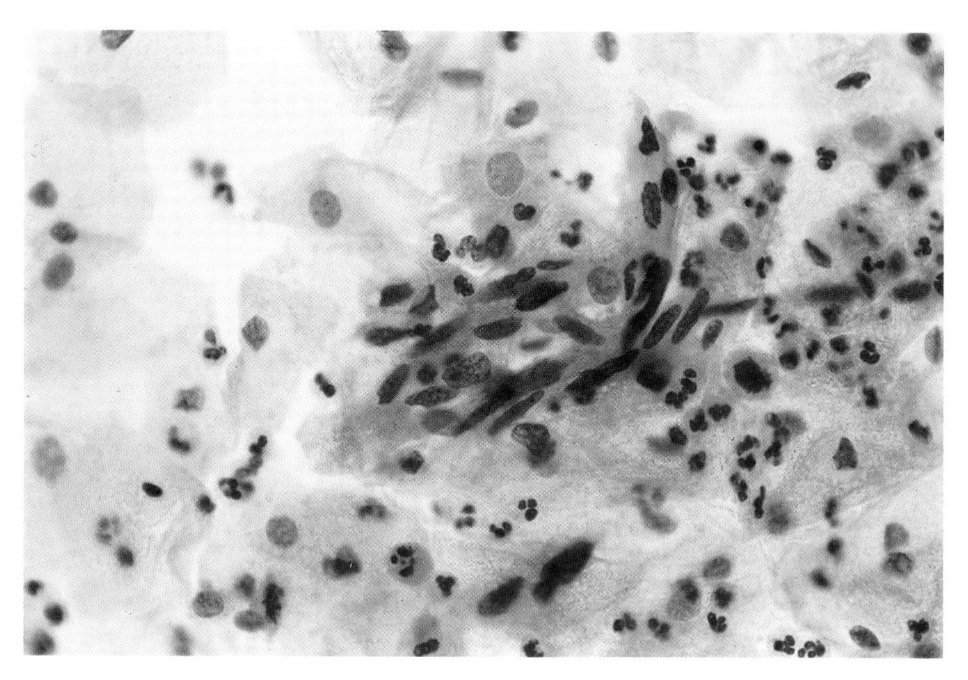

Figure 11.11 Cervical scrape cytological material from a microinvasive squamous carcinoma. Note the presence of fibre cells (Pap, ×630).

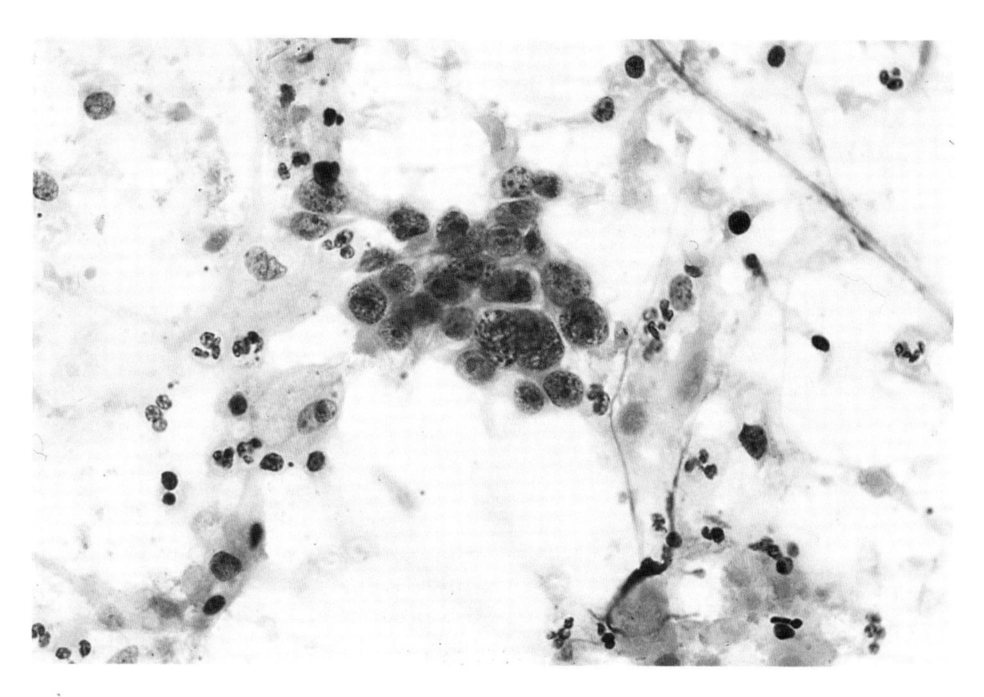

Figure 11.12 Cervical scrape from microinvasive squamous carcinoma. Undifferentiated malignant cells (Pap, ×630).

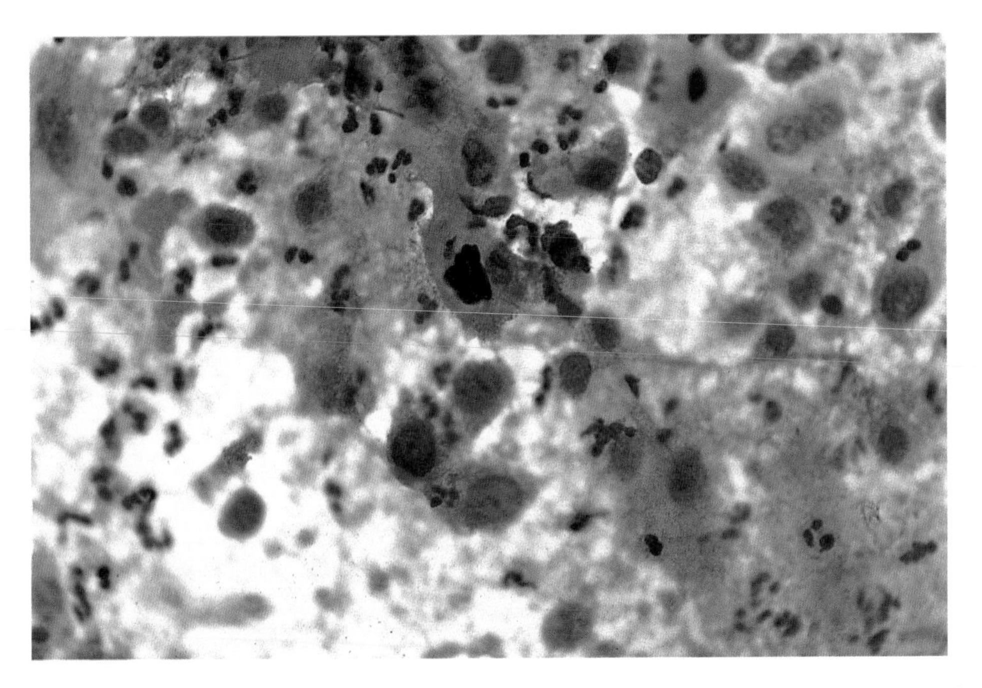

Figure 11.13 Same cervical smear as Figure 11.11 showing marked cellular pleomorphism. In some cells, there are changes suggestive of keratinization (Pap, ×630).

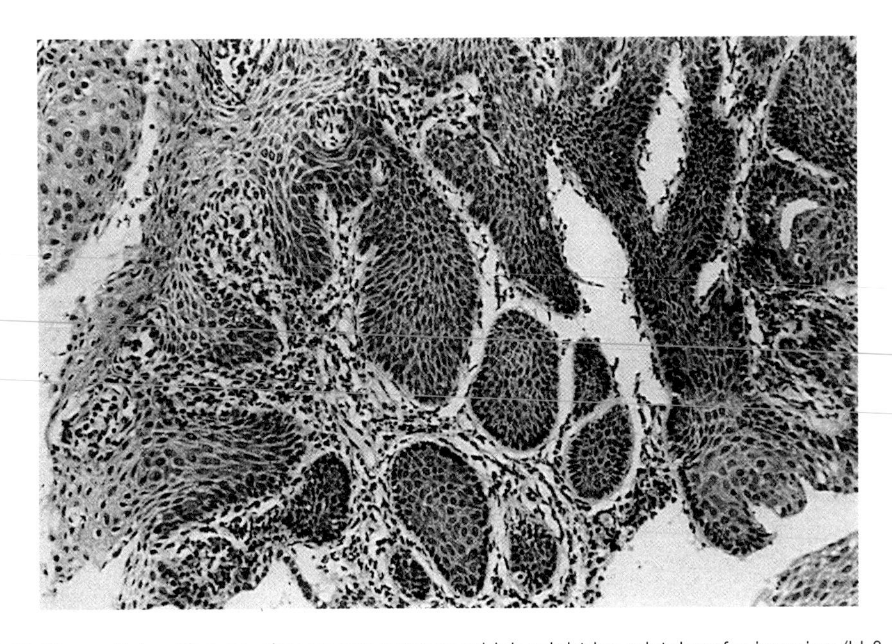

Figure 11.14 Tangential cutting causing an appearance which might be mistaken for invasion (H & E, ×190).

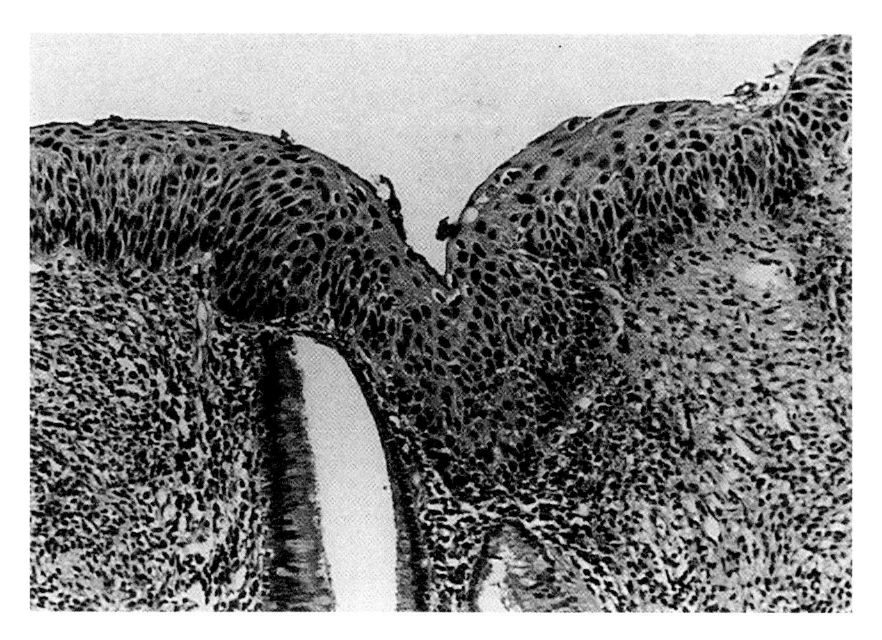

Figure 11.15 Non-invasive downward protuberance of lower margin of epithelium related to gland openings (H & E, ×190).

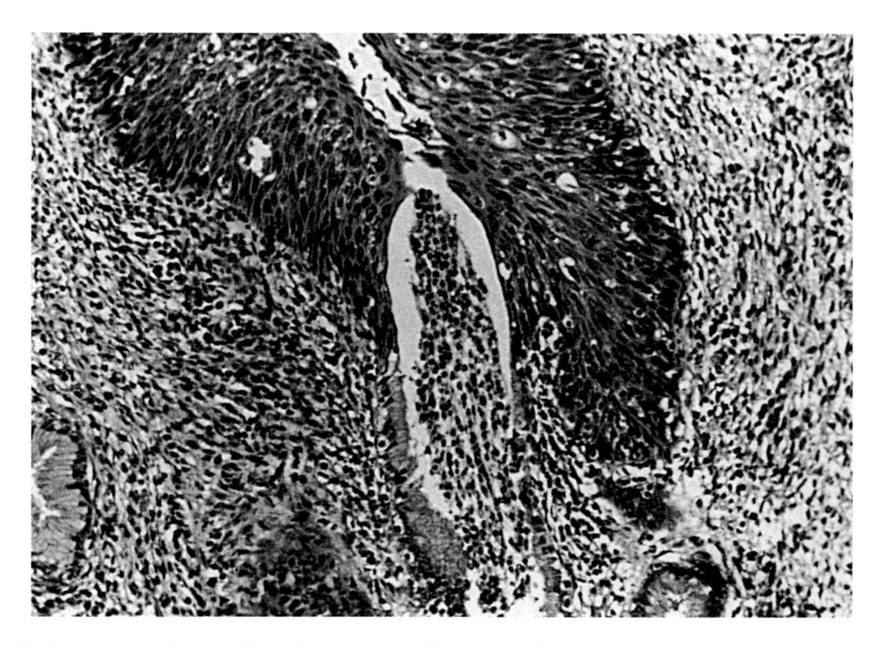

Figure 11.16 Deep extension of CIN 3 into glands (H & E, ×190).

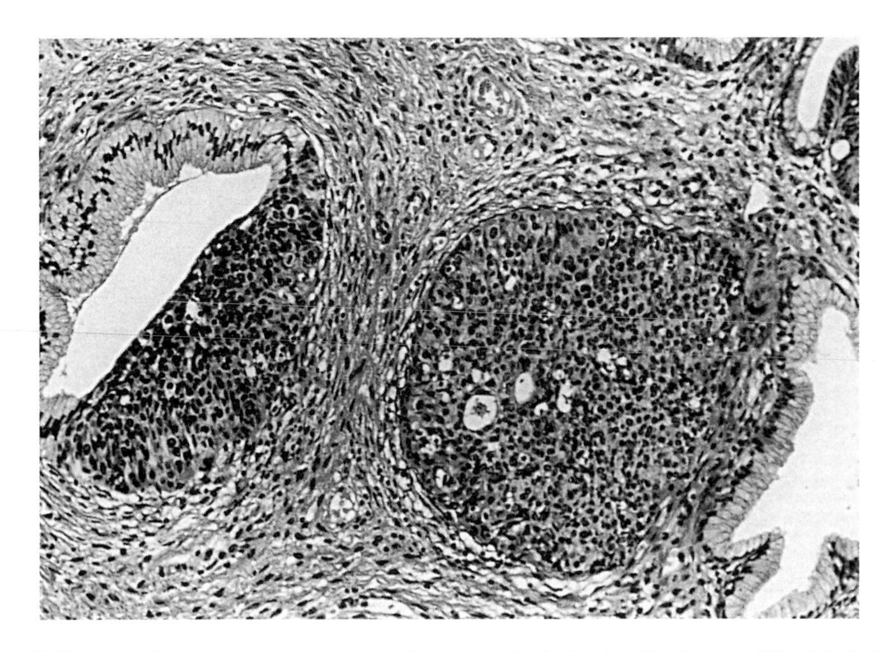

Figure 11.17 Partial and complete replacement of endocervical gland epithelium by CIN 3 (H & E, ×190).

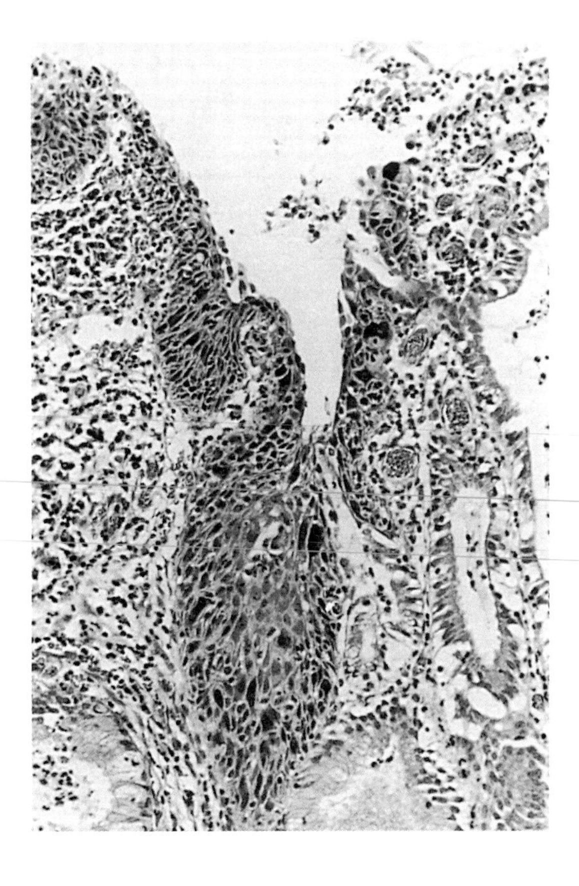

Figure 11.18 CIN 3 extending into gland mouths but showing bizarre plemorphism. The changes are not sufficient to be considered as questionable stromal invasion (H & E, × 190).

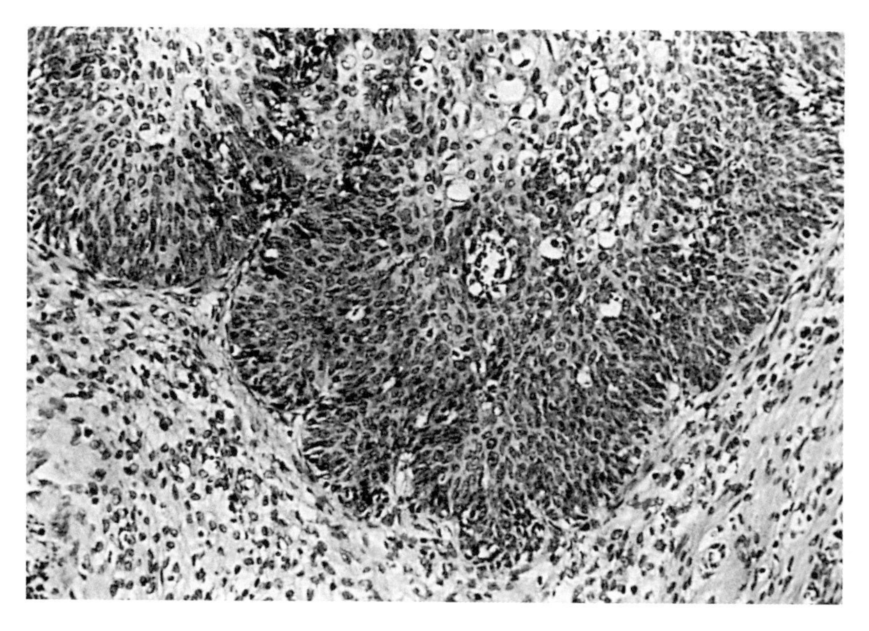

Figure 11.19 Downward extension of CIN 3 showing irregular protuberances, some lacking a basement membrane, interpreted as questionable stromal invasion (H & E, ×190).

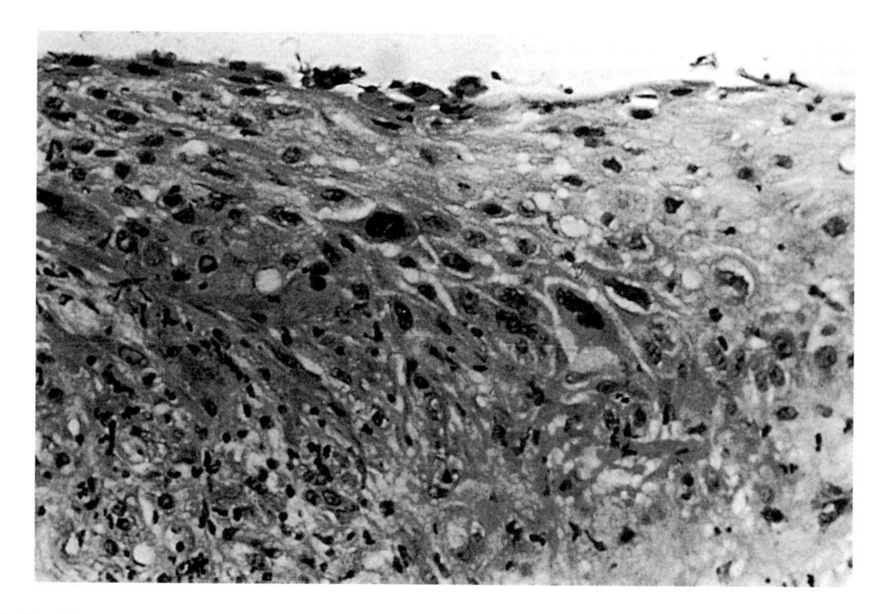

Figure 11.20 Absent basement membrane with altered pattern of basal cells in direction of increased differentiation and some inflammatory reaction in the underlying stroma, an appearance highly suggestive of early stromal invasion (H & E, ×310).

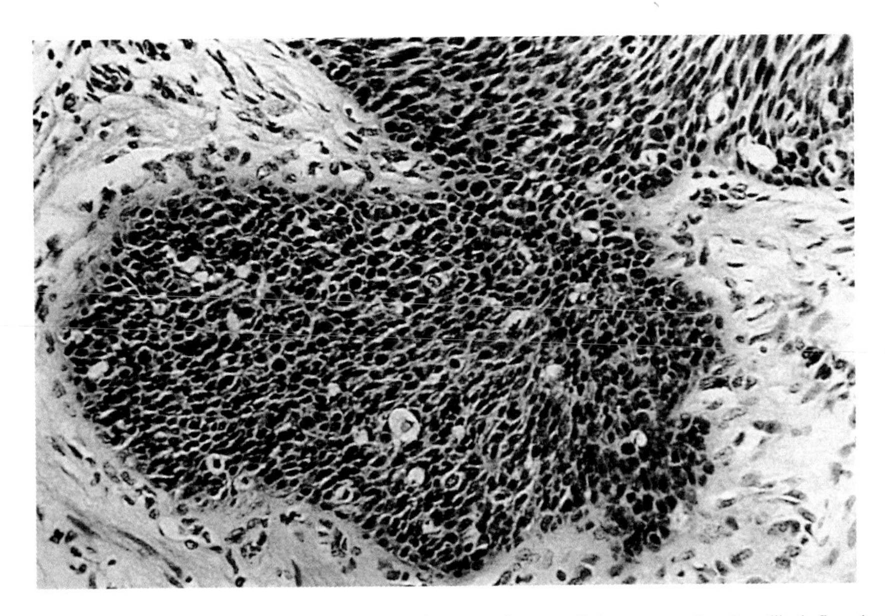

Figure 11.21 CIN 3 with protrusion into stroma and secondary protuberances having ill-defined margins with absent basement membrane. This appearance is indicative of microinvasion (H & E, ×310).

Figure 11.21, suspicion of microinvasion becomes a certainty.

The Society of Gynecological Oncologists in 1973 considered that a case of intraepithelial carcinoma with only questionable invasion should be regarded as intraepithelial carcinoma as far as clinical management is concerned (Creasman and Weed, 1979).

REFERENCES

Anderson, M. C. and Hartely, R. B. (1980) Cervical crypt involvement by intraepithelial neoplasia *Obstet. Gynecol.*, **55**, 546–550.

Anderson, M. C., Burghardt, E., Coppleson, J. W. M., Kolstadt, P., Richart, R. M. and Wade-Evans, T. (1982) in *Pre-clinical Neoplasia of the Cervix* (eds J. A. Jordan, F. Sharp and A. Singer), Royal College of Obstetricians and Gynaecologists, London, Appendix 1, p. 301.

Benson, W. L. and Norris, H. J. (1977) A critical review of the frequency of lymph node metastasis and death from microinvasive carcinoma of the cervix. *Obstet. Gynaecol.*, **49**, 632–638.

Boyce, J. G. *et al.* (1984) Vascular invasion in stage 1 carcinoma of the cervix. *Cancer*, **53**, 1175–1180.

Boyes, D. A., Worth, A. J. and Fidler, H. K. (1970) The results of treatment of 4389 cases of pre-clinical cervical squamous carcinoma. *J. Obstet. Gynaecol. Br. Commonwlth.*, **77**, 769–780.

Burghardt, E. (1973) *Early Histological Diagnosis of Cervical Cancer*. Saunders, Philadelphia, pp. 319–362.

Burghardt, E. (1982) Diagnostic and prognostic criteria in cervical micro-carcinoma. *Clin. Oncol.*, **1**, 323–333.

Creasman, W. T. and Weed, J. C. (1979) Microinvasive cancer versus occult cancer. *Int. J. Radiat. Oncol. Biol. Phys.*, **5**, 1871–1872.

Crissman, J. D., Makuch, R. and Budhraja, M. (1985) Histopathology grading of squamous cell carcinoma of the uterine cervix. An evaluation of 70 stage 1B patients. *Cancer*, **55** 1590–1596.

Ferenczy, A. (1982) Carcinoma and other malignant tumours of the cervix, in *Pathology of the Female Genital Tract*, 2nd edn (ed. A. Blaustein), Springer-Verlag, New York, Heidelberg, Berlin, pp. 184–222.

Kinney, W. K., Egorshin, E. V., Ballard, D. J. and Podratz, K. C. (1992) Long term survival and sequelae after surgical management of invasive cervical carcinoma diagnosed at the time of simple hysterectomy. *Gynecol. Oncol.*, **44**, 24–27.

Larsson, G., Alm, P., Gullberg, G. and Grundsell, M.

(1983) Prognostic factors in early invasive carcinoma of the uterine cervix. A clinical histopathological and statistical analysis of 343 cases. *Am. J. Obstet. Gynecol.*, **146**, 145–153.

Lohe, K. J. (1978) Early squamous cell carcinoma of the uterine cervix. III. Frequency of lymph node metastases. *Gynecol. Oncol.*, **6**, 51–59.

Maiman, M. A., Fruchter, R. G., Dimaio, T. M. and Boyce, J. G. (1988) Superficially invasive squamous cell carcinoma of the cervix. *Obstet. Gynecol.*, **72**, 399–403.

Mestwerdt, G. (1947), Probeexzision und Kolposkopie in der Fruhdiagnose des Portio Karinoms. *Zentralbl. Gynakäl.*, **4**, 326–332.

Nelson, J. H., Averette, H. F. and Richard, R. M. (1975) Detection, diagnostic evaluation and treatment of dysplasia and early carcinoma of the cervix. *Cancer*, **25**, 134–151.

Ng, A. B. P. and Reagan, J. W. (1969) Microinvasive carcinoma of the uterine cervix. *Am. J. Clin. Pathol.*, **52**, 511–529.

Poulsen, H. F., Taylor, C. W. and Sobin, L. H. (1975) *Histological Typing of Female Genital Tract Tumours*. World Health Organization, Geneva, pp. 15, 57–59.

Reinthaller, A., Tatra, G., Breitenecker, G. and Janisch, H. (1991) Prognostic factors in stage 1A-2B invasive cervix cancer after radical hysterectomy with special references to stroma reaction (in German). *Geburtshilfe Frauenheilkd.*, **51**, 809–813.

Roche, W. D. and Norris, H. J. (1975) Microinvasive carcinoma of the cervix: the significance of lymphatic invasion and confluent patterns of stromal growth. *Cancer*, **36**, 180–186.

Rubio, C. A., Söderberg, G. and Finhorn, N. (1974) Histological and follow-up studies in cases of micro-invasive carcinoma of the uterine cervix. *Acta Pathol. Microbiol. Sect. A*, **82**, 397–410.

Sano, T. and Veki, M. (1987) Stromal reactions to squamous cell carcinoma of the cervix. *Am. J. Obstet. Gynecol.*, **156**, 906–910.

Scully, R. E., Bonfiglio T. A., Kurman, R. J. *et al.* (1994) *Histological typing of female genital tract tumours.* World Health Organization International Histological Classification of Tumours, 2nd edn. Springer-Verlag, Berlin.

Sedlis, A., Sall, S., Tsukada, Y., Park, R., Manham, C., Shingleton, H. and Blessing, J. A. (1979) Microinvasive carcinoma of the uterine cervix: a clinical pathologic study. *Am. J. Obstet. Gynecol.*, **133**, 64–74.

Seski, J. C., Abell, M. R. and Morley, G. W. (1977) Microinvasive squamous carcinoma of the cervix: definition, histological analysis, late results of treatment. *Obstet. Gynecol.*, **50**, 410–414.

Tidbury, P., Singer, A. and Jenkins, D. (1992) The role of lesion size in invasion *Br. J. Obstet. Gynaecol.*, **99**, 583–586.

Ullery, J. C., Boutselis, J. G. and Botschner, A. C. (1965) Microinvasive carcinoma of the cervix. *Obstet. Gynecol.*, **26**, 866–875.

Walton Report (1976) Cervical cancer screening programs. Report of the Task Force of the Department of National Health and Welfare of Canada. *Can. Med. Assoc. J.*, **114**, 1003–1033.

12

MALIGNANT EPITHELIAL TUMOURS: CLASSIFICATION AND SQUAMOUS CELL CARCINOMA

12.1 CLASSIFICATION

Various types of malignant tumour occur in the cervix. The following categories are based on the WHO International Histological Classification of Tumours. Histological typing of female genital tract tumours (Scully *et al.*, 1994):

1. Squamous cell carcinoma (epidermoid carcinoma)
   - Keratinizing (well-differentiated and moderately differentiated)
   - Non-keratinizing (large cell and small cell types)
   - Special types:
     - Verrucous carcinoma
     - Warty (condylomatous) carcinoma
     - Papillary squamous carcinoma
     - Lymphoepithelioma-like carcinoma
     - Spindle cell (scirrhous) carcinoma
2. Adenocarcinoma
   - Mucinous adenocarcinoma
     - Endocervical type
       - Adenoma malignum (minimal deviation carcinoma)
       - Villoglandular papillary adenocarcinoma
     - Intestinal type
       - Signet-ring cell type
   - Endometrioid adenocarcinoma
   - Clear cell adenocarcinoma
   - Serous adenocarcinoma
   - Mesonephric adenocarcinoma
3. Other epithelial tumours
   - Adenosquamous carcinoma
     - Glassy cell carcinoma
     - Mucoepidermoid carcinoma
   - Adenoid cystic carcinoma
   - Adenoid basal carcinoma
   - Carcinoid-like carcinoma (argentaffinomas, neuroendocrine tumours)
   - Small cell (anaplastic) carcinoma
   - Undifferentiated carcinoma
   - Metastatic carcinoma

12.2 SQUAMOUS CELL (EPIDERMOID) CARCINOMA

Invasive squamous carcinoma is the most common malignant tumour of the cervix, accounting for 90% of primary carcinomas which arise in this organ. Most tumours are considered to arise from a focus of CIN originating in reserve cells in the transformation zone. A few may arise in basal cells of the original ectocervical epithelium, especially lesions close to the vagina (Burghardt, 1973). In some early invasive cancers, it is possible to identify an area of CIN at the edge of the lesion. The transition from CIN to invasive cancer may take many years. However, not all invasive carcinomas have a recognizable preinvasive stage; the clinical history suggests that some develop very rapidly indeed.

For many years squamous carcinomas were thought to represent a morphologically homogeneous group of tumours comprised exclusively of squamous cells. More recent studies have shown that only about 70% of the tumours are purely squamous (Buckley and Fox, 1989), the remaining 30% having a significant glandular component. This has been demonstrated by including a mucin stain as part of the routine histopathological examination of apparently squamous tumours.

The pattern of growth of squamous cell carcinoma is variable. It may be exophytic, having a polypoid or papillary pattern, or it may be flat or ulcerating. In all forms, the tumour sends out cords of cells into the stroma and surrounding tissue. These appear, in section, as islands of malignant cells surrounded by stroma. There is often an inflammatory infiltrate in the surrounding stroma composed mainly of lymphocytes and plasma cells.

The aggressiveness of squamous cancer depends to some extent on the age of the patient. Fenton *et al.* (1990) reviewed patients of 40 years and under, treated at the Institut Curie between 1970 and 1984. The overall 5-year survival rate was 75%. For the 36–40-year age group it was 85%, for the 30–35-year group it was 74%, and for those aged 29 years of less it was 67%. They concluded that the survival rate for women of 36 years or less, particularly those of 29 years or less, was significantly lower than for older women, especially if there was lymph node involvement. Buckley *et al.* (1988) had reached a similar conclusion, observing that the recognition of mucus secretion in a carcinoma and the detection of vascular permeation adjacent to the primary neoplasm identified the patients at greatest risk of having pelvic lymph node metastases.

12.2.1 Types

The WHO recognizes two common types of squamous cell carcinoma, namely keratinizing carcinoma in which keratin pearls are present and non-keratinizing carcinoma which lacks keratin pearls but may contain individually keratinized cells (Scully *et al.*, 1994) The keratinizing carcinomas may be well-differentiated or moderately differentiated (Buckley and Fox, 1989) and are composed of large cells. Non-keratinizing carcinomas may be of large cell or small cell type. Some of the large cell non-keratinizing tumours are moderately differentiated but virtually all the small cell tumours are poorly differentiated.

In one study, Chung *et al.* (1980) treated 87 previously untreated patients who had FIGO stage IB and IIA epidermoid cervical cancer: they reported an overall 2-year survival rate of 100% (20/20) for well-differentiated keratinizing carcinoma, 93% (76/82) for moderately differentiated keratinizing carcinoma, and 70% (14/20) for poorly differentiated non-keratinizing carcinoma. Although some authors have found no correlation between prognosis and tumour cell type (Gunderson *et al.*, 1974; Goellner, 1976; Beecham *et al.*, 1978) there are a number of reports that non-keratinizing small cell carcinomas have the worst prognosis (Wentz and Lewis, 1965; Fink and Denk, 1970; Reagan and Ng, 1973; Swan and Roddick, 1973; Sedlacek *et al.*, 1978). A very different picture is presented by Sugawa *et al.* (1982), particularly in patients given post-operative maintenance chemotherapy, who found the recurrence rate for large cell keratinizing carcinoma to be 41% (7/17), for large cell non-keratinizing 5% (3/58), and for small cell non-keratinizing 0% (0/25). After summarizing a number of studies, Wright *et al.* (1994) concluded that no histological grading or typing system had been shown to predict prognosis reliably.

Special types of squamous cell carcinoma include *verrucous carcinoma*, an exceptionally well-differentiated but insidiously progressive variant of keratinizing squamous cell carcinoma, with a characteristically blunt ('pushing') deeper margin; *warty (condylomatous) carcinoma*, morphologically similar to verrucous carcinoma but having a warty surface, cellular features of HPV infection and a deep margin that resembles typical squamous carcinoma (Kurman *et al.*, 1990); *papillary squamous carcinoma* which differs from warty carcinoma in that it has a papillary architecture and the papillary processes are lined by dysplastic cells (Randall *et al.*, 1986); *lymphoepithelioma-like carcinoma*, a circumscribed squamous carcinoma, densely infiltrated by lymphocytes and having a limited growth potential (Hamazaki *et al.*, 1968); and *spindle cell (scirrhous) carcinoma*, resembling a sarcoma histologically, with associated connective tissue proliferation (fibroplasia) and of low grade malignancy (Harris, 1982).

12.2.2 Cytological diagnosis of squamous carcinoma of the cervix

Some 90% of malignant lesions of the cervix can be detected by cytological examination of the cervical smear (Section 1.5). The main reason for failure of cytology to detect the tumour is the presence of blood, pus and necrotic debris in the smear which obscure the tumour cells. Tumours located deep in the cervical stroma may not be detected. Thus, a clinical suspicion of malignancy should always override a negative cytology report.

Well-differentiated keratinizing squamous carcinoma can be recognized from the cervical smear pattern with a high degree of accuracy. Less well-differentiated tumours are more difficult to type and it is not always possible to distinguish between invasive, microinvasive and intraepithelial neoplastic lesions from the cervical smear. However, there are certain factors that strongly suggest an invasive lesion and are an indication for urgent biopsy. The presence of highly keratinized abnormal squamous cells and undifferentiated tumour cells are all suggestive of an invasive lesion, especially if they are found in the presence of blood, pus and necrotic debris.

12.3 WELL-DIFFERENTIATED AND MODERATELY DIFFERENTIATED KERATINIZING SQUAMOUS CARCINOMA

The most characteristic feature of *well-differentiated* keratinizing squamous carcinomas is the presence of whorls of neoplastic squamous cells with large nuclei and keratinized cytoplasm forming the keratin pearls shown in Figures 12.1 and 12.2. Intercellular bridge formation is a more prominent feature of well-differentiated carcinoma of the cervix (Figure 12.3) than in the other grades of squamous carcinoma. Another characteristic is the presence of cells with clear cytoplasm (Figure 12.4) which contain glycogen giving a positive PAS reaction. These glycogen-containing cells are responsible for the misleading Schiller's iodine reaction described in Section 3.1. Mitotic figures may be sparse or frequent. Although the cells are usually fairly uniform, there may be considerable cellular pleomorphism (Figure 12.5). Keratohyaline granules may also be seen.

Moderately differentiated squamous carcinoma differs from the well-differentiated form in that keratinization is focal and less abundant. Epithelial pearls are scanty or absent The component cells show greater nuclear pleomorphism, with a higher nucleocytoplasmic ratio and more frequent mitotic figures (Figure 12.6). Intercellular bridge formation is less well defined and often difficult to demonstrate. Dyskeratosis (individual cell keratinization, see Figure 12.7) may be seen, the affected cells having conspicuous eosinophilic cytoplasm. These cells may occur singly or in small groups. Although it is generally considered that moderately differentiated keratinizing squamous carcinoma has a prognosis intermediate

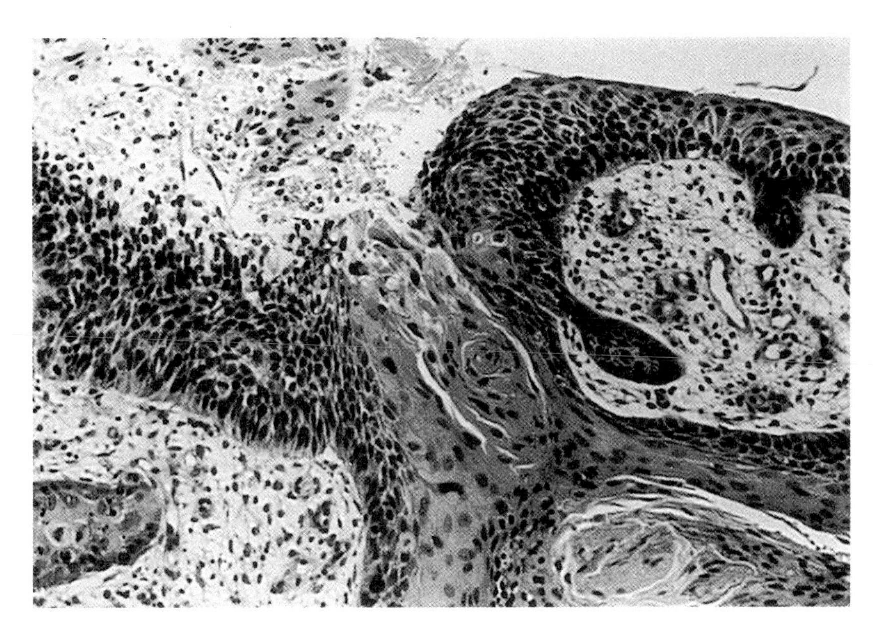

Figure 12.1 Well-differentiated squamous cell carcinoma. Keratin squames in the centre of the field are being extruded on to the surface (H & E, ×190).

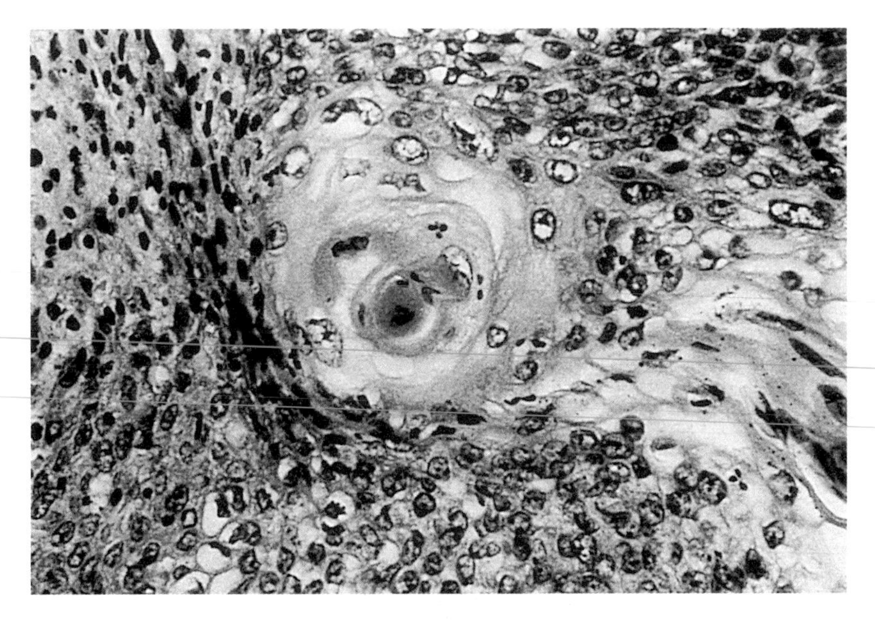

Figure 12.2 Cell nest with keratin pearl formation characteristic of well-differentiated squamous cell carcinoma (H & E, ×480).

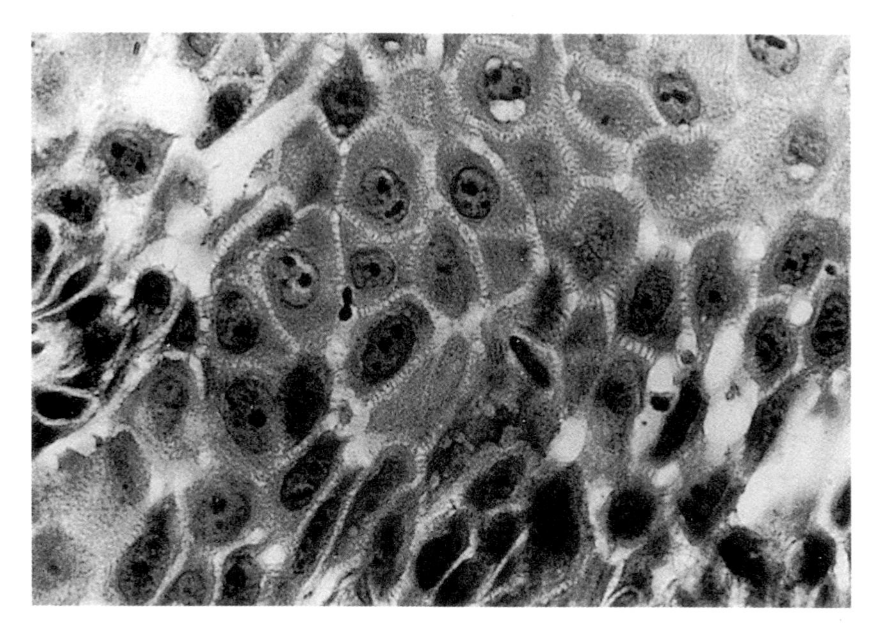

Figure 12.3 Prominent intercellular bridge formation in a well-differentiated squamous cell carcinoma (H & E, ×770).

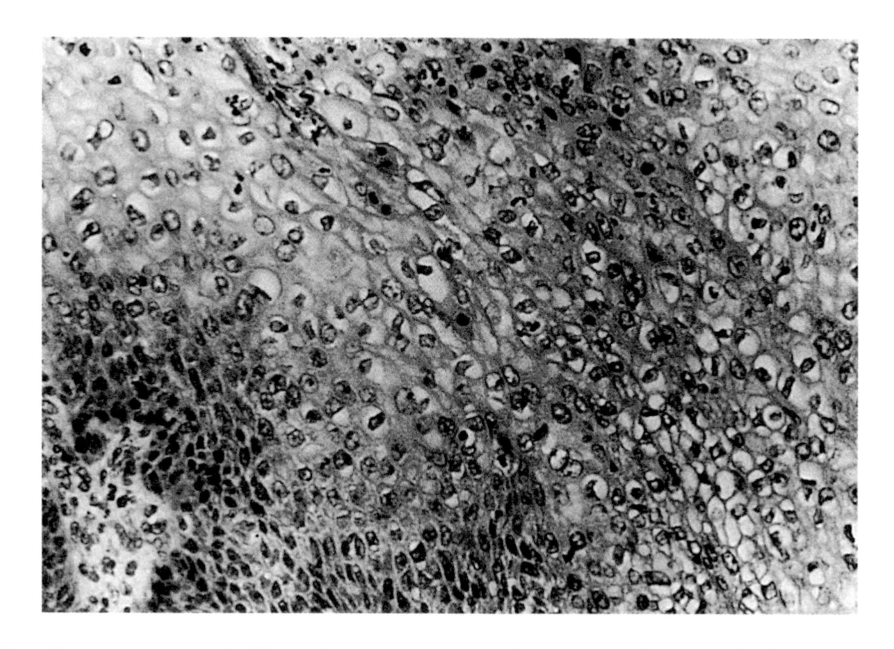

Figure 12.4 Clear cells in a well-differentiated squamous cell carcinoma (H & E, ×310).

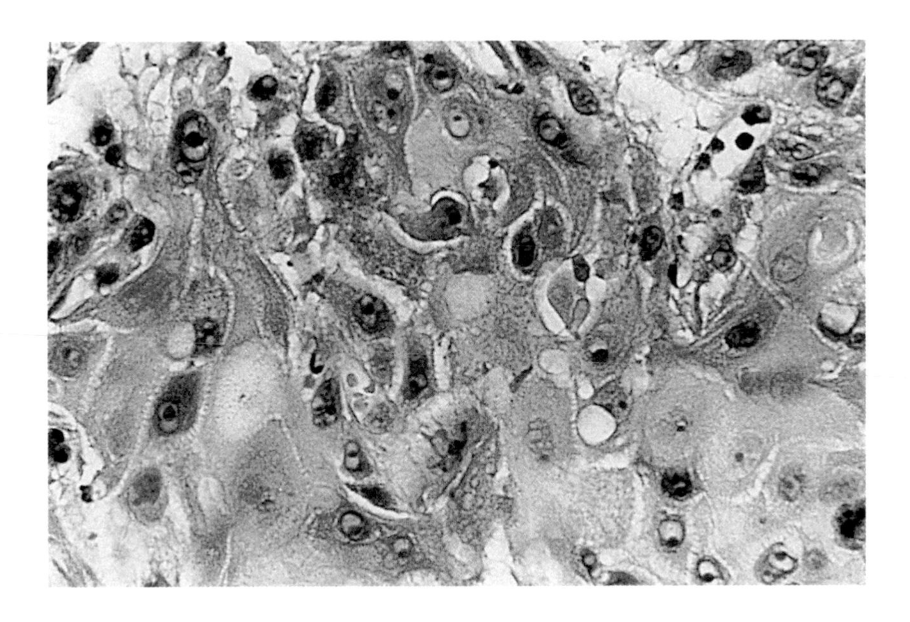

Figure 12.5 Pleomorphism in a well-differentiated squamous carcinoma (H & E, ×480).

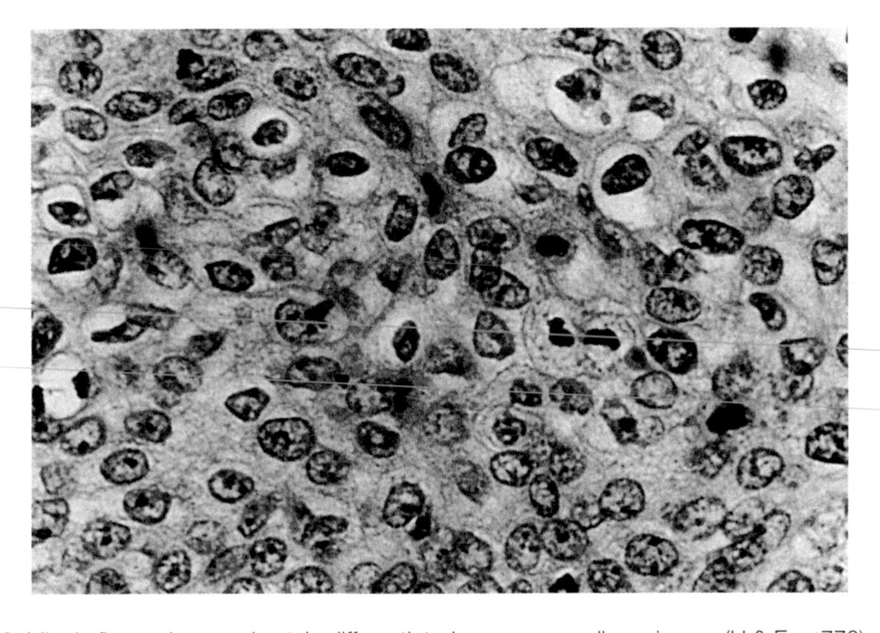

Figure 12.6 Mitotic figures in a moderately differentiated squamous cell carcinoma (H & E, ×770).

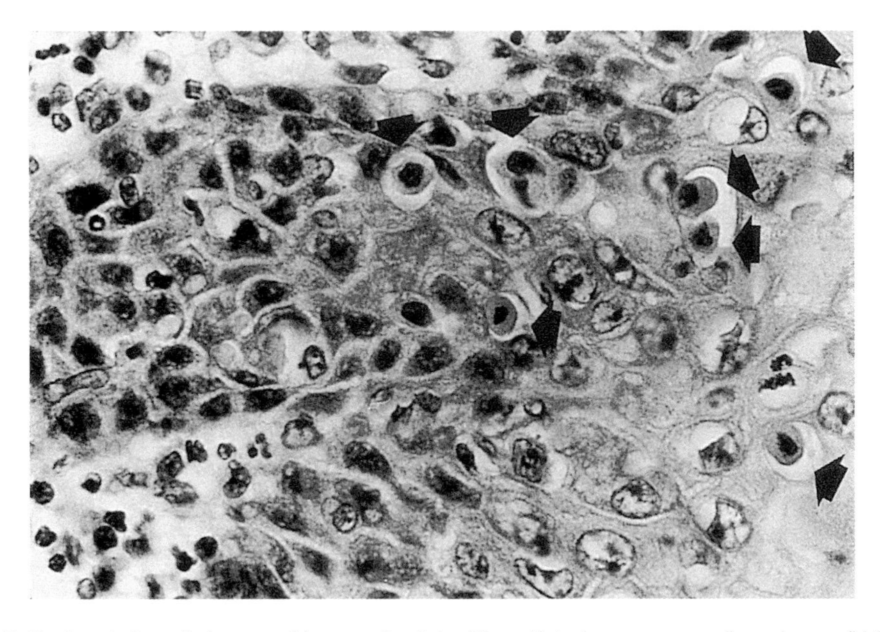

Figure 12.7 Dyskeratotic cells (arrowed) in a moderately differentiated squamous cell carcinoma (H & E, ×770).

between that of well and poorly differentiated lesions at the same stage, not all reported studies support this view (Section 12.2.1).

12.3.1 Cytology

The most striking finding in smears from patients with keratinizing squamous carcinoma is the presence of numerous abnormal squamous cells with highly keratinized orangeophilic cytoplasm (Figure 12.8) and irregular hyperchromatic nuclei (Figures 12.9 and 12.10). The cells may be single or in sheets and they often assume very bizarre forms. Spindle-shaped or tadpole-shaped cells are frequently seen (Figure 12.11(a)). Epithelial whorls and phagocytosis of tumour cells are common (Figure 12.11(a,b)). The nuclei of these cells are usually shrunken and pyknotic but may be very large indeed (Figure 12.12(a,b)) or multinucleated (Figure 12.12(c)). Karyorrhexis is frequently observed (Figure 12.13(a)) together with other degenerative changes such as vacuolation of nucleus or cytoplasm (Figure 12.13(b)). Solid-looking plaques of keratin studded with irregular, pyknotic nuclei (Figure 12.14) and highly keratinized anucleate squames are features of keratinizing carcinoma in smears and reflect the extensive dyskeratosis associated with this tumour.

In addition to the keratinized squames, clusters of cells or minibiopsies with varying amounts of basophilic or amphophilic cytoplasm (Figure 12.15(a)) may be seen, reflecting cells shed from non-keratinized portions of the tumour. These cells have irregular, nuclei with coarse chromatin. Hyperchromasia is a common but not inevitable features of these tumour cells. Usually the nuclei are rounded but spindle-shaped nuclei may also be seen. Occasionally, cells with large nucleoli (Figure 12.15(b)) or cells in mitosis may be seen. Phagocytosis of leucocytes by the tumour cells shown in Figure 12.13(c) is a rather unusual finding.

12.4 LARGE CELL NON-KERATINIZING SQUAMOUS CARCINOMA

The prevalence of this type of carcinoma in the USA has gradually increased from 38.5% of all

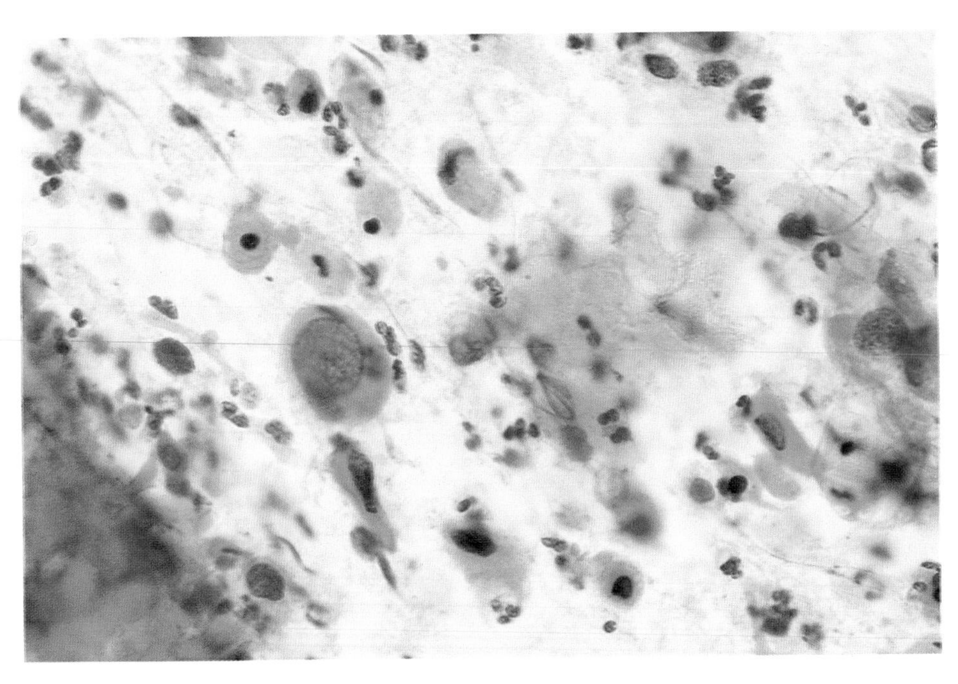

Figure 12.8 Keratinizing squamous carcinoma cells with dense orangeophilic cytoplasm (Pap, ×770).

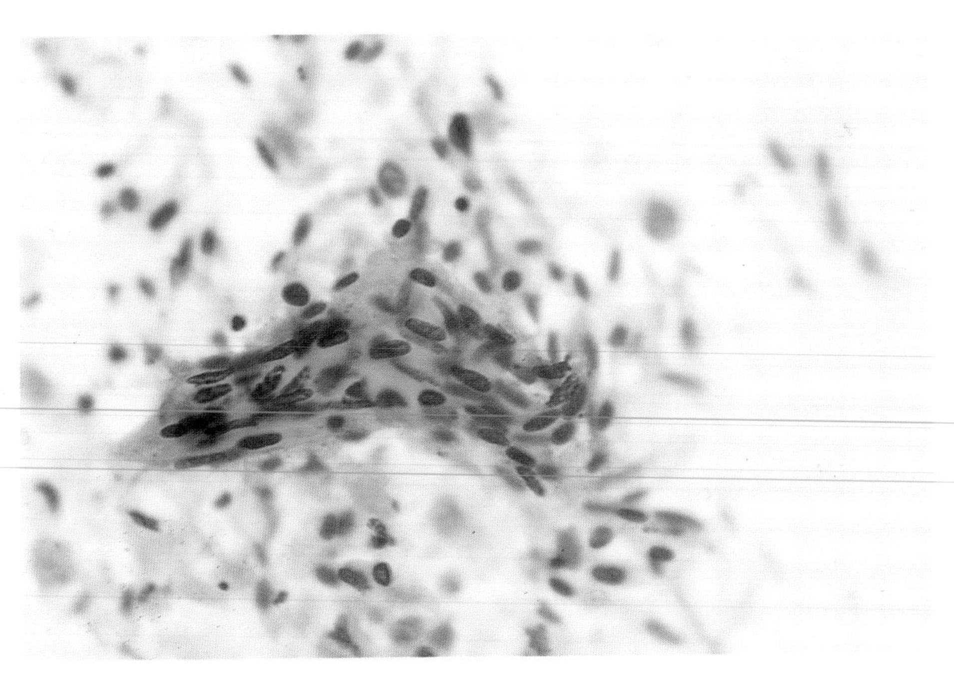

Figure 12.9 Cluster of cells with highly keratinized cytoplasm and spindle-shaped nuclei characteristic of keratinizing squamous carcinoma (Pap, ×630).

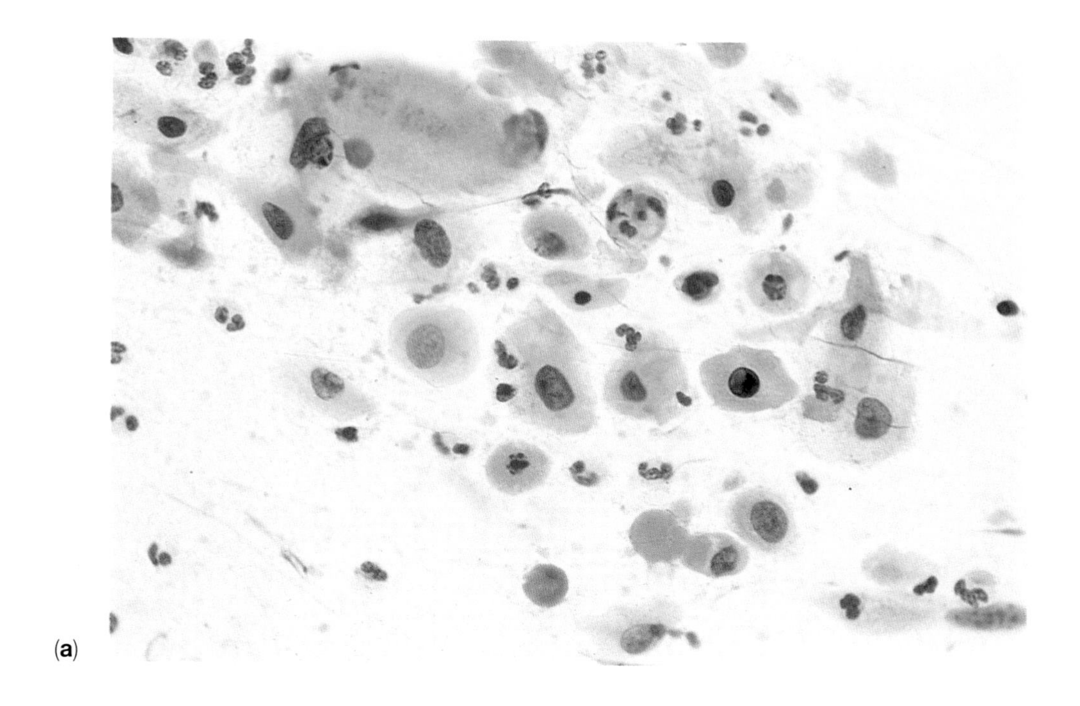

(**a**)

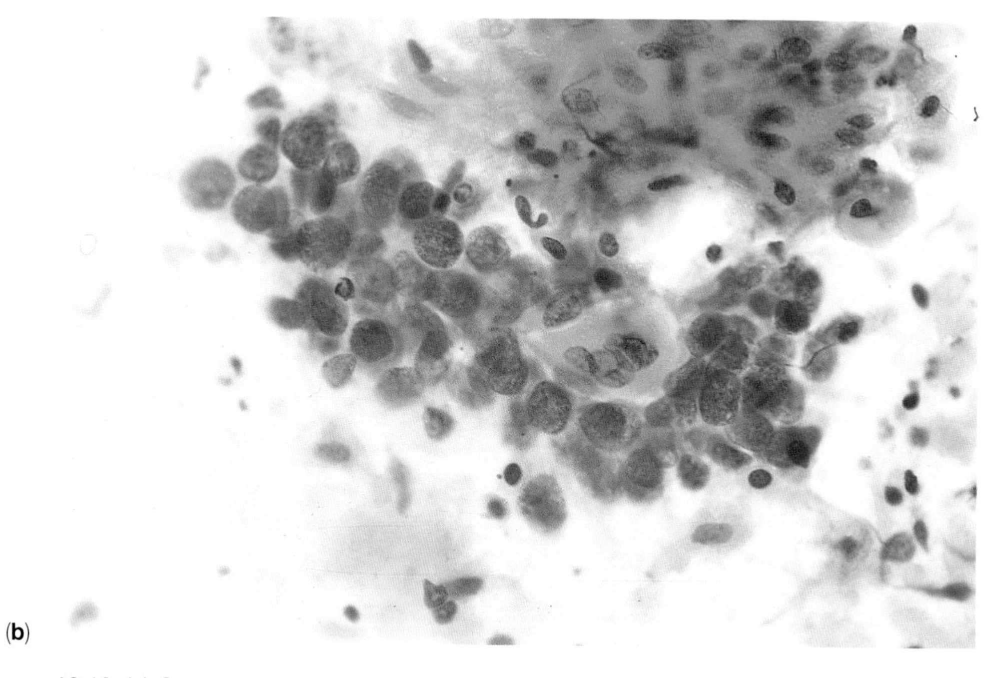

(**b**)

Figure 12.10 (**a**) Squames with keratinized cytoplasm and small irregular hyperchromatic nuclei in cervical smear from keratinizing squamous carcinoma (Pap, ×630). (**b**) Tumour cells from keratinizing squamous carcinoma shown in Figure 12.7. Note mixture of undifferentiated cells and well-differentiated cells (Pap, ×630).

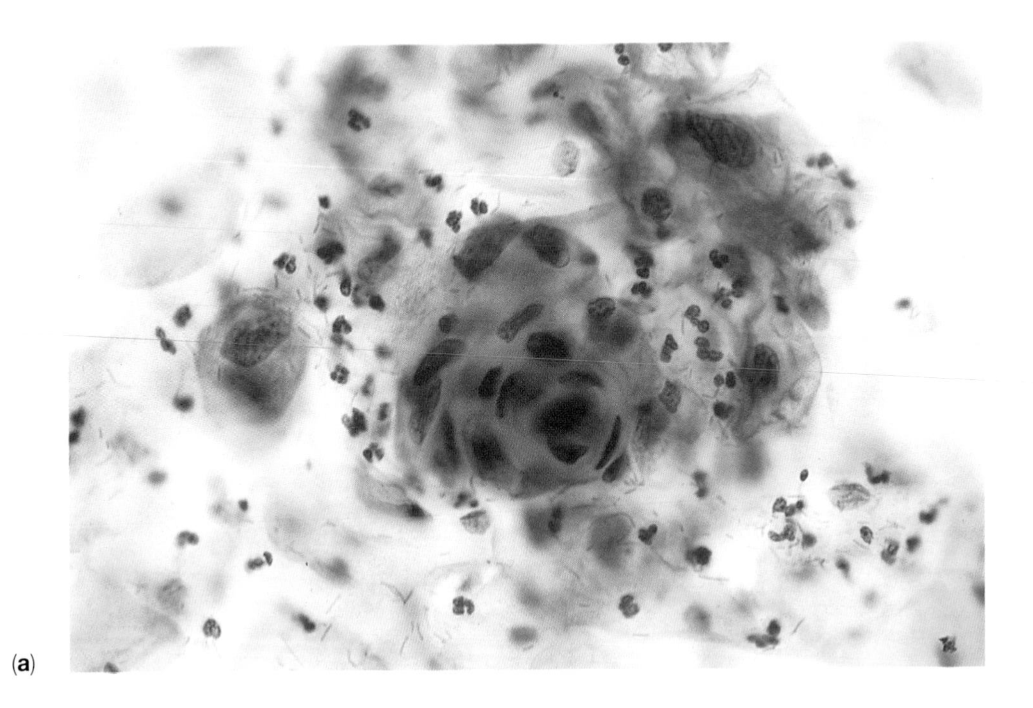

(**a**)

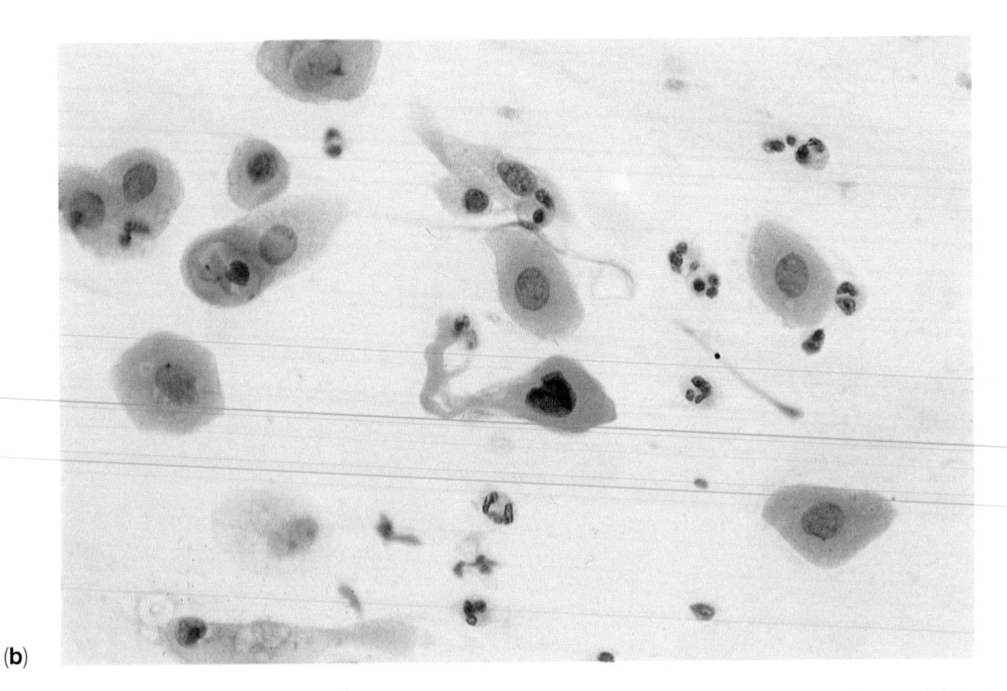

(**b**)

Figure 12.11 (**a** & **b**) Bizarre cells forms encountered in keratinizing squamous carcinoma. Note tadpole-shaped cells and epithelial whorl (Pap, ×630).

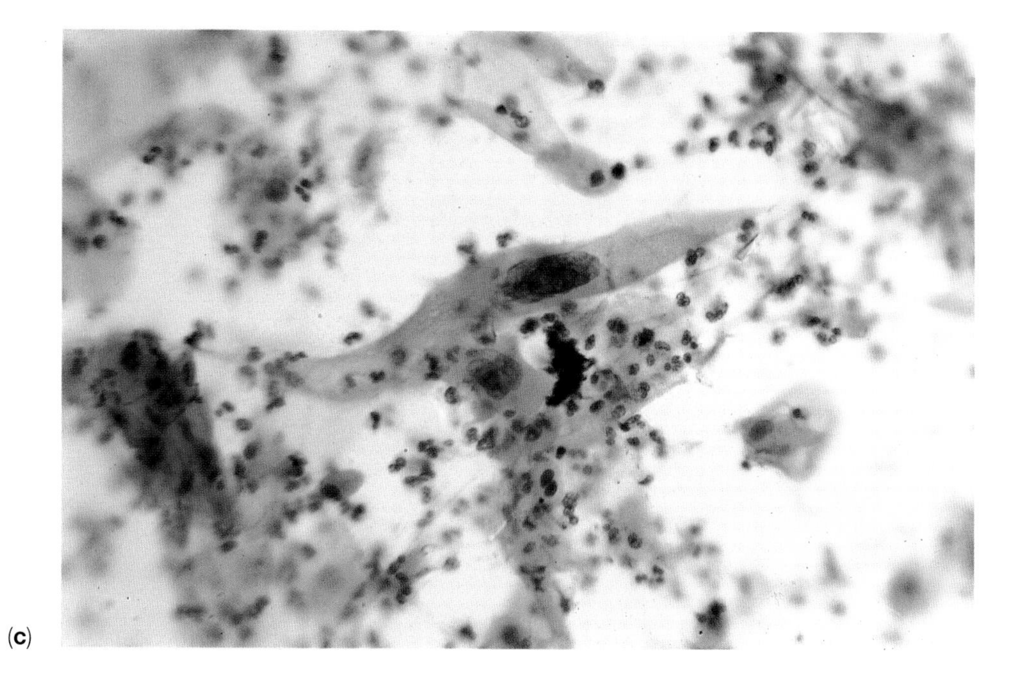

(**c**)

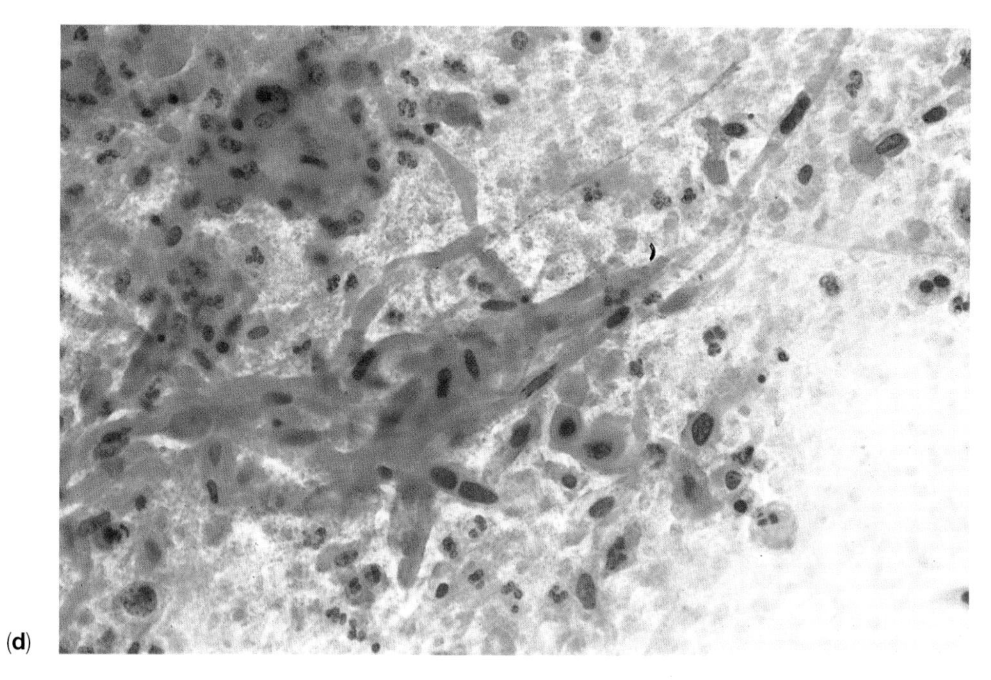

(**d**)

Figure 12.11 (**c** & **d**) Bizarre cell forms encountered in keratinizing squamous carcinoma. Note spindle-cells in invasive squamous carcinoma (Pap, ×630).

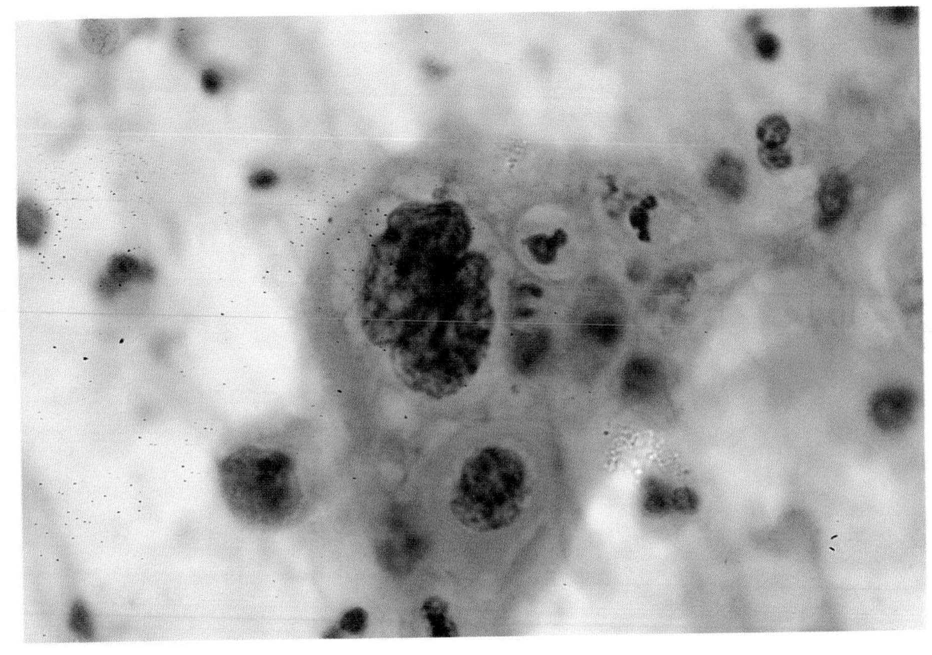

(e)

Figure 12.11 (**e**) Bizarre cell forms encountered in keratinizing squamous carcinoma. Note phagocytic tumour cell in invasive squamous carcinoma (Pap, ×630).

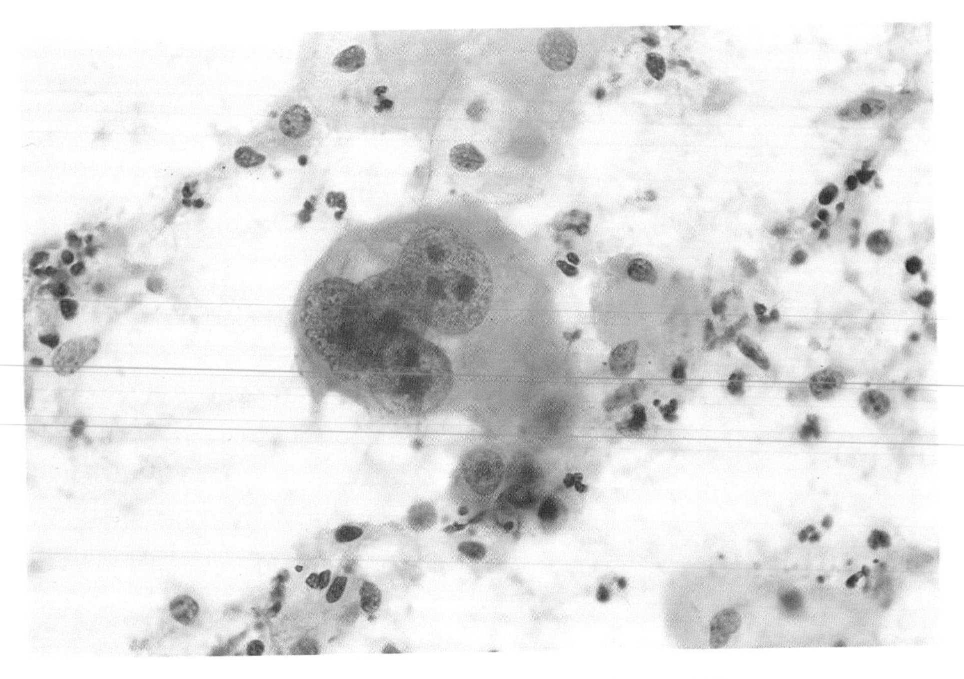

(a)

Figure 12.12 (**a**) Giant nuclei in keratinising squamous carcinoma (Pap, ×100).

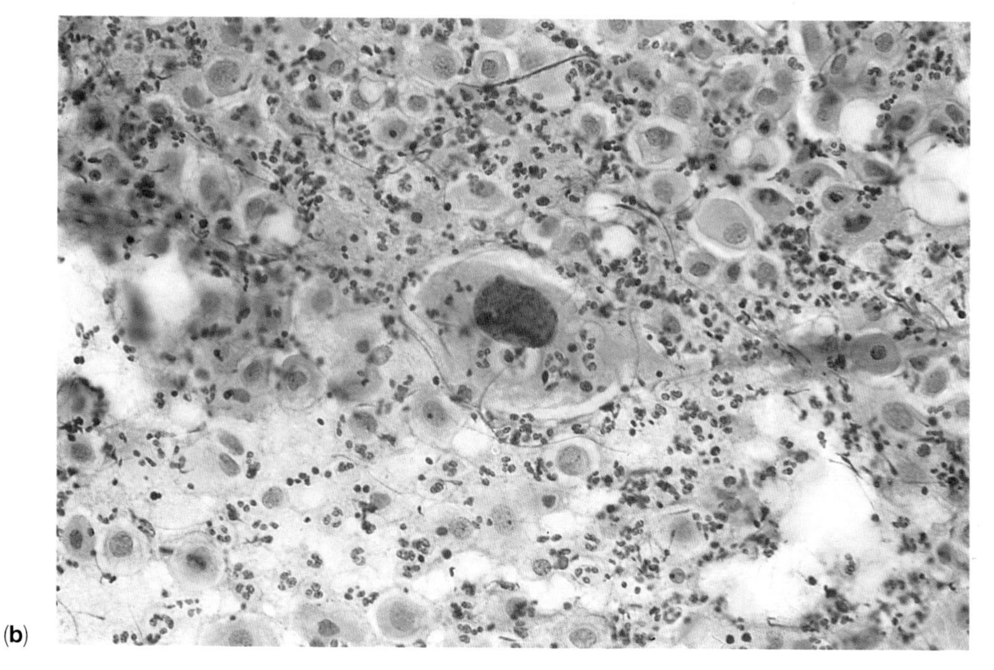

(**b**)

Figure 12.12 (**b**) Giant nuclei in keratinising squamous carcinoma (Pap, ×100).

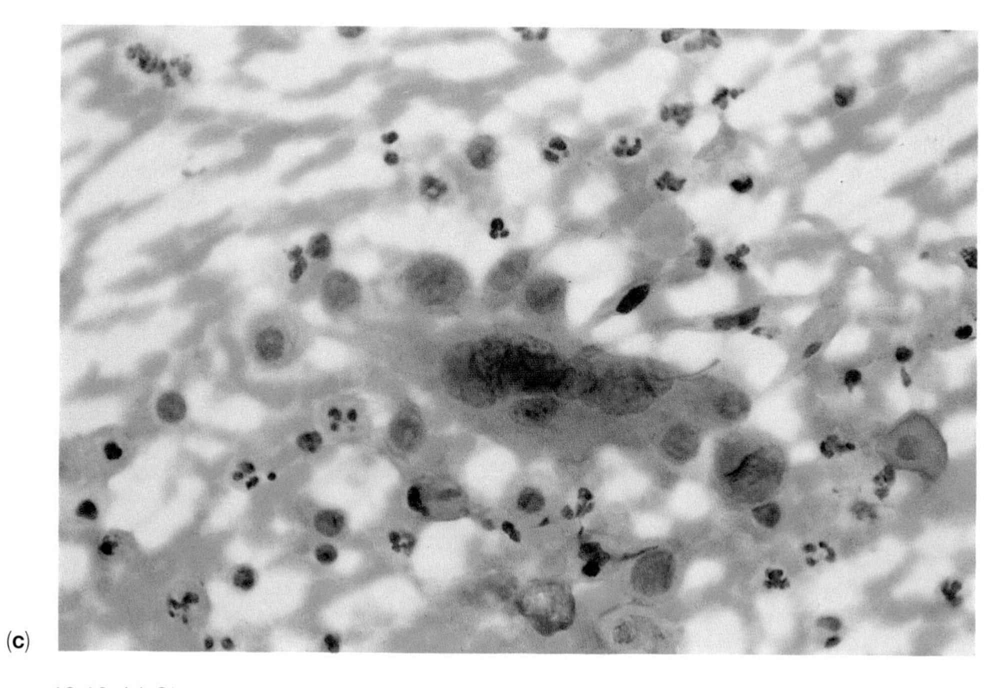

(**c**)

Figure 12.12 (**c**) Giant nuclei in multinucleate tumour cell (Pap, ×100).

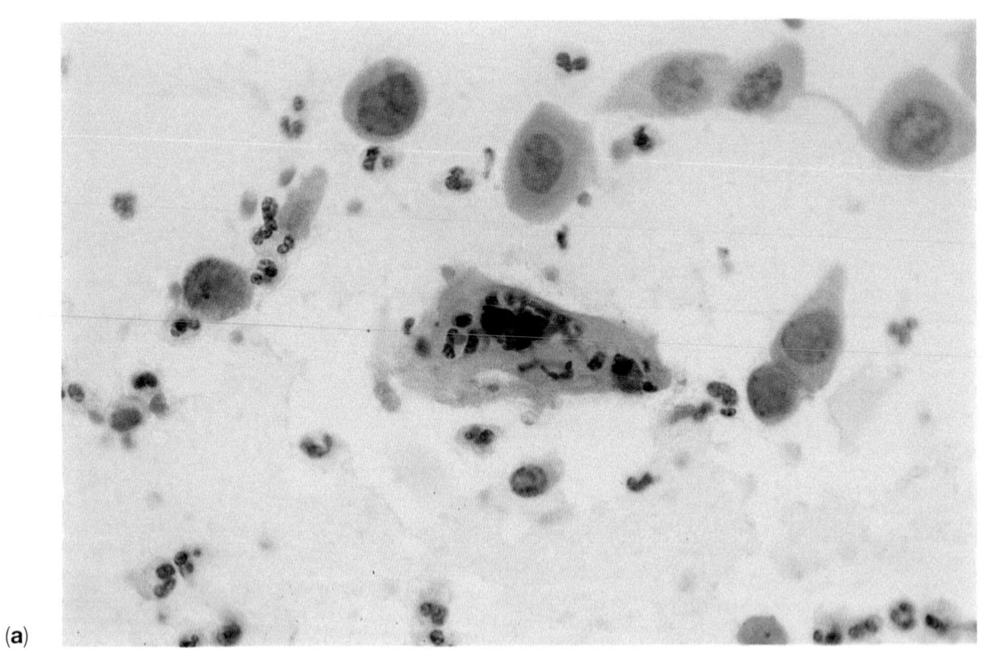

(**a**)

Figure 12.13 (**a**) Degenerative changes in cells in keratinizing squamous carcinoma. Note karyorrhexis (Pap, ×630).

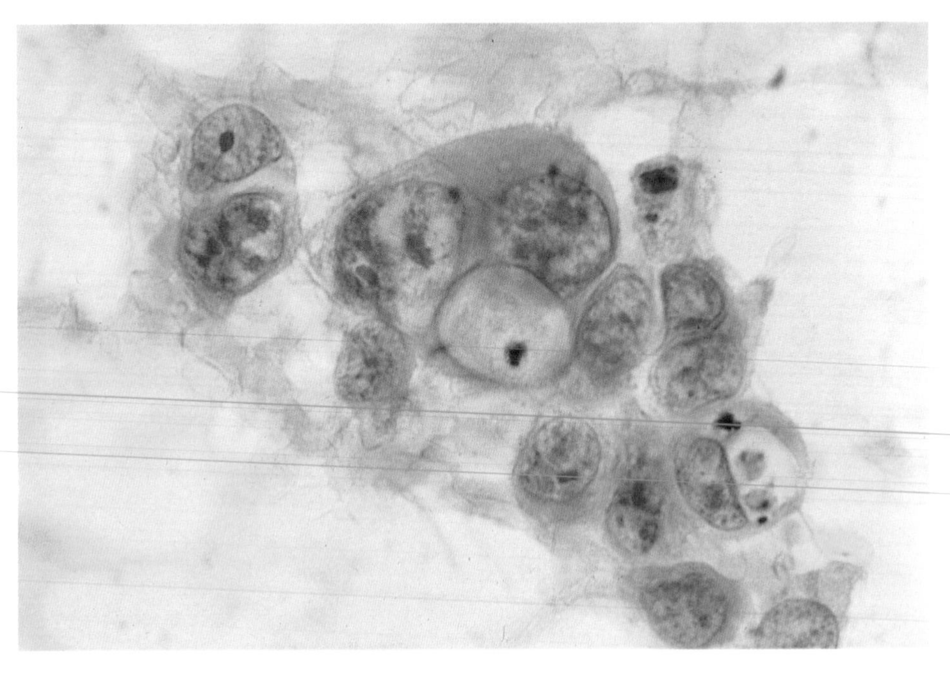

(**b**)

Figure 12.13 (**b**) Degenerative changes in cells with keratinizing squamous carcinoma. Note vacuolation (Pap, ×630).

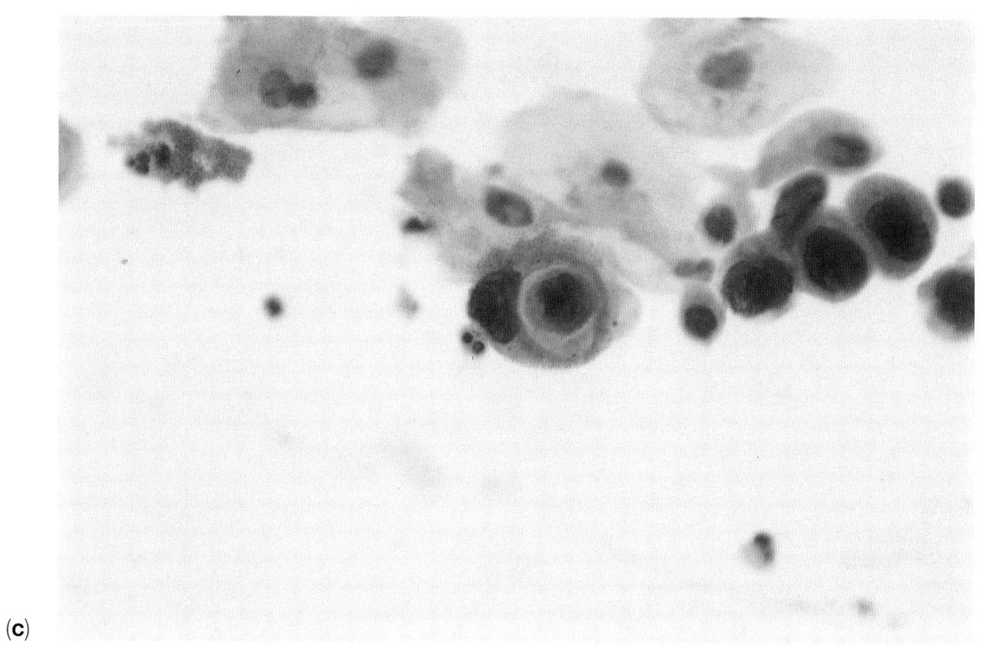

(c)

Figure 12.13 (**c**) Degenerative changes in cells in keratinizing squamous carcinoma. Note phagocytosis (Pap, ×630).

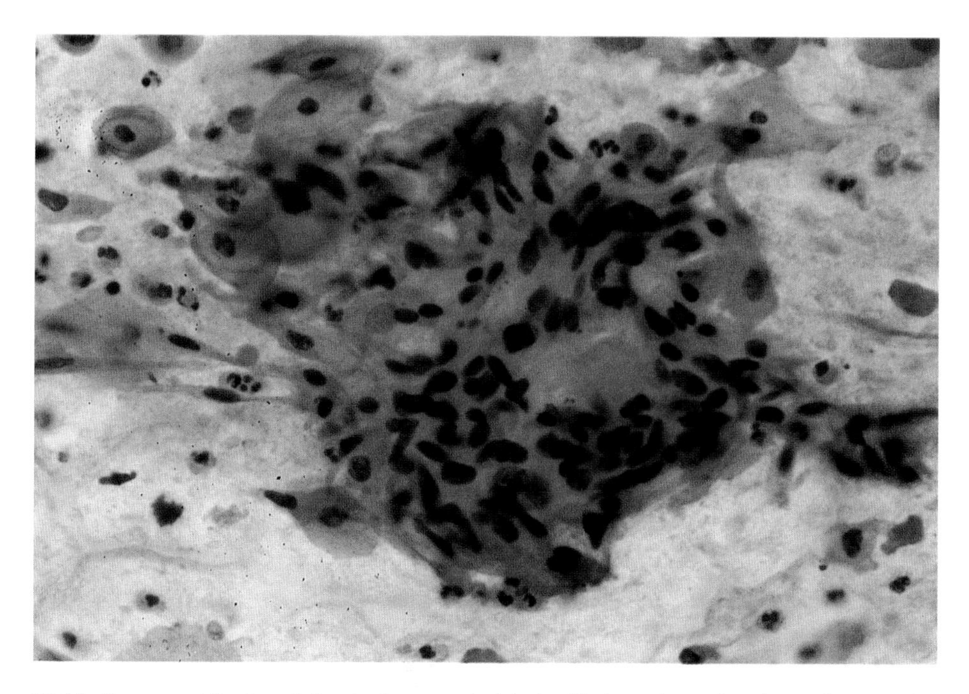

Figure 12.14 Orangeophilic keratinized plaques studded with irregular pyknotic or fragmenting nuclei. A common finding in keratinizing squamous cell carcinoma (Pap, ×630).

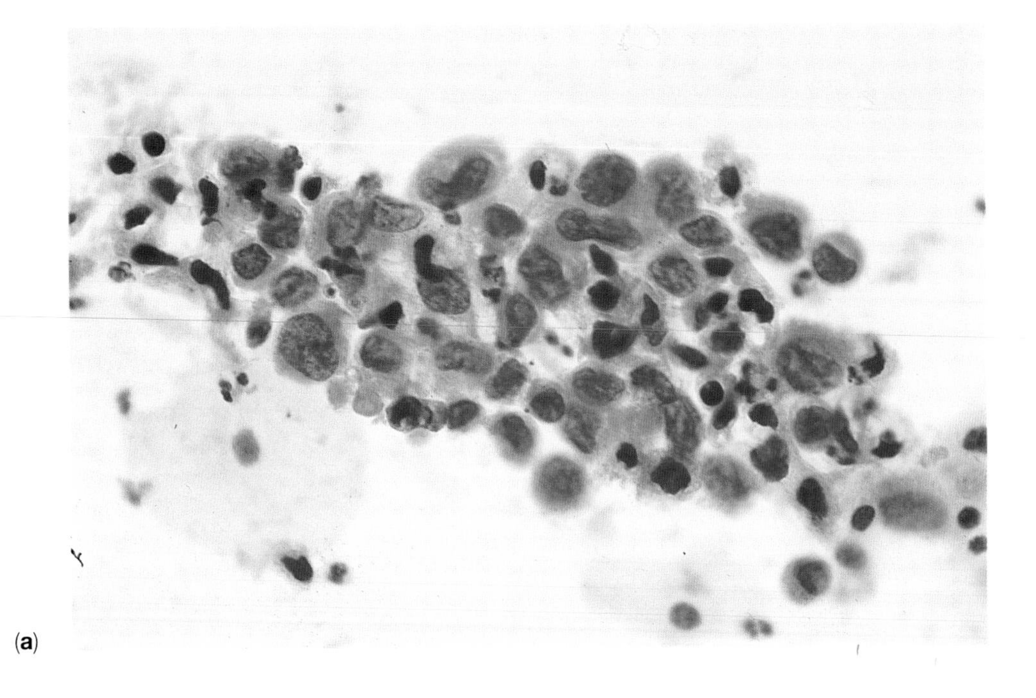

(a)

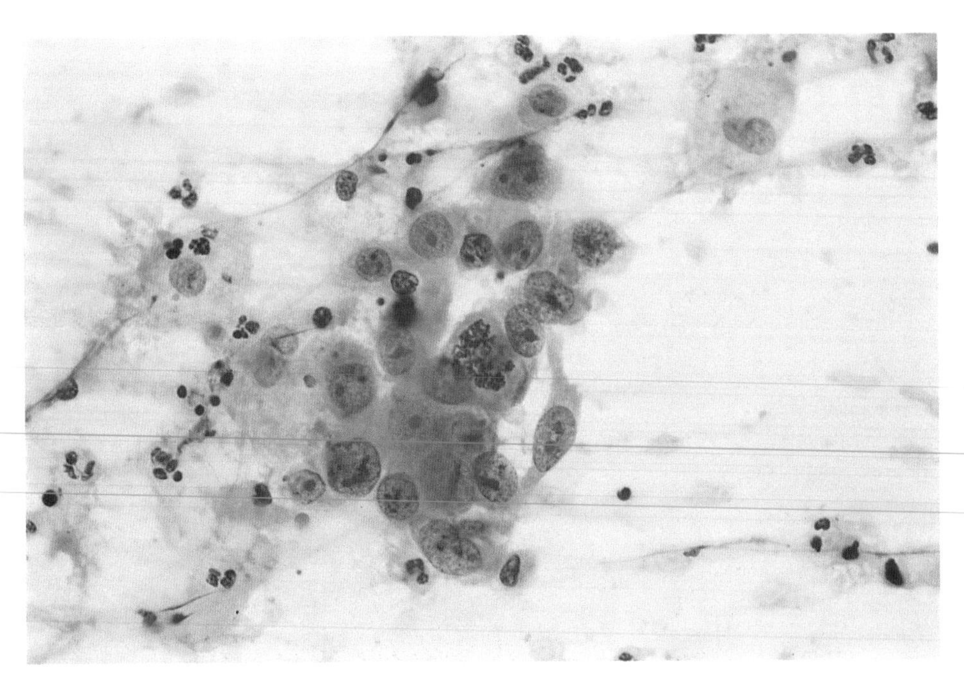

(b)

Figure 12.15 (**a**) Loose cluster of cells with basophilic cytoplasm and abnormal nuclei in keratinizing squamous carcinoma reflecting non-keratinizing areas of the tumour (Pap, ×630). (**b**) Cells with large nucleoli in keratinizing squamous cell carcinoma (Pap, ×630).

cervical squamous carcinomas in 1941–1945 to 72% in 1966–1970 (Reagan and Ng, 1973). These authors differ from most others in reporting a better 5-year survival rate (68%) for large cell non-keratinizing carcinoma than for large cell keratinizing carcinoma (42%). Histologically (Figure 12.16), the large polygonal cells of which the tumour is composed have a mainly eosinophilic cytoplasm. Keratinization is either absent or sparse, and little intercellular bridge formation is seen. Mitotic figures tend to be more numerous than in keratinizing carcinoma and may be quite frequent (Figure 12.17). The component cells are often fairly uniform (Figure 12.18) but malignant features such as nucleomegaly, pleomorphism, anisonucleosis, nuclear hyperchromatism, chromatin clumping and a high nucleocytoplasmic ratio may predominate (Figure 12.19). Cells may be arranged in moderate-sized aggregations or large sheets.

12.4.1 Cytology

The cells shed from non-keratinizing squamous carcinomas show much less pleomorphism than those shed from a keratinizing tumour. They may occur in large sheets or moderate-sized aggregates (Figure 12.20) but single cells are also seen. The hyperchromatic, irregular nuclei are surrounded by eosinophilic or basophilic cytoplasm which may be very scanty indeed (Figure 12.21). Mitotic figures are sometimes seen.

It is not unusual to detect highly keratinized cells in a cervical smear from a patient whose biopsy is reported as non-keratinizing

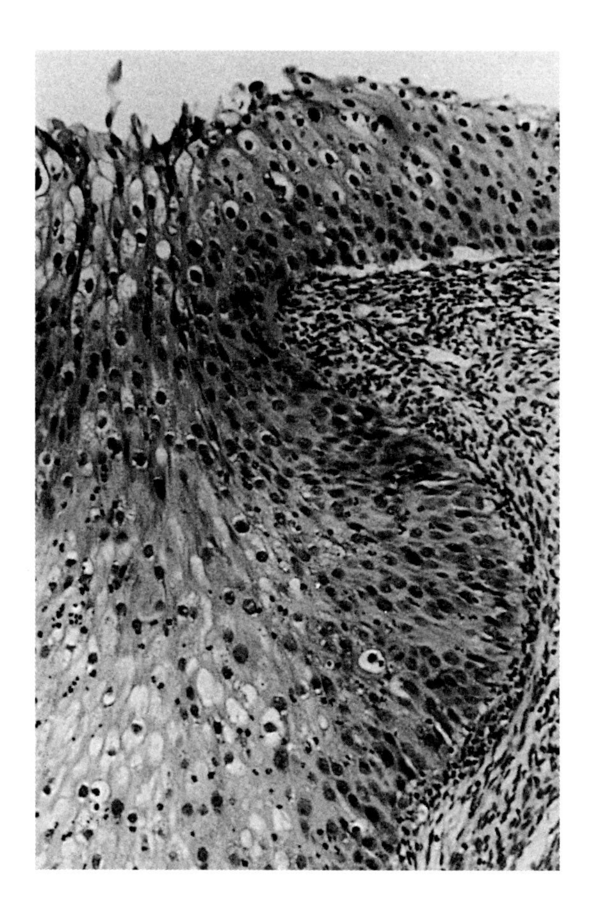

Figure 12.16 Large cell non-keratinizing squamous carcinoma. Most cells have an eosinophilic cytoplasm; some show vacuolation of varying degree (H & E, ×190).

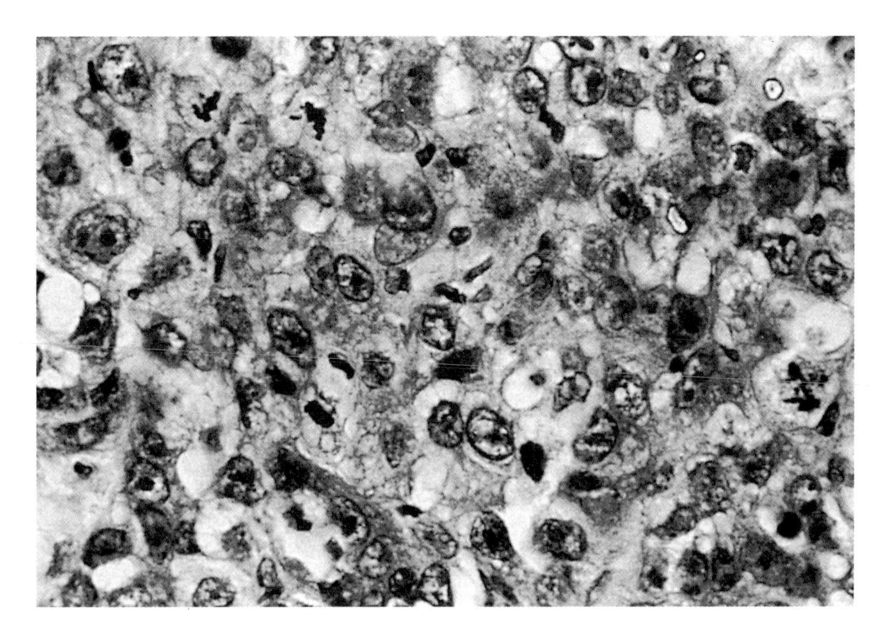

Figure 12.17 Mitotic figures in a large cell non-keratinizing squamous carcinoma (H & E, ×770).

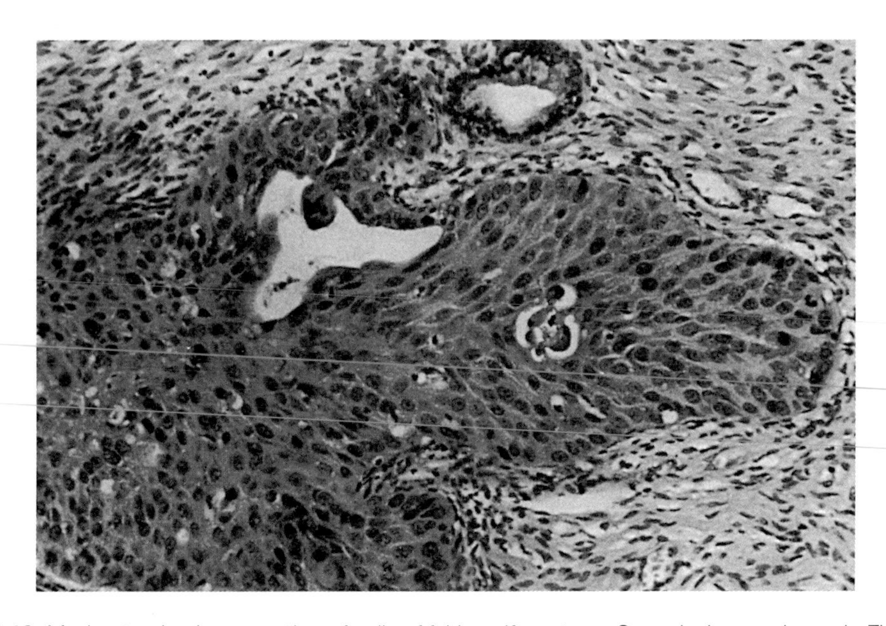

Figure 12.18 Moderate-sized aggregation of cells of fairly uniform type. Same lesion as shown in Figure 12.17 (H & E, ×190).

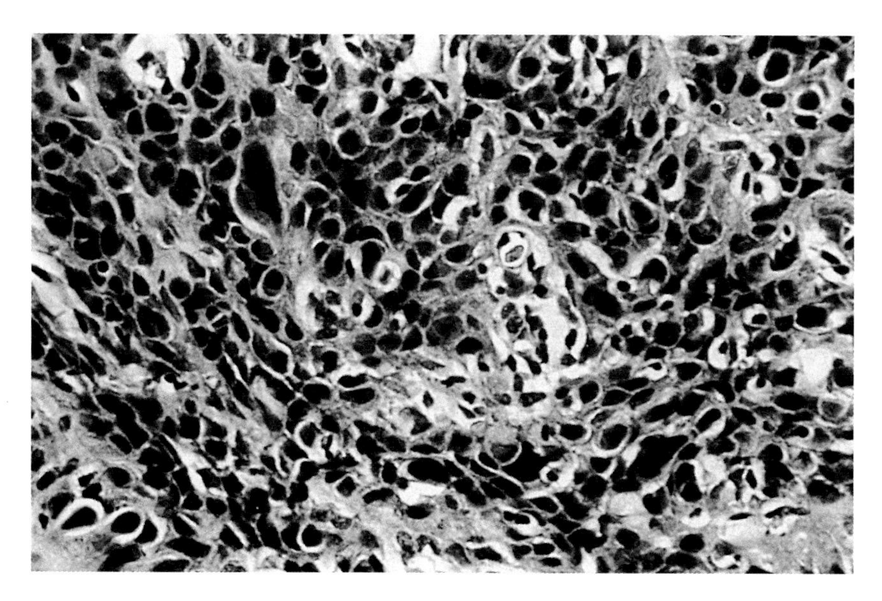

Figure 12.19 Pleomorphic large cell non-keratinizing squamous carcinoma (H & E, ×480).

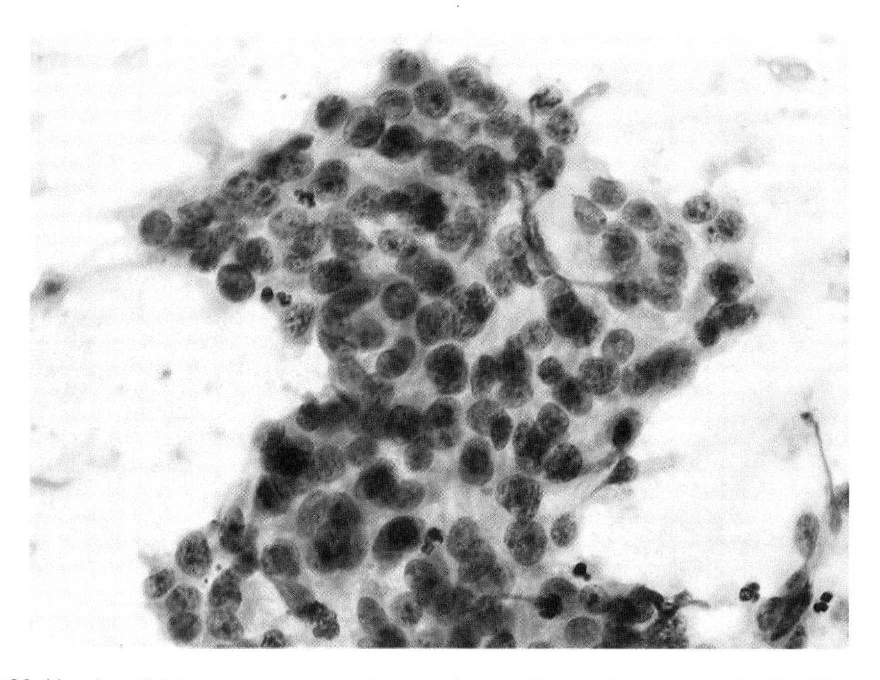

Figure 12.20 Non-keratinizing squamous carcinoma – large cell type. Aggregates of cells with amphophilic or basophilic cytoplasm and irregular hyperchromatic nuclei. Note similarity of cells to those in tumour shown in Figure 12.19 (Pap, ×630).

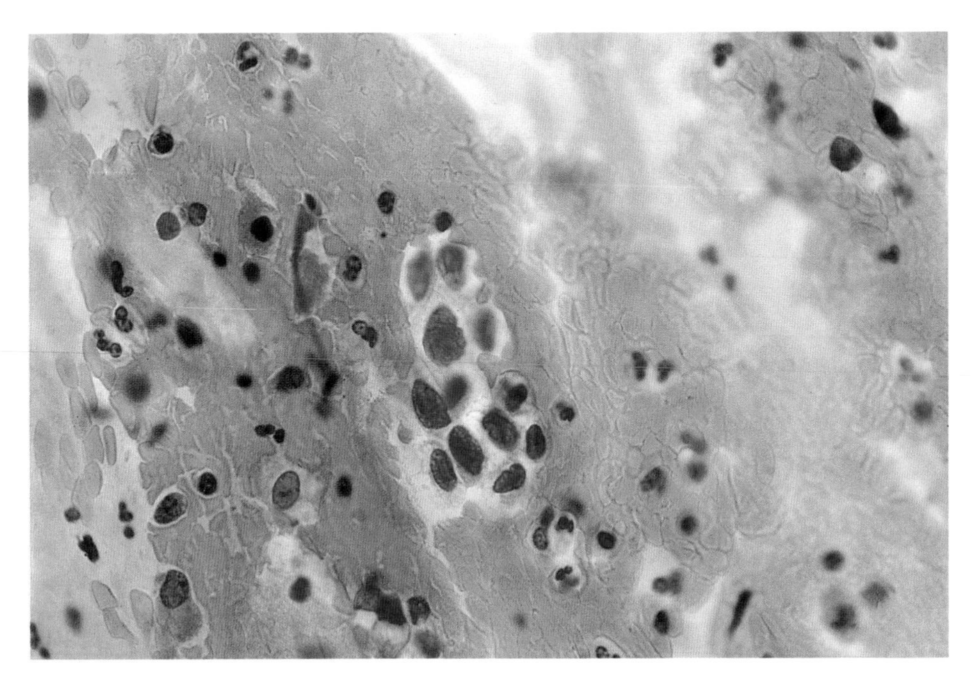

Figure 12.21 Clusters of malignant nuclei devoid of cytoplasm from an undifferentiated area of the tumour. Same case as shown in Figure 12.19 (Pap, ×630).

squamous carcinoma. Most squamous carcinomas have a mixed cell population and the keratinizing cells may be formed at the surface of a tumour which is classified histologically as non-keratinizing.

The cytological pattern of non-keratinizing squamous cell carcinoma may change after radiotherapy, as shown in Figure 12.22. Dense clusters of tumour cells with scanty cytoplasm and large hyperchromatic nuclei (Figure 12.22(a)) were found in the smear of a 74-year-old woman who was diagnosed as having a non-keratinizing large cell squamous carcinoma (Figure 12.22(b)) and treated by radiotherapy. Recurrence of the tumour was detected cytologically 6 years later. Sheets of tumour cells with abundant cytoplasm, large vesicular nuclei and irregular nucleoli (Figure 12.22(c)) were a prominent feature in the smear.

12.5 SMALL CELL NON-KERATINIZING SQUAMOUS CARCINOMA

This tumour type is composed of small cells with sparse cytoplasm having a high nucleo-cytoplasmic ratio, showing little or no evidence of keratin formation. The cells are arranged in fairly large aggregations or in a trabecular pattern (Figure 12.23). They have hyperchromatic nuclei, mainly oval in shape; many of the nuclei often appear opaque. Mitotic figures are frequent (Figure 12.24) and abnormal forms are often seen. Careful search may reveal evidence of squamous differentiation with rudimentary cell nest formation and/or dyskeratosis (Figure 12.25).

The proportion of squamous carcinoma of this type in the USA was found by Reagan and Ng (1973) to have fallen from 18.5% in

1941–1945 to 6.5% in 1966–1970. In view of the new WHO classification (Scully *et al.*, 1994) in which undifferentiated small cell carcinomas with neurosecretory or basaloid features are placed in different categories from small cell non-keratinizing squamous carcinomas, the proportion of the latter may have fallen still further. Of the various forms of squamous cervical carcinoma the small cell non-keratinizing type is generally considered to have the worst prognosis, but not all authors agree (Section 12.2.1).

12.5.1 Cytology

Smears from patients with ulcerating malignant lesions of the cervix frequently have a background of red blood cells, leucocytes and necrotic debris referred to as the 'malignant diathesis'. This may make identification of the tumour cells difficult, leading to a false-negative report. The identification of the small undifferentiated malignant cells which are characteristic of this tumour may be particularly troublesome and a confident diagnosis can only be made when small or large clusters of hyperchromatic nuclei showing marked anisonucleosis are seen (Figure 12.26). The chromatin structure is coarse and nucleoli are rarely seen.

12.6 SPECIAL TYPES OF SQUAMOUS CARCINOMA

12.6.1 Verrucous carcinoma

Verrucous carcinoma is an uncommon variant of well-differentiated keratinizing carcinoma (Figures 12.27 and 12.28) which is characterized by its large size, benign, warty appearance and slow, relentless, locally invasive growth. The tumour presents as an exophytic, cauliflower-like mass which may be fungating or ulcerated. Histologically, the lesion has features which are very similar to those found in condyloma acuminatum. It is composed of numerous papillary fronds (Figure 12.27) lined with several layers of benign-looking, well-differentiated epithelial cells with some surface keratinization (Figure 12.29). Unlike condyloma acuminatum, the papillae of the verrucous carcinoma sometimes lack a central connective tissue core (Figures 12.29 and 7.34). The component cells of the verrucous carcinoma usually lack the characteristic features of malignancy seen in the common keratinizing squamous carcinoma (Figure 12.30). Indeed, the diagnosis may be missed on a superficial biopsy unless the pathologist is aware of the gross appearance of the tumour (Anderson, 1985). Gilbert and Palladino (1966) observe that 'papillomas involving the cervix may be treacherous; their histologic appearances often belie their malignant potential'.

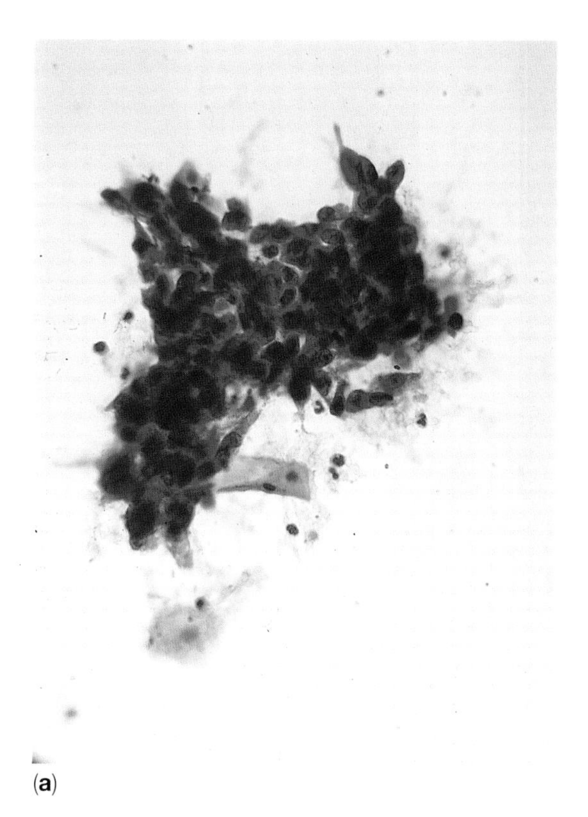

(**a**)

Figure 12.22 (**a**) Malignant cells in smear from woman aged 74, diagnosed as having non-keratinized large cell squamous carcinoma (Pap, ×630).

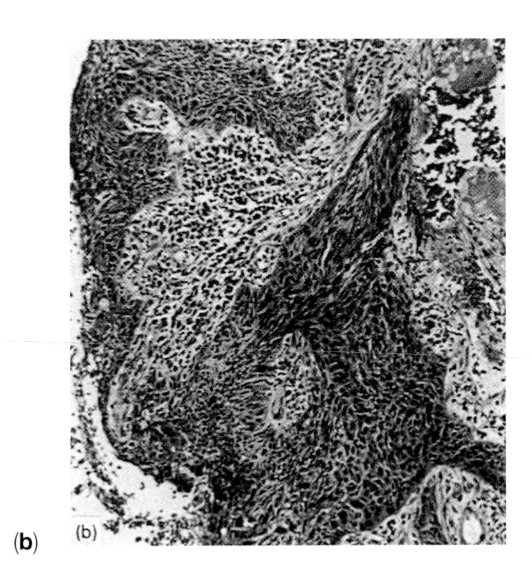

(**b**)

Figure 12.22 (**b**) Biopsy showing large cell non-keratinizing squamous carcinoma. Same case as shown in (**a**) (H & E, ×120).

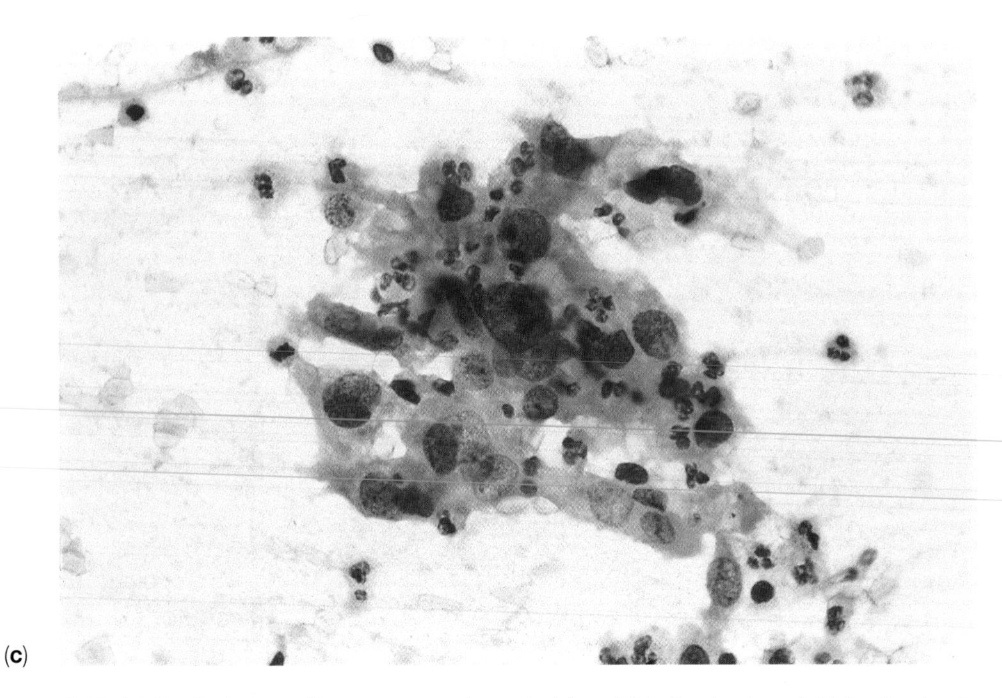

(**c**)

Figure 12.22 (**c**) Cervical smear. Same case as shown in (**a**) and (**b**) after treatment. Note sheets of malignant cells with vesicular nuclei (Pap, ×630).

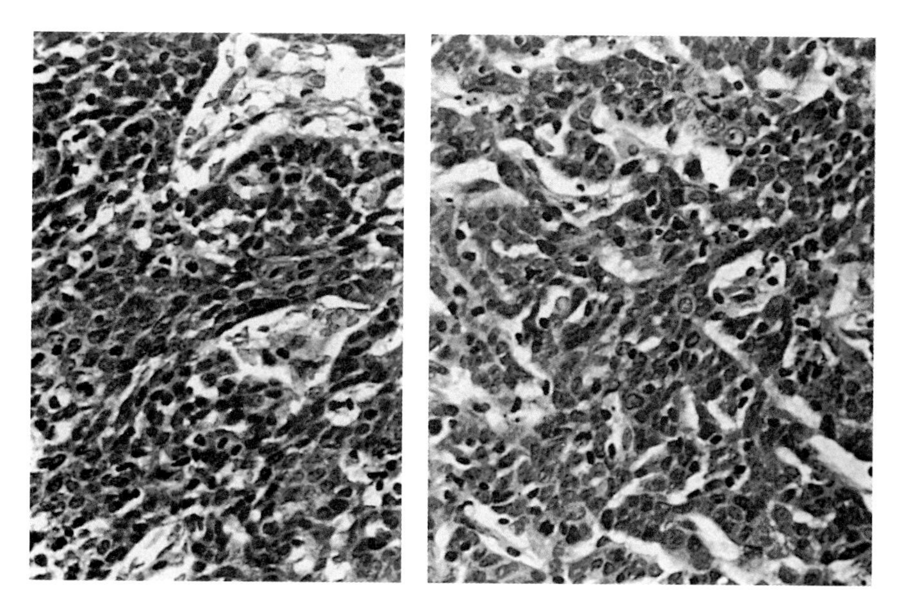

Figure 12.23 Small cell non-keratinizing carcinoma composed of cells with little cytoplasm arranged in fairly large aggregations or trabecular pattern (H & E, ×350).

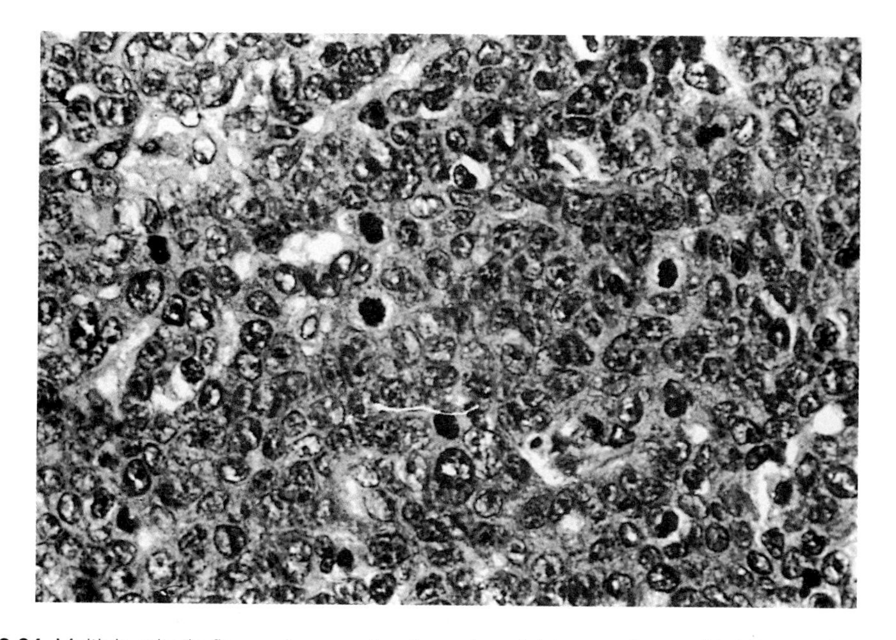

Figure 12.24 Multiple mitotic figures in a small cell non-keratinizing carcinoma (H & E, ×770).

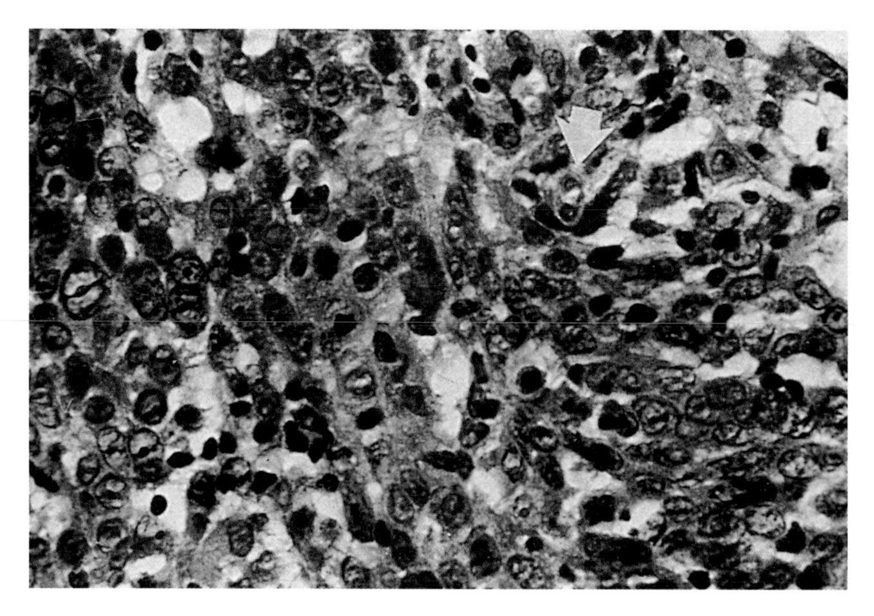

Figure 12.25 Rudimentary cell nest (arrowed) in a small cell non-keratinizing carcinoma (H & E, ×770).

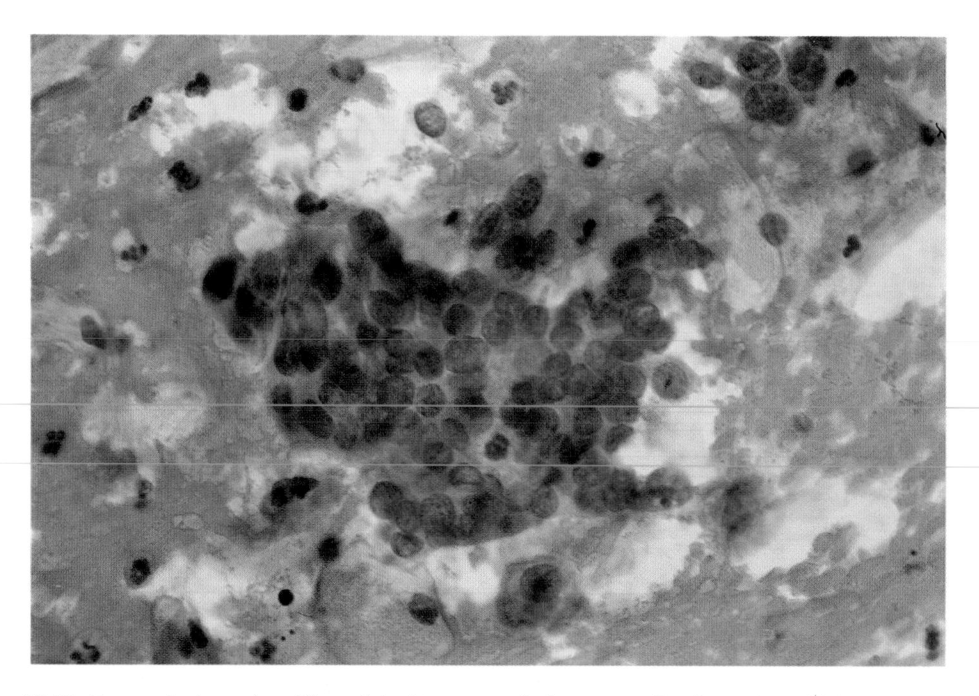

Figure 12.26 Dense clusters of undifferentiated tumour cells from a small cell non-keratinizing carcinoma. Note red blood cells and necrotic debris in this smear (Pap, ×630).

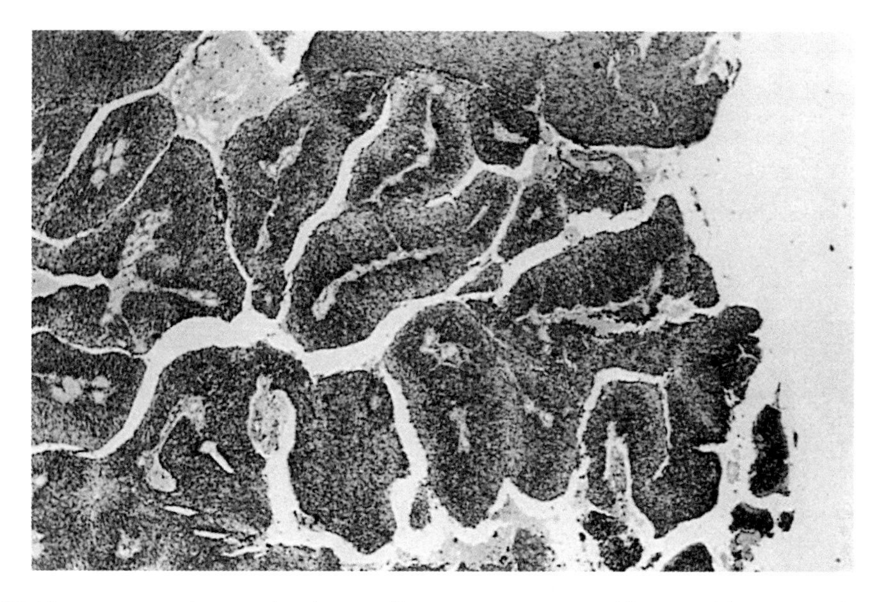

Figure 12.27 Verrucous carcinoma showing papillary processes resembling condyloma acuminatum (H & E, ×50).

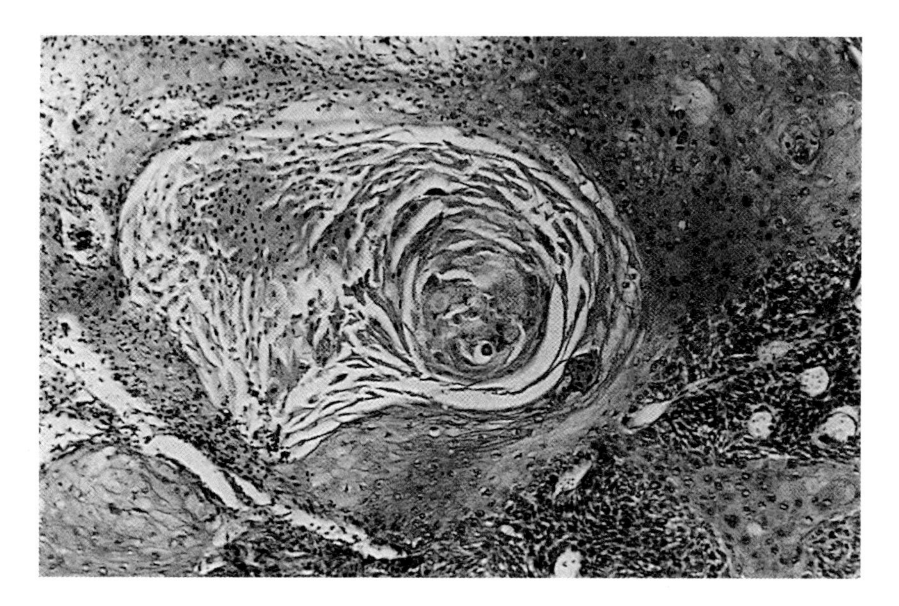

Figure 12.28 Cell nest with keratin pearl formation in a verrucous carcinoma (H & E, ×150).

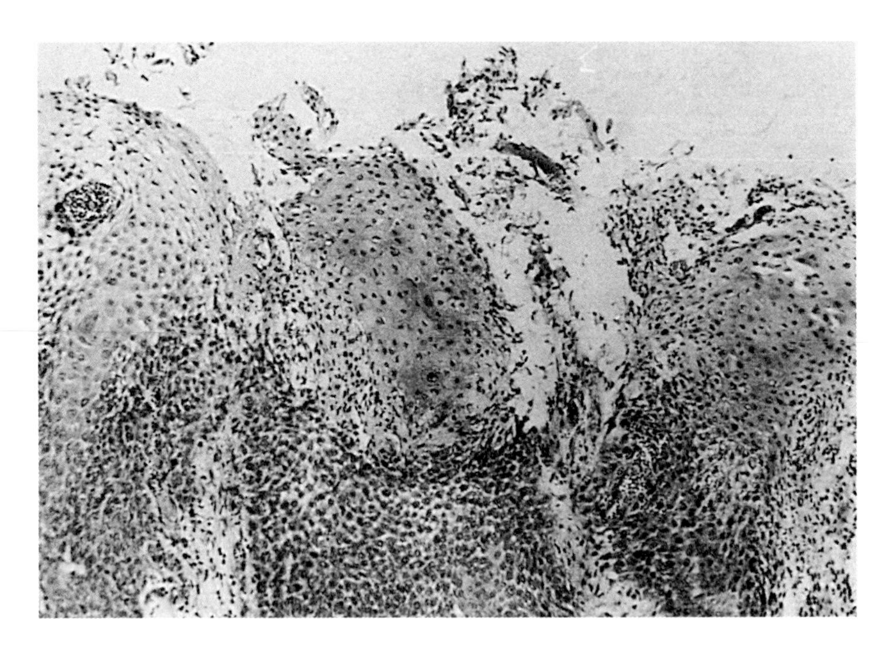

Figure 12.29 Epithelial papillae in a verrucous carcinoma. They lack the well-defined connective tissue core characteristic of a warty condyloma (H & E, ×120).

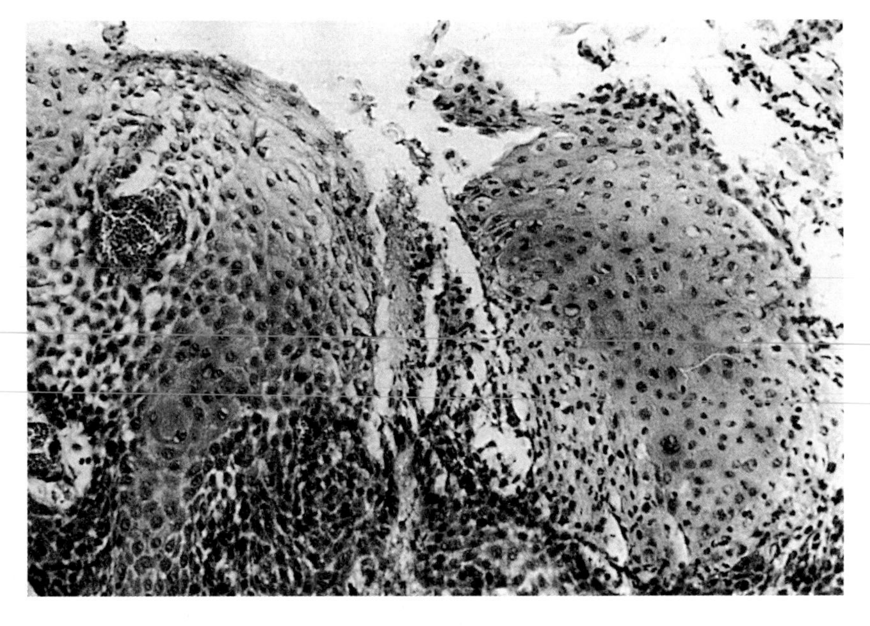

Figure 12.30 Bland cellular structure in a verrucous carcinoma giving the lesion a misleadingly benign appearance (H & E, ×190).

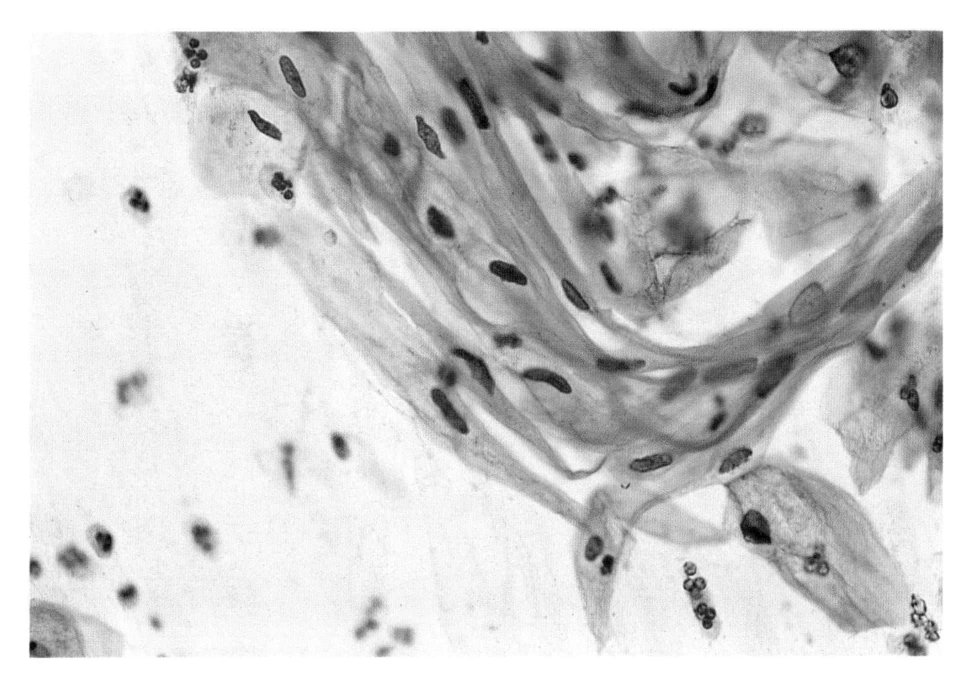

Figure 12.31 Verrucous carcinoma cells in cervical smear showing little evidence of malignancy. Same lesion as shown in Figures 12.30 and 12.32 (Pap, ×630).

The deep margins of the tumour tend to be bulbous, reflecting the fact that the tumour invades by pushing into the surrounding stroma (Anderson, 1985). At the junction of the tumour bulb and the stroma there is a considerable inflammatory reaction. Often, no clear evidence of invasion is detectable, the bland microscopic appearances being at variance with the relentless recurrence of the tumour even after apparently complete removal (Ferenczy, 1982). Correct diagnosis is essential as treatment is by wide local excision (Isaacs, 1976). According to Demian *et al.* (1973), and Kraus and Perez-Mesa (1966), verrucous carcinomas treated by radiotherapy are more likely to recur than those treated surgically. Lymph node involvement and lymph node metastases are rare (Lucas *et al.*, 1974).

In 1925, Buschke and Loewenstein drew attention to a group of slow-growing giant condylomata involving the prepuce of the penis which eroded and spread until they were indistinguishable clinically from an invasive carcinoma. Invasion however, was not seen histologically, nor were metastases noted (Judge, 1969). Subsequent reports confirmed that giant condylomata were not exclusively penile lesions and similar lesions involving the female genital tract were described. Kraus and Perez-Mesa (1966) reviewed 105 cases of giant condylomata of Buschke–Loewenstein type in the cervix, vulva or vagina and concluded that these tumours could not be distinguished from verrucous carcinoma. Lucas *et al.* (1974), in reviewing 16 examples of the latter tumour, came to the same conclusion.

The cytological diagnosis of verrucous carcinoma (Figures 12.31 and 12.32) is even more difficult than the histological diagnosis. The finding of cells showing features of an invasive carcinoma, as described in Section 12.3.1 or 12.4.1 is very much the exception. Koilocytotic

(a)

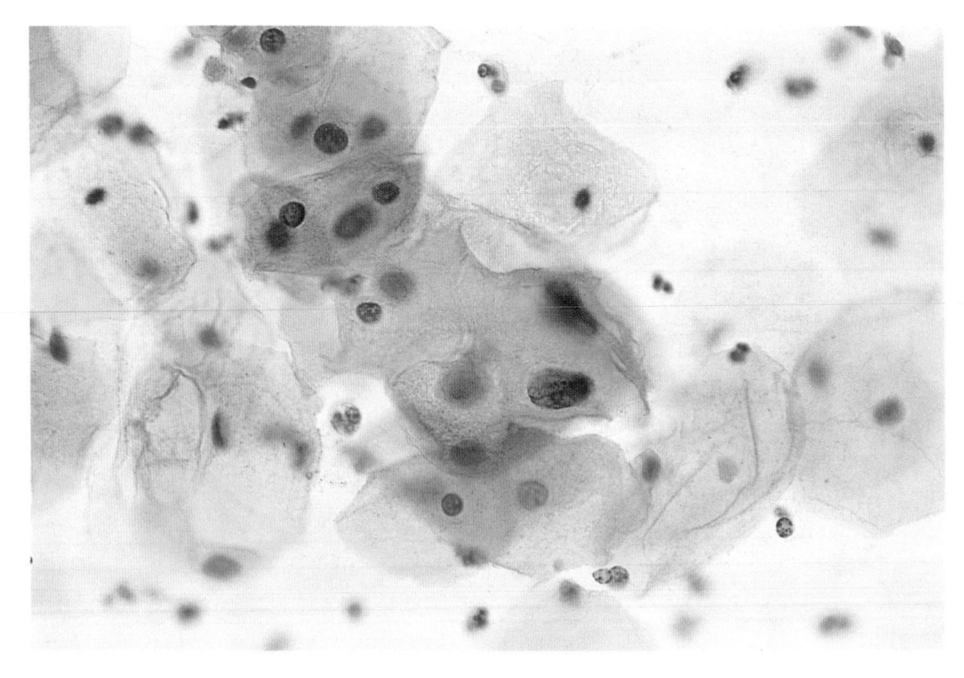

Figure 12.32 Verrucous carcinoma cells showing some malignant features. Same lesion as shown in Figures 12.29 and 12.30 (Pap, ×630).

atypia is likely to be present but an invasive carcinoma is unlikely to be suspected on cytological grounds.

12.6.2 Warty (condylomatous) carcinoma

The morphology of this lesion is similar to that of verrucous carcinoma from which it differs in having a deep margin which resembles that of typical squamous carcinoma, as in Figure 12.33. In addition, the component cells often show changes the changes of koilocytotic atypia with nuclear atypia and cytoplasmic clearing around the nucleus (Kurman *et al.*, 1990). It appears to be less aggressive than typical squamous carcinoma but clinical experience is limited (Wright *et al.*, 1994).

12.6.3 Papillary squamous carcinoma

Macroscopically, this tumour is usually exophytic, polypoid or wart-like (Randall *et al.*, 1989). Microscopically, unlike verrucous and warty carcinoma the papillary processes of the lesion are lined by dysplastic cells, and condylomatous changes are absent. The epithelial surface often bears a close histological resemblance to carcinoma *in situ*, for which it could be mistaken if the biopsy is too superficial. At the University of Virginia Medical Center it was found to constitute 1.6% of 365 consecutive squamous carcinomas (1968–1978). Its clinical course differed from that of typical squamous carcinoma in having a propensity for late metastases and recurrences. It is therefore important that any apparently *in situ* neoplasm having a papillary structure is completely excised.

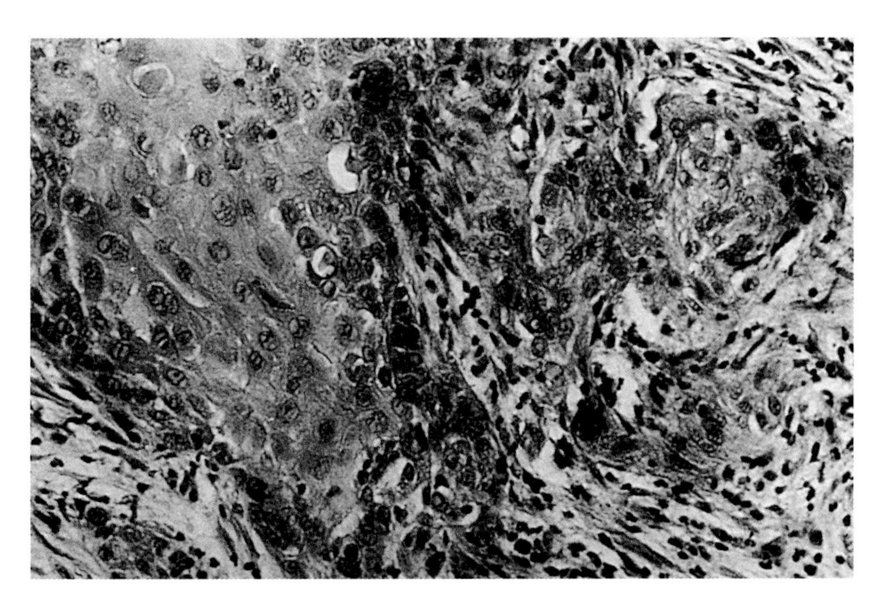

Figure 12.33 Invasive deep margin characteristic of warty (condylomatous) carcinoma (H & E, ×310).

12.6.4 Lymphoepithelioma-like carcinoma

This tumour was recognized as a separate entity by Hamazaki *et al.* (1968). Its relatively favourable prognosis was emphasized by Hasumi *et al.* (1977) who reported that it constituted 5.5% of 709 cases of cervical carcinoma, occurring in the age range of 30–61 years (mean age 43.5 years). Typically, it presents with vaginal bleeding (Halpin *et al.*, 1989). Examination reveals a superficially ulcerated tumour, usually arising within the endocervical canal, and on slicing, it is seen to be a circumscribed, slightly bulging, homogeneous greyish white growth (Hasumi *et al.*, 1977). Histologically, the tumour margin is sharply circumscribed with surrounding lymphocytic infiltration. The lesion is highly cellular and composed of anastomosing cords of cells separated by a loose fibrillary stroma which is densely infiltrated by lymphocytes. Eosinophils and plasma cells may also be seen (Halpin *et al.*, 1989).

The tumour cells are polygonal with large rounded or irregular nuclei showing remarkable uniformity of size (Mills *et al.*, 1985) and containing prominent nucleoli. The cytoplasm is variable in amount finely granular or flocculant and slightly eosinophilic. In places the lesion appears syncytial, but isolated tumour cells also occur. Mitotic figures are frequent usually not less that four per high-power field. No keratinization, intercellular bridge formation or gland formation is seen, but immunocytochemical foci of cytokeratin formation may be demonstrated in the tumour cells and leucocyte common antigen (LCA) in the lymphoid cells (Halpin *et al.*, 1989). The latter cells indicate a host response to the tumour which may well account for the low rate of metastasis, only two out of 39 (5%) in the series reported by Hasumi *et al.* (1977). Lymphoepithelioma-like carcinoma is radiosensitive, and radiation appears to be effective in eradicating localized low-stage disease (Mills *et al.*, 1985).

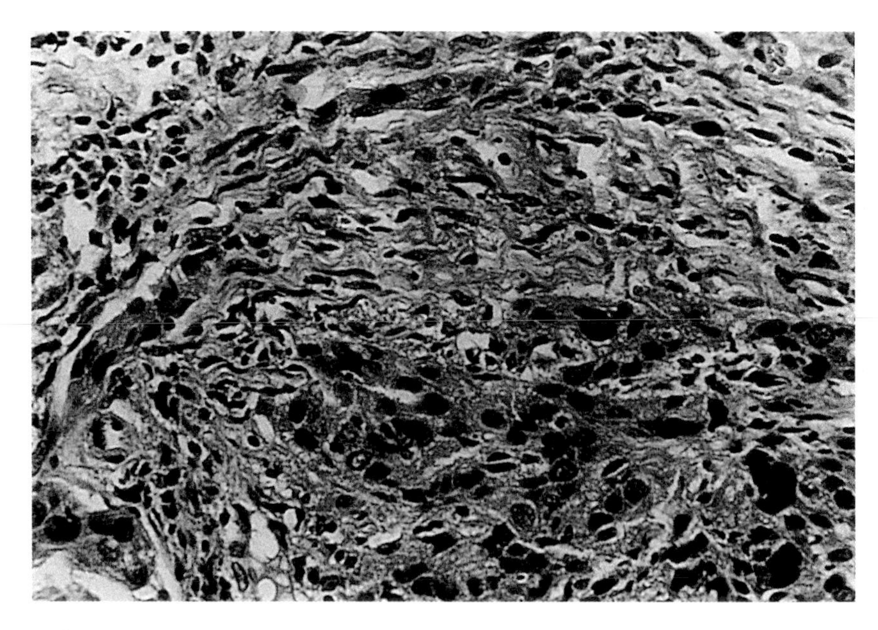

Figure 12.34 Spindle cell (scirrhous) variant of squamous carcinoma. A more recognizable squamous cell pattern can also be seen (H & E, ×480).

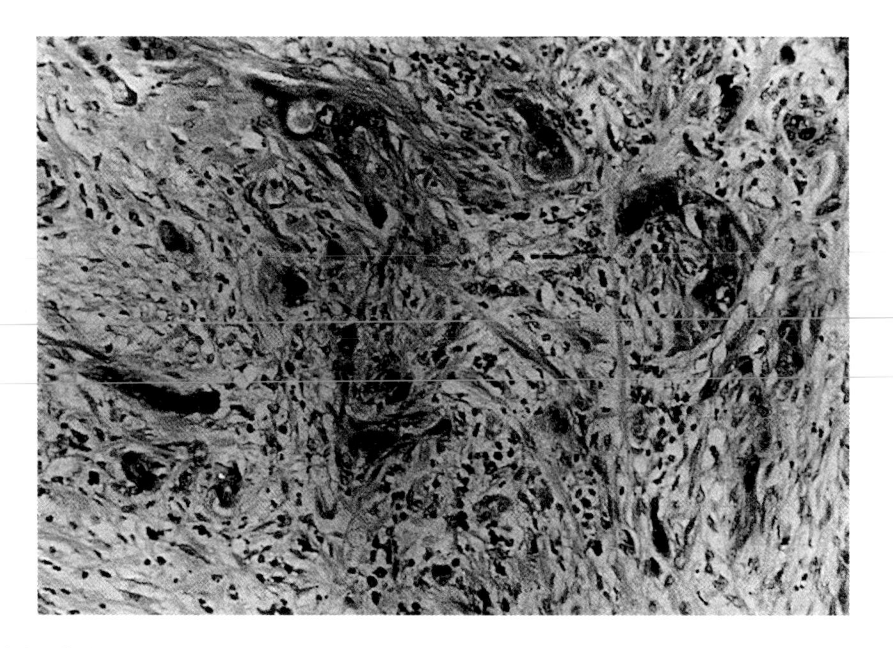

Figure 12.35 Spindle cell variant of squamous element in an adenosquamous carcinoma (H & E, ×190).

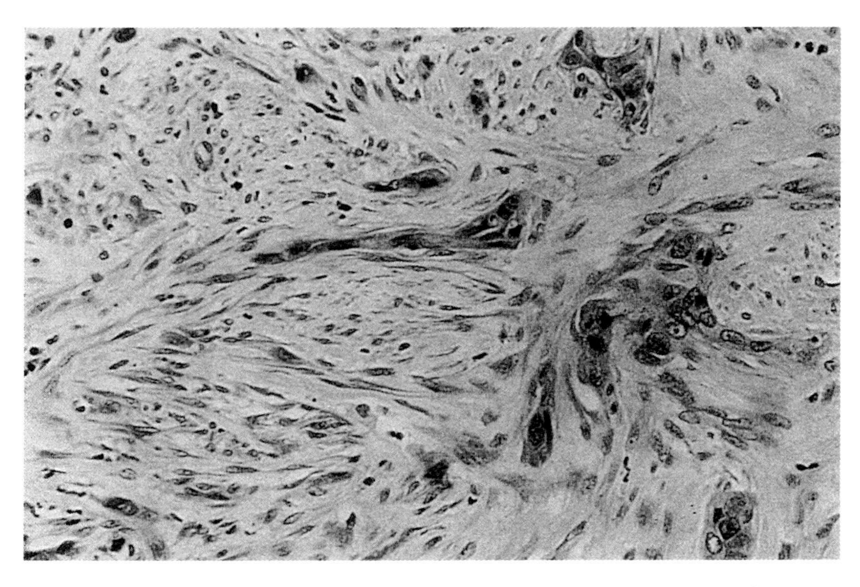

Figure 12.36 Fibroplasia. Stromal cells appear to be an integral part of the spindle cell tumour pattern. Same lesion as shown in Figure 12.35 (H & E, ×310).

12.6.5 Spindle cell (scirrhous) squamous carcinoma

This is a rare form of squamous cell carcinoma which is liable to be mistaken for spindle cell sarcoma. Careful search will usually reveal a part of the tumour which is more recognisably squamous (Figure 12.34). A spindle cell variant may also develop from the squamous element in adenosquamous carcinoma (Figure 12.35). The apparent discrepancy between the histologically aggressive-looking cells in spindle cell squamous carcinoma and the infrequency with which they appear to metastasize has been studied by Harris (1982) using electron microscopy. His findings indicate that the spindle cells are the result of partial or complete transformation of epithelial into mesenchymal cells. This could be the result of simple metaplasia or alternatively, and more probably, the result of cell fusion to form a hybrid. Cell fusion experiments have shown that a hybrid may retain the properties of both parents in a modified form. *In vivo* hybridization of transplanted A9H2 tumour cells and normal host mouse cells has shown a 200-fold reduction in tumorigenicity compared with A9H2 parent cells (Kao and Hartz, 1977).

This explanation appears to be consistent with our own observation that the 'stromal cells' surrounding the identifiably squamous spindle cells show a fibroplastic growth pattern suggesting that they are forming part of the tumour (Figure 12.36). Note that this spindle cell variant of squamous carcinoma is growing in the immediate proximity of fibrous connective tissue. Such tumours are unlikely to be present on the surface of the cervix and are consequently rarely available to provide desquamated cells in a cervical scrape preparation.

REFERENCES

Anderson, M. C. (1985) The pathology of cervical cancer. *Clin. Obstet. Gynecol.*, **12**, 87–119.

Beecham, J. B., Halvorsen, T. and Kolbenstvedt, A. (1978) Histologic classification, lymph node metastases and patient survival in stage IB cervical carcinoma. *Gynecol. Oncol.*, **6**, 95.

Buckley, C. H. and Fox, H. (1989) Carcinoma of the cervix, in *Recent Advances in Histopathology*, Vol. 14. Churchill Livingstone, Edinburgh, pp. 63–78.

Buckley, C. H, Beards, C. S. and Fox, H. (1988) Pathological prognostic indicators in cervical cancer with particular reference to patients under the age of 40 years. *Br. J. Obstet. Gynaecol.*, **95**, 47–56.

Burghardt, E. (1973) Early histological diagnosis of cervical cancer, in *Major Problems in Obstetrics and Gynaecology* (ed. E. A. Friedman), vol. 6, p. 43.

Buschke, A. and Loewenstein, L. (1925) Condylomata acuminata simulating cancer on penis. *Klin. Wochenschr.*, **4**, 1726–1728.

Chung, G. K., Stryker, J. A., Ward, S. P., Nahhas, W. A. and Mortel, R. (1980) Histologic grade and prognosis of carcinoma of cervix. *Obstet. Gynecol.*, **57**, 636– 642.

Demian, S. D. E., Bushkin, F. L. and Echevarria, R. A. (1973) Perineural invasion and anaplastic transformation of verrucous carcinoma. *Cancer*, **32**, 395–401.

Fenton, J., Chevret, S., Asselain, B. *et al.* (1990) Invasive cancer of the uterine cervix in young women: retrospective study of 236 cases. *Bull. Cancer*, **77**, 109–116.

Fink, F. M. and Denk, M. (1970) Cervical carcinoma: relationship between histology and survival following radiation therapy. *Obstet. Gynecol.*, **35**, 339– 343.

Gilbert, E. F. and Palladino, A. (1966) Squamous papillomas of the uterine cervix; review of the literature and report of a giant papillary carcinoma. *Am. J. Clin. Pathol.*, **46**, 115–211.

Goellner, J. R. (1976) Carcinoma of the cervix. Clinicopathologic correlation of 196 cases. *Am. J. Clin. Pathol.*, **66**, 775–785.

Gunderson, L. L., Weems, W. S., Herbertson, R. M. and Plenk, H. P. (1974) Correlation of histopathology with clinical results following radiation therapy for carcinoma of the cervix. *Am. J. Roentgenol. Radium Ther. Nucl. Med.*, **120**, 74–87.

Halpin, T. F., Hunter, R. E. and Cohen, M. B. (1989) Lymphoepithelioma of the uterine cervix. *Gynecol. Oncol.*, **34**, 101–105.

Hamazaki, K., Fujita, H. and Arata, T. (1968) Medullary carcinoma with lymphoid infiltration of the uterine cervix. *Jpn. J. Cancer Clin. (Gan No Rinsho)*, **14**, 787–792.

Hasumi, K., Sugano, H. and Sakamoto, G. (1977) Circumscribed carcinoma of the uterine cervix. with marked lymphocytic infiltration. *Cancer*, **39**, 2503–2507.

Harris, M. (1982) Spindle cell squamous carcinoma; ultrastructural observations. *Histopathology*, **6**, 197–210.

Isaacs, J. H. (1976) Verrucous carcinoma of the female genital tract. *Gynecol. Oncol.*, **4**, 259–269.

Judge, J. R. (1969) Giant condyloma acuminatum involving the vulva and rectum. *Arch. Pathol.*, **88**, 46–48.

Kao, F. T. and Hartz, J. A. (1977) Genetic and tumorigenic characteristics of cell hybrids formed in vivo between injected tumour cells and host cells. *J. Natl. Cancer Inst.*, **59**, 409–413.

Kraus, F. T. and Perez-Mesa, C. (1966) Verrucous carcinoma. *Cancer*, **19**, 22–38.

Kurman, R., Sasano, H., Koizumi, M. *et al.* (1990) Warty (condylomatous) carcinoma, in *Atlas of Tumour Pathology, Third Series, Fascicle 4. Tumours of the Cervix, Vagina and Vulva*. Armed Forces Institute of Pathology, Washington, DC, p. 200.

Lucas, W. E., Benirschke, K. and Lebher, T. B. (1974) Verrucous carcinoma of the female genital tract. *Am. J. Obstet. Gynecol.*, **119**, 435–440.

Mills, S. E., Austin, M. B. and Randall, M. E. (1985) Lymphoepithelioma-like carcinoma of the uterine cervix: a distinctive undifferentiated carcinoma with an inflammatory stroma. *Am. J. Surg. Pathol.*, **9**(12), 883–889.

Randall, M. E., Andersen, W. A., Mills, S. E. and Kim, J. A. C (1986) Papillary squamous cell carcinoma of the uterine cervix: a clinicopathologic study of nine cases. *Int. J. Gynecol. Pathol.*, **5**, 1–10

Reagan, J. W. and Ng, A. B. P. (1973) The Cellular Manifestations of Uterine Carcinogenesis, in *The Uterus* (eds H. J. Norris, A. T. Hertig and A. R. Abell), Int. Acad. Path. Monograph. Williams and Wilkins, Baltimore, pp. 320–347.

Scully, R. E., Bonfiglio, T., Kurman, R. J. *et al.* (1994) WHO International histological classification of tumours. Histological typing of female genital tract tumours. Springer-Verlag, Berlin, pp. 39–54.

Sedlacek, T. V., Mangan, C. E., Giuntoli, R. L. *et al.* (1978) Exploratory celiotomy for cervical carcinoma: the role of histologic grading. *Gynecol. Oncol.*, **6**, 138–144.

Sugawa, T., Yamagata, S. and Yamamoto, K. (1982) Chemotherapy for cancer of the cervix – current status and its evaluation. Asia–Oceania. *J. Obstet. Gynecol.*, **8**, 343–355.

Swan, D. S. and Roddick, J. W. (1973) A clinical-pathologic correlation of cell type classification for cervical cancer. *Am. J. Obstet. Gynecol.*, **116**, 666–670.

Wentz, W. B. and Lewis, G. C. (1965) Correlation of histologic morphology and survival in cervical cancer following radiation therapy. *Obstet. Gynecol.*, **26**, 228–232.

Wright., T. C., Ferenczy, A. and Kurman, R. J. (1994) Carcinoma and other tumours of the cervix, in *Blausteins Pathology of the Female Genital Tract*, 4th edn (ed. R. J. Kurman), Springer-Verlag, New York, pp. 274–326.

13

Adenocarcinoma *in situ* and glandular atypia of the cervix

13.1 ADENOCARCINOMA *IN SITU*

The term adenocarcinoma *in situ* (AIS) was introduced over 40 years ago by Friedell and McKay (1953), to describe changes in the endocervical epithelium which they considered to be the precursors of invasive adenocarcinoma of the cervix. These authors were not the first to recognize these changes. Burghardt (1973) quotes six reports of cervical intraepithelial adenocarcinoma which predate the report of Friedell and McKay by many years, that of Hauser (1894) being the earliest.

The evidence to support the concept that AIS is a precursor of invasive cancer is substantial. Examples of untreated AIS progressing to invasive adenocarcinoma have been recorded (Hopkins, 1988; Kashimura *et al.*, 1990). The changes of AIS are frequently found on the edge of an invasive adenocarcinoma. The average age of women with AIS is about 10 years less than that of women presenting with invasive glandular lesions (Qizilbash, 1975; Boon *et al.*, 1981; Brown and Wells, 1986) suggesting a long evolutionary period for the disease. Hopkins *et al.* (1988) compared the mean ages of women with AIS and women with invasive adenocarcinoma identified over the same period of time and found the mean age of 18 women with AIS was 37 whereas the mean age at presentation of 182 women with invasive adenocarcinoma was 47 years. Further evidence that AIS is a precursor of invasive adenocarcinoma is derived from the work of Boddington *et al.* (1976) who found abnormalities in cervical smears of six women several years before they were diagnosed as having invasive adenocarcinoma. However, there is considerable disparity in the reported incidence of AIS and invasive adenocarcinoma, indicating that if AIS is a frequent precursor of invasive adenocarcinoma, it is often overlooked (Christopherson *et al.*, 1979; Boon *et al.*, 1981).

13.1.1 Histology of adenocarcinoma *in situ*

Microscopically, the characteristic feature of AIS is replacement of the normal columnar epithelium of the endocervical canal and crypts by neoplastic cells with abnormal hyperchromatic nuclei (Figure 13.1). A sharp transition between normal and abnormal epithelium is frequently seen at low magnification due to the deeper staining of the neoplastic cells (Figures 13.1, 13.2 and 13.3(a)). The changes are often confined to the surface epithelium and the superficial endocervical crypts and are usually focal, involving only one quadrant of a cone biopsy. They are frequently found adjacent to an area of CIN (Figure 13.3(b)). Not infrequently, CIN is present in the surface epithelium whereas AIS may affect the crypts. The abnormal glandular cells usually maintain their columnar shape but exhibit anisonucleosis, nuclear pleomorphism and loss of polarity (Figure 13.3(c)). Multilayering of the nuclei (Figures 13.4 and 13.5) is frequently present, giving the epithelium a pseudostratified appearance. The basement membrane underlying the normal and neoplastic epithelium may be continuous (Figure 13.4), or it may be absent (Figure 13.2 and 13.3(c)) or ill-defined (Figure 13.5) Unlike CIN, the integrity of the basement membrane is not a useful guide to the presence of microinvasive or early invasive adenocarcinoma.

According to Scully *et al.* (1994) the epithelial lining of the glands is usually devoid of intracellular mucin and may resemble endometrial epithelium; in some cases the glands are lined by intestinal-type epithelium containing goblet cells and even argyrophyl cells. Jaworski *et al.* (1988) reviewed 72 cases of AIS and found that 41 were endocervical type, three were endometrioid type and 28 were of mixed pattern. The glands of AIS conform to the expected location of normal endocervical glands and do not extend deeper than the latter. They do not have a complex pattern and do not excite a desmoplastic stromal reaction (Scully *et al.*, 1994).

The cytoplasm in AIS is often more eosinophilic than normal and although mucin production may occur, it is considerably reduced. Cilia and brush borders may still be present (Figures 13.3(b) and 13.5). The mitotic index is also increased; the mitotic figures are usually normal and located toward the luminal aspect of the epithelium (Figure 13.6). Sometimes the epithelium is raised into papillary projections or intraluminal tufts (Figure 13.7), but overall the normal gland architecture is maintained (Figure 13.8(a)).

The WHO definition of adenocarcinoma *in*

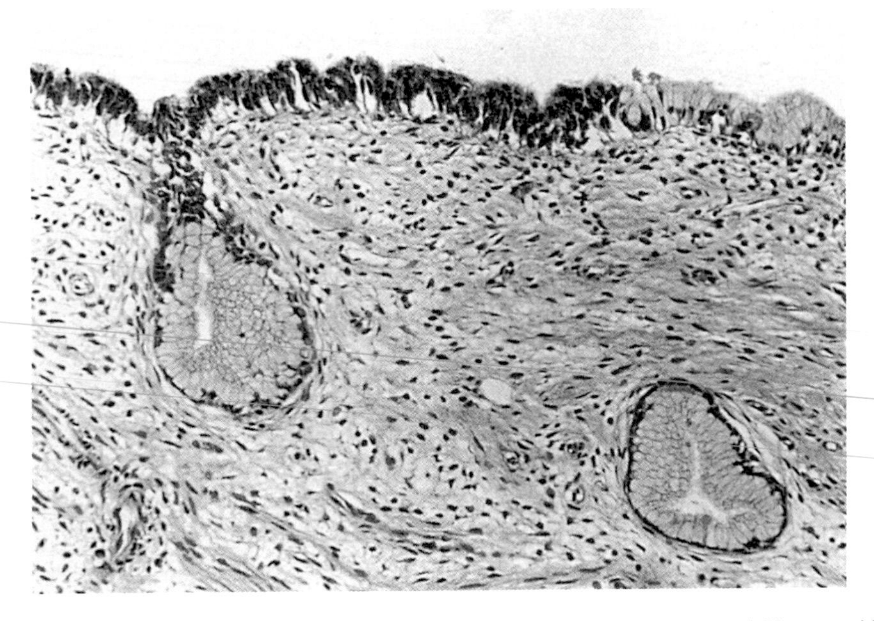

Figure 13.1 Adenocarcinoma *in situ* (AIS) of the endocervical canal in a woman aged 38 years. Normal architecture is maintained (H & E, ×190) (courtesy of Dr D.J.B. Ashley, Morriston Hospital, Swansea).

situ is 'a lesion in which normally situated glands are lined by cytological malignant glandular epithelium' (Scully *et al.*, 1994), to which they add: 'The epithelial lining of the glands is usually devoid of intracellular mucin and may resemble endometrial epithelium; in some cases the glands are lined by intestinal type epithelium containing goblet cells and even argyrophyl cells. The glands do not have a complex pattern and do not excite a desmoplastic reaction'.

The view of Brown and Wells (1986) that the concept of AIS should be extended to include budding and branching of the glands is unacceptable. If this concept is followed through to its natural conclusion, it becomes extremely difficult to distinguish AIS from invasive adenocarcinoma especially if deeper endocervical glands are involved or abnormal glands are found in more than one quadrant of a cone biopsy specimen. Yavner *et al.* (1990) found the basement membrane to be defective in clearly invasive lesions and also in branching and budding lesions, supporting the view that the latter findings are indicative of early invasion.

It is important to distinguish cervical adenocarcinoma *in situ* from non-neoplastic glands of endometrial type which occur quite commonly in the cervix. In both conditions affected glands are lined by cells with hyperchromatic nuclei. In normal glands of endometrial type, cells and nuclei are regularly arranged in contrast to the irregular, poorly polarized, often multilayered neoplastic epithelium which usually contains several mitotic figures. Other lesions from which AIS must be distinguished include papillary endocervicitis, tunnel clusters, microglandular hyperplasia, simple adenomatous hyperplasia, tubal metaplasia, intestinal metaplasia, endometriosis, Arias–Stella reaction of pregnancy and reactive atypia (Young and Clement, 1991). In reviewing all their relevant material from January 1987 to August 1989, Novotny *et al.* (1992) found that 66% (19 out of 29) of cytological diagnoses and 90% (19 out of 21) of histological diagnoses of AIS had to be revised to tubal metaplasia (Section 9.3)

In contrast to CIN lesions, AIS has no distinctive colposcopic pattern so the diagnosis of this lesion relies largely on histological examination of the transformation zone and endocervical canal.

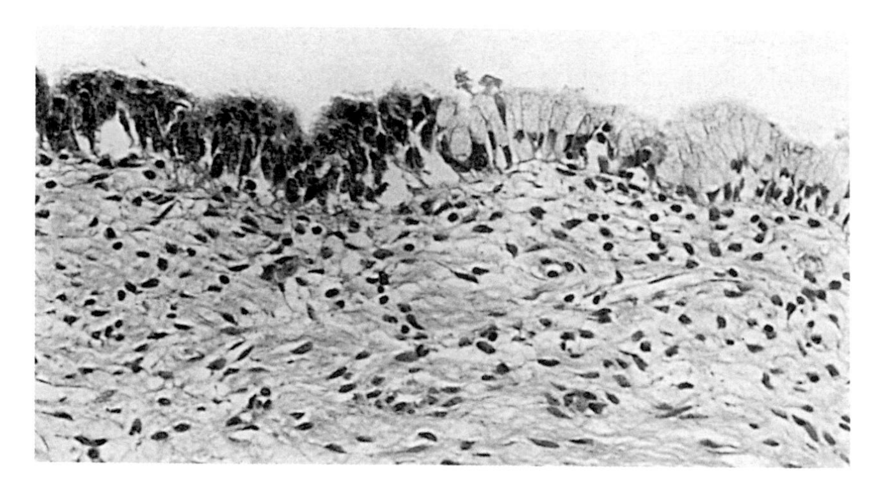

Figure 13.2 Characteristically sharp transition between normal endocervical epithelium (right) and AIS (left), the latter appearing hyperchromatic. No basement membrane can be seen deep to the AIS. Same lesion as shown in Figure 13.1 (H & E, ×310).

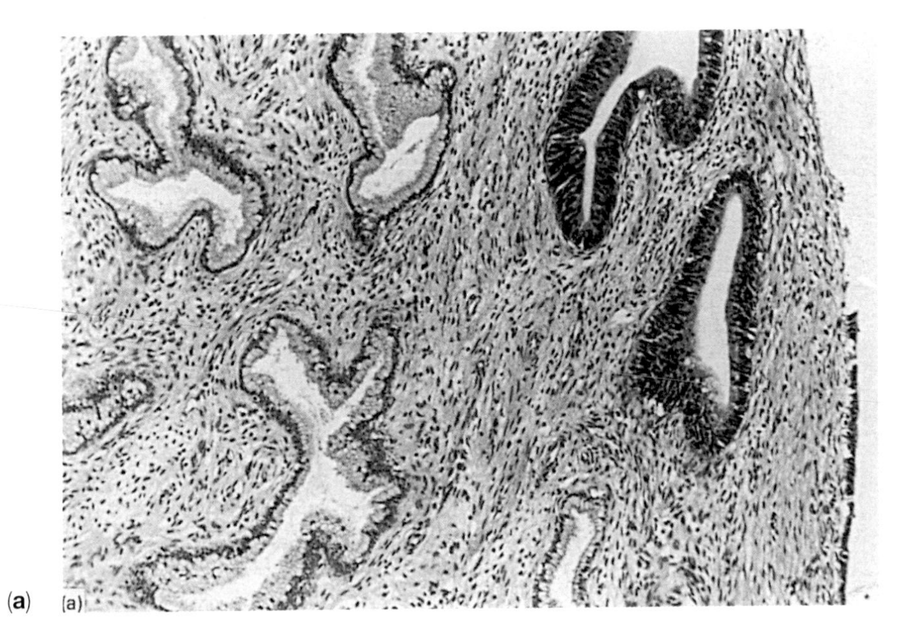

Figure 13.3 (**a**) At low magnification, the hyperchromatic AIS can be readily distinguished from the more deeply situated normal glands. Same case as shown in Figure 13.1 (H & E, ×120).

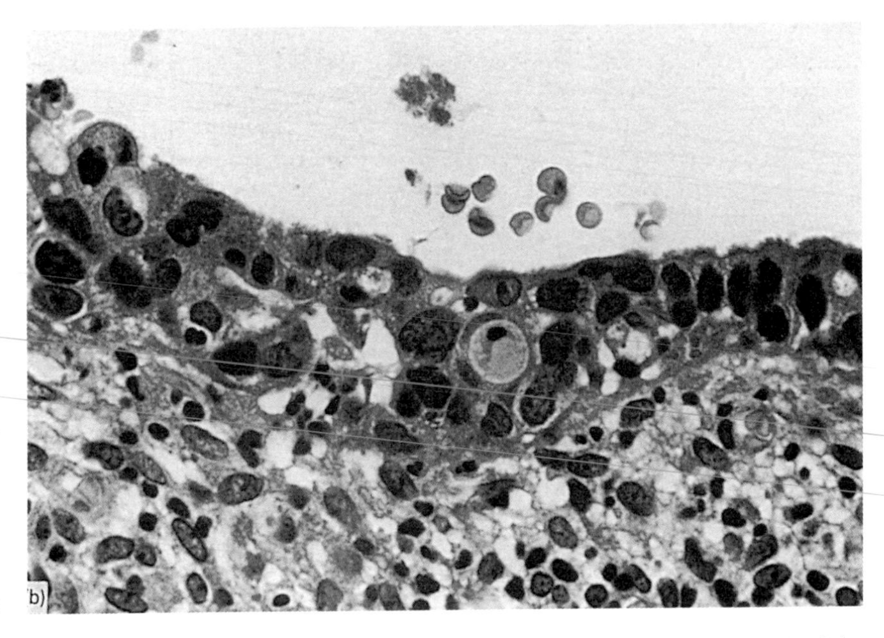

Figure 13.3 (**b**) Association of AIS with squamous carcinoma *in situ* (CIN 3). AIS is on the right and has ciliated columnar epithelium (H & E, 770) (courtesy of Dr D.J.B. Ashley).

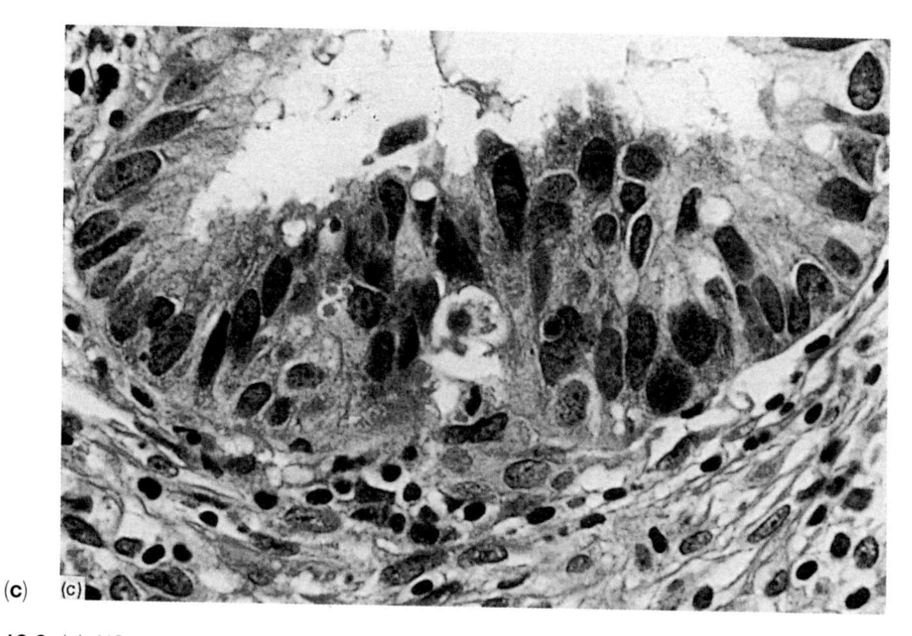

(**c**)

Figure 13.3 (**c**) AIS showing hyperchromatic nuclei with loss of polarization and absence of basement membrane. Same case as shown in (**b**) (H & E, ×770).

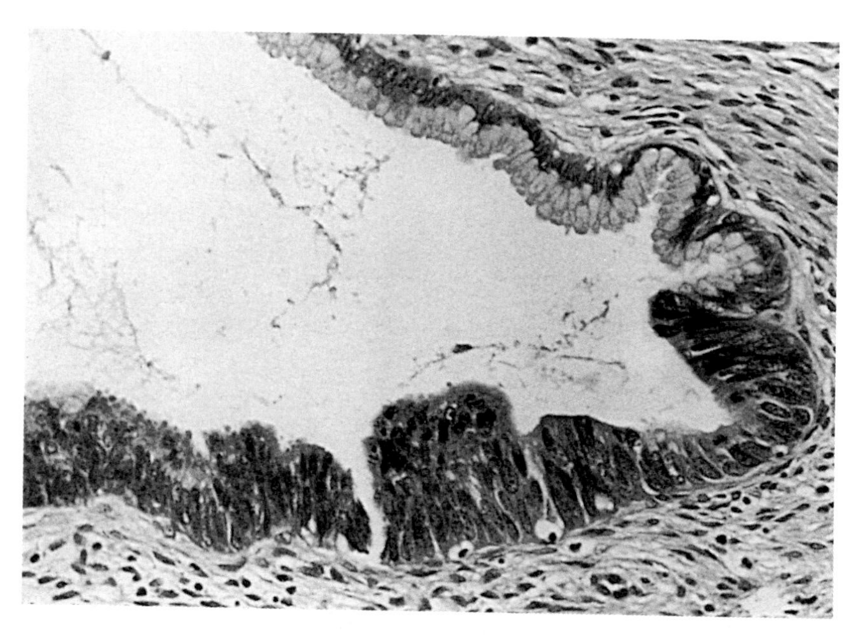

Figure 13.4 AIS involving an endocervical gland, showing enlarged, elongated, pseudostratified cells maintaining normal polarity. Same case as shown in Figure 13.1. Note continuity of basement membrane (H & E, ×310).

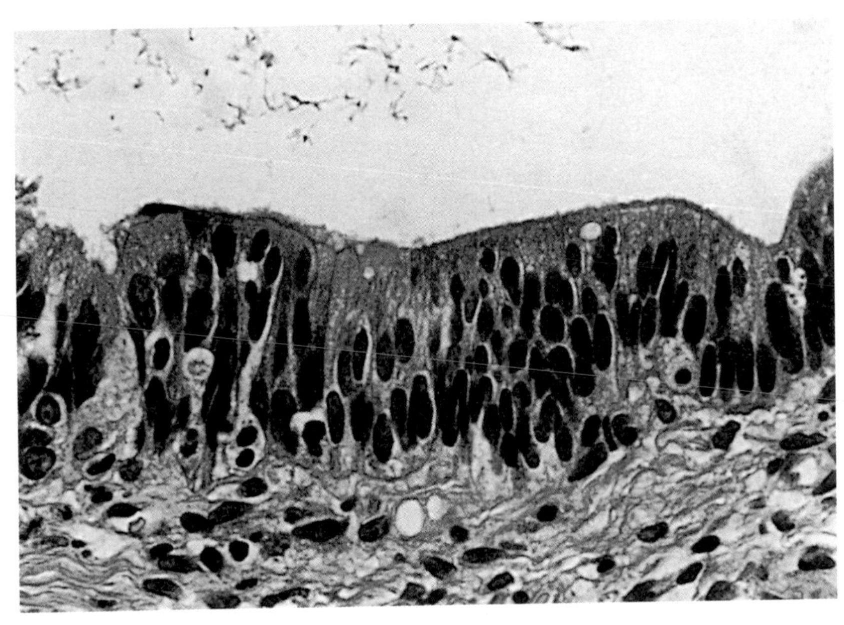

Figure 13.5 AIS with pseudostratified lining and ill-defined basement membrane in initial biopsy from a 33-year-old woman. Note brush border and cilia (H & E, ×770) (courtesy of Dr D.J.B. Ashley).

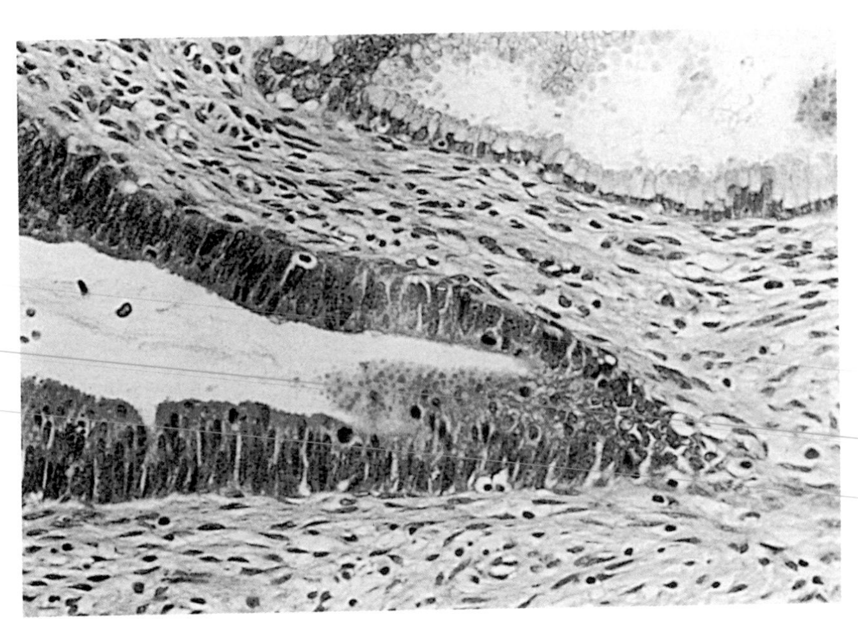

Figure 13.6 AIS of endocervical gland showing mitotic figures located towards the luminal aspect. Same case as shown in Figure 13.1 (H & E, ×310).

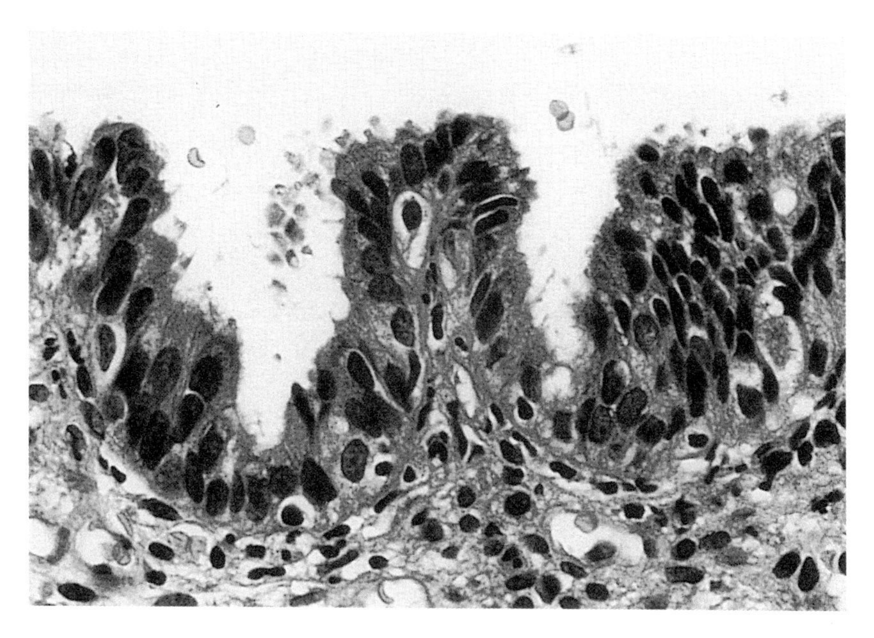

Figure 13.7 Papillary process formation in a zone of AIS. Same biopsy as shown in Figure 13.5 (H & E, ×770).

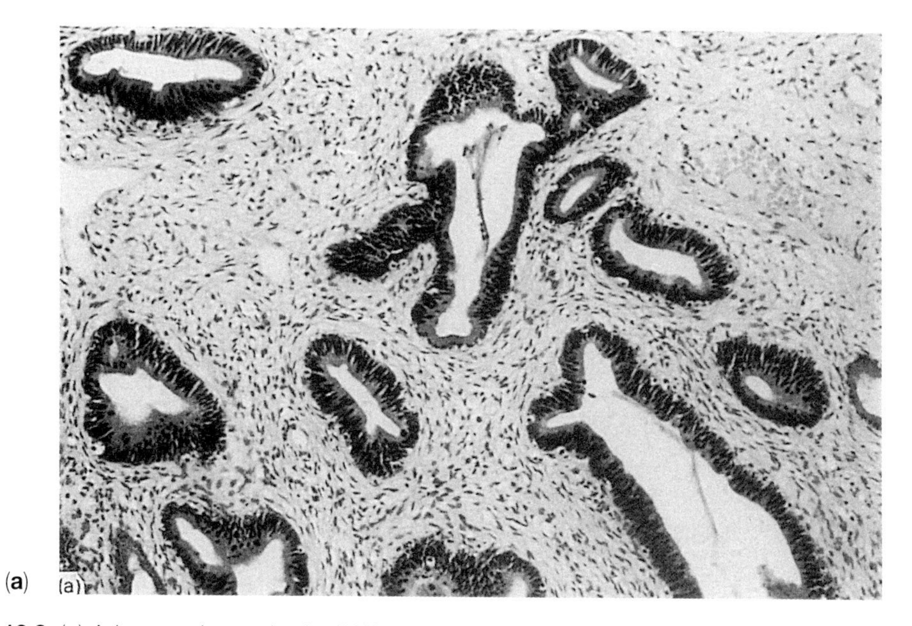

(**a**)

Figure 13.8 (**a**) Adenocarcinoma *in situ* (AIS) of cervix, showing extensive involvement of glandular epithelium in the neoplastic process while normal gland architecture is maintained (H & E, ×190).

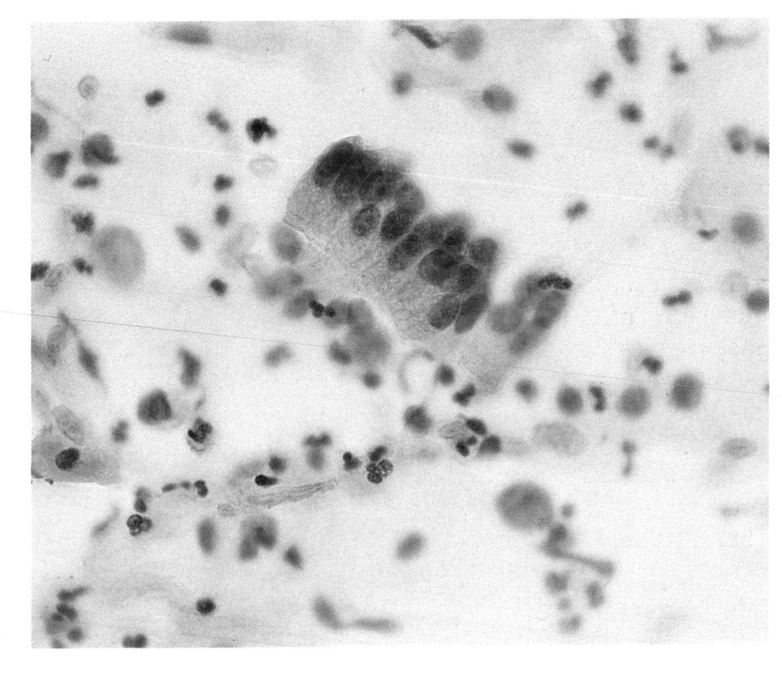

Figure 13.8 (**b**) Tissue fragments showing palisade from AIS in cervical smear. Note large, hyperchromatic basal nuclei (Pap, ×630).

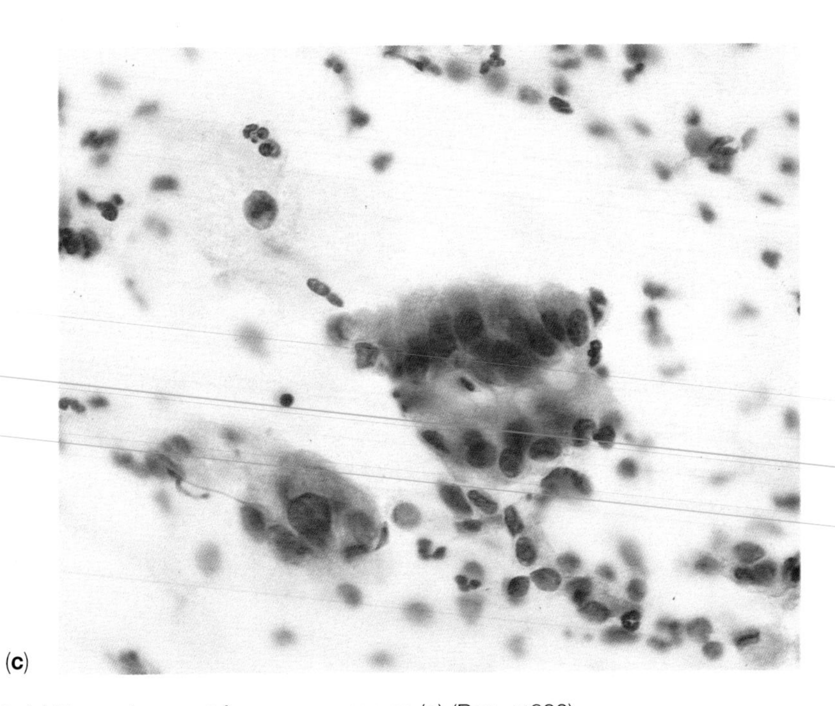

Figure 13.8 (**c**) Tissue fragment from same case as (**a**) (Pap, ×630).

13.1.2 Cytology of cervical adenocarcinoma *in situ*

A striking feature of AIS is the large number of endocervical cells in the smear. The morphology of the cells deviates only slightly from the normal and their glandular origin is readily recognizable (Figure 13.8(b,c)). The cells retain their columnar shape although the nucleocytoplasmic ratio may be slightly altered and the nuclei appear hyperchromatic (Figures 13.8–13.10). The cells may be aligned in a palisade or picket fence formation (Figure 13.8(b)) or a honeycomb pattern (Figure 13.9(b,c)). Many single endocervical cells are frequently scattered throughout the smears (Figure 13.10). The cells have a tendency to form gland-like rosettes (Figure 13.11) or loose clusters which may be either small (Figure 13.10) or large (Figure 13.12). Anisonucleosis is always present, but not marked, and nucleoli may be prominent (Figure 13.13). It is important to remember that the cytology of AIS is very similar to that found in early invasive endocervical adenocarcinoma (Figures 13.21 and 13.22) and a reliable diagnosis of AIS cannot be made from the smear. It is also worth noting that both CIN and AIS are not infrequently present in the same cervix so that the smear may contain both abnormal squamous and abnormal glandular cells.

In a colposcopic biopsy follow-up of 63 women whose cervical smears were reported as showing 'endocervical glandular atypia', Goff *et al.* (1992) found that two had invasive adenocarcinoma, five had AIS, six had squamous carcinoma, 25 had CIN, two had endometrial hyperplasia and seven had endocervical polyps. They concluded that in about half the cases with endocervical glandular atypia on cytology there was substantial cervical disease. Similarly, Taylor *et al.* (1993) found that atypical glandular cells in smears heralded high-grade CIN with sufficient frequency to warrant immediate colposcopy.

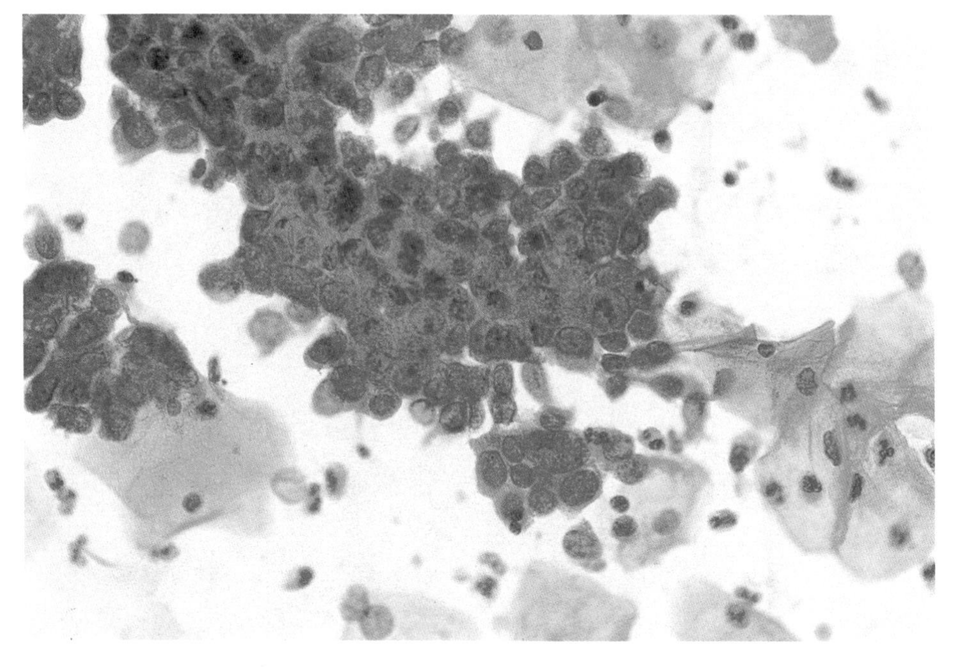

(**a**)

Figure 13.9 (**a**) Tissue fragment showing honeycomb pattern from AIS in cervical smear. Their shapes are consistent with an endocervical origin. Note nuclear hyperchromasia and anisonucleosis (Pap, ×630).

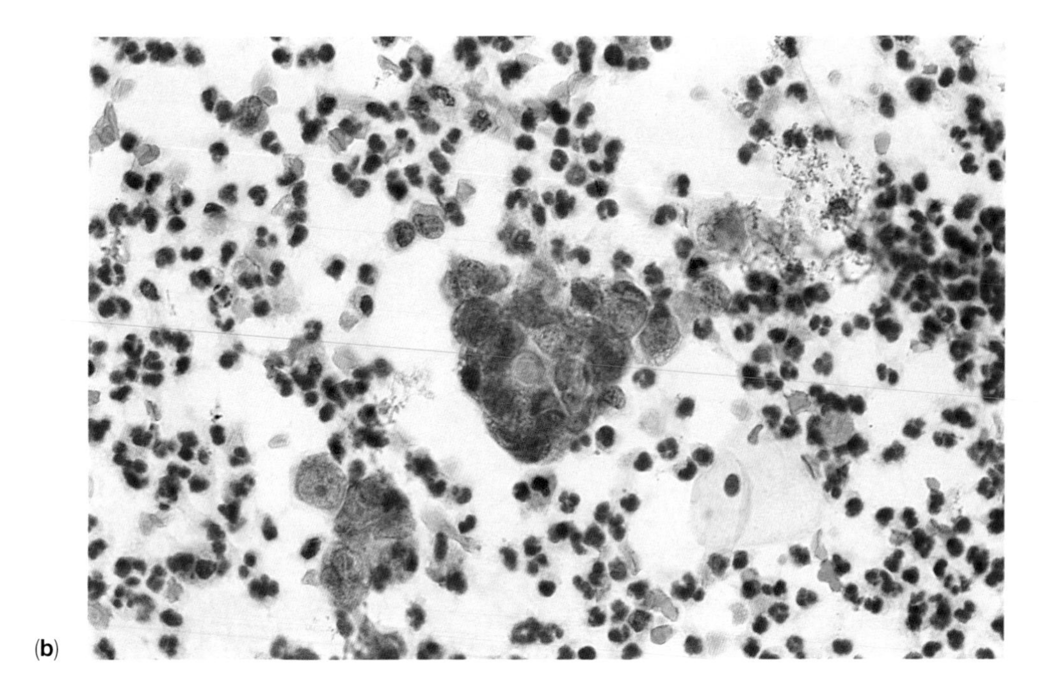

Figure 13.9 (**b**) Tissue fragment showing acinar formation (Pap, ×630).

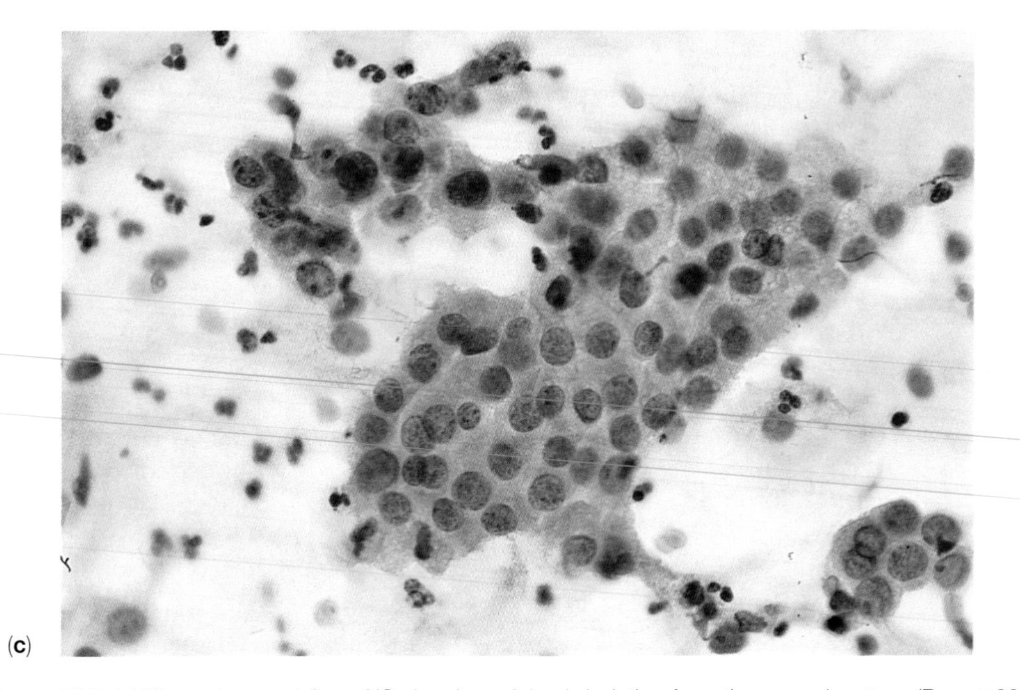

Figure 13.9 (**c**) Tissue fragment from AIS showing minimal deviation from the normal pattern (Pap, ×630).

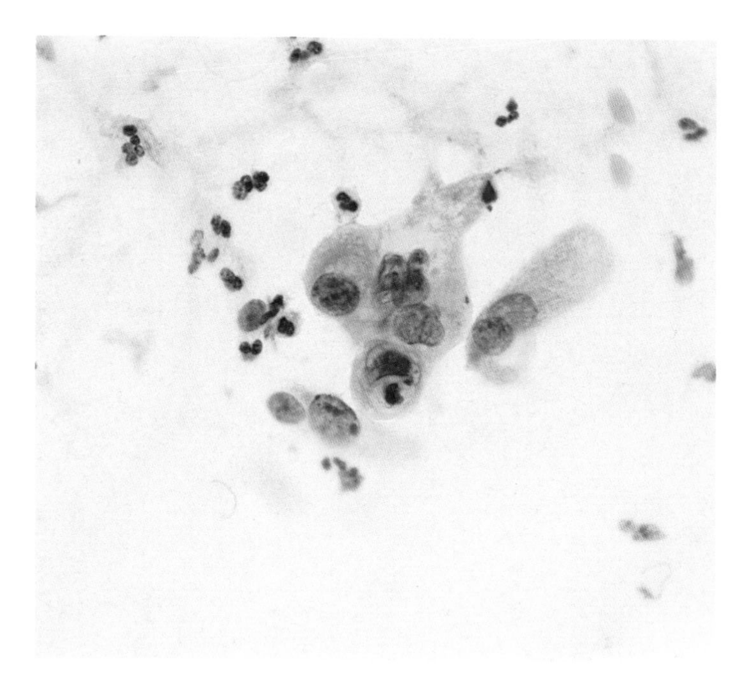

Figure 13.10 Discrete cell and small cluster from AIS retaining columnar cell morphology (Pap, ×630).

Figure 13.11 'Rosette' of cells from AIS. Same case as shown in Figure 13.5 (Pap, ×630).

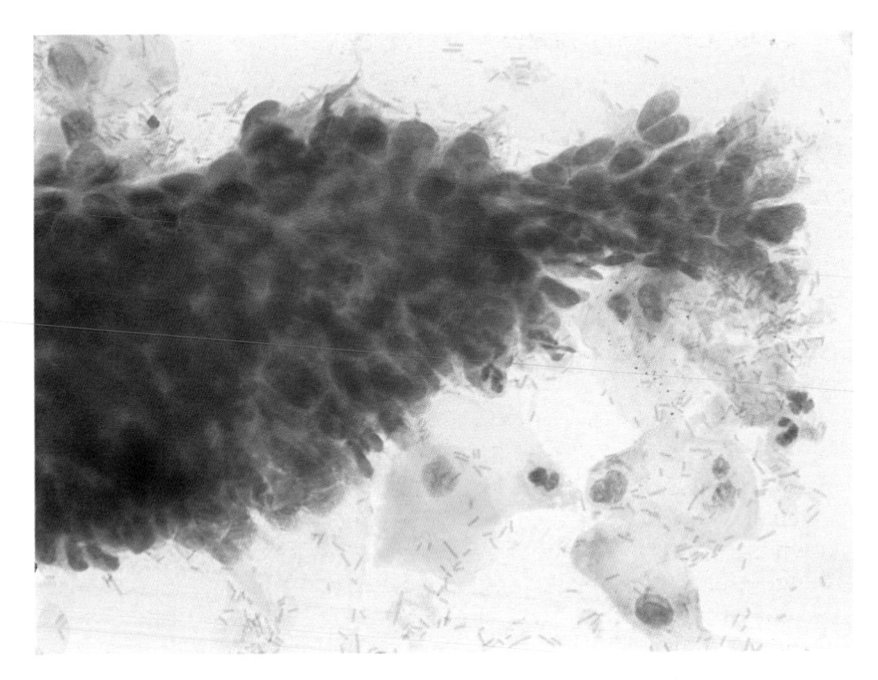

Figure 13.12 Cell cluster from AIS. Cells have enlarged hyperchromatic nuclei with chromatin clumping, slight anisonucleosis and increased nucleocytoplasmic ratio (Pap, ×630).

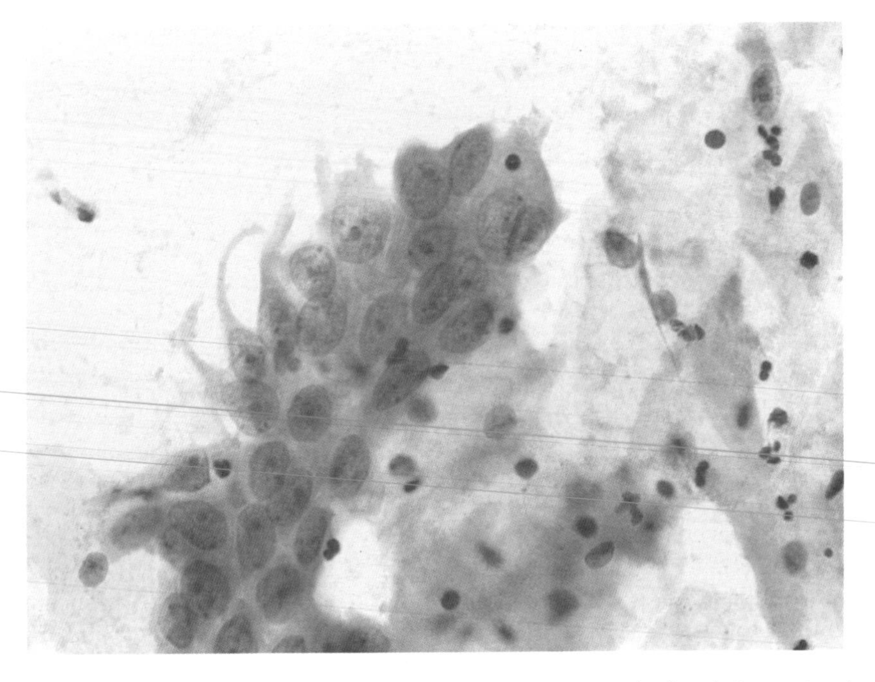

Figure 13.13 Exfoliated epithelium from AIS. The cells have prominent nucleoli and show a tendency to feathering (Pap, ×630).

13.1.3 AIS and glandular atypia of the cervix

Bousfield *et al.* (1980) observed that endocervical glands show a range of cellular atypia comparable with that seen in squamous cervical intraepithelial neoplasia and drew an analogy between preinvasive glandular lesions of the cervix and preinvasive squamous neoplasia. Attempts to characterize these glandular changes have been made using computed morphometry (Van Roon *et al.*, 1983) and histology (Brown and Wells, 1986) and grades of glandular atypia have been described. The lesions have been collectively described as cervical intraepithelial glandular neoplasia (CIGN) and Brown and Wells have identified low-grade and high-grade CIGN depending on the extent of nuclear stratification, mitotic activity, nuclear pleomorphism and irregularity of gland structure as exemplified in Figure 13.14.

Figure 13.14(a) is an example of a low-grade CIGN. The abnormal nuclei in the central gland show little pleomorphism and are limited to the basal half of the epithelium. Figure 13.14(b) is an example of high-grade CIGN with irregularity of the nuclei which occupy more than two-thirds the height of the epithelium. The significance of these histological patterns is not known, but the authors conclude that, in the present state of our knowledge, it is prudent to consider all grades of glandular atypia and AIS as a single lesion (CIGN) with a clinically similar significance.

Some authors advocate hysterectomy when the histological examination of a cone biopsy reveals AIS or glandular atypia on the grounds that the lesions may be multifocal and additional foci of disease may be located high in the canal (Buscema and Woodruff, 1984; Qizilbash, 1975). However, after reviewing 37 cases of AIS, Nicklin *et al.* (1991) found that the extent of AIS in women under 36 years was significantly less than the extent of AIS in older women. They concluded that women under the age of 36 years who are the ones most likely to wish to retain their fertility appear to have the lesions most amenable to conservative surgery. They considered that, providing the margins were uninvolved, conization was adequate treatment. The findings of Anderson and Arffman (1989) and Cullimore *et al.* (1992) lend support to this approach.

13.2 MICROINVASIVE ADENOCARCINOMA OF THE CERVIX

The difference between microinvasive and early (occult) invasive squamous carcinoma of the cervix has been defined with clarity and precision earlier in this book (Sections 11.1.1 and 11.3). Unfortunately, the distinction between microinvasive glandular neoplasia of the cervix and early invasive adenocarcinoma cannot yet be defined in such precise terms, as the dimensions of glandular lesions are not easily quantified. Despite this limitation, the term 'microinvasive adenocarcinoma' is not infrequently used.

Burghardt (1973) and Qizilbash (1975) applied the term microinvasive adenocarcinoma to those cervical lesions where, in addition to the replacement of the epithelium lining the endocervical canal and crypts by neoplastic cells, there is also budding and branching of affected glands (Figures 13.19 and 13.20). A major problem associated with this definition of microinvasive adenocarcinoma stems from the difficulty in distinguishing it from adenocarcinoma *in situ* on one hand, and early invasive (occult) adenocarcinoma on the other. This problem is clearly illustrated in Figures 13.15, 13.16 and 13.18–13.20. In the lesions shown in Figures 13.15 and 13.16, budding was confined to a few glands only, and there was doubt as to whether these minimal changes amounted to stromal invasion. In the lesions shown in Figures 13.18 and 13.19, budding and branching involved quite large areas of the gland circumference and the lesions may have progressed well beyond the microinvasive stage. In Figure 13.20, there is extensive field change and early

(**a**)

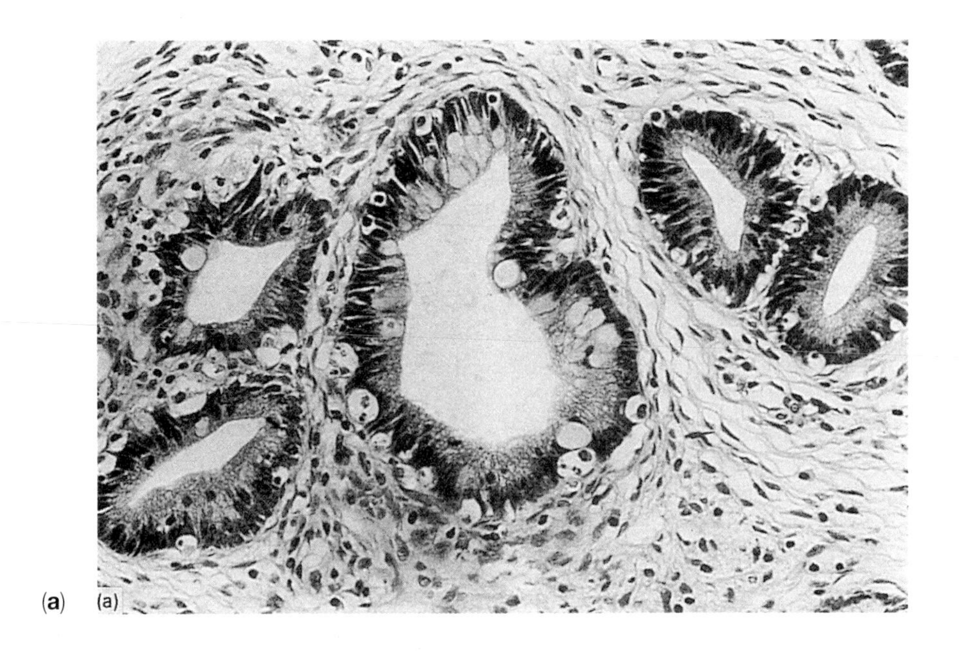

(**b**)

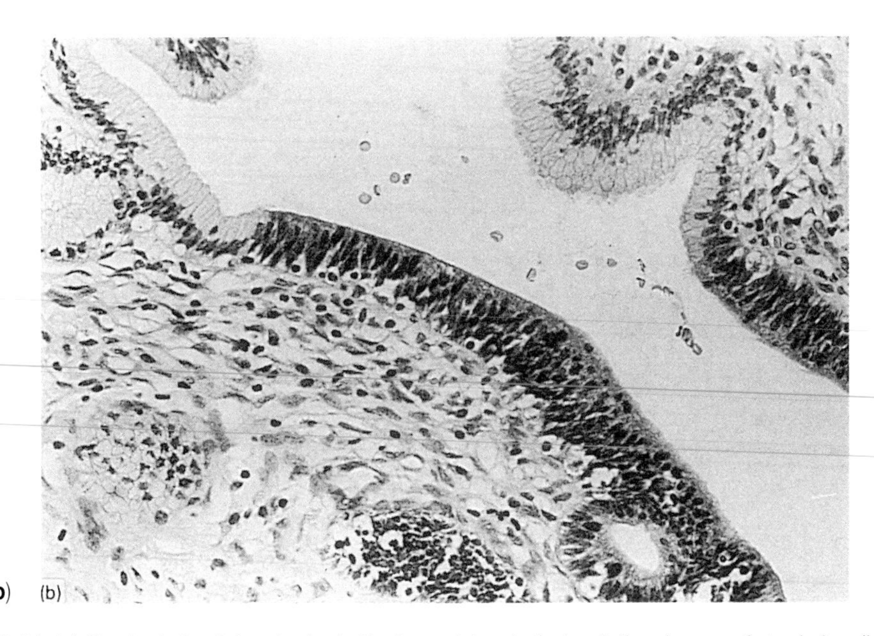

Figure 13.14 (**a**) Cervical glandular atypia. In the largest (central) gland, the degree of atypia is mild, the nuclei generally being limited to the basal half of the epithelium (H & E, ×480). (**b**) Sudden transition between normal epithelium and cervical glandular atypia of severe degree. Same biopsy as (**a**) (H & E, ×480).

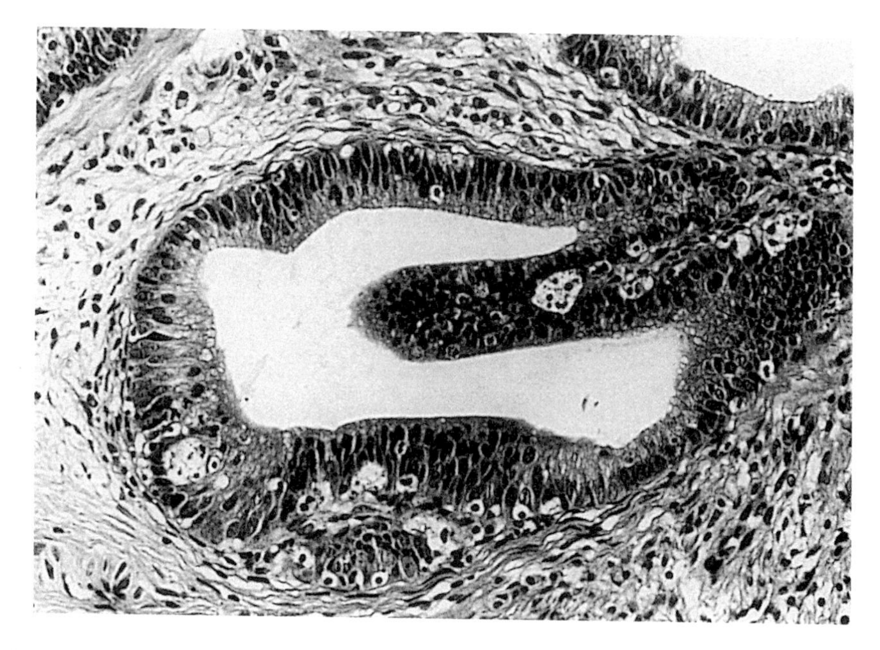

Figure 13.15 Abnormal superficial gland with single small bud. Stromal invasion is questionable (H & E, ×310).

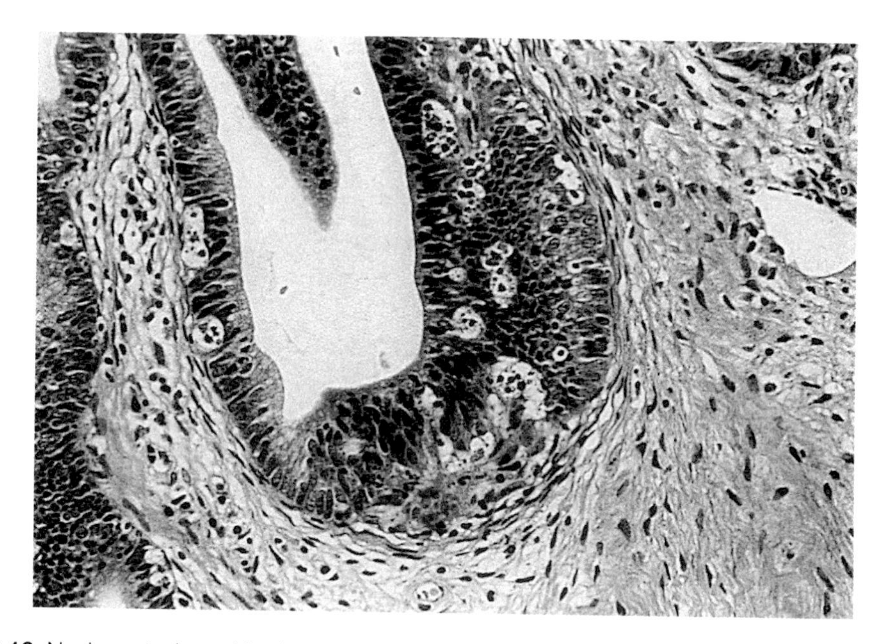

Figure 13.16 Nuclear atypia and budding involving superficial glands consistent with 'microinvasive adenocarcinoma' (H & E, ×310).

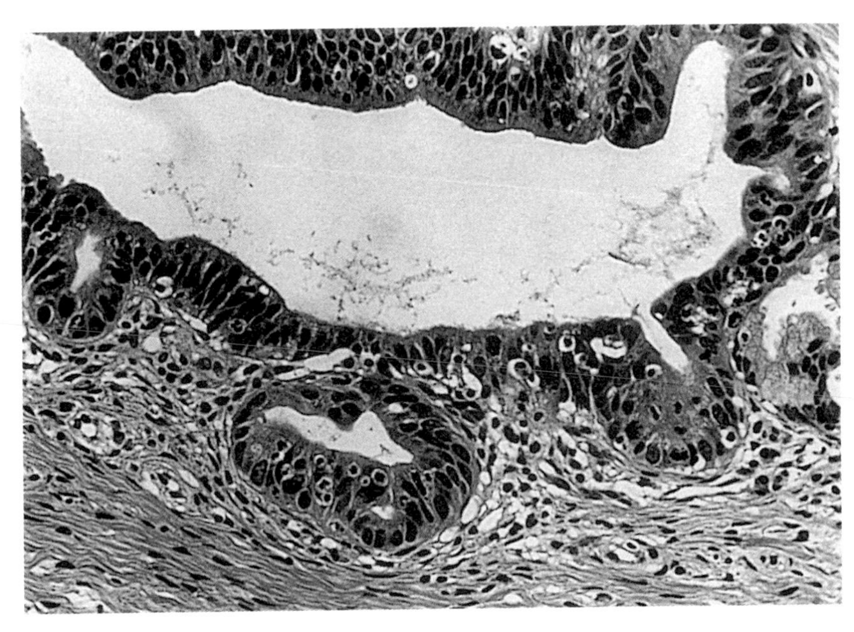

Figure 13.17 Nuclear atypia and budding involving superficial glands consistent with 'microinvasive adenocarcinoma'. Same cone biopsy as shown in Figure 13.18 (H & E, ×310).

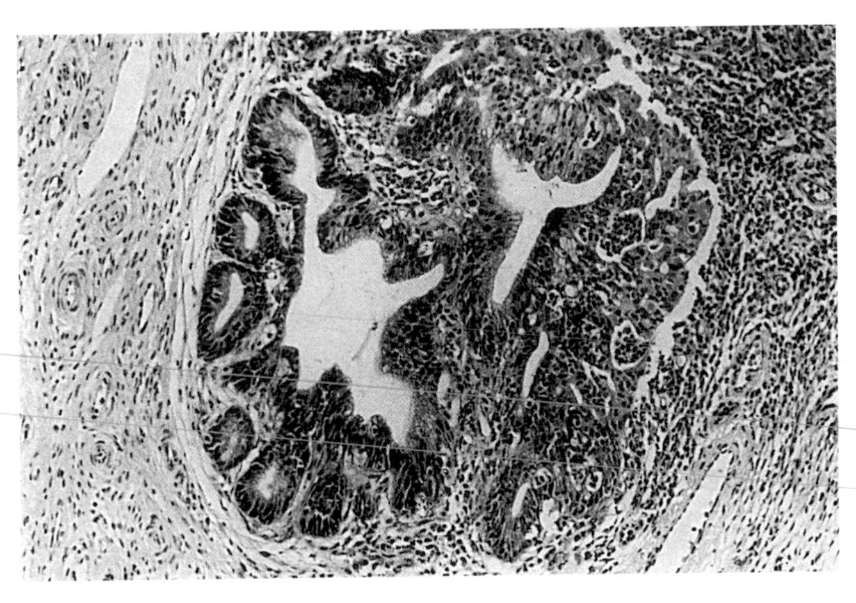

Figure 13.18 Nuclear atypia with extensive budding on right side of gland associated with cellular reaction in stroma. A precone biopsy from same case as Figure 13.5. As budding was confined to superficial glands, a diagnosis of 'microinvasive adenocarcinoma' could be considered (H & E, ×190).

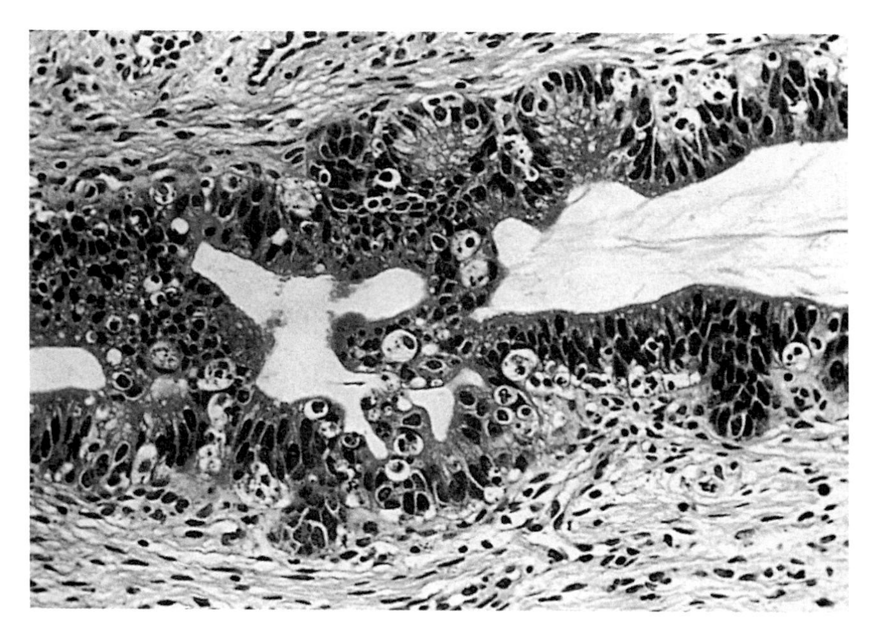

Figure 13.19 Atypical superficial gland with several buds in which cells are small and crowded together (tunnel clusters). Cone biopsy from case shown in Figure 13.5 consistent with 'microinvasive carcinoma' as defined by Burghardt (H & E, ×310).

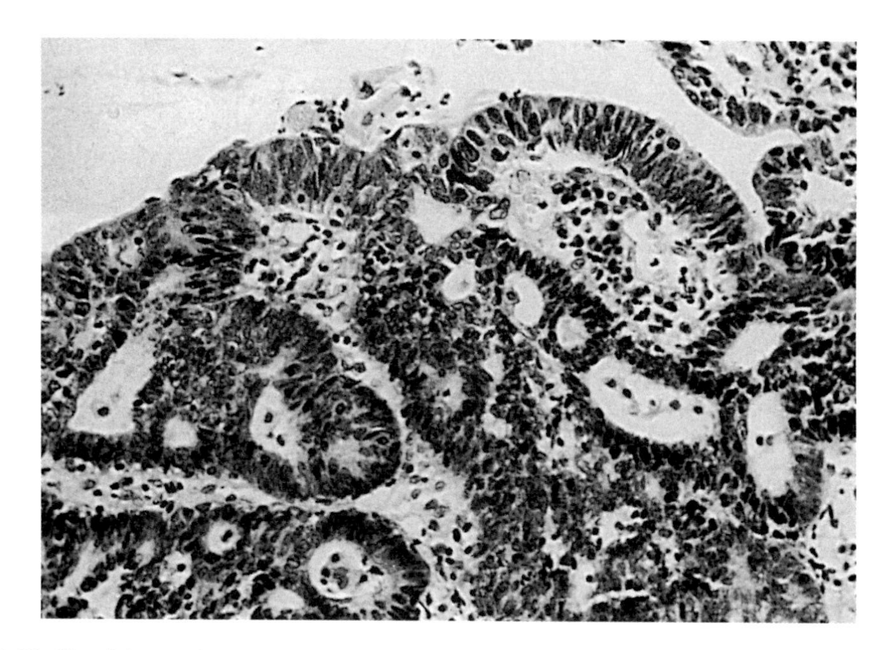

Figure 13.20 Glandular atypia occurring over a relatively large field involving many superficial glands in a cone biopsy from a 24-year-old woman. Such a lesion may have progressed well beyond the microinvasive stage (H & E, ×310).

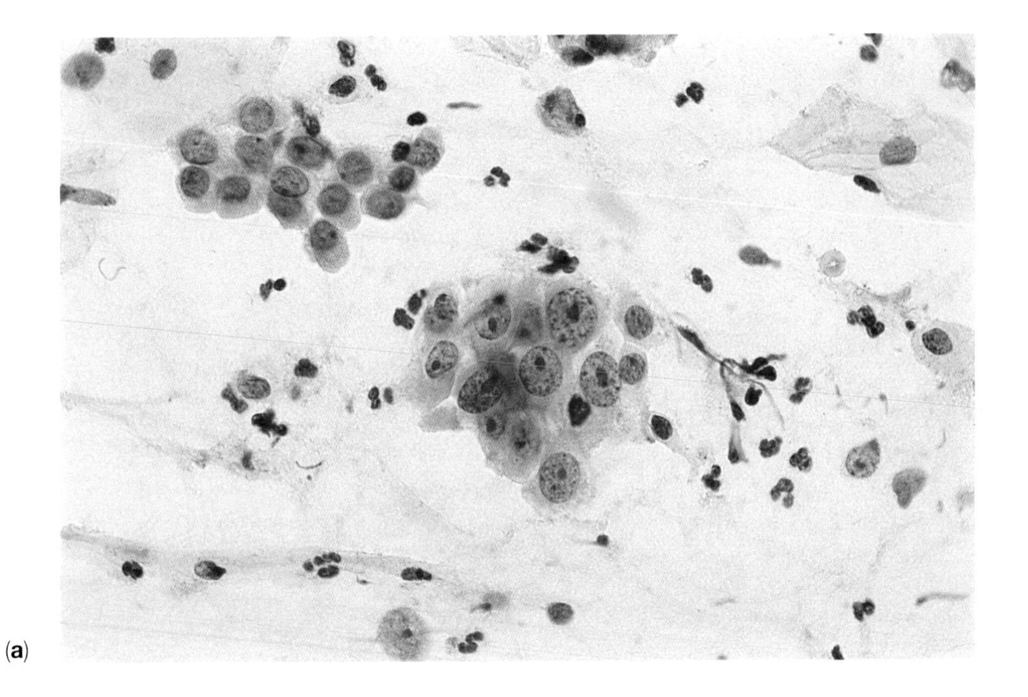

(**a**)

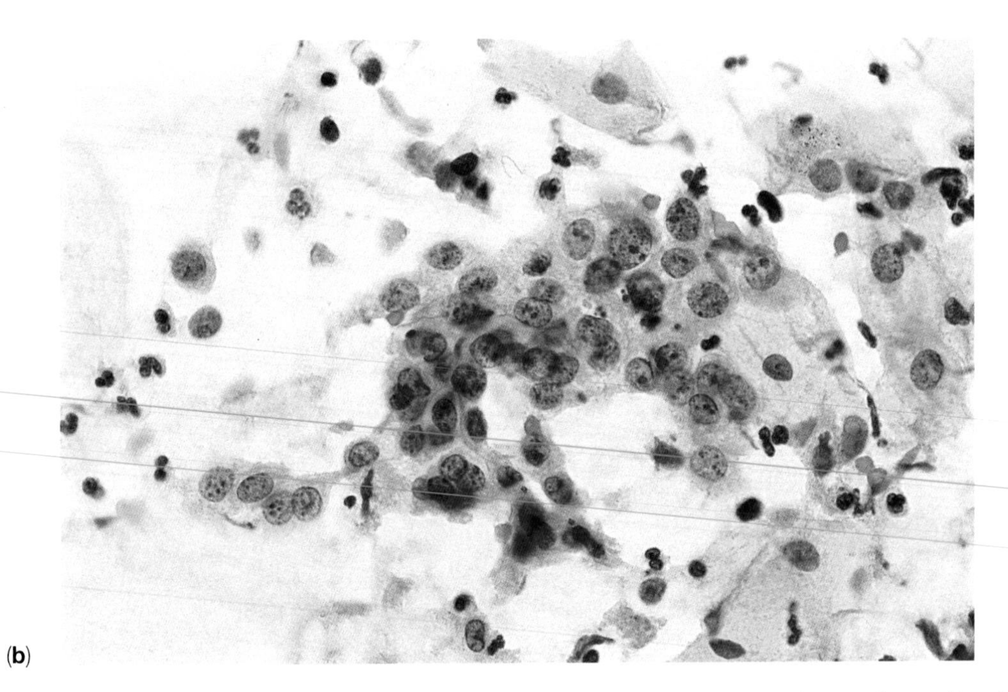

(**b**)

Figure 13.21 (**a,b**) Cells from lesion classified as 'microinvasive adenocarcinoma' according to Burghardt's definition (Pap, ×630).

invasive adenocarcinoma cannot be excluded, even though the glands did not invade deeply into the stroma.

It follows that the term microinvasive adenocarcinoma should be used with great caution, if at all. If there is evidence of lymphatic or blood vessel involvement it cannot be considered no matter how circumscribed the lesion. Teshima *et al.* (1985) use the term 'early stage adenocarcinoma' for invasive glandular lesions of 5 mm or less in depth (as in Figures 13.18–13.20). Of 22 such cases only one whose tumour was 3 mm in depth developed a pelvic recurrence after radical hysterectomy.

It should be noted that, occasionally, foci indistinguishable from squamous early stromal invasion arise from unequivocally glandular *in situ* malignant foci: Rollason *et al.* (1989) described two such cases and a third arising from a very early invasive adenocarcinoma, supporting the view that glandular and squamous *in situ* lesions are closely related possibly with a common origin in the subcolumnar reserve cell (Figures 13.21 and 13.22).

The cytology of the lesions described in this section cannot be reliably distinguished from AIS or early invasive adenocarcinoma (Figures 13.21 and 13.22). However, Lee *et al.* (1991) consider that the presence of large cells with irregular nuclei and uneven chromatin distribution (as in Figure 13.21) in smears containing no normal cells (as in Figure 13.22) helps to distinguish invasive adenocarcinoma from CIGN. They also find that 'feathering', rosettes, mitotic figures and very crowded nuclei with scanty cytoplasm helps to distinguish CIGN from benign conditions. Cervical endometriosis, tubal metaplasia and reactive changes due to a previous biopsy can be misdiagnosed cytologically as CIGN (Pacy *et al.*, 1988). So, occasionally, can microglandular endocervical hyperplasia (Figure 9.14).

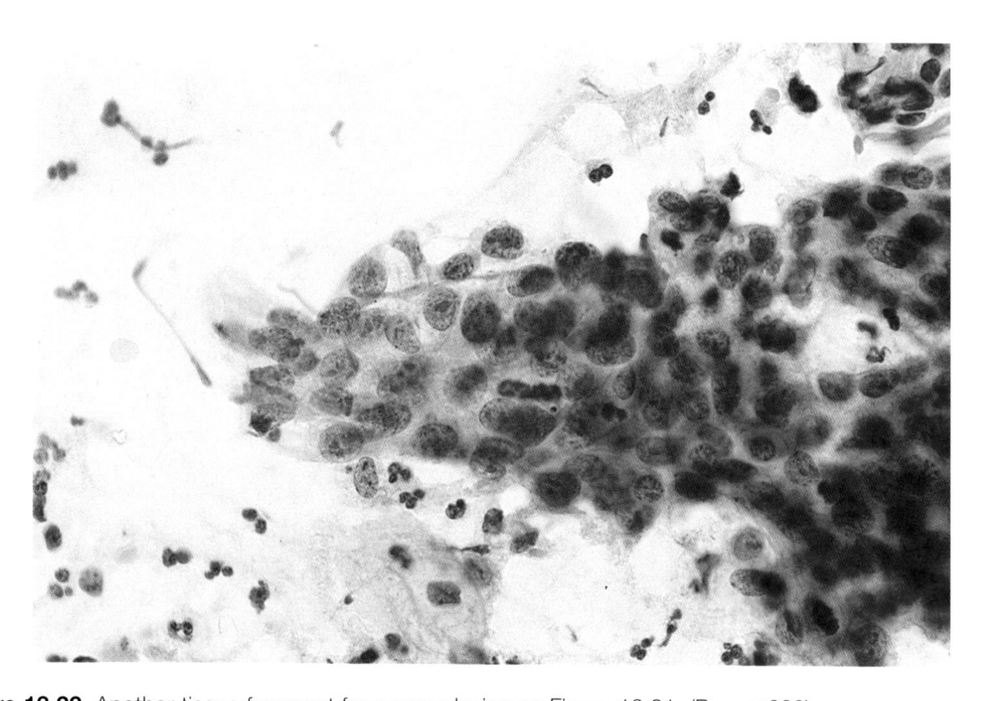

Figure 13.22 Another tissue fragment from same lesion as Figure 13.21. (Pap, ×630).

REFERENCES

Anderson, E. S. and Arfmann, E. (1989) Adenocarcinoma *in situ* of the uterine cervix: a clinico-pathologic study of 36 cases. *Gynecol. Oncol.*, **35**, 1–7.

Boddington, M. M., Spriggs, A. I. and Cowdell, R. H. (1976) Adenocarcinoma of the uterine cervix: cytological evidence of a preclinical evolution. *Br. J. Obstet. Gynaecol.*, **83**, 900–903.

Boon, M. E., Baak, J. P. A., Kurver, P. S. H., Overdiep, S. H. and Verdonk, G. W. (1981) Adenocarcinoma *in situ* of the cervix; an underdiagnosed lesion. *Cancer*, **48**, 768–773.

Bousfield, L., Pacey, F., Young, P., Krumins, I. and Osborn, R. (1980) Expanded cytological criteria for the diagnosis of adenocarcinoma of the cervix and related lesions. *Acta Cytol.*, **24**, 283–296.

Brown, L. J. R. and Wells, M. (1986) Cervical glandular atypia associated with squamous intraepithelial neoplasia: a premalignant lesion? *J. Clin. Pathol.*, **39**, 22–28.

Burghardt, E. (1973) Early histological diagnosis of cervical cancer, in *Major Problems in Obstetrics and Gynecology* (ed. E. A. Friedman), Saunders, Philadelphia, London, Toronto, Vol. 6, pp. 350–362.

Buscema, J. and Woodruff, J. D. (1984) The significance of neoplastic atypicalities in endocervical epithelium. *Gynecol. Oncol.*, **17**, 356–362.

Christopherson, W. M., Nealon, N. and Gray, L. A. (1979) Non-invasive precursor lesions of adenocarcinoma and mixed adenosquamous carcinoma of the cervix uteri. *Cancer*, **44**, 975–983.

Colgan, T. J. and Lickrish, G. M. (1990) The topography and invasive potential of cervical adenocarcinoma *in situ* with and without associated squamous dysplasia. *Gynecol. Oncol.*, **36**, 246–249.

Cullimore, J. E., Luesley, D. M., Rollason, T. P. *et al.* (1992) A prospective study of conisation of the cervix in the management of intraepithelial glandular neoplasia (CGIN A preliminary report. *Br. J. Obstet. Gynaecol.*, **99**, 314–318.

Friedell, G. H. and McKay, D. G. (1953) Adenocarcinoma *in situ* of the endocervix. *Cancer*, **6**, 887–897.

Goff, B. A., Atanasoff, P., Brown, E. *et al.* (1992) Endocervical glandular atypia in Papanicolaou smears. *Obstet. Gynecol.*, **79**, 101–104.

Hauser, G. (1894) Zur Histogenese des Krebses. *Virchows. Arch. Path. Anat.*, **134**, 482–498.

Hopkins, M. P., Roberts, J. A. and Schmidt, R. W. (1988) Cervical adenocarcinoma *in situ*. *Obstet. Gynecol.*, **71**, 842–844.

Jaworski, R. C., Pacey, N. F., Greenberg, M. L. and Osborn, R. A. (1988) The histologic diagnosis of adenocarcinoma *in situ* and related lesions of the cervix uteri. Adenocarcinoma *in situ*. *Cancer*, **61**(6), 1171–1181.

Kashimura, M., Shiohara, M., Oikawa, K. *et al.* (1990) An adenocarcinoma *in situ* that developed into invasive adenocarcinoma after 5 years. *Gynecol. Oncol.*, **36**, 128–133.

Nicklin, J. L., Wright, R. G. and Bell, J. R. (1991) A clinico pathological study of adenocarcinoma *in situ* of the cervix. The influence of cervical HPV infection and other factors and the role of conservative surgery. *Aust. N. Z. J. Obstet. Gynaecol.*, **31**(2), 170–183.

Novotny, D. B., Maygarden, S. J., Johnson, D. E. and Frable, W. J. (1992) Tubal metaplasia A frequent potential pitfall in the cytologic diagnosis of endocervical glandular dysplasia on cervical smears. *Acta Cytol.*, **36**, 1–10.

Qizilbash, A. H. (1975) *In situ* and microinvasive adenocarcinoma of the uterine cervix. A clinical, cytologic and histologic study of 14 cases. *Am. J. Clin. Pathol.*, **64**, 155–170.

Rollason, T. P., Cullimore, J. and Bradgate, M. G. (1989) A suggested columnar cell morphological equivalent of squamous cell carcinoma *in situ* with early stromal invasion. *Int. J. Gynecol. Pathol.*, **8**, 230–236.

Taylor, R. R., Guerrieri, J. P., Nash, J. D. *et al.* (1993) Atypical cervical cytology. Colposcopic follow-up using the Bethesda system. *J. Reprod. Med.*, **38**, 443–447.

Teshima, S., Shimosato, Y., Kishi, K. *et al.* (1985) Early stage adenocarcinoma of the uterine cervix Histopathologic analysis with consideration of histogenesis. *Cancer*, **56**(1), 167–172.

Van Roon, E., Boon, M. E., Kurver, P. J. H. and Baak, J. P. A. (1983) The association between precancerous columnar and squamous lesions of the cervix: a morphometric study. *Histopathology*, **7**, 887–896.

Yavner, D. L., Dwyer, I. M., Hancock, W. W. and Ehrmann, R. L. (1990) Basement membrane of cervical adenocarcinoma: an immunoperoxidase study of laminin and type IV collagen. *Obstet. Gynecol.*, **76**(6), 1014–1019.

Young, R. and Clement, P. B. (1991) Pseudoneoplastic lesions of the uterine cervix. *Semin. Diagn. Pathol.*, **8**, 234–249.

14
ADENOCARCINOMA

This chapter includes the various forms of malignant cervical epithelial tumour of glandular type, defined in the WHO classification as adenocarcinoma (Scully *et al.*, 1994). This comprises the following:

- Mucinous adenocarcinoma
  - Endocervical type
    - Adenoma malignum (minimal deviation adenocarcinoma)
    - Villoglandular papillary adenocarcinoma
  - Intestinal type
    - Signet-ring type
- Endometrioid adenocarcinoma
- Clear cell adenocarcinoma
- Mesonephric adenocarcinoma
- Serous adenocarcinoma

Although each type is histologically distinct, it is not uncommon for two or more histological forms to be present in a single tumour, including tumour types described in the preceding or following chapters. The frequent coexistence of glandular and squamous neoplasia in the cervix suggests that they may have aetiological factors in common. Thus, Sugimori *et al.* (1990) reported a uterus didelphys with adenocarcinoma in one cervix and squamous CIN in the other.

The cytological and histological studies by Boon *et al.* (1981) and the morphological and ultrastructural studies of Kudo *et al.* (1991) suggest a common histogenesis from the reserve cell. The indeterminate role of human papillomavirus (HPV) infection in squamous carcinogenesis, discussed in Section 10.6.4, is also evident in glandular lesions. Okagaki *et al.* (1989) found HPV DNA in 67% (14/21) of adenocarcinomas *in situ* and in a comparable proportion of invasive glandular lesions. Farnsworth *et al.* (1989), in studying 22 cases of AIS (including three with early invasion) found HPV messenger RNA expression in nearly 90%. Nagai (1990) found that 31% (10/32) of adenocarcinomas and 50% (4/8) of adenosquamous carcinomas were positive for HPV 18. However, Griffin *et al.* (1991) found evidence of HPV in only a small proportion of glandular lesions.

There has been speculation that oral contraceptives may be an aetiological factor in glandular neoplasia (Dallenbach-Hellweg, 1984) but Jones and Silverberg (1989) found no evidence that either oral contraceptives or microglandular endocervical hyperplasia were causally related to cervical adenocarcinoma.

Together, adenocarcinoma and adenosquamous carcinoma were thought to represent a relatively small fraction of all primary carcinomas of the cervix, in the order of 5–10%

(Cramer and Cutler, 1974). There is evidence that this proportion may be increasing (Davis and Moon, 1975; Gallup and Abell, 1977) and in 1990 it was estimated that adenocarcinomas and related tumours accounted for approximately 15% of cervical carcinomas (Young and Scully, 1990).

The use of mucin stains, strongly recommended in the diagnosis of cervical neoplasms, shows that many tumours thought to be squamous cell carcinomas are in fact adenosquamous carcinomas (Wells and Brown, 1986). This was confirmed by Shorrock *et al.* (1990) who, by using histochemical staining for mucins, found that out of 242 cases of cervical carcinoma, 64 initially diagnosed as squamous were found to be mucin-secreting. They also found that over a 10-year period there had been a significant increase in the proportion of carcinomas arising in patients of 45 years or younger with a disproportionate increase of adenocarcinomas and mucin-secreting subtypes in this age group.

Mucin-secreting carcinomas, even in the early stages, metastasize to lymph nodes more often than purely squamous lesions and are more frequent under the age of 40 (Buckley *et al.*, 1988). So the detection of mucus secretion and of vascular permeation adjacent to the primary growth identifies women at greatest risk of pelvic node metastases.

Hopkins and Morely (1991), in analysing 203 patients with adenocarcinoma and 756 with squamous carcinoma of the cervix, found that the 5-year survival rate for stage 1 cases was 90% for squamous lesions and 60% for adenocarcinoma. Survival rate was also influenced by tumour differentiation, nodal status, diabetes and cervical smear interval. Separate analysis of patients treated by radical hysterectomy revealed that the 5-year survival rate for patients with stage 2 carcinoma was 62% for squamous lesions and 47% for adenocarcinoma; for stage 3 carcinomas it was 36% for squamous lesions and 8% for adenocarcinomas. However, Grigsby *et al.* (1988) and Vesterinen *et al.* (1989) found no significant difference in the survival rate for the two types of cancer, except in stage 3.

Several different systems of classification have been proposed for the tumours (Abell, 1973; Tasker and Collins, 1974; Poulsen *et al.*, 1975; Shingleton *et al.*, 1981; Fu *et al.*, 1982).

We have used the current WHO classification (Scully *et al.*, 1994) throughout this chapter.

14.1 MUCINOUS ADENOCARCINOMA

This is an adenocarcinoma in which at least some of the cells contain a moderate to large amount of intracytoplasmic mucin (Scully *et al.*, 1994).

14.1.1 Endocervical type

This is characterized by mucin-containing cells which resemble those of the endocervix and is the most common type of adenocarcinoma in the cervix. It is believed to arise from the reserve cells related to the epithelium lining the endocervical canal (Kudo *et al.*, 1991) and occurs most commonly at the squamocolumnar junction (Teshima *et al.*, 1985; Matsukuma *et al.*, 1989). Two variants of the tumour are recognized: minimal deviation carcinoma (adenoma malignum) and papillary adenocarcinoma. The majority of the lesions are polypoid or ulcerative, although endophytic lesions are also found.

Histologically, endocervical adenocarcinomas are composed of branching and budding glands lined by neoplastic cells which closely resemble the glandular cells lining the endocervical canal. Three grades of endocervical adenocarcinoma are recognized: well-differentiated, moderately differentiated, and poorly differentiated forms (Scully *et al.*, 1994). The distinction is based on the tumour architecture and the morphology of the individual tumour cells.

In *well-differentiated endocervical adenocarcinoma* (Figures 14.1–14.3) the normal pattern of

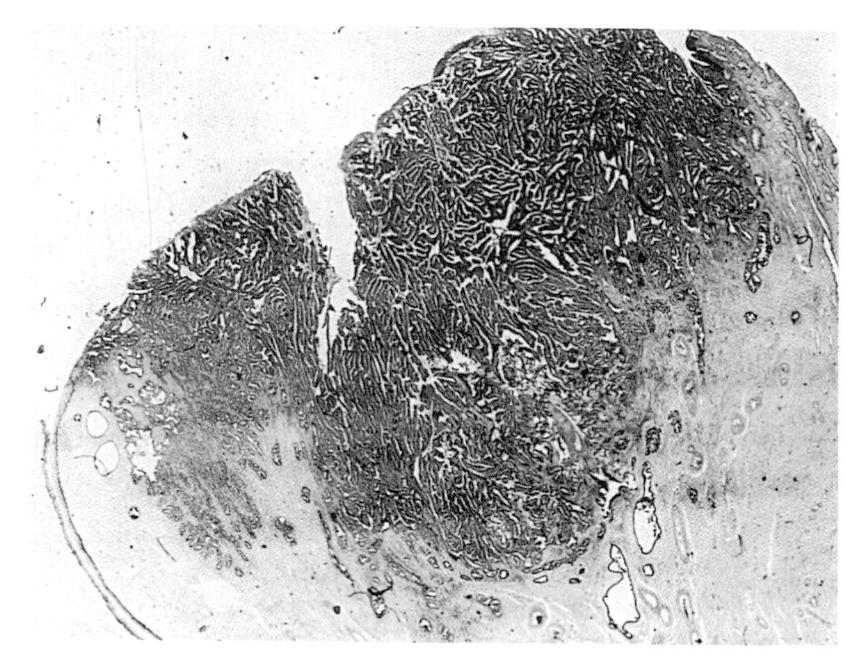

Figure 14.1 Well-differentiated endocervical adenocarcinoma. Focus of early invasive (occult) endocervical adenocarcinoma in a cervical biopsy. Note crowded, large irregular glands extending deep into the cervical stroma (H & E, ×7).

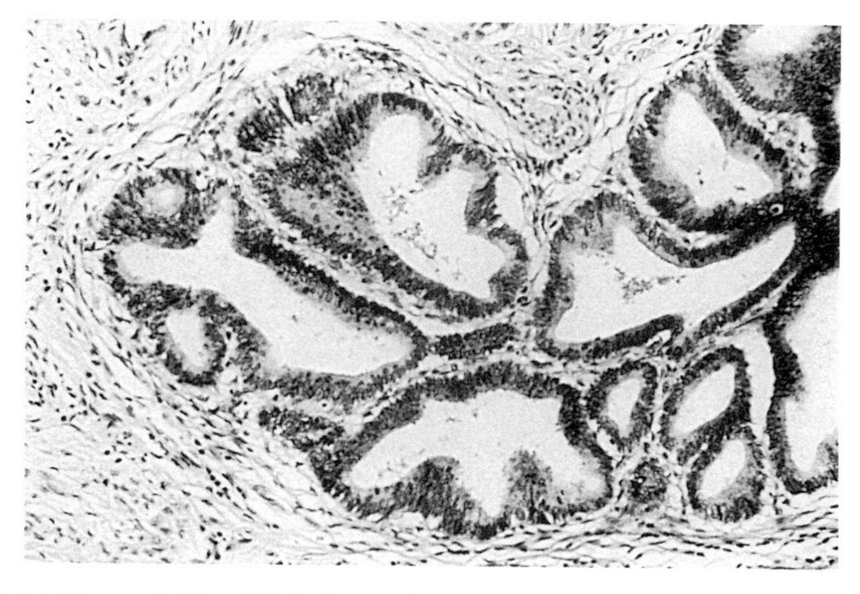

Figure 14.2 Acini from well-differentiated adenocarcinoma illustrated in Figure 14.1, showing crowded glands lined by abnormal endocervical cells (H & E, ×190).

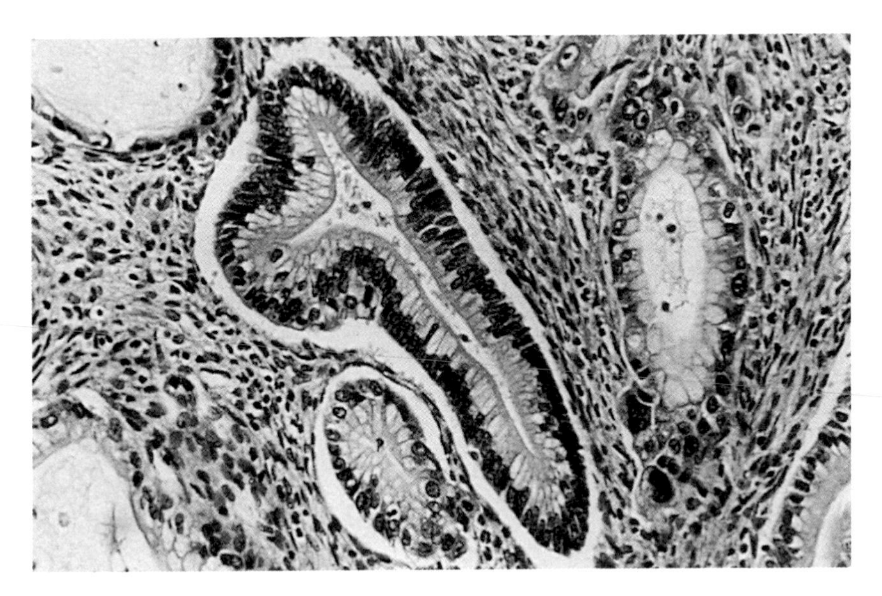

Figure 14.3 Well-differentiated endocervical adenocarcinoma. In addition to the near-normal endocervical cells there are a number of more pleomorphic forms (H & E, ×310).

endocervical glands may be maintained in many parts of the tumour, although areas of abnormal gland architecture will be seen. In these areas, the glands are often enlarged and irregular, and may be crowded, reduplicated and back-to-back (Figure 14.1).

The glands are lined by cells with a distinct columnar shape which show only a slight deviation from normal endocervical epithelium and are reminiscent of the cells seen in AIS (Figures 14.2 and 14.3) with marked nuclear hyperchromasia, cellular crowding and pseudostratification. Mucin secretion may be a prominent feature.

Moderately differentiated lesions (Figure 14.4) exhibit greater architectural abnormality of the glands which are more complex than those found in the well-differentiated form of endocervical carcinoma. The cells lining the glands are more pleomorphic, although cells resembling normal endocervical cells can still be recognized (Figure 14.5). Mucin secretion may be much less marked than in well-differentiated tumours. In general, the less well-differentiated the carcinoma, the less evidence there is of mucin secretion.

The cells in *poorly differentiated endocervical adenocarcinoma* show a tendency to form a solid growth pattern and gland formation is rare (Figures 14.6 and 14.7). Individual cells are barely recognizable as endocervical and show marked pleomorphism with frequent mitotic figures. Mucin secretion can only rarely be seen.

(a) Minimal deviation adenocarcinoma ('adenoma malignum')

This uncommon variant of well-differentiated endocervical adenocarcinoma has a poor prognosis (Zhang, 1989). Zhang's view that it grows rapidly and spreads early could be related to delay in diagnosis; the clinical features described were enlargement of the cervix and vaginal discharge, but since four of the nine cases died from their carcinomas he considered that these may have been late symptoms. Kudo *et al.* (1990) reported that in three out of four cases there was a diagnostic delay of between 6 months and 5 years.

Histologically, the tumour is so highly

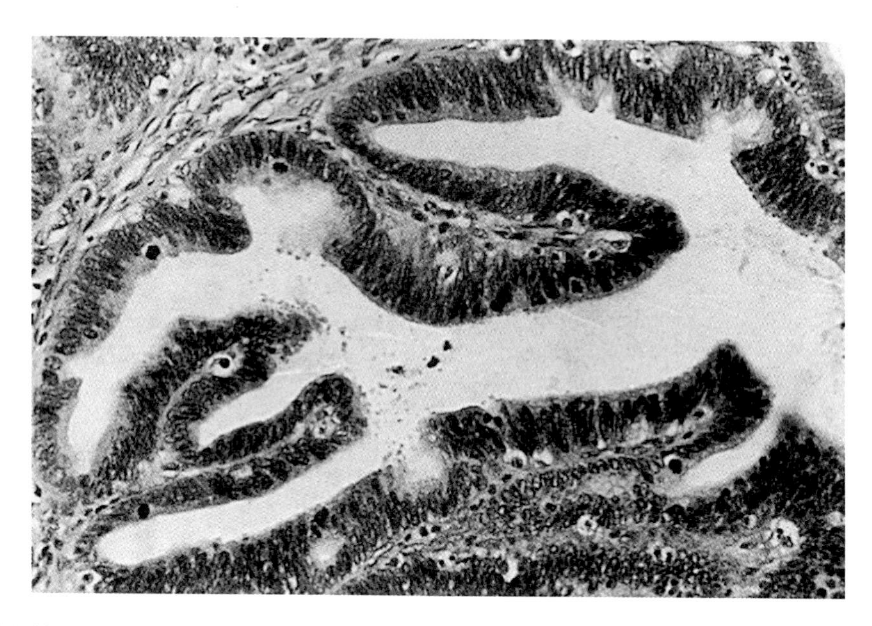

Figure 14.4 Moderately differentiated endocervical adenocarcinoma in a woman aged 50 years, showing greater complexity of glands (H & E, ×310).

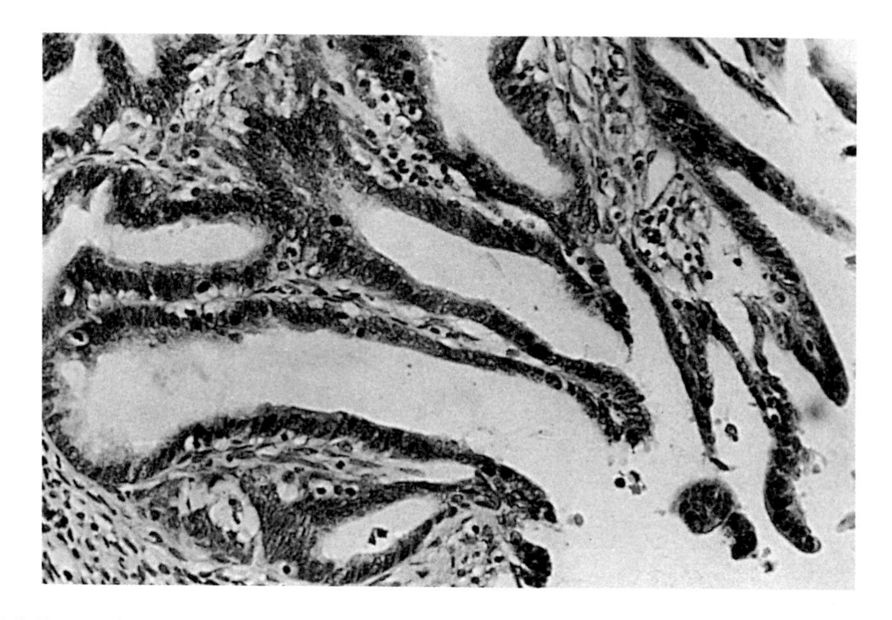

Figure 14.5 Focus of moderate differentiation in same adenocarcinoma as shown in Figure 14.4 (H & E, ×310).

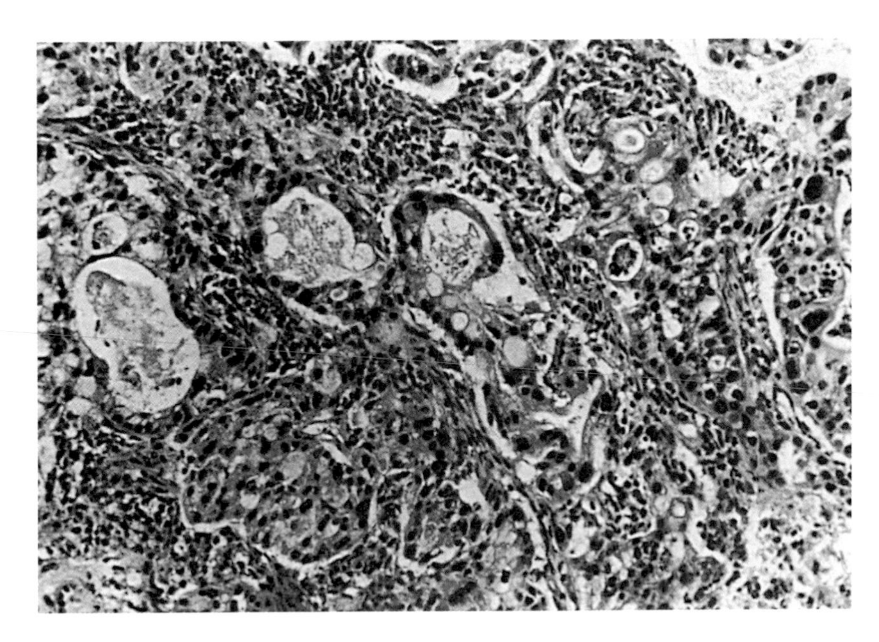

Figure 14.6 Poorly differentiated endocervical adenocarcinoma from a woman aged 51 years who died from carcinomatosis 5 months after hysterectomy. Gland formation and mucin secretion can still be identified in this field (H & E, ×240).

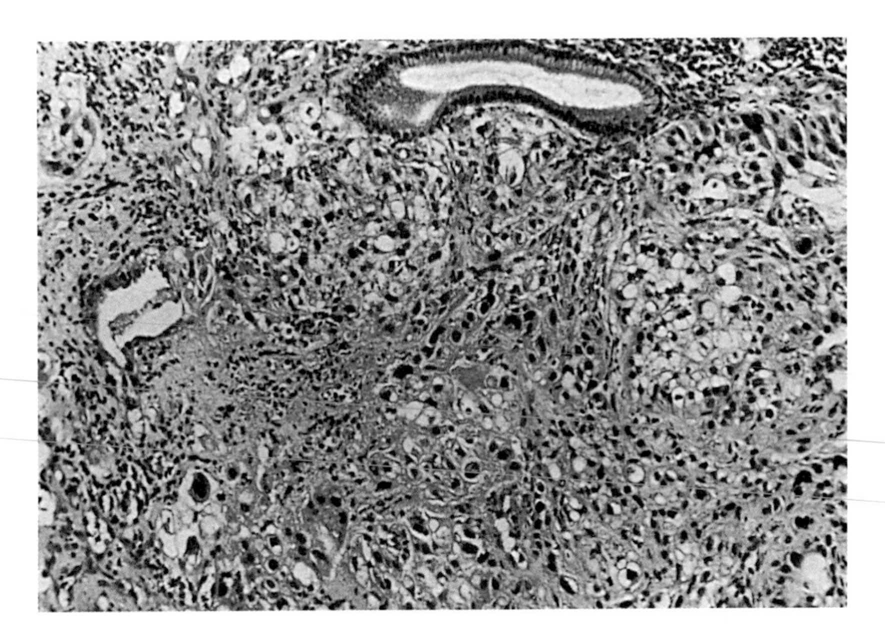

Figure 14.7 Poorly differentiated endocervical adenocarcinoma in which gland formation is scarcely recognizable. This is from the same section as that shown in Figure 14.6. Note normal endocervical gland at top of picture (H & E, ×190).

differentiated that the glandular architecture and cellular morphology is barely distinguishable from normal (Figure 14.8). There are however some features which should assist in its recognition. One is the presence of occasional mitotic figures in the cells lining the glands (Figures 14.9 and 14.10); another is the depth to which the glands extend into the cervical fibromuscular tissue. Normally, this depth is less than half the thickness of the cervical wall. In minimal deviation adenocarcinoma, the depth often exceeds two-thirds of the cervical wall (Figure 14.11). Other features are the presence of abnormally shaped glands (Figure 14.11), papillary process formation with enlargement of the columnar epithelial cells (Figure 14.12) and a tendency to bud into the stroma. Yamagata *et al.* (1991) described ballooned glands which ruptured with leakage of mucous material into the stroma. Gilks *et al.* (1989) found minor foci of less well-differentiated tumour in 58% (15/26) of cases. They also reported invasion of the cervical wall, an oedematous or desmoplastic stroma, vascular and perineural invasion (probably late in the natural history) and positive staining for CEA. Ultrastructurally, Kudo *et al.* (1990) noted intestinal metaplastic cells containing both secretory granules and microvilli with core filaments and rootlets, indicating a disorder of differentiation.

In view of the difficulty in making a histological diagnosis on a lesion which so closely resembles normal tissue, Bulmer *et al.* (1990) recommended the use of mucin histochemistry and immunohistochemistry. Their most consistent finding was cytoplasmic and luminal reactivity with antibodies to epithelial membrane antigens (Ep1 and HMFG 1) while normal cervix showed luminal labelling only. They also confirmed the report of Bates *et al.* (1985) who found that the sialomucin content of the tumour cells of adenoma malignum is increased compared with that of normal endocervical epithelium. Of clinical relevance is the occasional association of this tumour with Peutz–Jegher's syndrome (Gilks *et al.*, 1989; Podczaski *et al.*, 1991).

Although Silverberg and Hurt (1975) considered that adenoma malignum should have

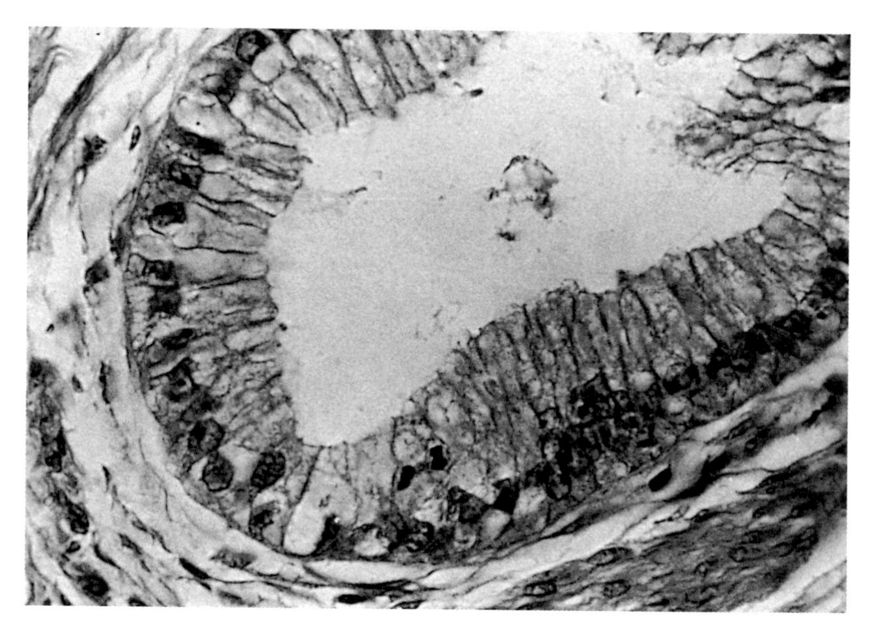

Figure 14.8 Minimal deviation adenocarcinoma ('adenoma malignum'). Essentially normal gland structure with minimal deviation of cell structure from normal. Same lesion as shown in Figures 14.9–14.12 (H & E, ×770).

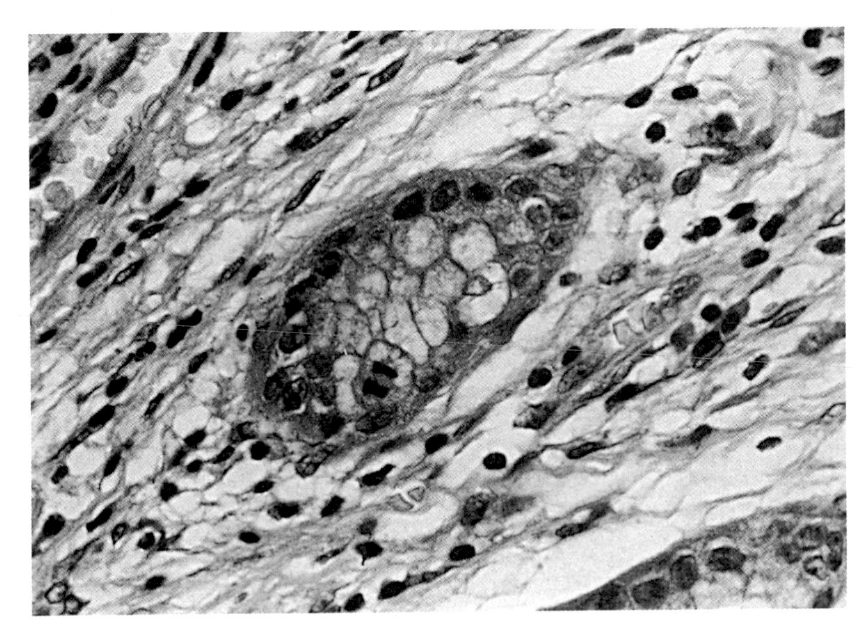

Figure 14.9 Minimal deviation adenocarcinoma. Small gland with mitotic figure situated on luminal aspect (H & E, ×770).

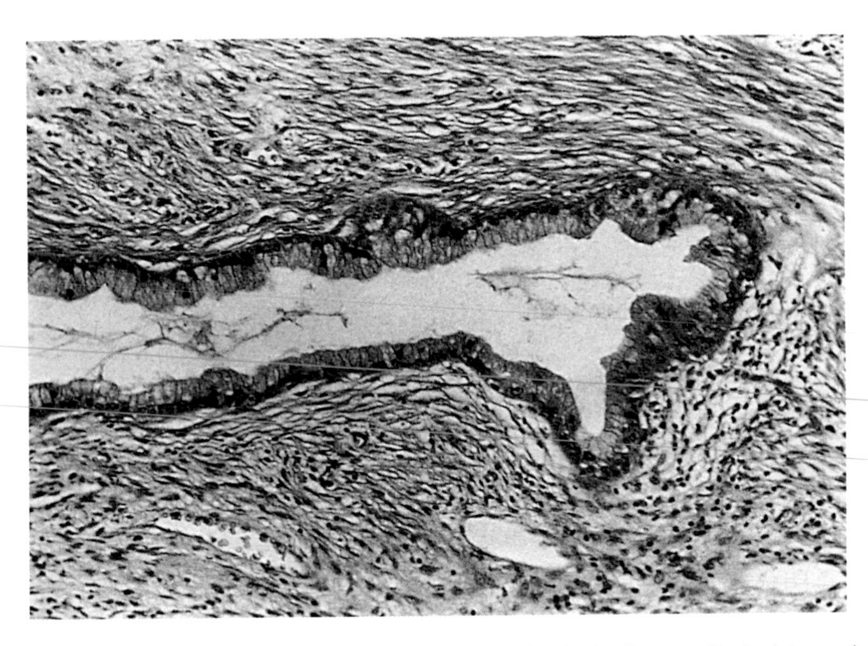

Figure 14.10 Minimal deviation adenocarcinoma with several mitotic figures situated towards the luminal surface (H & E, ×190).

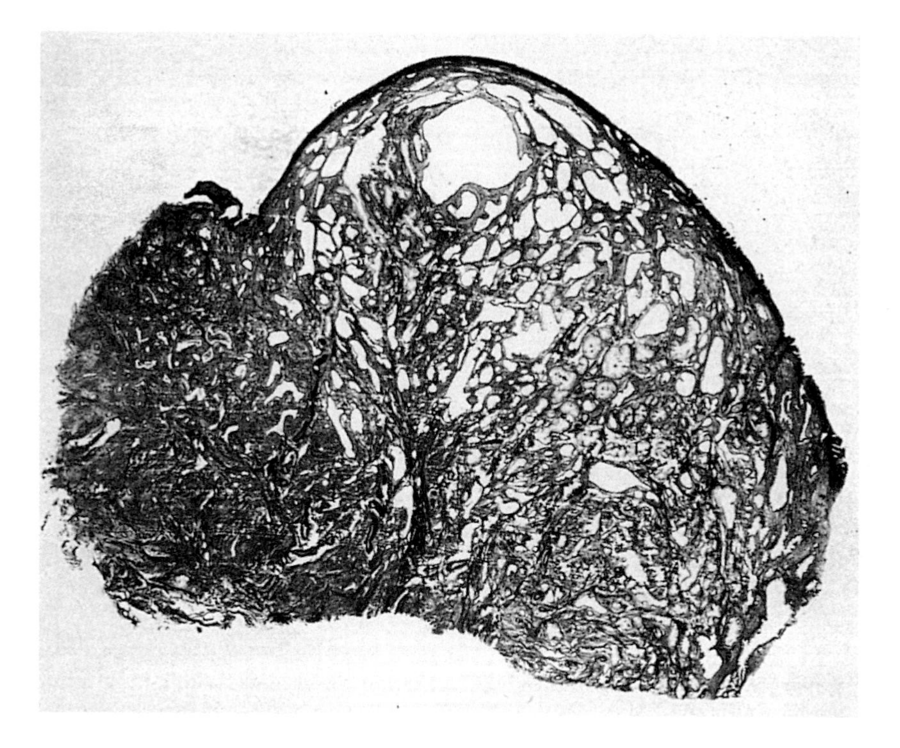

Figure 14.11 Minimal deviation adenocarcinoma of cervix extending deeply into the cervical wall (H & E, ×6).

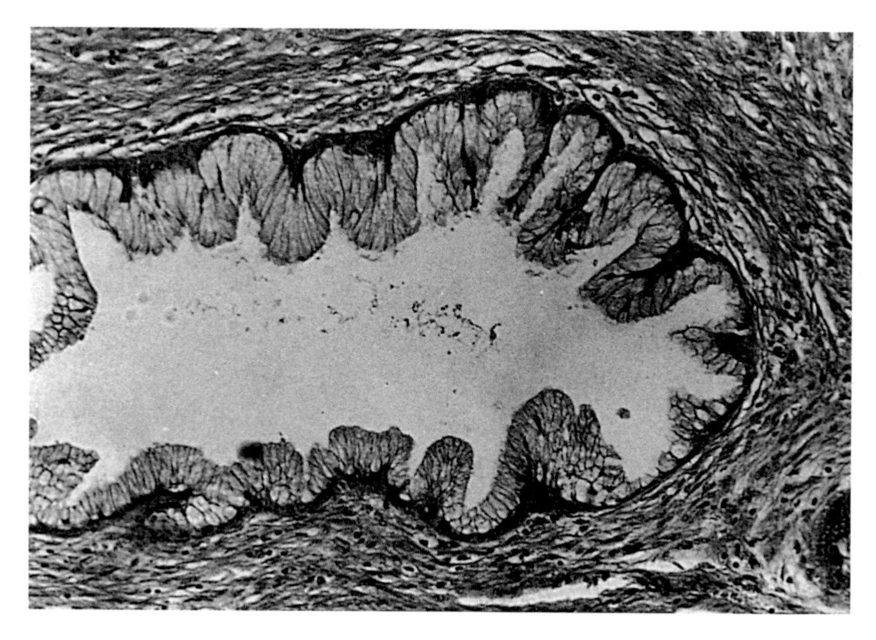

Figure 14.12 Minimal deviation adenocarcinoma. Gland containing papillary processes with elongated gland cells (H & E, ×190).

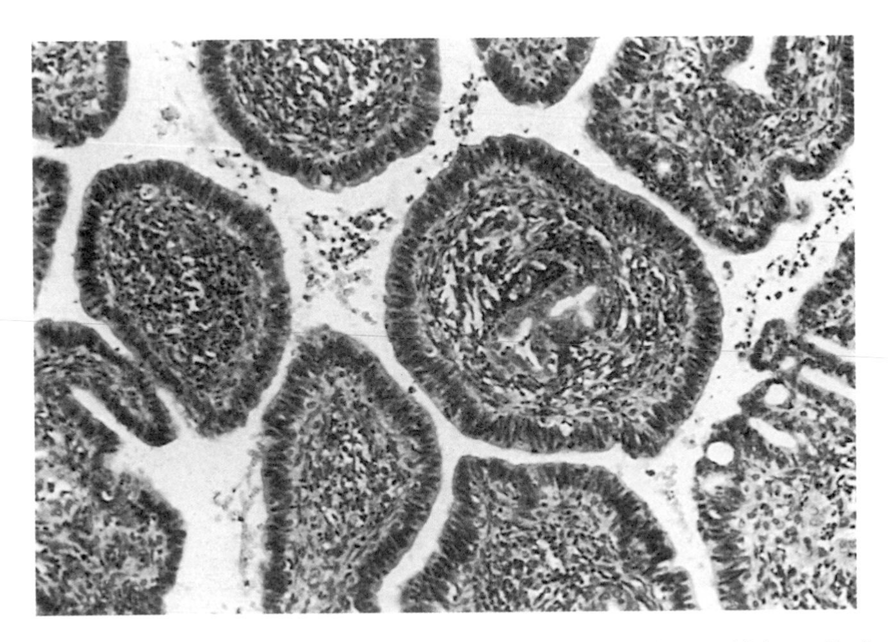

Figure 14.13 Well-differentiated papillary adenocarcinoma showing resemblance of lining epithelium to that of AIS. Note fibrous tissue core of the papillae. Same lesion as shown in Figure 14.14. The similarity of this tumour to papillary cervicitis is shown in Figure 8.8 (H & E. ×310).

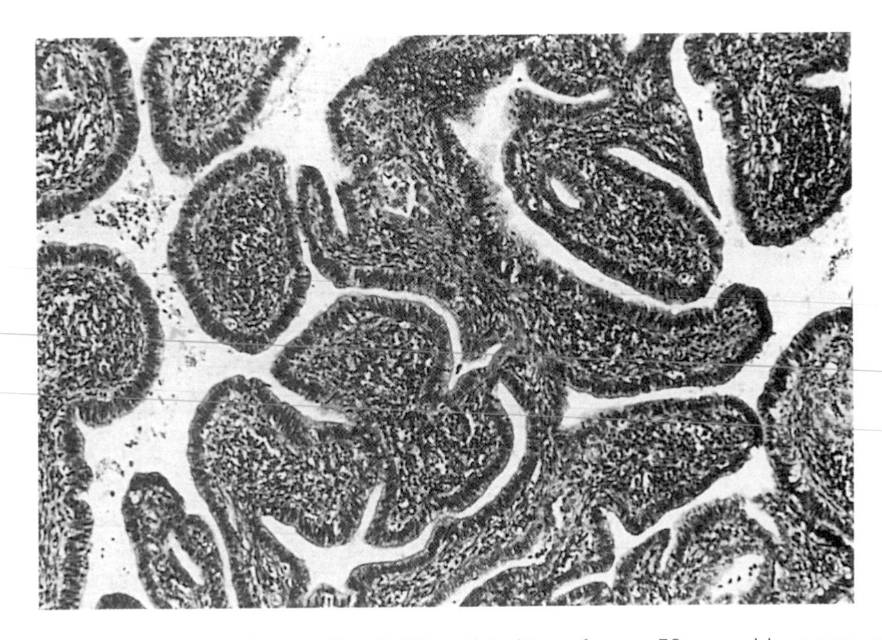

Figure 14.14 Papillary adenocarcinoma of well-differentiated type from a 50-year-old woman, arising in the endocervical canal close to the squamocolumnar junction (H & E, ×120).

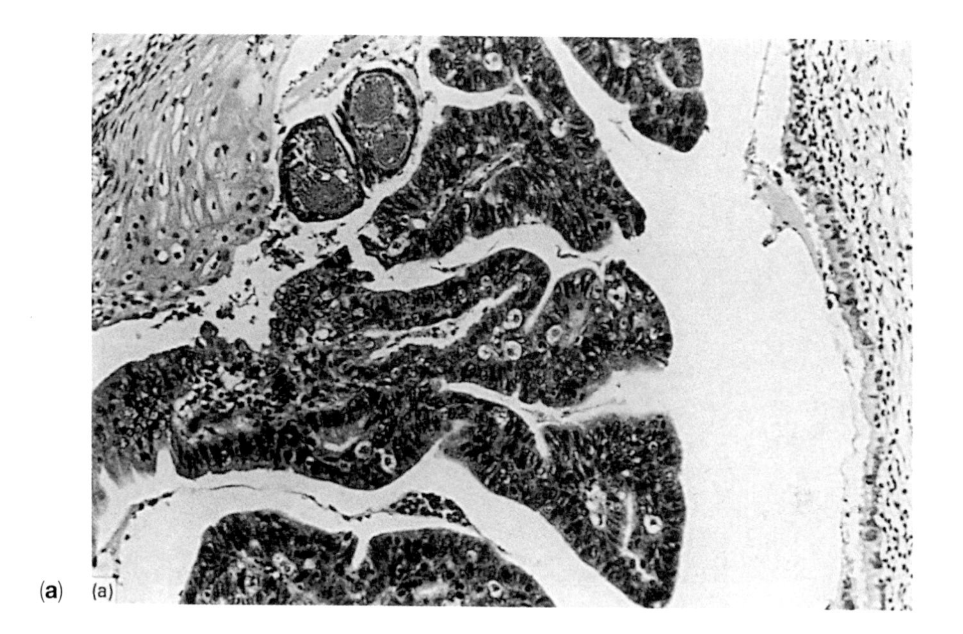

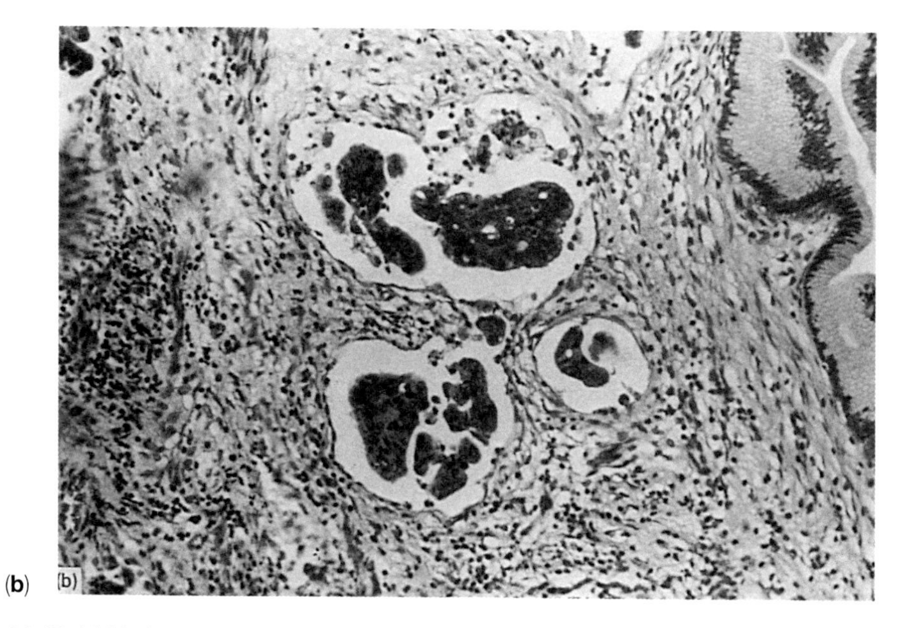

Figure 14.15 (**a**) Moderately differentiated papillary adenocarcinoma arising in a zone of MEH at the squamo-columnar junction in a woman aged 44 years. Although the tumour was quite small and apparently superficial, there was much lymphatic permeation in the cone biopsy specimen, shown in (**b**), and she died from carcinomatosis 2.5 years after radiotherapy and hysterectomy (H & E, ×190). (**b**) Lymphatic permeation by papillary adenocarcinoma shown in (**a**) (H & E, ×190).

a prognosis comparable with that of other well-differentiated adenocarcinomas, it is clear from the reports quoted above that the tumour has a high mortality rate which reflects the difficulty in early clinical and histological recognition.

Cytologically, it is extremely difficult to distinguish the cells of adenoma malignum from normal endocervical cells, although Kudo *et al.* (1990) state that the nuclei are somewhat more irregular in size and shape than normal columnar cells, with multilayered cell clusters arranged as honey combs, palisades or sheets with glandular openings.

(b) Villoglandular papillary adenocarcinoma

This is another variant of endocervical carcinoma which arises from the surface epithelium of the endocervix, usually at the squamocolumnar junction, and may be mistaken microscopically for a papillary erosion (Figure 8.8). It is composed of papillary fronds, each with a connective tissue core. Occasionally, psammoma bodies are present (Seltzer and Spitzer, 1983). Well-differentiated (Figures 14.13 and 14.14), moderately differentiated (Figure 14.15(a,b)), and poorly differentiated forms of the tumour may be seen.

The *well-differentiated* form is now recognized as a distinct histological entity which develops in young women (aged 23–54 years) and has an excellent prognosis with little likelihood of recurrence after excision (Young and Scully, 1989; Jones *et al.*, 1993) or of lymph node involvement (Hopson *et al.*, 1990). Characteristic features are its papillary exophytic structure and mild cytological atypia. Although invasion is usually present at the base of the tumour, the papillary component shows no desmoplastic stromal response and no evidence of infiltration. The epithelial cells lining the papillary processes resemble adenocarcinoma *in situ* (Fig 14.13) and often show pseudostratification. The deeper portions of the tumour are composed of branching tubular glands separated by a fibrous or fibromatous stroma which is fairly sharply demarcated from the adjacent cervical stroma. This tumour with its low grade of malignancy must be distinguished from papillary cervicitis (Figure 8.8), from less well-differentiated forms of papillary adenocarcinoma (Figure 14.15) which have a much worse prognosis, and also from typical endocervical adenocarcinoma with a minor papillary component.

14.1.2 Intestinal type of mucinous adenocarcinoma

This variant of mucinous adenocarcinoma is characterized by cells resembling carcinoma of the colon (Scully *et al.*, 1994), the cells having prominent brush borders, with many goblet cells and occasional Paneth cells and argentaffin cells. An adenocarcinoma *in situ* with similar features has been described by Lee and Trainer (1990). A typical feature of the invasive tumour is mucin production which may be so profuse that it is visible to the naked eye in the gross specimen. Microscopically, the tumour is composed of glands lined by mucin-producing cells which may show a high degree of pleomorphism (Figure 14.16). The glands may be so dilated with mucin that the cellular lining may be partially obscured (Figure 14.17). Occasionally, the mucin may spill over into the stroma to produce the pattern of 'myxoma cervix'. The tumour cells may contain little intracellular mucin, as in Figure 14.17, and show some pseudostratification (Wright *et al.*, 1994), the most reliable marker for intestinal differentiation is the demonstration of O-acetylated sialomucin (Fox *et al.*, 1988).

(a) Signet ring cell type

This is an undifferentiated mucocellular adenocarcinoma. The mucin, instead of being secreted outside the cell, as in the intestinal type, is relatively small in amount and retained within the cell wall, pushing the nucleus to one

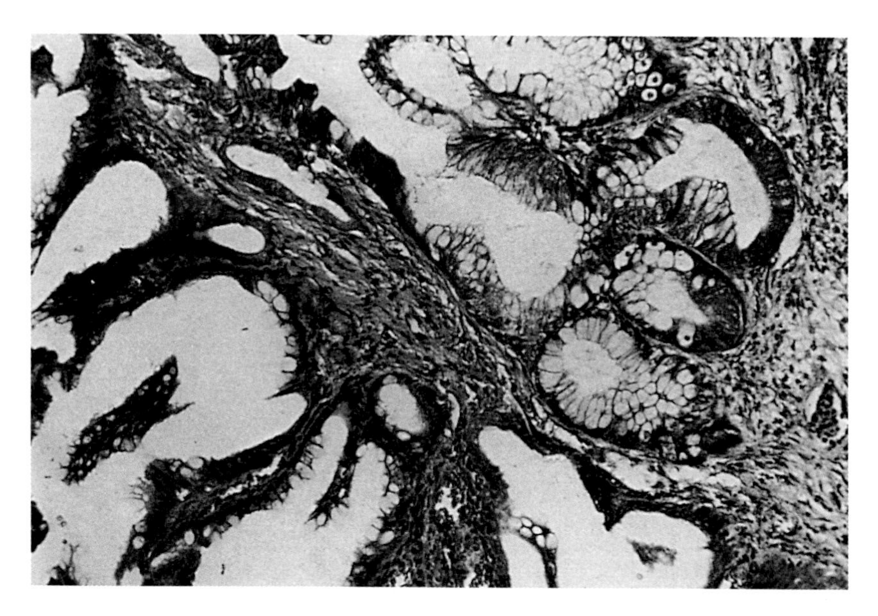

Figure 14.16 Mucinous variant of endocervical adenocarcinoma (intestinal type), moderately well-differentiated in a woman aged 31 years who died one year after hysterectomy and radiotherapy. Part of the tumour had an endometrioid structure (H & E, ×190).

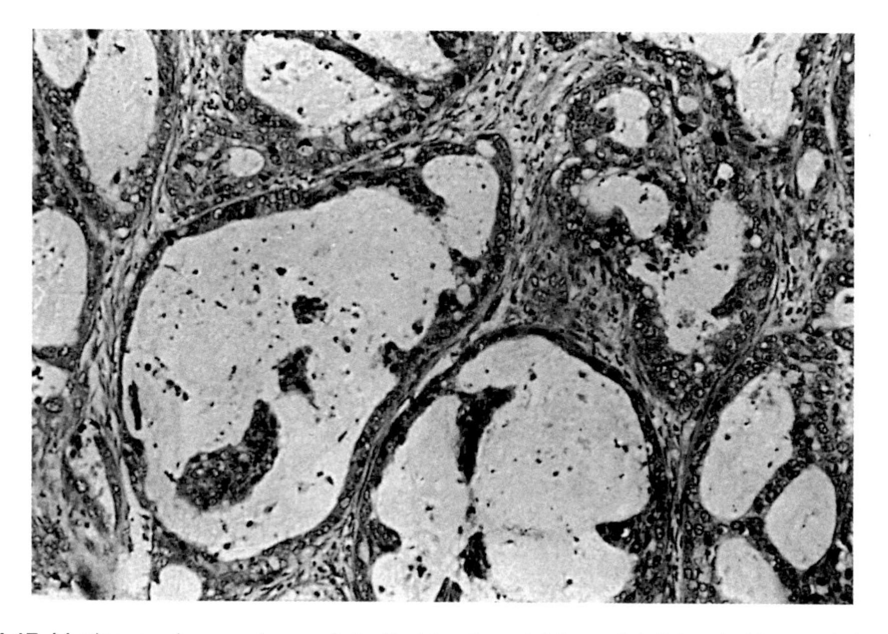

Figure 14.17 Mucinous adenocarcinoma (intestinal type) containing acini distended by mucin in which there appear to be islands of carcinoma tissue, from a woman aged 52 years, alive and well 2 years after hysterectomy and radiotherapy. Part of the tumour is adenosquamous carcinoma, shown in Figure 15.2 (H & E, ×190).

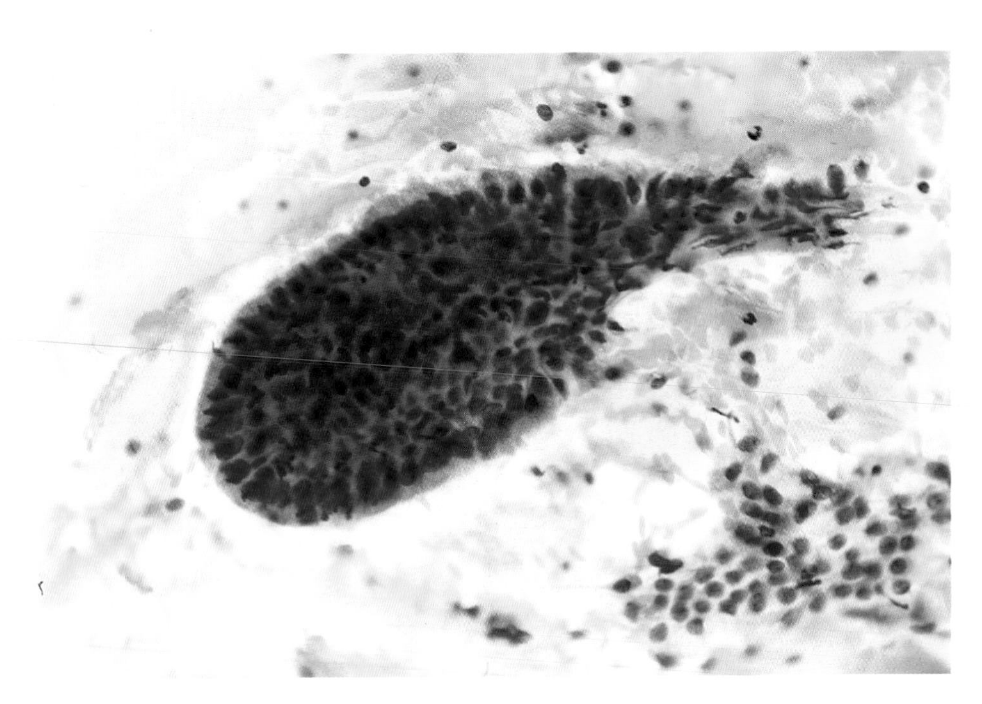

Figure 14.18 Cell clusters from well-differentiated invasive endocervical carcinoma. Note columnar shape of cells at edge of cluster (Pap, ×400).

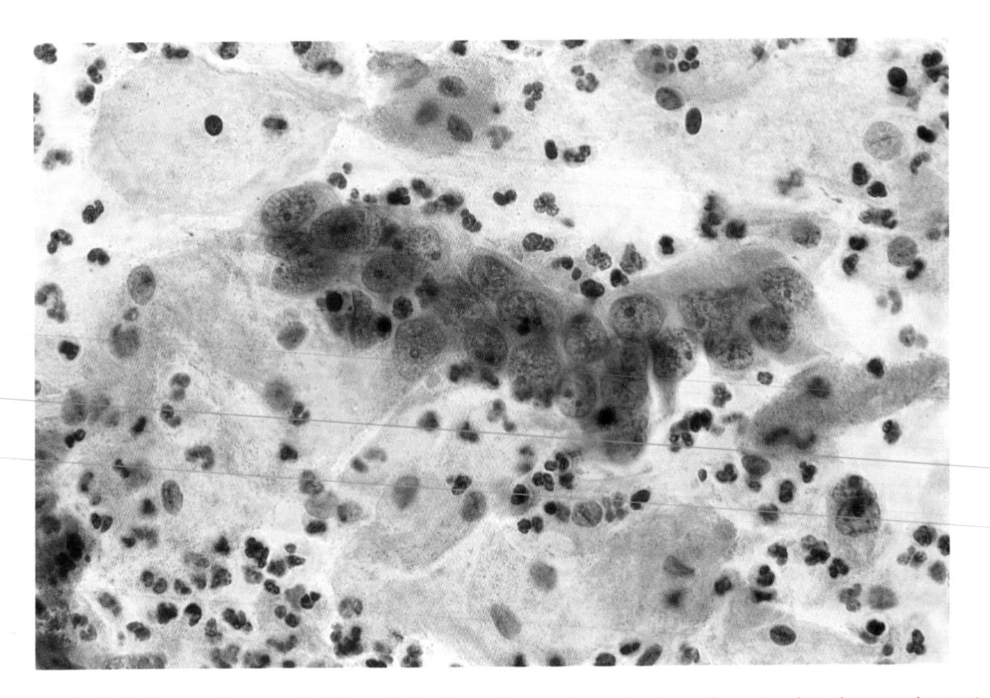

Figure 14.19 Cell clusters from well-differentiated invasive endocervical carcinoma, showing tendency to palisade formation (Pap, ×630).

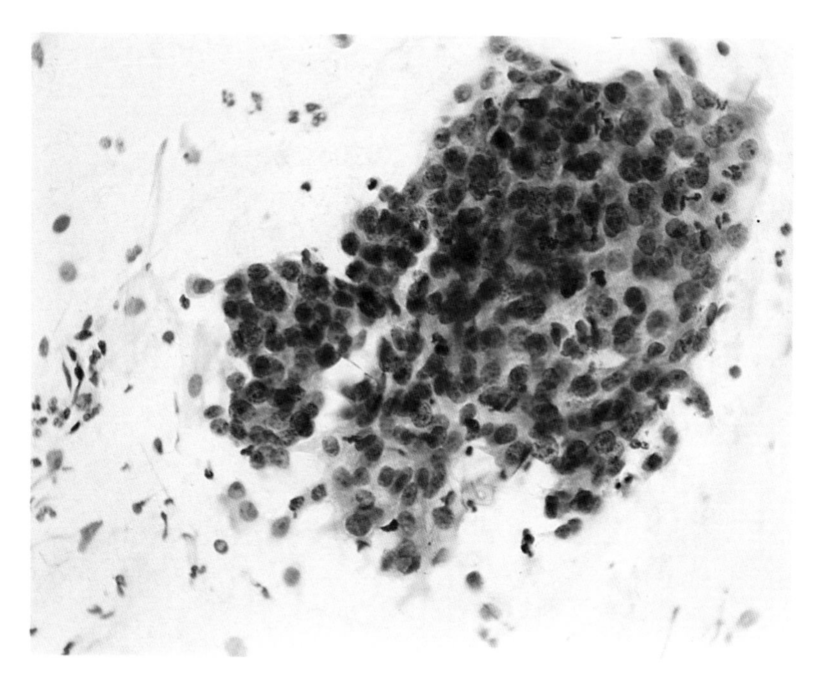

Figure 14.20 Dense cluster of cells with hyperchromatic nuclei from well-differentiated invasive endocervical carcinoma. Such a group on its own would be difficult to classify, although the rounded shape of the cluster suggests glandular origin (Pap, ×400).

side, giving the cell the microscopic appearance of a signet ring. The recognition of this lesion is greatly assisted by the use of mucin stains, such as alcian blue or mucicarmine which cause the tumour cells to stand out clearly, even when they are relatively sparse. Signet ring cell carcinoma rarely occurs in pure form and is usually associated with intestinal or endocervical adenocarcinoma (Wright *et al.*, 1994). Like most undifferentiated cancers, mucocellular carcinoma readily metastasizes, unless removed at an early stage.

14.1.3 Cytology of invasive endocervical adenocarcinoma

The cytological changes in smears from early invasive endocervical adenocarcinoma are often indistinguishable from those found in AIS (Section 13.1.2). In more advanced lesions the smears are often heavily bloodstained. In *well*-differentiated lesions, clusters of elongated cells, with an obvious columnar shape are usually identifiable (Figures 13.21, 14.18 and 14.19), although in some areas of the smears the nuclei are sometimes so densely packed and hyperchromatic that the glandular origin of the cluster is not obvious (Figure 14.20). In other clusters, there is often a tendency to gland formation or vacuolation (Figures 14.21 and 14.22) which makes tumour typing possible. Discrete tumour cells may be recognized as endocervical by their delicate vacuolated cytoplasm, vesicular eccentric nuclei and large nucleoli (Figure 14.23). Malignant cells shed from *moderately differentiated* (Figure 14.24) and *poorly differentiated endocervical carcinoma* (Figures 14.25 and 14.26) show more pleomorphism than the well-differentiated form. The cells may have huge nuclei, delicate chromatin structure, large nucleoli and

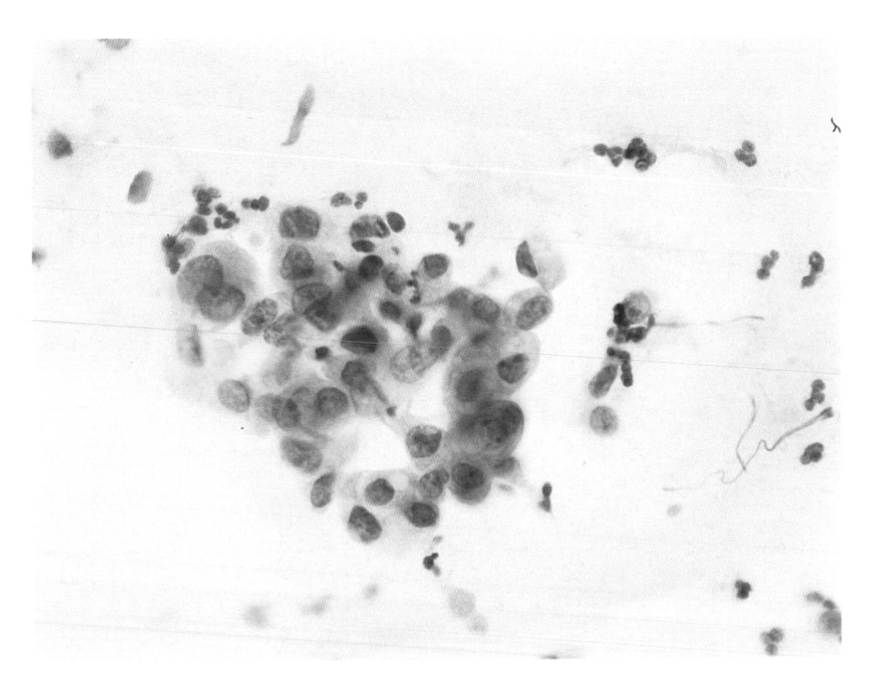

Figure 14.21 Cluster of cells from well-differentiated invasive endocervical carcinoma showing nuclear hyperchromasia and tendency to gland formation (Pap, ×630).

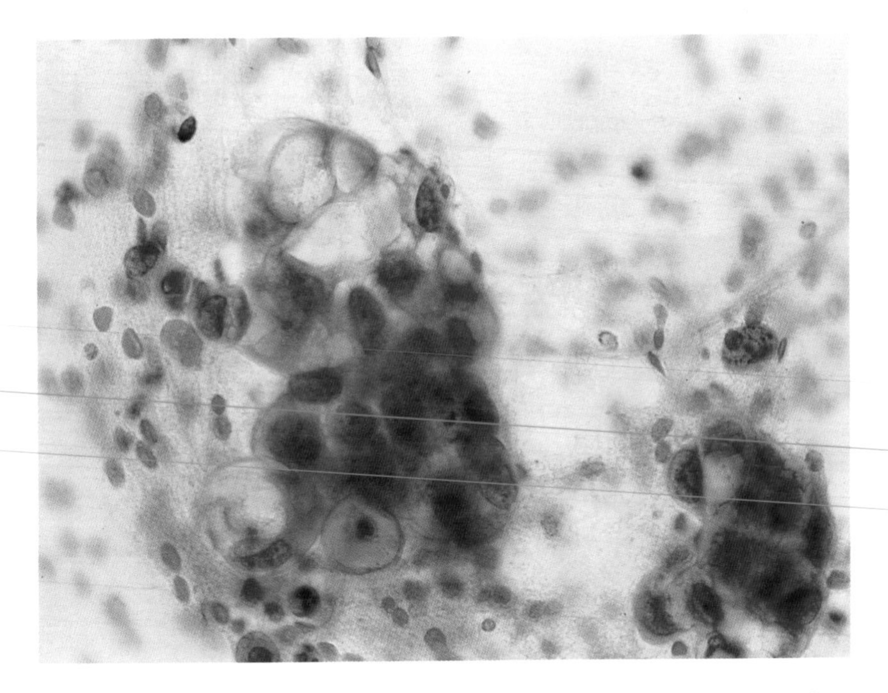

Figure 14.22 Cluster of carcinoma cells showing vacuolated cytoplasm characteristic of mucus-secreting adenocarcinoma (Pap, ×630).

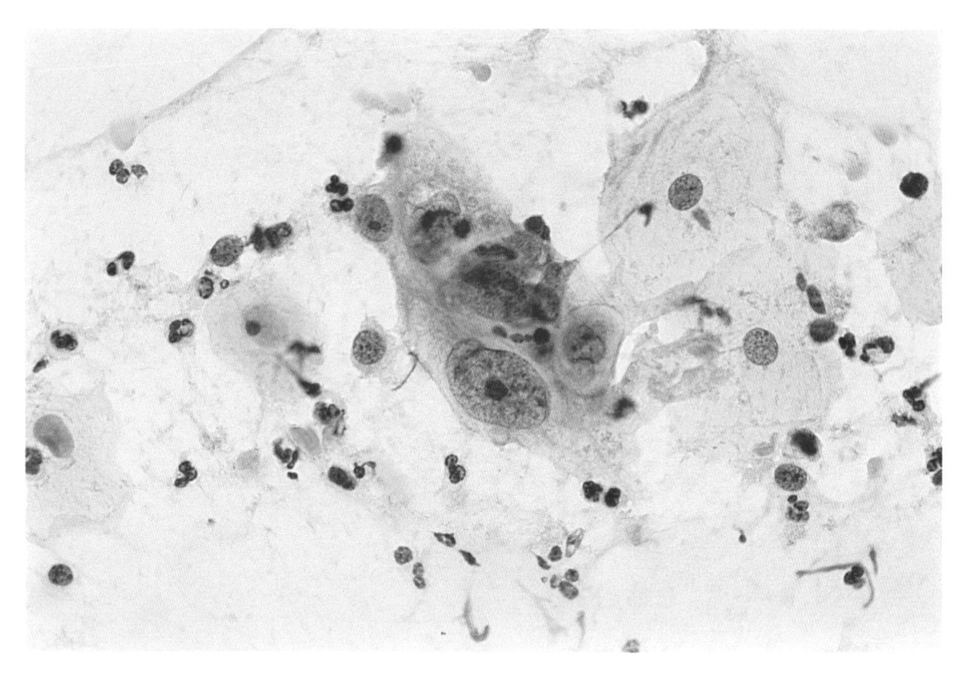

Figure 14.23 Small cluster of malignant cells from the same case as Figure 14.19. Note anisonuleosis, delicate vacuolated cytoplasm and large nucleoli (Pap, ×630).

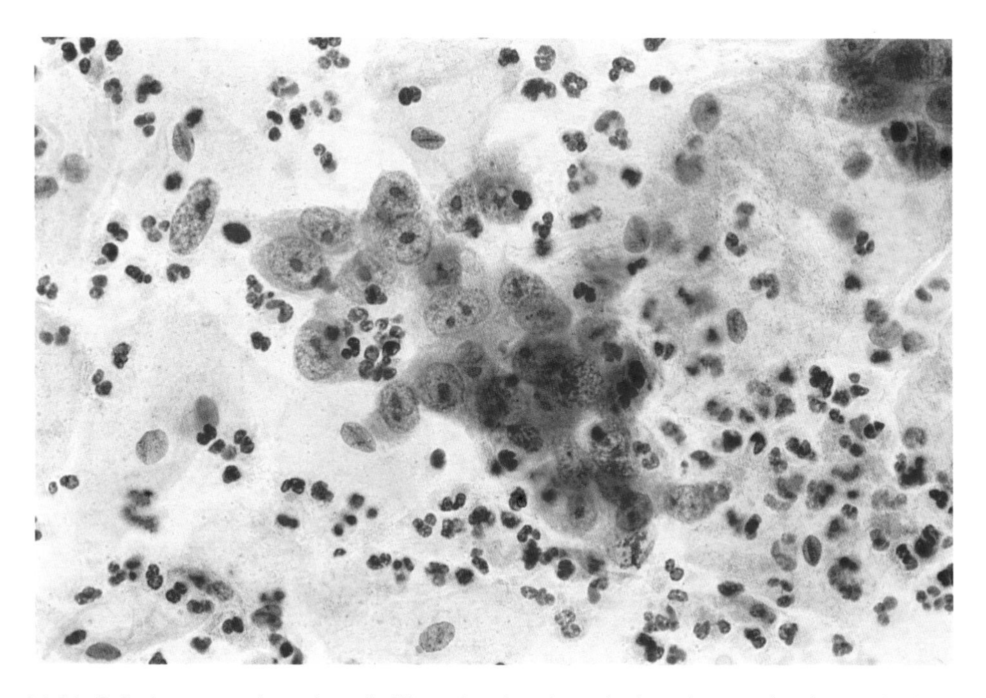

Figure 14.24 Cells from a moderately well-differentiated endocervical carcinoma showing marked anisonucleosis and prominent nucleoli (Pap, ×630).

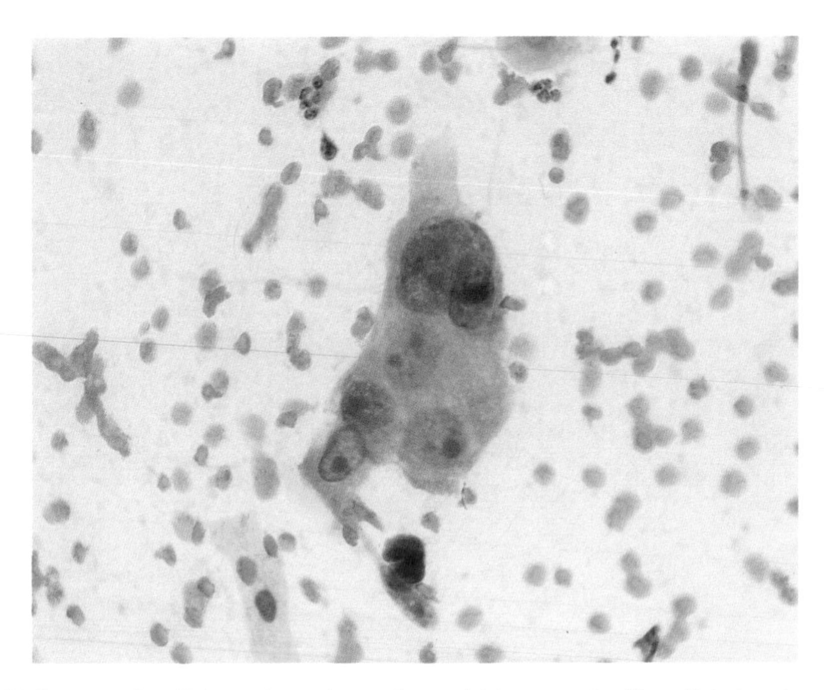

Figure 14.25 Malignant cells with large, hypochromatic, nuclei from poorly differentiated endocervical adenocarcinoma (Pap, ×630).

Figure 14.26 Malignant cell tissue fragments composed of densely packed nuclei with disorganised structure. Poorly differentiated endocervical carcinoma (Pap, ×630).

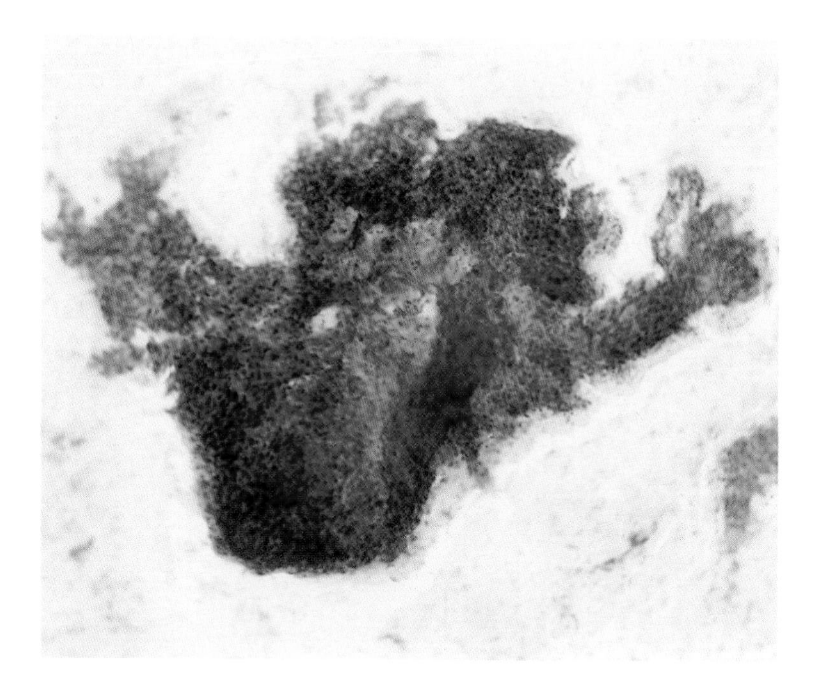

Figure 14.27 Tissue fragment from papillary adenocarcinoma (Pap, ×630).

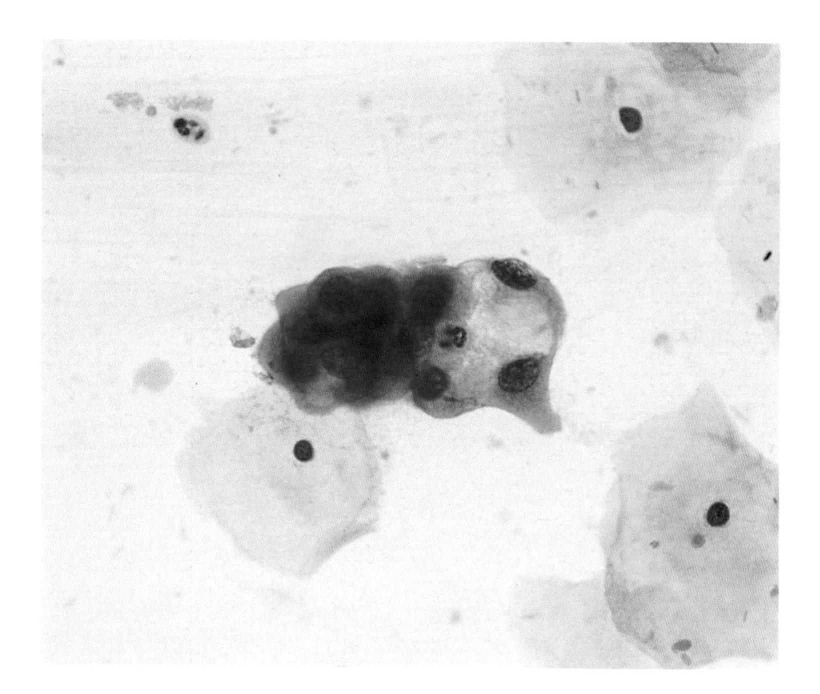

Figure 14.28 Discrete abnormal endometrial cells in smear from a patient with post-menopausal bleeding. An endometrial adenocarcinoma was diagnosed on curettings (Pap, ×630).

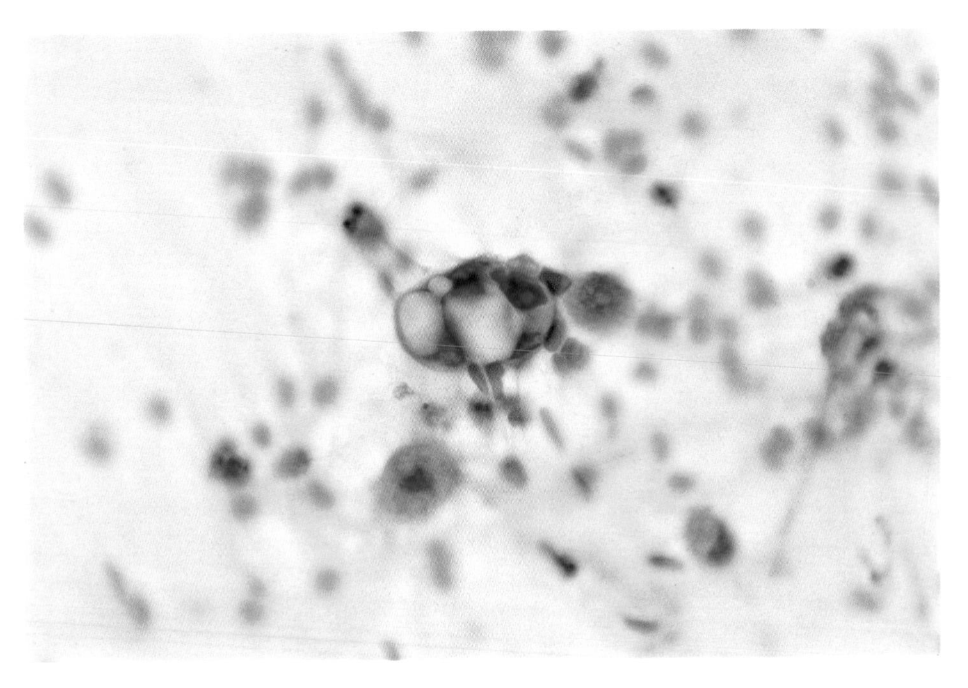

Figure 14.29 Abnormal vacuolated cells among parabasal cells in an atrophic smear. An endometrial adenocarcinoma was diagnosed on curettings (Pap, ×630).

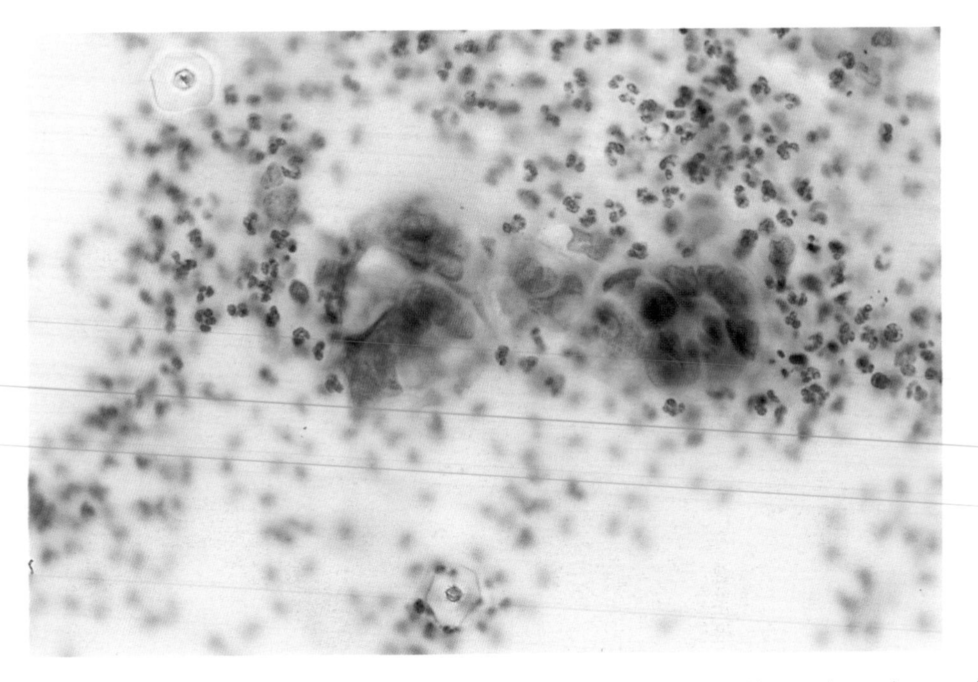

Figure 14.30 Necrotic tumour fragment in cervical smear from patient with ovarian adenocarcinoma (Pap, ×630).

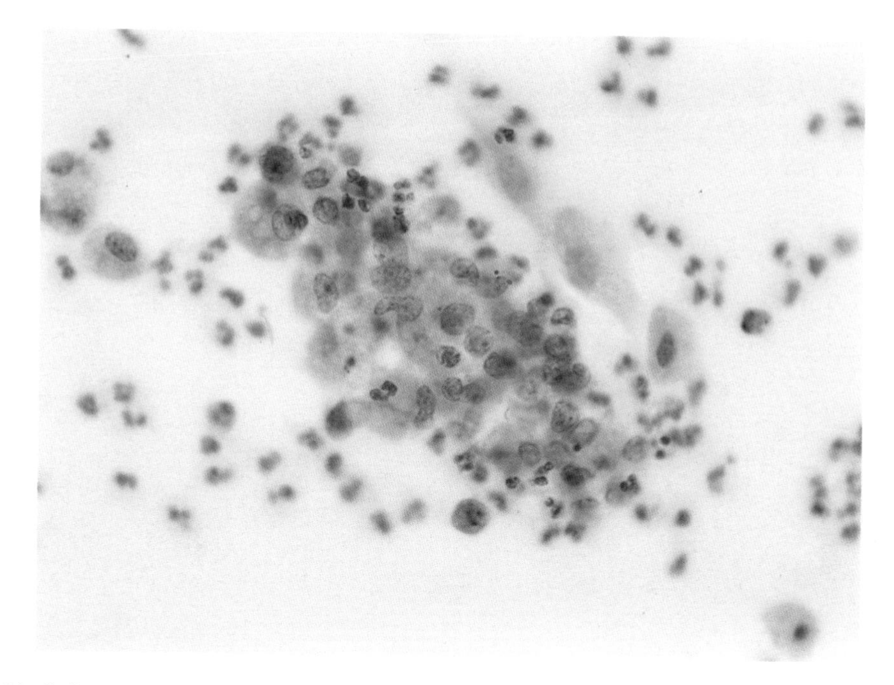

Figure 14.31 Collections of tumour cells in cervical smear from patient with endometrial adenocarcinoma (Pap, ×630).

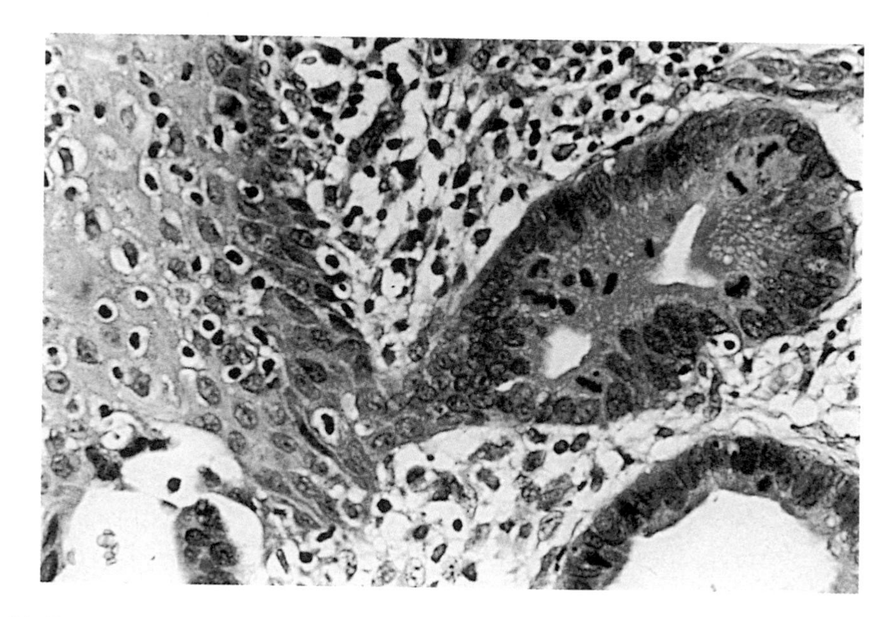

Figure 14.32 Endometrioid adenocarcinoma (minimal deviation) showing location at squamocolumnar junction. Same case as Figure 14.33 (H & E, ×480).

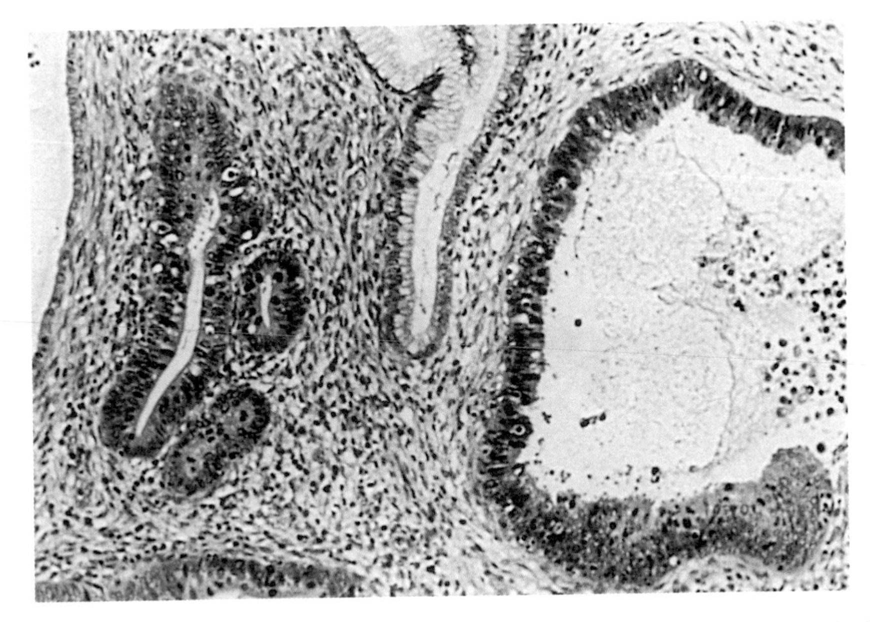

Figure 14.33 Minimal deviation endometrioid adenocarcinoma. The extent of the field change in this case was indicative of an invasive lesion (H & E, ×190).

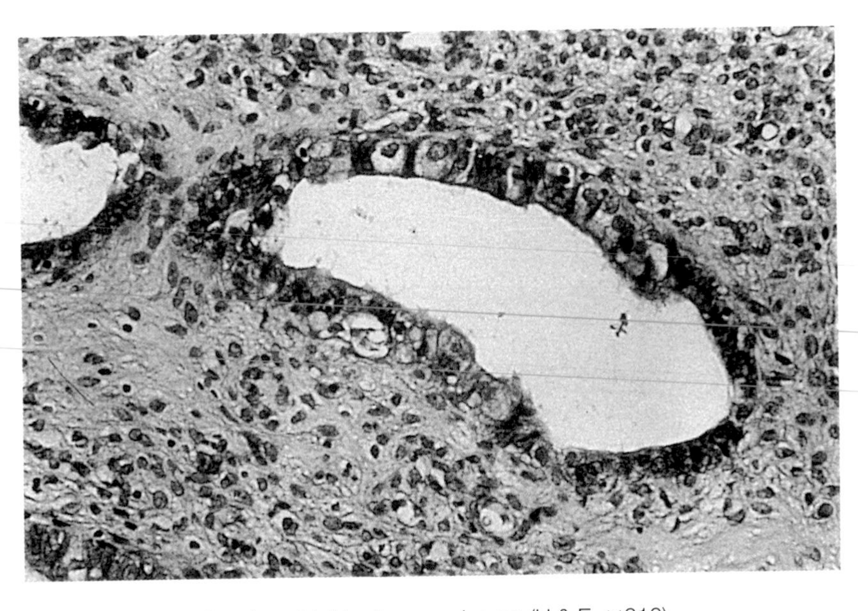

Figure 14.34 Well-differentiated endometrioid adenocarcinoma (H & E, ×310).

vacuolated cytoplasm (Figures 14.24 and 14.25), which distinguish them from squamous carcinoma cells. In very poorly differentiated lesions, the nuclei may be small and so densely packed that they mould into each other (Figure 14.26). Mucus secretion is not evident and the glandular origin of such a tumour would be difficult to ascertain cytologically. Villoglandular *papillary carcinoma* may be suspected if papillary clusters are found in the smear (Figure 14.27). Occasionally, a papillary frond with a connective tissue core may be seen. Ayer *et al.* (1991) predicted 70 cases of invasive adenocarcinoma cytologically, classifying them into well- and poorly differentiated adenocarcinomas along the lines described above. Close correlation with the histological findings were achieved.

14.2 ENDOMETRIOID ADENOCARCINOMA

This primary adenocarcinoma of the cervix has the histological characteristics of endometrial adenocarcinoma and the diagnosis will depend on the exclusion of a primary endometrial lesion. This distinction is difficult to make on a biopsy and diagnosis should be based on examination of a hysterectomy specimen (Poulsen *et al.*, 1975). Location of the tumour at the squamocolumnar junction is characteristic (Figure 14.32). Some examples of endometrioid carcinoma probably arise from an area of endometriosis in the cervix. Others may arise from cervical glands which have undergone endometrioid metaplasia (Section 9.3.2).

Four grades of endometrioid adenocarcinoma are recognized: minimal deviation, well-differentiated, moderately differentiated, and poorly differentiated. Examples of each are shown in Figures 14.32–14.36. The grading depends on the degree of resemblance of the tumour to normal endometrium. A well-differentiated adenocarcinoma is readily recognized as a tumour, although its glands resemble normal endometrium; a minimal deviation adenocarcinoma (adenoma malignum) mimics normal tissue and only careful scrutiny reveals it to be a tumour (Section 14.1.1(a)) In a series of five cases of minimal deviation endometrioid adenocarcinoma, Young and Scully (1993) reported that cervical neoplasia was recognized before biopsy in only one case. This patient had a cervical smear before biopsy which contained abnormal glandular epithelial cells; a comparable example is shown in Figure 14.37. Histological features that assist recognition are the shape, size and number of glands and their distribution deep in the cervical stroma (Figures 14.32 and 14.33).

Care must be taken to distinguish between endometrioid carcinoma arising in the cervix from a similar tumour arising in the uterus and extending into the cervix. With decreasing differentiation, it becomes increasingly difficult to identify the lesion as endometrioid although even a poorly differentiated lesion usually contains a more differentiated portion enabling correct identification to be made. Squamous metaplasia is a common finding in these tumours. The squamous cells often lack malignant features (Figure 14.38), the lesion then being known as adenoacanthoma. Although it has been the custom to consider such a lesion as being a separate and much less malignant tumour than adenosquamous carcinoma (Barrowclough and Jaarsma, 1980) it now appears probable that there are all gradations between the two. For convenience, adenosquamous carcinoma is considered separately (Section 15.1). However, it should be stressed that mixtures of one type of carcinoma with another are not uncommon in the cervix and adequate sampling is essential.

14.2.1 Distinction between primary adenocarcinoma of the cervix and cells shed from a primary carcinoma of the endometrium, fallopian tubes or ovary

In a cervical smear this may be difficult. To some extent, a distinction can be made on the number of tumour cells in the smear. Numerous well-preserved tumour cells will suggest

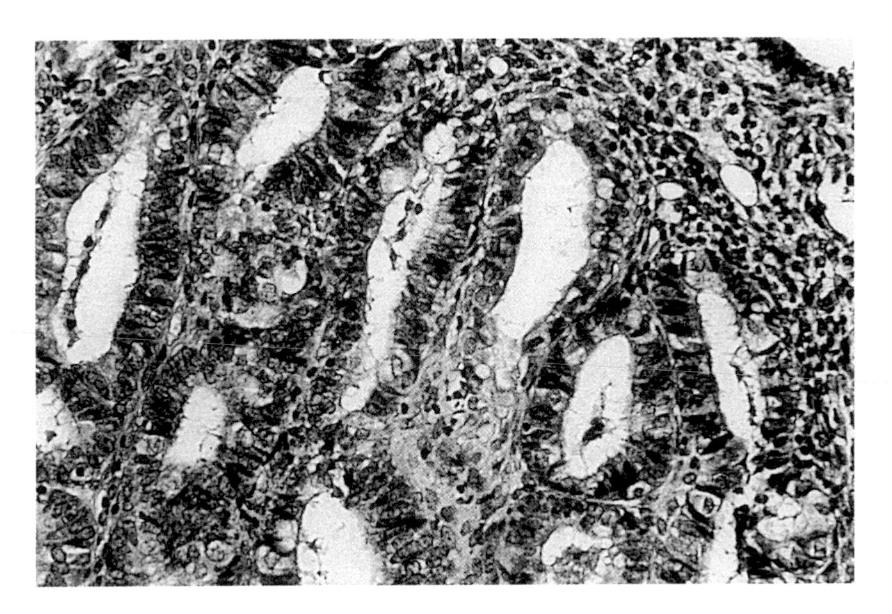

Figure 14.35 Moderately differentiated endometrioid adenocarcinoma from the same tumour as shown in Figure 14.34 (H & E, ×310).

Figure 14.36 Area of poorly differentiated endometrioid adenocarcinoma from the same tumour as shown in FIgure 14.34 (H & E, ×310).

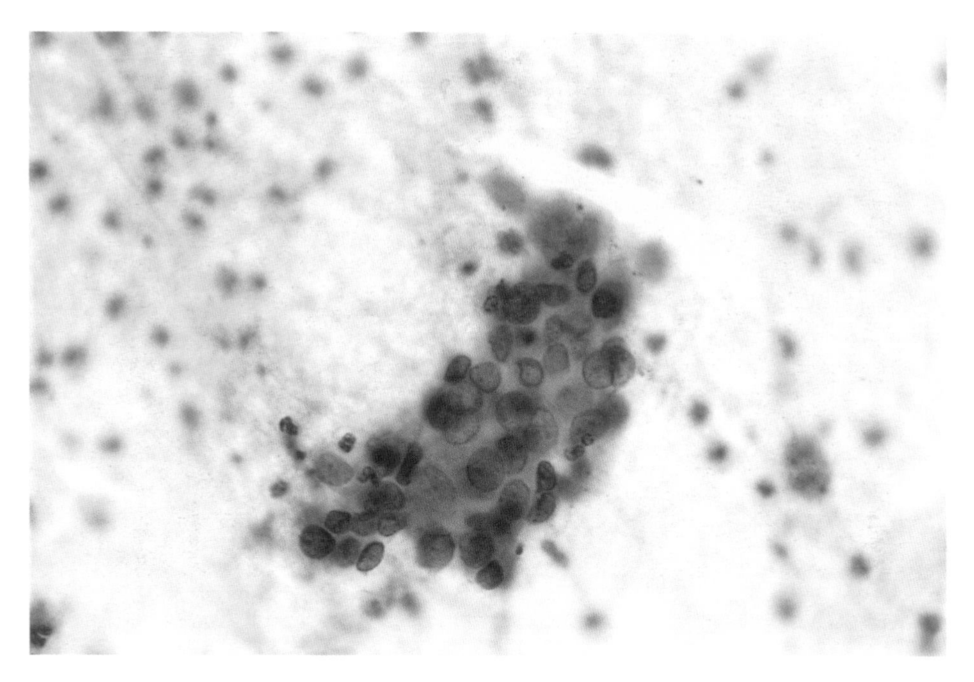

Figure 14.37 Cytological preparation from well-differentiated endometrioid adenocarcinoma shown in Figure 14.33. This can be recognized as glandular but probably not as endometrioid (Pap, ×630).

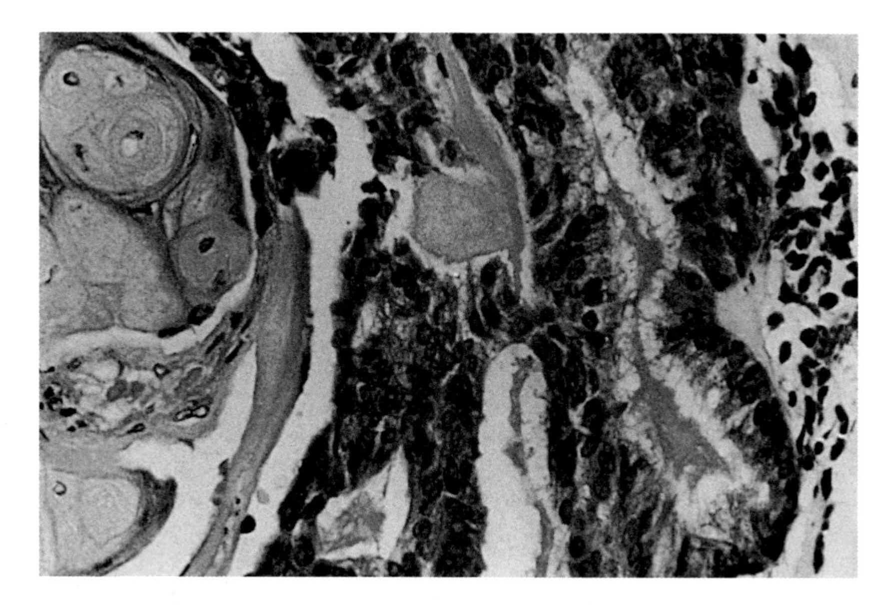

Figure 14.38 Squamous metaplasia in an endometrioid adenocarcinoma from a woman aged 69 years. The squamous cells appear to lack malignant features and such a lesion is often termed an adenoacanthoma (H & E, ×480).

the presence of cervical adenocarcinoma, whereas occasional discrete, poorly preserved glandular cells (Figures 14.28 and 14.29), scanty fragments of degenerating tumour tissue (Figure 14.30) and collections of necrotic cells in a smear that is otherwise free from blood and pus will suggest a tumour shed from the endometrial cavity or even from the ovary or fallopian tubes (Figures 14.30 and 14.31). The distinction between primary adenocarcinoma of the cervix and carcinoma of the breast, ovary, endometrium or other site metastatic to the cervix cannot be made with confidence on cytological grounds (Section 15.6).

14.3 CLEAR CELL ADENOCARCINOMA

This malignant tumour is an adenocarcinoma composed mainly of clear cells or hobnail cells arranged in solid, tubulocystic or papillary patterns or a combination of these patterns (Scully *et al.*, 1994). It was originally termed 'mesonephroid' because it was believed to be of mesonephric duct origin. There is now morphological and experimental evidence that the majority of clear cell tumours have in fact a müllerian (paramesonephric) origin (Forsberg, 1972; Puri *et al.*, 1977). Herbst *et al.* (1974) studied 170 such tumours in girls who had been exposed prenatally to diethylstilboestrol (DES); they showed that most tumours were superficially situated and appeared to be continuous with the endocervical epithelium. Of 55 clear cell carcinomas reported by Hanselaar *et al.* (1990), more than half had occurred in women exposed to DES *in utero*. These authors stress the importance of monitoring women at risk by palpation, colposcopy and both cervical and vaginal cytology in order to achieve early diagnosis.

Histologically, mesonephroid tumours are composed of cystic or tubular glands with a hobnail lining (Figure 14.39) with areas showing a papillary (Figure 14.40) solid or trabecular (Figure 14.41) structure. The 'clear cells' of a paramesonephric duct tumour are large and visible at fairly low magnification (Figures 14.42 and 14.43). Their cytoplasm contains abundant glycogen, giving a positive PAS reaction. The clear cells of a true mesonephric duct tumour are smaller and contain little glycogen and consequently their cytoplasm gives a negative PAS reaction although there is often PAS-positive material in the lumen of the tubules or cysts (Hart and Norris, 1972). Because the tumours of paramesonephric duct origin are superficially sited, the large, clear cells may be seen in cervical smears (Figures 14.44 and 14.45). Cells from the more deeply situated mesonephric duct tumours are unlikely to be exfoliated.

A link between the development of clear cell carcinoma of the cervix and vagina in young women and prenatal exposure to DES was first proposed in 1971 (Herbst *et al.*, 1971). Subsequent studies have shown that the risk of this cancer arising in the daughters of DES-exposed women is low being in the order of 1 per 1000 (Herbst and Anderson 1990). Burks *et al.* (1990) stress the importance of continuing to monitor DES-exposed women, observing that even after clear cell carcinoma has been treated they are still at risk: they quote a case where the first recurrence of the tumour occurred 17 years after initial therapy. In many cases where clear cell adenocarcinoma has been diagnosed, DES was not an aetiological factor. A number of non-neoplastic changes have been identified in the genital tract of DES-exposed daughters and these are described in Section 9.6.3.

14.4 MESONEPHRIC ADENOCARCINOMA

This tumour arises from mesonephric remnants in the lateral walls of the cervix and is characterized, in well-differentiated areas, by diffusely infiltrating small glands lined by mucin-free cuboidal epithelium with clear or granular cytoplasm. The glands are closely packed and contain eosinophilic, hyaline secretions in their lumens. Alternatively, the

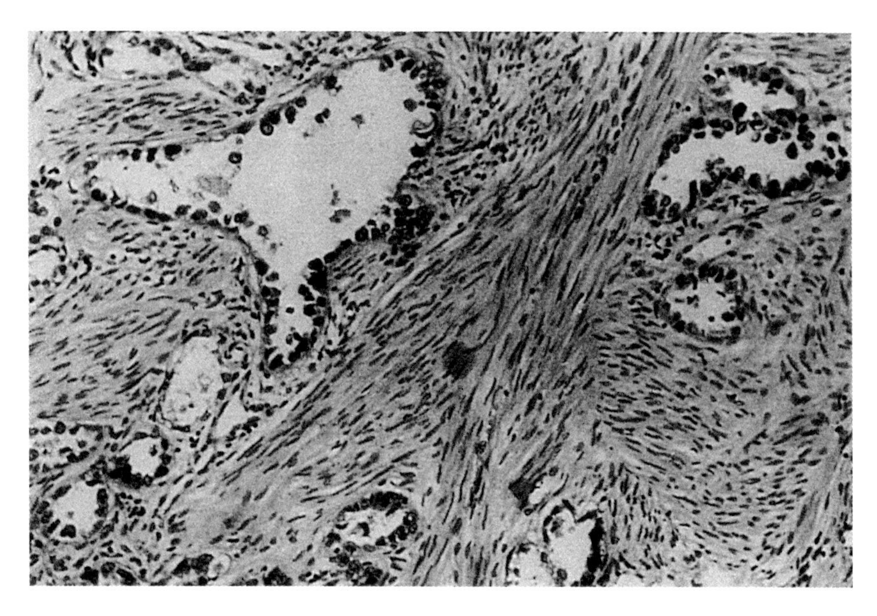

Figure 14.39 Tubules with hobnail lining cells in a clear cell carcinoma, probably of müllerian duct origin, occurring in a woman aged 23 years. She died four years after hysterectomy and radiotherapy (H & E, ×190).

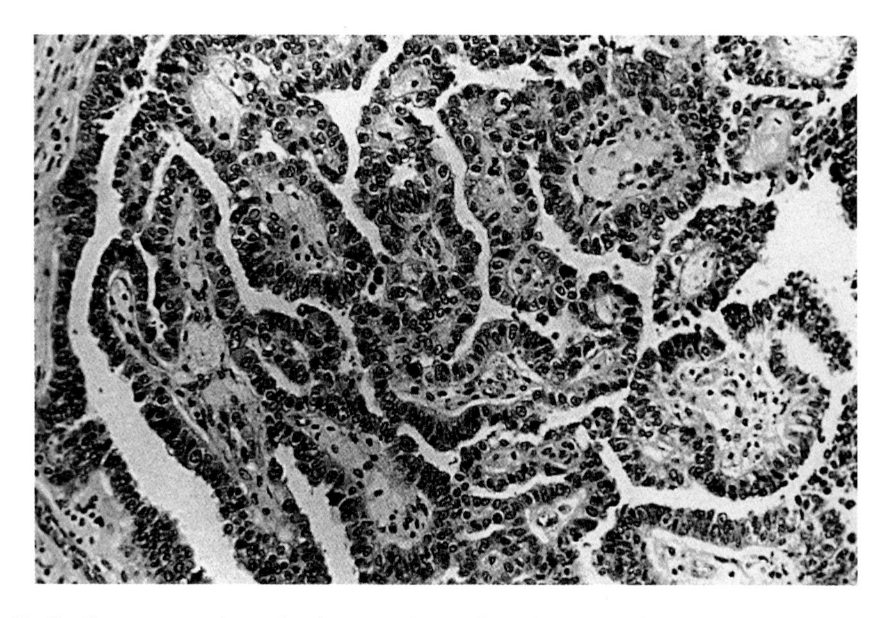

Figure 14.40 Papillary process formation in same clear cell carcinoma as shown in Figure 14.39 (H & E, ×190).

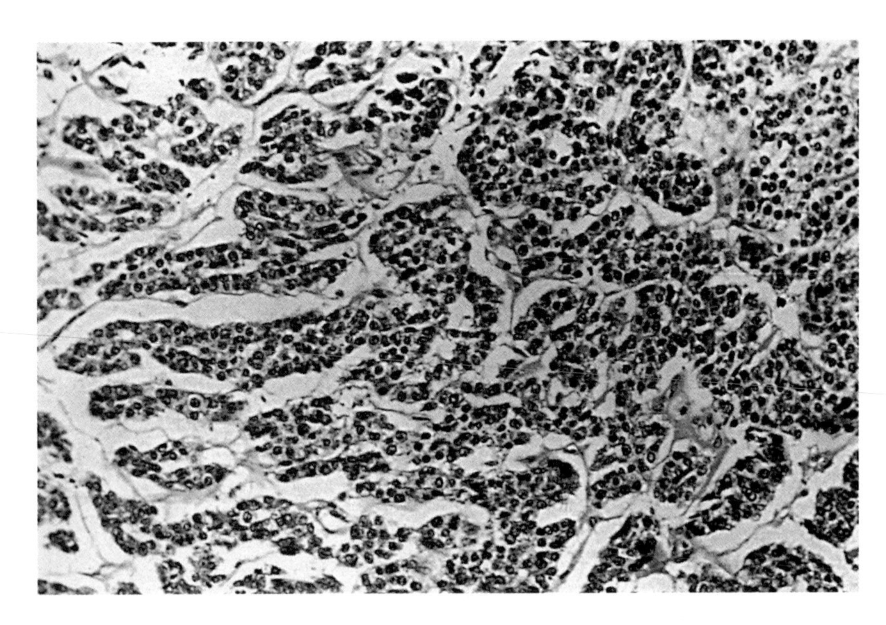

Figure 14.41 Solid nests and trabeculae of cells in same clear cell carcinoma as shown in Figure 14.39 (H & E, ×190).

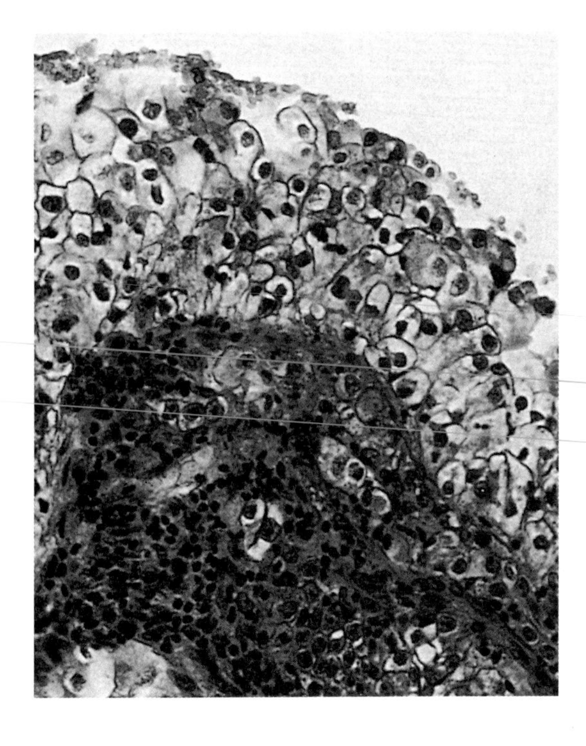

Figure 14.42 Clear cell carcinoma. Desquamated cells from this lesion are shown in Figure 14.44 (H & E, ×310).

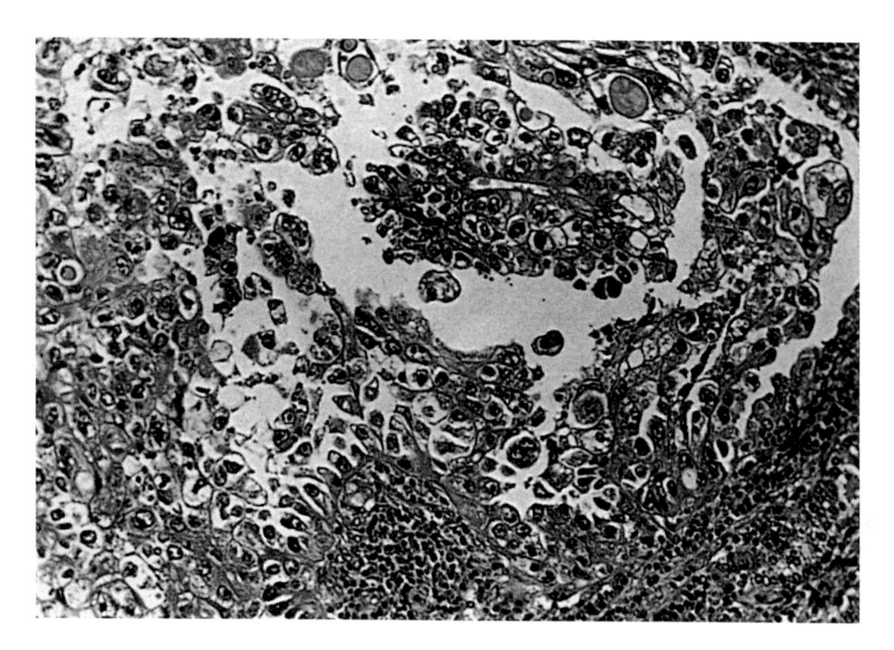

Figure 14.43 Clear cell carcinoma. Although the clear cells are less prominent in this tumour, the hobnail pattern and papillary process formation can be seen. The patient was 68 years old when biopsied, and died one year later in spite of radiotherapy (H & E, ×190).

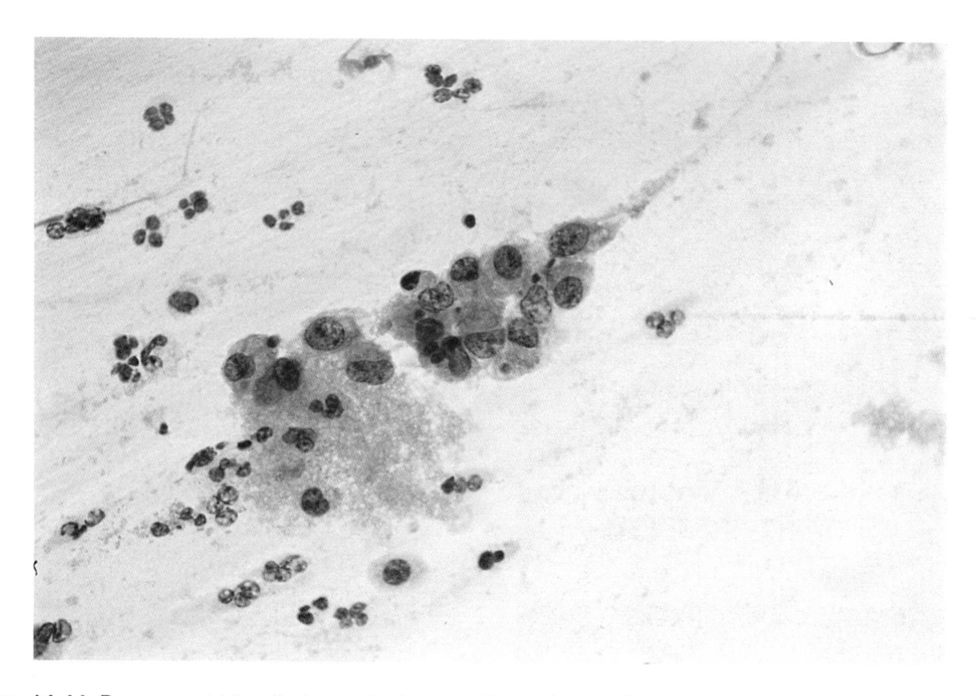

Figure 14.44 Desquamated cells in cervical smear from clear cell carcinoma shown in Figure 14.42 (Pap, ×630).

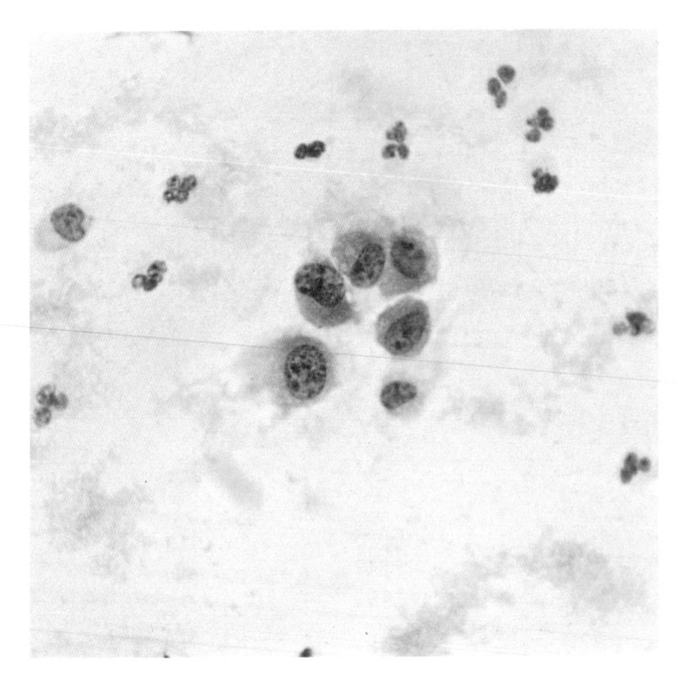

Figure 14.45 Desquamated cells in cervical smear from clear cell carcinoma. Note large nuclei with very irregular chromocentres (Pap, ×630).

tumour may present as large tubular glands resembling those of endometrioid adenocarcinoma (Scully *et al.*, 1994). The tumour is typically sited deeply in the cervical connective tissue with demonstrable continuity between mesonephric duct remnants and the tumour tissue. Such tumours are rare, although Hart and Norris (1972) reported that 3% of primary cervical adenocarcinomas were of this type. At one time they were confused with clear cell carcinoma.

All the examples of mesonephric hyperplasia and mesonephric adenocarcinoma examined histochemically by Lang and Dallenbach-Hellweg (1990) were positive for cytokeratin, including cytokeratin 8, and all were CEA-negative This contrasted with the observation of Valente and Susin (1987) who reported a case of mesonephric adenocarcinoma which was focally CEA-positive

The absence of lobular structure and the presence of cytologically malignant nuclei distinguishes mesonephric adenocarcinoma from mesonephric hyperplasia (Section 8.6; Wright *et al.*, 1994). A rare case of mesonephric adenocarcinoma arising in florid mesonephric hyperplasia has been described (Valente and Susin, 1987). The diagnosis of mesonephric adenocarcinoma should only be made if the tumour is deeply seated in the wall of the cervix, the endocervical epithelium is uninvolved and there is no history of DES exposure.

14.5 SEROUS (PAPILLARY SEROUS) ADENOCARCINOMA

This is a very rare tumour of the cervix. It is characterized by a complex pattern of papillae with cellular budding and the frequent presence of psammoma bodies (Scully *et al.*, 1994). Of the three cases described by Gilks and Clement (1992) two women aged 32 and 33 years presented with postcoital spotting and the third, aged 69 who was asymptomatic, was detected as a result of an abnormal cervical smear. Macroscopically the tumours were indistinguishable from typical endocervical adenocarcinoma. Histologically, their appearance closely resembled serous adenocarcinoma of the ovary. There were many papillary tufts with fibrovascular cores, covered by stratified cuboidal epithelial cells with eosinophilic cytoplasm which formed frequent small buds, some appearing detached. The tumour cells showed marked nuclear pleomorphism, with large and bizarre forms, often having prominent nucleoli. Although the three lesions were not deeply invasive, two had produced pelvic node metastases suggesting that they were more aggressive than the usual endocervical adenocarcinoma. It is important to distinguish these tumours from low-grade villoglandular papillary adenocarcinoma (Section 14.1.1.(b)). Occasionally, both tumour types may coexist. Metastasis from serous adenocarcinoma of the ovary or endometrium must also be excluded.

14.6 CLINICAL ASPECTS OF CERVICAL ADENOCARCINOMA

The mean age of patients presenting with invasive adenocarcinoma of the cervix varies with stage. Shingleton *et al.* (1981) reported the average age of 92 women with stage I disease to be 45. The average age of 28 women with more advanced lesions was 55 years. The women were on average five years younger than a comparable group of women with squamous carcinoma matched for stage.

Several authors claim that the prognosis associated with adenocarcinoma of the cervix is generally worse than that associated with squamous carcinoma of the cervix (Wheeless *et al.*, 1970; Abell, 1973; Gallup and Abell, 1977) and is related to histological type (Fu *et al.*, 1982). Shingleton *et al.* (1981), reviewed the survival data on 137 patients with pure and mixed adenocarcinomas of the cervix and concluded that histological type had no prognostic significance. This group also compared the clinical progress in patients with adenocarcinoma and adenosquamous carcinoma with a large group of patients with squamous carcinoma of the cervix matched by age, race, stage and treatment. They found that (contrary to earlier reports) neither adenocarcinoma nor adenosquamous carcinoma metastasize more frequently or more quickly, or are more radioresistant than squamous carcinoma. However, this is not the experience of all authors and Gallup *et al.* (1985), in a series of 127 patients with primary carcinoma of the cervix, found that there was a significant decrease in survival rate in patients with adenosquamous carcinoma compared with squamous carcinoma and adenocarcinoma. Prognosis in relation to mucus secretion, vascular permeation and clinical stage is considered towards the end of the introductory section of this chapter.

14.7 DIFFERENTIAL DIAGNOSIS OF CERVICAL ADENOCARCINOMA

Conditions which must be differentiated from cervical adenocarcinoma include microglandular endocervical hyperplasia, endocervical epithelial atypia associated 'with trauma or inflammation', the Arias–Stella pregnancy reaction affecting endocervical cells, metastatic adenocarcinoma from endometrium, breast, colon or other sites and cervical carcinoid.

14.7.1 Microglandular endocervical hyperplasia (MEH)

This is described in Section 9.2.1. It is a condition which has become quite common with the widespread use of contraceptives. Adenocarcinoma *in situ* (Figures 9.16 (a) and (b)) or invasive adenocarcinoma (Figure 14.15) may coexist with microglandular hyperplasia. Careful examination of the epithelium lining the hyperplastic glands will enable the correct diagnosis to be made.

14.7.2 The Arias–Stella reaction

This quite frequently occurs in endocervical epithelium during intrauterine and ectopic pregnancy (Figure 8.11) and could be mistaken for clear cell carcinoma. It is distinguished by fairly uniform nuclear enlargement, the absence of mitotic figures, the absence of glycogen and by not overlooking the possibility of pregnancy. Most clear cell carcinomas give a strongly positive reaction with PAS.

14.7.3 Metastatic carcinoma

Metastatic carcinoma varies in appearance according to the site of origin of the primary tumour (Section 15.7). An accurate history is invaluable for the accurate diagnosis of these lesions. Breast carcinoma (Figures 15.14 and 15.15) is one of the most common forms of metastatic cancer to affect the cervix and is usually distinguishable morphologically from any of the primary cervical cancers. *Endometrial carcinoma* (Figure 15.19), however, may be extremely similar to endometrioid carcinoma (Figures 14.32–14.38). Typically, alcian blue stains the primary endocervical adenocarcinoma quite strongly, although sometimes erratically, whereas extension of an endometrial carcinoma is usually alcian blue-negative. A poorly differentiated endocervical carcinoma may also give a negative staining reaction. Demonstration of carcinoembryonic antigen (CEA) by the immunoperoxidase technique assists in differentiation (Wahlstrom *et al.*, 1979; Agarwal and Sharma, 1990). CEA was identified in 80% of endocervical adenocarcinomas but in only 8% of endometrial adenocarcinomas. Cytological distinction between endometrial and cervical adenocarcinoma is discussed in Section 14.1.3.

Colonic carcinoma usually stains positively with alcian blue and is generally CEA antigen-positive. Morphologically, it may resemble cervical adenocarcinoma but the pattern of infiltration, by direct spread from rectum or colon, is usually distinctive. Occasionally, an *ovarian carcinoma* may produce a secondary deposit in the cervix (Figure 15.16) and its cells may be present in a cervical scrape (Figures 14.30 and 15.17).

14.7.4 Carcinoid tumours of the cervix (Figure 15.10)

These, if well-differentiated, may resemble adenocarcinomas and produce glandular acini scattered throughout the tumour. Differentiation is based on the rather characteristic carcinoid pattern and the demonstration of intracytoplasmic neurosecretory granules by histochemistry or electron microscopy (Section 15.4).

14.7.5 Cellular atypia associated with inflammatory change (Figure 7.3)

If due to inflammation, the cellular atypia is limited to the site of the inflammatory reaction. If atypicality extends beyond the inflammatory zone, neoplasia may be suspected.

REFERENCES

Abell, M. R. (1973) Invasive carcinomas of uterine cervix, in *The Uterus* (eds H. J. Norris, A. T. Hertig and M. R. Abell), *Int. Acad. Path. Monograph*, Williams and Wilkins, Baltimore, pp. 413–456.

Auersperg, N., Erber, H. and Worgh, A. (1972) Histologic variations among poorly differentiated invasive carcinomas of the human uterine cervix. *J. Natl. Cancer Inst.*, **51**, 1461–1477.

Agarwal, S. and Sharma, S. (1990) Localisation of carcinoembryonic antigen in uterine cervical neoplasia. *Ind. J. Med. Res.*, **92**, 452–455.

Ayer, B. S., Pacey, N. F. and Greenberg, M. L. (1991) The cytological features of invasive adenocarcinoma of the cervix uteri. *Cytopathology*, **2**, 181–191.

Barrowclough, H. and Jaarsma, K. W. (1980) Adenoacanthomas of the endometrium; a separate entity or a histological curiosity? *J. Clin. Pathol.*, **33**, 1064–1067.

Bates, C., Wells, M., Kingston, R., Bulmer, J. N. and Bird, C. C. (1985) Minimal deviation adenocarcinoma ('adenoma malignum') of the endocervix. A histochemical and immunohistochemical study of two cases. *J. Pathol.*, **145**, 71A.

Boon, M. E., Kirk, R. S. and Rietveld-Scheffers, P. E. M. (1981) The morphogenesis of adenocarcinoma of the cervix – a complex pathological entity. *Histopathology*, **15**, 565–577.

Buckley, C. H., Beards, C. S. and Fox, H. (1988) Pathological prognostic indicators in cervical cancer with particular reference to patients under the age of 40 years. *Br. J. Obstet. Gynaecol.*, **95**(1), 47–56.

Bulmer, J. N., Griffin, N. R., Bates, C. *et al.* (1990) Minimal deviation adenocarcinoma (adenoma malignum) of the endocervix: a histochemical and immunohistochemical study of two cases. *Gynecol. Oncol.*, **36**(1), 139–146.

Burks, R. T., Schwartz, A. M., Wheeler, J. E. and Antonioli, D. (1990) Late recurrence of clear-cell adenocarcinoma of the cervix: case report. *Obstet. Gynecol.*, **76**, 525–527.

Christopherson, W. M., Nealon, N. and Gray, L. A. (1979) Non-invasive precursor lesions of adenocarcinoma and mixed adenosquamous carcinoma of the cervix uterine. *Cancer*, **44**, 975–983.

Cramer, D. W. and Cutler, S. J. (1974) Incidence and histopathology of malignancies of the female genital organs in the United States. *Am. J. Obstet. Gynecol.*, **118**, 443–460.

Dallenbach-Hellweg, G. (1984) On the origin and histological structure of adenocarcinoma of the endocervix in women under 50 years of age. *Pathology Research and Practice*, **179**, 38–50.

Davis, J. R. and Moon, L. B. (1975) Increased incidence of adenocarcinoma of uterine cervix. *Obstet. Gynecol.*, **45**, 79–83.

Dougherty, C. M. and Cotton, N. (1964) Mixed squamous cell and adenocarcinoma of the cervix; combined adenosquamous and mucoepidermoid types. *Cancer*, **17**, 1132–1143.

Farnsworth, A., Laverty, C. and Stoler, M. H. (1989) Human papillomavirus messenger RNA expression in adenocarcinoma *in situ* of the uterine cervix. *Int. J. Gynecol. Pathol.*, **8**(4), 321–330.

Ferenczy, A. (1982) Carcinoma and other malignant tumours of the cervix, in *Pathology of the Female Genital Tract*, 2nd edn (ed. A. Blaustein), Springer-Verlag, New York, Heidelberg, Berlin, pp. 184–222.

Forsberg, J. G. (1972) Estrogen, vaginal cancer and vaginal development. *Am. J. Obstet. Gynecol.*, **113**, 83–87.

Fox, H., Wells, M., Harris, M. *et al.* (1988) Enteric tumours of the lower female genital tract: a report of three cases. *Histopathology*, **12**, 167–176.

Fu, Y. S., Reagan, J. W., Hsiu, J. G., Storaasli, J. P. and Wentz, W. B. (1982) Adenocarcinoma and mixed carcinoma of the uterine cervix I. A clinicopathologic study. *Cancer*, **49**, 2560–2570.

Gallup, D. G. and Abell, M. R. (1977) Invasive adenocarcinoma of the uterine cervix. *Obstet. Gynecol.*, **49**, 596–603.

Gallup, D. G., Harper, R. H. and Stock, R. J. (1985) Poor prognosis of patients with adenosquamous carcinoma of the cervix. *Obstet. Gynecol.*, **65**, 416–422.

Gilks, C. B. and Clement, P. B. (1992) Papillary serous adenocarcinoma of the uterine cervix: a report of three cases. *Mod. Pathol.*, **5**, 426–431.

Gilks, C. B., Young, R. H., Aguirre, P. *et al.* (1989) Adenoma malignum (minimal deviation adenocarcinoma) of the uterine cervix. A clinicopathological and immunohistochemical analysis of 26 cases. *Am. J. Surg. Pathol.*, **13**(9), 717–729.

Glücksmann, A. and Cherry, C. P. (1956) Incidence, histology and response to radiation of mixed carcinomas (adenoacanthomas) of the uterine cervix. *Cancer*, **9**, 971–979.

Griffin, N. R., Dockey, D., Lewis, F. A. and Wells, M. (1991) Demonstration of low frequency of human papillomavirus DNA in cervical adenocarcinoma and adenocarcinoma *in situ* by the polymerase chain reaction and *in situ* hybridisation. *Int. J. Gynecol. Pathol.*, **10**(1), 36–43.

Grigsby, P. W., Perez, C. A., Kuske, R. R. *et al.* (1988) Adenocarcinoma of the uterine cervix: lack of evidence for a poor prognosis. *Radiotherapy. Oncol.*, **12**(4), 289–296.

Hart, W. R. and Norris, H. J. (1972) Mesonephric adenocarcinoma of the cervix. *Cancer*, **29**, 106–113.

Herbst, A. L. and Anderson, D. (1990) Clear cell adenocarcinoma of the vagina and cervix secondary to intrauterine exposure to diethylstilbestrol. *Semin. Surg. Oncol.*, **6**, 343–346.

Herbst, A. L., Ulfelder, H. and Poskanzer, D. C. (1971) Adenocarcinoma of the vagina: association of maternal stilboestrol therapy with tumor appearance in young women. *N. Engl. J. Med.*, **284**, 878–881.

Herbst, A. L., Robboy, S. J., Scully, R. E. and Poskanzer, D. C. (1974) Clear cell adenocarcinoma of the vagina and cervix in young females: analysis of 170 registry cases. *Am. J. Obstet. Gynecol.*, **119**, 713–724.

Hopkins, M. P. and Morley, G. W. (1991) A comparison of adenocarcinoma and squamous cell carcinoma of the cervix. *Obstet. Gynecol.*, **77**(6), 912–917.

Hopson, L., Jones, H. A., Boyce, C. R. and Tarraza, H. M. (1990) Papillary villoglandular carcinoma of the cervix. *Gynecol. Oncol.*, **39**, 221–224.

Jones, M. W. and Silverberg, S. G. (1989) Cervical adenocarcinoma in young women: possible relationship to microglandular hyperplasia and use of oral contraceptives. *Obstet. Gynecol.*, **73**(6), 984–989.

Jones, M. W., Silverberg, S. G. and Kurman, R. J. (1993) Well-differentiated villoglandular adenocarcinoma of the uterine cervix: a clinico-pathological study of 24 cases. *Int. J. Gynecol. Pathol.*, **12**, 1–7.

Koss, L. G., Brannan, C. D. and Ashikari, R. (1970) Histologic and ultrastructural features of adenoid cystic carcinoma of the breast. *Cancer*, **26**, 1271–1279.

Kudo, R., Sagae, S., Kusanagi, T. *et al.* (1991) Morphology of adenocarcinoma *in situ* and microinvasive adenocarcinoma of the uterine cervix. A cytologic and ultrastructural study. *Acta Cytol.*, **35**(1), 109–116.

Lang, G. and Dallenbach-Hellwig, G. (1990) The histogenetic origin of cervical mesonephric hyperplasia and mesonephric adenocarcinoma of the uterine cervix studied with immunohistochemical methods. *Int. J. Gynecol. Pathol.*, **9**, 145–157.

Lee, K. R. and Trainer, T. D. (1990) Adenocarcinoma of the uterine cervix of small intestinal type containing numerous Paneth cells. *Arch. Pathol. Lab. Med.*, **114**, 731–733.

Matsukuma, K., Tsukamoto, N., Kaku, T. *et al.* (1989) Early adenocarcinoma of the uterine cervix; its histologic and immunohistologic study. *Gynecol. Oncol.*, **35**(1), 38–43.

Nagai, N. (1990) Molecular biologic study on the carcinogenesis of HPV in uterine cervical cancer and related lesions; analysis of HPV types 16, 18 E6/E7 gene mRNA (in Japanese). *Acta Obstet. Gynaecol. Jap.*, **42**(8), 823–833.

Okagaki, T., Tase, T., Twiggs, L. B. and Carson, L. F. (1989) Histogenesis of cervical adenocarcinoma with reference to human papillomavirus-18 as a carcinogen. *J. Reprod. Med.*, **34**(9), 639–644.

Podczaski, E., Kaminski, K. F., Pees, R. C. *et al.* (1991) Peutz–Jegher'syndrome with ovarian sex cord tumour with annular tubules and cervical adenoma malignum. *Gynecol. Oncol.*, **42**, 74–78.

Poulsen, H. E., Taylor, C. W. and Sobin, L. H. (1975) *Histological Typing of Female Genital Tract Tumours*, World Health Organisation, Geneva, pp. 15, 57–59.

Puri, S., Fenoglio, C. M., Richart, R. M. and Townsend, D. E. (1977) Clear cell carcinoma of cervix and vagina in progeny of women who received diethylstilboestrol; three cases with scanning and transmission electron microscopy. *Am. J. Obstet. Gynecol.*, **128**, 550–555.

Scully, R. E., Bonfiglio, T. A., Kurman, R. J. *et al.* (1994) *WHO International histological classification of tumours. Histological typing of female genital tract tumours.* Springer-Verlag, Berlin, pp. 39–54.

Seltzer, V. and Spitzer, M. (1983) Psammoma bodies in papillary adenocarcinoma of the cervix. *Int. J. Gynecol. Pathol.*, **2**, 216–221.

Shingleton, H. M., Bone, H., Bradley, D. H. and Soong, S. J. (1981) Adenocarcinoma of the cervix I: clinical evaluation and pathological features. *Am. J. Obstet. Gynecol.*, **139**, 799–814.

Silverberg, S. G. and Hurt, W. G. (1975) Minimal deviation adenocarcinoma ('adenoma malignum') of the cervix: a reappraisal. *Am. J. Obstet. Gynecol.*, **121**, 971–975.

Shorrock, K., Johnson, J. and Johnson, I. R. (1990) Epidemiological changes in cervical carcinoma with particular reference to mucin-secreting subtypes. *Histopathology*, **17**, 53–57.

Sugimori, H., Hachisuga, T., Nakamura, S. *et al.* (1990) Cervical cancers in uterus didelphys. *Gynecol. Oncol.*, **36**(3), 439–443.

Tasker, J. T. and Collins, J. A. (1974) Adenocarcinoma of the uterine cervix. *Am. J. Obstet. Gynecol.*, **18**, 344–348.

Teshima, S., Shimosato, Y., Kishi, K. *et al.* (1985) Early stage adenocarcinoma of the uterine cervix. Histopathologic analysis with consideration of histogenesis. *Cancer*, **56**(1), 167–172.

Valente, P. T. and Susin, M. (1987) Cervical adenocarcinoma arising in florid mesonephric hyperplasia: report of a case with immunocytochemical studies. *Gynecol. Oncol.*, **27**, 58–68.

Vesterinen, E., Forss, M. and Nieminen, U. (1989) Increase of cervical adenocarcinoma: a report of 520 cases of cervical carcinoma including 112 tumours with glandular elements. *Gynecol. Oncol.*, **33**(1), 3.

Wahlström, T., Lindgren, J., Korhonen, M. and Seppälä, M. (1979) Distinction between endocervical and endometrial adenocarcinoma with immunoperoxidase staining of carcinoembryonic antigen in routine histological tissue specimens. *Lancet*, **ii**, 1159–1160.

Wells, M. and Brown, L. J. C. (1986) Glandular lesions of the uterine cervix: the present state of our knowledge. *Histopathology*, **10**, 777–792.

Wheeless, C. R., Graham, R. and Graham, J. B. (1970) Prognosis and treatment of mucooepidermoid adenocarcinoma of the cervix. *Obstet. Gynecol.*, **35**, 928–932.

Wright, T. C., Ferenczy, A. and Kurman, R. J. (1994) Carcinoma and other tumors of the cervix, in *Blaustein's Pathology of the Female Genital Tract*, 4th edn (ed. R. J. Kurman), Springer-Verlag, New York, pp. 274–326.

Yamagata, S., Yamamoto, K. and Tsuchida, S. (1991) Clinicopathological, ultrastructural and immunohistochemical study on adenoma malignum of the uterine cervix. (in Japanese). *Acta Obstet. Gynaecol. Jap.*, **43**(1), 57–64.

Young, R. H. and Scully, R. E. (1989) Villoglandular papillary adenocarcinoma of the uterine cervix: a clinicopathologic study of 13 cases. *Cancer*, **63**, 1773–1779.

Young, R. H. and Scully, R. E. (1990) Invasive adenocarcinoma and related tumors of the uterine cervix. *Semin. Diagn. Pathol.*, **7**(3), 205–207.

Young, R. H. and Scully, R. E. (1993) Minimal-deviation endometrioid adenocarcinoma of the uterine cervix. A report of five cases of a distinctive neoplasm that may be misinterpreted as benign. *Am. J. Surg. Pathol.*, **17**, 660–665.

Zhang, J. M. (1989) Minimal deviation endocervical adenocarcinoma; clinical histologic and immunohistochemical study. *Chung-Hua Chung Liu Ysa Chih*, **11**(4), 275–277.

15

OTHER MALIGNANT EPITHELIAL TUMOURS

The following tumours are considered in this chapter:

Adenosquamous carcinoma
  Glassy cell carcinoma
  Mucoepidermoid carcinoma
Adenoid cystic carcinoma
Adenoid basal carcinoma
Carcinoid-like carcinoma (argentaffinoma, neuroendocrine tumour)
Small cell (anaplastic) carcinoma
Undifferentiated carcinoma
Metastatic carcinoma

15.1 ADENOSQUAMOUS CARCINOMA

This is a carcinoma composed of intermingled glandular and squamous cells. Shingleton *et al.* (1981) reported that one-third of their patients with adenocarcinoma had adenosquamous lesions. In addition, the use of mucin stains has revealed that many tumours thought to be of pure squamous type also have a glandular component (Wells and Brown, 1986). Mixed cell tumours of this type are thought to arise from undifferentiated subcolumnar reserve cells which have the capacity to differentiate into both squamous and glandular structures (Buckley and Fox, 1989). Adenosquamous carcinoma is to be distinguished from a collision tumour in which an adenocarcinoma and a squamous carcinoma arise in adjacent areas of the same cervix (Figure 15.1). An adenoacanthoma is an adenocarcinoma in which squamous metaplasia has occurred (Figure 14.38), but there is no clear dividing line between this and a well-differentiated adenosquamous carcinoma. The clinical implications of adenosquamous carcinoma are considered in Section 14.6.

The glandular component of adenosquamous carcinoma is typically of endocervical type but may be of endometrioid type, and must be more than a minor component to alter the diagnosis from squamous to adenosquamous carcinoma (Scully *et al.*, 1994). The squamous element may be composed of any of the squamous carcinoma types described in Chapter 12; hence a variety of histological patterns may be found. Of note is the example of adenosquamous carcinoma in which the squamous cells show spindle cell formation (Figure 15.2). The possible significance of the apparent stromal involvement in the tumour pattern and its apparently less malignant prognostic implications are discussed in Section 12.6.5. Part of this same tumour has the structure of a mucinous adenocarcinoma (Figure 14.17).

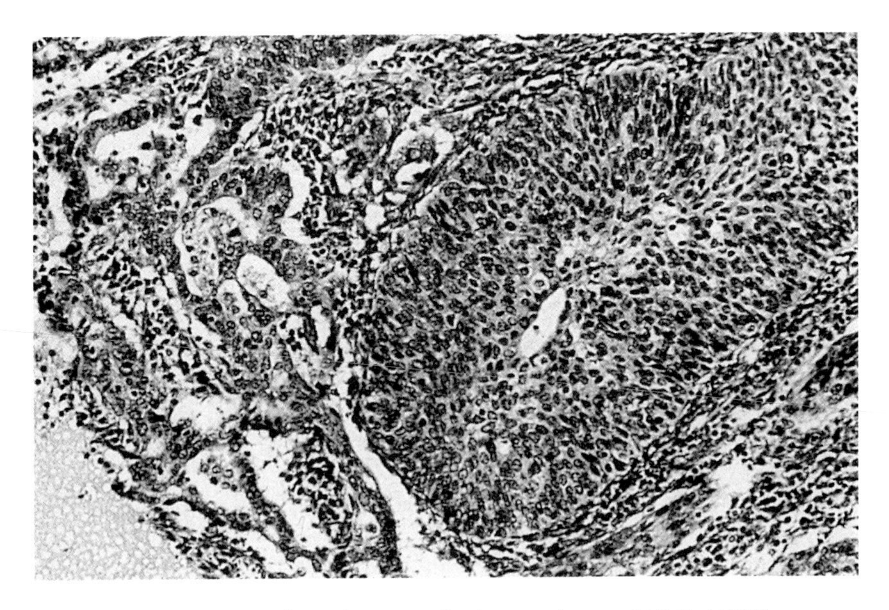

Figure 15.1 Collision of squamous cell carcinoma and adenocarcinoma (H & E, ×190).

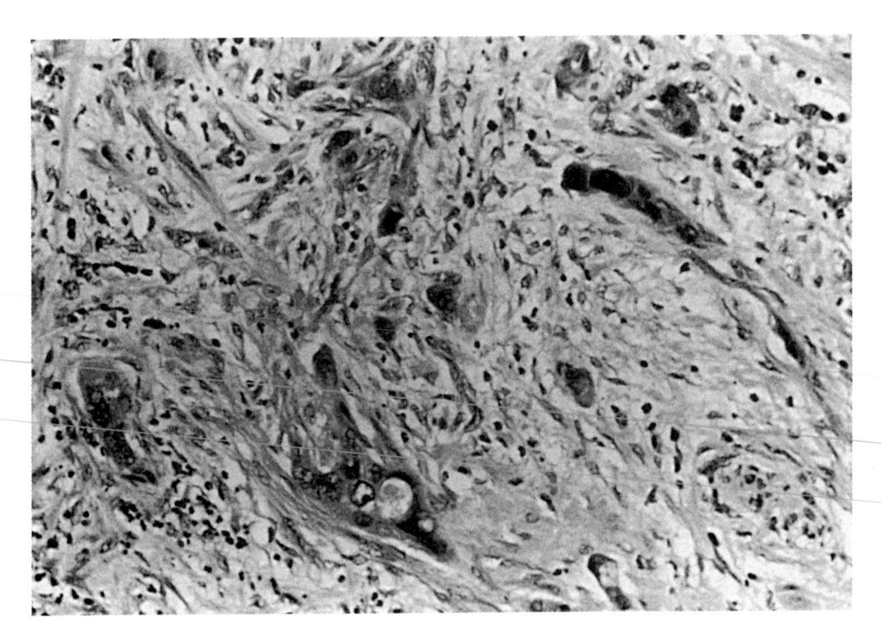

Figure 15.2 Adenosquamous carcinoma showing spindle cell differentiation. Part of the tumour is mucinous adenocarcinoma shown in Figure 14.17 (H & E, ×230).

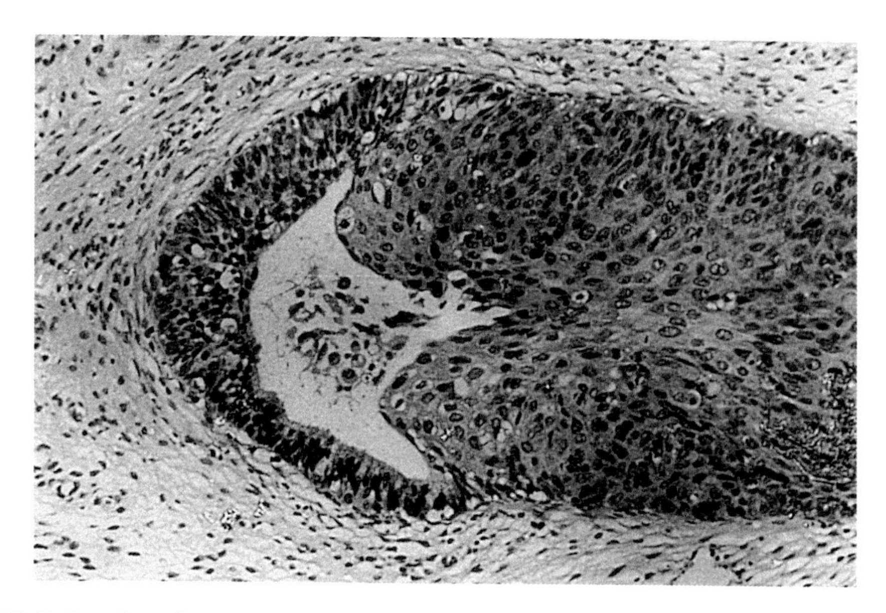

Figure 15.3 Preinvasive adenosquamous neoplastic lesion. The glandular elements are most clearly seen at the bottom left (H & E, ×190).

A preinvasive adenosquamous lesion has been described (Figure 15.3) in which cells showing squamous features such as intercellular bridge formation and cells showing glandular features can be identified. Such a lesion may give rise to squamous, glandular or adenosquamous carcinoma. Christopherson *et al.* (1979) consider that adenosquamous carcinoma, like squamous carcinoma, probably originates in subcolumnar reserve cells.

Three grades of invasive adenosquamous carcinoma are recognized: well-differentiated, moderately differentiated, and poorly differentiated (Figures 15.4 and 15.5(a,b)). A variant of the poorly differentiated form of the tumour is also known as glassy cell carcinoma (Figure 15.6).

The tumour shown in Figure 15.4 is an example of well-differentiated adenosquamous carcinoma. A glandular pattern can be recognized together with intervening sheets of cells showing definite intercellular bridge formation. The cells lining the glands show features of malignancy, as does the squamous cell component of the tumour. Moderately and poorly differentiated tumours are shown in Figure 15.5 (a,b). Occasional adenosquamous carcinomas show massive stromal infiltration by eosinophils which may be accompanied by eosinophilia in the peripheral blood (Gou, 1991). The cytological smear contains both squamous-looking and glandular tumour cells (Figure 15.5(c,d and e)).

15.1.1 Glassy cell carcinoma (Figure 15.6(a–c))

This is a variant of poorly differentiated adenosquamous carcinoma with a corresponding clinical course (Wright *et al.*, 1994). It is a very uncommon tumour which usually presents at a younger age than other types of cervical carcinoma. Of the 18 cases reported by Talerman *et al.* (1991) all but three were aged 34 years or less. Macroscopically, it is often large, producing a barrel-shaped deformity of the cervix. Histologically, it is composed of sheets of large cells with 'ground glass' cytoplasm having sharp cell borders in well-fixed

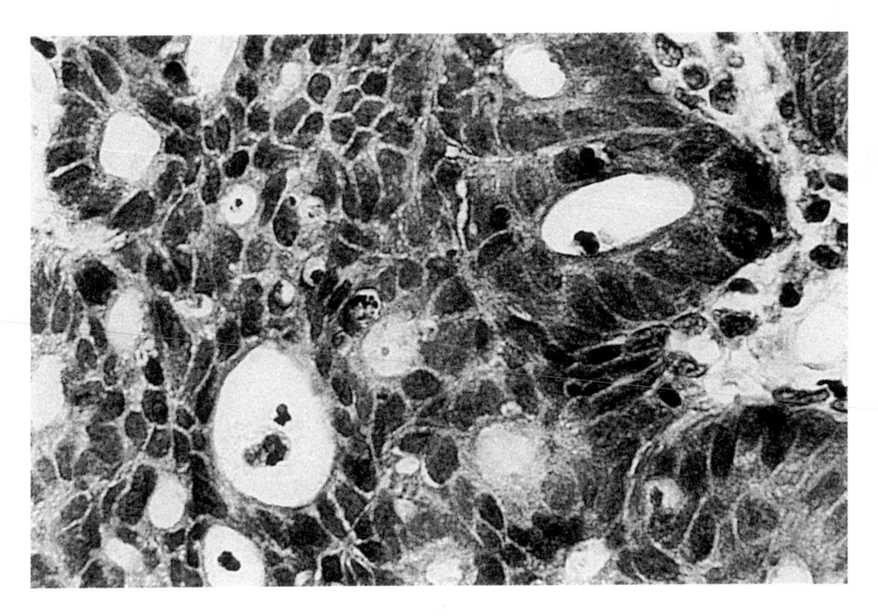

Figure 15.4 Adenosquamous carcinoma of well-differentiated type, showing clearly recognizable gland formation and squamous cells with intercellular bridge formation. Both the individual cells lining the glands and the squamous cells have malignant features. The patient was aged 45 years (H & E, ×770).

specimens and large nuclei with prominent nucleoli (Figure 15.6(a–b)) (Glucksmann and Cherry, 1956). Unlike squamous carcinoma, with which it might be confused, there is no dyskeratosis and no intercellular bridge formation, but an occasional keratin pearl, a poorly formed gland lumen or signet ring cells with intracellular mucin may be seen (Wright *et al.*, 1994) although Richard *et al.* (1981) did not detect any mucin in their cases. It is distinguished from paramesonephric clear cell carcinoma by the paucity of intracellular glycogen. When stained with PAS, the cytoplasm has a finely granular appearance. Cellular and nuclear pleomorphism is marked, tumour giant cells are frequent and mitotic activity is brisk (Talerman *et al.*, 1991). The stroma may be heavily infiltrated by lymphocytes, plasma cells and eosinophils. Electron microscopy may reveal abortive gland lumen formation, together with well-developed tonofilament desmosomal complexes, interdigitating microvilli and cytoplasmic microfilaments, supporting the view that the tumour is a variant of poorly differentiated adenosquamous carcinoma (Richard *et al.*, 1981; Talerman *et al.*, 1991). A cervical smear may contain tumour cells of the glassy cell type (Figure 15.6 (a–c)). The prognosis of glassy cell carcinoma is usually poor.

15.1.2 Mucoepidermoid carcinoma (Figure 15.7)

Mucoepidermoid carcinoma is defined as a tumour with the appearance of squamous carcinoma without any definite glandular pattern but with demonstrable intracellular mucin (Thelmo *et al.*, 1990). The mucin, which might not be suspected on routine haematoxylin and eosin staining, may be demonstrated by alcian blue, periodic acid–Schiff–diastase and mucicarmine. As well as occurring in signet ring, goblet or non-specific cells, the mucin may also collect in the intercellular spaces or fibrous stroma to form small or large lakes. The squamous pattern is usually that of a large cell non-keratinizing or

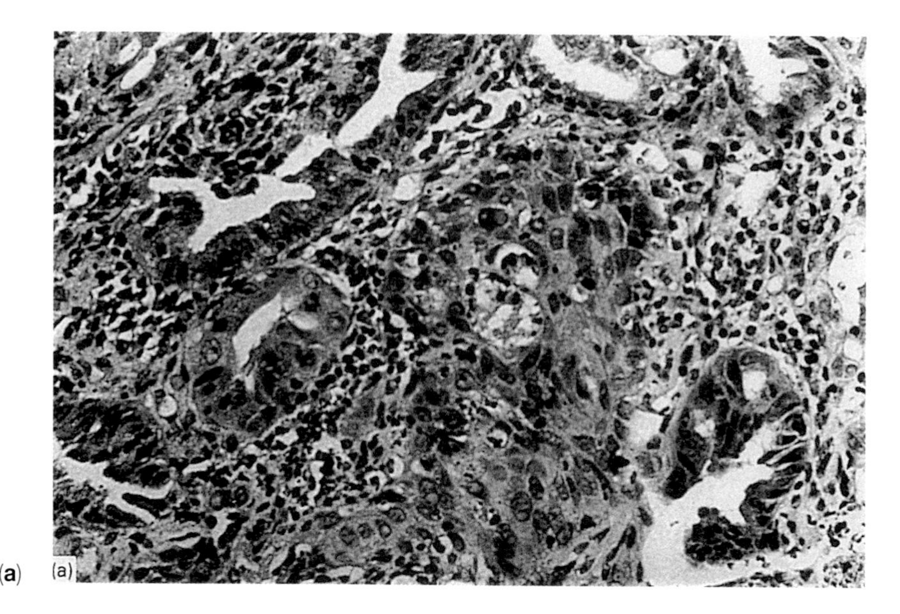

(**a**)

Figure 15.5 (**a**) Adenosquamous carcinoma showing moderate differentiation. Same tumour as shown in Figure 15.4. The glands are less regular and intercellular bridge formation is less easily seen. Although some malignant glandular cells and some squamous cells can be recognized, there are some cells whose direction of differentiation cannot be recognized (H & E, ×310).

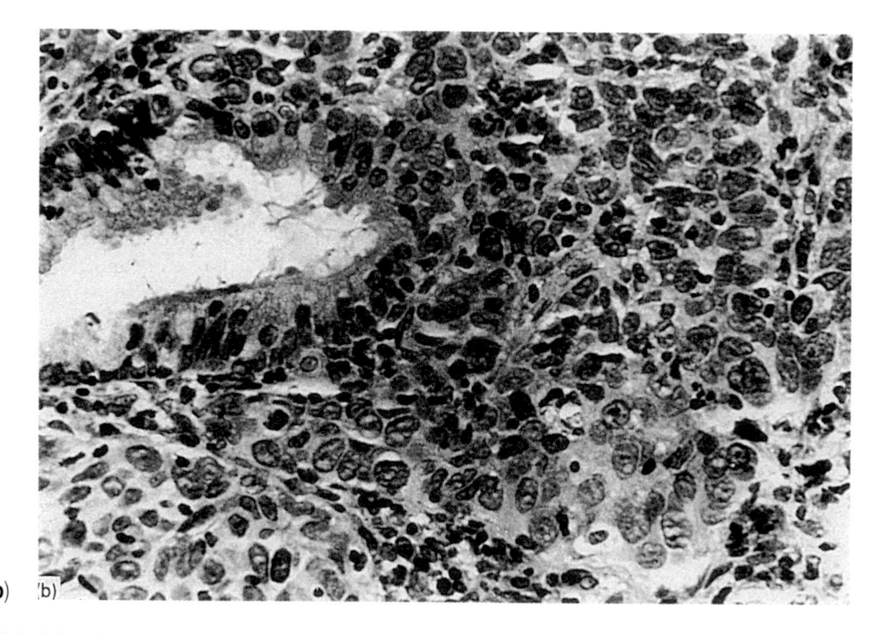

(**b**)

Figure 15.5 (**b**) Adenosquamous carcinoma showing moderate to poor differentiation from a woman aged 42 years. Glands can be recognized but intercellular bridge formation is difficult to detect (H & E, ×480).

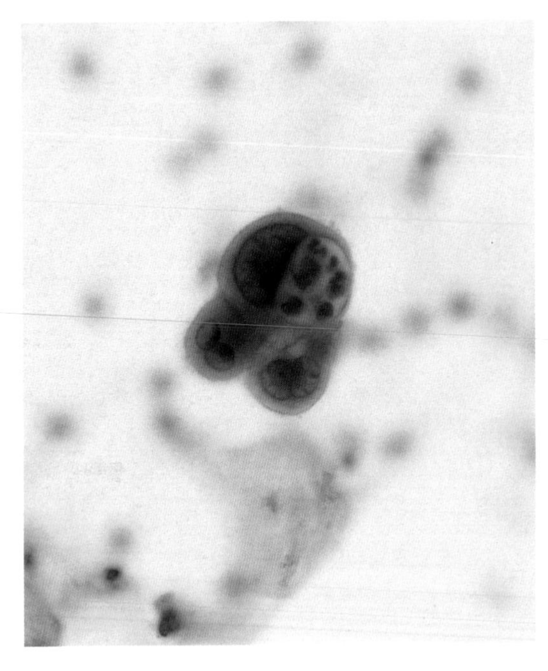

(**c**)

Figure 15.5 (**c–e**) Cytology of adenosquamous carcinoma. Note squamoid appearance of some tumour cells and glandular clusters in same smear (all Pap, ×630). (Courtesy of Dr Winifred Gray, John Radcliffe Hospital, Oxford).

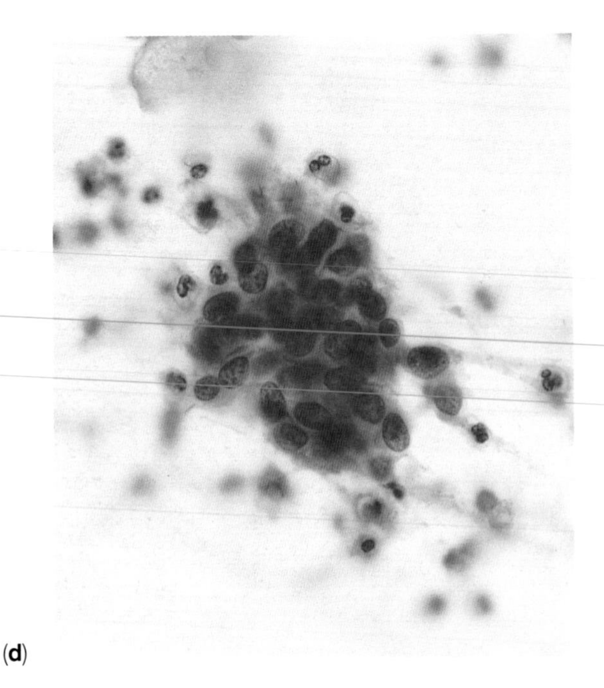

(**d**)

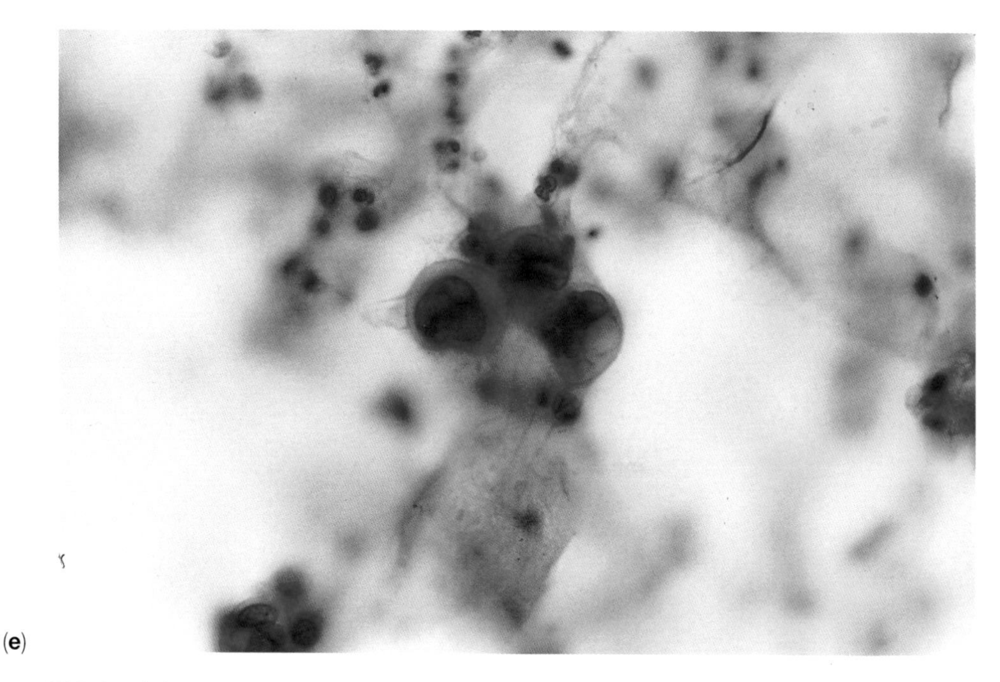

(e)

Figure 15.5 (**c–e**) Cytology of adenosquamous carcinoma. Note squamoid appearance of some tumour cells and glandular clusters in same smear (all Pap, ×630). (Courtesy of Dr Winifred Gray, John Radcliffe Hospital, Oxford).

focally keratinizing carcinoma (Buckley *et al.*, 1988) and the mucin-secreting areas lie either in the middle of tumour islands or in the centres of trabeculi. It may coexist with well-differentiated adenosquamous carcinoma (Dougherty and Cotton, 1964). In the series of 12 cases reported by Thelmo *et al.* (1990) the prevalence of nodal metastasis was 33%, compared with 14% for 265 cases of squamous carcinoma at the same stage (1B). This has prognostic implications and underlines the importance of using mucin stains as a routine on all cervical carcinomas (see also Section 14.6: Clinical aspects of cervical adenocarcinoma and adenosquamous carcinoma). A cervical scrape preparation from an adenosquamous or a mucoepidermoid carcinoma characteristically contains malignant cells of both glandular and squamous type.

15.2 ADENOID CYSTIC CARCINOMA

This is a rare cervical neoplasm with a characteristic cribriform pattern histologically similar to adenoid cystic carcinoma of the salivary glands but generally lacking the myoepithelial cell component of the latter (Scully *et al.*, 1994). It usually occurs in post-menopausal women but a few cases have been described in women under 40, the youngest being only 24 years old (King *et al.*, 1989). This pattern is due to the presence of cystic spaces containing hyaline eosinophilic material (Figure 15.8) surrounded by small basalar cells with pleomorphic nuclei and scanty cytoplasm. In addition to the cribriform pattern, much of the tumour is often composed of islands of basaloid cells (Figure 15.9). Although resembling their salivary counterparts histologically, the

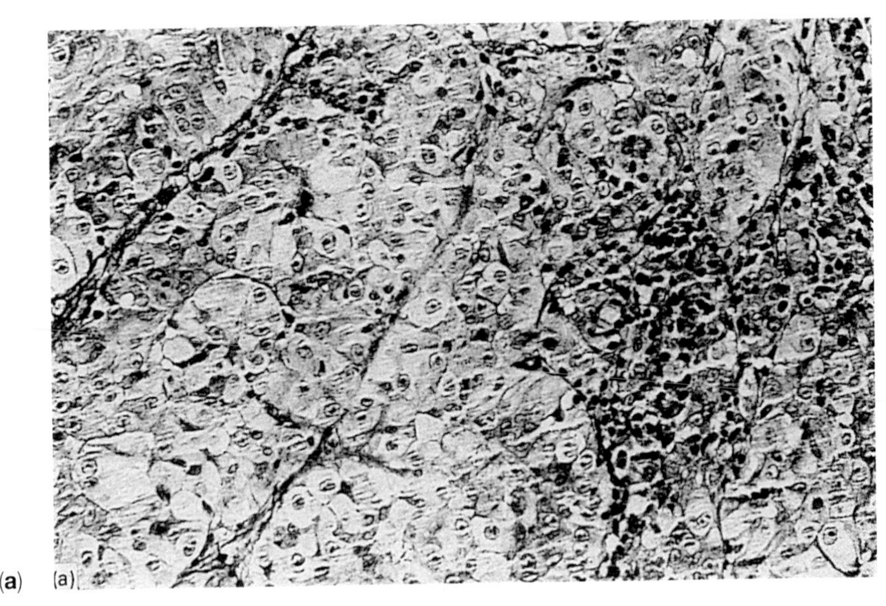

Figure 15.6 (**a**) Glassy cell carcinoma composed of large cells with ground-glass cytoplasm (H & E, ×180) (courtesy of Professor A. Ferenczy and Springer-Verlag).

Figure 15.6 (**b**) Glassy cell carcinoma showing glanularity of cytoplasm and prominent nucleoli. Same lesion as (**a**) (H & E, ×720).

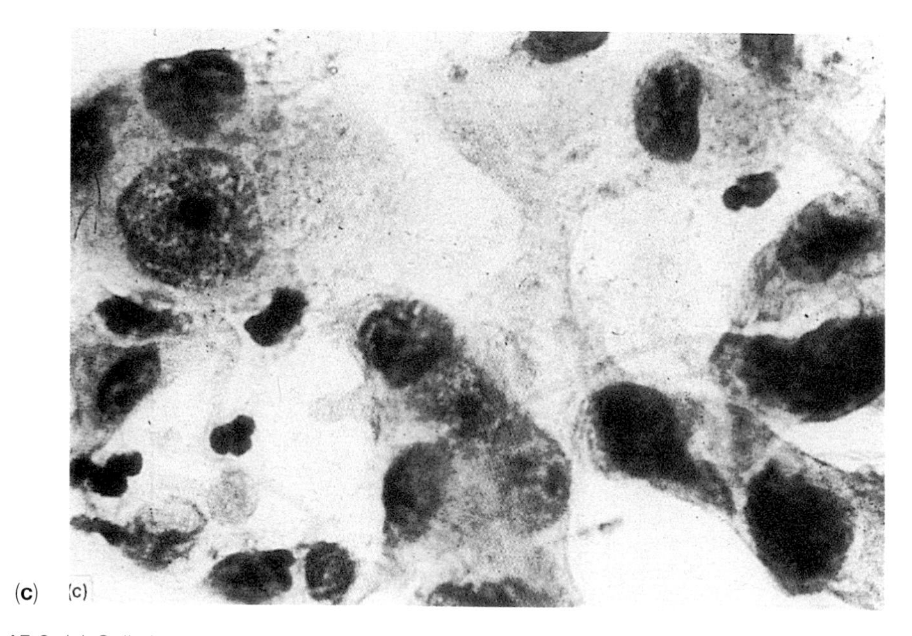

(c)

Figure 15.6 (**c**) Cells in cervical smear from glassy cell carcinoma showing similar features to (**b**) (Pap, ×630).

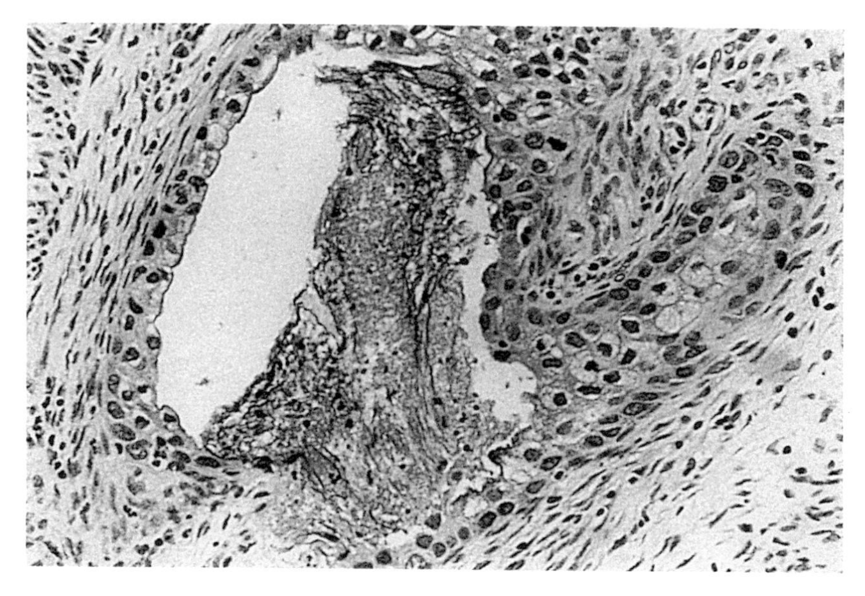

Figure 15.7 Mucoepidermoid carcinoma. Note cells with cytoplasmic vacuoles containing mucin droplets (H & E, ×310).

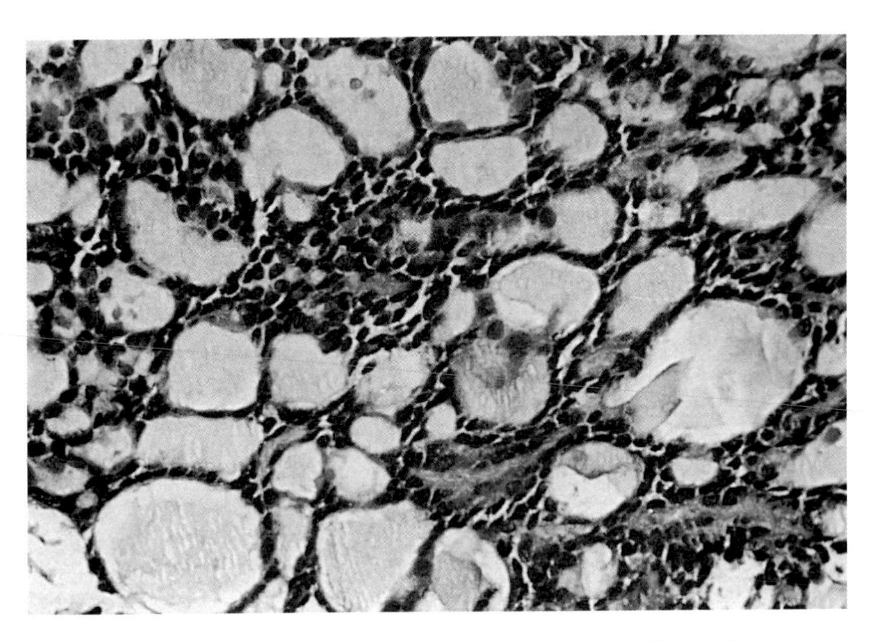

Figure 15.8 Adenoid cystic carcinoma of cervix showing cribriform pattern. The cystic spaces contain eosinophilic faintly fibrillar hyaline materials (H & E, ×480).

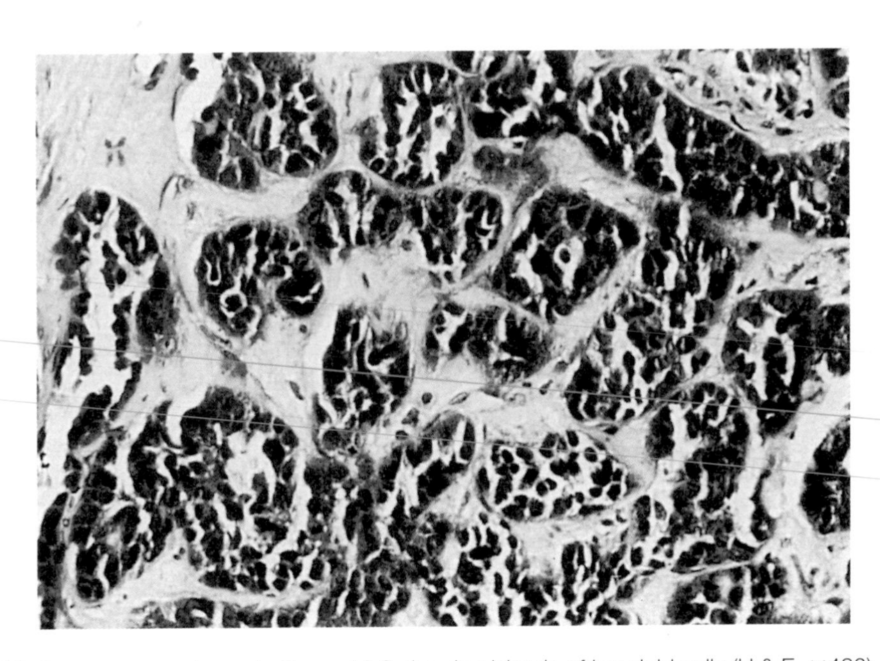

Figure 15.9 Same case as shown in Figure 15.8 showing islands of basaloid cells (H & E, ×480).

adenoid cystic carcinomas of the cervix differ in showing necrosis, a high mitotic rate and greater nuclear pleomorphism (Ferry and Scully, 1988). Histochemically, the hyaline material is mucicarmine-negative but strongly PAS-positive following diastase digestion (Mazur and Battifora, 1991). Pericellular areas, including the hyaline throughout the tumour are stained intensely by alcian blue, this staining being prevented by pretreatment with hyaluronidase. Ultrastructurally, the most striking feature is the presence of multiple layers of basal lamina encompassing clusters of tumour cells and forming pseudocysts which correspond with the fibrillar hyaline seen by light microscopy (Figures 15.8 and 15.9). Adenoid cystic carcinoma is usually associated with adenocarcinoma or squamous neoplasia (Fowler *et al.*, 1978), including CIN (SIL). Because of this frequent association, Buckley and Fox (1989) suggest that it may be regarded as a mixed tumour arising from multipotential reserve cells, the least differentiated areas having an adenoid cystic pattern and the better differentiated areas having either a squamous or an adenocarcinomatous pattern. Although relatively slow-growing, the tumour will progress and metastasize if not completely excised. Musa *et al.* (1985) reported that long-term survival occurred only if the treated patient had no parametrial involvement. No patient survived more than 5 years if there was obvious parametrial involvement. Radiotherapy only temporarily arrests its progress.

15.3 ADENOID BASAL CARCINOMA

This is a cervical carcinoma in which rounded, generally well-differentiated nests of basaloid cells show focal gland formation; central squamous differentiation may also be present (Scully *et al.*, 1994). It closely resembles adenoid cystic carcinoma histologically, from which it can be distinguished by the absence of cystic spaces containing hyaline eosinophilic material (with the corresponding absence of basal laminar material ultrastructurally) and the presence of smaller, less pleomorphic nuclei and fewer mitotic figures, consistent with its much less aggressive nature (Wright *et al.*, 1994). It consists of nests and cords of cells resembling the basal cells of the epidermis, in the centre of which there may be foci of squamous or glandular differentiation, or cystic spaces which may be filled with necrotic debris. The groups of tumour cell show prominent peripheral palisading with no significant stromal reaction (Daroca and Dhurandhar, 1980). Like adenoid cystic carcinoma it is frequently associated with CIN and is usually situated beneath, often arising from, CIN 3 (Ferry and Scully, 1988). Adenoid basal carcinoma is usually discovered as an incidental finding in operation specimens removed for an unrelated reason, reflecting its slow growth and low aggressiveness.

15.4 CARCINOID-LIKE CARCINOMA (ARGENTAFFINOMA, NEUROENDOCRINE TUMOUR)

This is a tumour resembling carcinoid tumours of the alimentary tract (Scully *et al.*, 1994). Although extremely rare, primary carcinoid tumour of the cervix is a well-recognized entity (Tateishi *et al.*, 1975; Albores-Saavedra *et al.*, 1976; Habib *et al.*, 1979; Wright *et al.*, 1994). *Well-differentiated* cervical carcinoid tumours only occasionally have the compactly arranged structure and scanty mitotic figures which characterize carcinoids of the appendix. Instead they tend to form trabeculae or solid nests of cells with glandular acini scattered throughout the tumour (Figure 15.10(a)). The latter contain diastase-resistant, PAS-positive material. The cells are usually round or polygonal and have a finely granular cytoplasm with round, oval or spindle-shaped nuclei. Mitotic figures may be numerous and vascular invasion appears to be frequent. Occasional findings have been foci of squamous metaplasia and amyloid formation in the stroma demonstrable by its dichroism and

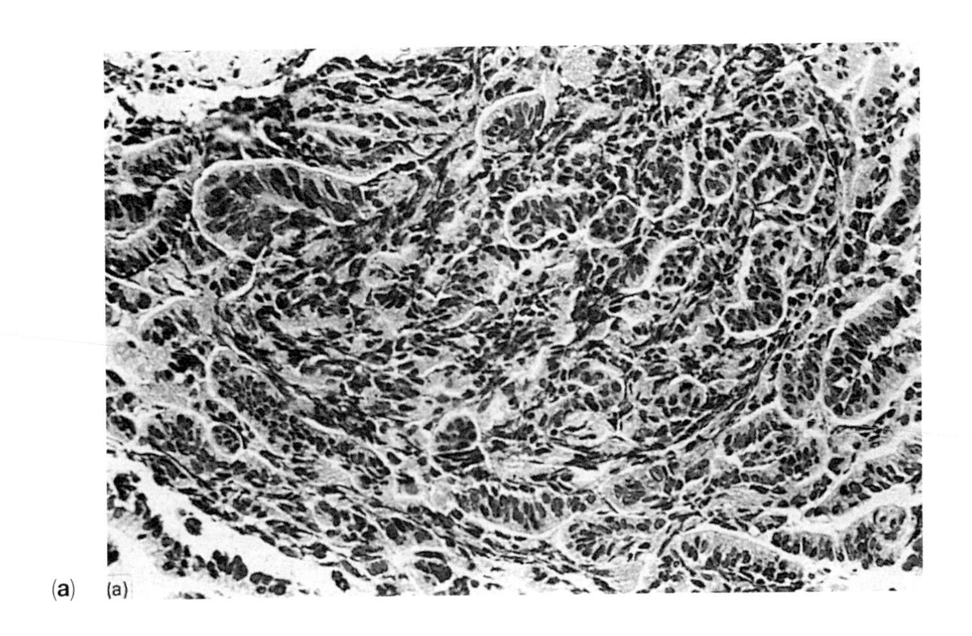

(a)

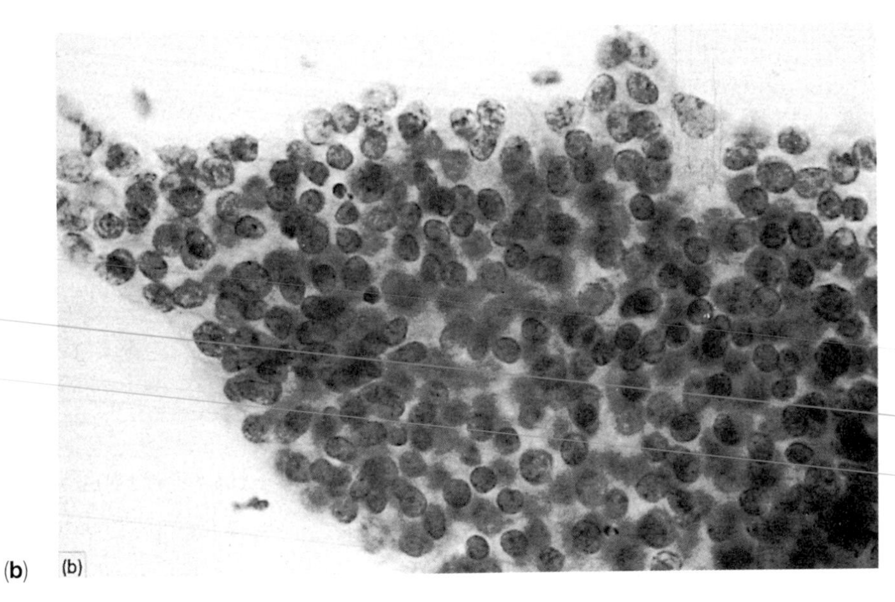

(b)

Figure 15.10 (**a**) Carcinoid tumour of cervix showing nests of tumour cells with occasional glandular acini (H & E, ×310). (**b**) Cluster of uniform cells from carcinoid tumour of cervix as shown in (**a**) (Pap, ×770).

staining by Congo red and crystal violet. Argyrophilic cytoplasmic neurosecretory granules demonstrable by Grimelius, stain and electron microscopy are usually present and the tumour has been termed 'argyrophil cell carcinoma' (Tateishi *et al.*, 1975). Fontana–Masson's stain for argentaffin granules is uniformly negative. The clinical course from diagnosis to death is of the order of 5–6 years (Albores-Saavedra *et al.*, 1976).

Immunohistochemically, many hormones such as ACTH, β-melanin-stimulating hormone, serotonin and histamine have been demonstrated (Matsuyama *et al.*, 1979) but to our knowledge, no examples of the carcinoid syndrome have been described with cervical carcinoid tumours. *Poorly differentiated* carcinoid tumours, which resemble oat cell carcinoma of the lung but still containing glandular acini, are now generally included with 'small cell carcinoma' (Section 15.5). Occasional examples of carcinoids occurring in association with other intraepithelial or invasive neoplasms have been described (Mullins and Hilliard, 1981; Stassart *et al.*, 1992). The case illustrated in Figure 15.10(a,b) was accompanied by an adenocarcinoma of endocervical type.

These tumours are rare and the chance of making the diagnosis on cytological material is small. Nevertheless, once the existence of the lesion is appreciated, it can be included in the differential diagnosis when compact clusters of fairly uniform cells are seen in a cervical smear (Figure 15. 10(b)).

15.5 SMALL CELL (ANAPLASTIC) CARCINOMA

Histologically, this closely resembles small (oat) cell carcinoma of lung (Scully *et al.*, 1994), characteristically showing nuclear moulding and crush artefacts. The diagnosis should be confined to those small cell carcinomas revealing little or no glandular or squamous differentiation, and not having the pattern of adenoid cystic or basal carcinoma. Small cell carcinoma invades the stroma diffusely in poorly defined trabeculae and clusters, unlike non-keratinizing squamous carcinoma which forms discrete nests. Auersperg (1972) found that many cervical carcinomas which appeared undifferentiated on light microscopy showed features of glandular or squamous differentiation using electron microscopy. Small cell carcinoma is to be distinguished from carcinoid-like carcinoma which has a more organoid structure, although immunohistochemically, 90% of the small cell carcinomas described by Stoler *et al.* (1991) showed some evidence of neurosecretory differentiation. They also found that 85% expressed HPV type 16 or 18 messenger RNA. Antibodies against leucocyte common antigen and neuroendocrine markers help to distinguish small cell carcinoma from lymphoproliferative disorders (Wright *et al.*, 1994). Of the 15 cases of small cell carcinoma reported by Gersell *et al.* (1988), 10 died as a direct result of their carcinomas and seven were associated with other forms of carcinoma (*in situ*, invasive or both). The histology and cytology of this tumour is shown in Figures 15.11(a–c).

15.6 UNDIFFERENTIATED CARCINOMA

This is a carcinoma that is not of small cell type and lacks evidence of glandular, squamous or other types of differentiation (Scully *et al.*, 1994). With the increasing use of special diagnostic techniques, e.g. immunohistochemistry and electron microscopy, the number of tumours included in this category is diminishing. Thus spindle cell carcinomas, including 'scirrhous carcinoma' (carcinoma with stromal fibroplasia) which were formerly classified as undifferentiated carcinomas are now included as a spindle cell variants of squamous or adenosquamous carcinoma; similarly glassy cell carcinomas are now considered to be a form of adenosquamous carcinomas. An example of a tumour which might be called undifferentiated carcinoma is shown in Figure 15.12. Auersperg *et al.* (1972) found that most

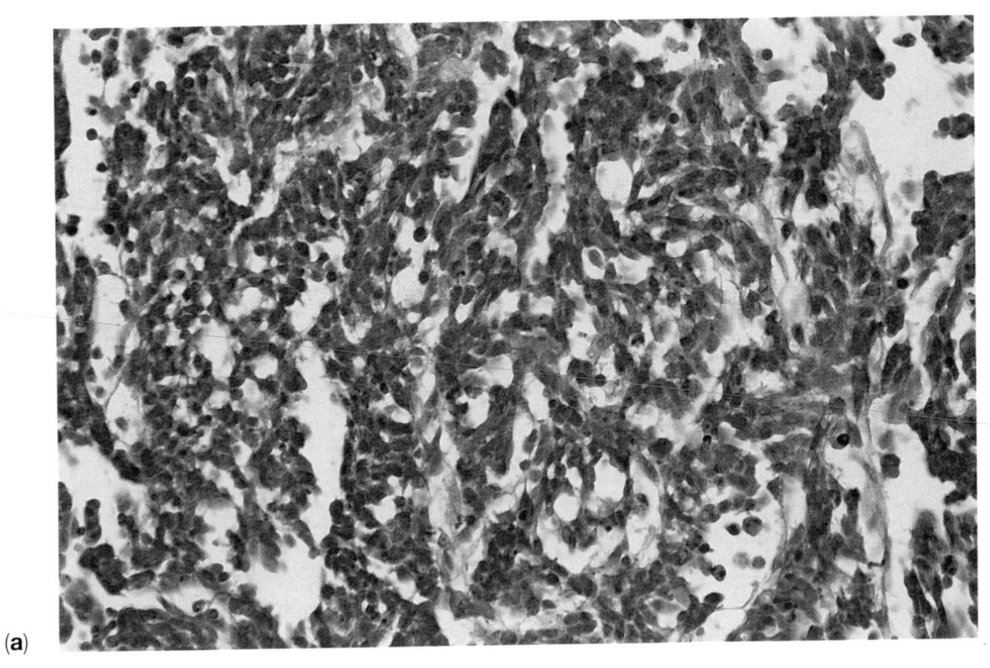

Figure 15.11 (**a**) Small cell anaplastic carcinoma of cervix. Note the close resemblance to small cell anaplastic (oat cell) carcinoma of lung (H & E, ×310).

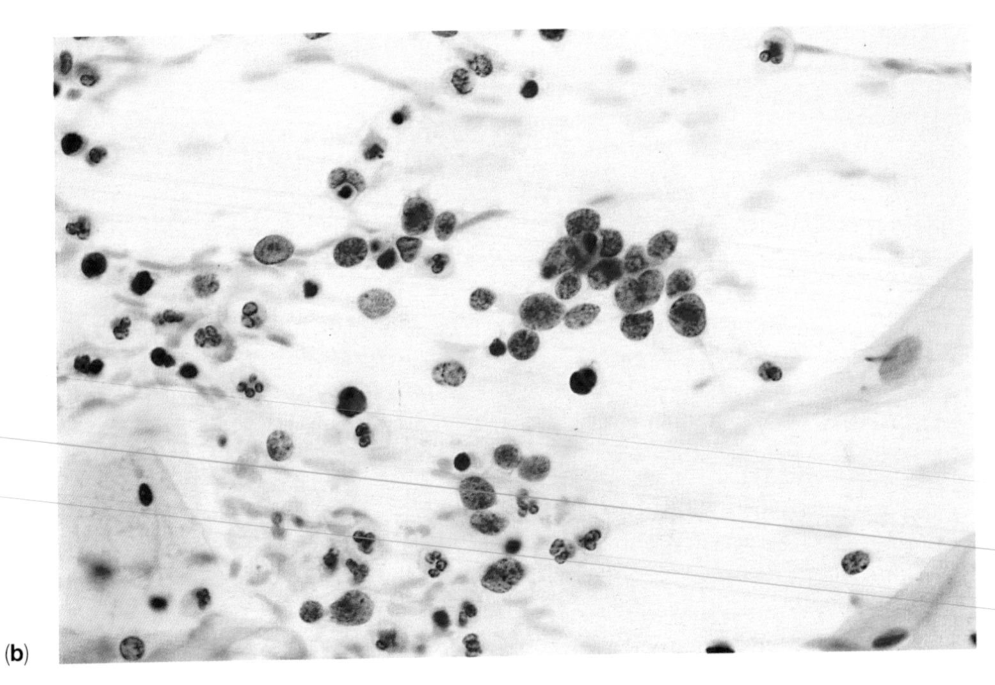

Figure 15.11 (**b**) Tumour cells from small cell anaplastic carcinoma. The cells are small and devoid of cytoplasm, and show nuclear moulding. Note the resemblance to small cell anaplastic (oat cell) carcinoma of lung (Pap, ×630).

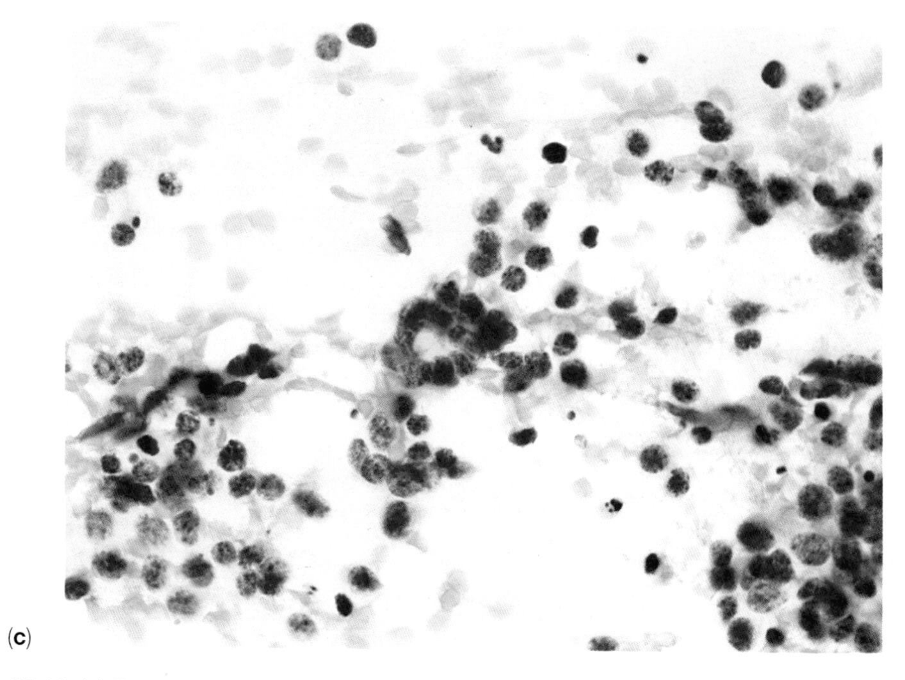

(c)

Figure 15.11 (**c**) Tumour cells from small cell anaplastic carcinoma. The cells are small and devoid of cytoplasm, and show nuclear moulding. Note the resemblance to small cell anaplastic (oat cell) carcinoma of lung (Pap, ×630).

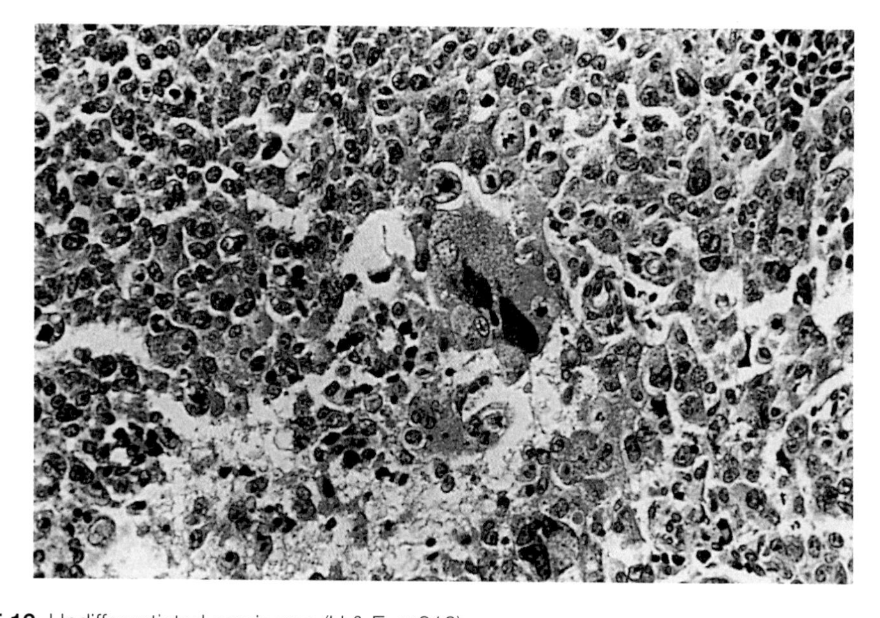

Figure 15.12 Undifferentiated carcinoma (H & E, ×310).

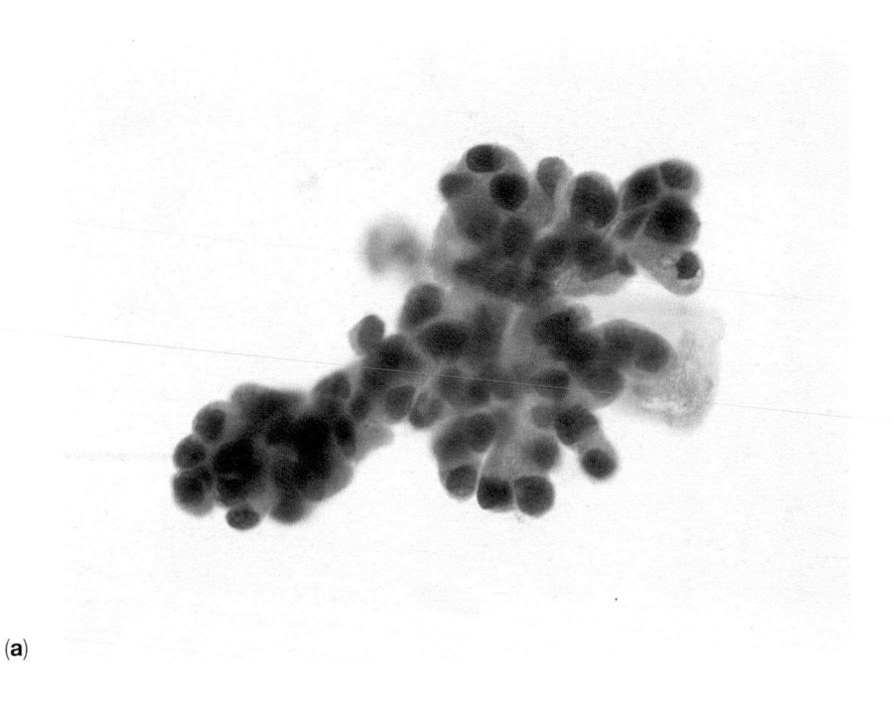

(a)

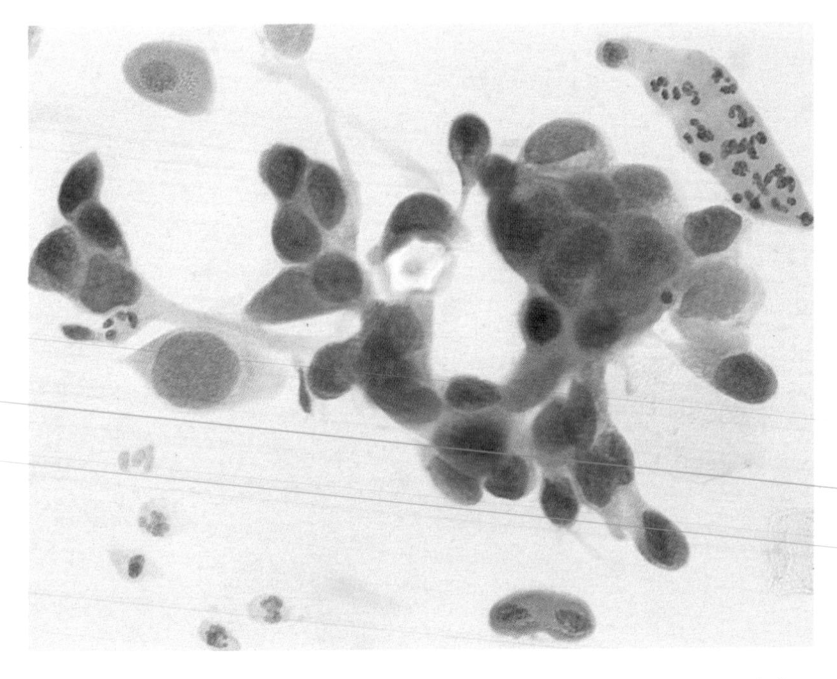

(b)

Figure 15.13 (**a,b**) Carcinoma of bladder cells in a cervical scrape. The cells are large and show some resemblance to urothelium (Pap, ×630).

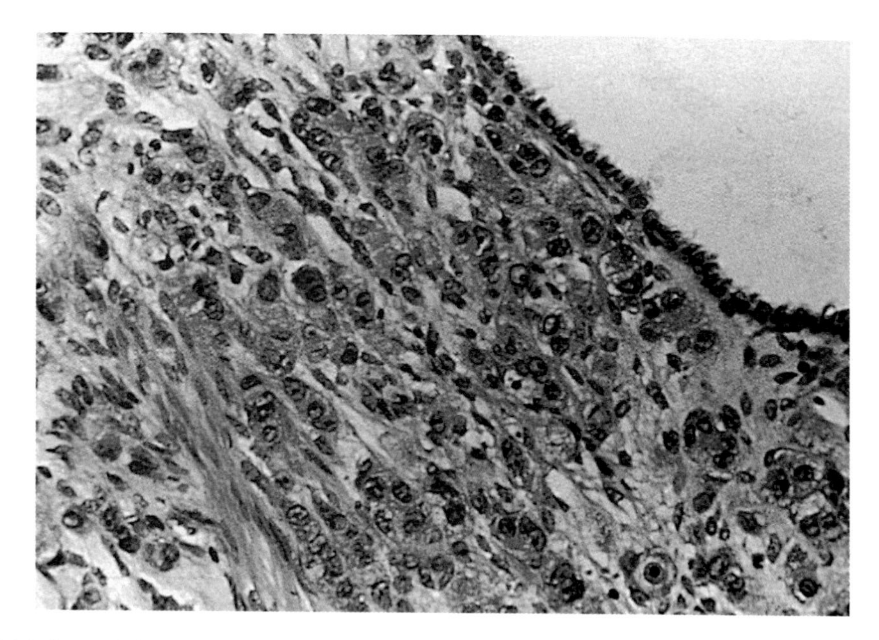

Figure 15.14 Deposit from breast carcinoma in cervix. The histological pattern resembles that of a primary breast carcinoma (H & E, ×310).

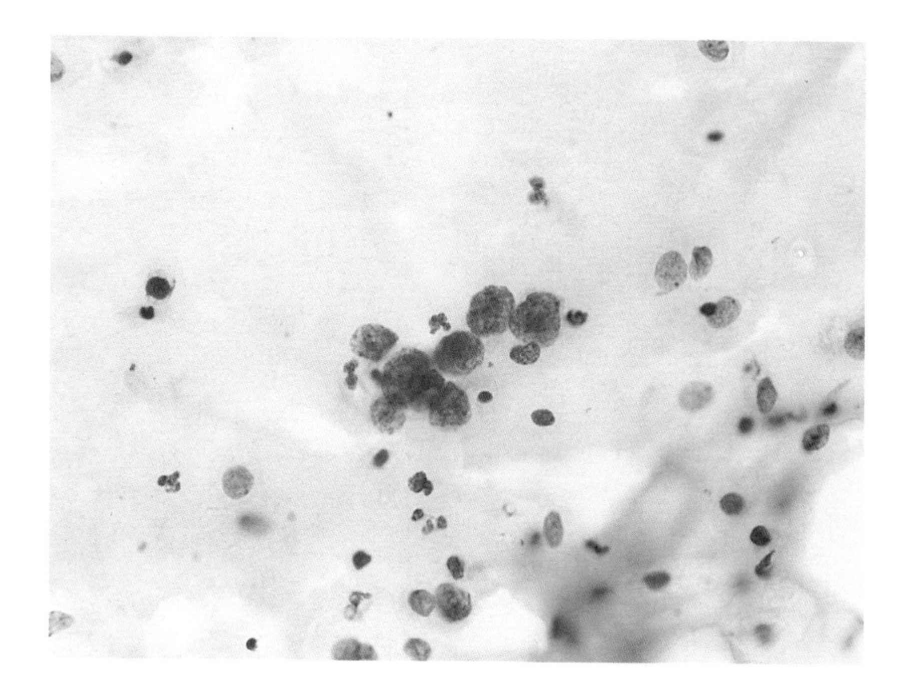

Figure 15.15 Breast carcinoma cells in a cervical scrape (Pap, ×630).

cervical carcinomas which appear undifferentiated by light microscopy showed features of glandular or squamous differentiation using electron microscopy.

15.7 METASTIC CARCINOMA OF THE CERVIX AND DIRECT SPREAD OF TUMOURS FROM OTHER SITES

Secondary tumours can involve the cervix from almost any primary site. *Endometrial carcinomas* are the most common, reaching the cervix either by direct spread, by vascular permeation, usually via the lymphatics, or by implantation. As a result of the latter, 'skip lesions' may develop in the endocervical canal, well away from the primary tumour. Such lesions have been described following dilatation and curettage (Fanning *et al.*, 1991). The differential diagnosis of primary cervical carcinoma, particularly of the endometrioid type from endometrial carcinoma may be very difficult. Not infrequently, primary cervical adenocarcinomas have a pattern which varies from one part of the tumour to another so that careful examination of all available material may enable a characteristically cervical pattern to be distinguished. The value of histochemistry and the demonstration of CEA antigen by immunohistochemistry is considered in Section 14.7.3.

Direct invasion of the cervix also occurs with *choriocarcinoma, carcinoma of the bladder, carcinoma of the rectum or pelvic colon, carcinoma and melanoma of the vagina* (Figure 16.21) and *sarcomas of the corpus uteri*. The cytological appearance of a bladder carcinoma involving the cervix is shown in Figure 15.13(a,b).

Metastases to the cervix from distant primaries are rare, the two most common being *breast* and *stomach*. A metastatic deposit of breast carcinoma is shown in Figure 15.14. Its histological pattern is usually distinctive, resembling a primary breast cancer in its structure and is generally unlikely to be confused with a primary cervical tumour. Its metastatic nature may also be suspected cytologically (Figure 15.15) if the possibility is borne in mind. A carcinomatous deposit from *ovary*, however (Figures 15.16 and 15.17), may resemble a papillary adenocarcinoma of cervix, a possibility which should be considered whenever the clinical or histological findings are unusual (Danielian, 1990). Fortunately, ovarian deposits in the cervix are very rare and when in doubt a primary cervical carcinoma is much more probable.

Other primary tumour sites which may metastasize to the cervix include gallbladder, pancreas and thyroid. Malignant melanomatous deposits can also occur (Figure 16.21(a,b)). Leukaemic infiltration of the cervix is not uncommon as a late manifestation (Lucia *et al.*, 1952). Secondary lymphomatous infiltration is occasionally seen (Lathrop, 1967); it is more common than primary lymphoma which it resembles histologically and cytologically (Section 16.10) but is distinguished clinically by the presence of a primary lesion elsewhere. Cervical deposits of granulosa cell carcinoma of ovary are rare (Figure 15.18).

It is not uncommon for metastatic carcinoma involving the cervix to be detected in cervical smears. Cells *exfoliated from the upper genital tract* may be found in a cervical scrape. Cells from an adenocarcinoma of the endometrium are relatively common but may easily be missed (Figures 14.29, 14.31 and 15.19). Papillary clusters of cells from metastatic deposit ovarian carcinoma can be seen in Figure 15.17. A metastatic deposit from a chorion carcinoma in a cervical smear is shown in Figure 15.20. The cytology of metastatic colon carcinoma is shown in Figure 15.21.

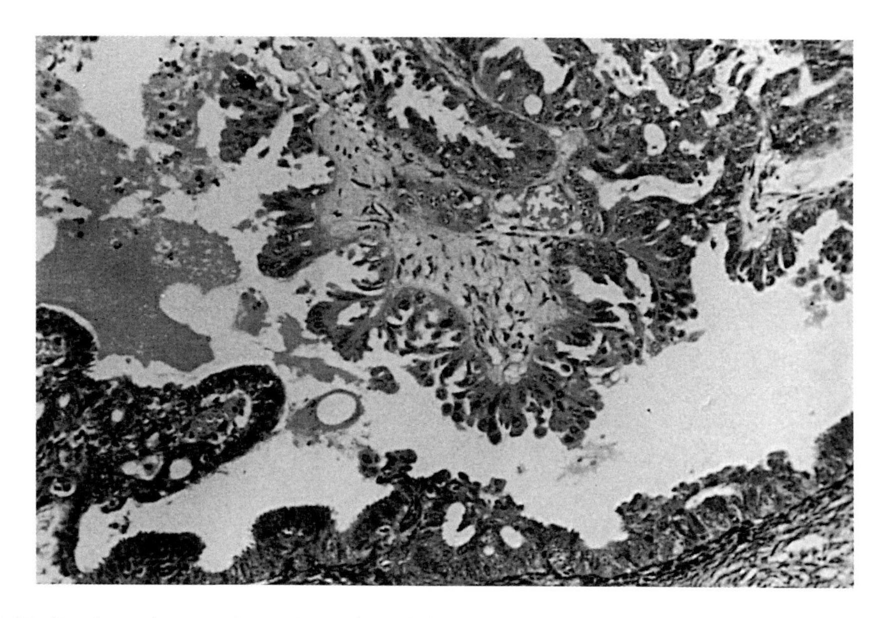

Figure 15.16 Ovarian adenocarcinomatous deposit in cervix, showing similarity to a primary papillary adenocarcinoma of cervix (H & E, ×190).

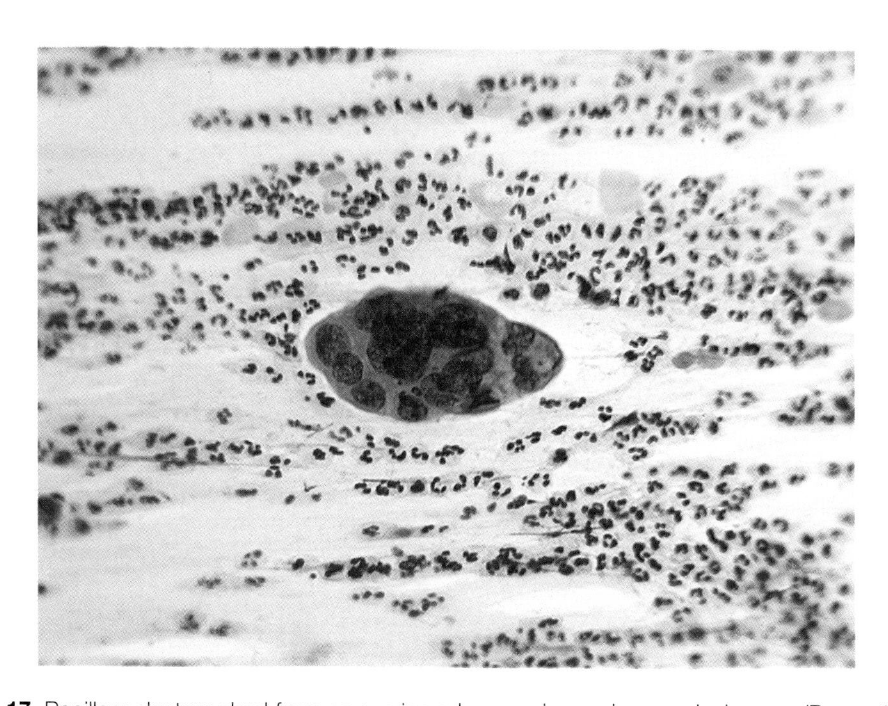

Figure 15.17 Papillary clusters shed from an ovarian adenocarcinoma in a cervical smear (Pap, ×630).

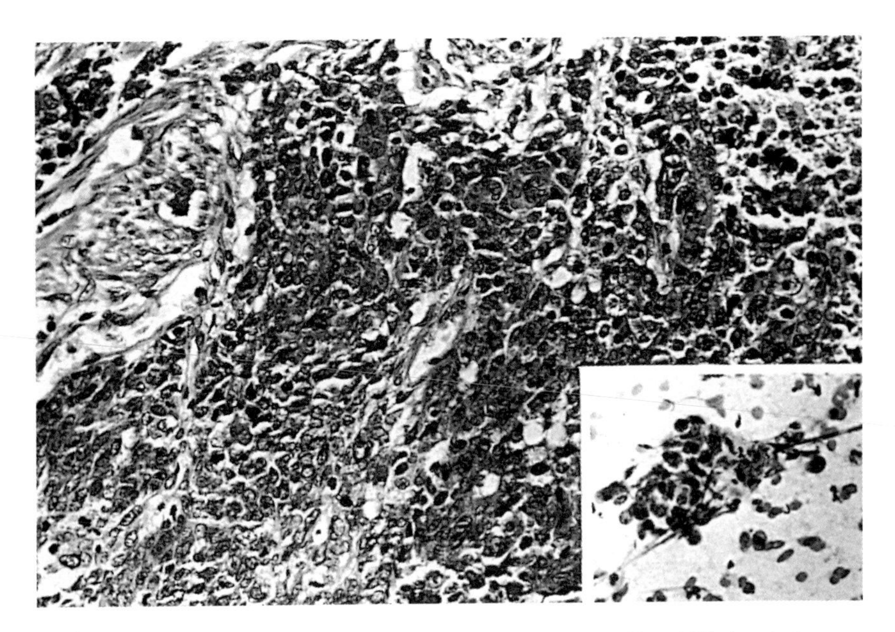

Figure 15.18 Cervical deposit of granulosa cell carcinoma of ovary (H & E, ×480). Inset shows granulosa cells in cervical smear from same case (Pap, ×480) (courtesy of Dr D.J.B. Ashley, Morriston Hospital, Swansea).

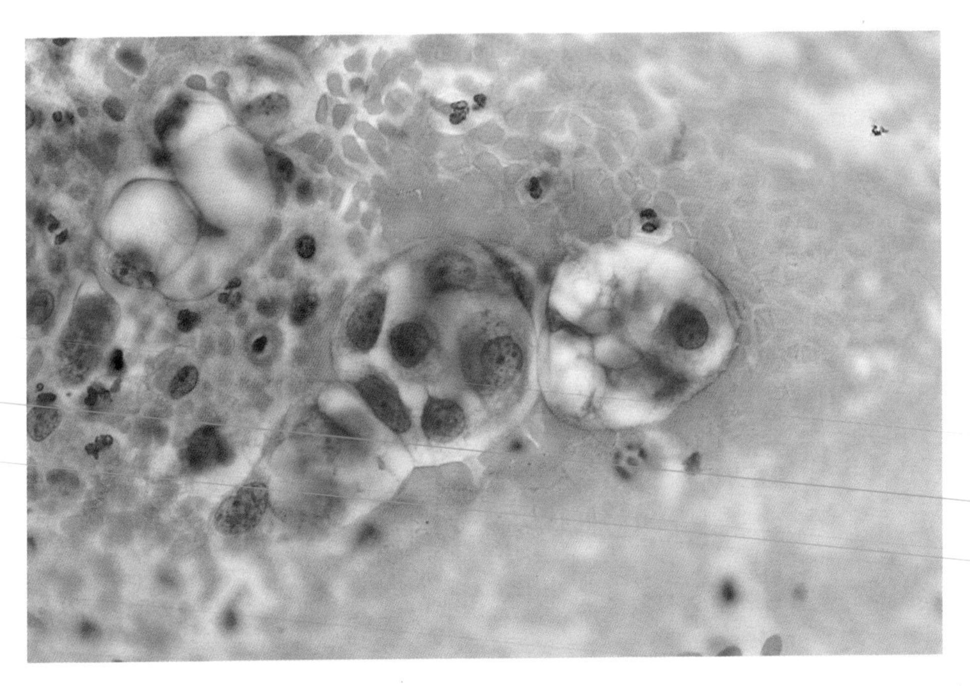

Figure 15.19 Malignant cell clusters from an endometrial carcinoma that has extended onto the cervix. Note the well-preserved appearance of the clusters (Pap, ×630).

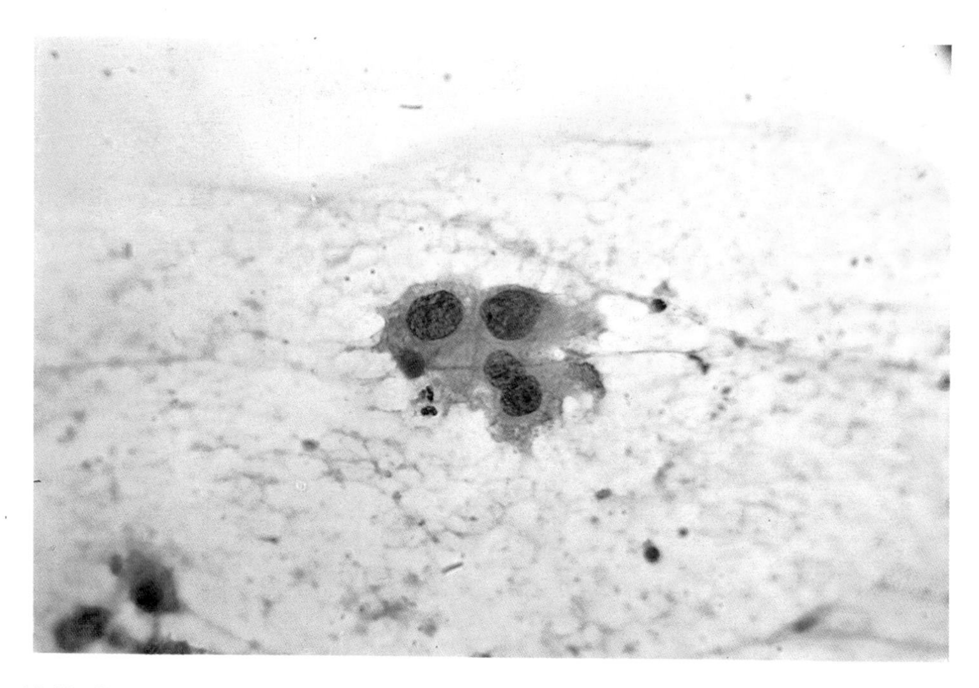

Figure 15.20 Cluster of malignant cells in a smear from a patient found to have chorion carcinoma (Pap, ×630).

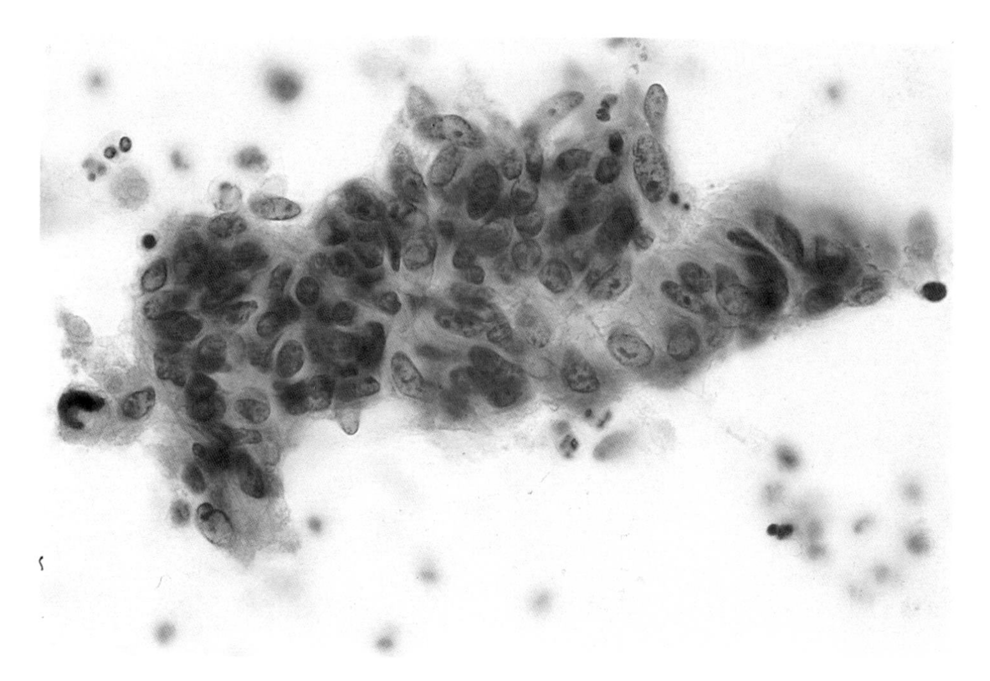

Figure 15.21 Metastatic colon carcinoma in a cervical smear (Pap, ×630). (Courtesy of Dr Psaropoulou).

REFERENCES

Albores-Saavedra, J., Larraza, O., Poucell, S. and Rodriguez-Martinez, H. A. (1976) Carcinoid of the uterine cervix. Additional observations on a new tumour entity. *Cancer*, **38**, 2328–2342.

Auersperg, N., Erber, H. and Worgh, A. (1972) Histologic variations among poorly differentiated invasive carcinomas of the uterine cervix. *J. Natl. Cancer Inst.*, **51**, 1461–1477.

Buckley, C. H. and Fox, H. (1989) Carcinoma of the cervix, in *Recent Advances in Histopathology, 14* (eds. P. P. Anthony and R. N. M. Macsween), Churchill Livingstone, Edinburgh, pp. 63–78.

Buckley, C. H., Beards, C. S. and Fox, H. (1988) Pathologic prognostic indicators in cervical cancer with particular reference to patients under the age of 40 years. *Br. J. Obstet. Gynaecol.*, **95**, 47–56.

Christopherson, W. M., Nealon, N. and Gray, L. A. (1979) Non-invasive precursor lesions of adenocarcinoma and mixed adenosquamous carcinoma of the uterine cervix. *Cancer*, **44**, 974–983.

Danielian, P. J. (1990) Ovarian metastatic carcinoma presenting as a primary cervical carcinoma. *Acta Obstet. Gynaecol. Scand.*, **69**, 265–266.

Daroca, P. J. and Dhurandhar, H. N. (1980) Basaloid carcinoma of uterine cervix. *Am. J. Surg. Pathol.*, **44**, 235–239.

Dougherty, C. M. and Cotton, N. (1964) Mixed squamous and adenocarcinoma of the uterine cervix. *Obstet. Gynecol.*, **45**, 79–83.

Fanning, J., Alvarez, P. M., Tsukada, Y. and Piver, M. S. (1991) Cervical implantation metastasis by endometrial adenocarcinoma. *Cancer*, **68**, 1335–1339.

Ferry, J. A. and Scully, R. E. (1988) 'Adenoid cystic' carcinoma and adenoid basal carcinoma of the uterine cervix. A study of 28 cases. *Am. J. Surg. Pathol.*, **12**, 134–144.

Fowler, W. C., Miles, P. A., Surwit, E. A. *et al.* (1978) Adenoid cystic carcinoma of the cervix. Report of 9 cases and a reappraisal. *Obstet. Gynecol.*, **52**, 337–342.

Gersell, D. J., Mazoujian, G., Mutch, D. G. and Rudloff, M. A. (1988) Small cell undifferentiated carcinoma of the cervix. A clinical, ultrastructural and immunocytochemical study of 15 cases. *Am. J. Surg. Pathol.*, **12**(9), 684–698.

Glucksmann, A. and Cherry, C. P. (1956) Incidence, histology and response to radiation of mixed carcinomas (adenocanthomas) of the uterine cervix. *Cancer*, **9**, 971–979.

Gou, K. J., Fujii, S. and Konishi, I. (1991) Clinicopathological study of cervical carcinoma massive infiltration of eosinophils (Jpn). *Acta Obstet. Gynaecol. Jap.*, **43**, 749–755.

Habib, A., Kaneko, M., Cohen, C. J. and Walker, G. (1979) Carcinoid of the uterine cervix. A case report with light and electron microscopy studies. *Cancer*, **43**, 535–538.

King, L. A., Talledo, O. E. and Gallup, D. G. (1989) Adenoid cystic carcinoma of the cervix in women under age 40. *Gynecol. Oncol.*, **32**, 26–30.

Matsuyama, M., Inoue, T., Aryoshi, Y. *et al.* (1979) Argrophyl carcinoma of the uterine cervix with ectopic production of ACTH, B-MSH, serotonin, histamine and amylase. *Cancer*, **44**, 1813–1823.

Mazur, M. T. and Battifora, H. A. (1982) Adenoid cystic carcinoma of the uterine cervix. Ultrastructure, immunofluorescence and criteria for diagnosis. *Am. J. Clin. Pathol.*, **77**, 494–500.

Mullins, J. D. and Hilliard, G. D. (1981) Cervical carcinoid ('argyrophil cell' carcinoma) associated with an endocervical adenocarcinoma: a light and ultrastructural study. *Cancer*, **47**, 785–790.

Musa, A. G., Hughes, R. R. and Coleman, S. A. (1985) Adenoid cystic carcinoma of the cervix: a report of 17 cases. *Gynecol. Oncol.*, **22**, 167–173.

Poulsen, H. E., Taylor, C. W. and Sobin, L. H. (1975) *Histological typing of female genital tract tumours.* WHO, Geneva, pp. 57–79.

Richard, L., Guralnick, M. and Ferenczy, A. (1981) Ultrastructure of glassy cell carcinoma of the cervix. *Diagn. Gynecol. Obstet.*, **3**, 31.

Scully, R. E., Bonfiglio, R. J. and Kurman, R. J. (1994) *WHO International histological classification of tumours. Histological typing of female genital tract tumours.* Springer-Verlag, Berlin, pp. 39–54.

Shingleton, H. M., Bone, H., Bradley, D. H. and Soong, S. J. (1981) Adenocarcinoma of the cervix, 1: clinical evaluation and pathological features. *Am. J. Obstet. Gynecol.*, **139**, 799–814.

Stassart, J., Crum, C. P., Yordan, C. L. *et al.* (1982) Argyrophilic carcinoma of the cervix: a report of a case with coexisting cervical intraepithelial neoplasia. *Gynecol. Oncol.*, **13**, 247–251.

Stoler, M. H., Stacey, M., Gersell, D. J. and Walker, A. N. (1991) Small-cell neuroendocrine carcinoma of the cervix. *Am. J. Surg. Pathol.*, **15**(1), 28–32.

Tateishi, R., Wada, A., Hayakawa, K. *et al.* (1975) Argyrophil carcinomas (apudomas) of the uterine cervix. *Virchows Arch, Pathol. Anat. Pathol.*, **366**, 257–273.

Talerman, A., Alenghat, E. and Okagaki, T. (1991) Glassy cell carcinoma of the uterine cervix. *Acta Pathol. Microbiol. Immunol. Scand. Suppl.*, **23**, 119–125.

Thelmo, W. L., Nicastri, A. D., Fruchter, R. *et al.*

(1990) Mucoepidermoid carcinoma of uterine cervix stage 1B. Long-term follow-up, histochemical and immunochemical study. *Int. J. Gynecol. Pathol.*, **9**, 316–324.
Wells, M. and Brown, L. J. C. (1986) Glandular lesions of the uterine cervix: the present state of our knowledge. *Histopathology*, **10**, 777–792.
Wright, T. C., Ferenczy, A. and Kurman, R. J. (1994) Carcinoma and other tumors of the cervix, in *Blaustein's Pathology of the Female Genital Tract*, 4th edn (ed. R. J. Kurman), Springer-Verlag, New York, pp. 274–326.

16

MESENCHYMAL AND MISCELLANEOUS TUMOURS

In this chapter the following tumours are considered:

Mesenchymal tumours
- Leiomyoma (fibromyoma)
- Adenomyoma and fibroadenoma
- Leiomyosarcoma
- Sarcoma botryoides (embryonal rhabdomyosarcoma)
- Stromal tumours, endocervical and endometrial
- Alveolar soft part sarcoma (ASPS)
- Mixed epithelial and mesenchymal tumours
  - Adenofibroma
  - Adenosarcoma
    - Malignant mesodermal mixed tumour (carcinosarcoma)
  - Other sarcomas

Miscellaneous tumours
- Pigmented lesions
  - Blue naevus
  - Malignant melanoma
- Malignant lymphoma
- Leukaemia and granulocytic sarcoma

16.1 LEIOMYOMA (FIBROMYOMA)

This is a benign neoplasm composed of smooth muscle cells with a variable amount of fibrous stroma (Scully *et al.*, 1994). It is the most common tumour of the uterus and about 8% of uterine leiomyomas occur in the cervix (Fluhman, 1961). They are usually solitary and as they enlarge, they cause distortion of the cervix, sometimes sufficient to cause obstruction of labour. An example of a small leiomyoma is shown in Figure 16.1. Its histological structure (Figure 16.2) is essentially similar to its counterpart, arising from the myometrium of the corpus uteri. It consists of bundles of smooth muscle fibres arranged in an interweaving whorled pattern. Zones of hyalinization are not uncommon (Figure 16.3). Cystic change, calcification and haemorrhage can also occur. Sometimes there is swelling and distortion of the muscle cells with pleomorphism of the nuclei. These changes occur especially in relation to pregnancy and progestogen therapy (Poulsen *et al.*, 1975). Most leiomyomas arise from the smooth muscle of the cervix but occasionally one arises from the muscular wall of a cluster of blood vessels; known as a 'vascular leiomyoma' type (Figure 16.4) it is often associated with a number of hyalinized arteries (Wright and Ferenczy, 1994). It may resemble a haemangioma but the latter is very rare in the uterus and tends to have numerous gaping vascular and capillary-lined spaces (Norris and Zaloudek, 1982).

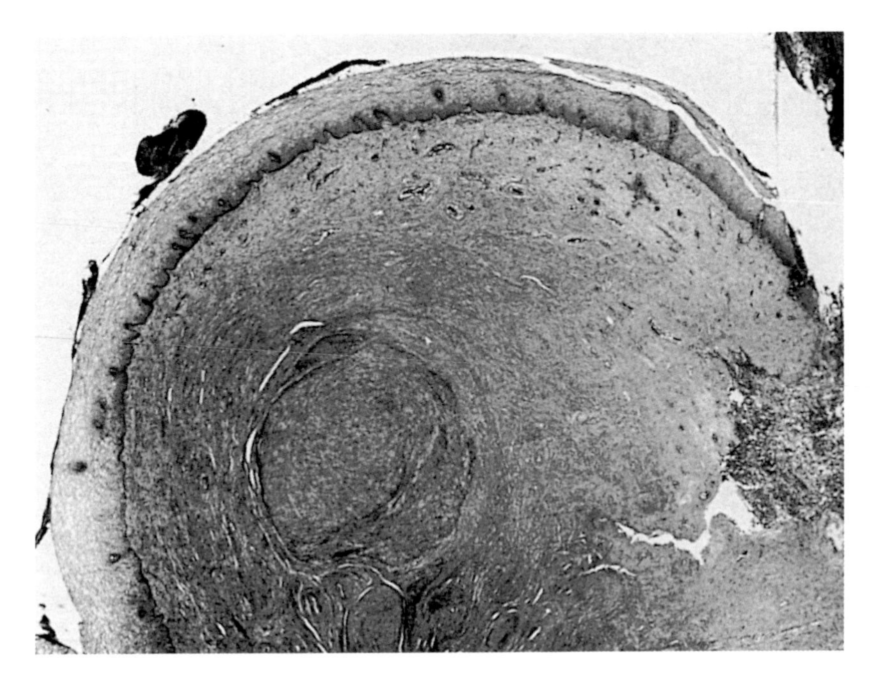

Figure 16.1 Cervical leiomyoma. Lesions causing symptoms are usually considerably larger than this (H & E, ×12).

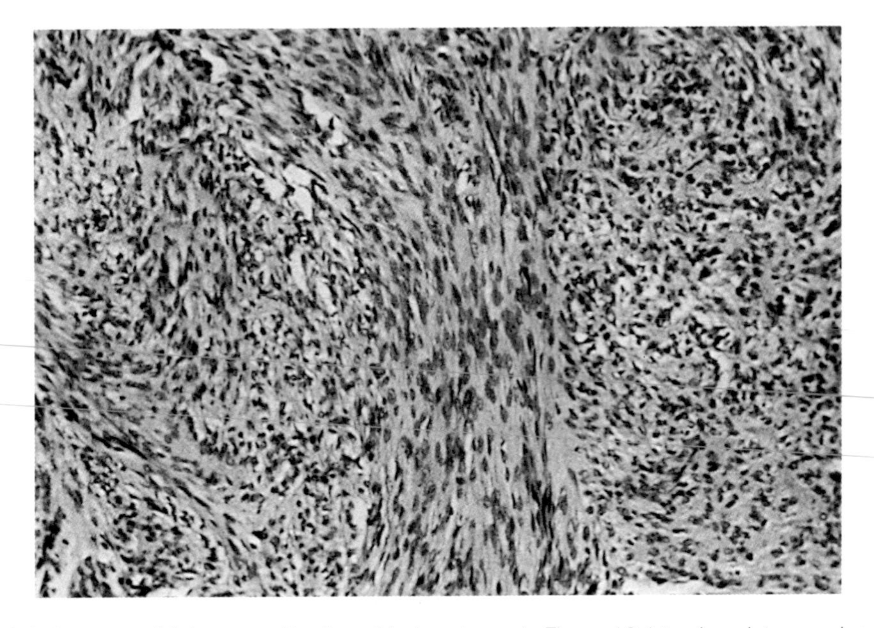

Figure 16.2 Leiomyoma. Higher magnification of lesion shown in Figure 16.1 to show interweaving bundles of smooth muscle (H & E, ×190).

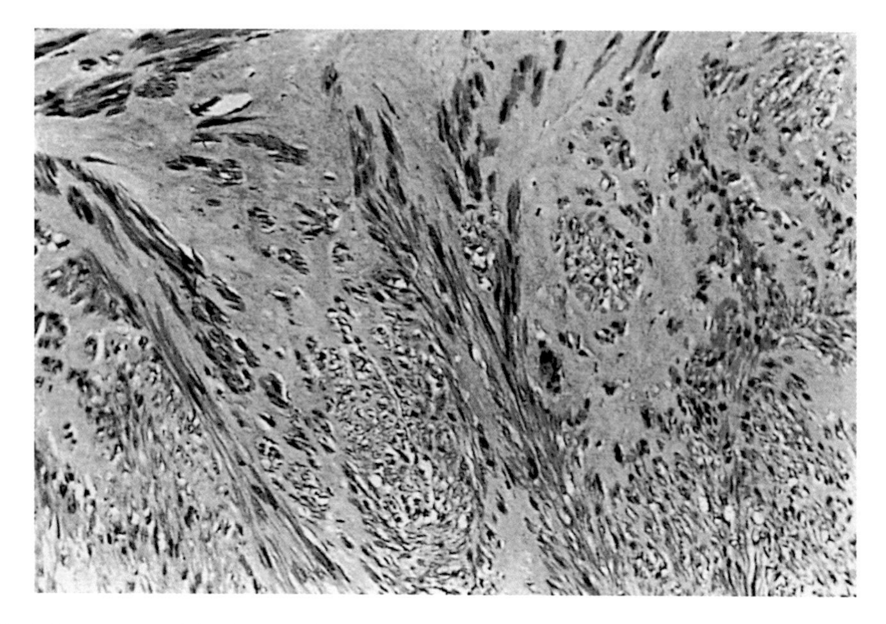

Figure 16.3 Zones of hyalinization in a leiomyoma (H & E, ×210).

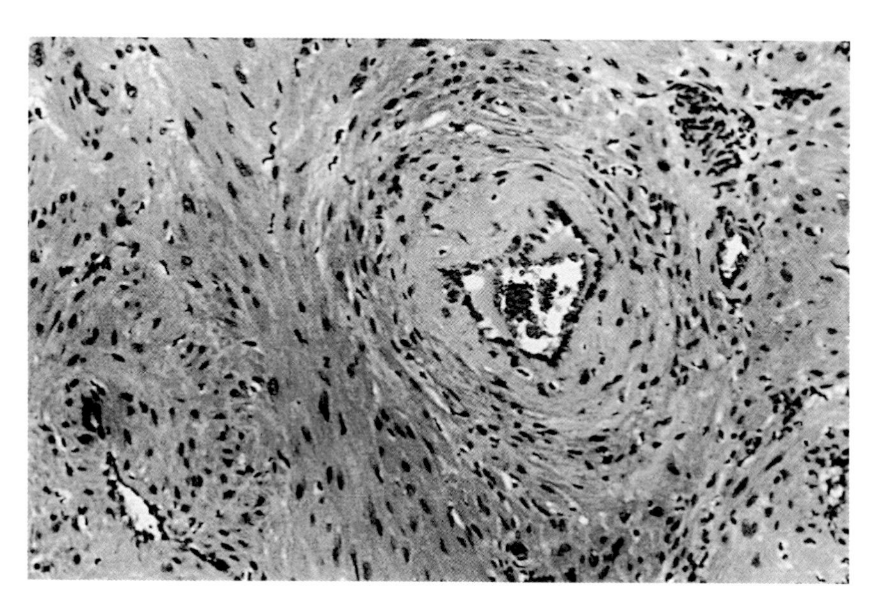

Figure 16.4 Part of a vascular leiomyoma showing continuity between the smooth muscle cells of the tumour and the muscular wall of the blood vessels (H & E, ×190).

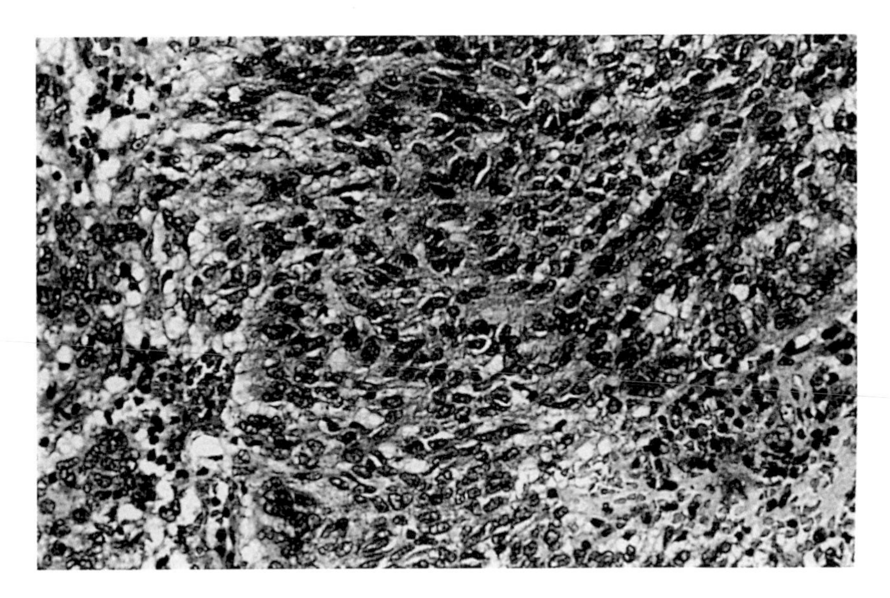

Figure 16.5 Cellular leiomyoma with densely packed plump muscle cells having prominent nuclei and ill-defined interweaving structure (H & E, ×610).

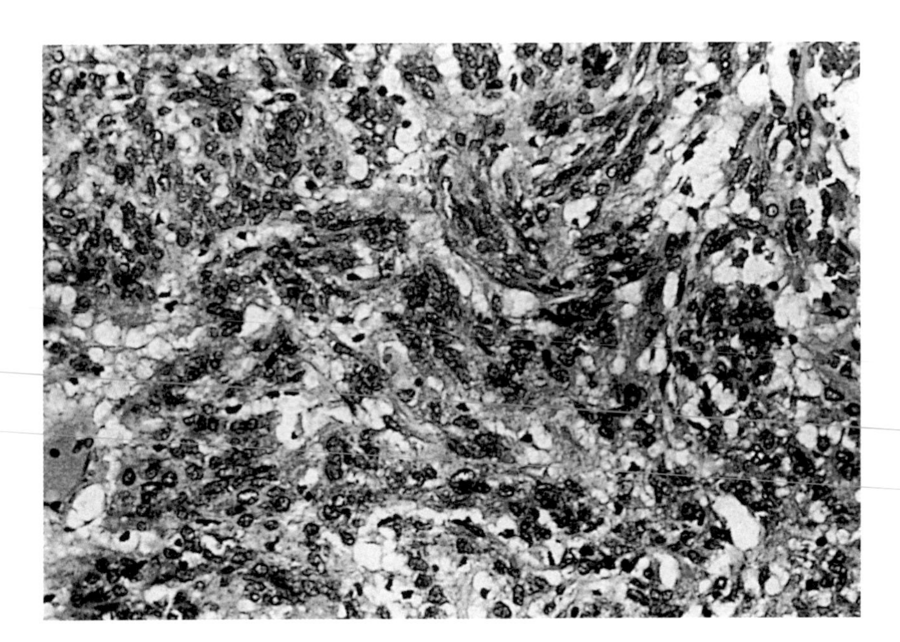

Figure 16.6 An atypical leiomyoma with large irregular nuclei, including some multinucleate forms. This is a variant of cellular leiomyoma occurring in same tumour as shown in Figure 16.6 (H & E, ×310).

There are several histological forms of leiomyoma which mimic – and must be distinguished from – leiomyosarcoma. The main forms are cellular, atypical and epithelioid leiomyomas. *Cellular leiomyoma* (Figure 16.5) is composed of rather plump muscle cells with prominent nuclei which tend to be densely packed together and it lacks a well-defined structure of interweaving muscle bundles. *Atypical leiomyoma* is a cellular leiomyoma in which many cells have large irregular nuclei, with scattered multinucleate forms (Figure 16.6). *Epithelioid leiomyoma* contains clusters or cords of cells which are often rounded or polygonal (Figure 16.7). Such a tumour in which the cells have an eosinophilic cytoplasm has been designated a leiomyoblastoma (Norris and Zaloudek, 1982). Epithelioid leiomyoma may also be of clear cell or plexiform pattern. In the clear cell form (Figure 16.8(a)) the nuclei are often displaced to the cell margin by the cytoplasmic vacuolation to produce a signet ring appearance. In the *plexiform tumour* or tumourlet (Patchefsky, 1970), there are little groups of cells which often form groups round capillaries (Figure 16.8(b)) but ultrastructurally have been found to contain the characteristic myofilaments and fibrils of a smooth muscle tumour (Nunez-Alonso and Battifora, 1979). These uncommon variants of leiomyoma may be regarded as patterns of tumour organization rather than specific types; more than one pattern may be present in one tumour (Figure 16.9). Even rarer variants are myxoid leiomyoma, in which there is abundant myxoid material between the smooth muscle cells, and lipoleiomyoma, in which there is an admixture of fat cells, smooth muscle and collagen; if the latter includes angiomatous tissue it becomes an angiolipoleiomyoma (Zaloudek and Norris, 1994). There may occasionally be a fatty component in epithelioid leiomyoma.

The feature which distinguishes most forms of leiomyoma from leiomyosarcoma is the absence or great sparsity of mitotic figures, <3 per 10 high-power fields (HPF) and usually only one in 20 to 50 HPF (Norris and Zaloudek,

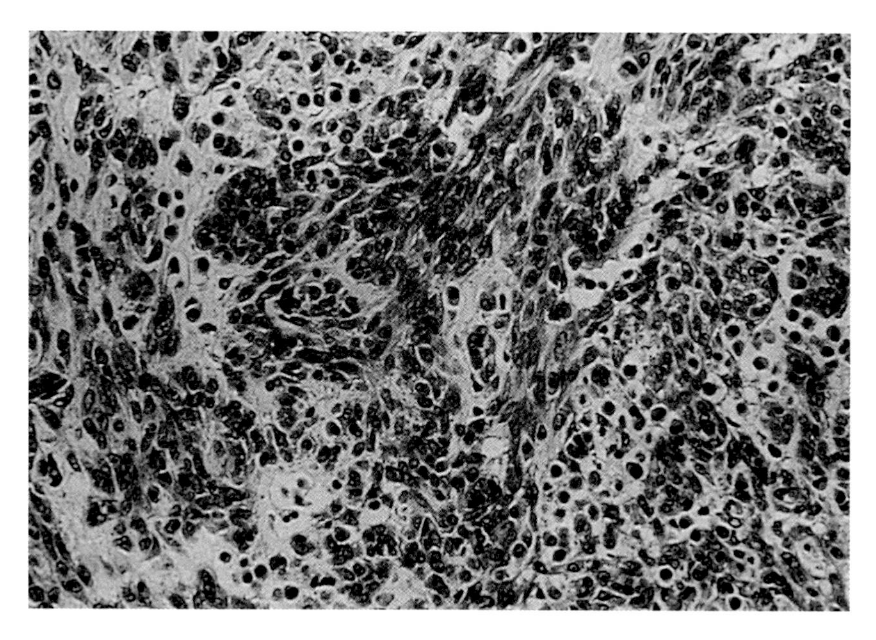

Figure 16.7 An epithelioid leiomyoma contains round or polygonal cells arranged in cords and clusters. An eosinopilic cytoplasm characterizes the subtype leiomyoblastoma (H & E, ×310).

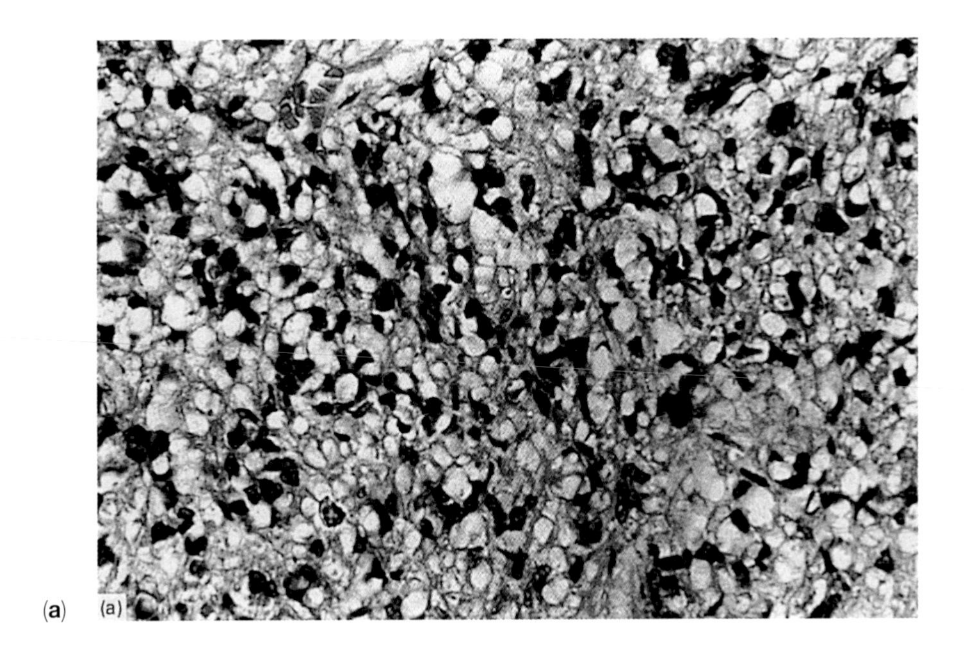

(**a**)

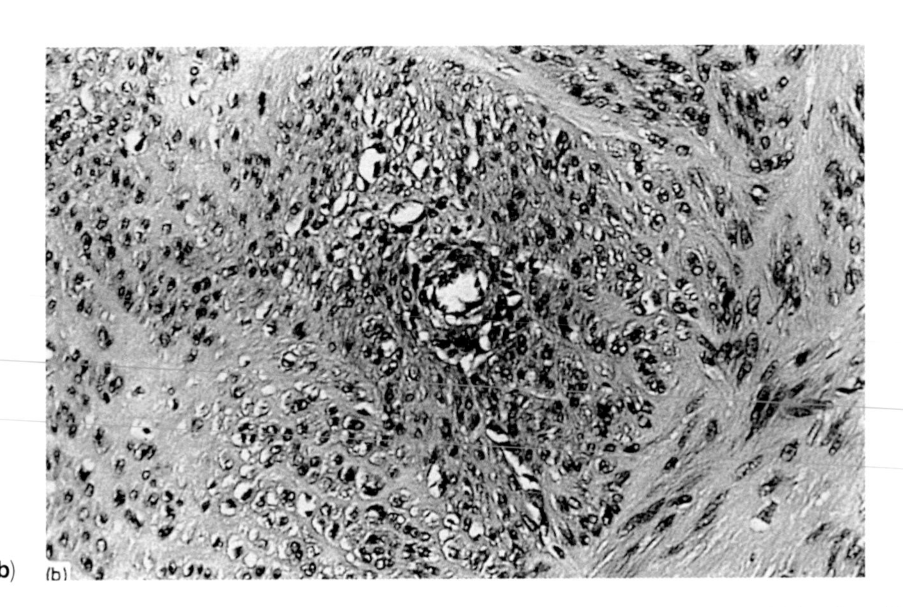

(**b**)

Figure 16.8 Epithelioid leiomyoma. (**a**) Clear cell type. The cells contain vacuoles which often displace the nucleus to the margin of the cell to give a signet ring appearance (H & E, ×580). (**b**) Plexiform type, showing leiomyoma cells clustered round capillaries (H & E, ×310).

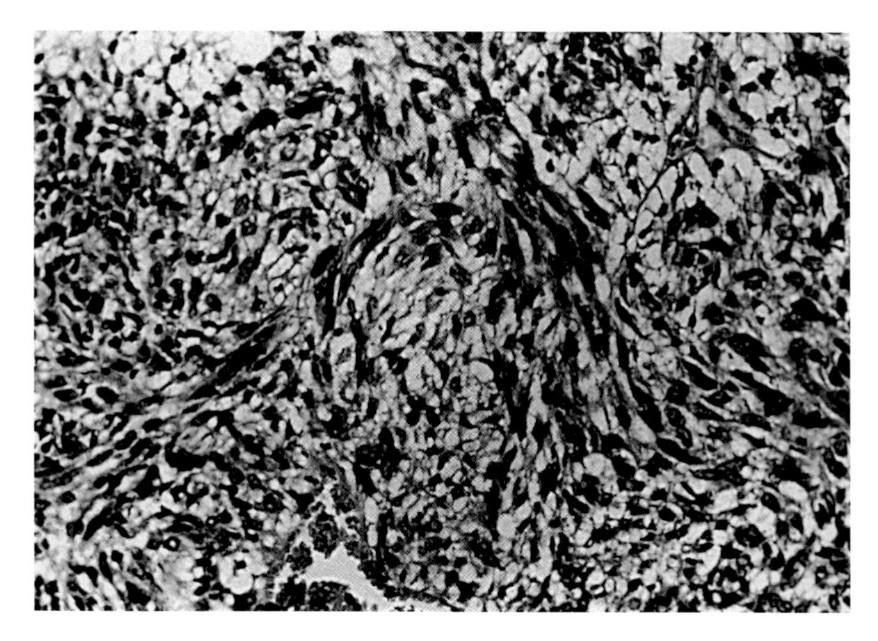

Figure 16.9 Variable pattern in a leiomyoma, that in the top right showing resemblance to some published examples of plexiform tumour. Same lesion as shown in Figure 16.8 (H & E, ×310).

1982). However, occasional leiomyomas in premenopausal women, known as 'mitotically active leiomyomas', clinically benign, have 5–15 mitotic figures per 10 HPF (Norris and Zaloudek, 1994). This diagnosis should not be made if the tumour contains atypical cells. Cellular atypicality in the absence of large numbers of mitotic figures or invasion is not indicative of malignancy (Corscaden and Singh, 1958). *Metastasizing leiomyoma* is apparently indistinguishable from a benign leiomyoma histologically and yet it gives rise to secondary deposits. Fortunately, it is extremely rare (see also 'Mitotically active leiomyomas', p. 403).

Usually leiomyomas do not reveal any evidence of their presence in cervical scrape preparations because of their submucosal situation. Occasionally, however, muscle fibres from a leiomyoma may be detected cytologically if the overlying mucosa becomes damaged and removed (Figure 16.10(a)).

16.2 ADENOMYOMA AND FIBROADENOMA

An adenomyoma is a benign tumour composed mainly of smooth muscle, resembling a leiomyoma but in addition containing glandular tissue. It is a rare lesion in the cervix. Wright and Ferenczy (1984) use the terms adenomyoma and fibroadenoma to describe benign tumours of smooth muscle and fibrous connective tissue respectively, each with intermingling mucin-secreting epithelium. It may be difficult to be certain whether the intermingling glands are cervical or endometrial (Figure 16.10(b))

Adenomyoma of endometrial type may present as a polypoid lesion at the external os. In a 'typical' lesion, the glands are of normal endometrial type. In an 'atypical polypoid adenomyoma' (Young *et al.*, 1986) the endometrial glands show varying degrees of hyperplasia, squamous metaplasia and atypicality up to

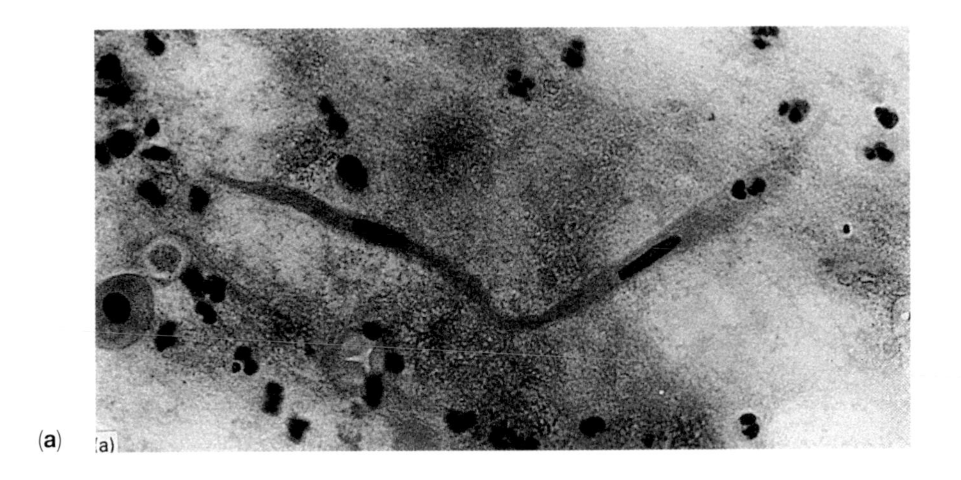

(**a**)

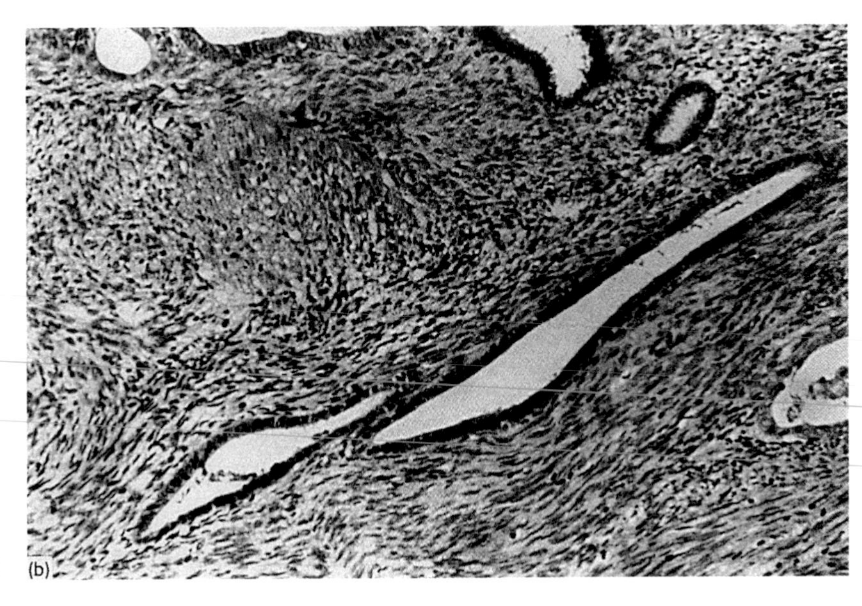

(**b**)

Figure 16.10 (**a**) Leiomyoma cells in a cervical scrape, showing characteristic spindle shape and elongated nucleus with a regular outline (Pap, ×770). (**b**) Cervical adenoymoma, consisting of bundles of smooth muscles with intermingling glands of indeterminate type (H & E, ×190).

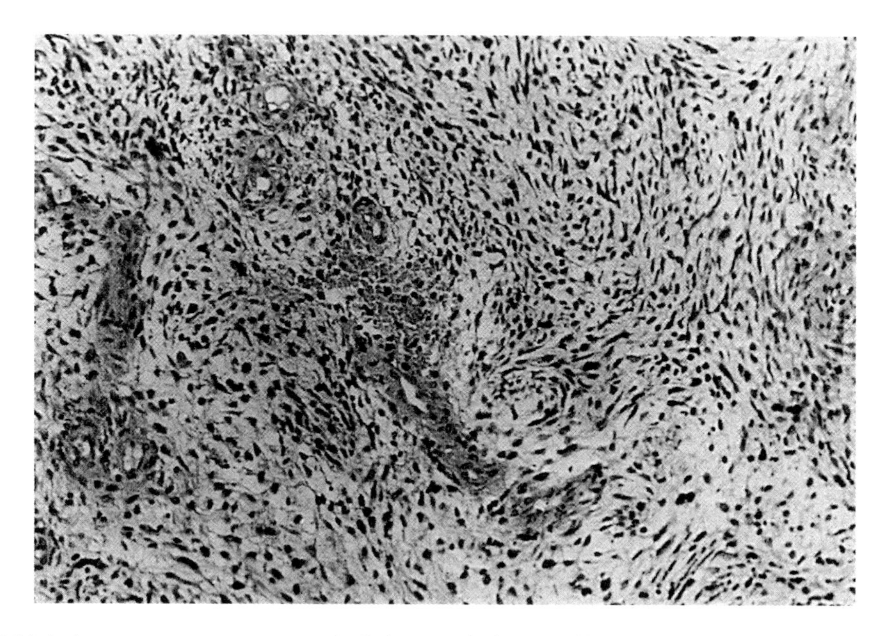

Figure 16.11 Leiomyosarcoma, composed of elongated pleomorphic cells and having a poorly defined interweaving pattern (H & E, ×190).

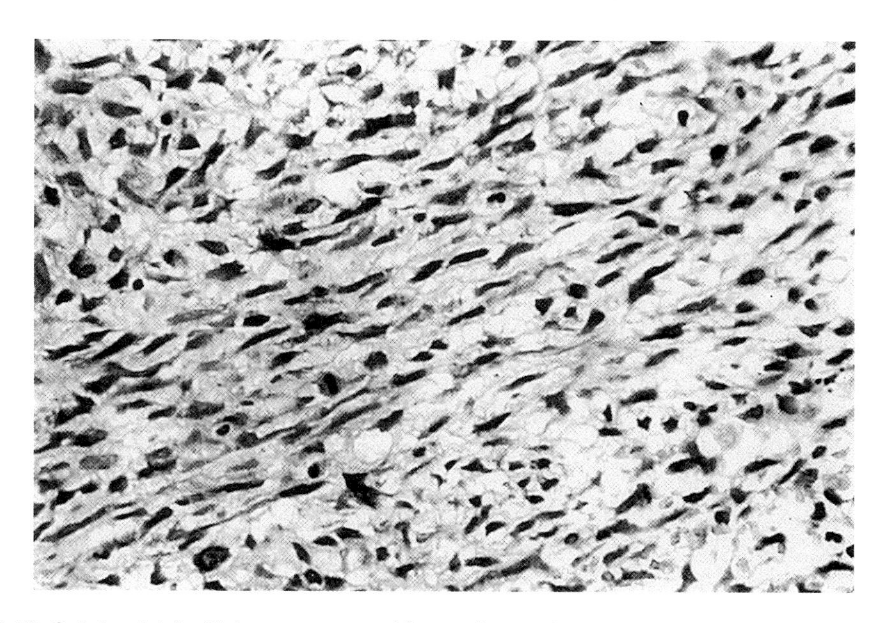

Figure 16.12 Cellular detail of leiomyosarcoma. Most cells are elongated and some show considerable pleomorphism. Same lesion as shown in Figure 16.11 (H & E, ×480).

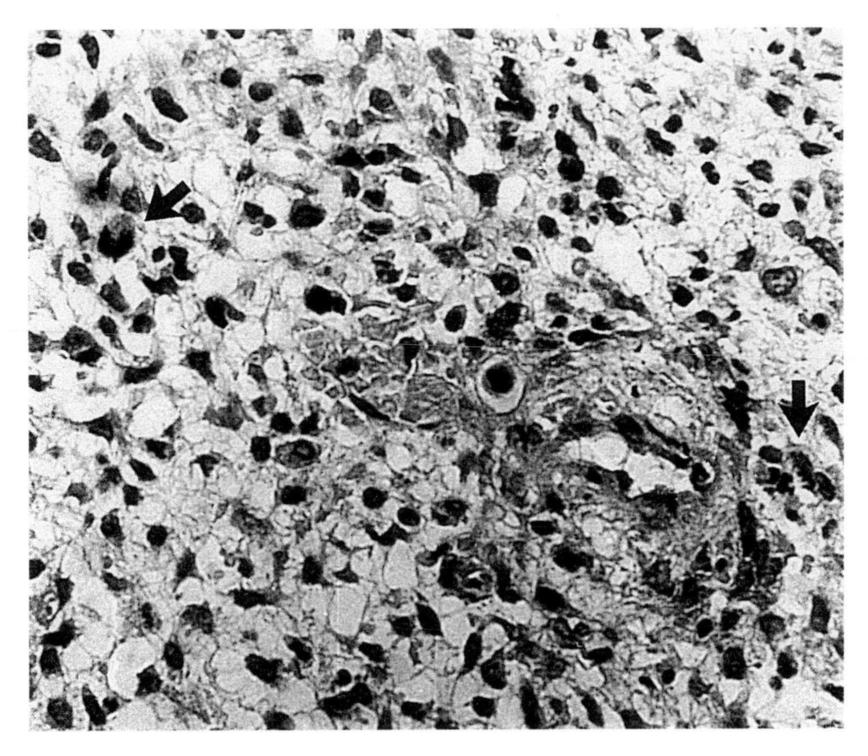

Figure 16.13 Mitotic figures (arrowed) in a leimyosarcoma provide the most reliable distinguishing feature from a leiomyoma. Same tumour as shown in Figures 16.11 and 16.12.

carcinoma *in situ* within neoplastic smooth muscle which occasionally is also atypical. Follow-up data suggest that the lesion is benign.

16.3 LEIOMYOSARCOMA

This malignant tumour of smooth muscle cells is the most common type of sarcoma to arise in the cervix, usually occurring between the ages of 40 and 60 years and a higher proportion of leiomyosarcomas than leiomyomas involve the cervix (Zaloudek and Norris, 1994). Typically, leiomyosarcoma is larger and softer that leiomyoma, having a less well-defined margin and often containing foci of haemorrhage or necrosis. It is usually composed of elongated pleomorphic cells (Figures 16.11 and 16.12) some of which have bipolar and multipolar mitotic figures (Poulsen *et al.*, 1975). The interweaving pattern, characteristic of most leiomyomas (Figure 16.2) is poorly defined. A diagnostic feature of leiomyosarcoma is the combination of high cellularity, nuclear atypicality and frequent mitotic figures (Figure 16.13), especially if some are abnormal. More than five mitotic figures per 10 HPF associated with moderate to marked cellular atypia, or more than 10 per 15 HPF whether or not cellular atypia is present should be considered indicative of leiomyosarcoma (Norris and Zaloudek, 1982, 1984). In low-grade leiomyosarcomas it may be necessary to search carefully for areas with increased mitotic activity (Figure 16.14(a)). The nuclei are usually hyperchromatic with chromatin clumping and prominent nucleoli. There is considerable cellular pleomorphism in poorly differentiated lesions. Multinucleated cells are often present, very occasionally with giant cells resembling osteoclasts (Darby *et al.*, 1975).

The number of giant cells does not correlate with survival. Many giant cells may be found in tumours with few mitotic figures and a benign course (Kempson and Bari, 1970).

Tumours with no cytological atypicality but with 5–15 mitotic figures per 10 HPF are designated 'mitotically active leiomyomas' (Norris and Zaloudek, 1994). Occasionally a smooth muscle tumour with less that five mitotic forms per 10 HPF which does not show cytological atypicality becomes invasive (Kempson and Bari, 1970). Atypical mitotic figures, necrosis, haemorrhage and evidence of tissue infiltration or vascular permeation may assist in identifying such lesions (but see also 'Metastasizing leiomyoma', p. 399).

Ultrastructurally, leiomyosarcomas differ from leiomyomas in having greater nuclear pleomorphism, free ribosomes and many mitochondria, but fewer and more disordered myofilaments and a greatly reduced extracellular collagenous stromal matrix (Ferenczy *et al.*, 1971; Böcker and Strecker, 1975; Norris and Zaloudek, 1982).

Cytologically, the desquamated cells may be spindle-shaped (Figure 16.14(b)). Such cells are more readily recognized as sarcomatous in origin than cells with round or oval nuclei, often arranged in clusters and lacking cytoplasm (Figure 16.14(c)) which are likely to be misinterpreted as carcinoma cells.

16.4 SARCOMA BOTRYOIDES (EMBRYONAL RHABDOMYOSARCOMA)

This highly malignant tumour is composed of cells with small, round to oval to spindle-shaped nuclei with inconspicuous nucleoli, some cells showing evidence of differentiation towards striated muscle cells (Scully *et al.*, 1994); characteristically it has a botryoid (grape-like) structure (Figure 16.15(a)). It is an extremely rare tumour, usually occurring in infants and young children but with a range extending from birth to 41 years (Zaino *et al.*, 1994). It has a polypoid structure with a loose, oedematous stroma which may not appear sarcomatous at first glance (Figure 16.15(b)). Closer examination reveals areas of increased cellularity often around blood vessels and also forming subepithelial dense cambial zones (Figure 16.15(c)). Many of the cells tend to be elongated with irregular nuclei and occasional mitotic figures. Skeletal muscle cells in varying stages of development may be recognized including occasional 'strap cells' with transverse striations (Figure 16.15(d)). In addition, cartilage and bone may be present. The surface of the neoplastic polyps is covered by normal stratified squamous epithelium, as in Figure 16.15(b), but frequently there are areas of ulceration. The tumour may spread via lymphatics or blood vessels but Hilgers *et al.* (1970) found that deposits were confined to the pelvis in 50% of autopsies on cases of vaginal botryoid sarcoma. They also found that 5-year survival, variously estimated as 1–35%, could be increased to nearly 50% by radical surgery, including pelvic exenteration. Multi-agent chemotherapy combined with surgery has greatly improved the chances of survival, in the order of 85% 3-year survival (Zaino *et al.*, 1994).

The diagnosis is difficult to make on cytological material except on the very rare occasion of finding a striated 'strap cell', particularly if the smear is from a child or young woman. In older women, this finding should suggest the diagnosis of malignant mesodermal mixed tumour (Section 16.7.3).

16.5 STROMAL TUMOURS

Both endocervical and endometrial stromal tumours can occur in the cervix.

16.5.1 Endocervical stromal sarcoma

The cells of this lesion closely resemble those of normal endocervical stroma and there is usually a rich reticular network (Abell and Ramirez, 1973). It arises from the mesenchyme

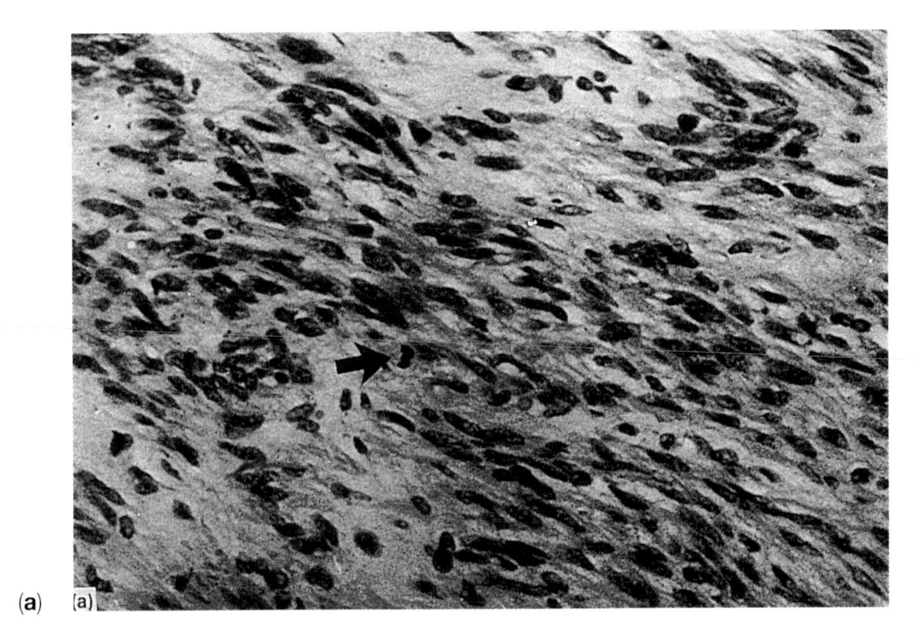

Figure 16.14 (**a**) In a low-grade leiomyosarcoma the detection of zones of increased mitotic activity may require careful searching. A pelvic recurrence was removed 7 years after hysterectomy (H & E, ×620).

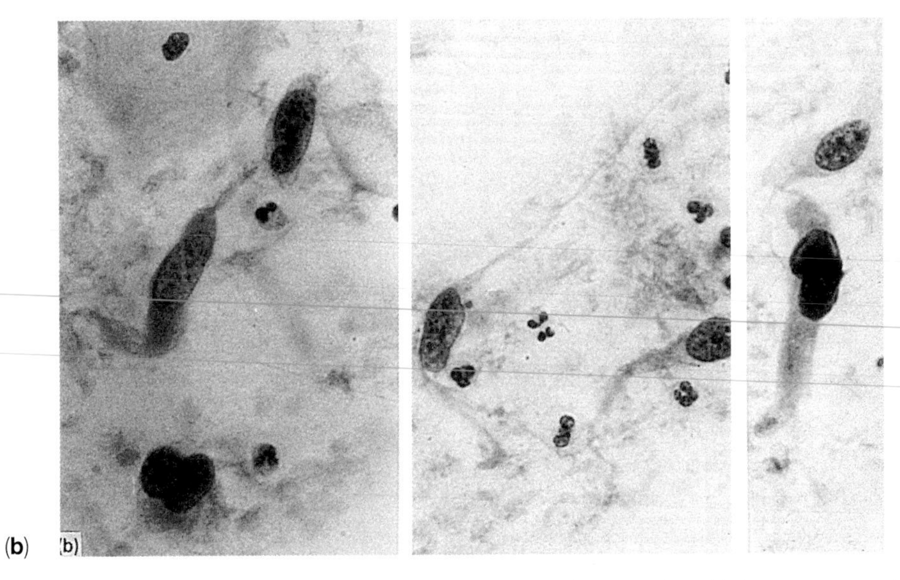

Figure 16.14 (**b**) Spindle-shaped sarcoma cells from a fairly well-differentiated leiomyosarcoma in a cervical smear (Pap, ×770).

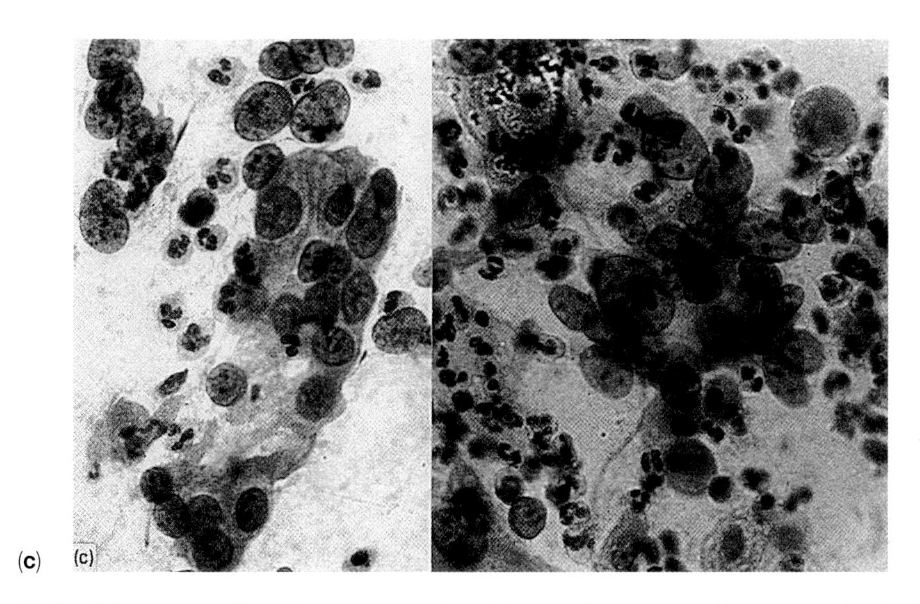

(c)

Figure 16.14 (**c**) Sarcoma cells with round or oval nuclei, many lacking cytoplasm, occurring in clusters and showing anisonucleosis (Pap, ×770).

of the substantia propria, like sarcoma botryoides (Section 16.4), from which it differs by the absence of striated muscle cells and by the fact that it usually occurs about the time of the menopause or later. The 12 cases reported by Abell and Ramirez (1973) all presented with irregular or post-menopausal bleeding. Nearly half were well-differentiated, of which only one had a recurrence and none died from the sarcoma. Six patients with the less well differentiated lesion died 2–24 months after diagnosis.

16.5.2 Endometrial stromal tumours

All endometrial stromal tumours are composed of cells identical to, or closely resembling, those of endometrial stroma (Norris and Taylor, 1966). They are subdivided as nodular or infiltrating according to the pattern of their margins. Those with well-circumscribed, 'pushing' margins, showing no infiltration of myometrium or lymphatics, are known as stromal nodules, and are benign. All 18 nodules described by Norris and Taylor (1966) were solitary and none recurred after removal. A small proportion (15%) contained epithelial islands with gland-like arrangements and, in addition, some (11%) contained endometrial glands.

Infiltrating tumours were divided into two groups on the basis of mitotic activity:

1. Those with less than 10 mitotic figures/10 HPF (20 cases).
2. Those with 10 or more mitotic figures/10 HPF (15 cases).

Those in group 1, considered to represent *endolymphatic stromal myosis* had a 100% 5-year survival rate in spite of occasional recurrences or spread to the peritoneum (Mazur and Askin, 1978). Those in group 2, considered to represent *stromal sarcoma*, had a 50% 5-year survival rate. The latter group are also characterized by anaplastic cells (Yoonessi and Harts, 1977), a

(**a**)

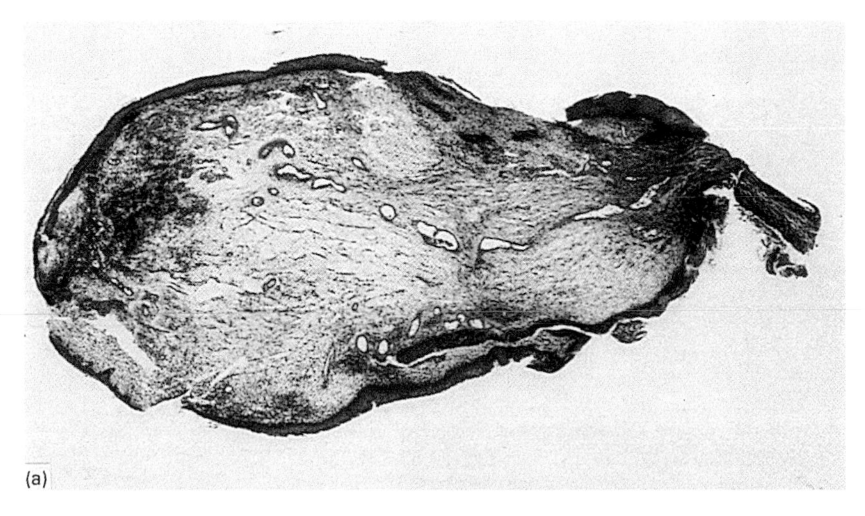

Figure 16.15 (**a**) Sarcoma botryoides. Low-power view of grape-like protuberance (H & E, ×20).

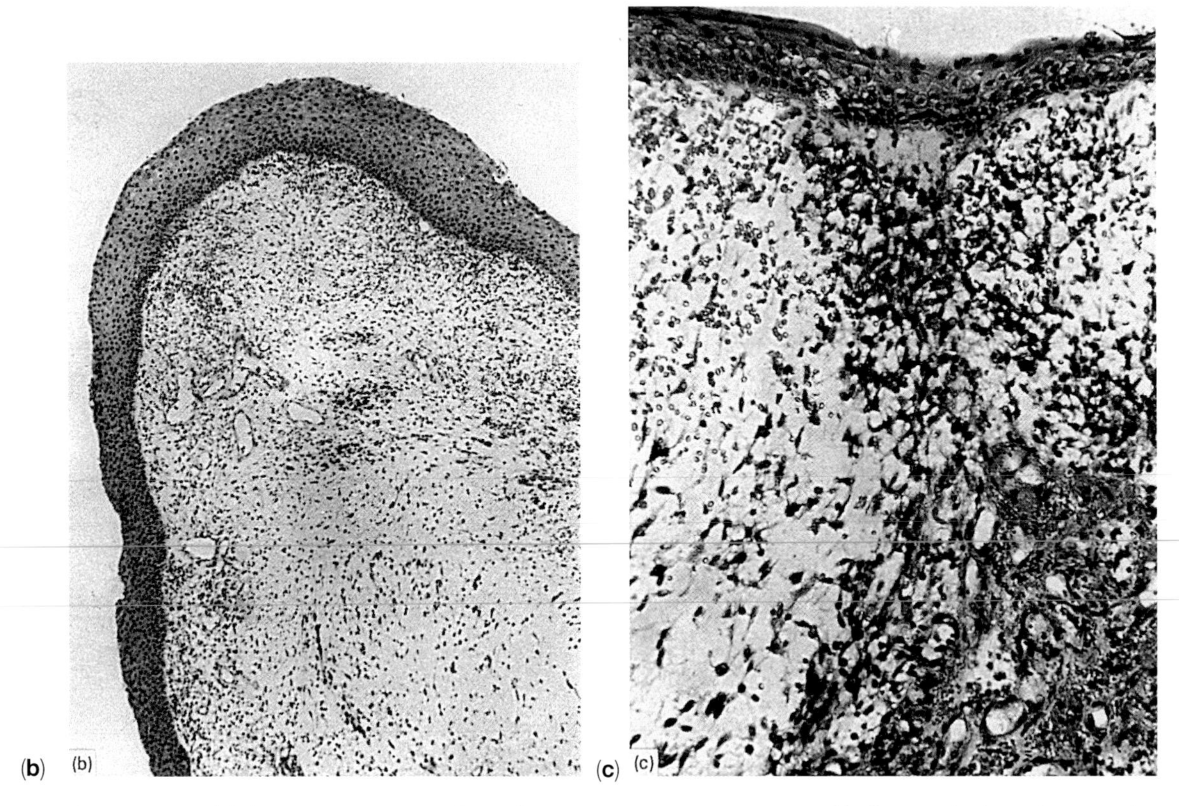

Figure 16.15 (**b**) Sarcoma botryoides. Polyp lined by stratified squamous epithelium enclosing a loose oedematous, bland-looking stroma. Same tumour as shown in (**a**) (H & E, ×80). (**c**) Subepithelial and perivascular zones of dense cellularity in a sarcoma botryoides. Same tumour as shown in (**a**) and (**b**) (H & E, ×190).

Figure 16.15 (**d**) Strap cells with transverse striations (arrowed) indicating that they are rhabdomyoblasts. Same tumour as shown in (**a**, **b** and **c**) (PTAH, ×770).

feature of differential diagnostic value when the mitotic figure rate is on the borderline.

16.6 ALVEOLAR SOFT PART SARCOMA (ASPS)

This sarcoma, which very rarely involves the cervix, is characterized microscopically by solid and alveolar groups of large epithelial-like cells with granular eosinophilic cytoplasm; most of the tumours contain intracytoplasmic PAS-positive, diastase-resistant, rod-shaped crystals (Scully *et al.*, 1994). The alveolar arrangement of the cells is made clearer by using silver impregnation for reticulin (Morimitsu *et al.*, 1993). Mitotic figures are few and the tumour tends to have a pushing border. Ultrastructurally, dense membrane-bound granules are present in the cytoplasm.

Immunohistochemically, the cytoplasm reacts positively with anti-desmin, anti-myoglobin and anti-neurone specific enolase. The growth of pelvic ASPS is slow but local and distant metastases may take place (Wilkinson, 1994). Lung is the most common site of distant metastasis.

16.7 MIXED EPITHELIAL AND MESENCHYMAL TUMOURS

The müllerian ducts are formed from mesenchyme and it is characteristic of mesenchymal tumours that they often retain the pluripotence of primitive mesenchymal tissue (Willis, 1967). This is reflected in this group of tumours in which both epithelial and connective tissues are neoplastic elements, varying in proportion and degree of abnormality. A tumour is designated 'homologous' if it is composed only of tissues normally found in the uterus, and 'heterologous' if it includes tissues foreign to the uterus, e.g. striated muscle, cartilage, osteoid (Figure 16.16) or fat. There is a gradation of aggressiveness from adenofibroma, at the benign end, through adenosarcoma, of relatively low-grade malignancy, to the highly malignant mesodermal (müllerian) mixed tumour, also designated carcinosarcoma (Norris and Zaloudek, 1994).

16.7.1 Adenofibroma

In this rare lesion, first described by Abell (1971), both epithelial and connective tissue elements are benign and the tumour is homologous. It occurs in menopausal or postmenopausal women. Macroscopically, it is a lobulated polypoid lesion, soft to firm in consistency, with a brownish cut surface (Zaloudek and Norris, 1994). Microscopically, it is composed of papillary processes having a compact fibrous tissue core lined by a single layer of columnar, cuboidal or flattened epithelium with foci of squamous metaplasia which extend into the branching clefts which are present. Varying types of epithelium – endocervical, endometrial or endometrial – may be found in the same lesion. The stroma is usually fibroblastic, but may also be endometrial; it may be uniformly cellular or hypocellular and collagenous. Mitotic figures are rare. If present in countable numbers, particularly if more than four per 10 HPF or if the stromal cells show atypicality, a diagnosis of adenosarcoma must be considered.

16.7.2 Adenosarcoma

This is a mixed tumour composed of benign or atypical epithelium of müllerian type and a malignant-appearing stroma (Scully *et al.*, 1994). It may be homologous or heterologous; both are of relatively low-grade malignancy. It can occur at any age from 14 to 69 years, the median being 57 years. It is usually a polypoid growth, often solitary but multiple papillary and polypoid masses may occur resembling the 'bunch of grapes' so characteristic of sarcoma botryoides. The cut surface is brown or greyish with scattered small cysts (Zaloudek and Norris, 1994). In about 25% of adenosarcomas there are foci of haemorrhage and necrosis. Histologically, the stroma is traversed by tubular glands or clefts and on the surface there are lobular protruberances with a lining epithelium which may be mucinous (Figure 16.17(a)), squamous or resemble inactive or regenerative endometrium. The epithelium may be hyperplastic or atypical and, in lesions where invasion has occurred, it participates in the invasion. The connective tissue component is usually a fibrosarcoma, but may be a small cell sarcoma showing some resemblance to endometrial stroma, often with subepithelial hypercellularity. Cellular atypicality may be mild, moderate or severe. It differs from papillary adenofibroma in having a more densely cellular stroma in which four or more mitotic figures per 10 HPF are found (Figure 16.17(b)). Heterologous tissues occur in 20–25% of cases, most commonly striated muscle and less frequently cartilage or fat.

Although considered to be of low-grade malignancy, 25–40% recur following hysterectomy. Distant metastases occur in about 5% of cases. About half the recurrent tumours show greater atypicality and mitotic activity than the primary tumour. Heterologous components may occur in the metastases but the epithelial element is often absent. The median interval between treatment and recurrence for all uterine adenosarcomas is 5 years (Norris and Zaloudek, 1982). It should be noted that grades intermediate between the classical adenosarcoma and mixed mesenchymal sarcoma also occur (H. Fox, personal communication, 1986).

16.7.3 Malignant mesodermal (müllerian) mixed tumour (carcinosarcoma)

The term 'carcinosarcoma' was originally reserved for mixed carcinoma and sarcoma with no heterologous elements, as in Figure 16.18. It is now generally used interchangeably with mesodermal (müllerian) mixed tumour (Scully *et al.*, 1994). It often contains heterologous tissues. The heterologous elements, e.g. rhabdomyosarcoma, chondrosarcoma or osteosarcoma are usually sufficiently well-differentiated for their nature to be readily identified (Figure 16.16).

Mesodermal mixed tumours usually occur after the menopause. These tumours (carcinosarcomas) may originate in the cervix but more commonly arise from the endometrium from which they may extend to involve the cervix. The risk of their occurrence is increased by previous radiotherapy, usually for carcinoma of the cervix (Thomas *et al.*, 1969; Fehr and Prem, 1974). Following radiotherapy, they tend to occur at a lower age with a median period between irradiation and tumour diagnosis of 16 years (Norris and Taylor, 1965). More recently the proportion of post-irradiation tumours has been reported to be lower (Salazar *et al.*, 1978). Metastases occur in the pelvic and para-aortic lymph nodes, the pelvic soft tissues, the vagina, the peritoneal surfaces of the upper abdominal cavity and the lungs (Zaloudek and Norris, 1994). If the heterologous tissue was chondrosarcoma and invasion minimal, Kempson and Bari (1970) considered that the condition was curable, but that the presence of osteoid, bone or rhabdomyoblasts indicated a highly malignant neoplasm with a fatal outcome (all the tumours in their study being large at the time of diagnosis).

Diagnosis is usually based on biopsy or curettage, although sampling errors may enable the full nature of the tumour to be recognized only after its resection. Cytologically the finding of sarcoma cells in a cervical scrape preparation from an older woman should suggest the possibility of a malignant müllerian mixed tumour, since other sarcomas are even more uncommon in adults and are seldom demonstrable cytologically (Barwick and LiVolsi, 1979).

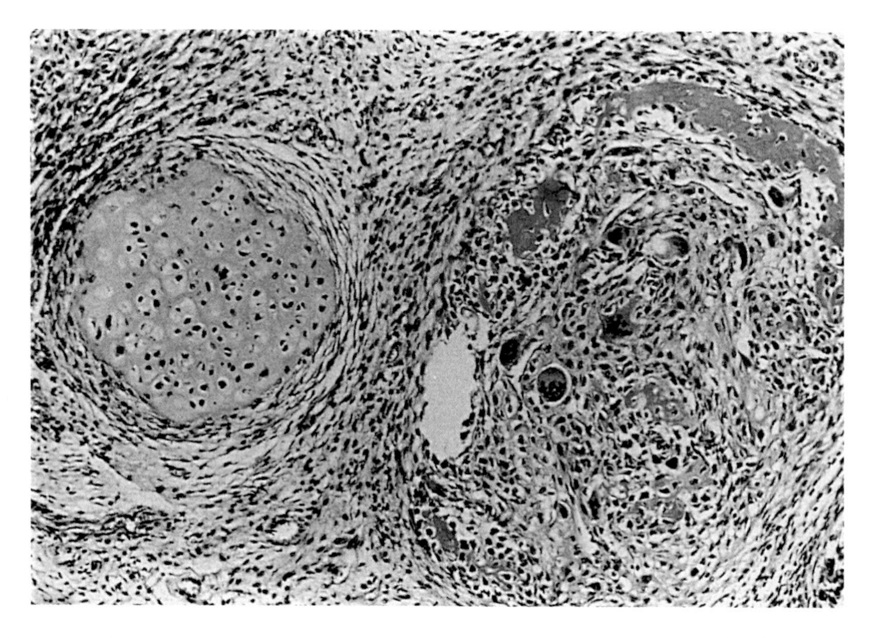

Figure 16.16 Heterologous malignant mesodermal mixed tumour (carcinosarcoma). Both cartilage and osteoid tissue can be seen (H & E, ×180) (courtesy of Dr A.S. Hill, Jessop Hospital for Women, Sheffield).

(**a**)

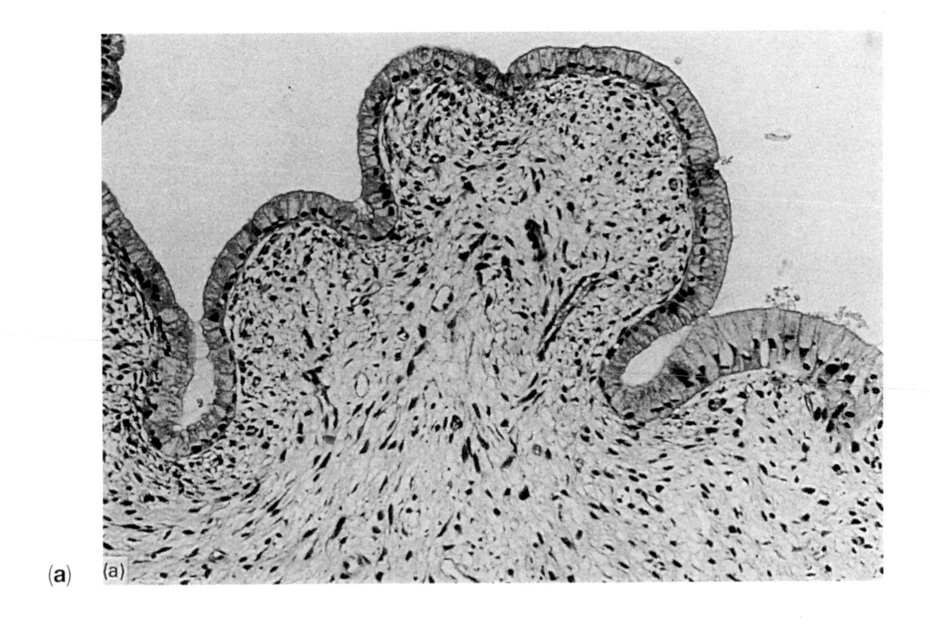

(**b**)

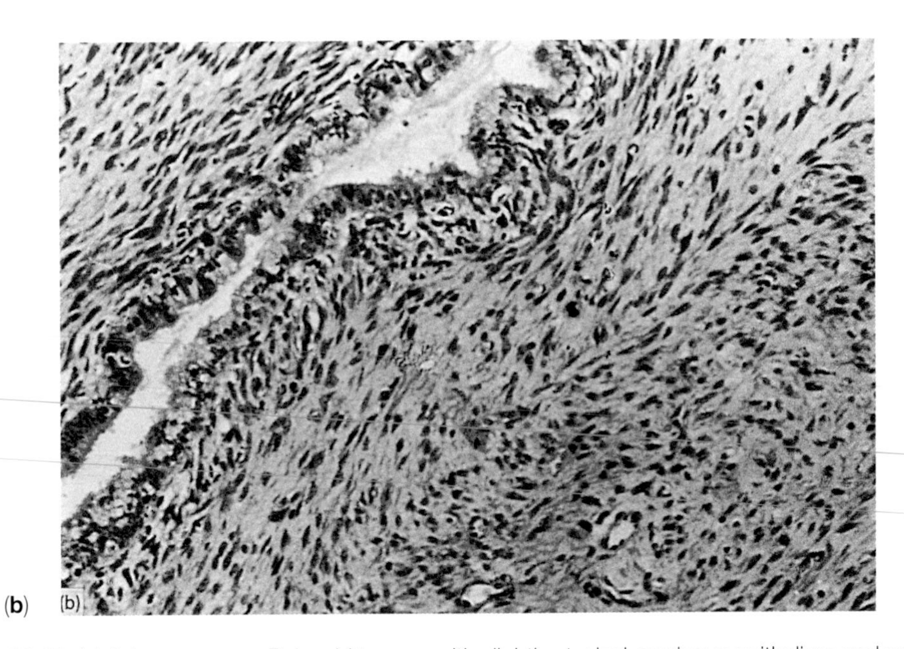

Figure 16.17 (**a**) Adenosarcoma. Polypoid tumour with slightly atypical mucinous epithelium and sarcomatous stroma (H & E, ×180) (courtesy of Dr A.S. Hill). (**b**) Adenosarcoma. The epithelium is of mucinous type. The fibrous stroma contained more than ten mitotic figures per 10 HPF (H & E, ×310). Same tumour as (**a**).

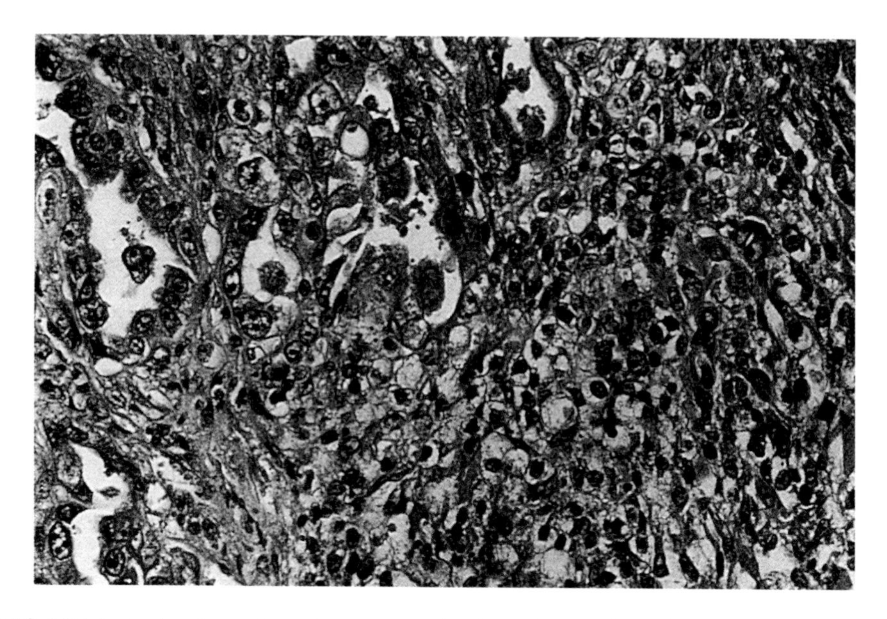

Figure 16.18 Histological pattern of carcinosarcoma (malignant mesodermal mixed tumour). Both glands and stroma show malignant features and show some resemblance to endometrium (H & E, ×310).

16.8 OTHER SARCOMAS

Some sarcomas are composed very largely of undifferentiated cells and much sectioning is needed, together with histochemical staining, to determine the histological type of the lesion. Many rare types have been described, including angiosarcoma and liposarcoma. If the tumour is so poorly differentiated as to defy classification it should be diagnosed as 'sarcoma unclassified' (Kempson and Bari, 1970).

16.9 PIGMENTED LESIONS

16.9.1 Blue naevus

Blue naevus of the cervix is seldom diagnosed. It is a symptomless benign lesion, visible as a small brown or blackish zone of discoloration deep to the endocervical canal lining, usually 5 mm in diameter or less (Jiji, 1971). This rare lesion is found in the lower endocervix and in endocervical polyps (Scully *et al.*, 1994).

Microscopically, it is composed of spindle-shaped cells containing abundant brown pigment, situated in the cervical stroma a short distance below the epithelium (Figure 16.19(a)). Many of the cells are long and thin with branching cytoplasmic processes, often with nuclei obscured by pigment. Because of its small size, the lesion is easily overlooked. Even if detected, it may be mistaken for a zone of siderosis, particularly when the pigment appears golden brown in colour. Its true nature is confirmed by a negative Perl's reaction for iron, positive staining with Masson–Fontana silver impregnation (Figure 16.20) and positive histochemical reactions for melanin with the ferrous ion technique of Lillie (1957).

The view of Goldman and Friedman (1967) that the lesion is probably the visceral analogue of the cutaneous blue naevus appears logical. Their suggestion that it may undergo malignant transformation requires substantiation, but would explain the occasional occurrence of primary malignant melanoma of the cervix (Jiji, 1971).

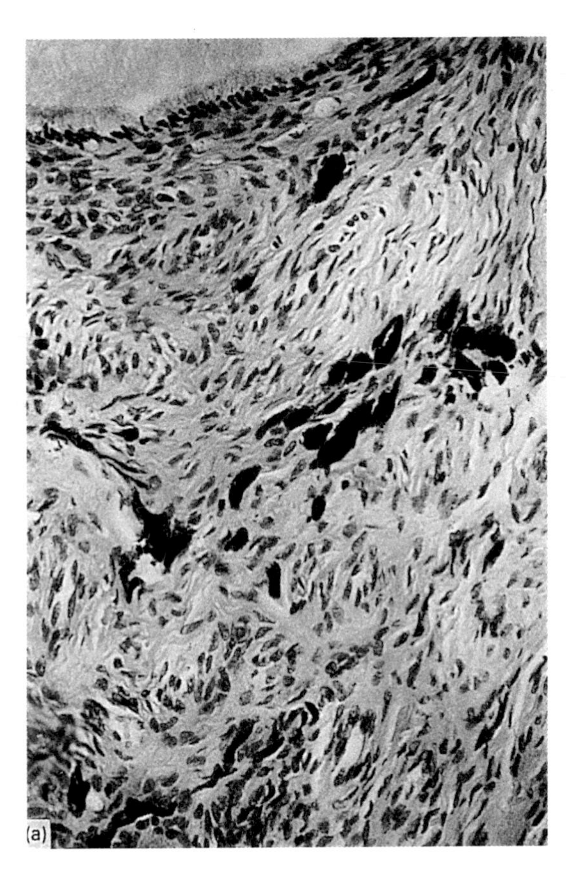

Figure 16.19 Blue naevus composed of spindle-shaped cells containing pigment which stained positively with Masson–Fontana stain (H & E, ×310). (Courtesy of Dr A.S. Hill).

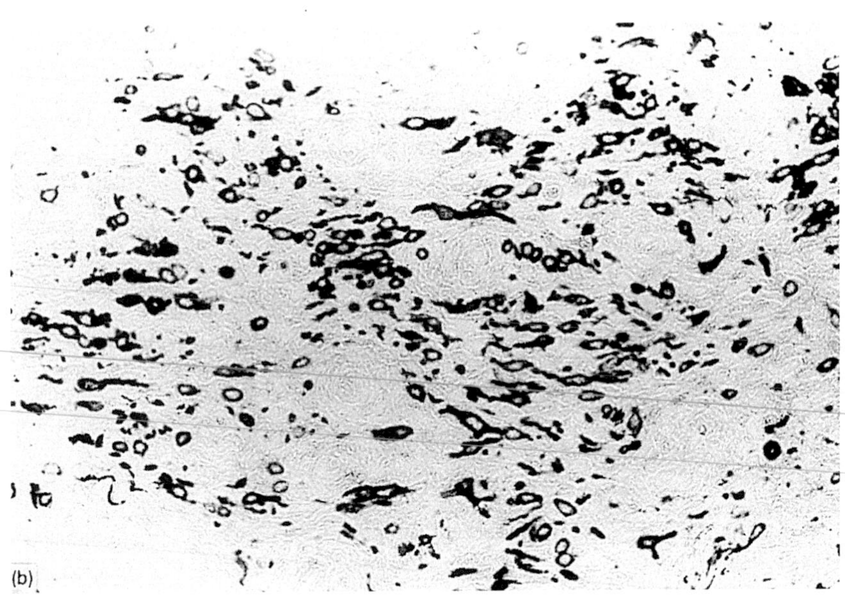

Figure 16.20 Blue naevus showing presence of melanin demonstrated by silver impregnation (Masson–Fontana, ×120). (Courtesy of Dr A. Levene, Royal Marsden Hospital, London).

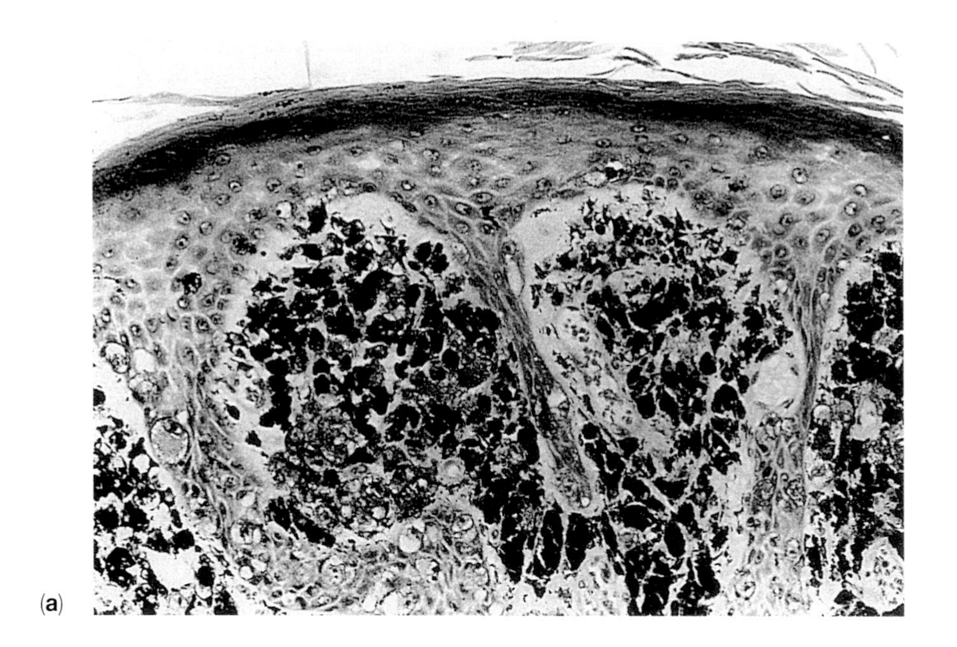

(**a**)

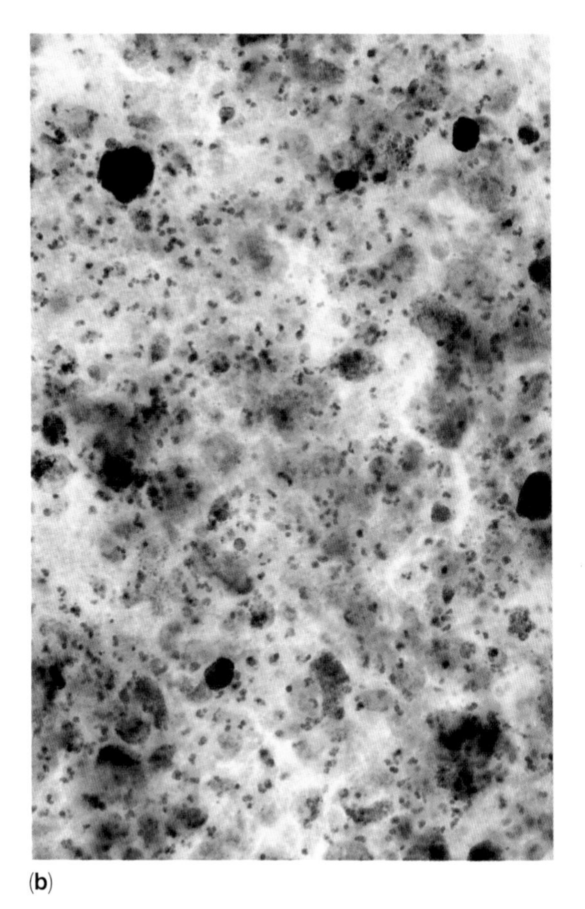

(**b**)

Figure 16.21 (**a**) Involvement of cervix by malignant melanoma extending from vagina. Melanoma cells are seen deep to and within the stratified squamous epithelium (H & E, ×480). (**b**) Pigment-laden malignant melanoma cells in a cervical smear (Pap, ×400).

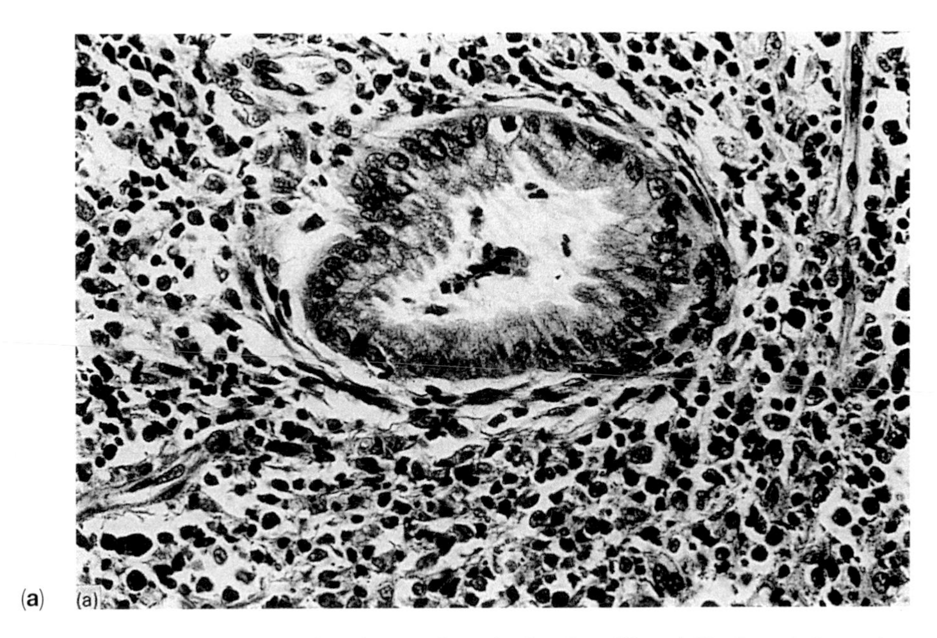

(a)

Figure 16.22 (**a**) Primary malignant lymphoma of cervix showing diffuse infiltration of stroma by mononuclear cells surrounding an intact gland (H & E, ×480).

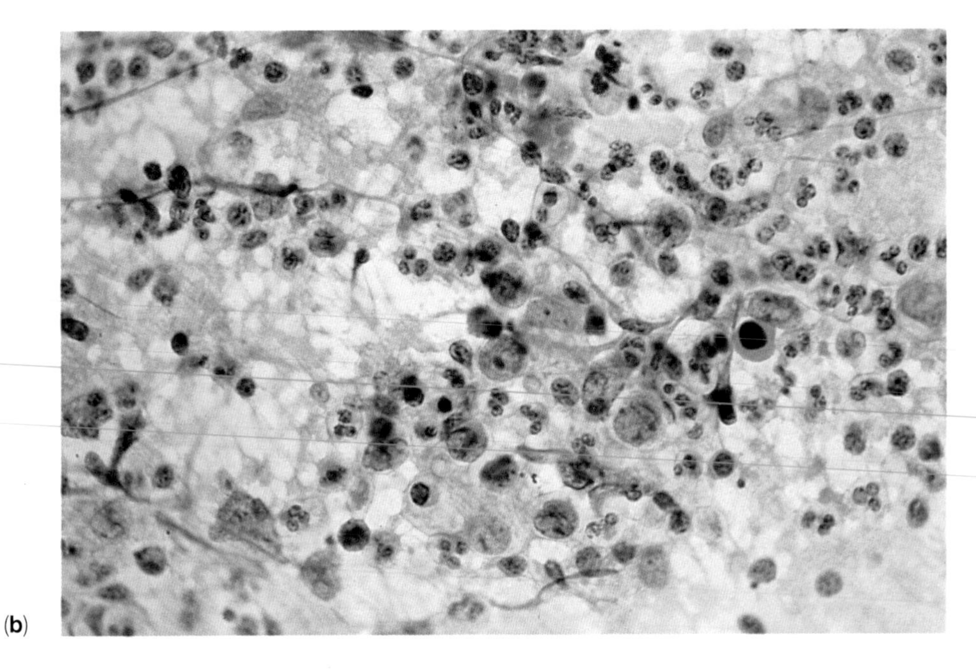

(**b**)

Figure 16.22 (**b**) Cervical smear from (**a**) showing lymphoma cells with prominent nucleoli (Pap, ×770).

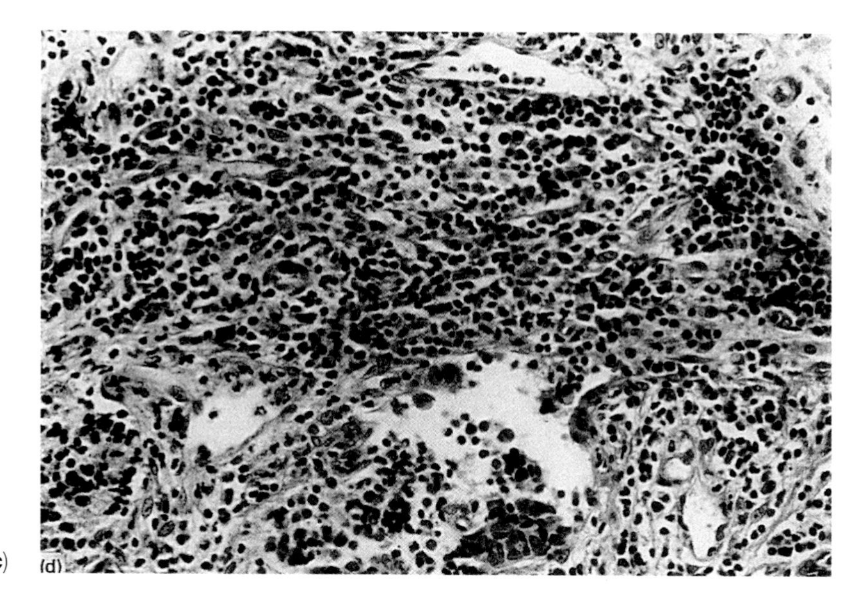

Figure 16.22 (**c**) Lymphoma-like infiltrate accompanying a cervical carcinoma. Small groups of carcinoma cells can be seen partially lining the tissue spaces near the lower edge (H & E, ×300). (Courtesy of Dr A.S. Hill).

16.9.2 Malignant melanoma

Occasional examples of primary malignant melanoma of the cervix have been recorded but it is an exceedingly rare lesion (Abell, 1961). It is very similar in appearance to primary malignant melanoma of the vagina which is slightly more common and may extend to involve the cervix (Figure 16.21(a)). The tumour described by Abell – and which he termed a melanoblastoma – invaded the parametrium, with lymphatic spread to regional lymph nodes and blood spread to more distant organs. He noted that 3.5% of uterine cervices had been found to possess melanin-containing cells which he interpreted as melanocytes of neurogenic origin. The prognosis for cervical malignant melanoma is quite poor, with 40% 5-year survival rate for stage 1 lesions and 14% for more advanced lesions (Wright *et al.*, 1994). Its diagnosis depends on the histological demonstration of junctional changes in the squamous epithelium and the absence of similar lesions elsewhere in the body. The cytological appearances of malignant melanoma are shown in Figure 16.21(b). The pigment-laden malignant cells obscure the other elements of the smear.

16.10 MALIGNANT LYMPHOMA

Malignant lymphoma is much less common as a primary lesion of the cervix than as a metastasis. The diagnosis of primary lymphoma should only be made if the lesion is confined to the region of the uterus and there has been no evidence of lymphoma elsewhere for several months (Fox and More, 1965). The age range of reported cases is 2 to 88 (median 43) years, usually presenting with vaginal bleeding or a pelvic tumour (Carr *et al.*, 1976). Characteristically, the cervix shows sessile or polypoid masses with a 'fish-flesh' appearance (Ferenczy, 1982).

Histologically, there is a diffuse infiltrate of mononuclear cells replacing the stroma without destroying the cervical epithelium (Figure 16.22(a)). The criteria for diagnosis are similar to those for lymphoma elsewhere. The infiltrating cells vary in type and maturity according to the type of lymphoma, often with prominent nucleoli and fairly frequent mitotic figures, and they lack the cohesiveness of an epithelial tumour. Muntz *et al.* (1991) reported 43 stage 1E lymphomas of cervix, of which approximately 90% were of intermediate grade, 70% being of the diffuse large cell type, and 20% of lower-grade follicular type. If the lesion ulcerates, lymphoma cells may be present in the cervical smear and provide the first clue to the diagnosis (Figure 16.22(b,c)). The *cytological* appearance of a cervical lymphoma is described by Carr *et al.* (1976) who found a monomorphic pattern of malignant cells smaller than the usual carcinoma cells, lying singly, some with bilobed nuclei and large nucleoli.

The main differential diagnosis is from the reactive infiltrate of chronic lymphocytic cervicitis (Section 7.2.1). A lymphoma-like infiltrate can occasionally accompany a carcinoma (Figure 16.22(d)). The essentially normal cells of a reactive infiltrate should be distinguishable from the abnormal cells of a lymphoma using a high-power objective; the reactive infiltrate is also less uniformly dense in most cases. The prognosis of patients with primary cervical lymphoma is considerably better than that with metastatic cervical lymphoma or other extranodal lymphomas of similar histologic type (Ferenczy, 1982). Most cases are localized to the cervix when first seen and can be cured by a combination of surgery and chemotherapy (Muntz *et al.*, 1991).

16.11 LEUKAEMIA AND GRANULOCYTIC SARCOMA

Rarely, leukaemia occurs as a primary tumour of the cervix (Scully *et al.*, 1994). Some patients with myeloid leukaemia present with cervical granulocytic sarcoma, an extramedullary tumour of malignant granulocytic precursor cells (Friedman *et al.*, 1992). It may accompany, precede, or signal relapse of acute myeloblastic leukaemia, or signal blastic transformation of chronic myeloproliferative disease, e.g. chronic myeloid leukaemia. As it is undifferentiated and uncommon it is often misdiagnosed on biopsy. Tissue imprint cytology with haematological staining, immunohistochemistry and histochemistry for detection of myeloid-specific substances, and electron microscopy aid in the diagnosis. Its early recognition and treatment, as for acute myeloid leukaemia, significantly improve an otherwise unfavourable prognosis.

REFERENCES

Abell, M. R. (1961) Primary melanoblastoma of the uterine cervix. *Am. J. Clin. Pathol.*, **36**, 248–255.

Abell, M. R. (1971) Papillary adenofibroma of the uterine cervix. *Am. J. Obstet. Gynecol.*, **110**, 990–993.

Abell, M. R. and Ramirez, J. A. (1973) Sarcomas and carcinosarcomas of the uterine cervix. *Cancer*, **31**, 1176–1192.

Albores-Saavedra, J., Larraza, O., Poucell, S. and Rodrigez-Martinez, H. A. (1976) Carcinoid of the uterine cervix. Additional observation on a new tumor entity. *Cancer*, **38**, 2328–2342.

Barwick, K. W. and LiVolsi, V. A. (1979) Malignant mixed müllerian tumors of the uterus. A clinicopathologic assessment of 34 cases. *Am. J. Surg. Pathol.*, **3**, 125–135.

Böcker, W. and Strecker, H. (1975) Electron microscopy of uterine leiomyosarcomas. *Virchows Arch. Pathol. Anat. Histol.*, **367**, 59–71.

Carr, I., Hill, A. S., Hancock, B. and Neal, F. E. (1976) Malignant lymphoma of the cervix uteri: histology and ultrastructure. *J. Clin. Pathol.*, **29**, 680–686.

Corscaden, J. A. and Singh, B. P. (1958) Leiomyosarcoma of the uterus. *Am. J. Obstet. Gynecol.*, **75**, 149–153.

Darby, A. J., Papadaki, L. and Beilby, J. O. W. (1975) An unusual leiomyosarcoma of the uterus containing osteoclast-like giant cells. *Cancer*, **36**, 495–504.

Daroca, P. J. and Dhurandor, D. (1980) Basaloid carcinoma of uterine cervix, *Am J. Surg., Pathol.* **4**, 235–239.

Fehr, P. E. and Prem, K. A. (1974) Malignancy of the uterine corpus following irradiation therapy for squamous cell carcinoma of the cervix. *Am. J. Obstet. Gynecol.*, **119**, 685–692.

Ferenczy, A. (1982) Carcinoma and other malignant tumours of the cervix. In *Pathology of the Female Genital Tract*, 2nd edn. (ed. A. Blaustein), Springer-Verlag, Berlin, pp. 184–222.

Ferenczy, A., Richart, R. M. and Olagaki, T. (1971) A comparative ultrastructural study of leiomyosarcoma, cellular leiomyoma, and leiomyoma of the uterus. *Cancer*, **28**, 1004–1018.

Fluhman, C. F. (1961) *The Cervix Uteri and its Diseases*, W. B. Saunders, Philadelphia, p. 245.

Fox, H. and More, J. R. S. (1965) Primary malignant lymphoma of the uterus. *J. Clin. Pathol.*, **18**, 723–728.

Friedman, H. D., Adelson, M. D., Elder, R. C. *et al.* (1992) Granulocytic sarcoma of the uterine cervix; literature review of granulocytic sarcomas of the female genital tract. *Gynecol. Oncol.*, **46**, 128–137.

Goldman, R. L. and Friedman, N. B. (1967) Blue nevus of the uterine cervix. *Cancer*, **20**, 210–214.

Habib, A., Kaneko, M., Cohen, C. J. and Walker, G. (1979) Carcinoid of the uterine cervix. A case report with light and electron microscopic studies. *Cancer*, **43**, 535–538.

Hilgers, R., Malkasian, G. D. and Soule, E. H. (1970) Embryonal rhabdomyosarcoma (botryoid type) of the vagina; a clinicopathologic review. *Am. J. Obstet. Gynecol.*, **107**, 484–501.

Jiji, V. (1971) Blue nevus of the endocervix (with) review of the literature. *Arch. Pathol.*, **92**, 203–205.

Kempson, R. L. and Bari, W. (1970) Uterine sarcomas. Classification, diagnosis and prognosis. *Hum. Pathol.*, **1**, 331–349.

Lathrop, J. C. (1967) Views and reviews: malignant pelvic lymphomas. *Obstet. Gynecol.*, **30**, 137–145.

Lillie, R. D. (1957) Ferrous ion uptake: a specific reaction of the melanins. *Arch. Pathol.*, **64**, 100–103.

Lucia, S. P., Mills, H., Lowenhaupt, E. and Hunt, M. L. (1952) Visceral involvement in neoplastic diseases of the reticulo-endothelial system. *Cancer*, **5**, 1193–1200.

Mazur, M. T. and Askin, F. B. (1978) Endolymphatic stromal myosis: unique presentation and ultrastructural study. *Cancer*, **42**, 2661–2667.

Morimitsu, Y., Tanake, H., Iwanaga, S. and Kojiro, M. (1993) Alveolar soft part sarcoma of the uterine cervix. *Pathol. Japon.*, **43**, 204–208.

Muntz, E. G., Ferry, J. A., Flynn, D. *et al.* (1991) Stage 1E Primary malignant lymphomas of the uterine cervix. *Cancer*, **68**, 2023–2032.

Norris, H. J. and Taylor, H. B. (1965) Postirradiation sarcomas of the uterus. *Obstet. Gynecol.*, **26**, 689–694.

Norris, H. J. and Taylor, H. B. (1966) Mesenchymal tumors of the uterus: a clinical and pathological study of 53 endometrial stromal tumors. *Cancer*, **19**, 755–766.

Norris, H. J. and Zaloudek, C. J. (1982) Mesenchymal tumors of the uterus, in *Pathology of the Female Genital Tract*, 2nd edn. (ed. A. Blaustein), Springer-Verlag, Berlin, pp. 352–392.

Norris, H. J. and Zaloudek, C. J. (1994) Mesenchymal tumors of the uterus, in *Blaustein's Pathology of the Female Genital Tract* (ed. R. J. Kurman), Springer-Verlag, New York, pp. 487–528.

Nunez-Alonso, C. and Battifora, H. A. (1979) Plexiform tumors of the uterus; ultrastructural study. *Cancer*, **44**, 1707–1714.

Patchefsky, A. S. (1970) Plexiform tumorlet of the uterus. Report of a case. *Obstet. Gynecol.*, **35**, 592–596.

Poulsen, H. E., Taylor, C. W. and Sobin, L. H. (1975) *Histological Typing of Female Genital Tract Tumors*. WHO, Geneva, pp. 55–73.

Salazar, O. M., Bonfiglio, T. A., Patten, S. F., Keller, B. E., Feldstein, M., Dunne, M. E. and Rundolph, J. H. (1978) Uterine sarcomas. Natural history, treatment and prognosis. *Cancer*, **42**, 1152–1160.

Scully, R. E., Bonfiglio, R. J. and Kurman, R. J. (1994) *WHO International histological classification of tumours. Histological typing of female genital tract tumours*. Springer-Verlag, Berlin, pp. 39–54.

Tateishi, R., Wada, A., Hayakawa, K., Hongo, J., Ishii, S. and Terekawa, N. (1975) Argyrophil cell carcinomas (apudomas) of the uterine cervix. *Virchows Arch. Pathol. Anat. Histol.*, **366**, 257–273.

Thomas, W. O., Harris, H. H. and Enden, J. A. (1969) Post-irradiation malignant neoplasms of the uterine fundus. *Am. J. Obstet. Gynecol.*, **104**, 209–219.

Willis, R. A. (1967) *Pathology of Tumours*, 4th edn., Butterworths, London, p. 659.

Wilkinson, E. J. (1994) Premalignant and malignant tumors of the vulva, in *Blaustein's Pathology of the Female Genital Tract* (ed. R. J. Kurman). Springer-Verlag, New York, pp. 87–129.

Wright, T. C. and Ferenczy, A. (1994) Benign diseases of the cervix, in *Blaustein's Pathology of the Female Genital Tract* (ed. R. J. Kurman). Springer-Verlag, New York, pp. 203–227.

Wright, T. C., Ferenczy, A. and Kurman, R. J. (1994) Carcinoma and other tumors of the cervix, in *Blaustein's Pathology of the Female Genital Tract* (ed. R. J. Kurman). Springer-Verlag, New York, pp. 279–326.

Yoonessi, M. and Hart, W. R. (1977) Endometrial stromal sarcomas. *Cancer*, **40**, 898–906.

Young, R. H., Treger, T. and Scully, E. (1986) Atypical polypoid adenomyoma of the uterus: a report of 27 cases. *Am. J. Clin. Pathol.*, **86**, 139–145.

Zaino, R. J., Robboy, S. J., Bentley, R. and Kurman, R. J. (1994), Diseases of the vagina, in *Blaustein's Pathology of the Female Genital Tract* (ed. R. J. Kurman). Springer-Verlag, New York, pp. 131–183.

Zaloudek, C. and Norris, H. J. (1994) Mesenchymal tumors of the uterus, in *Blaustein's Pathology of the Female Genital Tract* (ed. R. J. Kurman). Springer-Verlag, New York, pp. 487–528.

17

SCREENING FOR CERVICAL CANCER

17.1 EPIDEMIOLOGY OF CERVICAL CANCER AND ITS PREVENTION

Cervical cancer is the second most common cancer in women: world-wide cancer registry data indicate that about 500 000 new cases occur each year (Cancer Research Campaign Facts Sheet 22.1, 1995). The incidence varies from country to country, the highest incidence being recorded in Peru (54.6 cases per 100 000 female population) and the lowest incidence in Israel (2.2 cases per 100 000 female population) (Figure 17.1). The incidence in EC countries is shown in Figure 17.2.

The major risk factors for the disease are associated with sexual activity. Early age at first coitus and increasing number of sexual partners are primary risk factors (Rotkin, 1973). Parity, age at first pregnancy, prostitution, low socioeconomic status and history of sexually transmitted disease have all been shown to confer increased risk, but probably do so through their association with early age at first coitus and number of sexual partners (Rotkin, 1973; Harris *et al.*, 1980).

The strong link between sexual activity and cervical cancer has led to the hypothesis that a transmissible agent is involved in the pathogenesis of cervical cancer. Current research indicates that the human papillomaviruses have an important role to play in the carcinogenic process although the mechanism by which they exert their effect is not yet fully elucidated (Section 10.6.4).

Other risk factors that have been identified include use of oral contraceptives (Wright *et al.*, 1978; Vessey *et al.*, 1983), smoking (Harris *et al.*, 1980; Winkelstein *et al.*, 1990) and impaired immunological response. This has been convincingly shown in studies of renal allograft recipients (Kay *et al.*, 1970; Porecco *et al.*, 1975; Schneider *et al.*, 1983) and surveys of HIV-positive women (Bradbeer, 1987; Spurrett *et al.*, 1988). However, the increased risk of cervical neoplasia in immunosuppressed women appears to reflect their increased susceptibility to human papillomavirus infection (Sillman *et al.*, 1984) rather than a depleted lymphocyte count (Spurrett *et al.*, 1988). Occupation and promiscuity of the male partner have also been shown to affect the risk of the development of cervical cancer, leading to the concept of the high-risk male (Beral, 1974; Smith *et al.*, 1980; Skegg *et al.*, 1982; Zunzunqui *et al.*, 1986).

Invasive squamous cancer can be regarded as the end stage of the spectrum of diseases collectively known as cervical intraepithelial neoplasia (CIN). The concept that CIN is a precursor of invasive disease is based on evidence drawn from epidemiological, temporal

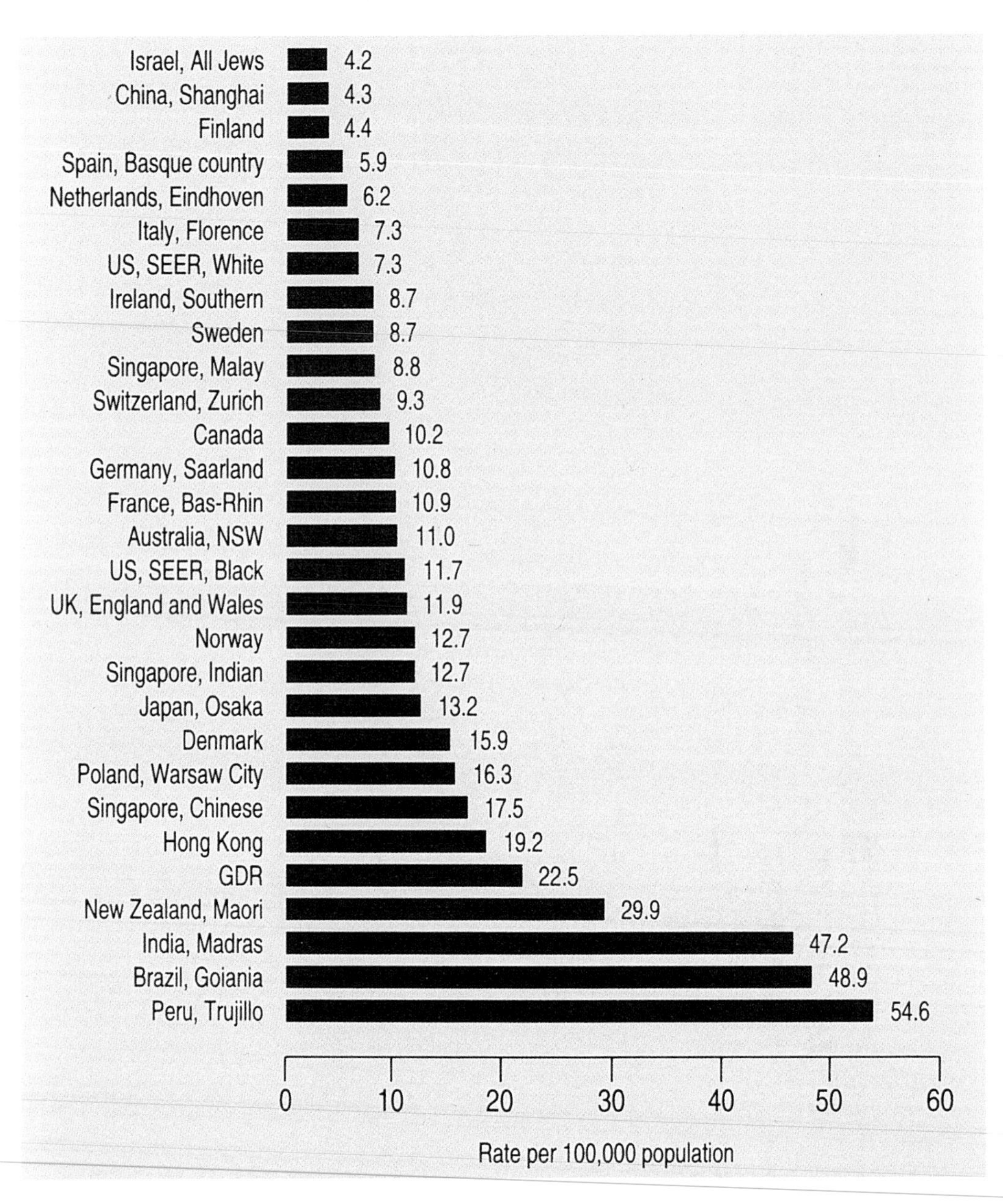

Figure 17.1 Cancer of the cervix uteri. World standardized incidence rates per 100 000 population 1983–1987+. +Dates vary for some populations. Cancer Research Campaign fact sheet 22.4, 1995.

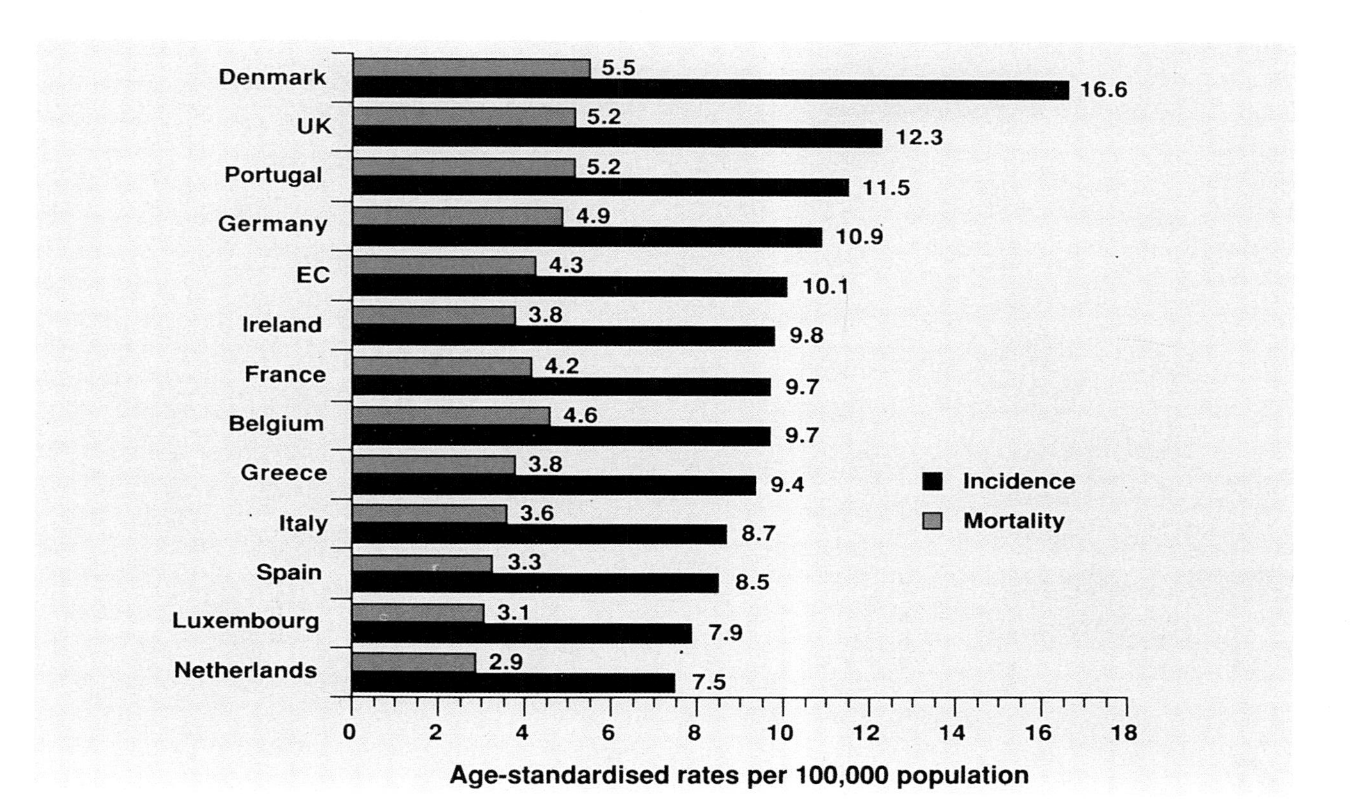

Figure 17.2 Cancer of the cervix uteri. Age-standardized incidence and mortality rates for 13 EC countries in 1990. Cancer Research Campaign fact sheet 12.3, 1994.

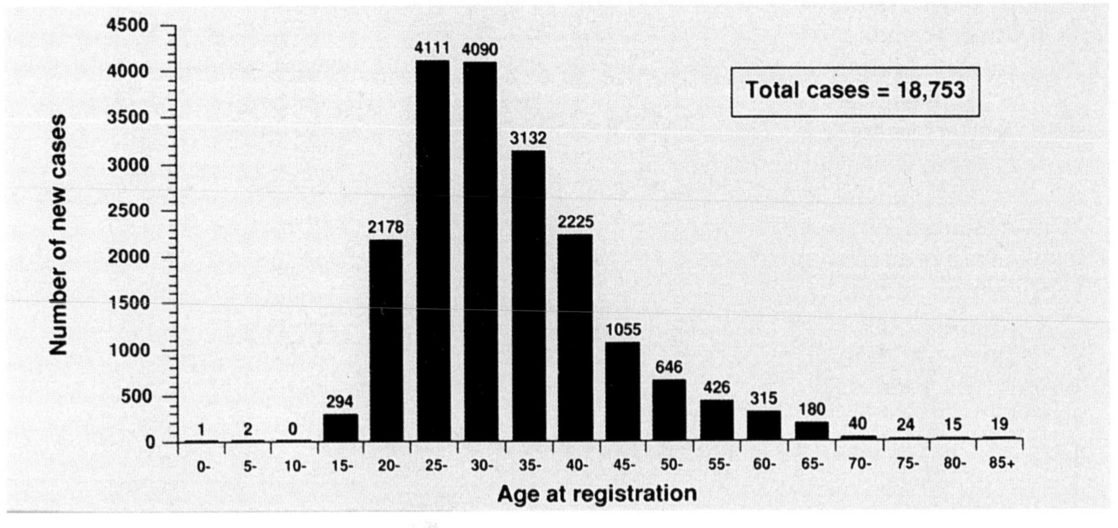

(**a**)

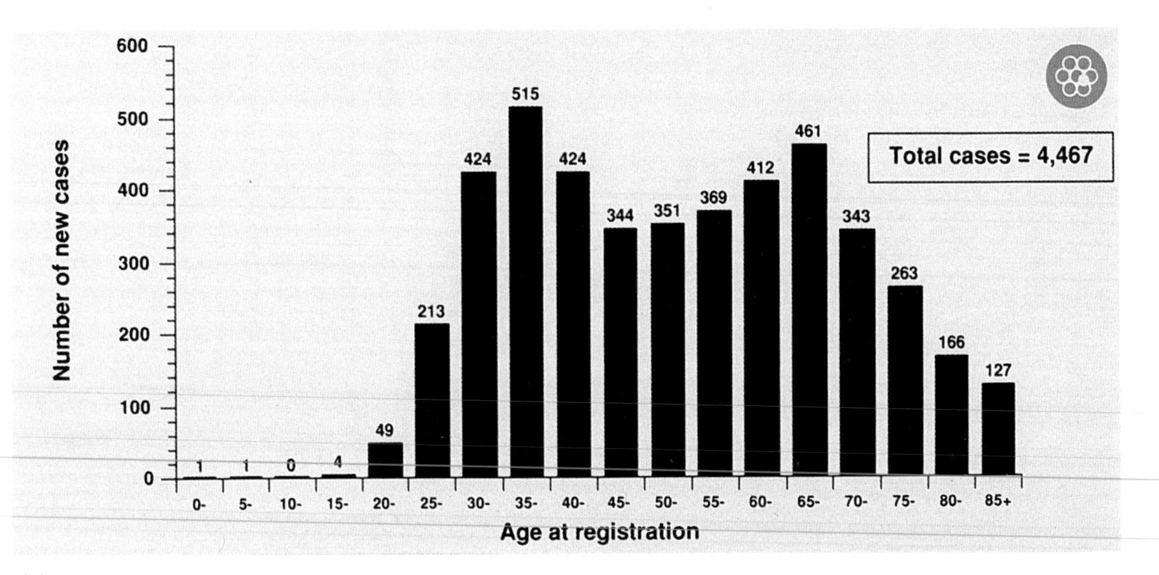

(**b**)

Figure 17.3 (**a**) New cases of carcinoma *in situ* of cervix (CIN 3) in England and Wales, 1988. (**b**) New cases of invasive carcinoma of cervix uteri, England and Wales, 1988. Cancer Research Campaign fact sheet 12.1, 1994.

and morphological studies as well as follow-up of untreated women with abnormal smears (Spriggs and Boddington, 1980; McIndoe *et al.*, 1984). Epidemiological studies have shown that the risk factors are the same for both preinvasive and invasive cervical cancer; temporal studies show that preinvasive cancer occurs most commonly about 10 or 15 years before the onset of invasive disease (Figure 17.3 (a,b)). The most convincing evidence of the link between invasive cervical cancer and CIN is derived from studies of those communities where screening has been successful, e.g. Finland, Iceland, British Columbia (see Chapter 1). In those communities where preinvasive disease has been identified and treated, there has been a significant reduction in the incidence and mortality from invasive cervical cancer.

This chapter addresses one of the key aspects of the screening programme which determine its success or failure, namely the accuracy of smear interpretation. Clearly, the effectiveness of screening is influenced by many factors. These include the ages and coverage of the target population, the interval between smears, the skill of the smear takers, and arrangements for follow-up and treatment of women with abnormal smears; these factors are discussed elsewhere (Coleman *et al.*, 1993). However, the question of diagnostic accuracy and the problems of consistency of smear interpretation are of key importance for pathologists and merit special attention in a handbook for pathologists.

17.2 ACCURACY OF CERVICAL CYTOLOGY

Errors of reporting cervical smears, particularly negative error (i.e. the failure to identify an abnormal smear on first examination), have been a cause for concern in the screening programme for many years. Contrary to what is considered now to be 'good screening practice' (Cochrane and Holland, 1971; Cuckle and Wald, 1984) the sensitivity and specificity of the Papanicolaou smear test was not established before it was introduced as a screening tool. Early attempts to establish the false-negative error rates were misleading as they were based solely on retrospective examination of biopsied cases (Spriggs, 1971). One of the first attempts to determine accurately the false-negative rate was made by Richart (1964). He reviewed the diagnosis offered on 635 Papanicolaou smears from 159 women with biopsy-proven dysplasia or carcinoma *in situ* and recorded a false-negative rate of 1.4%.

Yule (1973) used a different approach to record the false-negative rate. He identified 14 437 women who had had a smear taken as part of the population screening programme in the UK, and recalled them for a second smear within 3 months of taking the first smear. On the first round of screening, 143 women were found to have an abnormal smear. On the second round, 25 additional positive cases were found, giving a false-negative rate of 14.9%. Yule identified two causes for this high false-negative rate, namely 'screening error' and 'biological variability'. Screening error was due to failure of the screener to recognize the abnormal cells in the smear. Biological variability is a term used to describe failure of exfoliation of tumour cells and was assumed to explain those cases in which abnormal cells in the cervix are not transferred to the smear, despite the fact that every care is taken with the collection of the sample. Incomplete transfer of abnormal cells to smears is considered most likely to occur in cases of invasive cancer of the cervix and has been observed most frequently in smears from post-menopausal women (Husain *et al.*, 1974).

Later attempts to establish the false-negative rate were based on review of previous negative smears from women with a positive smear. Miller (1981) was one of the first to adopt this approach and reported a false-negative rate of 22% in the British Columbia screening programme. Attwood *et al.* (1985) reviewed the cervical smear history of women with invasive cancer of the cervix and observed that 16/26 smears (62%) reported initially as negative were found to contain abnormal cells on review. Bosch *et al.* (1992) reviewed the false-negative rates reported by 14 cytology laboratories who carried out a similar exercise and found that it ranged from 10% to 58%.

Ronco *et al.* (1996) estimated the sensitivity of cervical cytology in two ways. They invited experienced cytologists to examine and report on a test set of 61 smears and found that cytologists correctly interpreted 30 out of 34 abnormal smears under test conditions, giving a false-negative rate of 12%. They also determined the sensitivity of cytology by taking smears from 33 women immediately before cervical biopsy. Four smears from 19 women with biopsy-proven CIN contained no abnormal cells (21%). They assumed this finding to reflect sampling error, although it may have reflected biological variability of the tumour or the size of the lesion. Jarmulowicz *et al.* (1989) have shown that small lesions are less likely to be detected than larger lesions.

The examples cited above demonstrate that there are several different ways of expressing the sensitivity and the false-negative rate of cervical cytology and all have their limitations. Unfortunately, the values cited for the sensitivity of cervical cytology are often quoted without reference to the form of measurement used. For example, some authors include borderline smears and/or smears containing mildly dyskaryotic cells (suggestive of a low-grade lesion) among the category of smears classed as false-negative, whereas others include only smears showing severe dyskaryosis. Some authors include only biopsy-proven cases in their positive smear group while others may rely on the cytological findings only. Some use cervical scrapes, others use scrapes and endocervical brushings. Parameters should be carefully defined before comparing false-negative rates for cervical cytology.

Useful formulae for calculating the false-negative rate, sensitivity, specificity and positive predictive values of cervical cytology are shown in Figure 17.4.

Several authors have attempted to identify the reason for the errors of screening which leads to false-negative reporting (Reuter and Schenck, 1986). Two factors appear to influence performance – namely screener fatigue and the composition of the smear. Cervical screening is a tedious, repetitive task requiring intense and prolonged concentration by the screener as he/she proceeds to examine every cell in the smear. As each smear may contain up to 500 000 cells, it is hardly surprising that fatigue and loss of concentration ensue with potentially serious consequences for the patient.

There is evidence that certain smear patterns are more likely to result in a false-negative report than others. Roberts and Woodend (1993) studied the composition of 92 false-negative smears and found that they frequently contained minibiopsies of the cancerous epithelium which had not been recognized by the screener. Mitchell and Medley (1995) studied 71 false-negative smears and a corresponding number of correctly reported positive smears from women with CIN 3 and found that those smears that were correctly identified contained significantly more abnormal cells than those which were assigned false-negative reports. They found that when there were 50 or less abnormal cells in the smear the risk of a false-negative report being issued was 23.7 times greater than when there were 200 or more abnormal cells in the smear. Bosch *et al.* (1992) confirmed that smears that were erroneously reported as negative often contained few abnormal cells.

The risk of false-negative reporting was appreciated soon after the National Cervical Screening Programme was launched in the UK and resulted in the preparation of a set of laboratory procedures which minimized the risk of such errors (Husain *et al.*, 1974). The value of these quality control procedures and their place in laboratory practice are discussed in this chapter.

17.3 QUALITY ASSURANCE IN THE CYTOLOGY LABORATORY

Quality assurance in cytology involves the introduction of a set of planned and systematic

CALCULATION OF SENSITIVITY, SPECIFICITY, PREDICTIVE VALUE AND ACCURACY

TEST	OUTCOME +	OUTCOME -
+	a	b
-	c	d

a = True Positive (TP)
b = False Positive (FP)
c = False Negative (FN)
d = True Negative (TN)

$$\text{Sensitivity} = \frac{TP}{TP + FN} \times 100 \qquad \text{FN Rate} = \frac{FN}{TP + FN} \times 100$$

$$\text{Specificity} = \frac{TN}{TN + FP} \times 100 \qquad \text{FP Rate} = \frac{FP}{TN + FP} \times 100$$

$$\text{Positive Predictive Value} = \frac{TP}{TP + FP} \times 100$$

$$\text{Negative Predictive Value} = \frac{TN}{[TN + FN]} \times 100$$

$$\text{Accuracy} = \frac{TP + TN}{[(TP+FP) + (TN+FN)]} \times 100$$

Figure 17.4 Table for calculating the sensitivity and specificity of cervical cytology as a method of detecting CIN or invasive cervical cancer (NHSCSP publications No. 1, October 1995).

actions which will ensure optimal cytological practice. The outcomes can be measured as a single target figure or as a range of acceptable values. It is usual to describe quality assurance measures as being of two types: (i) those which apply within the laboratory and operate on a day-to-day basis (internal quality control measures); and (ii) those which are intermittently applied by an external peer group (external quality control measures). This division is somewhat artificial. In practice, quality assurance is irrevocably linked with training, education and continuing professional development, so that the detailed operation of quality assurance programmes will vary from laboratory to laboratory and need to be tailored to meet the special situations which apply in each laboratory

17.3.1 Internal quality assurance

Internal quality assurance refers to the procedures introduced by the staff in the laboratory to ensure that the results of screening are accurate. They include monitoring of specimens, reception, data entry, laboratory preparation, the accuracy of primary screening, staff/workload ratio, staff training and continuing professional development, and fail-safe procedures.

(a) Specimen reception

Alcohol-fixed cervical smears clearly labelled with the patient's name should be delivered to the laboratory in transport containers accompanied by a request form. The request form for smears from well women should contain the following information: married name, maiden name, date of birth, name and address of smear taker, details of sender, description of cervix, date of last menstrual period, parity, contraceptive use, hormone therapy and screening history (to include date and result of previous smear reports). The request forms accompanying smears from women with symptoms or signs suggestive of cervical cancer or who have been treated for preinvasive or invasive cancer should contain a clinical summary of the patient's condition.

Smears and request forms should be matched and the same laboratory number applied to both. Mismatches or broken smears or unlabelled smears should be recorded. Data entered by clerical staff should be audited at regular intervals.

(b) Laboratory preparation

Each batch of smears should be checked by senior staff for quality of stain and preparatory technique. Smears should be stained using the Papanicolaou method and there should be a written schedule of staining. Dehydration should be complete, nuclear chromatin should be clearly defined, and cytoplasmic staining should be transparent to permit examination of underlying cells. There should be good contrast between the superficial cells that contain keratin and the intermediate squames that do not. Coverslip size is extremely important as it determines the area of the smear screened which in turn will affect the opportunity to detect neoplastic cells. A minimum coverslip size of 40 mm is recommended. Preparation artefact should be avoided

(c) Primary screening

This should be undertaken by cytotechnologists who have completed a satisfactory period of training and passed a national or international test of competence to screen (Coleman *et al.*, 1993). It has been suggested that cytotechnologists should screen a minimum of 5000 smears under supervision before being allowed to sign out reports. Primary screeners should be supervised by more experienced screeners (supervisors) who are responsible for quality control of the primary screeners output. A model structure for flow of smears through the laboratory is shown in Figure 17.5.

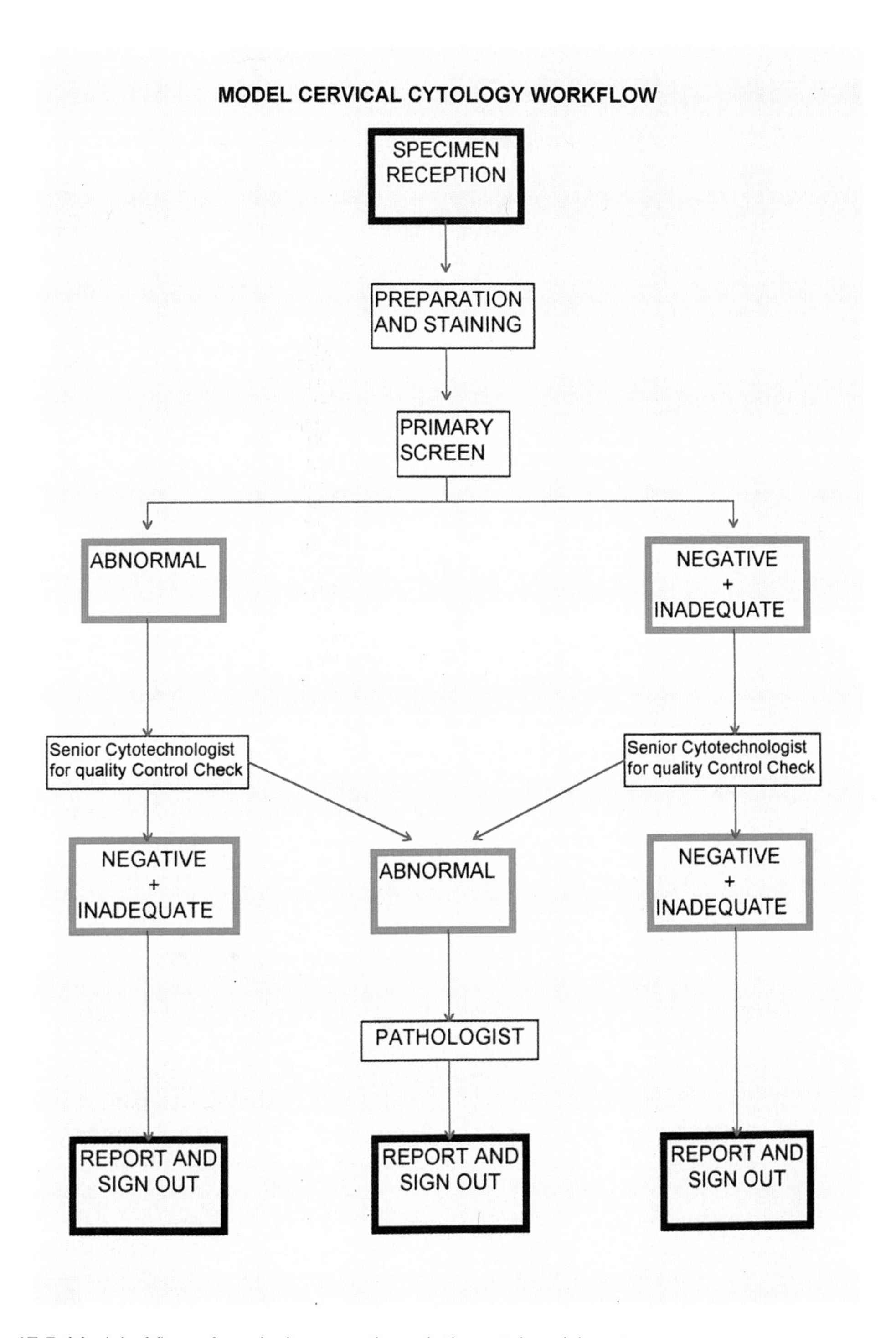

Figure 17.5 Model of flow of cervical smears through the cytology laboratory.

In many laboratories there is a written protocol for screeners regarding referral for a second opinion and certain categories of smears (e.g. borderline changes or worse, cases with symptoms or signs suggestive of cervical cancer, a past history of CIN or invasive cancer) are referred automatically for a pathologist's opinion. In the event of an abnormal smear being detected previous smears should be reviewed.

The assessment of the *adequacy* of the smear is an important task for the primary screener and the characteristics of an adequate smear are discussed in Sections 2.3 and 6.6.

Argument continues to rage regarding the significance of endocervical cells in the smear (Section 6.6). A compromise solution suggested by a working party sponsored by the NHS Cervical Screening Programme in the United Kingdom recommends that more than 80% of all cervical smears received should contain evidence of transformation zone sampling, i.e. endocervical cells and/or immature metaplastic cells (Quality Assurance Guidelines for the Cervical Screening Programme, NHS Publications No. 3, Sheffield, January 1996).

(d) Ensuring the accuracy of primary screening

One or more of the following quality control measures should be in place to reduce the risk of false-negative reporting

1. *Targeted double screening of negative cervical smears* from patients in selected categories, i.e. abnormal vaginal bleeding or a clinically suspicious cervix.

2. *Rescreening of 10% of a random sample of negative or inadequate smears.* This method of quality control has been widely practised by cytology laboratories in the United States for over 30 years in order to comply with the Clinical Laboratory Improvement Acts (CLIA) of 1967 and 1988. It has as its aims the detection of false-negative smears overlooked at the primary screening stage and the identification of cytotechnologists who are not competent to screen. Random rescreening has been severely criticized by several authors (Melamed, 1973; Jordan, 1992; Kaminsky *et al.*, 1995) who have demonstrated it to be a most inappropriate method of maintaining quality standards in cytology, as the likelihood of detecting an error of screening or identifying a cytotechnologist whose performance is substandard is very low indeed in view of the relatively low rate of abnormal smears in a well-woman population. Nevertheless, this approach has its advocates (Kreiger and Naryshkin, 1994) on the grounds that it makes the cytotechnologist 'quality conscious' and it is used in many cytology laboratories world-wide. Kreiger and Naryshkin reported that a study of random 10% rescreening in a large laboratory that processes over a quarter of a million smears annually revealed that the false-negative fraction, i.e. total false-negative smears per year expressed as a percentage of total positive smears per year, fell from 11% to 5% over 10 years. They were also able to assess the performance of individual screeners by this method.

3. *Seeding abnormal smears into the cytology workload.* This involves inserting known positive cases randomly among the cases to be screened. Although attractive in principle it is highly complicated in practice, and only a few laboratories have ever explored this approach. Bosch *et al.* (1992) introduced previously identified false-negative smears into the routine workload of five screeners who were unaware of the experiment and found that in only one of 25 cases were the abnormal cells recognized. When the experiment was repeated with the full knowledge of the screeners, all the false-negative smears were recognized as positive by three of the screeners, whereas the two remaining screeners continued to make errors of interpretation.

Hill (unpublished data, 1996) carried out a prospective study over a 17-month period which involved seeding smears with a high probability of being abnormal into the routine workload of the cytology laboratory. The cytotechnologists were unaware of the exercise. Some 235 smears were seeded of which 167 were abnormal on review. A total false-negative rate of 7.8% was identified for all grades of abnormality and a false-negative rate of 4.2% was calculated for smears showing moderate dyskaryosis (high-grade lesion) or worse. This is probably the most accurate estimate of false-negative rate in a true laboratory setting, although the laboratory itself was rather unusual in that the quality standards of the laboratory were very high: 40 % of negative smears were routinely double rescreened.

4. *Rapid review.* This is a relatively new approach to quality control which was developed in the UK. It involves supervisory staff performing a partial check under a low-power microscope objective (×10) of all cervical smears reported as negative or unsatisfactory by the primary screener. If the reviewer disagrees with the opinion of the primary screener, the smear is subjected to a full rescreen. In theory this approach has several advantages over the random rescreen of 10% of negative smears. It permits monitoring of laboratory performance on a continuous basis and provides timely feedback to the screeners. It also enables inaccurate reports to be amended before they are issued.

Initial studies of rapid review indicated that this method of quality control may be an improvement on 10% random rescreening. It was reported that a 30-second rapid rescreen could lead to a five-fold increase in the rate of detection of false-negative smears (Baker and Melcher, 1991; Faraker, 1993; Dudding, 1995). However, most of the initial studies were flawed and the real value of rapid review as a method of quality control has yet to be determined.

Baker *et al.* (1997a,b) used the AxioHOME microscope to evaluate rapid review as a method of quality assurance. The AxioHOME system is a microscope system which provides the cytotechnologist with feedback regarding his/her screening track and screening time. The system can be used to analyse slide coverage and time spent screening. From these data it is possible to determine screening rate.

The AxioHOME system was used to investigate slide coverage as performed by 15 cytotechnologists who were asked to review a test set of 22 slides comprising 18 negative slides, three positive slides and one unsatisfactory slide. They found that the sensitivity of rapid review ranged between 76.6% and 80%, depending on the method used. On several occasions abnormal cells were not identified by the screener even though they were included in the scanning track because the screening rate was too fast. On one occasion the scanning track failed to include the abnormal cells. Further research is in progress using the AxioHOME system to determine the characteristics of false-negative smears and to develop a method of checking which would minimize the risk of issuing an erroneous report in clinical laboratory practice.

5. *Review of previous negative cytology of women who present with a positive smear.* This has proved to be an instructive method of quality control and has been used by many authors to establish the laboratory false-negative rate. For adequate review of previous smears, the slides should be stored for a minimum of 10 years. The authors' experience of this form of quality control confirms the view that habituation and loss of attention on the part of the screener is a not uncommon reason for false-negative reports, as false-negative smears identified in this way not infrequently contain numerous dyskaryotic cells.

6. *Biopsy/smears correlation.* This form of quality control is an important part of every cytology laboratory's internal quality control scheme and laboratories are advised to keep their records in such a way that the smear history, biopsy and treatment of women with abnormal smears are available to the screener and the pathologist alike. Ideally, the biopsy and smear should be sent to the same laboratory,

but where this is not possible each cytopathology laboratory should ensure that they receive a copy of the biopsy report and vice versa. If there is a discrepancy between the cytological and histological findings, both cytopathologists and histopathologists should review all slides and the gynaecologists should be advised of their joint opinion of the case.

There are two problems which have to be borne in mind when interpreting the results of cytological/histological correlation. Firstly there is sampling error, both on the part of the smear taker and the colposcopist. Secondly, there is observer variability in the interpretation of the smears and biopsies

Observer variability has been studied by several groups (Evans and Sanerkin, 1970; Lambourne and Lederer, 1973; Evans *et al.*, 1974; Kern and Zivolich, 1977; Klinkhamer *et al.*, 1988, 1989; Ismail *et al.*, 1989, 1990; Robertson *et al.*, 1989; O'Sullivan *et al.*, 1994; Creagh *et al.*, 1995; Ciatto *et al.*, 1996) and show consistently that observer variability is at its lowest when pathologists are asked to report on biopsies or smears containing CIN 2/3 and at its greatest when experts are presented with low-grade lesions (borderline smears or biopsies)or inadequate smears. The distinction between borderline lesions and normal epithelium presents the greatest problems. Two of the studies have shown that interobserver variation can be reduced with training and experience (Klinkhamer *et al.*, 1989; O'Sullivan *et al.*, 1994).

Sampling error on the part of the colposcopist is much more likely to occur when punch biopsy rather than cone biopsy or laser loop excision biopsy is used to obtain material for histological correlation. The problem of sampling has been addressed by Ronco *et al.* (1996) and Giles *et al.* (1988). Giles *et al.* examined 200 asymptomatic women by colposcopy and cytology and found that the prevalence of CIN detected by cytology alone was 5%, whereas the prevalence of CIN detected by a combination of cytology and colposcopy rose to 11%. The difference in prevalence was mainly attributable to small, low-grade lesions of uncertain significance. Ronco *et al.* (1996) also found cytology less efficient than colposcopy for the diagnosis of low-grade lesions. Scrutiny of a wide range of cervical sampling devices using a study design involving meta-analysis of randomized and quasi-randomized studies of different spatulae and cervical brushes and swabs has shown that cervical neoplasia is most likely to be detected if an extended tip spatula or a combination of any spatula plus a Cytobrush or cotton swab or the Cervix brush are used to sample the cervix (Buntinx and Brouwers, 1996) (see also Section 2.1, Specimen collection).

17.3.2 External quality assurance

In order to obtain an overview of the quality of service provided by a cytology laboratory it is useful to obtain an independent assessment of the performance of a laboratory and compare the results against agreed national or international standards. In the United Kingdom and United States, laboratory performance is monitored externally in several ways – by proficiency testing, by accreditation, and by annual statistical returns.

(a) Proficiency testing

This scheme was introduced in 1968 in the United States to monitor the ability of medical and non-medical personnel to interpret cervical smears. It was introduced in response to a mandate from the US Department of Health that cytology laboratories obtain a valid permit and maintain minimum agreed standards of performance. The protocol for the test was devised and developed in the United States (Collins and Kaufmann, 1971) and introduced in the United Kingdom 20 years later. The scheme was designed to achieve an unbiased assessment by an independent external assessor of the performance of all grades of staff who reported on cervical smears. Briefly, Papanicolaou-stained cervical smears selected specifically for the purposes of the assessment are taken to each cytology laboratory participating in the scheme by a facilitator. Each member of staff in the laboratory who undertakes cervical screening is given 10 test slides and asked to report on them within 2 hours. The facilitator marks the test and informs the participants of the results. Tests are taken twice yearly. The overall laboratory performance in the test is compared with that of other laboratories in the region, usually on an anonymous basis.

Experience in the UK has shown that the scheme is useful at detecting unacceptable levels of performance. Personnel who fall below acceptable levels are not permitted to screen until they have obtained further training and demonstrated their competence to screen (Gifford and Coleman, 1994; Gifford *et al.*, 1997).

(b) Accreditation

This system of external audit employs a system of on-site inspection to ensure that a laboratory is able to provide a high-quality service. Inspectors are usually drawn from peers and carefully instructed in the assessment procedure. The accrediting body defines standards for organization and performance, and laboratories are assessed against those standards. On-site inspection ensures compliance with such standards. Usually, a certificate of accreditation is issued which is valid for one year.

(c) Annual statistical return

This form of external audit depends on the regular collection of information from the laboratory. For a cytology laboratory this will include information about annual workload (number of smears and/or number of women screened), staff/workload ratios, results of smear test, results of biopsies, interval to reporting, invasive cancer cases, invasive cancer deaths, etc. Annual returns are useful, provided that the information contained in the return is accurate and collected in a consistent manner; otherwise comparison with national standards will be invalid.

17.4 AUTOMATION OF CERVICAL SCREENING

Manual screening of cervical smears is a labour-intensive, time-consuming and tedious task requiring a high degree of skill and many years of experience for its perfection. There is little doubt that the introduction of fully automated or interactive automated screening systems would represent a significant advance in laboratory practice and enhance the scope for population screening in countries where screeners are in short supply and resources for training are scarce. At the time of writing, two automated systems are commercially available – namely the PAPNET system (developed by Neuromedical Systems, Inc.) and the AUTOPAP system (Neopath, Inc.) although several other systems are under development. Both PAPNET and AUTOPAP systems use computers to scan routinely prepared Papanicolaou-stained cervical smears for the presence of abnormal cells and have been shown in extensive trials to improve the accuracy of screening by reducing the risk of false-negative reporting. They both have the approval of the Federal Drug Agency in the United States for the interpretation of cervical smears for the purposes of quality control only.

17.4.1 The PAPNET system (Neuromedical Systems, Inc.)

Neuromedical Systems, Inc. was founded in 1989 in Suffern, New York. Australian and Asian headquarters were established in Hong Kong in November 1994 and European headquarters were established in Amsterdam soon after. They developed the PAPNET system which is an interactive automated cytology

Figure 17.6 The PAPNET system. (Courtesy Neuromedical Systems, Inc.).

testing system based on neural network technology and is designed to increase accuracy of a conventionally prepared cervical smear without introducing any new source of risk. The neural network in PAPNET represents a non-algorithmic branch of artificial intelligence and is ideally suited for recognizing patterns in natural scenes. Application of neural networks to cytology permits semi-automated screening without requiring specially prepared smears.

Stained cervical smears are bar coded by the cytology laboratory and shipped by special courier express to a scanning centre. On arrival, the smears are placed on a microscope stage using a robotic arm and scanned at two magnifications (× 10 and × 40) by an electronic camera. The camera scans the whole smear and extracts from each slide the 128 most suspect images (or tiles) which are stored on a compact disc. When the smears have been scanned they are returned to the laboratory with the disc to be reviewed by the cytotechnologist for the final diagnosis.

In the laboratory, the 128 images (tiles) are displayed on two pages (one showing single cells and one showing cell clusters) on a PAPNET review station (Figure 17.6). The smears are triaged by the cytotechnologist as 'negative' or 'review' on the basis of these images Slides triaged as 'negative' are dispatched without further investigation. Slides selected for 'review' are examined manually under the microscope and reported according to the microscopic findings.

PAPNET testing of over 16 000 presumed negative smears have shown that seven times more false-negatives can be identified using the PAPNET system than by manual re-examination of the same number of slides. Experience of over 2500 cases at St Mary's Hospital has confirmed the ability of the system to display abnormal cells with a high degree of accuracy. However, interpretation of the images is a skill which needs to be learned and experience with 400–500 smears is required before an experienced screener who is fully familiar with the microscopic appearances of cervical smears can be deemed competent to interpret the video images on the PAPNET screen. An initial training in their interpretation is provided by NSI.

17.4.2 AUTOPAP 300 and AUTOPAP QC (Neopath, Inc.)

Neopath, Inc. was founded in 1989 by Dr Alan Nelson and a technical team from the University of Washington Centre for Imaging Systems in Seattle, Washington. The initial product was the AUTOPAP 300 which is a classifying system for cervical smears. Cervical smears which are normal and adequate can be removed from the system. Remaining smears are screened manually in the normal way. The AUTOPAP 300 system comprises a high-speed video microscope, a field-of-view computer with associated image interpretation software, a touch sensor, a bar code reader, an input hopper, an output hopper and a communications interface, in a stand-alone instrument. This is installed in the laboratory and the company provides training on site for its use.

The cytotechnologist bar codes conventionally prepared cervical smears and loads them into the AUTOPAP slide trays, each of which holds eight slides. The slides are automatically processed by the AUTOPAP 300 which screens out the normal slides. The slides are ranked according to their likelihood of containing abnormal cells. The level of selection of smears for review is determined by the operator. Thus, the operator may decide to select 80% of the highest ranking smears for review when the system is used for primary screening or 20% when it is used for quality control purpose. Studies have shown that 80% of false-negative smears may be contained in this 20%. Up to 300 slides can be processed daily by the AUTOPAP 300.

Both AUTOPAP and PAPNET are still under development. A major drawback to their use in national screening programmes is the additional cost they impose on the screening programme when used as a quality control tool. In the UK there is special interest in the development of these systems for primary screening and the results of a multicentre trial of the PAPNET system for primary screening are awaited with interest.

For further information on automated screening, readers are referred to Coleman (1996), Mango (1994), Rosenthal *et al.* (1996), Colgan *et al.* (1995), Oewerkerk-Noordman *et al.* (1994), Boon *et al.* (1994) and Koss *et al.* (1994).

REFERENCES

Attwood, M. E., Woodman, C. B. J., Luesley, D. and Jordan, J. A. (1985) Previous cytology in patients with invasive carcinoma of the cervix. *Acta Cytol.*, **29**, 108–110.

Baker, A. and Melcher, D. H (1991) Rapid cervical cytology screening. *Cytopathology*, **2**, 299–301.

Baker, R. W., Wadsworth, J., Brugal, G. and Coleman, D. V. (1997a) An evaluation of rapid review as a method of quality control of cervical smears using the AxioHOME microscope. *Cytopathology*, **8**, 85–95.

Baker, R., Brugal, G. and Coleman, D. V. (1997b) Assessing slide coverage by cytoscreeners during the primary screening of cervical smears, using the AxioHOME microscope system. *Anal. Cell Path.*, **13**, 29–37.

Beral, V. (1974) Cancer of the cervix: a sexually transmitted infection? *Lancet*, **i**, 1037–1040.

Boon, M. E., Kok, L. P., Nygaard-Nielsen, M. *et al.* (1994) Neural network processing of cervical smears can lead to a decrease in diagnostic variability and an increase in screening efficacy: a study of 63 false negative smears. *Mod. Pathol.*, **7**, 957–961.

Bosch, M. M. C., Rietveld Scheffers, P. E. M. and Boon, M. E. (1992) Characteristics of false negative smears. *Acta Cytol.*, **36**, 711–716.

Bradbeer, C. (1987) Is infection with HIV a role factor for cervical intraepithelial neoplasia? *Lancet*, **2**, 1277–1278 (letter).

Buntix, F. and Brouwers, M. (1996) Relation between sampling device and detection of abnormality in cervical smears: a meta-analysis of randomised and quasi- randomised studies. *Br. Med. J.*, **313**, 1285–1290.

Ciatto, S., Cariaggi, M. P., Minuti, M. *et al.* (1996) Interlaboratory reproducibility in reporting inadequate cervical smears – a multicentre multinational study. *Cytopathology*, **7**, 386–390.

Cochrane, A. L. and Holland, W. W. (1971) Validation of screening procedures. *Br. Med. Bull.*, **27**, 3–8.

Coleman, D. V. (1996) Automation in cytology: a review of recent developments in the automated analysis of cervical smears. *J. Cell. Pathol.*, **1**, 3–12.

Coleman, D. V., Day, N., Douglas, G. *et al.* (1993) European guidelines for quality assurance in cervical cancer screening. *Eur. J. Cancer*, **29A** (Suppl. 4), S1–S31.

Collins, D. N. and Kauffman, W. (1971) New York State Computerised Proficiency Testing Program in Exfoliative Cytology Department. *Acta Cytol.*, **15**, 34–41.

Creagh, T., Bridger, J. E., Kupek, E. *et al.* (1995) Pathologist variation in reporting cervical borderline epithelial abnormalities and cervical intraepithelial neoplasia. *J. Clin. Pathol.*, **48**, 59–60.

Cuckle, H. and Wald, N. (1984) *Principles of Screening in Antenatal and Neonatal Screening* (ed. N. J. Wald), Oxford University Press, Oxford.

Dudding, N. (1995) Rapid rescreening of cervical smears: an improved method of quality control. *Cytopathology*, **6**, 95–99.

Evans, D. M. D. and Sanerkin, N. F. (1970) Cytology screening error rate, in *Cytology Automation* (ed. D. M. D. Evans), Livingstone, Edinburgh, pp. 5–13.

Evans, D. M. D., Shelley, G., Cleary, B. and Baldwin,

Y. (1974) Observer variation and quality control of cytodiagnosis. *J. Clin. Pathol.*, **27**, 945–950.

Faraker, C.A. (1993) Partial rescreening of all negative smears: an improved method of quality assurance in laboratories undertaking cervical cytology. *Cytopathology*, **4**, 47–50.

Gay, J. D., Donaldson, L. D. and Goellner, J. R. (1985) False negative results in cervical cytologic studies. *Acta Cytol.*, **29**, 1043–1046.

Gifford, C. and Coleman, D. V. (1994) Quality assurance in cervical cancer screening: results of a proficiency testing scheme for cytology laboratories in the North West Thames Region. *Cytopathology*, **5**, 197–206.

Gifford, C., Green, J. and Coleman, D. V. (1997) Evaluation of proficiency testing as a method of assessing competence to screen cervical smears. *Cytopathology*, **8**, 96–102.

Giles, J. A., Hudson, E., Crow, J., Williams, D. and Walker, P. (1988) Colposcopic assessment of the accuracy of cervical cytology screening. *Br. Med. J.*, **296**, 1099–1102.

Harri, R. W. C., Brinton, L. A. Cowdell, R. H. *et al.* (1980) Characteristics of women with dysplasia or carcinoma *in situ* of the cervix uteri. *Br. J. Cancer*, **42**, 359–369.

Hill, R. (1996) Development of a method for the direct measurement and continuous surveillance of the laboratory false negative cervical cytology screening/reporting rate and its application in a district laboratory. MSc Thesis, University of London.

Husain, O. A. N., Butler, E. B., Evans, D. M. D., McGregor, J. E. and Yule, R. (1974) Quality control in cervical cytology. *J. Clin. Pathol.*, **27**, 935–944.

Ismail, S. M., Colclough, A. B., Dinnen, J. S. *et al.* (1989) Observer variation in the histopathological diagnosis and grading of cervical intraepithelial neoplasia. *Br. Med. J.*, **298**, 717–710.

Ismail, S. M., Colclough, A. B., Dinnen, J. S. *et al.* (1990) Reporting cervical intraepithelial neoplasia (CIN): intra and inter pathologist variation and factors associated with disagreement. *Histopathology*, **16**, 371–376.

Jarmulowicz, M. R., Jenkins, D., Barton, S. E. *et al.* (1989) Cytological status and lesion size: a further dimension in cervical intraepithelial neoplasia. *Br. J. Obstet. Gynaecol.*, **96**, 1061–1066.

Jordan, S. W. (1992) Great expectations: cytology provisions of CLIA-88 and the role of professional societies. *Cytopathology Annual*, 235–247.

Kaminsky, F. C., Burke, R. J, Haberle, K. R. and Mullins, M. S. (1995) Rescreening policies in cervical cytology and their effect on detecting the truly positive patient. *Acta Cytol.*, **39**, 239–245.

Kay, S., Frable, W. J. and Hume, D. M. (1970) Cervical dysplasia and cancer in women on immunosuppression therapy for renal homotransplantation. *Cancer*, **26**, 1048–1052.

Kern, W. H. and Zivolich, M. R. (1977) The accuracy and consistency of the cytologic classification of squamous lesions of the uterine cervix. *Acta Cytol.*, **2**, 519–523.

Kinlen, L. J. and Spriggs, A. I. (1978) Women with positive cervical smears but without surgical intervention. *Lancet*, **ii**, 463–465.

Klinkhamer, P. J. J. M., Vooijs, G. P. and Haan, A. F. J. (1988) Intraobserver and interobserver variability in the diagnosis of epithelial abnormalities in cervical smears. *Acta Cytol.*, **32**, 794–800.

Klinkhamer, P. J. J. M., Vooijs, G. P. and Haan, A. F. J. (1989) Intraobserver and interobserver variability in the quality assessment of cervical smears. *Acta Cytol.*, **33**, 215–218.

Koss, L. G., Lin, E., Schreiber, K. *et al.* (1994) Evaluation of the PAPNET cytological screening system for quality control of cervical smears. *Am. J. Clin. Pathol.*, **101**, 220–229.

Kreiger, P. and Naryshkin, S. (1994) Random rescreening of cytologic smears: a practical and effective component of quality assurance programs in both large and small cytology laboratories. *Acta Cytol.*, **38**, 291–298.

Lambourne, A. and Lederer, H. (1973) Effects of observer variation in population screening for cervical cancer. *J. Clin. Pathol.*, **26**, 564–569.

McIndoe, W. A., McLean, M. R., Jones, R. W. and Mullins, P. R. (1984) The invasive potential of carcinoma *in situ* of the cervix. *Obstet. Gynecol.*, **64**, 451–458.

Mango, L. J. (1994) Computer assisted cervical cancer screening using neural networks. Cancer Letters, **77**, 155–162.

Melamed, M. R. (1973) Presidential address. *Acta Cytol.*, **17**, 285–288.

Melamed, M. R. and Flehinger, B. J. (1992) Editorial: Reevaluation of quality assurance in the cytology laboratory. *Acta Cytol.*, **36**, 461–465.

Miller, A. B. (1981) An evaluation of population screening for cervical cancer, in *Advances in Clinical Cytology* (eds L. G. Koss and D. V. Coleman), Butterworths, London, Chapter 3.

Mitchell, H. and Medley, G. (1995) Differences between Papanicolaou smears with correct and incorrect diagnoses. *Cytopathology*, **6**, 368–375.

O'Sullivan, J. P., Ismail, S. M., Barnes, W. S. F. *et al.* (1994) Interobserver variation in the diagnosis and grading of dyskaryosis in cervical smears: specialist cytopathologists compared with non specialists. *J. Clin. Pathol.*, **47**, 515–518.

Ouwerkerk-Noordam, E., Boon, M. E. and Beck, S. (1994) Computer assisted primary screening of cervical smears using the PAPNET method: comparison with conventional screening and evaluation of the role of the cytologist. *Cytopathology*, **5**, 211–218.

Porreco, R., Penn I., Droegemueller, W., Greer B. and Makowski, E. (1975) Gynecologic malignancies is immunosuppressed organ homograft recipients. *Obstet. Gynecol.*, **45**, 359–364.

Robertson, A. J., Anderson, J. M., Swanson-Beck, J. *et al.* (1989) Observer variability in histopathological reporting of cervical biopsy specimens. *J. Clin. Pathol.*, **42**, 231–238.

Reuter, B. and Schenck, U. (1986) Investigation of the visual cytoscreening of conventional gynaecologic smears. (ii) Analysis of eye movements. *Anal. Quant. Cytol. Histol.*, **8**, 210–218.

Richart, R. M. (1964) Evaluation of true false negative rate in cytology. *Am. J. Obstet. Gynecol.*, **89**, 723.

Robertson, J. H. and Woodend, B. (1993) Negative cytology preceding cervical cancer: causes and prevention. *J. Clin. Pathol.*, **46**, 700–702.

Ronco, G., Montanari, G., Aimone, V. *et al.* (1996) Estimating the sensitivity of cervical cytology: errors of interpretation and test limitations. *Cytopathology*, **7**, 151–158.

Rosenthal, D. L., Acosta, D. and Peters, R. K. (1996) Computer assisted rescreening of clinically important false negative cervical smears using the PAPNET testing system. *Acta Cytol.*, **40**, 120–126.

Rotkin, I. D. (1973) A comparison review of key epidemiological studies in cervical cancer related to current searches for transmissible agents. *Cancer Res.*, **33**, 1353–1357.

Schenck, U., Reuter, B. and Vohinger, R. (1986) Investigation of the visual cytoscreening of conventional gynecologic smears. (i) Analysis of slide movement. *Anal. Quant. Cytol. Histol.*, **8**, 35–44.

Schneider, V., Kay, S. and Lee, H. M. (1983) Immunosuppression as a high risk factor in the development of condyloma acuminata and squamous neoplasia of the cervix. *Acta Cytol.*, **27**, 220–224.

Sherman, M. E., Mango, L. J., Kelly, D. *et al.* (1994) PAPNET analysis of reportedly negative smears preceding the diagnosis of a high grade squamous intraepithelial lesion or carcinoma. *Mod. Pathol.*, **7**, 578–581.

Sillman, F., Stanek, A., Sedlis, A. *et al.* (1984) The relationship between human papillomavirus and lower genital intraepithelial neoplasia in immunosuppressed women. *Am. J. Obstet. Gynecol.*, **150**, 300–308.

Skegg, O. C. G., Gorwin, P. A., Paul, C. and Doll, R. (1982) Importance of the male factor in cancer of the cervix. *Lancet*, **2**, 581–584.

Smith, P. G., Kinlen, L. J., White, G. C., Adelstein, A. M. and Fox, A. J. (1980) Mortality of wives of men dying with cancer of penis. *Br. J. Cancer*, **41**, 422–428.

Spriggs, A. I. (1971) The cervical smear – its validity and reliability, in *The Early Diagnosis of Cancer of the Cervix* (ed. J. M. Rioggott), William Clowes and Sons Ltd, London, pp. 98–13.

Spriggs, A. I. and Boddington, M. M. (1980) Progression and regression of cervical lesions. *J. Clin. Pathol.*, **33**, 517–522.

Spurrett, B., Jones, D. S. and Stewart, G. (1988) Cervical dysplasia and HIV infection. *Lancet*, **1**, 239.

Vessey, M. P., Lawless, M., Macpherson, K. and Yeates, D. (1983) Neoplasia of the cervix uteri and contraception: a possible adverse effect of the pill. *Lancet*, **ii**, 930–934.

Winkelstein, W. (1977) Smoking and cancer of the uterine cervix: a hypothesis. *Am. J. Epidemiol.*, **106**, 257–259.

Winkelstein, W. (1990) Smoking and cervical cancer – current status: a review. *Am. J. Epidemiol.*, **131**, 945–957.

Wright, H., Vessey, M. P., Kenard, B., McPherson. K. and Doll, R. (1978) Neoplasia and dysplasia of the cervix uteri and contraception: a possible protective effect of the diaphragm. *Br. J. Cancer*, **38**, 273–279.

Yule, R. (1973) The prevention of cancer of the cervix by cytological screening of the population, in *Cancer of the Uterine Cervix* (ed. E. C. Easson), W. B. Saunders, London, pp. 11–25.

Zunzunqui, M. V., King, M. C., Coria, C. P. and Charlet, J. (1986) Male influence on cervical cancer risk. *Am. J. Epidemiol.*, **123**, 302–307.

INDEX

Page numbers in **bold** indicate tables and other set-out information; those in *italics* indicate illustrations.